Indoor Air Quality

Hermann Fromme

Indoor Air Quality

Occurrence and Health Effects of Contaminants

Springer

Hermann Fromme
Institute and Clinic for Occupational, Social and Environmental Medicine
LMU University Hospital, LMU Munich
Munich, Germany

ISBN 978-3-031-40077-3 ISBN 978-3-031-40078-0 (eBook)
https://doi.org/10.1007/978-3-031-40078-0

English translation of the 1st original German edition published by ecomed-Storck GmbH, Landsberg, 2021

This Springer imprint is published by the registered company Springer Nature Switzerland AG
The registered company address is: Gewerbestrasse 11, 6330 Cham, Switzerland

Paper in this product is recyclable.

I would like to dedicate this book with love to my son Fabian and my wife Ulla. Thank you very much for being with me.

Knowing is not enough; we must apply.
Willing is not enough; we must do.
(J.W. von Goethe)

Preface

Dear Reader,

we are all exposed to a wide range of influences from our living environment, with a broad spectrum of very different factors constantly affecting us. In addition to social and psychosocial factors, particular emphasis must be placed on the physical, chemical, and biological factors that can have a lasting and detrimental effect on our health. Against this background, the living environment is increasingly understood as a mutually influencing network of the most diverse factors in which we are integrated. Just how complex our living situation has now become is shown by the fact that around 100,000 chemicals are used on a large scale worldwide and we may also be exposed to many of them regularly. However, the aforementioned network is also subject to constant change, as new factors are added and old ones lose or disappear in importance. We must be aware of these developments as early as possible to be able to respond better to them with preventive measures.

People spend a large proportion of their lives indoors, so knowledge of this environmental compartment is particularly important. However, the significance of indoor spaces is often underestimated or not even adequately perceived. This is even though each adult takes about 22,000 breaths per day, inhaling approximately 11,000 liters of air. A large part of this takes place indoors. We must be aware that we will only be able to reliably assess health risks conveyed via indoor spaces if we significantly increase our knowledge in this area. So far, it is often still limited and incomplete.

Against this background, the book aims to present the current state of knowledge on indoor pollution in detail, to briefly describe health aspects and to show possibilities for a risk assessment. The focus is on chemical substances and physical and biological factors that occur more frequently indoors, are of particular health significance, or for which only limited information on their occurrence indoors is available to date. It also aims to help identify where "new" problems may arise in the future and where we lack data for a valid exposure and risk assessment as well as for a consequent risk management.

Munich, Germany Hermann Fromme

Contents

About the Author

Hermann Fromme is a professor at the Institute and Clinic for Occupational, Social and Environmental Medicine, University Hospital, Ludwig-Maximilians-University Munich (LMU), Germany. He started his professional career by studying social sciences at the University of Göttingen and completed his MD degree at the LMU and the Freie Universität Berlin. After holding various management positions in environmental health departments of the Berlin Senate Department for Social Affairs and Health, he was head of the Institute for Environmental Analysis and Human Toxicology in Berlin. Subsequently, Professor Fromme was head of the Department of Chemical Safety and Toxicology of the Bavarian Health and Food Safety Authority. His work focuses in particular on the identification of chemicals in the environmental media and in humans as well as on the performance of health-related risk assessments. This has resulted in numerous scientific articles, particularly in the fields of exposure science, indoor air contaminants, human biomonitoring, health effects, and regulatory aspects of pollutants.

Chapter 1
Indoor Environment: Background Information

Check for updates

Contents

Abstract Indoor environmental quality (IEQ), often shortened indoor air quality, is characterized by the spatial and temporal variability of pollutants, physical and biological factors in the various indoor spaces, as well as the thermal and structural conditions of the building, environmental influences, and also by the behavior and activities of the room users. This chapter provides an initial overview of basic aspects of the relationship between IEQ and health especially in schools. In addition, the so-called sick building syndrome is explained in more detail, which is a building-related health disorder that summarizes non-specific indoor-related health problems for both groups of people and individuals in commercial, public, and private buildings. Moreover, this chapter aims to present basic effects of indoor climate on our well-being indoors and discusses the effects of the dramatic climate change and its impact on human health.

Due to the very often high temporal and spatial variability of indoor conditions complex chemical formation and transformation processes occur in this microenvironment. This includes processes of deposition on the room surfaces and partly reaction with the surfaces, re-release to the room air, but also transformation of substances and new formation of the so-called secondary organic aerosols (SOA). Although these processes are not yet fully understood due to their diversity, they may be relevant to the exposure of the indoor room users.

It is of particular importance that indoor sampling and measurements should only ever be carried out with a clear question in mind and a defined measurement plan.

H. Fromme, *Indoor Air Quality*, https://doi.org/10.1007/978-3-031-40078-0_1

With this in mind, the different sampling techniques and more new passive sampler devices like the wristband technology were discussed in brief.

There is a large body of scientific publications, especially on the subject of ventilation of indoor spaces. The focus is on ventilation and air purification methods, but here only a first overview can be given and reference is made to more detailed literature.

Keywords Indoor Environmental Quality · IEQ · Health effects · Sick Building Syndrome · SBS · Indoor climate · Well-being indoors · Climate change · Heat waves · Secondary organic aerosols · SOA · Sampling techniques · VOC · SVOC · PM · Wristband · Air purification

Changing lifestyles and working habits have led to a situation in industrialized countries where up to 90% of our daily activities are spent inside buildings, where we can be exposed to a large number of chemical, physical, and biological factors, some of which are harmful to health. Uncontaminated indoor air is therefore of great importance to the health of the individual and the population. However, this is in contrast to our often incomplete knowledge of the exposure situation in this area of life. While we often have a large database with a high temporal and spatial resolution in outdoor air due to measuring stations, measuring programs, and research projects, this does not apply to indoor air. From not only a health protection point of view, but also a prevention point of view, this situation is unsatisfactory. In the past, this lack of reliable data has repeatedly led to measures being taken even though the technical basis for them was hardly available. Another drawback is the extensive lack of legally binding regulations on health protection and prevention from exposure in indoor living spaces, especially for public facilities such as schools and daycare centers. This is also surprising because it has long been known that indoor air quality in classrooms, for example, can adversely affect the well-being and health of children. Children are a particularly sensitive group because their lungs, immune systems, and detoxification capabilities are still developing. Also, due to, for example, their higher activity and breathing rate, they are relatively exposed to a higher dose of pollutants than adults. A crucial issue in maintaining and promoting the health of the population is therefore the air quality of our indoor spaces. For this reason, this problem is also highly relevant for political decision-makers.

Overall, the term "indoor space" is understood to mean very different types of rooms. In this paper, reference is made to the following definitions:

- Apartments with living rooms, bedrooms, craft rooms, sports rooms, basements, kitchens, and bathrooms.
- Workrooms or workplaces in buildings (e.g., offices, salesrooms) that are not subject to occupational health and safety controls concerning air pollutants.
- Public buildings such as hospitals, schools, kindergartens, gymnasiums, libraries, restaurants, theaters, cinemas, and other function rooms.
- Passenger compartments of motor vehicles and all means of public transport.

It should be noted here that for workrooms and industrial spaces the requirements according to the Workplace Ordinances must be considered. In principle, rooms in workplaces (such as offices) are considered to be indoor rooms in the sense of the above definition if the pollutants occurring there are not used directly as working substances.

1.1 Indoor Environmental Quality and Health

The central concept linking indoor air pollution and health is indoor environmental quality (IEQ). It is characterized not only by the spatial and temporal variability of gaseous and particulate pollutants, the physical factors (e.g., noise, light), the biological factors (e.g., molds), the thermal and structural conditions as well as the environmental influences resulting from the location of the object, but also by the behavior and activities of the users (e.g., cooking, heating). Against the background of these diverse influencing factors, it becomes understandable that the relationship between IEQ and health is also an extremely complex one and cannot be explained by one factor alone in any case. In the individual substance-related chapters, the most important health aspects are therefore briefly described and important literature references are given.

Several reviews have also examined in detail the state of scientific knowledge of the relationship between indoor environmental quality and health (Daisey et al. 2003; Hulin et al. 2012; Salthammer et al. 2016; Fisk 2018; Chithra and Shiva Nagendra 2018; Fisk et al. 2019; Tran et al. 2020; Manisalidis et al. 2020; Palacios et al. 2021; Sadrizadeh et al. 2022; Kumar et al. 2020). In addition, a National Institute for Health and Care Excellence (NICE) working group assessed study evidence in an extensive process that reviewed over 33,000 references on the topic of exposure to indoor pollutants and health effects (NICE 2020). Associations between individual and building characteristics and exposure situation with health effects were estimated. As a result, pollutants such as nitrogen dioxide, volatile organic compounds (VOC), particulate matter, and polycyclic aromatic hydrocarbons (PAH), as well as biological factors such as molds and inhalation allergens were found to be associated with various respiratory, cardiovascular, and neurological symptoms. However, the committee also clearly points out the limited power of the studies, particularly because of methodological limitations.

Basically, it must be taken into account that buildings and the people living and working in them can never be seen separately from their social references and living environment.

AlWaer et al. (2021), for example, introduced the term "unhealthy neighborhood syndrome" that illustrates how seemingly separate issues in urban design can help improve living conditions and health through an integrated program. Four key factors that influence human health in neighborhoods can be named: urban density, green space, movement (e.g., modal split), and housing.

1.1.1 Indoor Air Quality and Health in Schools

Overall, numerous studies have also shown that IEQ problems like moisture, mold damage, improper ventilation, and unsatisfactory temperature conditions in schools are associated with increased reporting of respiratory symptoms among pupils.

Particularly with the beginning of the 2000s, extensive studies have been conducted across countries in Europe in the school environment that have demonstrated a link between indoor air quality and various health effects, especially respiratory symptoms and mucosal irritation. For example, the HESE (Health Effects of School Environment) study, a multinational cross-sectional study conducted in five European countries in 2004–2005, included 654 children from 46 schools (HESE 2006). The multiple logistic regression model showed significant association between elevated CO_2 concentration (>1000 ppm per 100 ppm increase) and dry cough at night (OR = 2.99; 95% CI: 1.65–5.44) and rhinitis (OR = 2.07; 95% CI: 1.14–3.73) when adjusted for age, gender, passive smoke exposure at home, and PM_{10} (>50 μg/m^3). The School Environment and Respiratory Health of Children (SEARCH) project studied approximately 60 schools with around 6000 children from six European countries in 2006–2009 (Csobod et al. 2010). In addition, the 2011–2012 Schools Indoor Pollution and Health: Observatory Network in Europe (SINPHONIE) study examined a total of 5175 children and 1223 teachers attending 340 classrooms in 23 European countries (Csobod et al. 2014). Main findings are that the prevalence of diagnosed asthma, nasal allergy, and eczema was 8%, 9% and 17%, respectively, and that among all schoolchildren the commonest recent (<3 months) health event was a blocked nose (47%) followed by a runny nose, feeling cold or feverish, headache, feeling tired and sore throat (36%), as well as that 26% of the children felt better after returning home. In Finland, more than 12,000 students were also surveyed in 2017–2018 and associated with expert-assessed indoor-related health effects (Savelieva et al. 2020). This study shows that psychosocial factors and individual and allergic characteristics, in addition to school indoor air quality, were related to higher reporting of respiratory symptoms in administered questionnaires. But psychosocial factors explained more of the—low—variance between schools in symptom reporting than poor indoor air quality. These findings suggest that the role of other factors beyond IAQ should also be considered when interpreting symptom reporting in indoor air questionnaires.

In an extensive review WHO identified 90 substances by frequency of detection in public sections like schools and reduced the list during a prioritization process using WHO guidelines for indoor or ambient air quality, the most common chemical pollutants measured as well as expert judgment regarding toxicological or exposure potential. The final list of 30 substances was considered for the selection of priority chemicals for inclusion in the screening tool for assessment of health risks from combined exposure to multiple chemicals in indoor air of schools (see Table 1.1).

Table 1.1 Priority chemicals for inclusion in the screening tool for assessment of health risks in schools (WHO 2021)

N	Name	N	Name
1	Formaldehyde	17	Limonene
2	Nitrogen dioxide (NO_2)	18	n-decane
3	PM_{10}	19	Acetaldehyde
4	$PM_{2.5}$	20	BDE 47[+]
5	Benzene	21	Benzo[a]pyrene
6	Ozone (O_3)	22	Naphthalene
7	Carbon monoxide (CO)		o-xylene[a]
8	m,p-xylenes		d-limonene[a]
9	Styrene	23	Phenanthrene
10	Toluene	24	Trichloroethylene
11	1,4-dichlorobenzene	25	Diethyl phthalate (DEP)
12	Ethylbenzene	26	Diisobutyl phthalate (DiBP)
13	Butyl acetate	27	Di-n-butyl phthalate (DnBP)
14	Tetrachloroethylene	28	Galaxolide
15	α-pinene	29	SO_2
16	1,2,3-trimethylbenzene	30	Tonalide

[a]The two substances o-xylene and d-limonene are accounted, respectively, with m,p-xylenes and limonene; + a polybrominated diphenyl ether congener

1.1.2 Sick Building Syndrome

To describe the health effects of indoor environments in one term, the World Health Organization introduced the term sick building syndrome (SBS) in 1983. SBS refers to a building-related health disorder that summarizes non-specific indoor-related health problems for both groups of people and individuals in commercial, public, and private buildings (Tran et al. 2020). In contrast, the term Building-Related Illness (BRI) refers to scientifically well-defined health problems caused by indoor agents that can be identified, measured, and quantified (Seltzer 1994). These include humidifier fever, legionellosis, indoor-associated allergies (e.g., to house dust mites or molds), and indoor-associated malignancies, such as lung carcinoma caused by radon. Symptoms are caused by immunologic, infectious, toxic, or irritant mechanisms (Seltzer 1994).

In particular, evidence on the symptoms most frequently associated with SBS is available from studies in office buildings. In one of the first larger studies, Erdmann and Apte (2004) found statistically significant associations of mucous membrane and lower respiratory SBS symptoms with increasing CO_2. Using the 100-building 1994–1998 BASE (US EPA Building Assessment Survey and Evaluation) dataset the odds ratios for statistically significant associations of sore throat and wheeze symptoms with 100 ppm increases in CO_2 were 1.1 to 1.2, Moreover, the ProKlimA study conducted in Germany in 1994–2000 involved a comprehensive survey of approximately 5000 workers in 14 office buildings, as well as extensive indoor air measurements and medical examinations (Bischof et al. 2004). Although no proven link between indoor air pollutants and health effects could be found, 45% of women and 28% of men in the offices complained of mood disorders. The complaints were mainly expressed when staying in rooms with air-conditioning systems. The OFFICAIR study, conducted in eight European countries, surveyed 7441 people working in 167 office buildings between 2011 and 2012 (Bluyssen et al. 2016; Sakellaris et al. 2016). More than 37% of participants complained of indoor-related discomfort, particularly dry air (47%) and stuffy air (38%). A total of 38% of individuals complained about excessive indoor noise and 31% complained about indoor temperatures that were not comfortable. In a more recent evaluation on the OFFICEAIR project, Sakellaris et al. (2020) concluded that users in rooms with higher concentrations of xylene, ethylbenzene, a-pinene, d-limonene, styrene, formaldehyde, acrolein, propionaldehyde, hexanal, and ozone were more likely to report health symptoms. Licina and Yildirim (2021) studied a cohort of employees who transitioned from three non-WELL to three WELL-certified (ecological Standard given by the International Well Institute) office buildings and reported that the relative improvements owing to relocation into WELL buildings were attributed mostly to non-environmental factors, such as building and workplace cleanliness, colors and textures and furniture and that occupant satisfaction scores did not change during the first year of working in WELL buildings. The occurence of SBS symptoms was not different between two types of buildings, except for symptom of tiredness that was lower in WELL buildings.

Wolkoff (2013) summarized the available literature for the office environment and concluded that in general tiredness, dry, irritated or itchy eyes, and headache are among the top-three symptoms within a recall period of 30–90 days and that the outcome of the symptom reporting strongly depends on the study design, symptom frequency, recall period, context (work vs home), and season.

In a recent population-representative study from Sweden and Finland, a total of 8600 and 4606 people were surveyed in 2011–2012, respectively (Karvala et al. 2018). In this study, 4.8% of Swedes and 7.2% of Finns reported building-related symptoms. Notably, due to the ambiguous definition of SBS, prevalence rates appear to range widely (4.0–88%) (Srecec et al. 2005). In China, a total of 2134 pupils participated in the baseline investigation in 2010 and 1325 pupils stayed in the same classrooms until the follow-up in 2012 was recruited (Zhang et al. 2014). In conclusion, symptoms, as described for SBS, were commonly found, increased during the two-year follow-up period and PM_{10}, CO_2, and relative humidity were associated with an increase in the prevalence and incidence of SBS symptoms. Again, in another study of Chinese adults living in 165 residences relative humidity was a significant risk factor for SBS symptoms, especially for mucosal SBS symptoms and skin SBS symptoms (Hou et al. 2021). Using an online nationwide questionnaire survey involving 4996 participants, Suzuki et al. (2021) concluded that women (odds ratio (OR): 1.2), young individuals (OR: 2.4), and those with a history of allergies and mental illness (OR: 1.75) had a significant increased risk of SBS, while housing specifications designated by the manufacturer, for example,

TVOC values <400 μg/m^3 two weeks after completion, decreased the risk (OR = 0.77).

Moreover, investigating 1281 residents the prevalence of general, mucosal, and dermal symptoms were 14.4%, 9.5%, and 5.1%, respectively, in a Chinese study performed in 2016/17 (Shao et al. 2023). Newly built residential building, smaller floor area, wall paint peeling, dissatisfaction with indoor environment, indoor $PM_{2.5}$ concentration, indoor smoking, and less window opening hours per day were significant risk factors for residents' SBS symptoms in this study.

Even taking into account the methodological differences and limitations of the available studies, Redlich et al. (1997) assume the following core symptomatology of SBS:

- Mucous membrane irritation like eye irritation, throat irritation, cough.
- Neurotoxic effects like headaches, fatigue, lack of concentration.
- Respiratory symptoms like shortness of breath, cough, wheeze.
- Skin symptoms like rash, pruritus, dryness.
- Chemosensory changes like enhanced or abnormal odor perception and visual disturbances.

Very different factors are associated with the increased occurrence of SBS. Thus, associations with various chemical, physical, biological (e.g., mold due to moisture damage), building-related factors (e.g., ventilation, indoor humidity, cleaning), and also psychosocial factors (e.g., work stress, low social support), and personal factors (e.g., sex, reported allergy, tobacco smoking, stress load, high anxiety) have been described—not uniformly (e.g., Norbäck 2009, Lukcso et al. 2016, Azuma et al. 2018, Lu et al. 2015, Nakayama et al. 2019; Aziz et al. 2023). Of course, gender differences in the reporting of SBS symptoms need to be evaluated cautiously, as women often have to work under more unfavorable working conditions (Bullinger et al. 1999).

Nordin (2020) reviewed and discussed a biopsychosocial perspective of SBS including neurogenic inflammation, neural sensitization, classical conditioning, symptom misattribution and somatosensory amplification, and nocebo effect.

1.2 Indoor Climate, Comfort, and Climate Change

1.2.1 Indoor Climate and Comfort

Our well-being in indoor spaces depends on a variety of influences. These originate in the outdoor space (e.g., the noise), the building (e.g., narrow spaces), the individual interior design, the air quality, our personal feeling of being in the space, and also in the respective indoor climatic conditions. Important aspects that influence a comfortable indoor climate are indoor air temperature, indoor humidity, air velocity in the room, and thermal radiation of the enclosing surfaces (Bux and Kampmann 2019). In addition to these factors, physical activity, how we are dressed, our age, and whether we have pre-existing health conditions have an additional significant impact. All of these interact and should, as far as possible, always lead to a comfortable feeling in our very different living spaces such as the private home, public indoor spaces such as schools, daycare centers, and theaters.

Besides the room air temperature, the emission of thermal radiation from the surfaces of a room is important. It depends on the temperature, the building materials, and the color and texture of the room surfaces. Humans themselves react to unfavorable thermal conditions in a room with physiological adaptation reactions such as changes in peripheral blood circulation, increased perspiration, or cold shivering (van Marken Lichtenbelt et al. 2018).

Indoor humidity refers to the water vapor content of indoor air, which can be expressed either as relative humidity in percent or as absolute humidity in g water/kg air. In this context, high humidity levels of more than 12 g/kg of absolute humidity are perceived as unpleasantly humid. In this case, heat dissipation through sweating is limited. While, on the other hand, a very low indoor humidity of well below 30%, such as in the winter months, is perceived as dry air and often associated with health effects such as mucous membrane irritation (Wolkoff 2018b; Hou et al. 2021; Jones et al. 2022). Air velocity is a measure of the movement of air indoors, while increased air velocity is perceived as less disturbing during physical work, for example, because the body may still be able to dissipate heat through convection, even low air velocities below about 0.2 m/s in private living areas cause users to experience drafts (Bux and Kampmann 2019).

A comfortable indoor climate in living spaces, or thermal comfort, is achieved when the room occupant perceives the air temperature, air humidity, air velocity, and thermal radiation in his or her environment as optimal and wants it neither warmer nor colder (Bux and Kampmann 2019). Physiologically, the organism has to exert the least thermoregulatory effort in this case. Global and local comfort criteria can be used for assessment. Global comfort criteria are used to define average values of air temperature, velocities, indoor humidity, and ambient radiant temperature for a room based on the overall human thermal balance (Bux and Kampmann 2019). In contrast, local thermal comfort criteria include, for example, drafts, radiation asymmetries, and temperature stratification in the room, as well as warm/cold walls, floors, or ceilings. A simplified relationship can be shown for thermal comfort between room air temperature and relative humidity (see Fig. 1.1). It shows under which conditions a larger number of people perceive humidity and temperature as comfortable or still comfortable or already too dry/cold or already too humid/warm. Important sources of humidity in the room are listed as examples in Table 1.2. In addition, there is an increasing effort to develop more personalized heating and cooling systems through appropriate heat sensation models, taking into account small-scale factors (Vesela et al. 2017; Salehi et al. 2020). This is important because satisfactory working conditions can only be produced if there is a high level of acceptance by the users of an indoor space, such as an office. For example, various studies under laboratory and realistic conditions and in real objects showed that approval and comfort were always highest when users had the opportunity to actively influence the indoor climatic conditions themselves, such as through thermostats, individual adjustment

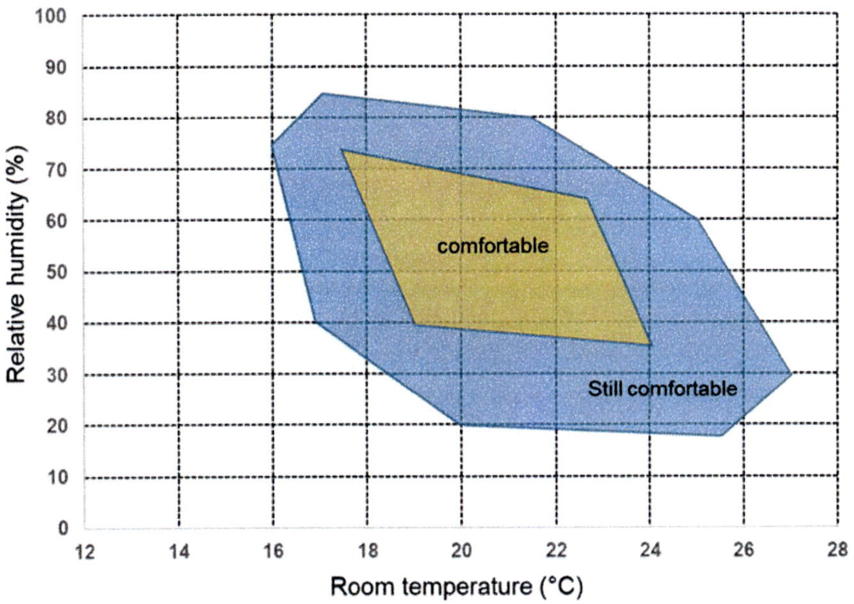

Fig. 1.1 Comfort diagram (according to Lenz 2005)

Table 1.2 Release of moisture in apartments (according to HMUEL 2012)

Activity	Moisture release
Potted plants	7–15 g/hour
Medium-sized rubber plant	10–20 g/hour
Drying laundry (4.5 kg spun laundry)	50–200 g/hour
Bath	approx. 1100 g/bath
Shower bath	approx. 1700 g/shower bath
Cooking	400,500 g/hour cooking time
Roast	approx. 600 g/hour cooking time
Dishwasher	approx. 200 g/wash
Washing machine	200–350 g/wash
Through humans	
Sleep	40–50 g/hour
Housework	approx. 90 g/hour
Strenuous activity	approx. 175 g/hour

options for ventilation, and windows that could be opened (Boerstra et al. 2015; Kwon et al. 2019).

Several recent reviews are available on topics such as thermal comfort (Kownacki et al. 2019; Malik et al. 2020), thermoregulation (Fu et al. 2016), indoor air movement (Zhu et al. 2015), and indoor humidity (Wolkoff 2018a, b).

1.2.2 Indoor Environment and Climate Change

In addition to the effects on environment, the human-induced climate change has various adverse direct and indirect effects on physical and mental health of humans globally, e.g. mediated through natural and human systems, including economic and social conditions and disruptions, extreme heat events with higher mortality and morbidity, increase of climate-related food-borne and water-borne diseases, increase of the incidence of vector-borne diseases, increase of animal and human diseases, including zoonoses in new regions, increase of the occurrence of diarrheal diseases due to increased rain and flooding, more trauma from weather and climate extreme events, loss of livelihoods and culture, and disruption of health services by extreme events such as floods (IPCC 2022).

Already in 2010, the Institute of Medicine (IOM) of the National Academy of Sciences established a committee on the effects of climate change on indoor air quality (IAQ) and public health (IOM 2011).

The IOM stated that climate change can affect IAQ and health in numerous ways:

- Changes in the outdoor concentrations of a pollutant due to alterations in atmospheric chemistry or other factors such as atmospheric circulation will affect indoor concentrations.
- Mitigation measures to reduce energy use in buildings could lead to systematically lower ventilation rates that would cause higher concentrations and exposures to indoor pollutants.
- Increased use of air conditioning, an expected adaptation measure, could exacerbate emissions of greenhouse gases and, if accompanied by reduced ventilation rates, increase the concentrations of pollutants emitted from indoor sources.

Three key findings derived from the committee's literature review underlie its conclusion that alterations in indoor environmental quality induced by climate change are an important public-health problem that deserves attention and action (IOM 2011):

- Poor indoor environmental quality is creating health problems today and impairs the ability of occupants to work and learn.
- There is inadequate evidence to determine whether an association exists between climate change-induced alterations in the indoor environment and any specific adverse health outcomes. However, available research indicates that climate change may make existing indoor environmental problems worse and introduce new problems by.
  - Altering the frequency or severity of adverse outdoor conditions that affect the indoor environment.
  - Creating outdoor conditions that are more hospitable to pests, infectious agents, and disease vectors that can penetrate the indoor environment.
  - Leading to mitigation or adaptation measures and changes in occupant behavior that cause or exacerbate harmful indoor environmental conditions.

– Opportunities exist to improve public health while mitigating or adapting to alterations in indoor environmental quality induced by climate change.

Moreover, Fisk (2015) and Vardoulakis et al. (2015) summarized the potential health consequences of changes in climate that affect indoor environments.

A particular problem is the transitions associated with global climate change. Increasingly occurring heat waves, which are also relevant for indoor environments, are of growing importance. For example, after the heat summer of 2003, increased mortality was demonstrated worldwide in various studies (Kohlhuber et al. 2006; Almeida et al. 2010; Benmarhnia et al. 2015; Arbuthnott and Hajat 2017; Murage et al. 2020; Mason et al. 2022). Following these heat waves, a corresponding heat warning system was developed and established in different countries as a forecast system. In addition, national and international behavioral recommendations have been published for vulnerable groups (e.g., elderly, persons with chronic diseases, young children) and in case of increased physical activity (Hajat et al. 2010; DWD 2020; NIHHIs 2022).

Since people, especially the less mobile and elderly, but also young children, spend most of their time indoors, knowledge about the climatic situation in a building is of great importance for assessing potential health risks during a hot spell (Walikewitz et al. 2018). For example, heat perception is strongly determined by the indoor air temperature in the afternoon (Franck et al. 2013). Compared to outdoor conditions, the indoor climatic situation is mainly dependent on the building fabric, the type of indoor use, and the ventilation (Pfafferott and Becker 2008). Estimates assume, taking into account the expected climatic development in Germany, that based on today's construction methods, an increase in indoor air temperature must be expected, which can quickly lead to overheating of indoor spaces during heat waves (Salthammer et al. 2018). Against this background, there is an urgent need to develop adaptation strategies that, on the one hand, reduce potential indoor health risks from climate change and, on the other hand, ensure that the impact of buildings on the climate is minimized.

1.3 Chemical Reactions in Indoor Air

Due to the very often high temporal and spatial variability of indoor conditions in terms of interior design, ventilation, UV light intensity, emission strength of pollutants, and exposure via outdoor air, complex chemical formation and transformation processes occur (Weschler and Carslaw 2018; Young et al. 2019; Abbatt and Wang 2020). Of particular importance is also the deposition of organic substances and particles on, for example, room surfaces, furnishings, clothing, and hair, as well as chemical reactions that subsequently take place (Milhem et al. 2020). Here, it became clear that chemical reactions of the gas and aerosol phase are also responsible for the new formation of the so-called secondary organic aerosols (SOA) and the modification of existing particles in indoor environments (summaries in

Weschler et al. 2006, Salthammer et al. 2002). Oxidants such as the hydroxyl and nitrate radicals, hydrogen peroxide, and ozone play a particular role in this process (Waring and Wells 2015). The most significant indoor source of ozone is the outdoor air, from where it is transported indoors by ventilation after photochemical formation in the atmosphere. Here, an indoor-to-outdoor ratio of about 0.1–0.8 is established (Zhang and Lioy 1994).

In particular, the formation of SOA by the reaction of ozone with monoterpenes and other unsaturated organic compounds, e.g., from detergents, indoor air fresheners, household products, and paints, has been observed in many experimental test chamber experiments (e.g., Wainman et al. 2000; Fan et al. 2003; Sarwar et al. 2004; Liu et al. 2004; Vartiainen et al. 2006; Destaillats et al. 2006; Aoki and Tanabe 2007; Waring et al. 2011; Wang et al. 2012; Lazaridis et al. 2015; Rösch et al. 2017; Pytel et al. 2020). A review, including potential health effects of the resulting downstream products, is provided by Rohr (2013). Recent test chamber experiments describe that various glycol ethers have the potential to form SOA (Li and Cocker 2018), and also benzo[*a*]pyrene can be converted by ozone into the corresponding—highly reactive—epoxides or epoxy diols (Zhou et al. 2017). During these reactions, not only new particle formation but also the formation of carbonyl compounds, such as formaldehyde and acetone, different organic acids, nitrogen compounds, hydrogen peroxide, and different radicals, such as the hydroxyl radical, were observed. Further reactions of the hydroxyl radical or ozone with other volatile organic compounds indoors (e.g., toluene, xylenes) have also been described (Shaughnessy et al. 2001; Fan et al. 2003; Wang et al. 2020). It is assumed that in the simultaneous formation of, e.g., hydrogen peroxide and the hygroscopic SOA is a mechanism that allows not only particle-bound H_2O_2, but also other irritant substances, to penetrate deeper into the respiratory tract without already reacting at the peripheral airway surfaces (Weschler et al. 2006). In general, increased particle growth appears to occur at higher relative humidity (Wainman et al. 2000). In addition, it has been observed that particle growth occurs even in indoor environments where no volatile precursors are present in the indoor air (Shaughnessy et al. 2001). This is explained by the reaction of ozone with unsaturated substances in indoor furnishing materials, such as linoleum coverings and carpets (Morrison and Nazaroff 2002).

Weschler and Shields (1999) first investigated SOA by exposing one of the two office rooms to different terpenes at a set ozone concentration (<5 to 45 ppb). Under realistic ozone and limonene concentrations, there was a mean mass increase of 2.5–5.5 $\mu g/m^3$ in a room with limonene compared to the control room. At the same time, an increase in particle number concentration by a factor of 10 was observed. The ozone concentration was the limiting factor in the formation of SOA in these experiments. The results of the aforementioned experiment are shown graphically in Fig. 1.2.

These results are confirmed by the investigation of apartments (Long et al. 2000; Hubbard et al. 2005) or test rooms furnished similar to living rooms (Singer et al. 2006; Rösch et al. 2017). While the Long et al. (2000) study determined the SOA effect after application of a pine oil-containing cleaner and an air freshener, Hubbard et al. (2005) examined the effect of ozone-forming air fresheners with simultaneous

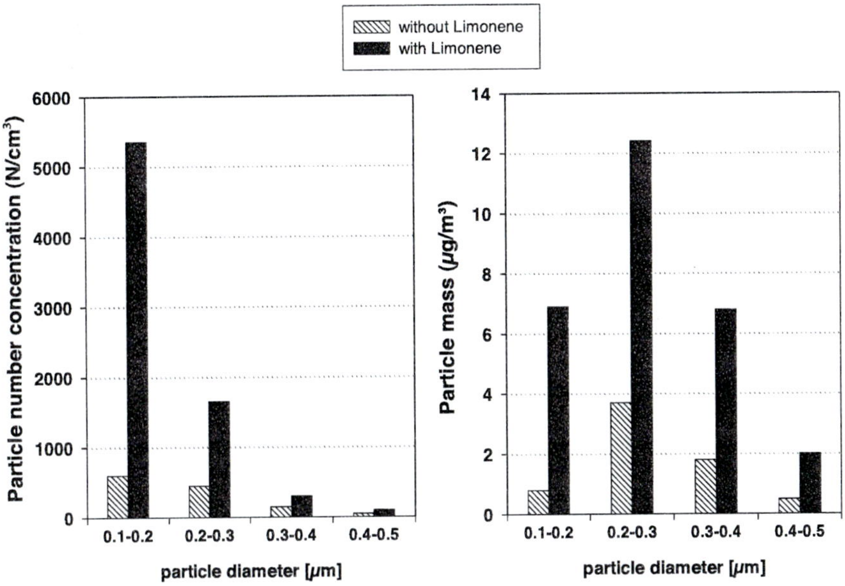

Fig. 1.2 Comparison of particle concentrations (left: number; right: mass, calculated from particle number concentrations in each size range) in an office with a limonene source and a room without (after Weschler and Shields 1999, Weschler 2003)

application of terpene-containing cleaners in three residential rooms in Texas. This study concludes that in the presence of ozone and terpenes, the number of particles in all size classes increases, and a shift from small to larger particle diameters is observed. Waring (2014), through modeling using measurement results from 143 California indoor environments, concluded that SOA account for a median of 6% of the total $PM_{2.5}$ in the room. Not only at higher ozone concentrations, but also with poor ventilation and higher occupant numbers such as in classrooms, the fraction can be much larger (Avery et al. 2019).

1.4 Humans as a Source

In addition to the complex interaction between—periodic—human activities such as smoking, cleaning, and cooking and indoor pollution load, the presence of man himself can also be a direct source. Household overcrowding, often a marker of poverty and social deprivation, on the one hand, increases exposure to risk factors and, on the other hand, leads to an aggravation of the indoor exposure situation (WHO 2018).

According to NAS (2022) typical types of direct human sources are:

- Substances excreted from skin (e.g., ammonia).
- Substances exhaled with breath (e.g., carbon dioxide, VOC).
- Desquamation of skin flakes and hair and their transfer onto room surfaces and dust.
- Skin and clothing-related biological emissions (e.g., viruses, bacteria).

This section presents the significance of skin or sebum and exhaled air emissions and their reactions with other reactive substances in indoor air. Sebum is a light yellow, lipid-rich fluid produced by the sebaceous glands and distributed on the skin surface. Although not fully understood, sebum is important for maintaining skin homeostasis, lubrication, thermoregulation, and protection from environmental stressors, pathogens, and contaminants by producing either proinflammatory or anti-inflammatory cytokines, chemokines, interleukins, pheromones, free fatty acids, and hormones (Géhin et al. 2023). Sebum can be found all over the body except for the palms of the hands and soles of the feet and the largest number of glands and the most sebum-rich parts of the body are the face, back, and upper chest (Géhin et al. 2023). Depending on the sampling techniques, the relative composition of sebum is 30–50% triacylglycerols/diacylglycerols, 15–30% free fatty acids, 12–20% squalene, 26–30% wax esters, 3–6% cholesterol esters, and 1.5–2.5% cholesterol (Picardo et al. 2009). Figure 1.3 shows the chemical structures of the main components of sebum, where squalene and wax esters are the very typical substances. Due to its composition, sebum, as the outer boundary layer of the body to the environment, can be a trap for lipophilic environmental contaminants.

The very reactive squalene of skin and clothing is able to react with indoor ozone and produces various compounds like aldehydes, ketones, carboxylic acids, and dicarbonyl species (Coffaro and Weisel 2022). First, Fruekilde et al. (1998) studied experimentally the reaction of foliage and epicuticular waxes and ozone and identified several emitted volatile compounds like 4-oxopentanal (4-OPA), 6-methyl-5-hepten-2-one (6-MHO), geranylacetone, and acetone. Further on, Wisthaler and Weschler (2010) performed different experiments as well as one with human subjects in a simulated office room under realistic ozone mixing ratios and found the following major volatile primary products produced from ozone reacting with squalene: acetone, 6-MHO, geranyl acetone. Secondary products derived after further chemical reactions were 4-OPA, 1,4-butanedial (succinic dialdehyde), 4-methyl-8-oxo-4-nonenal (4-MON), and 4-methyl-4-octene-1,8-dial (4-MOD). Overall they concluded that human occupant removes 10–25% of the ozone in the room and this process is matched by emissions into the air of volatile oxidation products originating from the skin. Additionally, Weschler (2016) gave an overview of the role of human occupants in indoor chemistry and summarized the different processes which can be observed in indoor air. Besides the aforementioned squalene ozone reaction he described the reaction of unsaturated fatty acids like cis-hexadec-6-enoic acid and cis-octadec-8-enoic acid of the sebum with ozone and the subsequent formation of decanal.

Breathing releases hundreds of substances, e.g., CO_2 (respiration), NO (catalyzed by nitric-oxide synthases; involved in vasodilation or neurotransmission), NH_3

Squalene

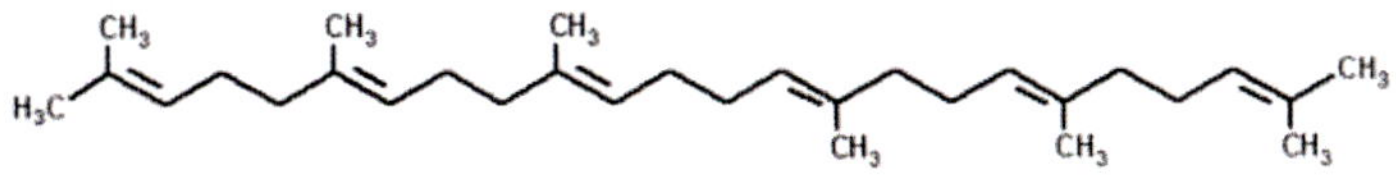

1-Palmitoyl-2-oleoyl-3-linoleoyl-rac-glycerol (as a triacylglycerol)

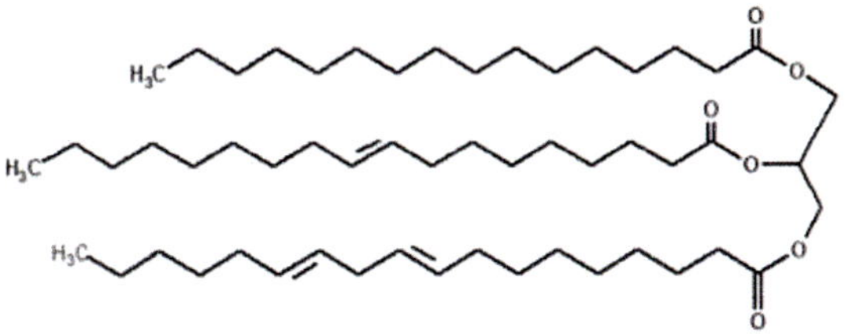

Lauryl palmitoleate (as a wax ester)

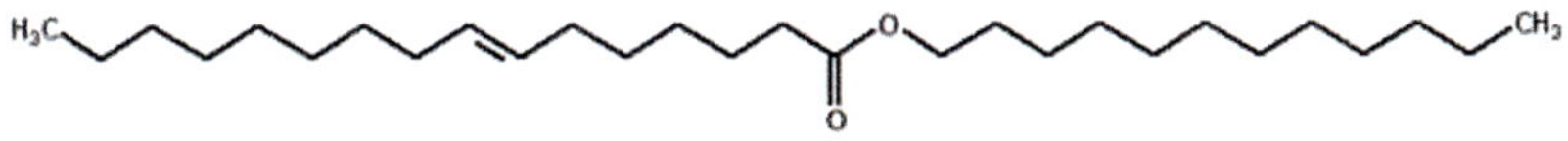

Linoleic acid (as a fatty acid)

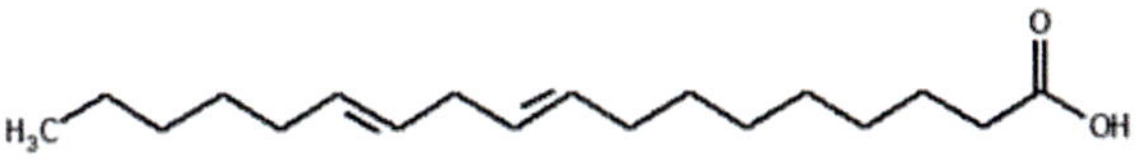

Fig. 1.3 Representatives of typical substance classes in sebum (modified from Géhin et al. 2023)

(protein metabolism), CH_4 (gut metabolism of carbohydrates), H_2 (gut bacterial metabolism of carbohydrates), H_2S (bacterial metabolism of thiol proteins), as well as hundreds of VOC (Agapiou et al. 2015). An overview of a variety of VOC emitted during exhalation and emitted by skin is available (Ruzsanyi et al. 2012; Mochalski et al. 2015; Roodt et al. 2018; Drabińska et al. 2021). Another field of work is the determination of a large number of volatile substances and their patterns in exhaled air in order to enable statements on diseases or the development of diseases. Recently, Issitt et al. (2022) summarized in a meta-analysis several previous reviews for lung, breast, and other cancers, irritable bowel diseases, diabetes, and liver disease and concluded that application of functional groups to biomarker discovery is helpful to better define the process of disease than single volatile markers. In another review Khan et al. (2023) presented an overview of cancer biomarkers in breath and other body fluids and of the analytical possibilities like biosensors.

Two studies determined overall VOC emission rates in cinema audiences (50–230 people per movie and around 83,000 in total) and in classrooms (volume = 670 m^3, air-exchange rate = 5 per hour, at least 17 occupants) and reported emission rates for children of 333 and 1060 µg/h/person for acetone, 357 and 329 µg/h/person for acetic acid, 189 and 187 µg/h/person for monoterpenes, and 252 and 114 µg/h/person for acetaldehyde (Stönner et al. 2018, Tang et al. 2016). In setting cinema the values are higher for adults with 419 µg/h/person for acetone, 201 µg/h/person for monoterpenes, and 221 µg/h/person for acetaldehyde, while lower levels of 205 µg/h/person were found for acetic acid.

1.5 Key Points for the Sampling of Indoor Pollutants

1.5.1 Volatile Organic Compounds (VOC)

Two sampling techniques, active and passive sampling, are used for the determination of VOC in indoor air. Sampling strategies and analytical methods are summarized in the ISO 16000 series, which applies to indoor environments such as dwellings, workrooms, or workplaces in buildings that are not subject to health and safety inspections regarding air pollutants (for example, offices, sales premises), public buildings (for example, hospitals, schools, kindergartens, sports halls, libraries, restaurants and bars, theaters, cinemas, and other function rooms), and also cabins of vehicles.

Passive samplers are systems that use a diffusion gradient to allow pollutants to diffuse from the sampled environmental medium into the collection phase of an enrichment medium over a longer period. Therefore, no pumping system or a power supply is needed. The following equation is used to calculate concentration (WHO 2020):

$$C\ (\mu g/m^3) = \frac{m\ (\mu g)}{Q\ \left(\frac{ml}{min}\right) x\ t(\min)}$$

where C is the concentration in indoor air, m is the mass of the pollutant, t is the sampling time, and Q is the sampling rate (known as diffusive uptake rate) that depends on the diffusion coefficient of the target pollutant, which is directly associated with the cross-sectional area of the opening of the sampler and inversely associated with the length of the diffusion zone of the device.

In active sampling, the air is aspirated by a sampling pump and the analyte is adsorbed on a suitable collection medium (e.g., activated carbon, silica gel, XAD, Tenax, Chromosorb). Here, the exact determination of the collected air volume is particularly important. Short-term measurements are possible with the active sampling technique. Personal or stationery collection systems can be used.

Subsequently, the VOC are desorbed from the collection medium in the laboratory by solvents or by thermodesorption and usually analyzed by gas chromatography (GC) or high-performance liquid chromatography (HPLC).

Indoor air measurements aim to obtain a comprehensive and, as far as possible, representative description of an exposure situation, which can subsequently be used as the basis for an assessment of indoor air quality. When performing measurements in indoor air the following basics always be ensured or required (Ad-hoc-AG 2014):

- The analysis is only carried out with a clear question and follows a defined measurement plan:
  - Why should be measured (reason)?
  - What is to be measured (substance/substance group)?
  - When should be measured (duration of sampling, sampling period)?
  - Where should measurements be taken (sampler location)?
  - How to measure (suitable analytical method)?
- The laboratory must have an established quality management (QM) system.
- The boundary conditions like activities in the room during sampling or climate parameters like temperature or ventilation are comprehensively documented.
- The final investigation report contains not only the measurement results, but also all information relevant for an evaluation and, if possible, an assessment.
- Other information important to the final evaluation, such as building characteristics, will be collected.

Two sampling techniques can be differentiated. Personal air sampling is a good way to determine the exposure of an individual by sampling with a portable air sampler near the breathing zone of the subject. Stationary indoor measurements should be taken at a height of 1.0–1.5 m above the floor at the reaching height of the room occupants and a distance of at least 1–2 m from walls. It may be necessary to investigate several sampling points in the room and usually also in a comparison room and/or in outdoor air.

Additionally, two measurement strategies can be differentiated:

- If an exposure and health assessment, e.g. comparison with reference values, is strived, sampling under conditions of use makes sense. In Germany for example, the following strategy is recommended (Ad-hoc-AG 2014): after intensive ventilation of approx. 3 minutes in winter and up to 10 minutes in summer, the windows and doors are closed and the measurement is carried out after one hour in the still closed room. If—as in schools—shorter usage cycles are present, the measurement strategy should be adapted. When examining rooms with a ventilation and air-conditioning system, the system must be operated for at least 3 hours under the usual operating conditions for the room before sampling.
- The second measurement strategy, which is intended in particular to provide additional information on sources in the room, is measurement under worst-case conditions. For this purpose, the room must first be thoroughly ventilated and

then kept closed for at least 8 hours (usually overnight). Sampling is then carried out in the unventilated room.

An overview of essential quality assurance and quality control (QA/QC) procedures is given by WHO (2020) and general requirements for the competence of testing laboratories by ISO/IEC 17025 (ISO/IEC 2017).

1.5.2 Semi-volatile Organic Compound (SVOC)

SVOC are present in the atmosphere in an equilibrium between particle and gas phases, which has a direct influence on the type of sampling and analysis. For example, artifacts due to adsorption and desorption processes must be taken into account during collection. Using passive sampler like polyurethane foam (PUF) disks some drawbacks have to be considered: calibration using a reference method, variability in uptake rates, and uncertainty in the ability to collect particle-associated chemicals (WHO 2020). Due to the limitations of passive sampling methods, SVOC like PAH were measured using, for example, low-volume sampler (LVS) active samplers that collected the air on a filter (e.g., quartz-fiber filter) which is often combined with a PUF disk. Usually, a size-selective inlet collecting PM_{10} or $PM_{2.5}$ particles was used.

1.5.3 Particulate Matter

Particulate airborne substances, when dispersed in the air, form colloidal systems with gases, also known as aerosols. The term includes both "solid" (dust) and "liquid" particles (mist). The environmental aerosol contains very different particles, which can vary considerably, especially in size, shape, and chemical composition. While the larger particles are predominantly responsible for the mass of the environmental aerosol, the number of particles and also the particle surface is almost exclusively determined by the ultrafine particles (UFP) or nanoparticles. This requires different measurement methods, which have been summarized, for example, in the review by Lowther et al. (2019). Different systems are used for sampling or measuring the particle mass, in which, for example, the air is sucked in using a volume-controlled pump and deposited on a filter according to the impaction principle, and the particle mass is then determined gravimetrically by differential weighing. Depending on the choice of sampling head with a size-selective inlet, different PM (Particulate Matter) fractions of the environmental aerosol (e.g., PM_{10}, PM_4, and $PM_{2.5}$) can be separated. Another principle would be the light scattering of particles, which is used as a method in continuously operating optical particle counters (so-called aerosol spectrometers). In this case, the determined concentration

Table 1.3 Definitions of mass-related particle fractions

Term	Definition
Common definitions in occupational health and safety	
Inhalable fraction	The mass fraction of all suspended particles inhaled through the mouth and nose
Extrathoracic fraction	The mass fraction of inhaled particles that does not penetrate beyond the larynx (E-dust)
Thoracic fraction	The mass fraction of inhaled particles that penetrates beyond the larynx (A-dust) (particles up to a maximum of approx. 50 μm[a]).
Alveolar fraction	The mass fraction of inhaled particles that penetrates into the non-ciliated airways (corresponds approximately to PM_4; particles up to a maximum of approx. 16 μm[a]).
Common definitions in environmental protection and health protection	
TSP	Total suspended particulates; total suspended particulate matter up to a particle diameter of approx. 70 μm[a].
PM_{10}	Particle fraction passing through a size-selective air inlet that has a separation efficiency of 50% for an aerodynamic diameter of 10 μm[a] (particles up to a maximum of approx. 20 μm[a]).
PM_4	Corresponds approximately to the alveolar fraction and the RPM
$PM_{2.5}$	Particle fraction passing through a size-selective air inlet that has a separation efficiency of 50% for an aerodynamic diameter of 2.5 μm[a] (particles up to a maximum of approx. 5–6 μm[a])

[a]Specified as aerodynamic diameter. This is the geometric diameter relative to a sphere of density 1 g/cm^3 having the same rate of descent in still or laminar flowing air

of the particles, the so-called number concentration, is converted into the particle mass by a mathematical algorithm.

In Table 1.3 as well as Fig. 1.4, diameter-related definitions of common dust fractions of occupational and environmental protection and the respective size distributions are compiled. Thus, the assessment of dust loads at workplaces is carried out according to the conventions of DIN EN 481, which distinguishes between two dust fractions with general dust limit values: E-dust and A-dust (DIN EN 1993). E-dust or inhalable fraction is defined as the mass fraction of all suspended particles up to 100 μm aerodynamic diameter that can be inhaled through the mouth and nose. Aerodynamic diameter is defined as the diameter of a sphere with the normalized density of 1 g/m^3 that has the same sinking velocity in still or laminar flowing air as the particles themselves. While particles below 5 μm are almost completely inhaled, inhalability decreases toward larger particles (non-inhalable fraction). E-dust can be subdivided into further dust fractions (e.g., thoracic or tracheobronchial fraction), depending on where it is deposited in the respiratory tract. A-dust or alveolar fraction is the part of the inhalable airborne dust that can penetrate into the alveoli.

The definition of the two dust fractions relevant to environmental protection is based on the EU Directive on ambient air quality and cleaner air in Europe (EU 2008). There, the conventions for the separation characteristics of $PM_{2.5}$ and PM_{10}, respectively, are established by specifying a defined separation efficiency of

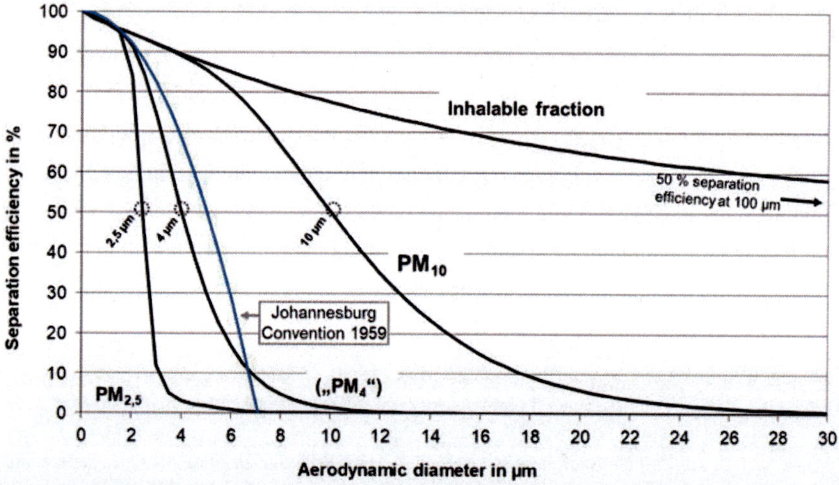

Fig. 1.4 Conventions of defined dust fractions according to DIN EN 481 (from Mattenklott and Höfert 2009)

50% of the particles at a certain aerodynamic diameter of 2.5 and 10 μm, respectively. The PM_{10} fraction essentially corresponds to the convention given in DIN EN 481 for the thoracic dust fraction.

In the context of the measurement of UFP or nanoparticles, the number concentration of particles is determined using the so-called condensation nuclei counters (CNC/CPC). Since the light absorption, which serves as a measurand, decreases strongly with the particle diameter, the very small particles are enlarged artificially (e.g., by the addition of butanol). To determine the particle size distribution, an electrical classifier (DMA, Differential Mobility Analyzer) can be connected upstream of the condensation nucleus counter. In this process, the particles contained in the sample gas stream are separated according to their electrical mobility following electrical charging. A combination of condensation core counter and electrical classifier is also called DMPS (Differential Mobility Particle Sizer) or SMPS (Scanning Mobility Particle Sizer).

1.5.4 New Passive Sampler Devices

In recent years, more and more passive sampler devices (PSD) have been developed, which allow the use in the context of extensive exposure studies, e.g. epidemiological studies or for the description and quantification of the human exposome. It is about wristband technology, that are now commercially available and have the general advantages of portability, low cost, easy installation and which are not perceived by the subjects to be disruptive in their lives. Therefore, they are particularly suitable for determining personal exposure, also in long-term studies.

Special requirements must be met by the sorptive material, which should have a linear (as possible) uptake, high capacity, and reproducible sampling behavior under typical deployment conditions (Fuentes et al. 2022). Wacławik et al. (2022) summarized the literature especially in relation to pre-analytical and analytical steps and comparison with other approaches.

Two comprehensive reviews are now available. First, Samon et al. (2022) provide a deep overview of silicon wristbands and compare it with other common exposure assessment tools, including biomarkers of exposure measured in biospecimens, and evaluate their utility in exposure assessments and epidemiological studies. Second, Fuentes et al. (2022) summarized under which conditions and quality measures (e.g., sampling and analysis) wristbands can support exposome studies. Overall, 47 studies using silicone wristbands to assess personal exposure and nine studies to assess personal exposure in conjunction with evaluation for health biomarkers were given by Samon et al. (2022). In the studies presented, not only numerous SVOC like PAH, PCB, PBDE, and other brominated flame retardants, organophosphate esters, pesticides, but also different VOC were measured.

Besides silicon wristband, however, other wearable passive sampler devices are also described. For example, Deng et al. (2022) used a combination of high-density polyethylene (HDPE) and thermoplastic polyurethane (TPU) as a sorbent material to collect PFAS. First, Lin et al. (2020) have developed a so-called fresh air wristband consisted of a commercially available triethanolamine-coated pad (looking like a watch) to collect NO_2 and a polydimethylsiloxane (PDMS) sorbent bar to sample VOC and SVOC. For example, DeLay et al. (2022) used this method in the comprehensive Venda Health Examination of Mothers, Babies, and their Environment (VHEMBE) birth cohort study in South Africa to quantify the exposure to PAH, organophosphate esters flame retardants, and pesticides.

- Samon et al. (2022) provide some helpful basic recommendations for the applications of silicone wristbands. Here in brief:
- Wristbands should be cleaned or conditioned predeployment then stored in airtight containers both before and after a participant has worn it for the sampling period.
- Quality control samples should be used for every step of the wristband deployment (e.g., predeployment cleaning blank, travel blank, extraction blank).
- Deployment of a silicone wristband has lasted anywhere between four hours and 30 days.
- Safe transport for up to seven days for some VOC and up to 30 days for SVOC is possible, but it is generally recommended to store wristbands at −20 °C.
- Chemical stability has been evaluated with little degradation in storage at −20 °C for up to 6 months among SVOC and up to 3 months for VOC.
- Wristbands have been used to evaluate some VOC exposures; however, researchers should keep in mind that more volatile compounds will likely desorb more quickly from the silicone and thus may not represent a completely accurate representation of personal exposures to VOC.

1.6 Measures to Improve Indoor Air Quality

To ensure good indoor air quality, Nazaroff (2013) formulated four basic principles:

- Prevention or minimization of uncontrolled emissions of indoors: Due to the variety of sources indoors, this is not always possible, but a more consistent avoidance of products (e.g., paints, cleaning agents, floor coverings) that could emit high levels of VOC into the indoor air can make a significant contribution to improving quality. From a precautionary point of view, products or materials with the Blue Angel, an environmental label in Germany, should be selected consistently. The necessity of the application should also always be critically questioned and application quantities reduced as far as possible. Furthermore, unnecessary indoor air pollution, such as that caused by smoking activities, must be avoided in any case.
- Avoiding moisture problems: Moisture problems indoors, even if the causal relationship is not fully understood, represent a health problem, for example, via mold. In addition to adequate ventilation, adapted to our behavior and activities, structural aspects in particular must be taken into account.
- Good ventilation: Often the simplest and most effective option is regular, adequate ventilation of indoor spaces, especially to prevent the accumulation of unavoidable emissions.
- Protection from outdoor air pollution: The importance of outdoor air as a possible source should not be underestimated. Particularly in the case of natural ventilation, noxious substances from the outside air, such as fine dust from motorized road traffic and also biological stressors, such as pollen, can enter indoor spaces. The most effective strategies in this case are, of course, often measures at the sources, e.g. through reduction techniques on motor vehicles and changes in our mobility, especially in urban living areas.

1.6.1 Ventilation

Ventilation is intended to bring outside air into buildings or rooms and distribute it there, as well as dilute substances in the indoor air or remove them to the outside. At the same time, it is intended to ensure well-being, comfort, and health in indoor spaces. There is a wide scientific literature on all aspects related to ventilation to which reference must be made at this point. This chapter can therefore only present the first keypoints. Several review papers have summarized results on the influence of ventilation on health effects (Seppänen et al. 1999; Wargocki et al. 2002; Sundell et al. 2011; Carrer et al. 2015; Fisk 2018).

Ventilation can be described in particular by the following factors (WHO 2009):

- Ventilation rate, which indicates the complete air exchange in the room per hour.
- Direction of air flow or flow through a room/building.
- Air distribution in the room or building.

Basically, a distinction must be made between natural ventilation (free ventilation), mechanical ventilation and hybrid ventilation (WHO 2009). In natural ventilation, air is transported from the outside to the inside depending on the climate, the temperature difference between inside and outside, the building conditions and the behavior of the users, as well as external conditions such as wind pressure and thermal buoyancy forces. In contrast, the term mechanical ventilation systems covers variously designed ventilation and air-conditioning systems that can be used centrally or decentrally for ventilation, temperature and humidity control, and air purification. According to DIN EN 16798, this is a combination of all the components required to provide a form of air treatment in which temperature is controlled, and possibly in conjunction with the control of ventilation, humidity and air purity (DIN EN 2017). This standard sets out principles for non-residential buildings, while DIN 1946–6:2019 sets out regulations for residential buildings (DIN 2019). In hybrid ventilation, mechanical ventilation systems are only used when natural ventilation is not sufficient.

Air Change Rate per hour (ACH), the air volume replaced within a defined space by ventilation and infiltration, is an important contributor to indoor air quality. Nazaroff (2021) performed a critical review of 40 studies and discuss in deep the various factors influencing ACH. He found a good estimate for the distribution is a geometric mean of 0.5 per hour and corresponding 10th, 25th, 50th, 75th, and 90th percentiles in such a distribution of 0.21, 0.31, 0.5, 0.80, and 1.2 per h, respectively.

Interesting starting points for linking ventilation requirements and comfort perception more closely could be seen in more personalized measures that enable both sufficient supply air and exhaust air close to the breathing zone in a targeted manner and, if necessary, adjustable by the user (Melikov 2016, 2020).

1.6.2 Air Purification

Although air purification technologies alone cannot ensure adequate indoor air quality, their use could be helpful in individual cases (Kelly and Fussell 2019). Various technical systems for cleaning the air from fine dust (particles) and volatile organic substances are available on the market (overview in: Luengas et al. 2015; Huang et al. 2016; US EPA 2018; González-Martín et al. 2021; Szczotko et al. 2022; Mata et al. 2022):

- Fibrous air filter media:

  Fibrous media air filters remove particles by capturing them onto fibrous filter materials and particle removal efficiency depends on a number of parameters including particle size, face velocity, filter thickness, filter porosity, filter fiber dimensions, dust loading conditions, and whether or not the media are modified

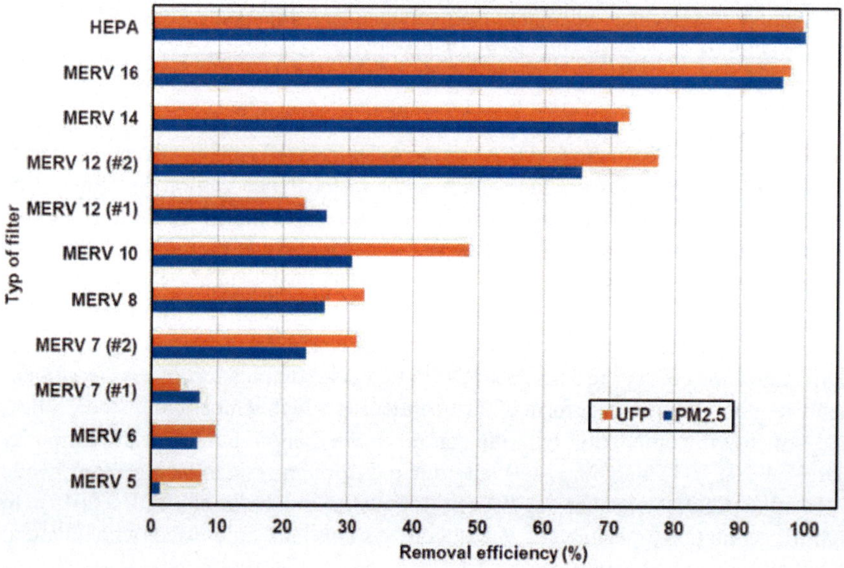

Fig. 1.5 Removal efficiency of ultrafine particles (UFP) and $PM_{2.5}$ for different types of filters (according to Azimi et al. 2014)

by the manufacturer to initially have an electrostatic charge on the fibers US EPA (2018). Due to different requirements for specific applications types of filter with different efficiencies are available on the market. The American Society of Heating, Refrigerating, and Air-Conditioning Engineers and the International Organization for Standardization have issued appropriate standards for evaluating filters (ASHRAE 2007, ISO 2016). Minimum Efficiency Reporting Value (MERV) ranging from MERV 1 to MERV 16 what means that the higher the MERV rating, the higher a filter's removal efficiency. The so-called HEPA (high efficiency particulate air) filters in particular have been used extensively for a long time and (depending on the filter class) also separate particles <0.3 μm in diameter, as well as bacteria, fungal and mold spores, with high efficiency. Azimi et al. (2014) estimated filter removal efficiency for UFP and $PM_{2.5}$ of outdoor origin using 194 outdoor particle size distributions and additionally using an infiltration factor to consider outdoor–indoor transport. They reported that although the MERV metric does not explicitly account for UFP or $PM_{2.5}$ removal efficiency, both tend to increase in efficiency with increasing MERV and are very high for HEPA filters (see Fig. 1.5). Moreover, outdoor particle size distributions characteristics and assumptions for particle density and size-resolved infiltration factors do not drastically impact estimates of filter removal efficiencies for $PM_{2.5}$, although the impacts are larger for UFP.

Many fibrous filter systems have U-shaped collection efficiencies, with high rates for particles <0.03 and > 3.0 μm. The filter media must be continuously renewed.

- Electronic filter systems: In particular, electrostatic precipitators (ESP) are used here, in which particles are charged in an electric field and deposited on an oppositely charged plate. Similarly, the so-called ion generators charge the air molecules and produce negative ions, whereby the particles are then deposited either on an oppositely charged plate or on other surfaces of the room.
- The third group is the so-called Ultraviolet Germicidal Irradiation systems (UVGI). In these, UV lamps are used to destroy or inactivate viruses, bacteria, fungal spores, and their fragments.
- For the removal of gaseous substances, methods such as absorption systems, photocatalytic oxidation, UV photolysis, ozonation, phytoremediation, and membrane separation have been described (Luengas et al. 2015; US EPA 2018). Absorption systems utilize the adsorptive properties of activated carbon, zeolites, silica gel, activated aluminum, mineral clay, or polymers. Such systems need to be continuously renewed. Their efficiency depends, among other things, on the substance. In photocatalytic oxidation, gaseous substances adsorb on surfaces coated with titanium dioxide, for example, and are then degraded by UV radiation (Costa et al. 2012; Sánchez et al. 2012; Tsang et al. 2019; Malayeri et al. 2019). This technique is not always effective in converting organic matter and unwanted by-products may also be produced (US EPA 2018). UV photolysis and ozonation aim to degrade microorganisms and gaseous substances by UV radiation and ozone, respectively (Wang et al. 2019).

Phytoremediation refers to techniques originating from soil remediation to achieve degradation of substances with the help of plants, for example (Pettit et al. 2018; Han and Ruan 2020). However, results to date in the field of indoor air purification are contradictory and the actual effectiveness of air purifiers is unknown under real conditions and needs further research (Thatcher et al. 2020; Szczotko et al. 2022; Matheson et al. 2023).

A review by the US Environmental Protection Agency saw significant reductions in PM_{10}, $PM_{2.5}$ and, in some cases, particle number concentrations in indoor air in 19 studies based on 20 intervention studies with filter systems (US EPA 2018). In contrast, only one of the three studies observed a significant decrease in allergen levels in air and dust. These results are also supported by other recent studies (Shao et al. 2017; Zhan et al. 2018; Cox et al. 2018; Brehmer et al. 2020; Sharma and Balasubramanian 2020; Park et al. 2021). Even though the aforementioned intervention studies showed an association between the use of air purifiers and individual health parameters, the US EPA estimates that the overall effects on health were rather limited. In addition, multiple interventions were often implemented simultaneously. Studies on the reduction of gaseous substances are limited in number and do not show consistent results (US EPA 2018). Studies with the human health endpoint in this case have not yet been published.

It must be seen as problematic that technologies such as ESP, ion generators, and UVGI in particular can form ozone or other substances of concern as a by-product and release it into indoor air (Waring et al. 2008; Jakober and Phillips 2008; Siegel 2016). There is also evidence that HEPA filters in portable cleaning equipment can be a source of endotoxins in indoor air (Niu et al. 2020). Other critical considerations are the costs of installation, maintenance, and possibly energy use. Many studies are also limited because they only target one pollutant or group of pollutants and attempt to measure its reduction. For example, Choe et al. (2022) measured continuously PM and CO_2 on weekdays over six months in 32 classrooms equipped with an air purifier with integrated high-performance filters in Korea in 2019. The air purifiers installed in the classroom reduced the indoor PM concentration by approximately 30%, however, as was to be expected, ventilation was still insufficient in most schools considering that the CO_2 concentration exceeded 1000 ppm.

1.7 Important Information Options on the Internet

The Internet has an outstanding importance as a source of information and thus a lot of data can be found there, which can be helpful for the assessment of indoor pollution. In the following sections, some links to general data sources are compiled as examples.

1.7.1 Toxicological Fact Databases

Hazardous substances database of the federal states (GdL): Basic information on harmful substances (http://www.gefahrstoff-info.de/).

ChemInfo/Information System on Chemicals/Common Substance Data Pool of the Federal Government and the Länder (GSBL): Basic information on pollutants (http://www.gsbl.de/).

Information System on Hazardous Substances (IGS): Basic Information on Pollutants (http://igsvtu.lua.nrw.de/igs_portal/index.htm).

Hazardous Substances Information System (GESTIS): Data bank of the Institute for Occupational Safety and Health of the German Social Accident Insurance (https://www.dguv.de/ifa/gestis/gestis-stoffdatenbank/index.jsp).

Hazardous substance information system chemical alkalis (GisChem) of the professional associations RCI and LOQHM: data sheets with basic information on hazardous substances (https://gischem.de/index.htm).

Hazardous Substances Information System of the Employer's Liability Insurance Association for the Construction Industry (WinGIS): Data sheets with basic information on hazardous substances in the construction industry (http://www.wingis-online.de/wingisonline/).

International Programme on Chemical Safety (INCHEM): Various World Health Organization and Food and Agriculture Organization publication series on chemicals, e.g. Environmental Health Criteria, Concise International Chemical Assessment Documents, Health and Safety Guides, International Chemical Safety Cards (https://www.who.int/ipcs/assessment/en/).

ECHA (European Chemicals Agency): Information platform for chemicals; classifications, labels, substance dossiers, etc. (https://echa.europa.eu/de/information-on-chemicals).

eChemPortsal of the OECD (Organization for Economic Co-operation and Development): Search option on, e.g., environmental behavior, chemistry and toxicology of substances (https://www.echemportal.org/echemportal/).

United Nations Environment Program (UNEP) Screening Information Datasets for Chemicals (SIDS): Compilations of physicochemical and toxicological data on various substances (http://www.chem.unep.ch/irptc/sids/OECDSIDS/sidspub.html).

INERIS: Chemical Substances Portal Toxicological Database of the Institut National de l'Environnement Industriel et des Risques, France (https://substances.ineris.fr/fr/).

eChemPortal: Metasearch in various European and international databases (http://webnet3.oecd.org/echemportal/).

National Library of Medicine (NLM) databases: Various databases, such as Hazardous Substances Database (HSDB) with metasearch via Toxnet (https://www.nlm.nih.gov/enviro/index.html).

Integrated Risk Information System (IRIS): Database of the US Environmental Protection Agency (US EPA): Comprehensive Substance Database (https://www.epa.gov/iris).

1.7.2 *Literature Search in Databases*

PubMed: query interface for Medline (https://pubmed.ncbi.nlm.nih.gov/). Especially important for literature in medicine and biological sciences.

Google Scholar: search in scientific journals and various internet sources (http://scholar.google.de/).

Web of Science: costly bibliographic search interface. Free access for employees of scientific institutions (https://clarivate.com/products/scientific-and-academic-research/research-discovery-and-workflow-solutions/webofscience-platform/).

Scopus: Another commercial bibliographic search platform (https://www.scopus.com/home.uri).

References

Abbatt JPD, Wang C (2020) The atmospheric chemistry of indoor environments. Environ Sci Processes Impacts 22:25

Ad-hoc-AG (2014) Determination and assessment of chemical contamination of the air of indoor workplaces (without activity with hazardous substances). Joint communication of the Working Group on Air Analyses of the Permanent Senate Commission for the Examination of Harmful Working Substances of the German Research Foundation and the German Working Group on Indoor Air Guidelines. Bundesgesundheitsbl 57:1002–1018. [in German]

Agapiou A, Amann A, Mochalski P, Statheropoulos M, Thomas CLP (2015) Trace detection of endogenous human volatile organic compounds for search, rescue and emergency applications. Trends Anal Chem 66:158–175

Almeida SP, Casimiro E, Calheiros J (2010) Effects of apparent temperature on daily mortality in Lisbon and Oporto, Portugal. Environ Health 9:12

AlWaer H, Speedie J, Cooper I (2021) Unhealthy neighbourhood "syndrome": a useful label for analysing and providing advice on urban design decision-making? Sustainability 13:6232

Aoki T, Tanabe S (2007) Generation of sub-micron particles and secondary pollutants from building materials by ozone reaction. Atmos Environ 41:3139–3150

Arbuthnott KG, Hajat S (2017) The health effects of hotter summers and heat waves in the population of the United Kingdom: a review of the evidence. Environ Health 16(Suppl 1):119

ASHRAE (American Society of Heating, Refrigerating and Air-Conditioning Engineers) (2007) Method of testing general ventilation air-cleaning devices for removal efficiency by particle size. ASHRAE Standard 52.2–2007. Atlanta, USA

Avery AM, Waring MS, DeCarlo PF (2019) Human occupant contribution to secondary aerosol mass in the indoor environment. Environ Sci Process Impacts 21:1301–1312

Azimi P, Zhao D, Stephens B (2014) Estimates of HVAC filtration efficiency for fine and ultrafine particles of outdoor origin. Atmos Environ 98:337–346

Aziz N, Adman MA, Suhaimi NS, Misbari S, Alias AR, Aziz AA, Lee LF, Khan MH (2023) Indoor air quality (IAQ) and related risk factors for sick building syndrome (SBS) at the office and home: a systematic review. IOP Conf Ser: Earth Environ Sci 1140:012007

Azuma K, Kagi N, Yanagi U, Osawa H (2018) Effects of low-level inhalation exposure to carbon dioxide in indoor environments: a short review on human health and psychomotor performance. Environ Int 121:51–56

Benmarhnia T, Deguen S, Kaufman JC, Smargiassia A (2015) Vulnerability to heat-related mortality – a systematic review, meta-analysis, and meta-regression Analysis. Epidemiology 26:781–793

Bischof W, Bullinger-Naber M, Kruppa B, Muller BH, Schwab R (2004) Exposures and health effects in office buildings. Fraunhofer IRB Verlag. ISBN 3-8167-6408-8. [in German]

Bluyssen PM, Roda C, Mandin C, Fossati S, Carrer P, de Kluizenaar Y, Mihucz VG, de Oliveira FE, Bartzis J (2016) Self-reported health and comfort in 'modern' office buildings: first results from the European OFFICAIR study. Indoor Air 26:298–317

Boerstra AC, te Kulve M, Toftum J, Loomans MGLC, Olesen BW, Hensen JLM (2015) Comfort and performance impact of personal control over thermal environment in summer: results from a laboratory study. Build Environ 87:315–326

Brehmer C, Norris C, Barkjohn KK, Bergin MH, Zhang J, Cui X, Teng Y, Zhang Y, Black M, Li Z, Shafer MM, Schauer JJ (2020) The impact of household air cleaners on the oxidative potential of $PM_{2.5}$ and the role of metals and sources associated with indoor and outdoor exposure. Environ Res 181:108919

Bullinger M, Morfeld M, von Mackensen S, Brasche S (1999) The sick-building-syndrome—do women suffer more? Zentralbl Hyg Umweltmed 202:235–241

Bux K, Kampmann B (2019) VIII–1.9 Indoor climate. In: Wichmann H-E, Fromme H (eds) Handbook of environmental medicine. Ecomed-Verlag, Landsberg, Germany. [in German]

Carrer P, Wargocki P, Fanetti A, Bischof W, De Oliveira FE, Hartmann T, Kephalopoulos S, Palkonen S, Seppänen O (2015) What does the scientific literature tell us about the ventilation and health relationship in public and residential buildings? Build Environ 94:273–286

Chithra VS, Shiva Nagendra SM (2018) A review of scientific evidence on indoor air of school building: pollutants, sources, health effects and management. Asian J Atmos Environ 12:87–108

Choe Y, Shin J-s, Park J, Kim E, Oh N, Min K, Kim D, Sung K, Cho M, Yang W (2022) Inadequacy of air purifier for indoor air quality improvement in classrooms without external ventilation. Build Environ 207:108450

Coffaro B, Weisel CP (2022) Reactions and products of squalene and ozone: a review. Environ Sci Technol 56:7396–7411

Costa A, Chiarello GL, Selli E, Guarino M (2012) Effects of TiO_2 based photocatalytic paint on concentrations and emissions of pollutants and on animal performance in a swine weaning unit. J Environ Manag 96:86–90

Cox J, Isiugo K, Ryan P, Grinshpun SA, Yermakov M, Desmond C, Jandarov R, Vesper S, Ross J, Chillrud S, Dannemiller K, Reponen T (2018) Effectiveness of a portable air cleaner in removing aerosol particles in homes close to highways. Indoor Air 28:818–827

Csobod E, Rudnai P, Vaskovi E (2010) School environment and respiratory health of children (Search). Regional Environmental Center for Central and Eastern Europe. https://www.isprambiente.gov.it/files/progetti/search-results-europe-rec.pdf

Csobod E, Annesi-Maesano I, Carrer P, Kephalopoulos S, Madureira J, Rudnai P, de Oliveira Fernandes E (2014) SINPHONIE. Schools indoor pollution and Health. Final Report, EUR 26738. https://publications.jrc.ec.europa.eu/repository/handle/JRC91160

Daisey JM, Angell WJ, Apte MG (2003) Indoor air quality, ventilation and health symptoms in schools: an analysis of existing information. Indoor Air 13:53–64

DeLay K, Lin EZ, Koelmel JP, Bornman R, Obida M, Chevrier J, Godri Pollitt KJ (2022) Personal air pollutant exposure monitoring in South African children in the VHEMBE birth cohort. Environ Int 170:107524

Deng Y, Wang F, Liu L, Chen D, Guo Y, Li Z (2022) High density polyethylene (Hdpe) and thermoplastic polyurethane (Tpu) wristbands as personal passive samplers monitoring per- and polyfluoroalkyl substances (PfAS) exposure to postgraduate students. Available at SSRN: https://doi.org/10.2139/ssrn.4263934

Destaillats H, Lunden MM, Singer BC, Coleman BK, Hodgson AT, Weschler CJ, Nazaroff WW (2006) Indoor secondary pollutants from household product emissions in the presence of ozone: a bench-scale chamber study. Environ Sci Technol 40:4421–4428

DIN (2019) DIN 1946-6:2019–12. Ventilation and air conditioning – Part 6: Ventilation for residential buildings – General requirements, requirements for design, construction, commissioning and handover as well as maintenance. Beuth Verlag, Berlin. https://www.beuth.de/en/standard/din-1946-6/314483915

DIN EN (1993) DIN EN 481:1993–09. Workplaces atmospheres; size fraction definitions for measurement of airborne particles

DIN EN (2017) DIN EN 16798-3. Energy performance of buildings – Ventilation for buildings – Part 3: For non-residential buildings – Performance requirements for ventilation and room-conditioning systems. Beuth Verlag, Berlin. https://www.beuth.de/en/standard/din-en-16798-3/265595146

Drabińska N, Flynn C, Ratcliffe N, Belluomo I, Myridakis A, Gould O, Fois M, Smart A, Devine T, Costello BL (2021) A literature survey of all volatiles from healthy human breath and bodily fluids: the human volatilome. J Breath Res 15(3)

DWD (Deutscher Wetterdienst) (2020) Heat warnings [in German]. https://www.dwd.de/DE/leistungen/hitzewarnung/hitzewarnung.html

Erdmann CA, Apte MG (2004) Mucous membrane and lower respiratory building related symptoms in relation to indoor carbon dioxide concentrations in the 100-building BASE dataset. Indoor Air Suppl 8:127–134

EU (2008) Directive 2008/50/EC of the European Parliament and of the Council on ambient air quality and cleaner air for Europe. https://www.ecolex.org/details/legislation/directive-200850ec-of-the-european-parliament-and-of-the-council-on-ambient-air-quality-and-cleaner-air-for-europe-lex-faoc080016/

Fan ZH, Lioy P, Weschler CJ, Fiedler N, Kipen H, Zhang JF (2003) Ozone-initiated reactions with mixtures of volatile orgbanic compounds under simulated indoor conditions. Environ Sci Technol 37:1811–1821

Fisk WJ (2015) Review of some effects of climate change on indoor environmental quality and health and associated no-regrets mitigation measures. Lawrence Berkeley National Laboratory. https://escholarship.org/content/qt6n39s7c6/qt6n39s7c6.pdf

Fisk WJ (2018) How home ventilation rates affect health: a literature review. Indoor Air 28:473–487

Fisk WJ, Chan WR, Johnson AL (2019) Does dampness and mold in schools affect health? Results of a meta-analysis. Indoor Air 29:895–902

Franck U, Kruger M, Schwarz N, Grossmann K, Röder S, Schlink U (2013) Heat stress in urban areas: indoor and outdoor temperatures in different urban structure types and subjectively reported well-being during a heat wave in the city of Leipzig. Meteorol Z 22:167–177

Fruekilde P, Hjorth J, Jensen NR, Kotzias D, Larsen B (1998) Ozonolysis at vegetation surfaces: a source of aceton, 4-oxopentanal, 6-methyl-5-hepten-2-one, and geranyl acetone in the troposphere. Atmos Environ 32:1893–1902

Fu M, Weng W, Chen W, Luo N (2016) Review on modeling heat transfer and thermo regulatory responses in human body. J Thermal Biol 62:189–200

Fuentes ZC, Schwartz YL, Robuck AR, Walker DI (2022) Operationalizing the exposome using passive silicone samplers. Curr Pollut Rep 8:1–29

Géhin C, Tokarska J, Fowler SJ, Barran PE, Trivedi DK (2023) No skin off your back: the sampling and extraction of sebum for metabolomics. Metabolomics 19:21

González-Martín J, Kraakman NJR, Pérez C, Lebrero R, Muñoz R (2021) A state-of-the-art review on indoor air pollution and strategies for indoor air pollution control. Chemosphere 262:128376

Hajat S, Sheridan SC, Allen MJ, Pascal M, Laaidi K, Yagouti A, Bickis U, Tobias A, Bourque D, Armstrong BG, Kosatsky T (2010) Heat-health warning systems: a comparison of the predictive capacity of different approaches to identifying dangerously hot days. Am J Public Health 100: 1137–1144

Han KT, Ruan LW (2020) Effects of indoor plants on air quality: a systematic review. Environ Sci Pollut Res 27:16019–16051

HESE (Health Effects of School Environment) (2006) Final scientific report prepared for the Health & Consumer Protection Directorate General. Siena, Italy. http://ec.europa.eu/health/ph_projects/2002/pollution/fp_pollution_2002_frep_04.pdf

HMUELV (Hessian Ministry for the Environment, Energy, Agriculture and Consumer Protection) (2012) Ventilation in residential buildings. Things to know about air exchange, and modern ventilation methods. Report of the Institut Wohnen und Umwelt. https://wirtschaft.hessen.de/sites/default/files/media/hmwvl/08_lueftung_im_wohngebaeude.pdf [in German]

Hou J, Sun Y, Dai X, Liu J, Shen X, Tan H, Yin H, Huang K, Gao Y, Lai D, Hong W, Zhai X, Norbäck D, Chen Q (2021) Associations of indoor carbon dioxide concentrations, air temperature, and humidity with perceived air quality and sick building syndrome symptoms in Chinese homes. Indoor Air 31:1018–1028

Huang Y, Ho S, Lu Y, Niu R, Xu L, Cao J, Lee S (2016) Removal of indoor volatile organic compounds via photocatalytic oxidation: a short review and prospect. Molecules 21:56

Hubbard HF, Coleman BK, Sarwar G, Corsi RL (2005) Effects of an ozone-generating air purifier on indoor secondary particles in three residential dwellings. Indoor Air 156:432–444

Hulin M, Simoni M, Viegi G, Annesi-Maesano I (2012) Respiratory health and indoor air pollutants based on quantitative exposure assessments. Eur Respir J 40:1033–1045

IOM (Institute of Medicine) (2011) Climate change, the indoor environment, and health. Washington, DC: The National Academies Press. https://doi.org/10.17226/13115

IPCC (The Intergovernmental Panel on Climate Change) (2022) Climate change 2022: impacts, adaptation and vulnerability. Sixth assessment report. https://www.ipcc.ch/report/ar6/wg2/

ISO (International Organization for Standardization) (2016) ISO 16890-1:2016: Air filters for general ventilation–Part 1: Technical specifications, requirements and classification system based upon particulate matter efficiency (ePM). December. Geneva, Switzerland

ISO/IEC (2017) ISO/IEC 17025. Testing and calibration laboratories. https://www.iso.org/ISO-IEC-17025-testing-and-calibration-laboratories.html

Issitt T, Wiggins L, Veysey M, Sweeney ST, Brackenbury WJ, Redeker K (2022) Volatile compounds in human breath: critical review and meta-analysis. J Breath Res 16(2)

Jakober C, Phillips T (2008) Evaluation of ozone emissions from portable indoor air cleaners: electrostatic precipitators and ionizers. Staff Technical Report. Sacramento, CA: California Air Resources Board. https://ww3.arb.ca.gov/research/indoor/esp_report.pdf

Jones ER, Cedeño Laurent JG, Young AS, Coull BA, Spengler JD, Allen JG (2022) Indoor humidity levels and associations with reported symptoms in office buildings. Indoor Air 32: e12961

Karvala K, Sainio M, Palmquist E, Claeson AS, Nyback MH, Nordin S (2018) Building-related environmental intolerance and associated health in the general population. Int J Environ Res Public Health 15:2047

Kelly FJ, Fussell JC (2019) Improving indoor air quality, health and performance within environments where people live, travel, learn and work. Atmos Environ 200:90–109

Khan H, Shah MR, Barek J, Malik MI (2023) Cancer biomarkers and their biosensors: a comprehensive review. Trends Anal Chem 158:116813

Kohlhuber M, Burckhardt F, Schindler P, Höller C, Beck H, Weber H, Fromme H (2006) Climate change in Bavaria. Health consequences and perspectives. https://www.lgl.bayern.de/gesundheit/arbeitsplatz_umwelt/projekte_a_z/doc/klima_gesundheit_lgl_2006.pdf. [in German]

Kownacki KL, Gao C, Kuklane K, Wierzbicka A (2019) Heat stress in indoor environments of Scandinavian urban areas: a literature review. Int J Environ Res Public Health 16:560

Kumar A, Kumar A, MMS C-P, Chaturvedi AK, Shabnam AA, Subrahmanyam G, Mondal R, Gupta DK, Malyan SK, Kumar SS, Khan SA, Yadav KK (2020) Lead toxicity: Health hazards, influence on food chain, and sustainable remediation approaches. Int J Environ Res Public Health 17:E2179

Kwon M, Remoy H, van den Dobbelsteen A, Knaack U (2019) Personal control and environmental user satisfaction in office buildings: results of case studies in the Netherlands. Build Environ 149:428–435

Lazaridis M, Serfozo N, Chatoutsidou SE, Glytsos T (2015) New particle formation events arising from painting materials in an indoor microenvironment. Atmos Environ 102:86–95

Lenz K (2005) Comfort and quality of living air. 3. Internationales Branchenseminar für Frauen 2005. https://www.forum-holzbau.com/pdf/lenz_behaglichkeit.pdf [in German]

Li L, Cocker DR (2018) Molecular structure impacts on secondary organic aerosol formation from glycol ethers. Atmos Environ 180:206–215

Licina D, Yildirim S (2021) Occupant satisfaction with indoor environmental quality, sick building syndrome (SBS) symptoms and self-reported productivity before and after relocation into WELL-certified office buildings. Build Environ 204:108183

Lin EZ, Esenther S, Mascelloni M, Irfan F, Godri Pollitt KJ (2020) The Fresh Air wristband: a wearable air pollutant sampler. Environ Sci Technol Lett 7:308–314

Liu XY, Mason M, Krebs K, Sparks L (2004) Full-scale chamber investigation and simulation of air freshener emissions in the presence of ozone. Environ Sci Technol 38:2802–2812

Long CM, Suh HH, Koutrakis P (2000) Characterization of indoor particle sources using continuous mass and size monitors. J Air Waste Manage Assoc 50:1236–1250

Lowther SD, Jones KC, Wang X, Whyatt D, Wild O, Booker D (2019) Particulate matter measurement indoors: a review of metrics, sensors, needs, and applications. Environ Sci Technol 53:11644–11656

Lu CY, Lin JM, Chen YY, Chen YC (2015) Building-related symptoms among office employees associated with indoor carbon dioxide and total volatile organic compounds. Int J Environ Res Public Health 12:5833–5845

Luengas A, Barona A, Hort C, Gallastegui G, Platel V, Elias A (2015) A review of indoor air treatment technologies. Rev Environ Sci Biotechnol 14:499–522

Lukcso D, Guidotti TL, Franklin DE, Burt A (2016) Indoor environmental and air quality characteristics, building-related health symptoms, and worker productivity in a federal government building complex. Arch Environ Occup Health 71:85–101
Malayeri M, Haghigat F, Lee C-S (2019) Modeling of volatile organic compounds degradation by photocatalytic oxidation reactor in indoor air: a review. Build Environ 154:309–323
Malik J, Bardhan R, Banerji P (2020) Rethinking indoor thermal comfort in the era of rebound and pre-bound effect for the developing world: a systematic review. Indoor Air 30:377–395
Manisalidis I, Stavropoulou E, Stavropoulos A, Bezirtzoglou E (2020) Environmental and health impacts of air pollution: a review. Front Public Health 20:14
van Marken Lichtenbelt WD, Pallubinsky H, te Kulve M (2018) Modulation of thermogenesis and metabolic health: a built environment perspective. Obes Rev 19(Suppl. 1):94–101
Mason H, King JC, Peden AE, Franklin RC (2022) Systematic review of the impact of heatwaves on health service demand in Australia. BMC Health Serv Res 28:960
Mata TM, Martins AA, Calheiros CSC, Villanueva F, Alonso-Cuevilla NP, Gabriel MF, Silva GV (2022) Indoor air quality: a review of cleaning technologies. Environments 9:118
Matheson S, Fleck R, Irga PJ, Torpy FR (2023) Phytoremediation for the indoor environment: a state-of-the-art review. Rev Environ Sci Biotechnol 22:249–280
Mattenklott M, Höfert N (2009) Dusts at workplaces and in the environment – comparison of definitions. Gefahrstoffe – Reinhaltung der Luft 69:127–129. [in German]
Melikov AK (2016) Advanced air distribution: improving health and comfort while reducing energy use. Indoor Air 26:112–124
Melikov AK (2020) COVID-19: reduction of airborne transmission needs paradigm shift in ventilation. Build Environ 186:107336
Milhem SA, Verriele M, Nicolas M, Thevenet F (2020) Does the ubiquitous use of essential oil-based products promote indoor air quality? A critical literature review. Environ Sci Pollut Res 27:14365–14411
Mochalski P, Unterkofler K, Teschl G, Amann A (2015) Potential of volatile organic compounds as markers of entrapped humans for use in urban search-and-rescue operations. Trends Anal Chem 68:88–10
Morrison GC, Nazaroff WW (2002) Ozone interactions with carpet: secondary emissions of aldehydes. Environ Sci Technol 36:2185–2192
Murage P, Kovats S, Sarran C, Taylor J, McInnes R, Hajat S (2020) What individual and neighbourhood-level factors increase the risk of heat-related mortality? A case-crossover study of over 185,000 deaths in London using high-resolution climate datasets. Environ Int 134:105292
Nakayama Y, Nakaoka H, Suzuki N, Tsumura K, Hanazato M, Todaka E, Mori C (2019) Prevalence and risk factors of pre-sick building syndrome: characteristics of indoor environmental and individual factors. Environ Health Prev Med 24:77
NAS (2022) Why Indoor Chemistry Matters. The National Academies Press, Washington, DC. https://nap.nationalacademies.org/catalog/26228/why-indoor-chemistry-matters
Nazaroff WW (2013) Four principles for achieving good indoor air quality. Indoor Air 23:353–356
Nazaroff WW (2021) Residential air-change rates: a critical review. Indoor Air 31:282–313
NICE (National Institute for Health and Care Excellence) (2020) Indoor air quality at home. NICE guideline (NG149). https://www.nice.org.uk/guidance/NG149
NIHHIS (2022) Heat.gov. National Integrated Heat Health Information System. US Department of Commerce. https://www.heat.gov/
Niu M, Shen F, Zhou F, Zhu T, Zheng Y, Yang Y, Sun Y, Li X, Wu Y, Fu P, Tao S (2020) Indoor air filtration could lead to increased airborne endotoxin levels. Environ Int 142:105878
Nordin S (2020) Mechanisms underlying nontoxic indoor air health problems: a review. Int J Hyg Environ Health 226:113489
Palacios J, Steele K, Tan Z, Zheng S (2021) Human health and productivity outcomes of office workers associated with indoor air quality: a systematic review. MIT Center for Real Estate Research Paper No. 21/14

Park HJ, Lee HY, Suh CH, Kim HC, Kim HC, Park YJ, Lee SW (2021) The effect of particulate matter reduction by indoor air filter use on respiratory symptoms and lung function: a systematic review and meta-analysis. Allergy Asthma Immunol Res 13:719–732

Pettit T, Irga PJ, Torpy FR (2018) Towards practical indoor air phytoremediation: a review. Chemosphere 208:960–974

Pfafferott J, Becker P (2008) Extension of the heat warning system to include indoor heat load prediction. Bauphysik 30:237–243. [in German]

Picardo M, Ottaviani M, Camera E, Mastrofrancesco A (2009) Sebaceous gland lipids. Dermatoendocrinology 1:68–71

Pytel K, Marcinkowska R, Zabiegała B (2020) Investigation of the dynamism of nanosized SOA particle formation in indoor air by a scanning mobility particle sizer and proton-transfer-reaction mass spectrometry. Molecules 25:2202

Redlich CA, Sparer J, Cullen MR (1997) Sick-building syndrome. Lancet 349:1013–1016

Rohr AC (2013) The health significance of gas- and particle-phase terpene oxidation products: a review. Environ Int 60:145–162

Roodt AP, Naudé Y, Stoltz A, Rohwer E (2018) Human skin volatiles: Passive sampling and GC × GC-ToFMS analysis as a tool to investigate the skin microbiome and interactions with anthropophilic mosquito disease vectors. J Chromatogr B Analyt Technol Biomed Life Sci 1097-1098:83–93

Rösch C, Wissenbach DK, Franck U, Wendisch M, Schlink U (2017) Degradation of indoor limonene by outdoor ozone: a cascade of secondary organic aerosols. Environ Pollut 226: 463–472

Ruzsanyi V, Mochalski P, Schmid A, Wiesenhofer H, Klieber M, Hinterhuber H, Amann A (2012) Ion mobility spectrometry for detection of skin volatiles. J Chromatogr B Analyt Technol Biomed Life Sci 911:84–92

Sadrizadeh S, Yao R, Yuan F, Awbi H, Bahnfleth W, Bi Y, Cao G, Croitoru C, de Dear R, Haghighati F, Kumarj P, Malayeri M, Nasiri F, Ruud M, Sadeghian P, Wargocki P, Xiong J, Yu W, Li B (2022) Indoor air quality and health in schools: a critical review for developing the roadmap for the future school environment. J Build Eng 57:10490

Sakellaris IA, Saraga DE, Mandin C, Roda C, Fossati S, de Kluizenaar Y, Carrer P, Dimitroulopoulou S, Mihucz VG, Szigeti T, Hänninen O, de Oliveira FE, Bartzis JG, Bluyssen PM (2016) Perceived indoor environment and occupants' comfort in European "modern" office buildings: The OFFICAIR Study. Int J Environ Res Public Health 13:444

Sakellaris I, Saraga D, Mandin C, de Kluizenaar Y, Fossati S, Spinazzè A, Cattaneo A, Mihucz V, Szigeti T, de Oliveira FE, Kalimeri K, Mabilia R, Carrer P, Bartzis J (2020) Association of subjective health symptoms with indoor air quality in European office buildings: the OFFICAIR project. Indoor Air 31:426–439

Salehi B, Ghanbaran AH, Maerefat M (2020) Intelligent models to predict the indoor thermal sensation and thermal demand in steady state based on occupants' skin temperature. Build Environ 169:106579

Salthammer T, Bednarek M, Fuhrmann F, Funaki R, Tanabe S-I (2002) Formation of organic indoor air pollutants by UV-curing chemistry. J Photochem Photobiol A 152:1–9

Salthammer T, Uhde E, Schripp T, Schieweck A, Morawska L, Mazaheri M, Clifford C, He C, Buonanno G, Querol X, Viana M, Kumar P (2016) Children's well-being at schools: impact of climatic conditions and air pollution. Environ Int 94:196–210

Salthammer T, Schieweck A, Gua J, Ameri S, Uhde E (2018) Future trends in ambient air pollution and climate in Germany – implications for the indoor environment. Build Environ 143:661–670

Samon SM, Hammel SC, Stapleton HM, Anderson KA (2022) Silicone wristbands as personal passive sampling devices: current knowledge, recommendations for use, and future directions. Environ Int 169:107339

Sánchez B, Sánchez-Muñoz M, Muñoz-Vicente M, Cobas G, Portela R, Suárez S, González AE, Rodríguez N, Amils R (2012) Photocatalytic elimination of indoor air biological and chemical pollution in realistic conditions. Chemosphere 87:625–630

Sarwar G, Olson DA, Corsi RL, Weschler CJ (2004) Indoor fine particles: the role of terpene emissions from consumer products. J Air Waste Manage Assoc 54:367–377

Savelieva K, Elovainio M, Lampi J, Ung-Lanki S, Pekkanen J (2020) Psychosocial factors and indoor environmental quality in respiratory symptom reports of pupils: a cross-sectional study in Finnish schools. BMJ Open 10(9):e036873

Seltzer JM (1994) Building-related illness. J Allergy Clin Immunol 94:351–361

Seppänen OA, Fisk WJ, Mendell MJ (1999) Association of ventilation rates and CO_2 concentrations with health and other responses in commercial and institutional buildings. Indoor Air 9: 226–252

Shao D, Dub Y, Liu S, Brunekreef B, Meliefste K, Zhao Q, Chen J, Song X, Wang M, Wang J, Xu H, Wu H, Wang T, Feng B, Shih-Chun Lung C, Wang X, He B, Huang W (2017) Cardiorespiratory responses of air filtration: A randomized crossover intervention trial in seniors living in Beijing Beijing Indoor Air Purifier StudY, BIAPSY. Sci Total Environ 603–604:541–549

Shao Z, Bi J, Yang J, Ma Z (2023) Indoor $PM_{2.5}$, home environmental factors and lifestyles are related to sick building syndrome among residents in Nanjing, China. Build Environ 235: 110204

Sharma R, Balasubramanian R (2020) Evaluation of the effectiveness of a portable air cleaner in mitigating indoor human exposure to cooking-derived airborne particles. Environ Res 183: 109192

Shaughnessy RJ, Mcdaniels TJ, Weschler CJ (2001) Indoor chemistry: ozone and volatile organic compounds found in tobacco smoke. Environ Sci Technol 35:2758–2764

Siegel JA (2016) Primary and secondary consequences of indoor air cleaners. Indoor Air 26:88–96

Singer BC, Coleman BK, Destaillats H, Hodgson AT, Lunden MM, Weschler CJ, Nazaroff WW (2006) Indoor secondary pollutants from cleaning products and air freshner use in the presence of ozone. Atmos Environ 40:6696–6710

Srecec D, Bas G, Weißbach W, Hornberg C, Dott W, Wiesmüller GA (2005) Prevalence of sick building syndrome reported in literature. Proc Indoor Air 2005

Stönner C, Edtbauer A, Williams J (2018) Real-world volatile organic compound emission rates from seated adults and children for use in indoor air studies. Indoor Air 28:164–172

Sundell J, Levin H, Nazaroff WW, Cain WS, Fisk WJ, Grimsrud DT, Gyntelberg F, Li Y, Persily AK, Pickering AC, Samet JM, Spengler JD, Taylor ST, Weschler CJ (2011) Ventilation rates and health: multidisciplinary review of the scientific literature. Indoor Air 21:191–204

Suzuki N, Nakayama Y, Nakaoka H, Takaguchi K, Tsumura K, Hanazato M, Hayashi T, Mori C (2021) Risk factors for the onset of sick building syndrome: a cross-sectional survey of housing and health in Japan. Build Environ 202:107976

Szczotko M, Orych I, Maka Ł, Solecka J (2022) A review of selected types of indoor air purifiers in terms of microbial air contamination reduction. Atmosphere 13:800

Tang X, Misztal PK, Nazaroff WW, Goldstein AH (2016) Volatile organic compound emissions from humans indoors. Environ Sci Technol 50:12686–12694

Thatcher A, Adamson K, Bloch L, Kalantzis A (2020) Do indoor plants improve performance and well-being in offices? Divergent results from laboratory and field studies. J Environ Psychol 71: 101487

Tran V, Park D, Lee YC (2020) Indoor air pollution, related human diseases, and recent trends in the control and improvement of indoor air quality. Int J Environ Res Public Health 17:2927

Tsang CHA, Li K, Zeng Y, Zhao W, Zhang T, Zhan Y, Xie R, Leung DYC, Huang H (2019) Titanium oxide based photocatalytic materials development and their role of in the air pollutants degradation: overview and forecast. Environ Int 125:200–228

US EPA (2018) Residential air cleaners. Portable air cleaners, Furnace and HVAC Filters, 3rd edn. EPA EPA 402-F-09-002. https://www.epa.gov/sites/production/files/2018-07/documents/residential_air_cleaners_-_a_technical_summary_3rd_edition.pdf

Vardoulakis S, Dimitroulopoulou C, Thornes J, Lai KM, Taylor J, Myers I, Heaviside C, Mavrogianni A, Shrubsole C, Chalabi Z, Davies M, Wilkinson P (2015) Impact of climate

change on the domestic indoor environment and associated health risks in the UK. Environ Int 85:299–313
Vartiainen E, Kulmala M, Ruuskanen TM, Taipale R, Rinne J, Vehkamäki H (2006) Formation and growth of indoor air aerosol particles as a result of d-limonene oxidation. Atmos Environ 40: 7882–7892
Vesela S, Kingma BR, Frijns AJ (2017) Local thermal sensation modeling – a review on the necessity and availability of local clothing properties and local metabolic heat production. Indoor Air 27:261–272
Wacławik M, Rodzaj W, Wielgomas B (2022) Silicone wristbands in exposure assessment: analytical considerations and comparison with other approaches. Int J Environ Res Public Health 19:1935
Wainman T, Zhang J, Weschler CJ, Lioy PJ (2000) Ozone and limonene in indoor air: a source of submicron particle exposure. Environ Health Perspect 108:1139–1145
Walikewitz N, Jänicke B, Langner M, Endlicher W (2018) Assessment of indoor heat stress variability in summer and during heat warnings: a case study using the UTCI in Berlin, Germany. Int J Biometeorol 62:29–42
Wang H, He C, Morawska L, McGarry P, Johnson G (2012) Ozone-initiated particle formation, particle aging, and precursors in a laser printer. Environ Sci Technol 46:704–712
Wang C, Lu S, Zhang Z (2019) Inactivation of airborne bacteria using different UV sources: performance modeling, energy utilization, and endotoxin degradation. Sci Total Environ 655: 787–795
Wang R, Huang Y, Cao G (2020) Heterogeneous oxidation of isoprene SOA and toluene SOA tracers by Ozone. Chemosphere 249:126258
Wargocki P, Sundell J, Bischof W, Brundrett G, Fanger PO, Gyntelberg F, Hanssen SO, Harrison P, Pickering A, Seppänen O, Wouters P (2002) Ventilation and health in nonindustrialindoor environments: report from a European Multidisciplinary Scientific Consensus Meeting (EUROVEN). Indoor Air 12:113–128
Waring MS (2014) Secondary organic aerosol in residences: predicting its fraction of fine particle mass and determinants of formation strength. Indoor Air 24:376–389
Waring MS, Wells JR (2015) Volatile organic compound conversion by ozone, hydroxyl radicals, and nitrate. Atmos Environ 106:382–391
Waring MS, Siegel JA, Corsi RL (2008) Ultrafine particle removal and generation by portable air cleaners. Atmos Environ 42:5003–5014
Waring MS, Wells JR, Siegel JA (2011) Secondary organic aerosol formation from ozone reactions with single terpenoids and terpenoid mixtures. Atmos Environ 45:4235–4242
Weschler CJ (2003) Indoor chemistry as a source of particles. In: Morawska L, Salthammer T (eds) Indoor environment. Airborne particles and settled dust. Wiley-VCH Verlag, Weinheim
Weschler CJ (2016) Roles of the human occupant in indoor chemistry. Indoor Air 26:6–24
Weschler CJ, Carslaw N (2018) Indoor chemistry. Environ Sci Technol 52:2419–2428
Weschler CJ, Shields HC (1999) Indoor ozone/terpene reactions as a source of indoor particle. Atmos Environ 33:2301–2312
Weschler CJ, Wells JR, Poppendieck D, Hubbard H, Pearce TA (2006) Workgroup report: indoor air chemistry and health. Environ Health Perspect 114:442–446
WHO (2009) Natural ventilation for infection control in health-care settings. Geneva, WHO Guideline. Switzerland. https://www.who.int/water_sanitation_health/publications/natural_ventilation/en/
WHO (2018) WHO housing and health guidelines. Licence: CC BY-NC-SA 3.0 IGO. https://www.who.int/publications/i/item/9789241550376
WHO (2020) Methods for sampling and analysis of chemical pollutants in indoor air. Supplementary publication to the screening tool for assessment of health risks from combined exposure to multiple chemicals in indoor air. https://www.euro.who.int/en/health-topics/environment-and-health/air-quality/publications/methods-for-sampling-and-analysis-of-chemical-pollutants-in-indoor-air

WHO (2021) Literature review on chemical pollutants in indoor air in public settings for children and overview of their health effects with a focus on schools, kindergartens and day-care centres. Supplementary publication to the screening tool for assessment of health risks from combined exposure to multiple chemicals in indoor air in public settings for children. WHO Europe. https://apps.who.int/iris/handle/10665/341467

Wisthaler A, Weschler CJ (2010) Reactions of ozone with human skin lipids: sources of carbonyls, dicarbonyls, and hydroxycarbonyls in indoor air. Proc Natl Acad Sci USA 107:6568–6575

Wolkoff P (2013) Indoor air pollutants in office environments: assessment of comfort, health, and performance. Int J Hyg Environ Health 216:371–394

Wolkoff P (2018a) Indoor air humidity, air quality, and health – an overview. Int J Hyg Environ Health 221:376–390

Wolkoff P (2018b) The mystery of dry indoor air – an overview. Environ Int 121:1058–1065

Young CJ, Zhou S, Siegel JA, Kahan TF (2019) Illuminating the dark side of indoor oxidants. Environ Sci Process Impacts 21:1229–1239

Zhan Y, Johnson K, Norris C, Shafer MM, Bergin MH, Zhang Y, Zhang J, Schauer JJ (2018) The influence of air cleaners on indoor particulate matter components and oxidative potential in residential households in Beijing. Sci Total Environ 626:507–518

Zhang J, Lioy P (1994) Ozone in residential air – concentrations, I/O ratios, indoor chemistry and exposures. Indoor Air 4:95–105

Zhang X, Li F, Zhang L, Zhao Z, Norback D (2014) A longitudinal study of sick building syndrome (SBS) among pupils in relation to SO_2, NO_2, O_3 and PM_{10} in schools in China. PLoS One 9: e112933

Zhou S, Yeung LWY, Forbes MW, Mabury S, Abbatt JPD (2017) Epoxide formation from heterogeneous oxidation of benzo[a] pyrene with gas-phase ozone and indoor air. Environ Sci Process Impacts 19:1292–1299

Zhu Y, Luo M, Ouyang Q, Huang L, Cao B (2015) Dynamic characteristics and comfort assessment of airflows in indoor environments: a review. Build Environ 91:5e14

Chapter 2
Basic Exposure Information and Special Exposure Situation

Contents

Abstract Exposure science has developed alongside epidemiology and toxicology as a field that aims to describe qualitatively and, in particular, quantitatively the contact of an individual or a population group with a chemical, physical, and biological stressor. Therefore, this chapter describes the basics of the exposure assessment process, its procedure and helpful sources to find data on human exposure factors like the US EPA Exposure Factors Handbook. Overall, the exposome is understood as the comprehensive—cumulative—description of the lifelong exposure history of individuals to changing exogenous and endogenous influences. In addition, the inclusion of human biomonitoring in exposure assessment and the importance of textiles as a source of indoor air pollution expands our scientific knowledge. Indispensable for understanding the health effects is the behavior of volatile and semi-volatile organic compounds, as well as particulate matter in the lungs. Many exposure and risk assessment strategies take their starting point from the exposure of sedimented dust indoors and its uptake through, e.g., hand-to-mouth contact. Nevertheless, some critical points like sampling conditions, type of analytical method, intake rates, bioaccessibility, and bioavailability have to be taken into account.

In addition, special important exposure situations such as the use of printers, open fires indoors, as well as the exposure inside aircraft cabins are presented.

H. Fromme, *Indoor Air Quality*, https://doi.org/10.1007/978-3-031-40078-0_2

Keywords Exposure assessment · Exposome · Indoor dust · Bioavailability · Printer · Aircraft cabin · Open fires

Historically, exposure science has developed alongside epidemiology and toxicology as a field that aims to describe qualitatively and, in particular, quantitatively the contact of an individual or a population group with a chemical, physical, and biological stressor from our living environment. Due to a large number of potentially problematic substances and the need to make scientifically validated statements ever more quickly, the US National Academy of Sciences already recommended a broadening and deepening of the term exposure in 2012 (NAS 2012). Accordingly, it sees exposure assessment as a scientific application that aims to predict impacts on human health, prevent them in advance if possible, or at least reduce their severity. It also aims to identify populations with high exposure and protect vulnerable individuals while assessing health and environmental risks. With this in mind, various approaches have been discussed on how to develop exposure science systematically and link it to epidemiology and toxicology (Krewski et al. 2010; Teeguarden et al. 2016; NAS 2017; Escher et al. 2017; Tan et al. 2018; Wambaugh et al. 2019; Bopp et al. 2019; Boyes and van Thriel 2020). On the exposure side, this is enabled by the concept of Aggregate Exposure Pathways (AEP), which sequentially represents all plausible and empirically supported relationships between a chemical or stressor from its source to its concentration at the target site of effect. Complementary to this, the concept of Adverse Outcome Pathways (AOP) is discussed in toxicology (Spinu et al. 2020). In this case, a continuum is described between the disruption of a biological system at the molecular level and the (adverse) (outcome) effects at the level of the whole organism, possibly also at the population level. These developments are particularly promoted by considerable improvements in analytical capabilities, which make it possible to reliably determine substances not only at very low concentrations, but also in large numbers.

2.1 Process of Exposure Assessment

In the context of exposure assessment, some methodological challenges have to be expected. For example, duration and frequency of exposure over time are important influencing factors, even though life and uniform exposure is often assumed in the assessment. Of course, this is often not the case; rather, exposure may exist only temporarily, possibly at varying levels. It must also be considered that exposure may occur during a particularly susceptible phase of human development, e.g., embryogenesis, or that there may be critical time windows for an effect. Another important consideration is the fact that we are not exposed to only one pollutant at a time and that this co-exposure, especially if there is an identical mechanism of action, can lead to possible additive or multiplicative effects.

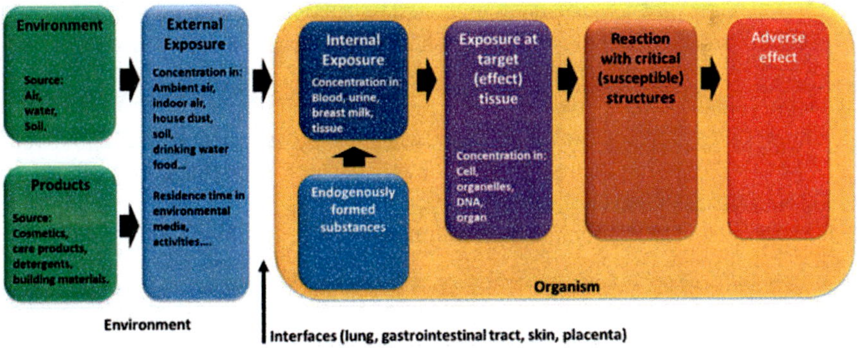

Fig. 2.1 Schematic representation of the exposure pathway

Figure 2.1 schematically illustrates the main steps in the development from the source of a substance to the health effect in terms of an Aggregate Exposure Pathway (AEP). Two major areas of sources can be distinguished: On the one hand, e.g. industrial processes, traffic, agriculture, and human activities lead to environmental pollution, whereby the substances generated are distributed between the different environmental media and can also change chemically there. On the other hand, the wide variety of products when used near or on humans also act as a source, such as cosmetics and personal care products, cleaning agents, odor improvers, furnishings, and building products. External exposure then involves determining the concentration of the pollutant in the various contact media of the organism. Direct measurement methods are available for this purpose, which can, for example, depict personal exposure, and indirect methods, in which exposure scenarios are considered. For this purpose, in addition to the concentration at the site of the respective exposure, knowledge of the source and its release characteristics, as well as data on the spatial and temporal residence of the target person or the target group in individual environmental areas, is required. As a third possibility, the external exposure can be recalculated based on measurements in body fluids such as blood or urine (so-called human biomonitoring), taking toxicokinetic data into account (see Sect. 2.3).

To perform an exposure assessment, it is necessary to consider and apply various human exposure factor data such as: anthropometric data (e.g., body weight, skin surface area, life expectancy), behavioral data (e.g., non-dietary ingestion rates, activity/time use patterns, consumer product use), physiological needs based data (e.g., water/food consumption rates, inhalation rates), and other data (e.g., building characteristics, ventilation rates). In an overview the Joint Research Centre (Reina et al. 2014) described exposure factor handbooks, guidance documents, and online databases from different countries like:

- The EU (https://data.jrc.ec.europa.eu/dataset/jrc-10114-10001).
- Germany (https://www.umweltbundesamt.de/en/www.uba.de/refxp).
- The USA (https://cfpub.epa.gov/ncea/risk/recordisplay.cfm?deid=236252).

- Canada (https://www.canada.ca/en/health-canada/services/chemical-substances/fact-sheets/canadian-exposure-factors-human-health-risk-assessments.html).
- Australia (https://www.health.vic.gov.au/publications/the-australian-exposure-factor-guidance-handbook).
- China (Only as book available under: https://www.sciencedirect.com/book/9780128031254/highlights-of-the-chinese-exposure-factors-handbook-adults).
- Japan (https://unit.aist.go.jp/riss/crm/exposurefactors/english_summary.html).
- Korea (open-access, http://creativecommons.org/licenses/by-nc/3.0/).

In principle, exposure assessment can most easily be performed deterministically using point estimates, such as a typical or probable percentile of exposure (e.g., the median), or high exposure, such as the 95th percentile of the distribution. In addition, it is possible to estimate exposure probabilistically by considering complete distribution data of the concentration of the target parameter and exposure factors, such as age, respiratory rate, or dust intake. The advantage of the second method is that it takes into account the full distribution rather than just individual characteristic values from the frequency distributions. This can lead to more realistic exposure estimates, preventing the mathematical accumulation of unfavorably chosen assumptions. On the other hand, the use of point estimates is a relatively easy method to implement in practice.

In addition, a distinction can be made between cumulative and aggregated exposure assessment. The term aggregated exposure assessment is understood to mean an overall consideration of all possible intake pathways for a single substance or substance group (Meek et al. 2011). Aggregated exposure estimates have the advantage that they can be used to determine the proportional importance of a source (e.g., indoor air) to other sources such as food. This is important, for example, for many persistent and low-volatile substances, since a large proportion of human exposure is via food intake. Several examples are available in the scientific literature so far (Fromme et al. 2007a, 2013; Tan et al. 2018; Clewell et al. 2020). Cumulative or combined exposure assessment, on the other hand, considers all substances with a common mechanism of action, taking into account interactions between substances (Kienzler et al. 2016).

If, in addition to the pollutant concentration, sufficient data are available on the conditions at the respective interface of the organism, the uptake into the organism can be estimated. Thus, to assess the total inhalation uptake, the concentration in the gas and particle phase in the—different—indoor and outdoor spaces, the residence times in the respective environmental compartment, the daily respiratory volume, and the uptake rates in the lungs could be considered. The total uptake (D_{Inhal}) over this pathway can be expressed by the following equation:

$$D_{\mathrm{Inhal}} = \frac{[(C_{\mathrm{Indoor}} \times V_{\mathrm{Indoor}}) + (C_{\mathrm{Outdoor}} \times V_{\mathrm{Outdoor}})] \times R_{\mathrm{Inhal}}}{\text{body weight}}$$

C_{indoor} (simplified here for central indoor living space) and C_{outdoor} are the concentrations in indoor air and outdoor air, $V_{\mathrm{indoor,}}$ and V_{outdoor} are the respiratory volume inhaled per day in these areas, and R_{inhal} is the uptake rate of the substance in the lungs upon inhalation. The respective respiratory volume can be calculated by taking into account the time spent indoors and outdoors.

After absorption into the organism, internal exposure can be measured as part of human biomonitoring. In this context, exposure monitoring is understood as the measurement of the concentration of pollutants or their metabolites in human test materials such as blood, urine, or saliva. This method can be used to make concrete statements about the overall exposure of the organism to a substance. It thus also provides a measure of the actual pollutant dose absorbed by the organism via all uptake pathways and also takes into account the individual characteristics above uptake, storage, metabolization, and excretion of the foreign substance in the human organism. Effect monitoring measures biological parameters that "react" to or indicate the effects of exposure to xenobiotics. It involves the quantitative recording of reactions and interactions of the foreign substance with enzymes, proteins, nucleic acids (e.g., enzyme inhibition in the context of lead poisoning or DNA adducts in the case of exposure to carcinogenic substances), and other functional biomolecules of the organism. In human biomonitoring, the measurement of the properties of certain genes or gene groups on the metabolism and toxicity of foreign substances is referred to as susceptibility monitoring. It should be noted that some contaminants, such as acetaldehyde and formaldehyde, not only reach humans from the environment but are also produced endogenously in the metabolism. In this case, this would have to be taken into account as a background exposure in the context of the assessment.

2.2 The Exposome

Human health is essentially dependent on genetic factors, environmental influences, and lifestyle factors, which can interact with each other. For example, environmental influences have the potential to affect variance in gene function through epigenetic mechanisms. Differences in environmental conditions are thought to account for 70–90% of chronic disease health risks (Rappaport and Smith 2010). The term exposome, introduced into the scientific discussion around 2005, is understood as the comprehensive—cumulative—description of the lifelong exposure history of individuals to changing exogenous and endogenous influences (Wild 2005, 2012; Turner et al. 2017; Hu et al. 2023). Here, three important domains can be distinguished, all of which together are intended to represent the changing background of exposure over time (see Fig. 2.2). The first, internal domain describes processes that substances from the environment experience in the human body or changes that occur in the organism itself, e.g., in the context of inflammation. The second, specific external area describes the direct environmental factors that affect humans, and the third, general external area describes the other influencing factors from the socio- and technosphere. It becomes understandable that an exact description of the complex exposome is very time-consuming and costly, since in part new—analytical—methods and data processing procedures have to be developed and used. Exemplary

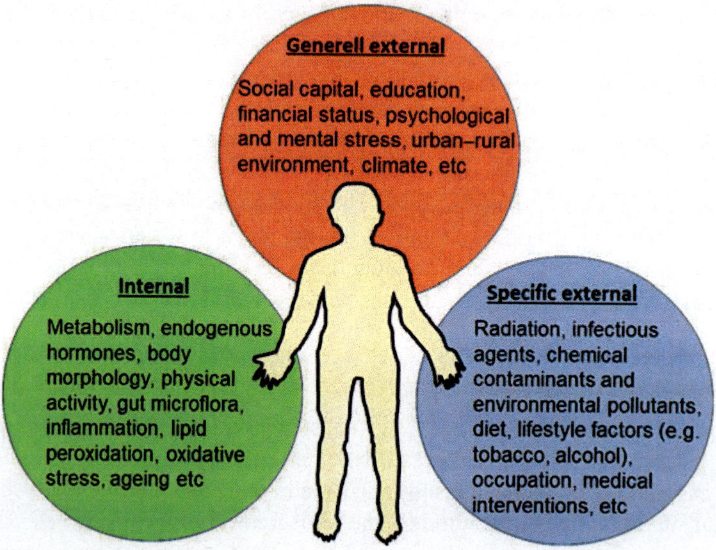

Fig. 2.2 Important aspects of the exposome (modified according to Wild 2012)

is a study from California in which small-scale changes in the biological and chemical exposome, among others, were tracked over time in 15 individuals over 890 days (Jiang et al. 2018). The goal was to map an interdependent network of influence and dependence that affects human health. An interesting extension of the exposome concept is discussed by Gudi-Mindermann et al. (2023) in that they describe the mechanisms by which social exposures are translated into health outcomes. In the process they integrated insights from health equity and environmental justice rersearch into the field of exposome research by providing the conceptual framework of the Social Exposome.

2.3 Importance of Human Biomonitoring in Exposure Assessment

A combination of exposome studies with human biomonitoring studies (HBM) also appears promising to make targeted use of the advantages of the two methods and to compensate for disadvantages (Dennis et al. 2017; Bocato et al. 2019). However, this will require developing more comprehensive approaches and, in particular, expanding the range of substances studied and moving from measurements at the one-time point to repeated measurements.

As indicated in Sect. 2.1, under certain conditions it is possible to back-calculate the total intake from the concentrations in blood or urine. Since the internal load always represents the integral of the intake over all possible pathways, this method cannot be used to differentiate between the individual pathways. This will be illustrated by two examples:

The first is based on a very important group of plasticizers, the phthalates. If the toxicokinetic data on human metabolism and, in particular, the characteristics of the excretion of the target substance are known and its metabolites are characterized, the total daily intake of the parent substances can be calculated based on the metabolite concentration excreted in the urine. The calculation can be made once on the based excreted urinary volume or creatinine-adjusted excretion. Based on the publications of Kohn et al. (2000) and Koch et al. (2007), the following formulas can be used to back-calculate the total intake:

$$Di\ [\mu g/kg\ KG/d] = \frac{C\ [\mu g/l] * UV_{norm}\ [l/kg\ bw/d]}{F_{UE}} * \frac{MW_{Di}}{MW_{Mono}}$$

$$Di\ [\mu g/kg\ KG/d] = \frac{C\ [\mu/g\ Crea] * CE\ [g/kg\ bw/d]}{F_{UE} * 1000} * \frac{MW_{Di}}{MW_{Mono}}$$

D_I is the total daily intake, C is the concentration of the metabolite in µg/l or µg/g creatinine, UV_{norm} the reference value for the excreted urinary volume, CE the daily creatinine excretion, F_{UE} the molar ratio between the proportion of phthalate metabolites excreted in the urine and the ingested amount of the respective diester, and MW the molecular weight of the diester or its monoester. For the determination of individual F_{UE}, toxicokinetic studies on the excreted concentration of metabolites in urine after oral administration of a known amount of a labeled phthalate are typically performed within 24 or 48 hours.

Another example is the INES study (Integrated Exposure Assessment Study), which examined duplicate samples of all—prepared—food consumed in adults and children every day for 7 consecutive days. In addition, urine or blood samples were measured daily to determine internal exposure and indoor air samples were analyzed to determine external exposure. This made it possible to relate exposure via air and food to total internal exposure for various contaminants and thus to determine the significance of individual pathways. Results have been published for phthalates and DEHA (Fromme et al. 2007b, 2013), PBDE (Fromme et al. 2009a), PFOS/PFOA (Fromme et al. 2007a), and PCB and PCDD/F (Fromme et al. 2009b, c).

Furthermore, Senerat et al. (2020) summarize the current evidence on the biomarkers which most frequently responded to variations in indoor air quality and found the most consistent association of IAQ with biomarkers for oxidative stress, coagulation, and inflammation. Advantages, informative value, and limitations of biomarker applicability require further in-depth scientific studies.

2.4 Importance of Textiles in Exposure Assessment

Clothing textiles can contain a wide range of chemical substances such as formaldehyde, glyoxal, organophosphorus flame retardants, organotin compounds, antimony compounds, perfluorinated and polyfluorinated substances, and pyrethroids, which are used as colorants, auxiliaries, and finishing agents to ensure, for example, a true-to-wash color, stable shape or a wrinkle-free garment. In addition, biocidal products such as insecticides, acaricides, or antimicrobial products can also be used to protect fibrous or polymerized materials. Of particular importance in this context is dermal uptake in the context of acute exposure when wearing new textiles with possibly higher migration rates from the product and chronic exposure during the period of use (Rovira and Domingo 2019).

In addition, the release of substances from textiles into the indoor air must also be considered. This can involve substances that are still present in the textiles after manufacture or substances that are subsequently adsorbed onto the textile fibers and then possibly released again (Salthammer et al. 2018; Licina et al. 2019). Studies on the effects on indoor air are only available to a limited extent so far.

For example, following cleaning of three textiles with perchloroethylene under realistic conditions, mean concentrations of approximately 0.5–1.4 mg/m^3 were measured in the bedroom closet where the cleaned garment hung on the first day, and elevated levels of approximately 0.06–0.14 mg/m^3 were also measured in the bedroom air (Tichenor et al. 1990). The authors observed a slow decrease of concentrations in the air and concluded that adsorptive surfaces of the room represent a sink for perchloroethylene. In another experiment, the emission of VOC from 5 foam mattresses was studied in test chambers of 2 and 3.2 m^3 at an air-exchange rate of 0.5/h (Hillier et al. 2003). After the conversion of the maximum concentrations to a standard room, the results were 219 µg/m^3 for the sum of siloxanes, 94 µg/m^3 for triethylenediamine, 70 µg/m^3 for toluene and dichloromethane, and 20 µg/m^3 for hexanal. Given the existing values, however, health effects are rather unlikely in this case.

In 80 German homes sampled in 1996–1998, all containing permethrin-finished woolen floor coverings, high levels of permethrin were found in <2 mm sieved house dust with a median of 9.7 mg/kg (90th percentile: 129 mg/kg), which was attributed to abrasion of the fiber (Berger-Preiss et al. 2002). In airborne dust, only 2.0 ng/m^3 (median) and 6.0 ng/m^3 (90th percentile) were determined, respectively. Human biomonitoring performed on the 139 apartment users also revealed only low urinary metabolite concentrations, which did not differ from the background of the general population. Licina et al. (2015) were able to show, using a human—clothed—manikin, that emissions in the breathing zone were highest when the source was located in the chest area of the clothing and that changes in position and lowering of room temperature also led to an increase in personal exposure. To simulate the uptake and release of particles, Licina and Nazaroff (2018) examined a garment in a test chamber. While larger particles (5.0–7.5 µm in diameter) were deposited on the clothing faster than smaller ones (0.5–1 µm), between 0.3 and 3%

could be released back into the air, depending on the intensity of movement of the textiles.

2.5 Indoor Residence Times

To be able to estimate exposure as correctly as possible, it is important to have not only measurement data in the individual environmental compartments but also information on the time spent in the respective environmental areas. This is all the more important since, for many pollutants, results are only available from stationary measurements in individual occupied areas. Data representing continuous, uninterrupted, personal exposure and thus enabling a better estimation of total exposure are still limited.

Table 2.1 summarizes the results of some comprehensive studies on time use. In Germany, representative data are available in a study performed 2003–2006 for 3- to 14-year-old children (Conrad et al. 2013), in which the children's parents were personally interviewed about a typical past week. While time spent indoors and outdoors decreases with age, it increases in other indoor spaces. Children of higher socioeconomic status spend less time outside than children of lower status. Somewhat older studies of children in the aforementioned age group reach similar conclusions (Brasche and Bischof 2005). In contrast, Canadian data in comparable age groups show longer time spent indoors and less time spent outdoors (Matz et al. 2015).

For adults, extensive data are reported from a study by Brasche and Bischof (2005), in which approximately 12,000 people in Germany were surveyed. The highest residence times in indoor living spaces were found to be 19.5 hours among persons over 64 years of age. Self-reported diary data (weekdays) are also available from the EXPOLIS study conducted in seven cities in Europe in 1996–2000, which found that 19- to 60-year-olds spent about 13.5 hours indoors, about 7.5 hours at work, and almost 2 hours outdoors (Schweizer et al. 2007). In Canada, adult residence times are again higher indoors and lower outdoors (Matz et al. 2014). One of the first, very comprehensive studies, was conducted US-wide in 1992–1994 as part of the US National Human Activity Pattern Survey (NHAPS) as a telephone survey (Klepeis et al. 2001). In addition, several smaller studies have attempted to more accurately assess personal time spent indoors using GPS (e.g., Wu et al. 2011; Tandon et al. 2013; Hu et al. 2016; Khawaja et al. 2019).

In a Belgian study with 16 participants, it was observed that exposure to black carbon was significantly higher among employees working in an external company than among people working at home, in particular, due to exposure in transport microenvironments (Dons et al. 2011). Although only about 8% of the time was spent on transportation, this resulted in about 31% of the total exposure. Using a dynamic time microenvironment activity (TMA) approach, Ramacher and Karl (2020) modeled the exposure situation in Hamburg. In this study, total exposure was 12% and 19% higher using TMA for NO_2 and $PM_{2.5}$, respectively, than when

Table 2.1 Median residence times of persons in different environmental areas in hours and minutes

Reference	Number	Age (years)	At home	At work	Other interiors	Outdoor space	Country, Year
Klepeis et al. (2001)[a]	9196	<5–>64	16:10	1:18	3:05	1:49	USA, 1992–1994
Schweizer et al. (2007)	1427	19–60	13:31	7:29	1:25	2:15	Europe, 7 cities, 1996–2000
Conrad et al. (2013)	1670	3–14	15:30	4:45		3:39	Germany, 2003–2006
	383	3–5	16:17	3:39		3:51	
	419	6–8	15:39	4:26		3:51	
	414	9–11	15:15	5:00		3:34	
	454	12–14	15:04	5:30		3:21	
Brasche and Bischof (2005)[a]	513	<7	17:36	–		–	Germany, 2000–2001
	1264	7–16	14:48	–		–	
	840	17–24	12:36	–		–	
	1216	25–34	13:18	–		–	
	1944	35–44	14:06	–		–	
	1683	45–54	14:24	–		–	
	1870	55–64	16:36	–		–	
	2467	>64	19:30	–		–	
Matz et al. (2015)[a]	217	1–4	17:42	3:42		1:48	Canada, 2010–2011
	316	5–11	17:36	4:18		1:48	
	216	12–19	16:42	5:00		1:30	
	1613	20–59	16:00	5:06		1:18	
	1049	>60	18:36	3:00		1:18	
Matz et al. (2014)	5011	25–64	16:44	3:30		0:30	Canada, 2010–2011

[a]Mean values

the only residential address was considered. The highest contribution to the total annual exposure was from car interiors with 40% (NO_2) and 33% ($PM_{2.5}$) and from cycling and bus use with 15% (NO_2) and 17% ($PM_{2.5}$). Subways and commuter trains have a share below 10%. Other studies also indicate that considering mobility characteristics of individuals is of particular importance for epidemiological studies (Molnár et al. 2006; Setton et al. 2011; Steinle et al. 2013; Cepeda et al. 2017; Zulauf et al. 2019; Borghi et al. 2021).

2.6 Fundamentals of the Behavior of Pollutants in the Lungs

Of central importance to understanding the health effects of airborne substances indoors is the structure and physiology of the respiratory tract, which each noxious agent encounters as a boundary layer in the organism after inhalation. Figure 2.3 shows graphically the different levels of the respiratory tract, which can be roughly divided into the extrathoracic, bronchial, and alveolar regions. In particular, the surface area of the alveoli, which in adults is approximately 70–80 m^2 and onto which approximately 10–20 m^3 of air separate per day, represents an important

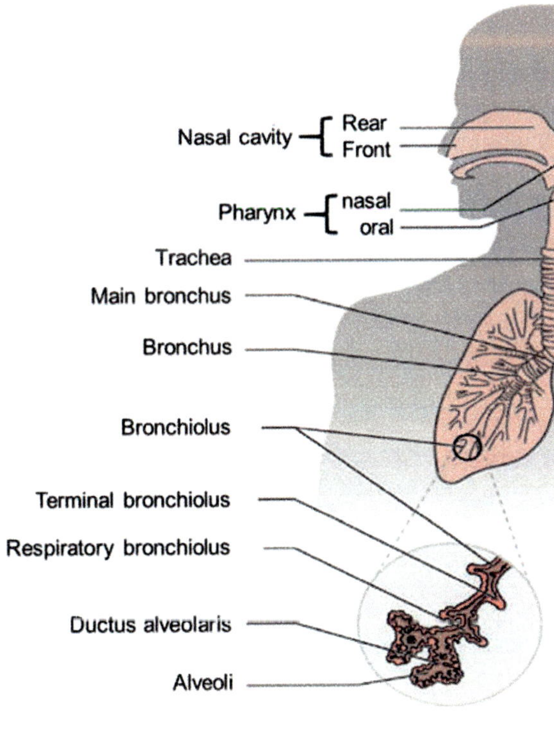

Fig. 2.3 Schematic structure of the lung (modified after Deng et al. 2019)

boundary area between the inside and the outside. To describe the behavior of volatile substances and aerosols in the different areas of the respiratory tract, e.g. in the context of inspiration and expiration, the site of local effect and deposition/absorption, in addition to results from animal experiments and experimental studies in volunteers, morphometric, digitally reconstructed respiratory tract models, e.g. based on imaging techniques such as high-resolution computed tomography, can be used (Ahookhosh et al. 2020; Islam et al. 2020).

In general, the following behavior of inhaled materials can be roughly differentiated (Ginsberg et al. 2010):

- Category 1 substances, such as chlorine, formaldehyde, and vinyl acetate, which have good water solubility or high reactivity, are predominantly absorbed in the extrathoracic region or react there with surfaces. Only at higher concentrations and/or increased physical exertion do they also increasingly reach deeper areas of the respiratory tract.
- Gases of category 2, such as ozone, with a lower reactivity and water solubility than substances of category 1, predominantly reach the bronchial and partly also the alveolar area. Local and systemic effects may occur in this case.
- Category 3 substances, such as chlorinated solvents, are less reactive and not water-soluble. They, therefore, reach the alveolar region and can subsequently be readily absorbed into the organism.

Because of the different physiological characteristics compared to adults and the toxicokinetic and toxicodynamic differences, children must be given special consideration in the context of exposure and risk assessment. Thus, the ventilation rate increases continuously from 3.6 m^3 in infants, 4.1 m^3 at the age of 3–6 months, 9.5 m^3 at the age of 2–3 years up to 20 m^3 in adults (Ginsberg et al. 2010). For example, children have a higher respiratory rate at rest of 400 ml/min × kg body weight compared with 150 ml/min × kg in adults due to their greater energy requirements, while the lung surface area is smaller at 3 m^2 (neonates) and 75 m^2 (adults) (Schneider et al. 2002). Normalized to body weight, however, a nearly identical ratio of 1 m^2/kg body weight is seen in all age groups. In addition, children have a 2- to 4-fold higher ratio of respiratory rate to surface area (in ml/min × cm^2) in the alveolar region (Ginsberg et al. 2010), while particle deposition models indicate a potential for greater particle dosimetry by approximately factor 3–10 in young children. However, these differences may not apply for all particle sizes and respiratory regions (Ginsberg et al. 2010).

2.6.1 Absorption of Volatile Organic Compounds (VOC)

The proportion of an inhaled substance that is absorbed in the lungs after inhalation of a volatile substance is the respiratory absorption factor. It depends on several factors, such as the physicochemical characteristics of the substance, the level of exposure, and individual physiological conditions (Beauchamp 2011). Mechanistic models for the uptake and distribution of volatile organic compounds show a dependence on substance-specific water solubility and reactivity. For example, formaldehyde, which is very reactive and highly soluble, is already absorbed in the upper tracheobronchial or nasal region, whereas substances with lower solubility, such as acrolein and acetaldehyde, penetrate deeper into the lung and are absorbed in the alveolar region (Asgharian et al. 2012). Considering the anatomical differences of the upper respiratory tract and the different respiration rates of seven children and adults, there were no significant differences in the local absorption rate of formaldehyde in the upper respiratory tract for children compared to adults (Ginsberg et al. 2005; Garcia et al. 2009).

A larger number of study results are available for aromatic hydrocarbons in particular, but these were carried out at very high concentrations that are not to be expected under environmental conditions today (Huang et al. 2016). It was only with improved chemical analysis, known as proton transfer reaction time-of-flight mass spectrometry, that it was possible to perform online measurements at low concentration ranges. When seven subjects were exposed to BTX aromatics at concentrations of approximately 2.0–3.2 $\mu g/m^3$ (benzene) and 3.1–14.8 $\mu g/m^3$ (toluene), a mean respiratory absorption factor of 55% for benzene, 56% for toluene, and 67% for ethylbenzene and xylene were observed (Huang et al. 2016). In a follow-up study, inhaled and exhaled air was analyzed for aromatics five times each in 50 subjects, and mean respiratory absorption factors were found to be 28% for benzene, 63% for toluene, and 67% for other C_8 aromatics (Huang et al. 2017). In the inhaled air, the concentrations for benzene, in this case, were 0.6–10.6 $\mu g/m^3$ and for toluene 1.3–33.4 $\mu g/m^3$. There were no differences between the sexes.

2.6.2 *Deposition of Semi-volatile Organic Compounds (SVOC)*

Wei et al. (2018) describe in their review the few available results of in vitro methods (with different synthetic lung fluids and extraction methods) for the determination of absorption rates (bioaccessibility) and of publications in which bioavailability was determined, usually in rats. For definitions of terms, see also Sect. 2.8.4. Resorption availability varied between 25 and 85% for B[*a*]P and B[*k*]Flu and was below 1.3% for the 5 other PAH. Experiments on inhalation bioavailability were also carried out mainly with B[*a*]P or pyrene and showed values between 20 and 65% for B[*a*]P, with dependence on the level of the concentrations and the diameter of the particles. Again, very different methods were used. A mechanistic model for estimating the resorption availability of gaseous- and particle-bound SVOCs was developed by a French research group (Wei et al. 2020). Under the model assumptions of this study, for phthalates, the resorption availability after inhalation ranged from 62 to 66% for DEHP and DiNP, from 93 to 98% for DnBP, DiBP, and DEP, and from 71 to 79% for PBDE and 78% for B[*a*]P.

2.6.3 *Particle Deposition*

The location and level of deposition of particles in the respiratory tract depend on particle size and are influenced by respiratory tract morphology, respiratory mechanics (frequency, depth, respiratory flow), particle geometry, and chemical–physical characteristics of the particles (Deng et al. 2019). Figure 2.4 shows deposition rates for the main regions of the respiratory tract (nasopharyngeal, tracheobronchial, alveolar) based on a predictive computational model from the International Commission on Radiological Protection (ICRP 1994). It is assumed that particles are not inhaled as aggregates, in which ultrafine particles are usually present, but as single particles of defined size via the nose. Particles are predominantly deposited by three mechanisms (Löndahl et al. 2014):

- Impaction, in which particular particles of approx. >3 μm impact with high velocity due to mass inertia on a surface, in particular at branching points of the bronchi.
- Sedimentation, in which smaller (approx. >0.3–3 μm) and slower particles are deposited on respiratory surfaces by the action of gravitational forces.
- Diffusion, in which very small particles (approx. <0.3 μm) take on changing directions due to Brownian molecular motion through interaction with the gas molecules in the air and can thus be deposited on a surface with a certain probability. However, in this case, a large proportion of the particles is also exhaled again.

A current compilation of studies on the deposition of micro-and nanoscale particles can be found in Islam et al. (2020). Löndahl et al. (2014) already summarized the results of about 40 studies on the deposition of particles <300 nm. In particular, they discuss the manifold methodological problems in the context of study performance.

While particles with a diameter >10 μm are deposited almost exclusively in the nasal region in all age classes, deposition in the extrathoracic region increases with increasing airflow and particle diameter for particles of 0.5–10 μm but decreases with age (Cheng et al. 1995; Golshahi et al. 2013; Xi et al. 2014). In general, it must be taken into account that the differences between individuals can be large. Compared to adults, children's exposure is about a factor of 2 higher when normalized to the respiratory surface (Ginsberg et al. 2004).

Inhaled ultrafine or nanoscale particles with a diameter of <100 nm are deposited in the respiratory tract predominantly by diffusion processes, whereas electrostatic precipitation occurs only if the particles are electrically charged (Oberdörster et al. 2005; Kleinstreuer et al. 2008). Deposition of these particles is observed in all three main regions of the respiratory tract. Particles with a diameter of 1 nm are deposited to 90% in the nasopharyngeal compartment, to about 10% in the tracheobronchial region, and almost not at all in the alveolar region (Oberdörster et al. 2004). Particles of 5 nm, on the other hand, show a uniform deposition of 30% in all regions, while 20-nm particles accumulate to more than 50% in the alveoli (ICRP 1994). The

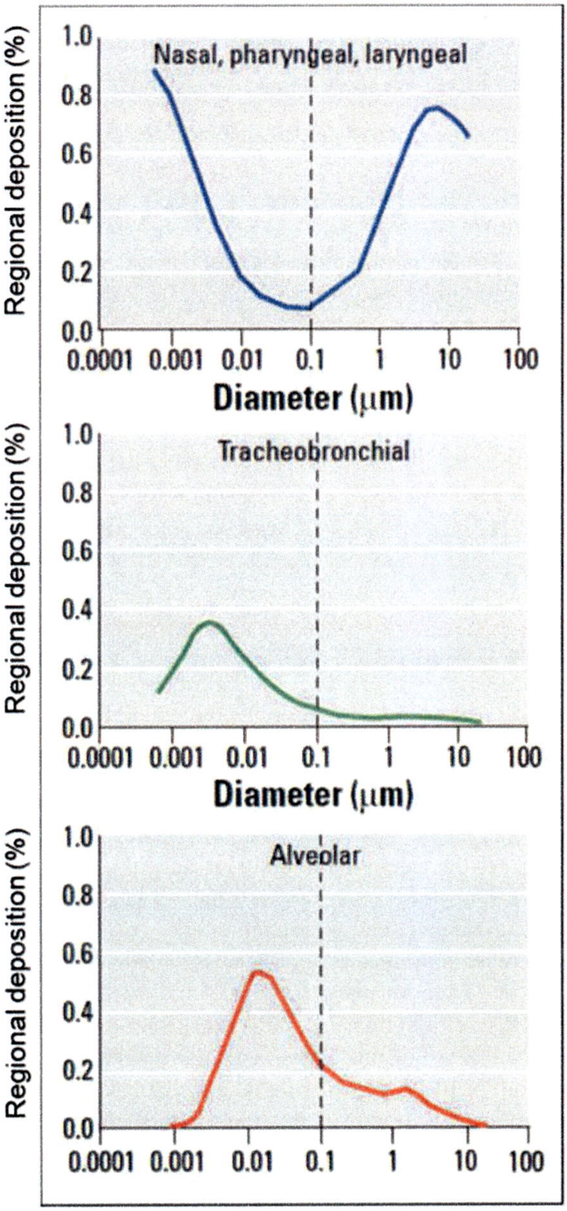

Fig. 2.4 Proportion of fractional deposition of particles in the areas of the human respiratory tract during normal nasal breathing (modified after Oberdörster et al. 2005)

different deposition efficiencies are decisive for the potentially toxic effects of inhaled particles of different sizes for their distribution into extrapulmonary organs.

A first study suggests that the total deposition of nanoparticles (50 nm diameter; 10,000 particles/cm^3) in the lungs was significantly higher, mean 13% vs. 7%, in 19 COPD patients compared to 19 non-smoking lung healthy individuals (Aaltonen et al. 2018). In contrast, for diesel engine particles with diameters of 20–50 nm and concentrations of $<10^5$–10^6 particles/cm^3, deposition was lower in COPD patients compared to healthy individuals (Löndahl et al. 2012). In a study of 128 children, a high deposition of 58% was observed upon inhalation of an urban particle mix (20 nm diameter: 6600 particles/cm^3) (Guo et al. 2020). Consideration must be given to the high interindividual variance in particle deposition, which ranged from 1–7% (50 nm), 6–24% (75 nm), and 14–44% (100 nm) in 7 healthy subjects (Jakobsson et al. 2016).

Fibrous particles represent a special case, for which only limited data on deposition after inhalation are available so far (see Asgharian et al. 2018). In this case, inhalability, shape, and orientation in the air stream, and density of fibers are key determinants of behavior in the respiratory tract. One way to model the deposition/retention of fibers in the lungs could be to form an equivalent diameter, which is calculated taking into account the fiber geometry.

2.7 Basics of Absorption Via the Skin

Absorption into the organism can occur via the skin during direct contact with products such as clothing, cosmetics, and personal care products, toys, or cleaning agents. Transdermal absorption depends, among other things, on the concentration level of the substance and its chemical structure, water solubility, octanol-water partition coefficient, exposure time, the respective formulation of the substance, and the thickness and integrity of the skin. Again, skin permeation studies have been conducted using animal skin models, predominantly rodents (rat, mouse and guinea pig), and human ex vivo skin (Olkowska and Gržinić 2022). In general, for phthalates the results obtained using animal or human skin seems comparable, but it is necessary to investigate the metabolization of phthalates on the skin, which can drastically impact the permeation rates.

Another important issue is the possible uptake of gaseous indoor air pollutants via the skin. While for many substances findings from animal experiments and experimental studies in humans on the uptake after application of a substance to the skin are available, they are very limited for uptake directly from the air. Weschler and Nazaroff (2012), considering a transdermal coefficient and different modeling assumptions, estimated dermal versus inhalation uptake (D/I) for about 80 indoor air pollutants. Based on these theoretical considerations, they arrive at a D/I ratio of >10 for parabens, low-molecular-weight phthalates, synthetic musks, and o-phenylphenol, and a D/I of 1–10 for texanol, α-terpineol, 4-oxopentanal, chlorpyrifos, linalool, and 2-butoxyethanol (ethylene glycol monobutyl ether). Weschler et al. (2015) exposed 6 men over 6 h to a mean of 114–140 μg/m^3 DnBP and 233–291 μg/m^3 DEP, respectively, once to estimate dermal uptake alone and once separately for inhalation and dermal uptake. While direct dermal airborne uptake was 4.0 μg/(μg/m^3) for DEP and 3.1 μg/(μg/m^3) for DnBP, inhalation uptake was 3.8 μg/(μg/m^3) (DEP) and 3.9 μg/(μg/m^3) (DnBP), respectively. As a result, the study shows that significant dermal uptake directly from the air must be expected for these two chemicals. This is also true for substances with comparable physicochemical characteristics. Based on these results, Lorber et al. (2016) conclude that the sum of inhalation and dermal uptake relative to the total uptake of DnBP in the USA is approximately 25%. Based on a toxicokinetic model and urinary excretion of the specific metabolite, they conclude that the theoretically derived transdermal models overestimate DnBP intake by a factor of 1.1–1.4 compared to cumulative metabolite excretion. When subjects were exposed only via inhalation, no difference between model and measurement could be found. Comparable results were obtained in another study, in which a certain overestimation was also observed for DnBP intake, but not for DEP intake (Morrison et al. 2016). One thing to consider in the aforementioned studies is the large variance in individual results, varying by a factor of 4 (DEP) and 6 (DnBP) for dermal uptake between individuals.

Another study attempted to model dermal uptake of various phthalates by wipe sampling of body surfaces (Gong et al. 2014). The authors estimate that approximately 10–20% of the total exposure may be due to dermal uptake after exposure via air.

Pelletier et al. (2017) emphasize performing uncertainty analyses as part of exposure assessments. For calculations of gaseous SVOC, they conclude that the variability of direct dermal uptake from the air depends primarily on the variability or uncertainty of indoor air concentrations and only secondarily on the uncertainties related to physicochemical parameters (e.g., the octanol/air partition coefficient).

2.8 Dust as a Basis for Risk Evaluation

The examination of indoor dust for pollutants has been firmly established for a long time and is carried out on a large scale. The essential aim here is to obtain an initial overview of the pollution situation indoors and to identify any previously unrecognized sources or undetected substances. On the other hand, various uncertainties have to be taken into account when evaluating the results, which will be discussed later in this chapter. In a comprehensive review of about 3000 articles and 355 chemicals, Dong et al. (2019) compiled the priority substances detectable in dust. Figure 2.5 shows the organic substances with the highest weighted medians. Among them, phthalates, organophosphate flame retardants, and chlorinated kerosenes are found in particular, in addition to fatty acids. Also, many metals can be quantified in higher contents in dust. In a meta-analysis, Mitro et al. (2016) had already published a compilation of data on indoor dust exposure to 45 priority chemicals and quantified inhalation and dermal uptake in addition to oral uptake, based on the physicochemical characteristics of the individual substances. In this case, too, phthalates were of greatest importance.

Although there is no universal definition of the term indoor dust, today it is understood to mean the dust deposited on the surfaces of an indoor space. It is a complex and very heterogeneous mixture of particles, including pollutants adsorbed

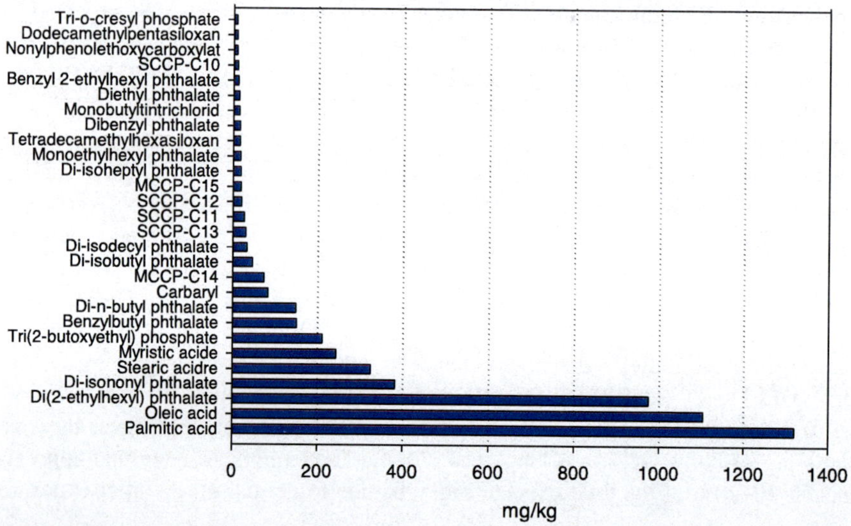

Fig. 2.5 Ranking of organic substances in house dust (modified after Dong et al. 2019) (SCCP: short-chain chlorinated paraffins; MCCP: medium-chain chlorinated paraffins)

to them, from a wide variety of natural and anthropogenic sources. It thus also represents the "pollutant memory" of a room. Indoor dust can consist, for example, of particles from the outdoors (sedimented airborne particulate matter, entrained soil particles) and from indoor combustion processes (e.g., from kitchens and furnaces), as well as fibers from textiles and furnishings, abrasion from building materials and interior furnishings, organic material from room occupants (e.g., hair, dander), and biological material such as pollen, bacteria, fungi, and insects, or their constituents and excretory products. However, the quantitative proportions of these potential dust constituents can vary considerably. For example, Butte and Walker (1994), studying 318 dust samples, conclude that the organic content in the samples can vary from <5% to >95%. In summarizing studies up to 1998, Oomen and Lijzen (2004) assume that the proportion of soil particles in house dust can range from 8% to >80%.

Hand-to-mouth contact is a major contributor to indoor dust intake. It is part of the natural play instinct of children and can account for a quantitative share of their exposure. For example, children up to 18 months of age exhibit this behavior on average every 20 minutes, and 18–36 months old children as often as every 5 minutes (Smith and Norris 2003). Results of two meta-analyses showed that hand-to-mouth frequency was highest in toddlers with 15–26/per hour and then decreased in older children (Xue et al. 2007, 2010). Against this background, a higher house dust intake has to be expected especially in toddler and crawling age.

2.8.1 Basic Aspects of an Assessment of Dust

It is not possible to estimate substance intake and, as a consequence, possible health risks from house dust concentrations alone, since a variety of uncertainties at different levels strongly influence the results. The following factors are of importance here:

- Sampling conditions: e.g., from which surface in which room the dust is taken and with which system? Is the dust vacuumed old or new? Is the dust bag of a vacuumer used?
- Analytics: e.g., whether the dust is sieved or only large dust particles such as stones and hair are removed. Is the sample homogeneous? Which dust fraction is ultimately analyzed (2 mm or 63 μm fraction)? How are values below the limit of quantification handled?
- The exposure scenarios chosen: Are children or adults, a single or prolonged exposure considered? Are unusual behaviors with possibly high dust intake considered? What quantity of dust is assumed to be ingested by an individual?
- Age and personal behavior of the group of people under consideration: e.g., play behavior, frequency of hand-to-mouth activities, indoor residence times, personal cleaning habits, indoor cleaning.
- Absorption pathways and conditions: e.g. behavior of substances in the gastrointestinal tract. What proportion is actually absorbed? With which toxicological values or health endpoints should the absorbed dose be compared?

Against this background and the resulting large variance of the results, a health-related evaluation of pollutant concentration in indoor dust is questionable. However, a consideration in the context of the source search and for the delimitation of the question can be meaningful.

In the following, some important key points of the aforementioned uncertainties will be presented in more detail.

2.8.2 Intake Rates for Dust

Moya and Phillips (2014) summarized the methodology and results of methods to quantify soil and dust ingestion rates: the tracer element method, the biokinetic model comparison method, and the activity pattern method.

Uptake levels were estimated based on tracer studies using substances such as titanium, aluminum, silicon, or vanadium. These are substances that, on the one hand, are ubiquitously distributed in the soil, but, on the other hand, are not systemically absorbed after oral intake and are thus completely excreted with the feces. Quantification in the soil and the feces can then be used to calculate the amount added to the soil if other sources of the tracer (e.g., titanium from toothpaste) can be reliably excluded. In this context, a variety of study results have been

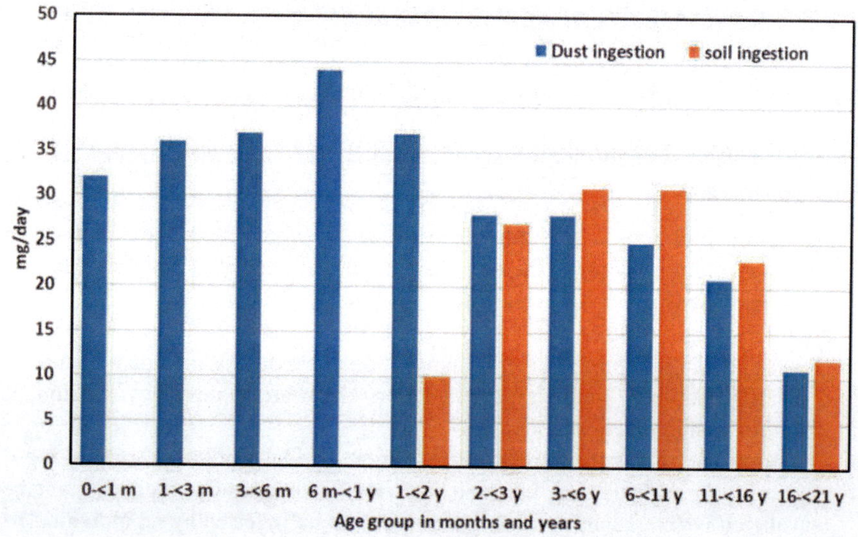

Fig. 2.6 Annual mean daily total dust and soil ingestion rate (mg/day) (modified after Özkaynak et al. 2022)

described in the scientific literature (e.g., Calabrese et al. 1989; Clausing et al. 1987; Davis et al. 1990; Van Wijnen et al. 1990; Bothe 2004).

Furthermore, biokinetic models compare the internal exposure of a substance in body with predictions from a biokinetic model that includes exposures to toxicants in air, food, water, soil, and dust via ingestion, inhalation, and dermal routes (Moya and Phillips 2014). An example is the estimate of soil uptake rates of lead (Von Lindern et al. 2003).

In addition, the amount of added soil material was calculated by some authors via model calculations considering parameters such as particle adhesion to body surfaces, dust load, event duration, contact duration with surfaces, particle transfer from room surfaces to the skin, etc. (Özkaynak et al. 2011; Wilson et al. 2013). For example Özkaynak et al. (2022) developed their model for age groups previously not studied and including new exposure scenarios and described results of annual means of total daily dust and soil ingestion rates by age group (see Fig. 2.6).

Several reviews have summarized results of dust intake rates (Cohen Hubal et al. 2000; Klenow et al. 2016; US EPA 2017).

In terms of dust, the US Environmental Protection Agency, in its Exposure Factors Handbook, after evaluating all available scientific studies, arrives at a daily dust intake as a central tendency (or as an upper percentile) of 20 mg/day (40 mg/day) at ages <6 months, 40 mg/day (100 mg/day) at ages 6 ≤ 12 months, 30 mg/day (100 mg/day) at ages 2 ≤ 6 years, 30 mg/day (100 mg/day) at ages 6 ≤ 12 years, and 20 mg/day (60 mg/day) at ages 12 years and older (US EPA 2017). However, the agency specifically notes that the confidence of the soil intake rates is low due to very limited study quality and large uncertainties. It also cautions that the estimates for all age groups are based on the assumption that 55% of total exposure is via house dust and 45% via soil.

In addition, some children show unusual behavior by ingesting larger amounts of soil (pica behavior, geophagy). In these cases, soil ingestion rates can be significantly higher, up to 5–7 g/day (Calabrese and Stanek 1992). The US EPA uses a ground intake rate of 1 g/day for these behaviors (US EPA 2017). However, overall data are limited and the prevalence of these behaviors in the population is unknown.

2.8.3 *Significance of the Particle Size Distribution*

The particle size distribution of the house dust is also very variable. Since different methods of sample preparation are used, e.g. sieving to a defined particle fraction, the question arises as to what influence these measures can have on the measurement result. Although study results are not always consistent, the majority tend toward higher concentrations the smaller the particles examined (Ruby and Lowney 2012). One explanation could be the relatively larger surface area of small particles compared to larger particles. For example, Gustafsson et al. (2022b) observed a surface area of 2.5 m^2/g in the 3.7 μm fraction of 32 vacuum cleaner bags dust samples and only 0.8 m^2/g in the 1000–2000 μm fraction. Nevertheless, Melymuk et al. (2022) analyzed different dust size fractions for carbon and PAH content and observed an irregular distributions across size fractions which did not follow the theoretical understanding of partitioning to size fractions and suggests caution in interpretating. The unclear separation of fibers and fine particles adhering to fibers may be contributing to this effect and again the unknown contributions from polymer particles and fibers.

In 28 offices in China 45%, 29%, and 26% of the total mass was found in the >2000 μm, 200–50 μm, and <50 μm particle fractions, respectively (Cao et al. 2013). In this study, the highest content of 18 brominated flame retardants was found in the fraction of particles with an aerodynamic diameter of about 20 μm. This was supported by Gustafsson et al. (2018), they found 41% of the particle mass in the fraction >2000 μm in 32 samples from Swedish vacuum cleaner bags. In contrast, in another study of 15 offices, no concentration differences of organophosphate flame retardants were found in the 5 collected fractions of dust between 50 and 2000 μm (Guo et al. 2019). Meanwhile, Liu et al. (2018) found the highest PAH concentration in the <43–2000 μm fraction in 6 fractions of office dust with diameters of <43–2000 μm. For B[*a*]P, for example, 824 ng/g in the fraction <43 μm, but only 57 ng/g in the fraction from 200–2000 μm. For phthalates, no clear result was obtained in this study.

About metals in indoor dust, the analysis of 376 samples in Australia revealed arsenic, copper, lead, and zinc in particular in the 90–150 μm fraction, whereas chromium and manganese were detected in the sieve fraction <45 μm (Doyi et al. 2020). Other studies found evidence that lead occurs particularly in smaller sieve fractions (Bright et al. 2006; Kempken and Butte 2010; Ikegami et al. 2014).

Choate et al. (2006) were able to show, albeit for soil particles, that over 90% of soil particles adhering to hands had a diameter below 250 μm and about 70% below 63 μm. Another group of studies also described that almost exclusively particles below 250 μm were found on the skin and the maximum of the particle size distribution was for particles with a diameter of 20–30 μm (Cao et al. 2012, 2013).

2.8.4 Bioaccessibility and Bioavailability

To assess the amount of substance in dust that has the possibility of being absorbed systemically into the organism after oral intake, the bioaccessibility or, even better, the bioavailability of the substance must be known.

Two important definitions exist that must be distinguished:

- Bioaccessibility (BA) is the amount of a soil-bound contaminant that can be mobilized in the gastrointestinal tract for subsequent absorption after oral intake.
- In contrast, the term bioavailability (BV) refers to the rate and proportion of a pollutant—in this case—supplied orally that is available systemically in the organism, i.e. after absorption from the gastrointestinal tract. Often, only the relative bioavailability (BV_{rel}) can be determined, in which the uptake of a substance from the house dust matrix is compared relative to the uptake of an orally supplied standard preparation in animal experiments or volunteers. For this purpose, the blood concentration of the target substance is measured over time and the AUC (area under the curve) is determined. The AUC is the integral of the area under the plasma level-time curve, which is proportional to the amount of contaminant occurring in the systemic circulation. To determine the BV_{rel}, the AUC of the amount of pollutant orally supplied with the dust is then compared with the AUC of a standard solution of the same pollutant.

For the German Soil Protection Ordinance, for example, a method for determining the bioaccessibility of organic and inorganic pollutants from contaminated soil materials has been standardized and validated for some pollutants in animal experiments in DIN 19738:2017-06 (DIN 2004; Barkowski et al. 2016). The aim was to obtain data on the mobilizable pollutant content as realistically as possible using an in vitro test system mimicking the human gastrointestinal tract using synthetic digestive juices. For house dust, however, this systematic, methodical information is lacking. As with soil, the BA of pollutants in dust is influenced, on the one hand, by the substance properties (e.g., lipophilicity, molecular weight, dissociation, binding strength to particles) and, on the other hand, by the specific conditions in the gastrointestinal tract (e.g., pH in the different sections, residence times, type and composition of other foods).

In a review paper, Raffy et al. (2018) summarize study results on the bioaccessibility (BA) of organic pollutants between 2011 and 2017. The following weighted medians (ranges) for contaminant classes are described in this work: PAH 7% (2–17%), phthalates 13% (1–81%), PBDEs 37% (11–71%), organophosphate flame retardants 55% (8–103%), PCB 74% (3–92%). Recent studies also show considerable differences within a substance group. For example, for PAH in Chinese dust samples from offices, a BA of 6.3% was found for benz[*a*]anthracene and 88% for acenaphthene (Liu et al. 2018). The highest BA was observed in the largest particle fraction of 200–2000 μm. In an in vitro digestion test, BA of phthalates from house dust ranged from 10.2% for DEHP to 32% for DMP (Kang et al. 2012).

For organophosphate flame retardants, the mean BA in 17 US dust samples (<53 μm) was 80% (Fang and Stapleton 2014). In 17 Chinese office dust samples (50–2000 μm), mean BA for TCP was 24% and ranged from 86–100% for TCEP, TCiPP, and TBOEP (Guo et al. 2019). For flame retardants, mean BA in 17 US dust samples (<53 μm) ranged from 30% (BDE 209) to 75% (BDE 47) for individual PBDE, 50% for EHTBB, and 25% for BEH-TEBP (Fang and Stapleton 2014). For the three isomers of HBCDD, only a mean RV of 12% (α-HBCDD), 16% (β-HBCDD), and 19% (γ-HBCDD) were reported in 7 dust samples (Garcia-Alcega et al. 2016).

For pyrethroids, a mean BA of 32% (cypermethrin), 46% (permethrin), 50% (deltamethrin), and 51% (cyfluthrin) were found in 6 dust samples that had been sieved to <250 μm (Wang et al. 2018). When proteins or carbohydrates were added to the test system at the same time as the dust samples, BA also increased.

In the investigation of 200 house dust samples from northern Germany, which were sieved to <63 μm, mean BA of between 44% (PCB 153) and 90% (PCB 180) were found for PCB, 49% for permethrin, and between 31% and 56% for organochlorine pesticides when milk was added at the same time (Ertl and Butte 2012). Also in this study, the BA of all substances investigated was lower when no milk had been added at the same time.

For metals, Turner (2011) summarized study results before 2007 that showed bioaccessibility of lead in house dust between 25 and 80%. The authors themselves found 30% (stomach model only) and 18% (stomach, intestine model) for lead and 53% and 42% for cadmium. In 376 house dust samples from Australia, the mean BA of lead ranged from 42% to 62% (Doyi et al. 2020). Higher BA of about 40% (arsenic), 80% (lead), and about 95% (cadmium), and 73% (arsenic), 63% (lead), and 68% (cadmium) were determined by studies from France and China (Glorennec et al. 2012; Ma et al. 2020).

While the database on relative bioavailability (BV_{rel}) of soils is very large, such studies with indoor dust have rarely been performed. For example, Huwe et al. (2008) studied the absorption and bioaccumulation of various PBDE in four male rats over 21 days. BV_{rel} ranged from <5% (BDE 209) to approximately 70% (BDE 47, BDE100). In contrast, a very high BV_{rel} of 26–96% was found for BDE 209 in a Chinese mouse study in which 6 different house dust samples were administered (Pan et al. 2016). The authors explain the high BV_{rel}, which was also determined for the parallel observed absorption availability, with the specific delivery form in the oil.

In a German study, 5 house dust samples (<63 μm sieved) were administered orally with the feed to 7 piglets (Plichta et al. 2019). The mean BV_{rel}, in this case, was 43% for DEHP, 11% for DINP, and 26% for DINCH. Differences between individual dust samples were small in this study. In an initial study in human subjects, 11 participants ingested 6 grams of house dust (Plichta et al. 2022). From the examination of the excreted metabolites, a mean bioavailability of 51% for DEHP, 26% for DiNP and 19% for DINCH could be calculated in this study.

Gustafsson et al. (2022a) administered orally and by inhalation residential dust samples (diameter > 4 μm) spiked with PFOA to 10 rats and found that PFOA is readily bioavailable and has a rapid systemic distribution following an inhalation or oral exposure.

For 8 house dust samples (<100 μm sieved) fed orally to mice, BV_{rel} ranged from 45–91% for phenanthrene, 19–72% for fluoranthene, 15–56% for pyrene, and 32–75% for benzo[*a*]pyrene (Kang et al. 2018).

In an initial study, Freeman et al. (1995) reported a mean BV_{rel} for arsenic of 19% in female monkeys administered house dust from the vicinity of an industrial plant. Li et al. (2014b) found a mean BV_{rel} of 51% for arsenic when 12 dust samples (<150 μm sieved) were orally administered to female mice, although this varied from 22 to 86% for individual samples. The BV_{rel} of lead ranged from 29 to 60% in another Chinese experiment with mice fed 12 dust samples (Li et al. 2014a). The Chinese research group also found a mean BV_{rel} for arsenic of 17% (9–37%), for lead of 25% (11–34%), and for cadmium of 46% (28–68%) in the mouse model (Zhao et al. 2018). For lead and cadmium in dust, the BV_{rel} here was lower than that calculated for intake from food (rice). Using 3 dust samples from the USA, higher values were found than in the aforementioned studies, with RV_{rel} of 41%, 64%, and 100% (Sowers et al. 2021). The authors attribute this to the use of lead-based paints, which contained high levels of readily absorbable cerussite (white lead ore) and hydrocerussite.

2.9 Special Indoor Exposure Situations

2.9.1 Use of Laser Printers and Photocopiers

Possible sources of particles during the printing process include the toner and the paper, but also components of the printer such as lubricants or additives in the plastics (Gu et al. 2020). For example, modern toners of laser printers consist of synthetic resins, magnetizable metal oxides, pigments, and various additives (Gu et al. 2020).

When laser printers and photocopiers are operated indoors like offices a significant increase in the number of ultrafine particles can be observed (Lee and Hsu 2007; Kagi et al. 2007; Destaillats et al. 2008; Schripp et al. 2008; Wensing et al. 2008; Betha et al. 2011; Tang et al. 2012; Wang et al. 2012; Castellano et al. 2012; Setyawati et al. 2020). For example, in a study of 110 laser printers, the devices were observed to emit between 3.4×10^8 and 1.6×10^{12} particles per minute, with an aerodynamic diameter of 34 nm (Scungio et al. 2017). The level of emissions can vary greatly depending on the type of printer, the number of printed pages, and the specific ventilation conditions (Betha et al. 2011).

Further studies could show that laser printer particles are almost exclusively secondary organic aerosols (SOA), which are formed by condensation and nucleation from the low-volatile organic compounds emitted during the printing operation, while toner particles are hardly found (Gu et al. 2020).

Various studies have been conducted on the possible health effects of laser printer emissions summarized by Gu et al. (2020). They concluded that health effects are likely to be low according to the current state of knowledge. Of note in this context is a controlled exposure study in which 52 subjects were exposed to laser printer emissions in a test chamber for 75 minutes (Karrasch et al. 2017). There were 23 healthy subjects, 14 subjects with mild but stable asthma, and 15 subjects with a history of symptoms related to laser printers. Exposure was blinded once to 3000 particles/cm^3 (background) and then at 100,000 particles/cm^3. There were no coherent effect patterns in lung function, bronchial and alveolar NO, cytokines in serum and nasal secretions concerning the aforementioned subgroups. Overall, the small changes were classified as not clinically significant. Also about various psychological or cognitive effects, no difference was seen between the two exposure scenarios in the study, however, the group that already reported anamnestic effects in the vicinity of laser printers showed a higher psychological susceptibility to adverse reactions (Herbig et al. 2018).

2.9.2 *Use of 3D Printers*

A new source in office interiors, possibly also in homes, can be 3D printers. These three-dimensional printing devices use the so-called FFF (fused filament fabrication) or FDM (fused deposition modeling) technology. In this process, a thermoplastic material is melted and deposited layer by layer onto a moving support plate to form a three-dimensional element. Acrylonitrile butadiene styrene (ABS), polylactide (PLA), polycarbonate (PC), and rigid polyethylene (HDPE, high-density polyethylene) are commonly used as thermoplastic materials (Azimi et al. 2016).

2.9.2.1 Health Aspects

There are hardly any studies on the possible health effects of the emitted particles so far. With regard to the effects in experimental animals, only two studies are available to date. Farcas et al. (2020) exposed male Sprague Dawley rats to a single concentration (240 µg/m³ with a geometric mean particle mobility diameter of 85 nm) generated by ABS emissions from a commercial desktop 3-D printer. Rats were exposed to ABS emissions or control air for 4 hours per day, 4 days/week for five exposure durations (1, 4, 8, 15, and 30 days). Overall, minimal transient signs of pulmonary and systemic toxicity like increased number of macrophages, higher amounts of IFN-γ and IL-10 in the lavage fluid, but neither pulmonary oxidative stress responses nor histopathological changes in the airways were observed. In another study four male Sprague Dawley rats placed in a chamber that were exposed to printer emitted particles of on average 72 μg/m^3 (mean diameter: 50 nm; mean PNC: 497000 particles/cm^3) 5 h per day over 21 days (Carll et al. 2020). Compared to controls an increase in cardiovascular risk by augmenting sympathetic influence, impairing ventricular performance and repolarization, and inducing hypertension and arrhythmia were seen.

Gümperlein et al. (2018) exposed 26 healthy volunteers for one hour to 1,600,000 particles/cm^3 (mean) from a 3D printer (with ABS) or to a low-emitting printer (with PLA). There were no acute effects in terms of inflammation, but only a small increase in exhaled NO with ABS printing compared to PLA printing, which could be interpreted as a sign of a first inflammatory response. The same working group exposed 28 young subjects with seasonal allergic rhinitis to 3D printer emissions from PLA or ABS over two hours (Würzner et al. 2022). Compared with control data functional allergen response was not different between PLA and ABS, the diffusing capacity to inhaled nitric oxide slightly declined after both exposures, and overall the authors concluded that results do not provide solid evidence that short-term exposures to 3D printer emissions elicit health effects in subjects with allergic rhinitis that may be considered as clinically relevant.

2.9.2.2 Emission and Exposure in the Indoor Air

Typically, the particle number concentration (PNC) in the room increases significantly after the start of printing, then persists during the printing operation, and drops at the end (Azimi et al. 2016; Steinle 2016; Seeger et al. 2018; Ding et al. 2020; Bernatikova et al. 2021). Approximately 10^8 to 10^{11} particles are emitted per minute during operation (Stephens et al. 2013; Azimi et al. 2016; Vance et al. 2017). The number and diameter of the emitted particles depend on the type of plastic, its color, and the printing temperature (Yi et al. 2016; Stabile et al. 2017). Investigating 44 marketable filament products in a test chamber, the total number of emitted particles varied by approximately four orders of magnitude (10^9 to 10^{13}), indicating that origin of polymers, manufacturer-specific additives, and undeclared impurities have a strong influence (Tang and Seeger 2022).

Two studies in which HEPA filters were used in 3D printers were able to show that particle emissions as PNC could be reduced by up to >99% (Kwon et al. 2017; Katz et al. 2020). However, since printers are not fully enclosed devices, small emissions that can reach indoor air in addition to the filter must be expected (Katz et al. 2020). Electron-optical investigations using TEM (transmission electron microscopy) showed compact carbon aggregates and chain-like particle structures when operated without a filter system and using ABS (Katz et al. 2020). In another study, containment of a 3D printer, e.g. full enclosed over the printer, significantly reduces the emissions, especially if the enclosure is equipped with ventilation (Viitanen et al. 2021).

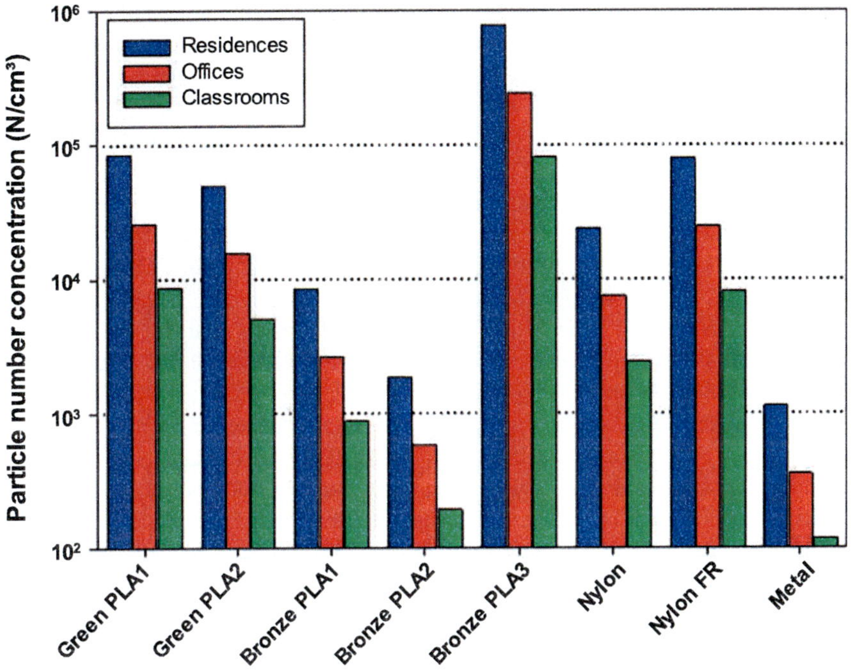

Fig. 2.7 Estimated particle number concentration in three indoor spaces using different filaments for printing (modified after Zhang et al. 2023) (Room volume 30.6, 28.2, and 231 m^3 for office, residence, and school; air-exchange rate: 0.68, 0.23, and 0.82 per hour for office, residence, and school)

While no significant increase in PNC is to be expected in large, adequately ventilated office spaces, an increase in mean PNC from 960 particles/cm^3 to 3200 particles/cm^3 could be determined in a small, poorly ventilated space (Steinle 2016). Similar results were obtained by Zontek et al. (2017) and García-Gonzáleza and López-Pola (2022). Poikkimäki et al. (2019) also modeled indoor air concentrations in a 40 m^3 living room and an air-exchange rate of 0.5/hour. They calculated mean levels when using ABS of ~30,000 particles/cm^3 and ~1000 particles/cm^3 for PLA, respectively. Zhang et al. (2023) analyzed particle size distribution (7 nm –10 μm) in a test chamber using three different 3D printers and three pure PLA filaments with green and bronze coloring, two composite PLA filaments with bronze powder, two composite nylon filaments, and one composite metal filament. After that, PNC was estimated for a typical residential room, an office, and a school classroom (see Fig. 2.7). The modeled estimation of PNC in the three indoor spaces showed high indoor concentrations only for residences and printing operations with PLA filaments, and especially high when PLA filaments with an 80% bronze content were used.

Currently, a so-called portable 3D pen is being very actively marketed with the argument that it promotes creativity, especially among children. Like a 3D printer the 3D-pen operated with the FFF technology and the extruder nozzle of a 3D-pen is moved manually by the user. Kim and Lee (2022) tested the emissions of two pens in a 2.5 m^3 chamber and the more realistic exposure in a 197 m^3 classroom and found high median PNC for ABS, PLA, and PCL of $4.2 \times 10^5 - 1.2 \times 10^6$, $5.4 \times 10^4 - 1.1 \times 10^5$, and $3.4 \times 10^4 - 4.6 \times 10^4$ particles/cm^3, respectively, about 20 cm from the point of origin of the emission. It remains questionable whether such a toy is really recommendable for children without protective measures.

In addition to particulate emissions, the emission of VOC must also be taken into account. For example, Azimi et al. (2016) found high emissions of caprolactam when using nylon filaments, styrene when using ABS and HIPS (high impact polystyrene), and lactide when using PLA in five 3D printers operated with different plastic materials. Comparable results were also reported by Floyd et al. (2017), Zhang et al. (2019), and Wojtyła et al. (2020) in test chamber studies. TVOC emissions, depending on printing temperature, plastic material, printer type, and material color, varied on average from 147 µg/h (polyvinyl alcohol) to 1660 µg/h (nylon filaments) (Davis et al. 2019). From these data, the authors calculated TVOC indoor air concentrations with mean values of 30–256 µg/m^3 for a typical living room or 3–26 µg/m^3 for a school classroom. García-Gonzáleza and López-Pola (2022) quantified VOC during seven printing tests using PLA and ABS in a 113 m^3 room and found low levels of limonene (1.0–1.2 µg/m^3), styrene (0.7–1.2 µg/m^3), toluene (2.0–5.2 µg/m^3), naphthalene (0.8–0.9 µg/m^3), and benzene (0.8–1.8 µg/m^3). Again, in room with a volume of 264 m^3 and with three fresh air inlets (no window) equipped with two 3D printers relatively low means of 30 µg/m^3 for formaldehyde, 8.4 µg/m^3 for acetaldehyde, 6.3 µg/m^3 for acrolein, 0.4 µg/m^3 for benzene, 13.0 µg/m^3 for toluene, and 2.7 µg/m^3 fore styrene were reported (Kim et al. 2022).

Because metals in the form of elemental particles, salts, metal oxides, and organometallics have been added into polymers for a variety of functional and aesthetic purposes at relatively high concentrations of >5% wt, Tedla et al. (2022) gave an overview of trace metal contents of various commercial filaments. Overall, common metal emissions detected in ABS and PLA include Cu, Al, Fe, Zn, Ca, Mg, and Na, in ABS emissions Cr and Ni, in PLA emissions Ti, and in PC emissions Cr, Al, Fe, Ni, Ca, and Na. Using a chamber and printing with seven filaments, modeled concentrations in a typical residence were estimated to 34 and 298 ng/m^3 for aluminum, 0.22 ng/m^3 for cadmium (only one result), and 0.06–0.48 ng/m^3 for lead (Zhang et al. 2022a). Silicon was in this case the element with the highest concentrations of 82–3930 ng/m^3.

2.9.2.3 Recommendations

UL Chemical Safety announces that ANSI/CAN/UL 2904 "Standard Method for Testing and Assessing Particle and Chemical Emissions from 3D Printers" (ANSI 2019) has been published. It contains measurement and assessment protocols for the emission of particles and volatile chemicals from various 3D printers, printing media and printing applications, and applies to freestanding 3D printers.

The Indoor Air Hygiene Commission at the German Environment Agency has published guidance on dealing with 3D printers for domestic use (IRK 2020). In particular, the commission has recommended the following:

- Always ventilate sufficiently and, especially if the printer is used frequently, operate it in a separate room with good ventilation.
- To avoid unnecessary exposure to emissions, a spatial distance should be maintained during the printing process.
- The manufacturer's instructions for operation must be strictly observed.
- Professional and commercially operated 3D printers are not suitable for home use.
- 3D printers do not belong in the hands of children.

2.9.3 *Open Fires*

The building sector is responsible for 9.8 Gt CO_2 equivalent emissions by upstream emissions from power generation and commercial heat, as well as emissions, for instance, by gas and coal boilers, and cooking and lighting devices that burn kerosene, biomass, and other fuels (Lamb et al. 2021). On a global scale 2.6 billion people are without access to clean cooking fuels and technologies (IEA 2021). Nevertheless, among low- and middle-income countries, the use of gaseous fuels rose steadily from 36% in 2000 to 51% in 2019, overtaking unprocessed biomass fuels (wood, crop waste, and dung) as the predominant type of cooking fuel. This represents 6% of the total greenhouse gases emissions.

On the other side, cooking and heating with fossil fuel materials (household air pollution) result in significant indoor air exposure (e.g., CO, PM, and VOC) of persons and subsequent severe health outcomes (Bruce et al. 2015; Odo et al. 2021; Aithal et al. 2021; Younger et al. 2022). WHO concluded in 2014 that inefficient and highly polluting stoves for cooking and heating resulting in some four million premature deaths annually among children and adults from respiratory and cardiovascular diseases, and cancer (WHO 2014).

During cooking activities in 112 (56 biomass, 23 kerosene, and 33 liquified petroleum gas (LPG)) households in Southern India TVOC concentrations were 4617 μg/m^3, 2839 μg/m^3, and 1639 μg/m^3 while total BTEX (e.g., benzene, toluene, ethylbenzene, m,p-xylene, o-xylene and styrene) concentrations were 139 μg/m^3, 121 μg/m^3 and 245 μg/m^3 among biomass, LPG, and kerosene using households, respectively (Puttaswamy et al. 2021). Median benzene levels were 67.5 μg/m^3 (kerosene), 21.5 μg/m^3 (LPG), and 19.6 μg/m^3 (biomass). Univariate analysis showed that season, type of food cooked, cooking type, and kitchen ventilation significantly influence indoor VOC levels among biomass fuel users but not in kerosene fuel users. The type of cooking (i.e., boiling or steaming) and kitchen configuration (i.e., moderate ventilation) significantly lowered total VOC levels by 3 to 4 times among LPG fuel households.

Balakrishnan et al. (2013) monitored 24-h $PM_{2.5}$ concentrations in 617 rural households in India on a cross-sectional basis between November 2004 and March 2005 and found in solid cook fuel using households means (medians) of 163 μg/m^3 (106 μg/m^3) in the living area and 609 μg/m^3 (472 μg/m^3) in the kitchen area. Fuel type, kitchen type, ventilation, geographical location, and cooking duration were significant predictors of $PM_{2.5}$ concentrations. Additionally, they summarized the results of 18 studies in the Americas, Africa, and Asia in 1999–2011 with $PM_{2.5}$ concentrations in a wide range between 37 and 1354 μg/m^3 (24-hour means).

In a recent US study during the winter median indoor $PM_{2.5}$ was 30 μg/m^3 in 10 Alaska homes, 29 μg/m^3 in 23 Navajo Nation homes, and 16 μg/m^3 in 59 Montana homes using wood stoves for heating (Walker et al. 2021).

In China, 2073 complete 24-hour personal measurements were performed at 17 villages in 2015–2016 and geometric means of 108 μg/m^3 in the heating season and 65 μg/m^3 in the non-heating season were reported (Lee et al. 2021). Outdoor $PM_{2.5}$ was the dominant variable for explaining within-participant variance in exposure (16%), while household fuel use (8%) and smoking status (27%) explained the most between-participant variance in this study. In another Chinese study 477 indoor air samples were collected in two rural and one urban area in 2017–2018 (Chan et al. 2021). Overall, the mean $PM_{2.5}$ levels were two to three fold higher in the cool than warm season and in rural than urban sites. Again, an additional study reported high $PM_{2.5}$ means of 135 μg/m^3 (winter) and 47.1 μg/m^3 (summer) in kitchen and 113 μg/m^3 (winter) and 50.8 μg/m^3 (summer) in living rooms in 81 homes in rural China (Huang et al. 2022). Fuel type can significantly affect $PM_{2.5}$ concentrations in the kitchen, with greater $PM_{2.5}$ pollution associated with wood combustion than electricity.

As part of the Prospective Urban and Rural Epidemiological (PURE) cohort, the PURE-AIR study was done in 120 rural communities in eight countries (Bangladesh, Chile, China, Colombia, India, Pakistan, Tanzania, and Zimbabwe) in 2017–2019 and gravimetric 48-hour $PM_{2.5}$ measurements were obtained from 2541 kitchens and from 998 individuals (Shupler et al. 2020). Figure 2.8 shows a concentration gradient across primary cooking fuels and nearly equivalent average personal exposure to kitchen exposure ratios of 0.79 for men and 0.82 for women.

First in 2006 WHO published annual interim air quality targets (IT-1: 35 μg/m^3; IT-2: 25 μg/m^3; IT-3: 15 μg/m^3; IT-4: 10 μg/m^3) for $PM_{2.5}$, in order to provide incremental steps for air quality improvement in settings where baseline levels are high (or very high) and short-term achievement of the guideline levels might be unrealistic in practice (WHO 2021). WHO stated that at 35 μg/m^3 the predicted relative risk of acute lower respiratory infections is 1.29 reaching only 1.02 at the concentration of 10 μg/m^3.

Pope et al. (2021) summarized the results of 50 studies from Africa, Asia, South and Latin America and concluded that $PM_{2.5}$ kitchen concentrations, burning biomass more cleanly through improved combustion stoves (ICS) with (63% reduction) or without (−52%) wth venting through flue or chimney. Forced-draft combustion

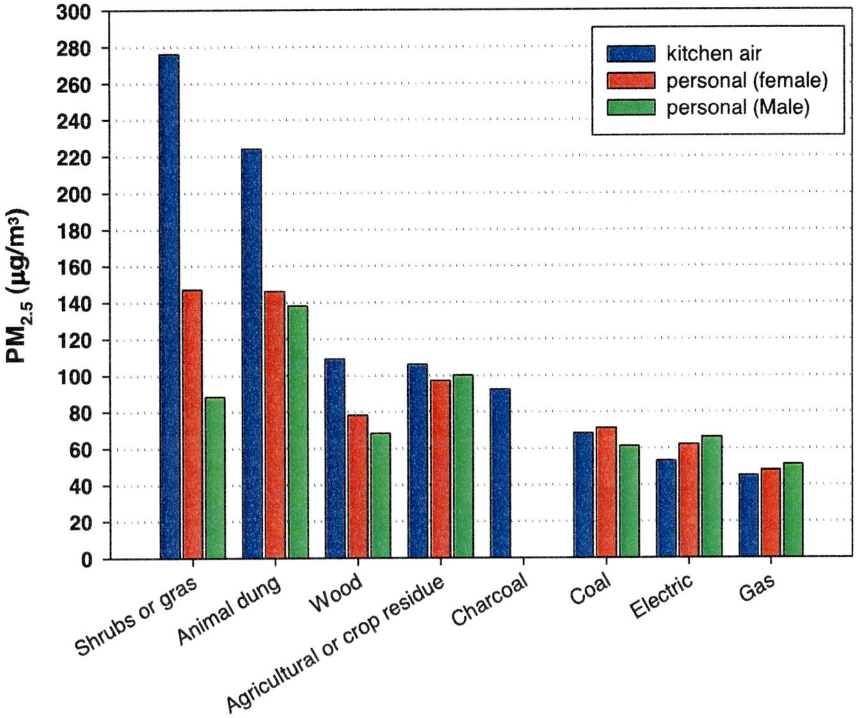

Fig. 2.8 $PM_{2.5}$ concentrations in indoor air by cooking fuel type using stationary and personal (for females and males) measurements (modified after Shupler et al. 2020)

(−50%) was less effective than cooking with clean fuels including ethanol (−83%), LPG (−83%), and electricity (−86%). Only studies of clean fuels consistently achieved post-intervention kitchen $PM_{2.5}$ levels at or below the health-based WHO interim target level 1 of 35 µg/m^3.

As an alternative to the fireplace stove, the so-called ethanol stoves operated with methylated spirits or burning gels are used indoors, on the one hand for decoration, but also to generate heat, on the other. Various studies in test rooms and living spaces have investigated indoor air pollution with combustion products (Schripp et al. 2014; Höllbacher et al. 2017; Habarta et al. 2018). Mean CO_2 levels ranged from 1082 to 3050 ppm and formaldehyde and acetaldehyde concentrations ranged from 62 to 117 µg/m^3 and 108 to 1440 µg/m^3, respectively, depending on the fuel type in the three studies. Operating ethanol stoves did not result in a significant increase in PM_{10} or $PM_{2.5}$ levels, but did result in mean particle number concentrations in the Habarta et al. (2018) and Schripp et al. (2014) studies of 91,000 particles/cm^3 and 4100–411,000 particles/cm^3 for methylated spirits and 136,000 particles/cm^3 and 37,000–224,000 particles/cm^3 for gels, respectively. In these studies, maximum NO_2 levels of 1270 and 2933 µg/m^3 as well as 780 and 771 µg/m^3 were found for spirits and burning gels, respectively.

2.9.4 *Aircraft Cabins*

The total number of passengers carried on scheduled air services continuously increased to 4.5 billion in 2019, before the restrictive measures to prevent the spread of COVID-19 pandemic let to a collapse in passenger numbers (ICAO 2022). Nevertheless, the market shows first sign of recovery.

In general, the indoor environment of commercial aircraft is characterized by high daytime solar load, very low humidity, reduced air pressure, high density of occupants with limited space, high air-exchange rates, continuous noise, and a high load of furnishings (Mayer et al. 2022).

To supply fresh air, to pressurize the aircraft fuselage, and to control temperature pressurized outside air is taken from the compressor stage of the engine (bleed air), cooled down and mixed with cabin air (AIRBUS 2021). The cabin air, taken from the under-floor area of the cabin and filtered, is then mixed with 40–60% outside air (AIRBUS 2021). An example of an aircraft air generation and distribution system is given in Fig. 2.9. Most aircraft are equipped with High Efficient Particle Absorption (HEPA) filter used to filter the recirculation air. In an experimental study on 23 HEPA filters with glass fiber media that are used in different commercial airliner models the tested filters had a median filtration efficiency of >99.97% for particles with a diameter of 0.3–0.5 µm (Zhang et al. 2022b).

Overall, the ventilation system leads to an air-exchange rate greater than 20 per hour. Additionally, auxiliary power units (APU) are smaller turbine engines that

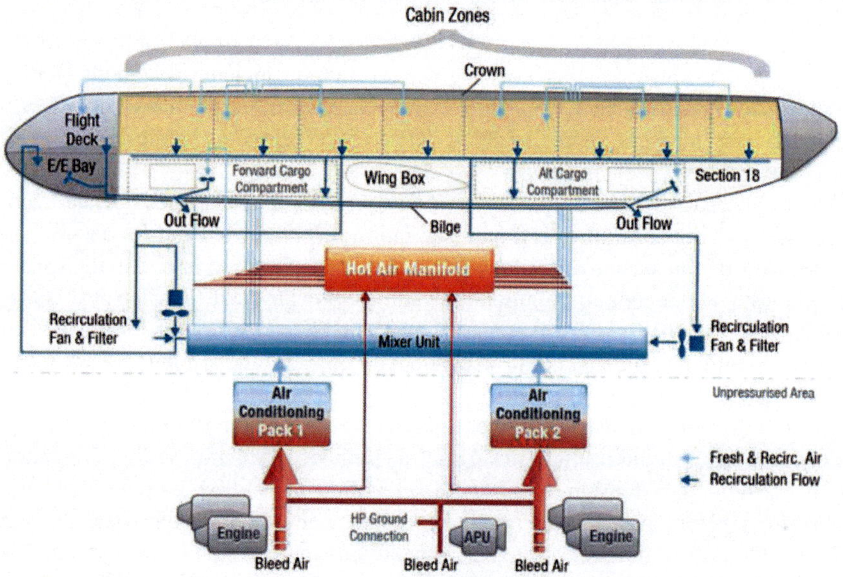

Fig. 2.9 Example of an aircraft air generation and distribution system (Airbus 2021)

provide power to an aircraft during approach, while taxiing, or positioned at the gate, unless an alternative source of electricity and preconditioned air is made available.

2.9.4.1 Health Aspects

In the literature and by authorities, unusual changes in cabin air quality, referred to as odor, smoke or fume events are repeatedly described when noticeable odors are perceived or the air was contaminated by trace amounts of oil, hydraulic fluid, or other agents from the main engines or the APU. Between 2012 and 2017, a total of 438 events were reported to the US Federal Aviation Administration (FAA), with APU accounting for 18%, while engine air and engine oil accounted for only 0.31 and 0.38% of incidents, respectively (Mayer et al. 2022). However, the measurements of cabin air has not identified a substance or mixture in sufficient levels to fully explain the symptomology of health effect (Rosenberger et al. 2016; Wolkoff et al. 2016; Schuchardt et al. 2017; Rosenberger 2018).

For example, Bekö et al. (2015) reported the health effects of ozone on 83 flights. On a basis of a questionnaire, 46% of passengers reported at least one symptom related to the eyes or dry mouth, 26% of them dry mouth or lips, 22.1% dry eyes, 18.9% nasal stuffiness, and between 5 and 10% having itchy eyes; headache; dry, irritated, sore throat and runny nose or sneezing. Medium and long duration flights were significantly associated with more symptoms compared to short flights. Nevertheless, other factors like stress, work control and satisfaction, seat comfort, baseline health (prior to the flight) and history of atopy are important influencing factors.

In a recent study, 559 subjects rated comfort, health symptoms, and sleepiness multiple times and heart rate variability was measured using a full and half-occupied aircraft segment in a low-pressure tube during a 4-h simulated flight under four different ventilation conditions (Herbig et al. 2023). Effect sizes showed that occupation impacts passengers' responses much more than increasing recirculation air rate. Only if an aircraft is more or less fully occupied decreasing health and well-being with increasing recirculation air rate. Overall, the close proximity of other people seems to increase stress and exacerbate reactions to air quality.

For commercial aircraft, various authorities such as the FAA, European Aviation Safety Agency (EASA), and the Civil Aviation Administration of China (CAAC) have issued some minimum requirements for indoor air quality. All three authorities demand 50 ppm for CO, 5000 ppm for CO_2, and 100 ppb for ozone (Mayer et al. 2022). Minimum ventilation rates are 3.5–4.7 l/s per person (0–8000 ft. pressure) in the USA and China and 4.7 l/s per person (min. outside) in Europe.

2.9.4.2 Occurrence in Cabin Air

Mayer et al. (2022) give an overview of indoor cabin quality and a summary of existing studies. In a comprehensive review, Chen et al. (2021) summarize the VOC,

SVOC, ozone, CO_2, and NO_2 measurements from a total of 2251 flights from 47 studies conducted in aircraft cabins between 1967 and 2019. The results of some VOC with the highest concentrations are shown in the Table 2.2 and those of some SVOC and naphthalene are summarized in the Table 2.3. The mean VOC concentrations most frequently appearing were toluene (15 μg/m^3), benzene (5.9 μg/m^3), ethylbenzene (2.3 μg/m^3), formaldehyde (5.4 μg/m^3), acetaldehyde (6.4 μg/m^3), and limonene (24 μg/m^3), while the SVOC with high concentrations and high frequency of detection were tributyl phosphate (TiBP) (495 ng/m^3), tris (chloroisopropyl) phosphate (TCPP) (506 ng/m^3), and trichloroethylene (TCE) (483 ng/m^3). Additionally, Yin et al. (2022) characterized VOC in 56 commercial flights and found six key substances (formaldehyde, benzene, tetrachloroethylene, trichloromethane, 1,2-dichloroethane, and naphthalene) that merit priority attention under normal flight conditions.

For studies that corrected measurements for changes in pressure a mean value of 1315 ppm CO_2 (range: 410–874 ppm) was reported (Chen et al. 2021). Mean relative humidity (RH) level was 16%, with minimum RH levels from 0.9% to 15% and maximum RH levels from 13% to 77%.

Chen et al. (2021) found 11 studies in which PM was measured in generally low concentrations. Furthermore, Yu et al. (2021) found mean $PM_{2.5}$ and PM_{10} concentrations of 0.15–2.7 μg/m^3 and 0.26–3.2 μg/m^3, respectively, in four domestic flights (2–3 hours) in China in 2019. Low concentrations in cabin air are also reported for ultrafine particles. UFP were measured on 55 flights in 2008–2010, with a maximum 15-minute concentration across flights ranging from 1 to 312,000 particles/cm^3 depending on latitude. The mean UFP for flights with lower latitude routes was 1104 particles/cm^3 compared to a mean of 55 particles/cm^3 for flights with higher latitude routes (Spengler et al. 2012). Again, the cabin air of 83 flights was examined and low median concentrations (of the mean values of the flights) of 64 particles/cm^3 (0.1–22,052 particles/cm^3) were found (Bekö et al. 2015).

During a typical commercial flights at altitudes of 5490 to 12,500 meters, tropospheric ozone (around 25 to 900 ppb) enters the cabin from the outside via the ventilation and is deposited on the surfaces or reacts with other pollutants (Chen et al. 2021). Chen et al. (2021) summarize 21 studies between 1967 and 2018 from which a mean value of 70 μg/m^3 and minimum levels from 0 to 43 μg/m^3 and maximum levels from 21 to 589 μg/m^3 can be derived, respectively. Using a computational fluid dynamics-based model to evaluate ozone distribution suggested that the ozone concentration was non-uniform in the aircraft cabin and that the ozone removal proportion of the human body was found to be 89.3%, followed by material (10.4%) and air reactions (<1%) (He et al. 2022). Thus, surface removal plays a major role in ozone removal and passengers are the main ozone sink.

For example, in a report prepared for the European Aviation Safety Agency, Schuchardt et al. (2017) measured ozone in 61 flights with aircrafts using bleed air systems and 8 flights with an aircraft using an alternative no-bleed air supply system in 2015 to 2016. The median concentrations (95th percentiles) in the cabin were 11 μg/m^3 (30 μg/m^3) and 6 μg/m^3 (15 μg/m^3).

Table 2.2 VOC with the highest medians in indoor cabin on non-smoking flights in μg/m^3 (modified from Chen et al. 2021)

Compound	CAS No.	Min.	Medium	Max.	Number of studies	Number of flights
Ethanol	64-17-5	81	82	3009	2	77
1-Propanol	71-23-8	0.6	81	81	1	69
Isoalkanes C_{14}–C_{20}		25	62	62	1	69
1,2-Propanediol	57-55-6	11	45	45	1	69
Acetonitrile	75-05-8	19	19	27	1	69
Cyclopentasiloxane	294-40-6	9.8	18	18	1	69
Acetone	4468-52-4	0.5	16	49	4	139
Decanal	112-31-2	2.7	15	36	4	186
Isopropyl alcohol	67-63-0	3.5	13	13	2	132
Limonene	138-86-3	1.4	12	276	6	339
Toluene	108-88-3	2.5	12	123	7	402
Acetic acid	64-19-7	1.1	12	16	3	88
Menthol	15,356-70-4, 491-02-1	1	12	12	2	83
Isoprene	78-79-5	0.8	9	14	3	183
6-methyl-5-hepten-2-one/ 6-MHO	129,085-68-3	0.2	8.5	16	4	122
1-Hexanol,2-ethyl-	103-09-3	4.8	7.7	12	1	14
Dimethylformamide	68-12-2	0	7.7	7.7	2	120
Dioctyl ether	629-82-3	0.4	6.4	6.4	1	69
Formaldehyde	50-00-0	2.7	5.9	7.1	4	248
Perfluoro derivatives		0	5.5	5.5	1	69
Nonanal	124-19-6	1.9	5.4	24	5	191
Acetaldehyde	75-07-0	5.2	5.3	7.7	3	240
Benzoic acid	65-85-0	3.3	5.3	5.3	1	69
1,3-Butanediol	107-88-0	0.4	5.2	5.2	1	69
Ethyl acetate	141-78-6	3.9	4.9	16	3	183
2-Phenoxyethanol	122-99-6	1	4.6	4.6	1	69
2-Ethyl-1-hexanol/2-Ethylhexanol	104-76-7	2.9	4	5.9	2	120
5,9-Undecandien-2-one-6,10-dimethyl	3796-70-1	0.1	3.9	3.9	1	69
Perchloroethylene	127-18-4	0.6	3.8	16	4	197
Hexanoic acid	142-62-1	3.8	3.8	6.2	1	69
Hydroxyacetone	116-09-6	0.3	3.3	3.3	1	69
Octanal	124-13-0	1.3	2.9	10	3	135
Butanone/2-butanone	78-93-3	1.2	2.9	2.9	2	152
Hexanal	66-25-1	1.7	2.8	14	4	234
Tetradecane	629-59-4	2.1	2.6	2.6	1	69
1-Butanol	71-36-3	0.9	2.4	2.4	1	69
2,2,4,4,6,8,8-Heptamethyl nonane	4390-04-9	0.6	2.4	2.4	1	69

(continued)

Table 2.2 (continued)

Compound	CAS No.	Min.	Medium	Max.	Number of studies	Number of flights
Methoxy-bis-1,2′--dipropane-1,2-diol ether		0	2.4	2.4	1	69
Heptanal	111-71-7	0.7	2.3	4.6	2	121
2-Ethylhexyl salicylate	118-60-5	0.3	2.3	2.3	1	69
Undecane	1120-21-4	0	2.2	13	5	239
Butyl acetate	123-86-4	0.7	2.2	2.2	1	69
Isovaleraldehyde	3842-03-3	2.1	2.1	2.2	1	108
Octanoic acid	124-07-2	1.4	2.1	2.1	1	69
Benzaldehyde	100-52-7	0	2	14	5	247
Eucalyptol	470-82-6	0.5	2	2	1	69
2-Methylpentane	107-83-5	0.1	2	2	1	63

In terms of biological factors, samples collected in the cabin air showed geometric mean concentrations for the sum of mVOC (excluding butanols) of 3192 ng/m^3 (290–32,460 ng/m^3), 3.7 times higher than in a random sample of 92 homes (Fu et al. 2013). Endotoxin was measured in 32 air samples collected during four commercial aircraft flights and geometric mean air concentrations of 1.5 EU/m^3 with significant differences within aircraft type (geometric means from 0.53–5.3 EU/m^3) were reported (Hines et al. 2003). Airborne endotoxin levels were not significantly related to either carpet or seat dust endotoxin levels. A total of 513 airborne bacterial samples were collected during 12 flights and geometric mean levels of 136 total colony-forming units per cubic meter were observed (McKernan et al. 2008). Total bacterial concentrations observed during flight were significantly lower than for boarding and deplaning in this study.

The exposure of aircraft cabins to viruses is described in Sect. 6.5.3.

2.9.4.3 Occurrence in Cabin Dust

Endotoxin was measured in 64 dust samples collected during four commercial aircraft flights and mean concentrations (range) in carpet and chair dust of 447 EU/mg (281–624 EU/mg) and 140 EU/mg (132–663 EU/mg) were reported (Hines et al. 2003). Mean endotoxin levels were significantly higher in carpet dust than in seat dust. Dust was collected by vacuum cleaner from cabin seats from 18 flights and geometric mean concentrations of cat (Fel d1) and dog (Can f1) allergens in cabins with textile seats were 5359 ng/g and 6067 ng/g, respectively, while in cabins having leather seats the GM were only 107 ng/g and 227 ng/g (Fu et al. 2013).

Allen et al. (2013) analyzed dust from 19 commercial aircraft parked overnight for flame retardants by vacuuming the carpet. Median values were 950 n/g for BDE 47, 950 ng/g for BDE 99, 495,000 ng/g for BDE 209, 7600 ng/g for sum HBCD, and

Table 2.3 Examples of some organophosphates, PAH, and pesticides in indoor cabins of aeroplanes with no events in ng/m^3 (modified from Chen et al. 2021)

Compound	CAS No.	Min	Max	Mean
Organophosphates				
Tris (chloro-isopropyl) phosphate (TCPP)	13,674-84-5	23	9977	506
Tributyl phosphate (TBP)	126-73-8	<LOD	1610	92
Triisobutyl phosphate (TiBP)	126-71-6	<LOD	9100	495
Tris (butoxy-ethyl) phosphate (TBEP)	78-51-3	<LOD	642	71
Tris (chloroethyl) phosphate (TCEP)	115-96-8	<LOD	324	15
Diphenyl-2-ethylhexyl phosphate (DPEHP)	1241-94-7	<LOD	282	15
Triphenyl phosphate (TPP)	115-86-6	1	119	8.7
Tris (1,3-dichloro-isopropyl) phosphate (TDCPP)	13,674-87-8	<LOD	49	7.7
Tricresyl phosphates (TCP)	1330-78-5	<LOD	14,900	35
Tri-o-cresyl phosphate (ToCP)	78-30-8	<LOD	22,800	50
Tri-m-cresyl phosphate (TmCP)	563-04-2	<LOD	428	4.4
Tri-p-cresyl phosphate (TpCP)	563-04-2	<LOD	57	2.1
Tri-mmp-cresyl phosphate (T-mmp-CP)		<LOD	691	6.5
Tri-mpp-cresyl phosphate (T-mpp-CP)		<LOD	339	4.2
Tris (ethyl-hexyl) phosphate (TEHP)	78-42-2	<LOD	88	3.5
Polycyclic aromatic hydrocarbons				
Naphthalene	91-20-3	<LOD	49,100	1241
Acenaphthene	83-32-9	<LOD	24	5.7
Phenanthrene	85-01-8	<LOD	21	4.9
Fluorene	86-73-7	<LOD	12	3.0
Pyrene	129-00-0	<LOD	15	2.6
Retene	483-65-8	<LOD	2.0	1.4
Acenaphthylene	208-96-8	<LOD	3.3	0.8
Fluoranthene	206-44-0	<LOD	1.9	0.5
Anthracene	120-12-7	<LOD	1.1	0.3
Pesticides				
trans-Permethrin	61,949-77-7	<LOD	2.0	1.5
cis-Permethrin	61,949-76-6	<LOD	0.9	0.9
Hexachlorobenzene	118-74-1	<LOD	2.3	0.2

LOD limit of detection

330 and 110 ng/g for anti-Dechlorane Plus and syn-Dechlorane Plus, respectively. Tris(1,3-dichloro-isopropyl)phosphate (TDCPP) and bis-(2-ethylhexyl)-tetrabromophthalate (TBPH) were found at median levels of 2100 ng/g and 640 ng/g, respectively. Higher concentrations compared to flats were found for BDE 209, probably due to the use of flame retardants in the floor coverings.

2.9.5 *Rubber Granules and Indoor Synthetic Turf*

In addition to combustion-related emissions from motorized road traffic, emissions from tire wear and brake abrasion are increasingly coming into focus. Using passive samples from different sites near highly frequented roads in Germany, single-particle analysis with electron microscopy showed that 89% of all analyzed particles were derived from traffic-related sources like road wear (39%), tire wear (33%), and brake wear (17%) (Sommer et al. 2018). In light of this, these emissions of particles and adsorbed chemicals to them pose a consistent inhalation risk to humans in areas exposed to motorized traffic and result in deposition to vegetation and water and soil surfaces (Johannessen et al. 2022). Furthermore, particles can also enter indoor spaces with the environmental aerosol.

In addition, in recent years there has been intensive discussion of another exposure situation, which is connected with the increasing recycling of tires, which is quite desirable from an ecological point of view. Rubber granules contain a wide variety of substances depending on the material from which they are produced or recycled and were often used as infill material in artificial turf and sportscenter and playground surfaces (ECHA 2017). The most common material used for these granules are end-of-life vehicle tires which made typically using styrene butadiene rubber. Based on industry estimates, the quantity of end-of-life tires rubber infill that is used on European sport fields is about 80,000 to 130,000 tons per year with about 30% in synthetic turf, 24% in sports and children playgrounds, and 24% in molded objects (ECHA 2017). Various volatile and semivolatile substances in particular PAH, phthalates, VOC, antioxidants as well as metals can be detected in such granules (ECHA 2017). For example, Armada et al. (2022a) collected 78 crumb rubber samples from football pitches in 17 countries worldwide and found medians of 15 μg/g for the 16 US EPA priority PAH, 28 μg/g for DEHP, and 2132 μg/g for DINP.

In addition to the classic synthetic antioxidants that are also widely used in rubber production, such as benzothiazole and its derivatives, other antioxidants are increasingly being investigated in indoor spaces. These include phenolic antioxidants like guanidine derivatives and amino antioxidants like N-(1,3-dimethylbutyl)-N′-phenyl-p-phenylenediamine (6-PPD). The chemical structures of the before mentioned substances are given in Fig. 2.10. Some important key points on these substances in brief:

Benzothiazole (BTH, CAS No.: 95-16-9) and its derivatives like 2-hydroxybenzothiazole (2-OH-BTH, CAS No. 934-34-9) are thermally stable compounds used primarily as a vulcanization accelerators in the manufacture of rubber, as fungicides in paper and leather manufacturing, as photosensitizers in photography, and as corrosion inhibitors in antifreeze formulations (Liao et al. 2018).

Guanidine derivatives like 1,3-diphenylguanidine (1,3-DPG, CAS No., 102-06-7) have been widely used as vulcanization accelerators in rubber industry and depending on the specific application, the concentration in rubber may vary from

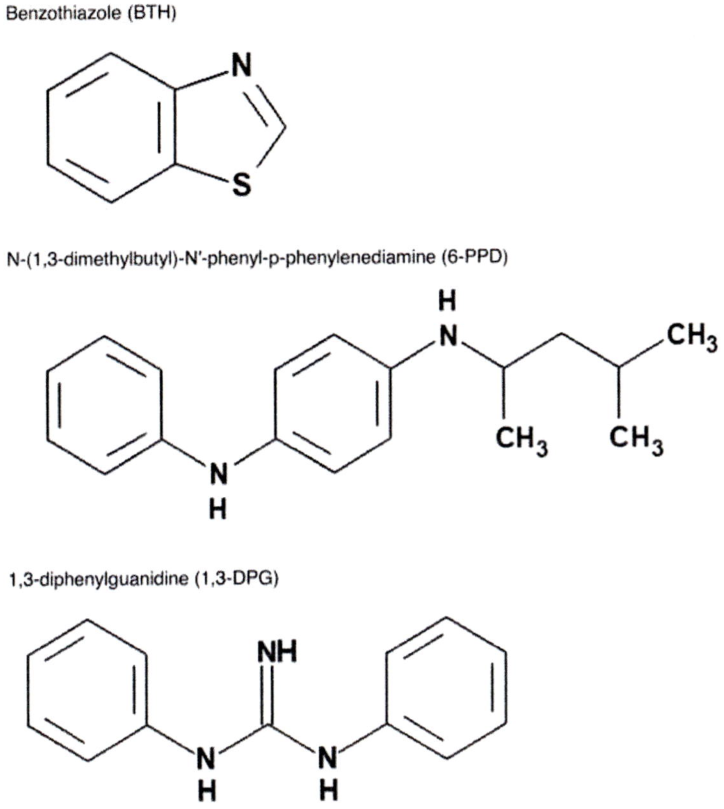

Fig. 2.10 Chemical structures of three common antioxidants

0.25% to 2.0% by weight (NTP 1995). It is manufactured and imported to the EU in amounts of 1000–10,000 tons per year (ECHA 2017).

6-PPD (CAS No.: 793-24-8) is mainly used for protecting rubber against aging (e.g., in vehicle tires). In China, 6-PPD contributed the largest proportion (~54%) among all rubber antioxidants, and its annual output reached 0.2 million tons in 2020 (Zhang et al. 2022d). 6-PPD can migrate to the surface of rubber products, was released into the atmosphere, and will be rapidly degraded by ozone or other photooxidants to, e.g., 6-PPD quinone (OECD 2004).

2.9.5.1 Benzothiazoles in Outdoor and Indoor Air and in Facilities with Synthetic Turf

Overall, without a special emission situation, only low concentrations of BTH are found in the outdoor air. Thus, Naccarato et al. (2021) found BTH in only one of five PM_{10} samples collected over six days in Italy, with a level of 0.027 ng/m^3. BTH and 2-OH-BTH were also found in three Chinese cities in 2018/19 with means from 0.034–0.056 ng/m^3 and 0.164–0.416 ng/m^3, respectively, with 2-OH BTH as the predominant compound accounted for 50–80% of the total concentrations (Liao et al. 2021).

For the first time, in the gas and particulate phase, a median level of 41.9 ng/m^3 for BTH (range: 4.7–1233 ng/m^3) and 2.5 ng/m^3 for 2-OH-BTH (range: 0.18–872 ng/m^3) were found in 26 indoor residential spaces in the USA in 2014 (Wan et al. 2016). For the sum of benzothiazoles in this study, the median in the particulate fraction was 6.2 ng/m^3 and in the gas phase 32.7 ng/m^3. Significantly higher median concentrations were found in 7 car interiors with 146 ng/m^3 of BTH and 3.5 ng/m^3 for 2-OH-BTH. In contrast, very high BTH levels of around 11,000 to 14,000 ng/m^3 were observed in one indoor synthetic turf crumb rubber fields in the USA (Simcox et al. 2011). Overall, the very limited indoor air measurements available indicate that higher concentrations must be expected in indoor spaces where rubber is increasingly used. In a study collected air samples on the edge of 17 outdoor sport fields in six European countries a median of 1500 ng/m^3 (95th percentile: 6900 ng/m^3) was reported (Schneider et al. 2020). On the contrary, using an active air sampling device with Tenax TA as a sorbent, Armada et al. (2022b) reported low levels of 65.1 ng/m^3 for BTH in the air of one indoor playground for children.

A health assessment of BTH is difficult due to the limited toxicological data available. Based on an oral animal study, the German Committee on Indoor Guide Values (AIR) has set a provisional guideline value I (precautionary value) of 15 µg/m^3, at which level sensory irritative effects are also not to be expected (AIR 2018).

2.9.5.2 Other Chemicals in Indoor Air Facilities with Synthetic Turf

In a study collected air samples on the edge of 17 outdoor sport fields in six European countries no increased concentrations of metals (aluminum, cobalt), 16 US EPA priority PAH (95th percentile: 1798 ng/m^3), or other semi-volatile substances in air samples were seen compared to background level, but some VOC were slightly increased with 95th percentiles of 18 µg/m^3 for methyl isobutyl ketone, 31 µg/m^3 for tert-butylamine, 0.4 µg/m^3 for 2-heptanone, 1.5 µg/m^3 for cyclohexanone (Schneider et al. 2020).

Measurement results of other, actually expected substances, in indoor facilities where granular material from tires used are only available to a very limited extent. First results of a total of four measurements in two sports halls with synthetic turf were carried out in Norway in 2005 and showed concentrations for ∑PAH of 6.5–174 ng/m^3, for B[*a*]P < 0,01–1.15 ng/m^3, and for dibutylphthalate 15.1–53.7 ng/m^3 (Dye et al. 2006).

In the course of collecting two personal samples in one indoor facility with crumb rubber turf and one outdoor control site higher maximums compared to outdoor air were reported for acetone, acrolein, bromoform, cyclohexane, heptane, methyl ethyl ketone, methyl isobutyl ketone, and methylene chloride (Simcox et al. 2011;

Ginsberg et al. 2011). Several PAH were ten-fold higher in one air sample on indoor turf than background, including 1-methylnaphthalene, 2-methylnaphthalene, fluorene, naphthalene, and pyrene or were found indoors only, namely, acenaphthene, acenaphthylene, fluorene, naphthalene, and 2,6-dimethylnaphthalene. For example, 113 ng/m^3 naphthalene was found indoors and 4.5–16.9 ng/m^3 in outdoor air. Using an active air sampling device with Tenax TA as a sorbent, Armada et al. (2022b) reported low levels of 12 ng/m^3 for sum of 11 PAH, 17.4 ng/m^3 for diisobutyl phthalate, 17.9 ng/m^3 for dibutyl phthalate, and 2.05 ng/m^3 for butylated hydroxytoluene in the air of one indoor playground for children.

2.9.5.3 Antioxidants in Residences

In 158 dust samples collected from homes and offices in 2006–2012, median concentrations of ∑BTH in 4 countries could be found to be 2000 ng/g in Korea, 1290 ng/g in the USA, 605 ng/g in Japan, and 857 ng/g in China (Wang et al. 2013). In Spain in 2020/21, a median concentration of 470 ng/g (range: <170–3230 ng/g) was reported in 16 dust samples sieved to 75 μm (Núñez et al. 2022). Even with unfavorable assumptions, the authors assume that the hazard index is well below 1 in the context of dust absorption.

Li and Kannan (2023) collected 216 indoor dust samples from 11 countries worldwide in 2011–2014 and analyzed three guanidine derivatives. The overall median concentration of sum of 1,3-DPG,3-di-o-tolylguanidine, and 1,2,3-triphenylguanidine in house dust was 150 ng/g (range: 2.9–11,000 ng/g), respectively, and 1,3-DPG accounted for 87 ≥ 97% of the sum concentrations of the three compounds. In a previous study from Northern California in 2015/16, 38 dust samples were analyzed and a higher median of 3218 ng 1,3-DPG/g (95th percentile: 9659 ng/g) were reported (Shin et al. 2020). In a second study that used dust samples from China, Australia, USA, and Vietnam 2017/19, high medians of 1,3-DPG in the range of 305 ng/g in Vietnam to 11,400 in the USA were also found (Tan et al. 2021). Nevertheless, the authors concluded that the estimated intake of target antioxidants by toddlers via dust ingestion, even under the high exposure scenario, was determined to be 2−4 orders of magnitude lower than the reference doses of selected antioxidants.

6-PPD has not been measured indoors, but results are now available from a recent Chinese study that determined the transformation product 6-PPD quinone. Dust samples were collected in 16 residential bedrooms and 17 vehicles in 2021 and medians of 7.7 ng/g for 6-PPD quinone (range: 0.97–26.1 ng/g) and 38.4 ng/g (range: 19.7–71.4 ng/g), respectively, were found (Zhang et al. 2022d). Further results are not yet available, but measurement results in the outdoor air prove a low exposure to 6-PPD and 6-PPD quinone. A total of 81 $PM_{2.5}$ ambient air samples were collected in six Chinese cities in 2018/19 and 6-PPD and 6-PPD quinones were found with medians from 0.9 to 8.4 pg/m^3 (range: 0.02–487 pg/m^3) and 1.7 to 6.7 pg/m^3 (range: 0.1–84 pg/m^3), respectively (Zhang et al. 2022c).

References

Aaltonen HL, Jakobsson JK, Diaz S, Zackrisson S, Piitulainen E, Löondahl J, Wollmer P (2018) Deposition of inhaled nanoparticles is reduced in subjects with COPD and correlates with the extent of emphysema: proof of concept for a novel diagnostic technique. Clin Physiol Funct Imaging 38:1008–1014

Ahookhosh K, Pourmehran O, Aminfar H, Mohammadpourfard M, Sarafraz MM, Hamishehkar H (2020) Development of human respiratory airway models: a review. Eur J Pharmaceutical Sci 145:105233

AIR (2018) Indoor air guide values for benzothiazole. German Committee on Indoor Guide Values (AIR). Bundesgesundheitsbl Gesundheitsforsch Gesundheitsschutz 63:1304–1310. [in German]

Airbus (2021) Cabin air quality – key to a comfortable flight. How to make an aircraft breathe. https://aircraft.airbus.com/en/newsroom/news/2021-01-cabin-air-quality-key-to-a-comfortable-flight

Aithal SS, Gill S, Satia I, Tyagi SK, Bolton CE, Kurmi OP (2021) The effects of household air pollution (HAP) on lung function in children: a systematic review. Int J Environ Res Public Health 18:11973

Allen JG, Stapleton HM, Vallarino J, McNeely E, McClean MD, Harrad SJ, Rauert CB, Spengler JD (2013) Exposure to flame retardant chemicals on commercial airplanes. Environ Health 16:17

ANSI (American National Standards Institute) (2019) ANSI/CAN/UL 2904. Standard method for testing and assessing particle and chemical emissions from 3D printers. https://chemicalinsights.org/wp-content/uploads/2019/05/3DPrint_Standard_Brief_Version-2.pdf

Armada D, Llompart M, Celeiro M, Garcia-Castro P, Ratola N, Dagnac T, de Boer J (2022a) Global evaluation of the chemical hazard of recycled tire crumb rubber employed on worldwide synthetic turf football pitches. Sci Total Environ 812:152542

Armada D, Celeiro M, Dagnac T, Llompart M (2022b) Green methodology based on active air sampling followed by solid phase microextraction and gas chromatography-tandem mass spectrometry analysis to determine hazardous substances in different environments related to tire rubber. J Chromatogr A 1668:462911

Asgharian B, Price OT, Schroeter JD, Kimbell JS, Singal M (2012) A lung dosimetry model of vapor uptake and tissue disposition. Inhal Toxicol 24:182–193

Asgharian B, Price OT, Kuempel ED, Jarabek AM (2018) Dosimetry of inhaled elongate mineral particles in the respiratory tract: the impact of shape factor. Toxicol Appl Pharmacol 361:27–35

Azimi P, Zhao D, Pouzet C, Crain NE, Stephens B (2016) Emissions of ultrafine particles and volatile organic compounds from commercially available desktop three-dimensional printers with multiple filaments. Environ Sci Technol 50:1260–1268

Balakrishnan K, Ghosh S, Ganguli B, Sambandam S, Bruce N, Barnes DF, Smith KR (2013) State and national household concentrations of $PM_{2.5}$ from solid cookfuel use: results from measurements and modeling in India for estimation of the global burden of disease. Environ Health 12:77

Barkowski D, Krüger G, Machtolf M, Derz K, Küchler F (2016) Studies on the resorption availability of organic and inorganic pollutants for the further update of Annex 1 of the BBodSchV. TEXTE 15/2016. Bericht im Auftrag des Umweltbundesamtes. https://www.umweltbundesamt.de/sites/default/files/medien/378/publikationen/texte_15_2016_untersuchungen_zur_resorptionsverfuegbarkeit.pdf [in German]

Beauchamp J (2011) Inhaled today, not gone tomorrow: pharmacokinetics and environmental exposure of volatiles in exhaled breath. J Breath Res 5:037103

Bekö G, Allen JG, Weschler CJ, Vallarino J, Spengler JD (2015) Impact of cabin ozone concentrations on passenger reported symptoms in commercial aircraft. PLoS One 10:e0128454

Berger-Preiss LK, Leng G, Idel H, Sugiri D, Ranft U (2002) Indoor pyrethroid exposure in homes with woollen textile floor coverings. Int J Hyg Environ Health 205:459–472

Bernatikova S, Dudacek A, Prichystalova R, Klecka V, Kocurkova L (2021) Characterization of ultrafine particles and VOCs emitted from a 3D Printer. Int J Environ Res Public Health 18:929
Betha R, Selvam V, Blake DR, Balasubramanian R (2011) Emission characteristics of ultrafine particles and volatile organic compounds in a commercial printing center. J Air Waste Manag Assoc 61:1093–1101
Bocato MZ, Ximenez JPB, Hoffmann C, Barbosa F (2019) An overview of the current progress, challenges, and prospects of human biomonitoring and exposome studies. J Toxicol Environ Health Part B 22:131–156
Bopp SK, Kienzler A, Richarz A-N, van der Linden SC, Paini A, Parissis N, Worth AP (2019) Regulatory assessment and risk management of chemical mixtures: challenges and ways forward. Crit Rev Toxicol 49:174–189
Borghi F, Spinazzè A, Mandaglio S, Fanti G, Campagnolo D, Rovelli S, Keller M, Cattaneo A, Cavallo DM (2021) Estimation of the inhaled dose of pollutants in different micro-environments: a systematic review of the literature. Toxics 12:140
Bothe M (2004) Quantification of ingestion of soil by children Series of publications Reactor Safety and Radiation Protection. Report for the BMU – 2004-647. Bonn: VKTA, Verein für Kernverfahrenstechnik und Analytik Rossendorf eV. [in German]
Boyes WK, van Thriel C (2020) Neurotoxicology of nanomaterials. Chem Res Toxicol 33:1121–1144
Brasche S, Bischof W (2005) Daily time spent indoors in German homes – baseline data for the assessment of indoor exposure of German occupants. Int J Hyg Environ Health 208:247–253
Bright DA, Richardson GM, Dodd M (2006) Do current standards of practice in Canada measure what is relevant to human exposure at contaminated sites? I: a discussion of soil particle size and contaminant partitioning in soil. Hum Ecol Risk Assess 12:591–605
Bruce N, Pope D, Rehfuess E, Balakrishnan K, Adair-Rohani H, Dora C (2015) WHO indoor air quality guidelines on household fuel combustion: strategy implications of new evidence on interventions and exposureerisk functions. Atmos Environ 106:451–457
Butte W, Walker G (1994) Sense and nonsense of house dust tests – the pros and cons. House dust as a measurement parameter for detecting indoor contamination with permethrin, pentachlorophenol and lindane. In: Kommission Reinhaltung der Luft im VDI und DIN (Eds.), Indoor Air Pollution. VDI-Report No. 1122. Düsseldorf, VDI-Verlag. [in German]
Calabrese EJ, Stanek ES (1992) Distinguishing outdoor soil ingestion from indoor dust ingestion in a soil pica child. Regul Toxicol Pharmacol 15:83–85
Calabrese EJ, Barnes R, Stanek 3rd EJ, Pastides H, Gilbert CE, Veneman P, Wang XR, Lasztity A, Kostecki PT (1989) How much soil do young children ingest: an epidemiological study. J Reg Tox Appl Pharm 10:123–127
Cao ZG, Yu G, Chen YS, Cao QM, Fiedler H, Deng SB, Huang J, Wang B (2012) Particle size: a missing factor in risk assessment of human exposure to toxic chemicals in settled indoor dust. Environ Int 49:24–30
Cao Z-G, Yu G, Chen Y-S, Liu C, Liu K, Zhang T-T, Wang B, Deng S-B, Huang J (2013) Mechanisms influencing the BFR distribution patterns in office dust and implications for estimating human exposure. J Hazard Mater 252–253:11–18
Carll AP, Salatini R, Pirela SV, Wang Y, Xie Z, Lorkiewicz P, Naeem N, Qian Y, Castranova V, Godleski JJ, Demokritou P (2020) Inhalation of printer-emitted particles impairs cardiac conduction, hemodynamics, and autonomic regulation and induces arrhythmia and electrical remodeling in rats. Part Fibre Toxicol 17:7
Castellano P, Canepari S, Ferrante R, L'Episcopo N (2012) Multiparametric approach for an exemplary study of laser printer emissions. J Environ Monit 14:446–454
Cepeda M, Schoufour J, Freak-Poli R, Koolhaas CM, Dhana K, Bramer WM, Franco OH (2017) Level of ambient air pollution according to mode of transport: a systematic review. Lancet Public Health 2:e23–e34
Chan KH, Xia X, Ho KF, Guo Y, Kurmi OP, Du H, Bennett DA, Bian Z, Kan H, McDonnell J, Schmidt D, Kerosi R, Li L, Lam KBH, Chen Z (2021) Regional and seasonal variations in

household and personal exposures to air pollution in one urban and two rural Chinese communities: a pilot study to collect time-resolved data using static and wearable devices. Environ Int 146:106217

Chen R, Fang L, Liu J, Herbig B, Norrefeldt V, Mayer F, Fox R, Wargocki P (2021) Cabin air quality on non-smoking commercial flights: a review of published data on airborne pollutants. Indoor Air 31:926–957

Cheng Y-S, Smith SM, Yeh H-C, Kim D-B, Cheng K-H, Swift DL (1995) Deposition of ultrafine aerosols and thoron progeny in replicas of nasal airways of young children. Aerosol Sci Technol 23:541–552

Choate LM, Ranville JF, Bunge AL, Macalady DL (2006) Dermally adhered soil: 1. Amount and particle-size distribution. Integr Environ Assess Manage 2:375–384

Clausing P, Brunekreef B, van Wijnen JH (1987) A method for estimating soil ingestion by children. Int Arch Occup Environ Health 59:73–82

Clewell RA, Leonard JA, Nicolas CI, Campbell JL, Yoon M, Efremenko AY, McMullen PD, Andersen ME, Clewell 3rd HJ, Phillips KA, Tan YM (2020) Application of a combined aggregate exposure pathway and adverse outcome pathway (AEP-AOP) approach to inform a cumulative risk assessment: a case study with phthalates. Toxicol In Vitro 66:104855

Cohen Hubal EA, Sheldon LS, Burke JM, McCurdy TR, Berry MR, Rigas ML, Zartarian VG, Freeman NC (2000) Children's exposure assessment: a review of factors influencing Children's exposure, and the data available to characterize and assess that exposure. Environ Health Perspect 108:475–486

Conrad A, Seiwert M, Hünken A, Quarcoo D, Schlaud M, Groneberg D (2013) The German Environmental Survey for Children (GerES IV): reference values and distributions for time-location patterns of German children. Int J Hyg Environ Health 21:25–34

Davis S, Waller P, Buschbom R, Ballou J, White P (1990) Quantitative estimates of soil ingestion in normal children between the ages of 2 and 7 years: populationbased estimates using aluminium, silicon, and titanium as soil tracer elements. Arch Environ Health 45:112–122

Davis AY, Zhang Q, Wong JP, Weber RJ, Black MS (2019) Characterization of volatile organic compound emissions from consumer level material extrusion 3D printers. Build Environ 160: 106209

Deng Q, Deng L, Miao Y, Guo Y, Li Y (2019) Particle deposition in the human lung: health implications of particulate matter from different sources. Environ Res 169:237–245

Dennis KK, Marder E, Balshaw DM, Cui Y, Lynes MA, Patti GJ, Rappaport SM, Shaughnessy DT, Vrijheid M, Barr DB (2017) Biomonitoring in the era of the exposome. Environ Health Perspect 125:502–510

Destaillats H, Brett RLM, Singer C, Hodgson AT, McKone TE (2008) Indoor pollutants emitted by office equipment: a review of reported data and information needs. Atmos Environ 42:1371–1388

DIN (2004) DIN 19738 (2004–07). Soil quality – Absorption availability of organic and inorganic pollutants from contaminated soil material. Beuth-Verlag, Berlin. https://www.beuth.de/de/norm/din-19738/72163151

Ding S, Wan MP, Ng BF (2020) Dynamic analysis of particle emissions from FDM 3D printers through a comparative study of chamber and flow tunnel measurements. Environ Sci Technol 54:14568–14577

Dong T, Zhang Y, Jia S, Shang H, Fang W, Chen D, Fang M (2019) Human indoor exposome of chemicals in dust and risk prioritization using EPA's ToxCast Database. Environ Sci Technol 53:7045–7054

Dons E, Int Panis L, van Poppel M, Theunis J, Willems H, Torfs R, Wets G (2011) Impact of time–activity patterns on personal exposure to black carbon. Atmos Environ 45:3594–3602

Doyi INY, Strezov V, Isley CF, Yazdanparast T, Taylor MP (2020) The relevance of particle size distribution and bioaccessibility on human health risk assessment for trace elements measured in indoor dust. Sci Total Environ 733:137931

Dye C, Bjerke A, Schmidbauer N, Mané S (2006) Measurement of air pollution in indoor artificial turf halls. Report Number NILU OR 3. https://www.nilu.com/wp-content/uploads/dnn/03-2006-eng.pdf

ECHA (2017) An evaluation of the possible health risks of recycled rubber granules used in synthetic turf sports fields. Annex XV report. https://echa.europa.eu/documents/10162/13563/annex-xv_report_rubber_granules_en.pdf/dbcb4ee6-1c65-af35-7a18-f6ac1ac29fe4

Ertl H, Butte W (2012) Bioaccessibility of pesticides and polychlorinated biphenyls from house dust: in-vitro methods and human exposure assessment. J Expo Sci Environ Epidemiol:1–10

Escher BI, Hackermüller J, Polte T, Scholz S, Aigner A, Altenburger R, Böhme A, Bopp SK, Brack W, Busch W, Chadeau-Hyam M, Covaci A, Eisenträger A, Galligan JJ, Garcia-Reyero N, Hartung T, Hein M, Herberth G, Jahnke A, Kleinjans J, Klüver N, Krauss M, Lamoree M, Lehmann I, Luckenbach T, Miller GW, Müller A, Phillips DH, Reemtsma T, Rolle-Kampczyk-U, Schüürmann G, Schwikowski B, Tan YM, Trump S, Walter-Rohde S, Wambaugh JF (2017) From the exposome to mechanistic understanding of chemical-induced adverse effects. Environ Int 99:97–106

Fang M, Stapleton HM (2014) Evaluating the bioaccessibility of flame retardants in house dust using an in vitro tenax bead-assisted sorptive physiologically based method. Environ Sci Technol 48:13323–13330

Farcas MT, McKinney W, Qi C, Mandler KW, Battelli L, Friend SA, Stefaniak AB, Jackson M, Orandle M, Winn A, Kashon M, LeBouf RF, Russ KA, Hammond DR, Burns D, Ranpara A, Thomas TA, Matheson J, Qian Y (2020) Pulmonary and systemic toxicity in rats following inhalation exposure of 3-D printer emissions from acrylonitrile butadiene styrene (ABS) filament. Inhal Toxicol 32:403–418

Floyd EL, Wang J, Regens JL (2017) Fume emissions from a low-cost 3-D printer with various filaments. J Occup Environ Hyg 14:523–533

Freeman GB, Schoof RA, Ruby MV, Davis AO, Dill JA, Liao SC, Lapin CA, Bergstrom PD (1995) Bioavailability of arsenic in soil and house dust impacted by smelter activities following oral administration in cynomolgus monkeys. Fundam Appl Toxicol 28:215–222

Fromme H, Schlummer M, Möller A, Gruber L, Wolz G, Ungewiß J, Böhmer S, Dekant W, Mayer R, Twardella D (2007a) Exposure of an adult population to perfluorinated substances using duplicate diet portions and biomonitoring data. Environ Sci Technol 41:7928–7933

Fromme H, Gruber L, Schlummer M, Wolz G, Boehmer S, Angerer J, Mayer R, Liebl B, Bolte G (2007b) Intake of phthalates and di(2-ethylhexyl)adipate: results of the Integrated Exposure Assessment Survey based on duplicate diet samples and biomonitoring data. Environ Int 33: 1012–1020

Fromme H, Körner W, Shahin N, Wanner A, Albrecht M, Boehmer S, Parlar H, Mayer R, Liebl B, Bolte G (2009a) Human exposure to polybrominated diphenyl ethers (PBDE), as evidenced by data from duplicate diet study, indoor air, house dust, and biomonitoring in Germany. Environ Int 35:1125–1135

Fromme H, Shahin N, Boehmer S, Albrecht M, Parlar H, Liebl B, Mayer R, Bolte G (2009b) Dietary intake of non-dioxin-like polychlorinated biphenyls (PCBs) in Bavaria. Results from INES (Integrated Exposure Assessment Survey). Das Gesundheitswesen 71:275–280. [in German]

Fromme H, Albrecht M, Boehmer S, Büchner K, Mayer R, Liebl B, Wittsiepe M, Bolte G (2009c) Intake and body burden of dioxin-like compounds in Germany. The INES Study. Chemosphere 76:1457–1463

Fromme H, Gruber L, Schuster R, Schlummer M, Kiranoglu M, Bolte G, Völkel W (2013) Phthalate and di-(2-ethylhexyl) adipate (DEHA) intake by German infants based on the results of a duplicate diet study and biomonitoring data (INES 2). Food Chem Toxicol 53:272–280

Fu X, Lindgren T, Guo M, Cai GH, Lundgren H, Norbäck D (2013) Furry pet allergens, fungal DNA and microbial volatile organic compounds (MVOCs) in the commercial aircraft cabin environment. Environ Sci Process Impacts 15:1228–1234

Garcia GJM, Schroeter JD, Segal RA, Stanek J, Foureman GL, Kimbell GS (2009) Dosimetry of nasal uptake of water-soluble and reactive gases: a first study of interhuman variability. Inhal Toxicol 21:607–618

Garcia-Alcega S, Rauert C, Harrad S, Collins CD (2016) Does the sourcemigration pathway of HBCDs to household dust influence their bio-accessibility? Sci Total Environ 569–570:244–251

García-Gonzáleza H, López-Pola T (2022) Health and safety in 3D printing. Int J Occup Environ Saf 6:14–25

Ginsberg G, Hattis D, Sonawane B (2004) Incorporating pharmacokinetic differences between children and adults in assessing children's risks to environmental toxicants. Toxicol Appl Pharmacol 198:164–183

Ginsberg GL, Foos BP, Firestone MP (2005) Review and analysis of inhalation dosimetry methods for application to children's risk assessment. J Toxicol Environ Health A 68:573–615

Ginsberg G, Foos B, Dzubow RB, Firestone M (2010) Options for incorporating children's inhaled dose into human health risk assessment. Inhal Toxicol 22:627–647

Ginsberg G, Toal B, Simcox N, Bracker A, Golembiewski B, Kurland T, Hedman C (2011) Human health risk assessment of synthetic turf fields based upon investigation of five fields in Connecticut. J Toxicol Environ Health A 74:1150–1174

Glorennec P, Lucas JP, Mandin C, Le Bot B (2012) French children's exposure to metals via ingestion of indoor dust, outdoor playground dust and soil: contamination data. Environ Int 45: 129–134

Golshahi L, Vehring R, Noga MI, Finlay WH (2013) In vitro deposition of micrometer-sized particles in the extrathoracic airways of children during tidal oral breathing. J Aerosol Sci 57: 14–21

Gong M, Zhang Y, Weschler CJ (2014) Measurement of phthalates in skin wipes: estimating exposure from dermal absorption. Environ Sci Technol 48:7428–7435

Gu J, Karrasch S, Salthammer T (2020) Review of the characteristics and possible health effects of particles emitted from laser printing devices. Indoor Air 30:396–421

Gudi-Mindermann H, White M, Roczen J, Riedel N, Dreger S, Bolte G (2023) Integrating the social environment with an equity perspective into the exposome paradigm: A new conceptual framework of the Social Exposome. Environ Res 233:116485

Gümperlein I, Fischer E, Dietrich-Gümperlein G, Karrasch S, Nowak D, Jörres RA, Schierl R (2018) Acute health effects of desktop 3D printing (fused deposition modeling) using acrylonitrile butadiene styrene and polylactic acid materials: an experimental exposure study in human volunteers. Indoor Air 28:611–623

Guo H, Zheng X, Ru S, Sun R, Mai B (2019) Size-dependent concentrations and bioaccessibility of organophosphate esters (OPEs) in indoor dust: a comparative study from a megacity and an e-waste recycling site. Sci Total Environ 650:1954–1960

Guo L, Salimi F, Wang H, Hofmann W, Johnson GR, Toelle BG, Marks GB, Morawska L (2020) Experimentally determined deposition of ambient urban ultrafine particles in the respiratory tract of children. Environ Int 145:106094

Gustafsson Å, Krais AM, Gorzsás A, Lundh T, Gerde P (2018) Isolation and characterization of a respirable particle fraction from residential house-dust. Environ Res 161:284–290

Gustafsson Å, Wang B, Gerde P, Bergman Å, Yeung LWY (2022a) Bioavailability of inhaled or ingested PFOA adsorbed to house dust. Environ Sci Pollut Res Int 29:78698–78710

Gustafsson Å, Bergman Å, Weiss JM (2022b) Estimated daily intake of per- and polyfluoroalkyl substances related to different particle size fractions of house dust. Chemosphere 303:135061

Habarta C, Sysoltseva M, Eckert S, Fembacher L, Frenzen A, Wolf J, Winterhalter R, Scheu C, Fromme H (2018) Ethanol fireplaces – effects on indoor air quality. Gefahrstoffe-Reinh Luft 78: 375–382. [in German]

He J, Yin Y, Pei J, Sun Y, Liu Z, Chen Q, Yang X (2022) A model to evaluate ozone distribution and reaction byproducts in aircraft cabin environments. Indoor Air 32:e13178

Herbig B, Jorres RA, Schierl R, Simon M, Langner J, Seeger S, Nowak D, Karrasch S (2018) Psychological and cognitive effects of laser printer emissions: a controlled exposure study. Indoor Air 28:112–124

Herbig B, Norrefeldt V, Mayer F, Reichherzer A, Lei F, Wargocki P (2023) Effects of increased recirculation air rate and aircraft cabin occupancy on passengers' health and well-being – results from a randomized controlled trial. Environ Res 216(Pt 4):114770

Hillier K, Schupp T, Carney I (2003) An investigation into VOC emissions from polyurethane flexible foam matresses. Cell Polym 22:237–259

Hines CJ, Waters MA, Larsson L, Petersen MR, Saraf A, Milton DK (2003) Characterization of endotoxin and 3-hydroxy fatty acid levels in air and settled dust from commercial aircraft cabins. Indoor Air 13:166–173

Höllbacher E, Ters T, Rieder Gradinger C, Srebotnik E (2017) Emissions of indoor air pollutants from six user szenarios in a model room. Atmos Environ 150:389–394

Hu M, Li W, Li L, Houston D, Wu J (2016) Refining time-activity classification of human subjects using the global positioning system. PLoS One 11:e0148875

Hu H, Liu X, Zheng Y, He X, Hart J, James P, Laden F, Chen Y, Bian J (2023) Methodological challenges in spatial and contextual exposome-health studies. Crit Rev Environ Sci Technol 53:827–846

Huang Z, Zhang Y, Yan Q, Zhang Z, Wanga X (2016) Real-time monitoring of respiratory absorption factors of volatile organic compounds in ambient air by proton transfer reactiontime-of-flight mass spectrometry. J Hazard Mater 320:547–555

Huang Z-H, Zhang Y-L, Yan Q, Wang Z-Q, Zhang Z, Wang X-M (2017) Decreased human respiratory absorption factors of aromatic hydrocarbons at lower exposure levels: the dual effect in reducing ambient air toxics. Environ Sci Technol Lett 4:463–469

Huang Y, Wang J, Chen Y, Chen L, Chen Y, Du W, Liu M (2022) Household $PM_{2.5}$ pollution in rural Chinese homes: levels, dynamic characteristics and seasonal variations. Sci Total Environ 13:153085

Huwe JK, Hakk H, Smith DJ, Diliberto JJ, Richardson V, Stapleton HM, Birnbaum LS (2008) Comparative absorption and bioaccumulation of polybrominated diphenyl ethers following ingestion via dust and oil in male rats. Environ Sci Technol 42:2694–2700

ICAO (International Civil Aviation Organization) (2022) The World of Air Transport in 2019. https://www.icao.int/annual-report-2019/Pages/the-world-of-air-transport-in-2019.aspx

ICRP (International Commission on Radiological Protection) (1994) Human respiratory tract model for radiological protection. Publication 66. International Commission on Radiological Protection. 24. Pergamon Press, Oxford, United Kingdom, pp. 272 Annals of ICRP

IEA (International Energy Agency) (2021) Tracking SDG 7: The energy progress report. https://www.irena.org/publications/2021/Jun/Tracking-SDG-7-2021

Ikegami M, Yoneda M, Tsuji T, Bannai O, Morisawa S (2014) Effect of particle size on risk assessment of direct soil ingestion and metals adhered to children's hands at playgrounds. Risk Anal 34:1677–1687

IRK (2020) Handling and use of additive manufacturing processes ("3D printers") in private households. Statement of the Indoor Air Hygiene Commission at the German Environment Agency. Bundesgesundheitsbl 63:370–371. [in German]

Islam MS, Paul G, Ong HX, Young PM, Gu YT, Saha SC (2020) A review of respiratory anatomical development, air flow characterization and particle deposition. Int J Environ Res Public Health 17:380

Jakobsson JKF, Hedlund J, Kumlin J, Wollmer P, Löndahl J (2016) A new method for measuring lung deposition efficiency of airborne nanoparticles in a single breath. Sci Reports 6:36147

Jiang C, Wang X, Li X, Inlora J, Wang T, Liu Q, Snyder M (2018) Dynamic human environmental exposome revealed by longitudinal personal monitoring. Cell 175:277–291

Johannessen C, Liggio J, Zhang X, Saini A, Harner T (2022) Composition and transformation chemistry of tire-wear derived organic chemicals and implications for air pollution. Atmos Pollut Res 13:101533

Kagi N, Fujii S, Horiba Y, Namiki N, Ohtani Y, Emi H, Tamura H, Kim YS (2007) Indoor air quality for chemical and ultrafine particle contaminants from printers. Build Environ 42:1949–1954

Kang Y, Man YB, Cheung KC, Wong MH (2012) Risk assessment of human exposure to bioaccessible phthalate esters via indoor dust around the Pearl River Delta. Environ Sci Technol 46:8422–8430

Kang Y, Zeng B, Man YB, Liu J, Yang Y, Li S, Situ K, Xiong W, Zeng L, Zhang Q, Luo J, Pan W, Jiang F, Wong MH (2018) Comparison of sorption kinetics of PAHs by sorptive sinks and caco-2 cell and the correlation between bioaccessibility and bioavailability of PAHs in indoor dust. Sci Total Environ 645:170–178

Karrasch S, Simon M, Herbig B, Langner J, Seeger S, Kronseder A, Peters S, Dietrich-Gümperlein G, Schierl R, Nowak D, Jörres RA (2017) Health effects of laser printer emissions: a controlled exposure study. Indoor Air 27:753–765

Katz EF, Goetz D, Wang C, Hart JL, Terranova B, Taheri ML, Waring MS, DeCarlo PF (2020) Chemical and physical characterization of 3D printer aerosol emissions with and without a filter attachment. Environ Sci Technol 54:947–954

Kempken B, Butte W (2010) Concentrations of lead, cadmium, manganese and zinc in fractions of household dust. Gefahrst Reinh Luft 70:98–102. [in German]

Khawaja I, Woodfield L, Collins P, Benkwitz A, Nevill A (2019) Exploring children's physical activity behaviours according to location: a mixed-methods case study. Sports 7:240

Kienzler A, Bopp SK, van der Linden S, Berggren E, Worth A (2016) Regulatory assessment of chemical mixtures: requirements, current approaches and future perspectives. Regul Toxicol Pharmacol 80:321–334

Kim D, Lee K (2022) Characteristics of ultrafine particles emitted from 3D-pens and effect of partition on children's exposure during 3D-pen operation. Indoor Air 32:e12978

Kim B, Shin JH, Kim HP, Jo MS, Kim HS, Lee JS, Lee HK, Kwon HC, Han SG, Kang N, Gulumian M, Bello D, Yu IJ (2022) Assessment and mitigation of exposure of 3-D printer emissions. Front Toxicol 18:817454

Kleinstreuer C, Zhang Z, Donohue JF (2008) Targeted drug-aerosol delivery in the human respiratory system. Annu Rev Biomed Eng 10:195–220

Klenow S, Heiland A, Heinemeyer G (2016) Pollutants in house dust: improving health assessment by determining the actual dust intake of children and adults. Report commissioned by the German Environment Agency. https://www.bfr.bund.de/de/schadstoffe_im_hausstaub__verbesserung_der_gesundheitlichen_bewertung_durch_ermittlung_der_tatsaechlichen_staubaufnahme_von_kindern_und_erwachsenen__schadstoffe_hausstaub_-193057.html. [in German]

Klepeis NE, Nelson WC, Ott WR, Robinson JP, Tsang AM, Switzer P, Behar JV, Hern SC, Engelmann WH (2001) The National Human Activity Pattern Survey (NHAPS): a resource for assessing exposure to environmental pollutants. J Expo Anal Environ Epidemiol 11:231–252

Koch HM, Müller J, Angerer J (2007) Determination of secondary, oxidised di-iso-nonylphthalate (DINP) metabolites in human urine representative for exposure to commercial DINP plasticizers. J Chromatogr B 847:114–125

Kohn MC, Parham F, Masten SA, Portier CJ, Shelby MD, Brock JW, Needham LL (2000) Human exposure estimates for phthalates. Environ Health Perspect 108:A440–A442

Krewski D, Acosta D, Andersen M, Anderson H, Bailar JC, Boekelheide K, Brent R, Charnley G, Cheung VG, Green S, Kelsey KT, Kerkvliet NI, Li AA, McCray L, Meyer O, Patterson RD, Pennie W, Scala RA, Solomon GM, Stephens M, Yager J, Zeise L (2010) Toxicity testing in the 21st century: a vision and a strategy. J Toxicol Environ Health B Crit Rev 13:51–138

Kwon O, Yoon C, Ham S, Park J, Lee J, Yoo D, Kim Y (2017) Characterization and control of nanoparticle emission during 3D printing. Environ Sci Technol 51:10357–10368

Lamb WF, Wiedmann T, Pongratz J, Andrew R, Crippa M, Olivier JGJ, Wiedenhofer D, Mattioli G, Al Khourdajie A, House J, Pachauri S, Figueroa M, Saheb Y, Slade R,

Hubacek K, Sun L, Kahn Ribeiro S, Khennas S, de la Rue du Can S, Chapungu L, Davis SJ, Bashmakov I, Dai H, Dhakal S, Tan X, Geng Y, Gu B, Minx J (2021) A review of trends and drivers of greenhouse gas emissions by sector from 1990 to 2018. Environ Res Lett 16:07300

Lee C-W, Hsu D-J (2007) Measurements of fine and ultrafine particles formation in photocopy centers in Taiwan. Atmos Environ 41:6598–6609

Lee M, Carter E, Yan L, Chan Q, Elliott P, Ezzati M, Kelly F, Schauer JJ, Wu Y, Yang X, Zhao L, Baumgartner J (2021) Determinants of personal exposure to $PM_{2.5}$ and black carbon in Chinese adults: a repeated-measures study in villages using solid fuel energy. Environ Int 146:106297

Li ZM, Kannan K (2023) Occurrence of 1,3-diphenylguanidine, 1,3-di-o-tolylguanidine, and 1,2,3-triphenylguanidine in indoor dust from 11 countries: implications for human exposure. Environ Sci Technol Apr 3

Li HB, Cui XY, Li K, Li J, Juhasz AL, Ma LQ (2014a) Assessment of in vitro lead bioaccessibility in house dust and its relationship to in vivo lead relative bioavailability. Environ Sci Technol 48: 8548–8455

Li HB, Li J, Juhasz AL, Ma LQ (2014b) Correlation of in vivo relative bioavailability to in vitro bioaccessibility for arsenic in household dust from China and its implication for human exposure assessment. Environ Sci Technol 48:13652–13659

Liao C, Kim U-J, Kannan K (2018) A review of environmental occurrence, fate, exposure, and toxicity of benzothiazoles. Environ Sci Technol 52:5007–5026

Liao X, Zou T, Chen M, Song Y, Yang C, Qiu B, Chen ZF, Tsang SY, Qi Z, Cai Z (2021) Contamination profiles and health impact of benzothiazole and its derivatives in PM2.5 in typical Chinese cities. Sci Total Environ 755(Pt 2):142617

Licina D, Nazaroff WW (2018) Clothing as a transport vector for airborne particles: chamber study. Indoor Air 28:404–414

Licina D, Melikov A, Sekhar C, Tham KW (2015) Transport of gaseous pollutants by convective boundary layer around a human body. Sci Technol Built Environ 21:1175–1186

Licina D, Morrison GC, Bekö G, Weschler CJ, Nazaroff WW (2019) Clothing-mediated exposures to chemicals and particles. Environ Sci Technol 53:5559–5575

Liu L, He R, Cui X, Ma LQ (2018) Impact of particle size on distribution, bioaccessibility, and cytotoxicity of polycyclic aromatic hydrocarbons in indoor dust. J Hazard Mater 357:341–347

Löndahl J, Swietlicki E, Rissler J, Bengtsson A, Boman C, Blomberg A, Sandström T (2012) Experimental determination of the respiratory tract deposition of diesel combustion particles in patients with chronic obstructive pulmonary disease. Part Fibre Toxicol 9:30

Löndahl J, Möller W, Pagels JH, Kreyling WG, Swietlicki E, Schmid O (2014) Measurement techniques for respiratory tract deposition of airborne nanoparticles: a critical review. J Aerosol Med Pulmonary Drug Deliv 27:229–254

Lorber M, Weschler CJ, Morrison G, Bekö G, Gong M, Koch HM, Salthammer T, Schripp T, Toftum J, Clausen G (2016) Linking a dermal permeation and an inhalation model to a simple pharmacokinetic model to study airborne exposure to di(n-butyl) phthalate. J Expo Sci Environ Epidemiol 00:1–9

Ma J, Li Y, Liu Y, Wang X, Lin C, Cheng H (2020) Metal(loid) bioaccessibility and children's health risk assessment of soil and indoor dust from rural and urban school and residential areas. Environ Geochem Health 42:1291–1303

Matz CJ, Stieb DM, Davis K, Egyed M, Rose A, Chou B, Brion O (2014) Effects of age, season, gender and urban-rural status on time-activity: Canadian human activity pattern survey 2 (CHAPS 2). Int J Environ Res Public Health 11:2108–2124

Matz CJ, Stieb DM, Brion O (2015) Urban-rural differences in daily time-activity patterns, occupational activity and housing characteristics. Environ Health 14:88

Mayer F, Fox R, Space D, Bezold A, Wargocki P (2022) IAQ in commercial air transportation. In: Zhang Y, Hopke PK, Mandin C (eds) Handbook of indoor air quality. Springer

McKernan LT, Wallingford KM, Hein MJ, Burge H, Rogers CA, Herrick R (2008) Monitoring microbial populations on wide-body commercial passenger aircraft. Ann Occup Hyg 52:139–149

Meek ME, Boobis AR, Crofton KM, Heinemeyer G, Van Raaij M, Vickers C (2011) Risk assessment of combined exposure to multiple chemicals: a WHO/IPCS framework. Regul Toxicol Pharmacol 60:S1–S14

Melymuk L, Rozárka Jílková R, Kolář M, Svobodová P, Vrana B, Hilscherová K (2022) Questioning the appropriateness of sieving for processing indoor settled dust samples. J Environ Expo Assess 1:15

Mitro SD, Dodson RE, Singla V, Adamkiewicz G, Elmi AF, Tilly MK, Zota AR (2016) Consumer product chemicals in indoor dust: a quantitative meta-analysis of U.S. studies. Environ Sci Technol 50:13611–13611

Molnár P, Johannesson S, Boman J, Barregård L, Sällsten G (2006) Personal exposures and indoor, residential outdoor, and urban background levels of fine particle trace elements in the general population. J Environ Monit 8:543–551

Morrison GC, Weschler CJ, Bekö G (2016) Dermal uptake directly from air under transient conditions: advances in modeling and comparisons with experimental results for human subjects. Indoor Air 26:913–924

Moya J, Phillips L (2014) A review of soil and dust ingestion studies for children. J Expo Sci Environ Epidemiol 24:545–554

Naccarato A, Tassone A, Martino M, Elliani R, Sprovieri F, Pirrone N, Tagarelli A (2021) An innovative green protocol for the quantification of benzothiazoles, benzotriazoles and benzosulfonamides in PM_{10} using microwave-assisted extraction coupled with solid-phase microextraction gas chromatography tandem-mass spectrometry. Environ Pollut 285:117487

NAS (2017) Using 21st century science to improve risk-related evaluations. National Academies Press, Washington, DC, USA. https://www.ncbi.nlm.nih.gov/books/NBK424978/

NAS (National Academies of Sciences, Engineering, and Medicine) (2012) Exposure science in the 21st century: a vision and a strate. National Academies Press, Washington, DC, USA. http://www.nap.edu/catalog.php?record_id=13507

NTP (National Toxicology Program) (1995) NTP technical report on toxicity studies of 1,3-diphenylguanidine (CAS No. 102-06-7) administered in feed to F344/N rats and B6C3F mice. NIH publication 95–3933. https://ntp.niehs.nih.gov/ntp/htdocs/st_rpts/tox042.pdf?utm_source=direct&utm_medium=prod&utm_campaign=ntpgolinks&utm_term=tox042

Núñez M, Fontanals N, Borrull F, Marcé RM (2022) Multiresidue analytical method for high production volume chemicals in dust samples, occurrence and human exposure assessment. Chemosphere 301:134639

Oberdörster G, Sharp Z, Atudorei V, Elder A, Gelein R, Kreyling W, Cox C (2004) Translocation of inhaled ultrafine particles to the brain. Inhal Toxicol 16:437–445

Oberdörster G, Oberdörster E, Oberdörster J (2005) Nanotoxicology: an emerging discipline evolving from studies of ultrafine particles. Environ Health Perspect 113:823–839

Odo DB, Yang A, Knibbs LD (2021) A systematic reviewand appraisal of epidemiological studies on household fuel use and its health effects using demographic and health surveys. Int J Environ Res Public Health 18:1411

OECD (2004) SIDS initial assessment report for SIAM 18. N-(1,3-dimethylbutyl)-N′-phenyl-1,4-phenylenediamine. https://hpvchemicals.oecd.org/UI/handler.axd?id=5e1a446c-5969-479c-9270-7ced8726952e

Olkowska E, Gržinić G (2022) Skin models for dermal exposure assessment of phthalates. Chemosphere 295:133909

Oomen AG, Lijzen JPA (2004) Relevancy of human exposure via house dust to the contaminants lead and asbestos. Report RIVM 711701037/2004. http://rivm.openrepository.com/rivm/bitstream/10029/9070/1/711701037.pdf

Özkaynak H, Xue J, Zartarian VG, Glen G, Smith L (2011) Modeled estimates of soil and dust ingestion rates for children. Risk Anal 31:592–608

Özkaynak H, Glen G, Cohen J, Hubbard H, Thomas K, Phillips L, Tulve N (2022) Model based prediction of age-specific soil and dust ingestion rates for children. J Expo Sci Environ Epidemiol 32:472–480

Pan W, Kang Y, Zeng L, Zhang Q, Luo J, Wong MH (2016) Comparison of in vitro digestion model with in vivo relative bioavailability of BDE-209 in indoor dust and combination of in vitro digestion/Caco-2 cell model to estimate the daily intake of BDE-209 via indoor dust. Environ Pollut 218:497–504

Pelletier M, Bonvallot N, Ramalho O, Blanchard O, Mandin C, LeBot B, Mercier F, Glorennec P (2017) Dermal absorption of semivolatile organic compounds from the gas phase: Sensitivity of exposure assessment by steady state modeling to key parameters. Environ Int 102:106–113

Plichta V, Völkel W, Fembacher L, Spolders M, Wöckner M, Aschenbrenner B, Schafft H, Fromme H (2019) Bioavailability of phthalate and DINCH® plasticizers, after oral administration of dust to piglets. Toxicol Lett 314:82–88

Plichta V, Völkel W, Fembacher L, Wöckner M, Nowak D, Fromme H (2022) The oral bioavailability of di-2-ethylhexyl phthalate (DEHP), di-isononyl phthalate (DiNP) and di-(isononyl)-cyclohexane-1,2-dicarboxylate (DINCH®) in house dust. Toxicol Lett 355:82–87

Poikkimäki M, Koljonen V, Leskinen N, Närhi M, Kangasniemi O, Kausiala O, Maso MD (2019) Nanocluster aerosol emissions of a 3D printer. Environ Sci Technol 53:13618–13628

Pope D, Johnson M, Fleeman N, Jagoe K, Duarte R, Maden M, Ludolph R, Bruce N, Shupler M, Adair-Rohani H, Lewis J (2021) Are cleaner cooking solutions clean enough? A systematic review and meta-analysis of particulate and carbon monoxide concentrations and exposures. Environ Res Lett 16:083002

Puttaswamy N, Natarajan S, Rao Saidam S, Mukhopadhyay K, Sadasivam S, Sambandam S, Balakrishnan K (2021) Evaluation of health risks associated with exposure to volatile organic compounds from household fuel combustion in southern India. Environ Adv 4:100043

Raffy G, Mercier F, Glorennec P, Mandin C, Le Bot B (2018) Oral bioaccessibility of semi-volatile organic compounds (SVOCs) in settled dust: a review of measurement methods, data and influencing factors. J Hazard Mater 352:215–227

Ramacher MOP, Karl M (2020) Integrating modes of transport in a dynamic modelling approach to evaluate population exposure to ambient NO_2 and $PM_{2.5}$ pollution in urban areas. Int J Environ Res Public Health 17:2099

Rappaport SM, Smith MT (2010) Epidemiology. Environment and disease risks. Science 330:460–461

Reina V, Kephalopoulos S, Zenié A, Borges T, del Río Martín A, Radovnikovic A, Stecca L (2014) Harmonising and integrating existing exposure factors systems world-wide. Feasibility study report. https://publications.jrc.ec.europa.eu/repository/bitstream/JRC91832/lbna26915enn.pdf

Rosenberger W (2018) Effect of charcoal equipped HEPA filters on cabin air quality in aircraft. A case study including smell event related in-flight measurements. Build Environ 143:358–365

Rosenberger W, Beckmann B, Wrbitzky R (2016) Airborne aldehydes in cabin-air of commercial aircraft: measurement by HPLC with UV absorbance detection of 2,4-dinitrophenylhydrazones. J Chromatogr B 1019:117–127

Rovira J, Domingo JL (2019) Human health risks due to exposure to inorganic and organic chemicals from textiles: a review. Environ Res 168:62–69

Ruby MV, Lowney YW (2012) Selective soil particle adherence to hands: Implications for understanding oral exposure to soil contaminants. Environ Sci Technol 46:12759–12771

Salthammer T, Zhang Y, Mo J, Koch HM, Weschler CJ (2018) Assessing human exposure to organic pollutants in the indoor environment. Angew Chem Int Ed Engl 57:12228–12263

Schneider K, Gerdes H, Hassauer M, Oltmanns J, Schulze J (2002) Consideration of the child risk group in the derivation of health-related environmental standards. Report within the framework of the "Environment and Health" action program. Funding reference (UFOPLAN) 201 61 215. https://www.apug.de/archiv/pdf/kinderempfindlichkeit.pdf [in German]

Schneider K, de Hoogd M, Haxaire P, Philipps A, Bierwisch A, Kaiser E (2020) ERASSTRI – European Risk Assessment Study on Synthetic Turf Rubber Infill – Part 2: Migration and monitoring studies. Sci Total Environ 718:137173

Schripp T, Wensing M, Uhde E, Salthammer T, He C, Morawska L (2008) Evaluation of ultrafine particle emissions from laser printers using emission test chambers. Environ Sci Technol 42: 4338–4343

Schripp T, Salthammer T, Wientzek S, Wensing M (2014) Chamber studies on nonvented decorative fireplaces using liquid or gelled ethanol fuel. Environ Sci Technol 48:3583–3590

Schuchardt S, Bitsch A, Koch W, Rosenberger W (2017) CAQ Preliminary cabin air quality measurement campaign; 2017. https://www.easa.europa.eu/sites/default/files/dfu/EASA%20CAQ%20Study%20Final%20Report_21.03.2017.pdf

Schweizer C, Edwards RD, Bayer-Oglesby L, Gauderman WJ, Ilacqua V, Jantunen MJ, LAI H, Nieuwenhuijsen M, Künzli N (2007) Indoor time–microenvironment–activity patterns in seven regions of Europe. J Expo Sci Environ Epidemiol 17:170–181

Scungio M, Vitanza T, Stabile L, Buonanno G, Morawska L (2017) Characterization of particle emission from laser printers. Sci Total Environ 586:623–630

Seeger S, Brodner D, Jacobi T, Rasch F, Rothardt M, Wilke O (2018) Emissions of fine and ultrafine particles and volatile organic compounds from different filament materials operated on a low-cost 3D printer. Gefahrst Reinh Luft 3:79–87

Senerat AM, Manemann SM, Clements NS, Brook RD, Hassett LC, Roger VL (2020) Biomarkers and indoor air quality: a translational research review. J Clin Transl Sci 5:e39

Setton E, Marshall JD, Brauer M, Lundquist KR, Hystad P, Keller P, Clouthier-Fisher DP (2011) The impact of daily mobility on exposure to traffic-related air pollution and health effect estimates. J Expo Sci Environ Epidemiol 21:42–48

Setyawati MI, Singh D, Krishnan SPR, Huang X, Wang M, Jia S, Goh BHR, Ho CG, Yusoff R, Kathawala MH, Poh TY, Ali NABM, Chotirmall SH, Aitken RJ, Riediker M, Christiani DC, Fang M, Bello D, Demokritou P, Ng KW (2020) Occupational inhalation exposures to nanoparticles at six Singapore printing centers. Environ Sci Technol 54:2389–2400

Shin HM, Moschet C, Young TM, Bennett DH (2020) Measured concentrations of consumer product chemicals in California house dust: Implications for sources, exposure, and toxicity potential. Indoor Air 30:60–75

Shupler M, Hystad P, Birch A, Miller-Lionberg D, Jeronimo M, Arku RE, Chu YL, Mushtaha M, Heenan L, Rangarajan S, Seron P, Lanas F, Cazor F, Lopez-Jaramillo P, Camacho PA, Perez M, Yeates K, West N, Ncube T, Ncube B, Chifamba J, Yusuf R, Khan A, Hu B, Liu X, Wei L, Tse LA, Mohan D, Kumar P, Gupta R, Mohan I, Jayachitra KG, Mony PK, Rammohan K, Nair S, Lakshmi PVM, Sagar V, Khawaja R, Iqbal R, Kazmi K, Yusuf S, Brauer M (2020) Household and personal air pollution exposure measurements from 120 communities in eight countries: results from the PURE-AIR study. Lancet Planet Health 4:e451–e462

Simcox N, Bracker A, Ginsberg G, Toal B, Golembiewski B, Kurland T, Hedman C (2011) Synthetic turf field investigation in Connecticut. J Toxicol Environ Health A 74:1133–1149

Smith SA, Norris B (2003) Reducing the risk of choking hazards: mouthing behaviour of children aged 1 month to 5 years. Inj Control Saf Promot 10:145–154

Sommer F, Dietze V, Baum A, Sauer J, Gilge S, Maschowski C, Gieré R (2018) Tire abrasion as a major source of microplastics in the environment. Aerosol Air Qual Res 18:2014–2028

Sowers TD, Nelson CM, Diamond GL, Blackmon MD, Jerden ML, Kirby AM, Noerpel MR, Scheckel KG, Thomas DJ, Bradham KD (2021) High lead bioavailability of indoor dust contaminated with paint lead species. Environ Sci Technol 55:402–411

Spengler JD, Vallarino J, McNeely E, Estephan H (2012) In-flight/on-board monitoring: ACER's component for ASHRAE 1262, Part 2. Airliner cabin environmental research (ACER) program. https://www.faa.gov/data_research/research/med_humanfacs/cer/media/In-FlightOnboardMonitoring.pdf

Spinu N, Cronin MTD, Enoch SJ, Madden JC, Worth AP (2020) Quantitative adverse outcome pathway (qAOP) models for toxicity prediction. Arch Toxicol 94:1497–1510

Stabile L, Scungio M, Buonanno G, Arpino F, Ficco G (2017) Airborne particle emission of a commercial 3D printer: the effect of filament material and printing temperature. Indoor Air 27: 398–408

Steinle P (2016) Characterization of emissions from a desktop 3D printer and indoor air measurements in office settings. J Occup Environ Hyg 13:121–132
Steinle S, Reis S, Sabel CE (2013) Quantifying human exposure to air pollution – moving from static monitoring to spatio-temporally resolved personal exposure assessment. Sci Total Environ 443:184–193
Stephens B, Azimi P, El Orch Z, Ramos T (2013) Ultrafine particle emissions from desktop 3D printers. Atmos Environ 79:334–339
Tan Y-M, Leonard JA, Edwards S, Teeguarden J, Paini A, Egeghy P (2018) Aggregate exposure pathways in support of risk assessment. Curr Opin Toxicol 9:8–13
Tan H, Yang L, Huang Y, Tao L, Chen D (2021) "Novel" synthetic antioxidants in house dust from multiple locations in the Asia-Pacific region and the United States. Environ Sci Technol 55: 8675–8682
Tandon PS, Saelens BE, Zhou C, Kerr J, Christakis DA (2013) Indoor versus outdoor time in preschoolers at child care. Am J Prev Med 44:85–88
Tang CL, Seeger S (2022) Systematic ranking of filaments regarding their particulate emissions during fused filament fabrication 3D printing by means of a proposed standard test method. Indoor Air 32:e13010
Tang T, Hurraß J, Gminski R, Mersch-Sundermann V (2012) Fine and ultrafine particles emitted from laser printers as indoor air contaminants in German offices. Environ Sci Pollut Res Int 19: 3840–3849
Tedla G, Jarabek AM, Byrley P, Boyes W, Rogers K (2022) Human exposure to metals in consumer-focused fused filament fabrication (FFF)/3D printing processes. Sci Total Environ 814:152622
Teeguarden JG, Tan Y-M, Edwards SW, Leonard JA, Anderson KA, Corley RA, Kile ML, Simonich SM, Stone D, Tanguay RL, Waters KM, Harper SL, Williams DE (2016) Completing the link between exposure science and toxicology for improved environmental health decision making: the aggregate exposure pathway framework. Environ Sci Technol 50:4579–4586
Tichenor BA, Sparks LE, Jackson MD, Guo Z, Mason MA, Plunkett CM, Rasor SA (1990) Emissions of perchloroethylene from dry cleaned fabrics. Atmos Environ Part A 24A:1219–1229
Turner A (2011) Oral bioaccessibility of trace metals in household dust: a review. Environ Geochem Health 33:331–341
Turner MC, Nieuwenhuijsen M, Anderson K, Balshaw D, Cui Y, Dunton G, Hoppin JA, Koutrakis P, Jerrett M (2017) Assessing the exposome with external measures: commentary on the state of the science and research recommendations. Annu Rev Public Health 38:215–239
US EPA (2017) Soil and Dust Ingestion. Update for Chapter 5 of the Exposure Factors Handbook. EPA/600/R-17/384F. Washington, DC, USA. https://www.epa.gov/sites/production/files/2018-01/documents/efh-chapter05_2017.pdf
Van Wijnen JH, Clausing P, Brunekreef B (1990) Estimate soil ingestion by children. Environ Res 51:147–162
Vance ME, Pegues V, Van Montfrans S, Leng W, Marr LC (2017) Aerosol emissions from fuse-deposition modeling 3D printers in a chamber and in real indoor environments. Environ Sci Technol 51:9516–9523
Viitanen AK, Kallonen K, Kukko K, Kanerva T, Saukko E, Hussein T, Hämeri K, Säämänen A (2021) Technical control of nanoparticle emissions from desktop 3D printing. Indoor Air 31: 1061–1071
Von Lindern I, Spalinger S, Petroysan V, von Braun M (2003) Assessing remedial effectiveness through the blood lead: soil/dust lead relationship at the Bunker Hill Superfund Site in the Silver Valley of Idaho. Sci Total Environ 303:139–170
Walker ES, Noonan CW, Semmens EO, Ware D, Smith P, Boyer BB, Erdei E, Hopkins SE, Lewis J, Belcourt A, Ward TJ (2021) Indoor fine particulate matter and demographic, household, and wood stove characteristics among rural US homes heated with wood fuel. Indoor Air 31:1109–1124

Wambaugh JF, Bare JC, Carignan CC, Dionisio KL, Dodson RE, Jolliet O, Liu X, Meyer DE, Newton SR, Phillips KA, Price PS, Ring CL, Shin H-M, Sobus JR, Tal T, Ulrich EM, Vallero DA, Wetmore BA, Isaacs KK (2019) New approach methodologies for exposure science. Curr Opin Toxicol 15:76–92

Wan Y, Xue J, Kannan K (2016) Benzothiazoles in indoor air from Albany, New York, USA, and its implications for inhalation exposure. J Hazard Mater 311:37–42

Wang H, He C, Morawska L, McGarry P, Johnson G (2012) Ozone-initiated particle formation, particle aging, and precursors in a laser printer. Environ Sci Technol 46:704–712

Wang L, Asimakopoulos AG, Moon HB, Nakata H, Kannan K (2013) Benzotriazole, benzothiazole, and benzophenone compounds in indoor dust from the United States and east Asian countries. Environ Sci Technol 47:4752–4759

Wang J, Lin K, Taylor A, Gan J (2018) In vitro assessment of pyrethroid bioaccessibility via particle ingestion. Environ Int 119:125–132

Wei W, Bonvallot N, Gustafsson A, Raffy G, Glorennec P, Krais A, Ramalho O, Le Bot B, Mandin C (2018) Bioaccessibility and bioavailability of environmental semivolatile organic compounds via inhalation: a review of methods and models. Environ Int 113:202–213

Wei W, Ramalho O, Mandin C (2020) Modeling the bioaccessibility of inhaled semivolatile organic compounds in the human respiratory tract. Int J Hyg Environ Health 224:113436

Wensing M, Schripp T, Uhde E, Salthammer T (2008) Ultra-fine particles release from hardcopy devices: sources, real-room measurements and efficiency of filter accessories. Sci Total Environ 15:418–427

Weschler CJ, Nazaroff WW (2012) SVOC exposure indoors: fresh look at dermal pathways. Indoor Air 22:356–377

Weschler CJ, Bekö G, Koch HM, Salthammer T, Schripp T, Toftum J, Clausen G (2015) Transdermal uptake of diethyl phthalate and di(n-butyl) phthalate directly from air: experimental verification. Environ Health Perspect 123:928–934

WHO (2014) WHO indoor air quality guidelines: household fuel combustion. Geneva, Switzerland. https://www.who.int/publications/i/item/9789241548885

WHO (2021) WHO global air quality guidelines Particulate matter ($PM_{2.5}$ and PM_{10}), ozone, nitrogen dioxide, sulfur dioxide and carbon monoxide. https://www.who.int/publications/i/item/9789240034228

Wild CP (2005) Complementing the genome with an "exposome": the outstanding challenge of environmental exposuremeasurement in molecular epidemiology. Cancer Epidemiol Biomark Prev 14:1847–1850

Wild CP (2012) The exposome: from concept to utility. Int J Epidemiol 41:24–32

Wilson R, Jones-Otazo H, Petrovic S, Mitchell I, Bonvalot Y, Williams D, Richardson GM (2013) Revisiting dust and soil ingestion rates based on hand-to-mouth transfer. Hum Ecol Risk Assess 19:158–188

Wojtyła S, Klama P, Śpiewak K, Baran T (2020) 3D printer as a potential source of indoor air pollution. Int J Environ Sci Technol 17:207–218

Wolkoff P, Crump DR, Harrison TC (2016) Pollutant exposures and health symptoms in aircrew and office workers: is there a link? Environ Int 87:74–84

Wu X, Bennett DH, Lee K, Cassady DL, Ritz B, Hertz-Picciotto I (2011) Longitudinal variability of time-location/activity patterns of population at different ages: a longitudinal study in California. Environ Health 10:80

Würzner P, Jörres RA, Karrasch S, Quartucci C, Böse-O'Reilly S, Nowak D, Rakete S (2022) Effect of experimental exposures to 3-D printer emissions on nasal allergen responses and lung diffusing capacity for inhaled carbon monoxide/nitric oxide in subjects with seasonal allergic rhinitis. Indoor Air 32:e13174

Xi J, Si X, Zhou Y, Kim JW, Berlinski A (2014) Growth of nasal and laryngeal airways in children: implications in breathing and inhaled aerosol dynamics. Respir Care 59:263–273

Xue J-P, Zartarian VG, Moya J, Freeman N, Beamer P, Black K, Tulve N, Shalat S (2007) A meta-analysis of children's hand-to-mouth frequency data for estimating nondietary ingestion exposure. Risk Anal 27:411–420

Xue J-P, Zartarian VG, Tulve N, Moya J, Freeman NCG, Auyeung W, Beamer P (2010) A meta-analysis of children's object-to-mouth frequency data for estimating non-dietary ingestion exposure. J Expo Sci Environ Epidemiol 20:536–545

Yi J, LeBouf RF, Duling MG, Nurkiewicz T, Chen BT, Schwegler-Berry D, Virji MA, Stefaniak AB (2016) Emission of particulate matter from a desktop three-dimensional (3D) printer. J Toxicol Environ Health A 79:453–465

Yin Y, He J, Pei J, Yang X, Sun Y, Cui X, Lin CH, Wei D, Chen Q (2022) Influencing factors of carbonyl compounds and other VOCs in commercial airliner cabins: on-board investigation of 56 flights. Indoor Air 31:2084–2098

Younger A, Alkon A, Harknett K, Jean Louis R, Thompson LM (2022) Adverse birth outcomes associated with household air pollution from unclean cooking fuels in low- and middle-income countries: a systematic review. Environ Res 204(Pt C):112274

Yu N, Zhang Y, Zhang M, Li H (2021) Thermal condition and air quality investigation in commercial airliner cabins. Sustainability 13:7047

Zhang Q, Pardo M, Rudich Y, Kaplan-Ashiri I, Wong JPS, Davis AY, Black MS, Weber RJ (2019) Chemical composition and toxicity of particles emitted from a consumer-level 3D printer using various materials. Environ Sci Technol 53:12054–12061

Zhang Q, Weber RJ, Luxton TP, Peloquin DM, Baumann EJ, Black MS (2022a) Metal compositions of particle emissions from material extrusion 3D printing: emission sources and indoor exposure modeling. Sci Total Environ 25:160512

Zhang X, Liu J, Liu X, Liu C, Chen Q (2022b) HEPA filters for airliner cabins: state of the art and future development. Indoor Air 32:e13103

Zhang Y, Xu C, Zhang W, Qi Z, Song Y, Zhu L, Dong C, Chen J, Cai Z (2022c) p-Phenylenediamine antioxidants in $PM_{2.5}$: the underestimated urban air pollutants. Environ Sci Technol 56:6914–6921

Zhang Y-J, Xu T-T, Ye D-M, Lin Z-Z, Wang F, Guo Y (2022d) Widespread N-(1,3-dimethylbutyl)-N′-phenyl-p-phenylenediamine quinone in size-fractioned atmospheric particles and dust of different indoor environments. Environ Sci Technol Lett 9:420–425

Zhang Q, Weber RJ, Luxton TP, Peloquin DM, Baumann EJ, Black MS (2023) Metal compositions of particle emissions from material extrusion 3D printing: emission sources and indoor exposure modeling. Sci Total Environ 860:160512

Zhao D, Wang JY, Tang N, Yin DX, Luo J, Juhasz AL, Xiang P, Li HB, Ma LQ (2018) Coupling bioavailability and stable isotope ratio to discern dietary and non-dietary contribution of metal exposure to residents in mining-impacted areas. Environ Int 120:563–571

Zontek TL, Ogle BR, Jankovic JT, Hollenbeck SM (2017) An exposure assessment of desktop 3D printing. J Chem Health Saf 24:15–25

Zulauf N, Dröge J, Klingelhöfer D, Braun M, Oremek GM, Groneberg DA (2019) Indoor air pollution in cars: an update on novel insights. Int J Environ Res Public Health 16:2441

Chapter 3
Volatile Organic Compounds and Very Volatile Organic Compounds

Contents

Abstract In this chapter, the large group of volatile organic compounds (VOC) like formaldehyde, other aldehydes, glycol ethers, isothiazolinone, methyl methacrylate, emerging PFAS, oximes, and naphthalene as well as the very volatile organic compounds (VVOC) are presented individually in detail. In addition to a brief description of the chemical–physical characteristics and the use of the substance, an overview of main health risks and, in particular, references to further scientific literature are given. In the foreground, however, are extensive presentations of the occurrence of the substances in indoor air and in sedimented dust.

Keywords VOC · VVOC · Formaldehyde · Glycol ethers · Emerging PFAS

For outdoor air, measurement results of many substances are often available in larger numbers. In contrast, knowledge about air quality in homes and especially in publicly accessible indoor spaces is patchy in many areas. This is the case, for example, in the common rooms of daycare centers or the classrooms of schools and similar facilities. This is even though most of the population in Western countries spends most of their time indoors. Children, young people, and trainees spend around 30–50% of their daytime in schools or comparable public facilities. Here, they can be exposed to various substances and other factors influencing their health. Indoor public spaces are also of great importance in the context of preventive

H. Fromme, *Indoor Air Quality*, https://doi.org/10.1007/978-3-031-40078-0_3

Table 3.1 Classification of chemical substances in the interior (according to Seifert 1999)

Designation	Abbreviation	Boiling point (°C)
Very volatile organic compounds	VVOC	<0 to 50–100
Volatile organic compounds	VOC	50–100 to 240–260
Semi-volatile organic compounds	SVOC	240–260 to 380–400
Particulate organic matter	POM	>380

concepts. On the one hand, a large number of people can be reached at the same time with appropriate measures and, on the other hand, these are often sensitive age groups of the population who have a right to special protection of their health.

In the meantime, various organic and inorganic compounds can be detected in indoor air. Since the volatility of organic substances increases with the boiling point, a general classification of volatile and non-volatile substances according to this criterion was proposed by a WHO working group (see Table 3.1). Volatile organic compounds (VOC) are substances present in gaseous form with a boiling point range of approx. 50–100 °C to approx. 240–260 °C. This group includes substances with very different boiling points and very different chemical and physical characteristics. If the boiling point is lower, the term VVOC (very volatile organic compounds) is used, but the definition is not yet clear. With decreasing volatility, the group of semi-volatile compounds (SVOC) follows. They occur indoors predominantly in particle-bound form. According to this classification, the dust fractions or particulate air pollutants (POM, Particulate Organic Matter) are at the end.

In addition, subgroups such as the so-called mVOC (microbial volatile organic compounds) can be distinguished in indoor air. In this case, it is a large number of organic compounds that can be formed by microorganisms (e.g., molds) and released into the indoor air. In addition, various substances can be grouped as oVOC (odor active volatile organic compounds). These are odor-active volatile organic compounds that can still be perceived even in very low concentrations.

3.1 Volatile Organic Compounds (VOC)

The very extensive group of VOC comprises thousands of carbon containing individual substances, including methane (CH_4), which has a particularly strong impact on the climate. All other VOC are therefore also referred to as non-methane VOC (NMVOC). In the following, only the common term VOC is used, without taking methane into account.

In outdoor air, VOCs have multiple sources such as technical processes with incomplete combustion (e.g., motor vehicle exhaust), evaporative losses from fuels, fugitive emissions from industrial processes and applications (e.g., when using solvents), and biological processes such as plant metabolism, decay, and degradation processes. For example, Yang et al. (2022) reviewed the literature on VOC source analyses published from 2015 to 2021 using highly time-resolved monitoring data from ambient air in 8 countries and China (responsible for 84% of the studies). Overall, taking into account significant differences between the cities/regions studied, this study concluded that vehicle emissions and fuel evaporation were the primary contributors to the ambient VOC in the world. The average contribution in other countries and China for vehicle emissions was 23.9% and 25.1%, for fuel evaporation 22.4% and 18.5%, for coal combustion 12.0% and 28%, for industrial emissions 22.0% and 18.0%, and for solvent 23.0% and 15.5%, and for biogenic emissions 16.8% and 8.8%, respectively.

The sources of VOC in indoor environments are diverse and subject to significant change over time, e.g., due to formulation changes in consumer products. Thus, significant changes in the composition of VOC have been observed in recent decades (Weschler 2009). While aromatic and chlorinated solvents were dominant in the past, today terpenes and glycols are more likely to be observed in indoor air. The following major groups of sources can be named:

- Buildings: emissions from building materials and furnishings such as flooring, wood materials, and curtains; ventilation systems; bioaerosols after, e.g., moisture damage.
- Activities: application of paints, varnishes, and adhesives; smoking activities; operation of open fires; cleaning products; personal care products/cosmetics; indoor air fresheners; hobby activities; operation of technical equipment and office machines.
- By the human being himself: exhalation and body exhalations; clothing.
- Outdoor air: combustion processes such as motor vehicle traffic and domestic fires as well as from industrial processes; emissions from textile cleaning.
- Soil/groundwater: emissions of VOC via soil passage from contaminated sites/ soil contamination.

In a recent review of the occurrence of chemicals in European residences, of the 65 individual VOC commonly found in this microenvironments, 52 (80%) were identified to have sources associated with building and construction materials (e.g., brick, wood products, adhesives and materials for flooring installation), 41 (63%) linked with consumer products (passive, electric, and combustible air fresheners, hair sprays, deodorants), 19 were associated with candle and incense burning, and nine (14%) were associated with space heating (Halios et al. 2022).

In addition, complex chemical reactions occur indoors in the gas and aerosol phases and the formation of the so-called secondary organic aerosols (SOA), which can consist of different, newly formed organic substances (Uhde and Salthammer 2009). See also Sect. 1.3.

3.1.1 *Health Aspects*

Due to a large number of individual substances, the effects of VOC are complex and, depending on the level of exposure, range from mild disturbances of well-being to toxic effects. In this context, an association is repeatedly made between the level of VOC concentration indoors and rather unspecific symptoms such as fatigue, headaches, reduced performance, and irritative effects on mucous membranes of the eye and upper respiratory tract. In assessment practice, however, it is often difficult to attribute the symptoms or illnesses described to VOC or a single substance.

Epidemiological studies on the relationship between TVOC (total volatile organic compounds) levels in indoor environments and health effects did not provide a consistent picture and showed methodological limitations (Andersson et al. 1997; Seifert 1999; Maung et al. 2022). In addition, several population-based studies attempted to find an association between indoor or outdoor VOC concentrations and the development or worsening of asthma. Five recent reviews see a partial relationship in this regard, but all conclude that there is little evidence for a pronounced association (Heinrich 2011; Tagiyeva and Sheikh 2014; Nurmatov et al. 2015; Kanchongkittiphon et al. 2015; Rufo et al. 2016).

Due to the large number of VOC, it would make sense to prioritize the individual substances according to the significance of their health effects. The following is a first example. Halios et al. (2022) reviewed the occurrence and toxicological properties of 65 VOC and identified the 17 most health relevant VOC, according to their adverse-effect endpoints (i.e., respiratory, irritation of the upper airway system and eyes, cardiovascular, neurological, and carcinogenic) after inhalation. Figure 3.1 shows the substances and their weighted average geometric mean values in indoor air showing effects in at least one up to five of the afore mentioned categories. However, it should not be forgotten that, of course, for each individual substance the respective level of carcinogenic and non-carcinogenic risk is of great importance.

3.1.2 *Total Volatile Organic Compounds (TVOC)*

In addition to the determination of individual VOC, the sum of the concentrations of the individual volatile organic compounds, the so-called TVOC value (Total Volatile Organic Compounds), can be used for a first and simplified assessment of indoor air quality (Mølhave et al. 1997). By definition, this is the sum of all substances that elute analytically in the retention range between n-hexane and n-hexadecane on a non-polar separation column using the following equation:

$$\mathrm{TVOC} = S_{\mathrm{id}} + S_{\mathrm{un}}$$

where S_{id} the sum of the concentrations of the identified compounds in mg/m^3 and S_{un} the sum of concentrations of unidentified VOC in mg/m^3 using the response factor of toluene.

Controlled experiments with human subjects have attempted to establish a relationship between TVOC levels and health effects.

At first, Mølhave et al. (1986) conducted the first controlled test chamber experiments with a VOC mixture consisting of 22 individual substances on

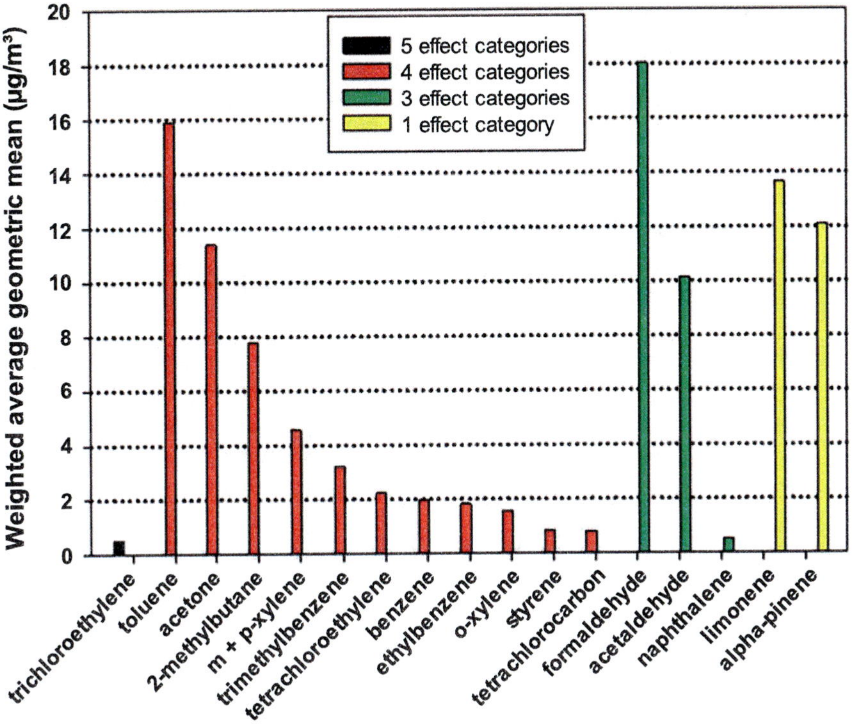

Fig. 3.1 Ranking of the most relevant VOC according to their association with the number of adverse-effect categories (modified after Halios et al. 2022)

62 healthy volunteers. These were exposed to a concentration of 0, 5, or 25 mg/m^3 for 2.75 hours. The test subjects complained of poor indoor air quality, odors, and dry mucous membranes at concentrations as low as 5 mg/m^3 and reported more irritative symptoms. The study was repeated with 16 subjects in the USA at a concentration of 25 mg/m^3 or filtered air (Koren et al. 1992). While various psycho-neurological performance tests remained unchanged, subjects here also complained more of headaches, faster fatigability, and eye and throat irritation. In another experimental study, both healthy subjects and subjects reporting complaints of sick building syndrome (SBS) showed more subjective symptoms, but the effects were more pronounced in the SBS group (Kjærgaard et al. 1991).

Nevertheless, Gminski et al. (2011a) exposed twenty-four healthy non-smokers to clean air or emissions from oriented strand boards (OSB) for 2 h under controlled conditions in a 48 m^3 test chamber at three different time points and chemosensory irritation, exhaled nitric oxide (eNO) concentration, eye blink frequency, lung function, and subjective perception of irritation of eyes, nose, and throat were examined before, during, and after exposure. Total VOC exposure concentrations reached 8900 $\mu g/m^3$ for the fresh OSB panels and emissions consisted predominantly of a-pinene, Δ3-carene, and hexanal. No irritating or pulmonary effects were seen and all the subjective ratings of discomfort were at a low level. Only the ratings for smell of emissions increased significantly during exposure in comparison with clean air. In a second experiment of the same working group, 15 healthy non-smokers were exposed for 2 h under controlled conditions to TVOC levels of 5000 μg/m^3, 8000 μg/m^3, and 13,000 μg/m^3 emitted from pinewood panels in a 48 m^3 test chamber (Gminski et al. 2011b). Again, the emissions consisted predominantly of α-pinene and Δ3-carene. No concentration-dependent effects before or after exposure to the emissions were measured with respect to sensory irritation, pulmonary function, eNO, and eye blink frequency. Only the odor of the emissions was perceived by the study subjects, rated as being closer to "pleasant" than to "unpleasant."

In any case, it must be taken into account that the TVOC concept does not include the possibility of interactions between the many compounds and no toxicological information of the single substances are considered (Mølhave 2003). Therefore, the indicator is not an official recommendation and no definitive conclusions should be made based on this indicator alone, but he is a reasonable screening tool needed for exposure evaluation, source identification, and indoor air quality evaluation. A recent overview of the limitations and advantages of the TVOS concept is given by Salthammer (2022).

Table 3.2 summarizes the results of some recent studies on TVOC concentrations in indoor air. For example, in the first representative study by the German Federal Environment Agency in 2003–2006, a median of 300 μg/m^3 and the 95th percentile of 1100 μg/m^3 were observed. The second study 2014–2017 showed comparable values of 270 μg/m^3 and 950 μg/m^3, respectively (UBA 2008; Bethke et al. 2021). Within the summary of measurement results of different German research institutes from 2006–2012, a median of 360 μg/m^3 (95th percentile: 1975 μg/m^3) was found for TVOC in 146 dwellings (Hofmann et al. 2014).

3.1.3 *Occurrence in the Air of Residences*

A large number of studies are available in the scientific literature that has investigated VOC or individual volatile organic compounds in residential air. Several recent reviews summarize the published study results (Cometto-Muñiz and Abraham 2015; Paciência et al. 2016; Shrubsole et al. 2019; Vardoulakis et al. 2020; Mannan and Al-Ghamdi 2021; You et al. 2022). Figure 3.2 shows an example of the substances with the highest concentrations in indoor air identified in the work of Cometto-Muñiz and Abraham (2015).

In Germany, study results have been published by private research laboratories in the years 2006–2012 (Hofmann et al. 2014). In addition, results from two representative studies by the German Environment Agency are available. In the first, the air of 555 children's rooms was investigated in 2003–2006 as part of the Children's Environmental Survey, and in the second, 639 indoor rooms were investigated in 2014–2017 as part of the German Environmental Health Survey (GerES V) (UBA

Table 3.2 Indoor TVOC values in μg/m³

Reference	Quantity	Location	Median	Minimum-Maximum	Country, Year
Raw et al. (2004)	876	Residences	202	15–3360	UK, 1997–99
UBA (2008)	555	Residences	300	n.s.-3000	Germany, 2003–06
Kotzias et al. (2009)	111	Public buildings	49[a]	8–282	EU-wide, 2003–08
	60	Schools/daycare centers	41[a]	9–193	EU-wide, 2003–08
Hofmann et al. (2014)	146	Residences	430	n.s.-1975[b]	Germany, 2006–12
	271	Schools	470	n.s.-2600[b]	Germany, 2006–12
Langer and Bekö (2013)	157	Houses	180	–	Sweden, 2011
Fromme et al. (2016)	63	Daycare centers	179	30–748	Germany, 2012
Langer et al. (2015)	21	Passive houses	272	n.s.-1550[c]	Sweden, 2012–13
	20	Conventional houses	145	n.s.-650[c]	Sweden, 2012–13
Yang et al. (2020)	169	Energy efficient houses	212	n.s.-2292	Switzerland, 2013–16
Bethke et al. (2021)	639	Residences	270	99[d]-6700	Germany, 2014–17
Chen et al. (2017)	31	Residences	1270	290–3060	China
Sun et al. (2022b)	75	Residences	5	5–1850	China, 2017–21
Pei et al. (2020)	251	Residences	–	104–1151	China
Heeley-Hill et al. (2021)	60	Residences	370[e]/426[f]	–	UK, 2019

n.s. not specified
[a]Arithmetic mean
[b]95th percentile
[c]Estimated from figure in publication
[d]Tenth percentile
[e]Summer
[f]Winter

2008; Bethke et al. 2021). The substances with the highest medians in the latter study are ethyl acetate (16.6 μg/m³), decamethylcyclopentasiloxane (14.0 μg/m³), and limonene (12.0 μg/m³). The statistical characteristic values for the individual substances from the two aforementioned studies are summarized in Table 3.3 Again, 61 VOC were quantified in a study conducted in 622 homes (2246 measurements through the whole year) in 2006–2008 (Rösch et al. 2014). In this study the highest medians were reported for α-pinene (15.5 μg/m³), limonene (13.0 μg/m³), and toluene (8.1 μg/m³). Alvarez-Vaca et al. (2022) reported medians of 4.5 μg/m³ for

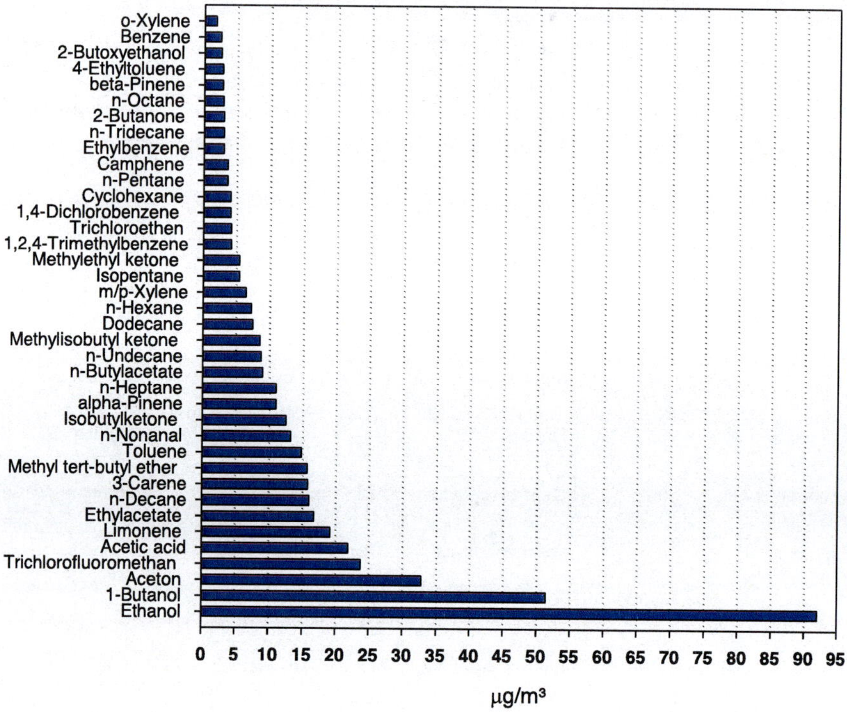

Fig. 3.2 Ranking of median VOC concentrations in the air of homes/schools (modified after Cometto-Muñiz and Abraham 2015)

α-pinene, 4.8 µg/m^3 for butyl acetate, 7.0 µg/m^3 for toluene, and 10 µg/m^3 for limonene in a study in 159 residences in Luxembourg from 2014 to 2019.

In a very comprehensive Canadian study, which examined 3524 residential interiors nationwide, the following individual substances could be found with the highest geometric mean values in 2012–2013: Limonene 23.9 µg/m^3, decamethylcyclopentasiloxane 9.6 µg/m^3, toluene 8.2 µg/m^3, and α-pinene 6.7 µg/m^3 (Li et al. 2019). In the annual course of VOC concentrations in indoor air, significant differences can be observed for the individual substances. For some substance classes, the time course over the year based on the aforementioned Canadian study by Li et al. (2019) is shown as an example in Fig. 3.3. For example, aromatic and aliphatic hydrocarbons show higher levels not only in the winter period, but also in June and July. In an area of unconventional natural gas exploitation by hydraulic fracturing, significantly higher VOC concentrations, namely acetone, 2-methyl-2-propanol, chloroform, 1,4-dioxane, and decanal were reported in Canada (Caron-Beaudoin et al. 2022). Highest medians were 28.2 µg/m^3 for limonene, 20.4 µg/m^3 for hexenal, 20.1 µg/m^3 for decamethylcyclopentasiloxane, and 12.5 µg/m^3 for nonanal. In residential indoor air in 99 homes in southeast Louisiana, 29 VOC were targeted in cross-sectional analyses using a 48-h sampling

Table 3.3 Volatile organic substances in the air of German residences (in μg/m³)

Substance	Children's Environmental Survey 2003–06 ($N = 555$) (UBA 2008)				German Environmental Health Study 2014–17 ($N = 639$) (Bethke et al. 2021)			
	% ≥ LOQ	Median	95th percentile	Maximum	% ≥ LOQ	Median	95th percentile	Maximum
Alkanes								
n-Heptane	63	1.4	22.8	414	54	1.3	13.7	150
n-Octane	54	1.1	10.3	69.4	41	<1.0	5.1	55.0
n-Nonan	47	<1.0	12.1	66.4	48	<1.0	7.9	85.0
n-Decan	60	1.4	14.9	108	70	1.1	11.0	93.0
n-Undecan	60	1.4	14.8	135	77	1.4	12.0	230
n-Dodecane	54	1.1	7.9	186	63	1.0	6.4	33.0
n-Tridecan	42	<1.0	4.2	44.0	53	<1.0	2.8	32.0
n-tetradecane	85	1.7	5.4	55.4	74	1.1	3.8	18.0
n-pentadecane	63	1.2	3.7	21.1	56	0.8	2.7	5.0
n-Hexadecane	46	<1.0	2.3	9.7	64	0.9	2.7	76.0
n-Heptadecane	23	<1.0	2.0	6.3	12	<1.0	1.0	85.0
n-Octadecane	7	<1.0	2.0	3.5	3	<1.0	<1.0	150
Cyclohexane	73	2.5	39.1	456	45	<1.0	6.2	33.0
Methylcyclohexane	52	1.1	26.5	400	45	<1.0	16.7	64.0
Aromatics								
Benzene	75	1.8	7.7	61.0	54	1.1	3.6	29.0
Toluene	100	13.5	57.6	2400	99	4.7	27.0	160
Ethylbenzene	68	1.4	6.8	40.8	70	0.9	5.0	30.0
m-, p-xylene	91	3.2	16.0	200	95	1.9	15.0	93.0
o-Xylene	57	1.2	5.5	47.8	55	0.7	6.0	39.0
Isopropylbenzene	7	<1.0	1.3	3.7	4	<1.0	<1.0	7.0
n-Propylbenzene	18	<1.0	2.6	14.4	18	<1.0	1.7	21.0

(continued)

Table 3.3 (continued)

Substance	Children's Environmental Survey 2003–06 ($N = 555$) (UBA 2008)				German Environmental Health Study 2014–17 ($N = 639$) (Bethke et al. 2021)			
	% ≥ LOQ	Median	95th percentile	Maximum	% ≥ LOQ	Median	95th percentile	Maximum
2-Ethyltoluene	14	<1.0	2.3	13.6	–	–	–	–
3-Ethyltoluene	39	<1.0	5.5	41.2	15	<1.0	3.4	21.0
4-ethyltoluene	14	<1.0	2.6	19.7	17	<1.0	2.3	30.0
1,2,3-Trimethylbenzene	20	<1.0	2.9	12.3	22	<1.0	2.8	19.0
1,2,4-Trimethylbenzene	68	1.5	10.3	58.8	65	0.8	8.8	50.0
1,3,5-Trimethylbenzene	19	<1.0	2.9	24.3	19	<1.0	1.9	120
Styrene	41	<1.0	4.8	32.0	58	0.9	4.4	14.0
Naphthalene	7	<1.0	1.2	4.9	18	<1.0	1.1	10.0
4-Phenylcyclohexene	0				0			
Halogenated compounds								
1,1,1-Trichloroethane	5	<1.0	<1.0	14.3				
Trichloroethene	4	<1.0	<1.0	64.3				
Perchloroethene	7	<1.0	1.4	28.0				
1,4-Dichlorobenzene	1	<1.0	<1.0	11.3				
Oxygenated compounds								
Ethyl acetate	98	9.3	70.8	785	84	16.6	204	1200
Butyl acetate	88	4.1	30.7	214	97	4.9	31.8	150
1-Methoxy-2-propanol acetate	23	<1.0	3.6	93.5				
Methyl ethyl ketone	6	<7.5	9.2	139				
Methyl isobutyl ketone	27	<1.0	2.6	17.3				
1-Butanol	98	5.4	17.6	71.6				
Isobutanol	9	<3.5	4.9	40.7	54	1.2	10.0	51.0
2-Methoxyethanol	5	<1.0	1.2	14.1				

2-ethoxyethanol	10	<1.0	1.5	7.1	0			
2-Butoxyethanol	61	1.4	10.3	117	6	<1.0	6.1	13.0
2-Butoxyethyl acetate	–	–	–	–	0			
2-Butoxyethoxyethanol	33	<1.0	6.0	35.5	0			
2-Phenoxyethanol	36	<1.0	3.7	11.5	26	<1.0	7.4	12.0
1-Methoxy-2-propanol	64	1.5	8.4	86.1	45	<1.0	12.0	91.0
1-Butoxy-2-propanol	47	<1.0	12.8	126	50	<1.0	41.6	81.0
1-Phenoxy-2-propanol	2	<1.0	<1.0	3.1	0			
2-ethyl-1-hexanol	85	2.6	11.4	67.0				
Dipropylene glycol monobutyl ether	18	<1.0	3.2	35.0	2	<1.0	<1.0	12.0
2,2,4-trimethyl-1,3-pentanediol monoisobutyrate (Texanol®)	27	<1.0	2.8	126	3	<1.0	<1.0	76.0
2,2,4-trimethyl-1,3-pentanediol diisobutyrate (TXIB®)	59	1.2	5.5	62.1	73	1.2	11.0	25.0
Terpenes								
α-Pinene	99	9.8	67.6	800	99	6.4	42.0	200
β-Pinene	57	1.2	8.3	47.8	70	1.8	10.8	21.0
Lime	93	11.5	103	400	97	12.0	100	1100
Δ-3-Caren	75	2.6	22.7	336	83	2.3	26.3	150
Longifolene	16	<1.0	1.8	8.0	0			
TVOC[a]		300	1100	3000		270	950	6700

[a] Total Volatile Organic Compounds, sum of all VOC in the range from n-hexane to n-hexadecane, in toluene equivalent units
LOQ limit of quantification

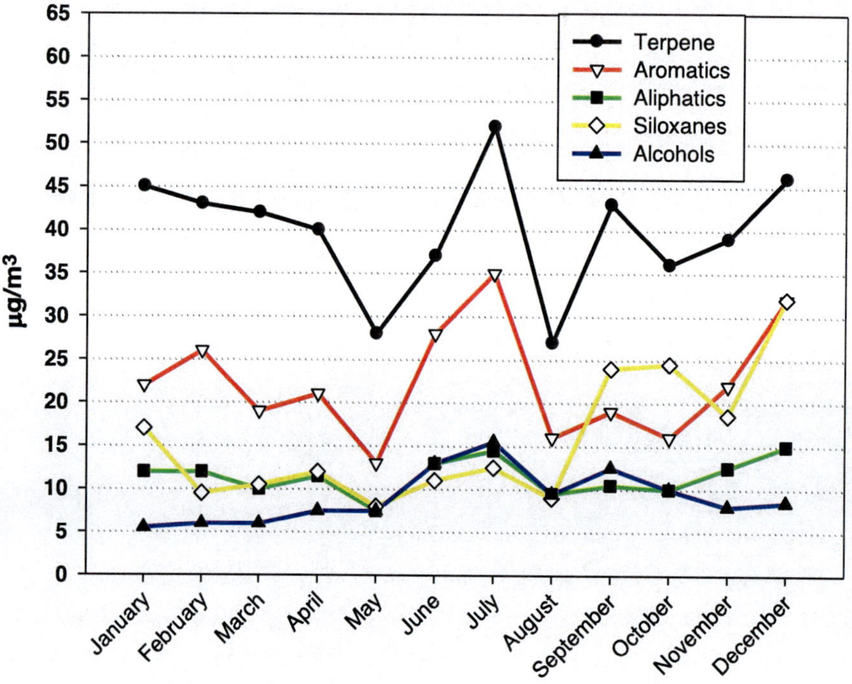

Fig. 3.3 Annual variation of VOC groups in indoor air in Canada (modified after Li et al. 2019)

period with medians of 2.1 µg/m^3 for m,p-xylenes, 4.7 µg/m^3 for α-pinene, 1.1 µg/m^3 for benzene, 4.9 µg/m^3 for toluene, 0.74 µg/m^3 for ethylbenzene, and 4.1 µg/m^3 for pentane (Wickliffe et al. 2020).

The homes of 5017 randomly-selected participants in the Japan Environment and Children's Study (JECS) performed in two sampling periods between 2014–2017 showed for aromatic VOC the highest medians of 11 and 9.7 µg/m^3 for toluene and 3.7 and 3.5 µg/m^3 for ethylbenzene (Jung et al. 2021). Liu et al. (2022) performed a systematic review of field measurements of eight target VOC (benzene, toluene, xylenes, acetaldehyde, p-dichlorobenzene, butadiene, trichloroethylene, and tetrachloroethylene) in residences, offices, and schools in China from 2000 to 2021. Very high levels were reported in residences and offices with medians of 24.2 µg/m^3 and 46.5 µg/m^3 for benzene and 40.9 µg/m^3 and 83.9 µg/m^3 for toluene, respectively. The authors concluded that indoor pollution of benzene, toluene, and xylenes has been more serious in China than in other countries and found a declining trend from 2000 to 2021.

3.1.4 Occurrence in the Air of Public Facilities

The exposure situation in schools and daycare centers is summarized in a WHO review (WHO 2021).

In Europe, very extensive studies have also investigated VOC (or selected VOC) in school classrooms. For example, the Schools Indoor Pollution and Health: Observatory Network in Europe (SINPHONIE) study examined a total of 340 classrooms in 23 European countries in 2010–2012 (Csobod et al. 2014) and the Transnational Adaption Actions for Integrated Indoor Air Quality Management Project (InAirQ) studied for 5 consecutive days 64 primary schools in Czech Republic, Hungary, Italy, and Slovenia in 2017–2018 (Szabados et al. 2021). In two German studies, a total of 273 rooms in schools and daycare centers in Schleswig-Holstein and 165 classrooms of Bavarian schools were investigated in 2005 (Ostendorp et al. 2009; Fromme et al. 2008). The study of 63 daycare centers in 3 German states revealed the following 95th percentiles: Limonene (93.8 μg/m^3), decamethylcyclopentasiloxane (D5) (56.5 μg/m^3), 2-(2-ethoxyethoxy) ethanol (55.5 μg/m^3), 1-methoxy-2-propanol (38,0 μg/m^3), 1,2-propanediol (36.9 μg/m^3), α-pinene (29.8 μg/m^3), ethyl acetate (29.5 μg/m^3), and methyl ethyl ketone (26.2 μg/m^3) (Fromme et al. 2016). The median TVOC level in this study was 179 μg/m^3 and the 95th percentile was 617 μg/m^3. A total of 10 facilities exceeded the TVOC target value of 300 μg/m^3 or the hygienically safe range. As part of the summary of measurement results from various private measurement institutes from 2006–2012, TVOC with a median of 470 μg/m^3 (95th percentile: 2600 μg/m^3) was found in a total of 271 classrooms (Hofmann et al. 2014). The Finnish Institute of Occupational Health analyzed a total number of samples 3583 from schools and 727 from kindergartens collected from 20,210 to 2019 and found relatively low 95th percentiles in the two sites of 2.0 and 2.0 μg/m^3 for benzene, 4.0 and 5.0 μg/m^3 for toluene, 4.0 and 3.0 μg/m^3 for limonene, and 9.0 and 12.0 μg/m^3 for decamethylcyclopentasiloxane (Wallenius et al. 2022).

Another study examines 16 VOC in 144 classrooms in 37 conventional and high-performance elementary schools in the USA in 2015/16 and observed a median TVOC of 29.3 μg/m^3 (maximum: 197 μg/m^3) in all classrooms (Zhong et al. 2017). Aromatics, alkanes, and terpenes were the major compounds detected. The highest medians were 3.5 μg/m^3 for d-limonene, 2.2 μg/m^3 for n-hexane, 1.3 μg/m^3 for methyl cyclohexane, and 0.7 μg/m^3 for toluene. The within-school variance exceeded that between schools, and no major differences were found beween conventionalk and high performance buildings.

Yoon et al. (2011) measured TVOC in 71 classrooms at 17 Korean preschools and reported geometric means in urban indoor air of 433 μg/m^3 (range: 73–1928 μg/m^3) and 233 μg/m^3 (62–1103 μg/m^3) in rural indoor air. The geometric mean concentration of benzene in urban samples of 6.6 μg/m^3 was higher than in rural with 4.0 μg/m^3.

3.2 Aldehydes

3.2.1 *Formaldehyde*

In the large group of aldehydes, formaldehyde (methanal) represents the structurally simplest aldehyde. Figure 3.4 shows its chemical structure. Formaldehyde (CAS No. 50-00-0) is a colorless, flammable, highly reactive, pungent-smelling gas (WHO 2010). It is used on a large scale in industry in a variety of processes. Formaldehyde is used at industrial and professional level as (ECHA 2019): Material for the production of FA-based resins such as urea-formaldehyde (UF) resins, melamine-formaldehyde (MF) resins, phenol-formaldehyde (PF) resins, polyacetal resins, polyurethanes (MDI); use as an intermediate in the chemical synthesis; use as reagent and bactericidal agent used in healthcare applications such as tissue preservation, embalming fluids in autopsy rooms and pathology departments, disinfectant in operating rooms, vaccines, animal medicines; use as preservative, biocidal and cleaning agent in food applications.

In the environment, formaldehyde is released from natural and anthropogenic sources. In indoor air, formaldehyde can be formed during combustion processes such as tobacco smoking. In addition, emissions from wood-based materials such as particleboard, in which it is used as a binder, and emissions from insulation materials such as (urea)-formaldehyde in-situ foams (so-called UF in-situ foams) are relevant sources (Salthammer 2019a, b). In addition, emissions from products such as paints, varnishes, adhesives, disinfectants, and cosmetics can pollute indoor air. Secondary formation from the oxidation of, e.g., alkenes has also been described indoors. In this context, Pibiri et al. (2020), for example, were able to show that surface coatings of particleboard, e.g., with latex paints, can reduce formaldehyde emissions by more than 70–90%, even in the long term. A major problem in the context of health assessment of formaldehyde must be seen in the temporary and intermittent active sources that can lead to high indoor concentrations (Salthammer 2019a, b).

In Germany, coated and uncoated wood-based materials (particleboard, blockboard, veneer panels, and fiberboard) may only be placed on the market if it has previously been proven by an examination in a test room that the compensatory

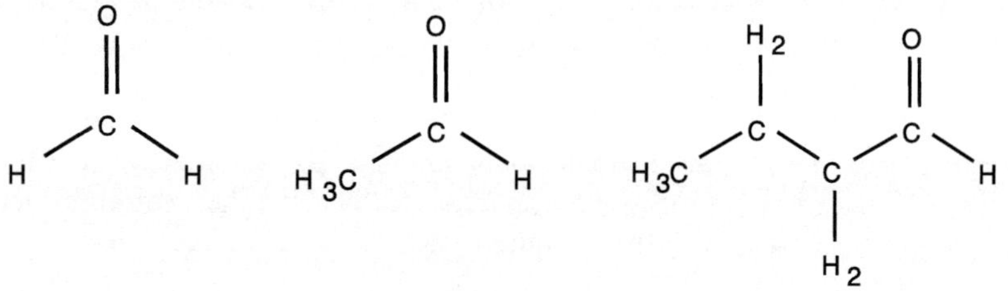

Fig. 3.4 Chemical structure of some aldehydes; left: methanal (formaldehyde), middle: ethanal (acetaldehyde), right: butanal (butyraldehyde)

concentration of formaldehyde in the air caused by it does not exceed 0.1 ppm (approx. 124 μg/m^3) (ChemVerbotsV 2017). In addition, detergents, cleaning and care products with a mass content of more than 0.2% formaldehyde may not be placed on the market. It should be noted that, in contrast to the aforementioned test chamber conditions, wood-based materials are often installed coated indoors.

The U.S. Department of Housing and Urban Development has set standards for formaldehyde emissions in manufactured housings of less than 0.2 ppm for plywood and 0.3 ppm for particle board. The US EPA considers purchasing composite wood products certified as compliant with ANSI/HPVA HP-1-2009 (for hardwood plywood), ANSI A208.1–2009 (for particleboard), or ANSI A208.2-2009 (for medium-density fiberboard). These standards include limits on formaldehyde emissions. On July 7, 2021, the final Formaldehyde Emissions from Composite Wood Products Regulations were published in Canada and recommended the following emission standards: for hardwood plywood 0.05 ppm; for particleboard 0.09 ppm; for medium-density fiberboard 0.11 ppm; for thin medium-density fiberboard 0.13 ppm; and for a laminated product 0.05 ppm.

3.2.1.1 Health Aspects

There are numerous reviews on the health effects of inhaled formaldehyde (e.g., WHO 2010; ECHA 2012; Kanchongkittiphon et al. 2015; Nielsen et al. 2017; AIR 2016a; b; Bernardini et al. 2020; Hartwig et al. 2020; La Torre et al. 2023). Formaldehyde, due to its water solubility and high reactivity, is absorbed after inhalation almost exclusively in the nasal region and only in a small proportion in the pharynx and upper bronchial region (Schroeter et al. 2014). The focus of indoor exposure is therefore on effects in this anatomical region. In particular, these are sensory irritant effects of the receptors of the trigeminal nerve at the eye and in the nasal mucosa and only at high concentrations, due to saturation effects of degrading enzymes, cytotoxic effects in the nose and throat. After chronic inhalation of formaldehyde, various studies in rats showed nasal cell carcinomas, while results of epidemiological studies on carcinogenic effects were inconclusive.

Since 2004, the International Agency for Research on Cancer has classified formaldehyde as a human carcinogen (Group 1) (IARC 2006). At that time, the organization saw sufficient evidence for this effect in animal experiments and in the context of epidemiological studies (regarding tumors in the nasopharyngeal area and leukemias). In 2014, the European Chemicals Agency classified formaldehyde in category 1B (substance likely to be carcinogenic in humans; classification is predominantly based on evidence in animals) in Annex VI according to EC Regulation 1272/2008. With regard to mutagenicity, formaldehyde is classified in category 2 (substance of concern for humans because it may cause heritable mutations in germ cells of humans).

Other tumor localizations have also been repeatedly discussed in the scientific literature. Whether formaldehyde can also lead to cancer in the upper respiratory tract in humans cannot be clearly deduced from the available epidemiological studies on employees (AIR 2016a). Kwak et al. (2020) also found no evidence for an increase in risk of lung cancer in a recent review. Furthermore, recent meta-analyses of epidemiological studies did not show associations between formaldehyde exposure and non-Hodgkin's lymphoma (NHL) (Catalani et al. 2019) or acute myeloid leukemia (AML) (Allegra et al. 2019).

3.2.1.2 Occurrence in the Air of Residences

Formaldehyde is one of the substances that have been very frequently investigated in indoor air for decades. The different sampling and analytical methods are summarized in a review (Dugheri et al. 2021).

In an overview of studies published in the 1990s and 2000s, mean values between 9.3 and 46.1 μg/m^3 have been described (Sarigiannis et al. 2011). Recent reviews on the occurrence of formaldehyde in indoor air have been published by Shrubsole et al. (2019) and Salthammer (2019b). Again, the results of 33 studies were summarized by Vardoulakis et al. (2020) with concentrations ranging from 7.5 μg/m^3 in Quebec City, Canada, to 134 μg/m^3 in new homes (first year) across various cities in Japan.

In France, a median (90th percentile) of 19.7 μg/m^3 (39 μg/m^3) was determined for formaldehyde in the air of 567 dwellings in 2002–2005 (Langer et al. 2016) and a median of 27 μg/m^3 (10.7–47.7 μg/m^3) in bedrooms and of 23 μg/m^3 (9.6–37.2 μg/m^3) in living rooms in 10 Spanish dwellings in 2014 (Rovira et al. 2016). Within the scope of two German studies performed in 2003–2006 and 2014–2017, respectively, representative data on the occurrence of formaldehyde in indoor residential air were collected for Germany. Thus, a median of 23.5 μg/m^3 and a 95th percentile of 47.7 μg/m^3 were observed for formaldehyde in the first study and concentrations in a comparable range were observed in the second study with 24.9 μg/m^3 and 55.1 μg/m^3, respectively (Birmili et al. 2022). As part of the summary of measurement results from various German private research institutes from 2006–2012, formaldehyde was found in 95% of the 979 samples with a median of 38.0 μg/m^3 (95th percentile: 125 μg/m^3) (Hofmann et al. 2014). In 2013–2016, an investigation of 160 energy-efficient homes in Switzerland revealed 14 μg/m^3 as the median or 50 μg/m^3 as the maximum formaldehyde (Yang et al. 2020). Wallner et al. (2015) conducted an extensive study in 2010–2013 examining 62 energy-efficient homes (with mechanical ventilation) and 61 with natural ventilation and found no evidence of poorer indoor air quality in the mechanically ventilated homes. The medians (95th percentiles) were 27 and 22 μg/m^3 (53 and 67 μg/m^3) in the control rooms and 40 and 31 μg/m^3 (67 and 59 μg/m^3) in the energy-efficient indoor rooms during the two measurement periods. Alvarez-Vaca et al. (2022) reported a median (90th percentile) of 24.9 μg/m^3 (55.5 μg/m^3) for formaldehyde in a study in 159 residences in Luxembourg from 2014 to 2019.

The homes of 5017 randomly-selected participants in the Japan Environment and Children's Study (JECS) performed in two sampling periods between 2014–2017 showed for formaldehyde medians of 11 and 15 μg/m^3 (Jung et al. 2021). On the contrary, in 31 Chinese living rooms a very high median of 130 μg/m^3 (range:

20–370 μg/m^3) was reported (Chen et al. 2017). In a nationwide study in China, 68,324 samples from residences collected since 1986 were summarized (Fang et al. 2022). The median concentrations of formaldehyde in newly renovated residences have decreased from 175 μg/m^3 in 2001–2005 to 153 μg/m^3 in 2011–2015, respectively, while the corresponding median formaldehyde concentration in residences that were renovated beyond one year was 39 μg/m^3 in 2011–2015.

3.2.1.3 Occurrence in the Air of Public Facilities

In the AIRMEX study (European Indoor Air Monitoring and Exposure Assessment), which investigated 186 public buildings and schools in 11 cities across Europe in 2003–2008, a median of 14.1 μg/m^3 (95th percentile: 31.5 μg/m^3; range: 8.2–50 μg/m^3) could be observed for formaldehyde (Geiss et al. 2011). In addition, the Schools Indoor Pollution and Health: Observatory Network in Europe (SINPHONIE) study examined a total of 340 classrooms in 23 European countries in 2010–2012 (Csobod et al. 2014) with a median of 12 μg/m^3 (range: 1.0–66 μg/m^3). Moreover, Szabados et al. (2021) investigated in the Transnational Adaption Actions for Integrated Indoor Air Quality Management Project (InAirQ) for 5 consecutive days 64 primary schools in Czech Republic, Hungary, Italy, and Slovenia in 2017–2018. They found a median value 8.0 μg/m^3 (95th percentile: 17.2 μg/m^3; range: 2.2–33.9 μg/m^3).

French daycare centers (N = 28) showed median formaldehyde concentrations of 10.9 and 15.0 μg/m^3 in the cold and warm seasons, respectively (Roda et al. 2011). The level was significantly higher in the warm season. Villanueva et al. (2018) studied a total of 18 facilities in rural and 18 in urban areas of Spain in 2013. They found formaldehyde with medians of 27.7 and 20.4 μg/m^3 in school classrooms and 35.3 and 14.4 μg/m^3 in daycare centers, respectively. The Finnish Institute of Occupational Health analyzed 938 samples from schools and 68 from kindergartens collected from 20,210 to 2019 and found medians (95th percentiles) of 3.0 μg/m^3 (14 μg/m^3) and 4.0 μg/m^3 (20 μg/m^3) (Wallenius et al. 2022).

In a nationwide study in China, 577 samples from schools collected since 1986 were summarized (Fang et al. 2022). The median concentrations of formaldehyde in newly renovated schools have decreased from 228 μg/m^3 in 2001–2005 to 94 μg/m^3 in 2011–2015, respectively, while the corresponding median formaldehyde concentrations in schools that were renovated beyond one year was 35 μg/m^3 in 2011–2015. Moreover, within the framework of an epidemiological study Sun et al. (2022a) sampled air in 75 residences in China from 2017 to 2021 and reported a median level of 20 μg/m^3 (range: 5–790 μg/m^3).

3.2.1.4 Occurrence in Occupational Settings

In many workplaces where it is used as an intermediate in the production of some plywood adhesives, fertilizer, paper, and urea-formaldehyde resins in industrial processes, sometimes high inhalation and dermal exposure must be expected

(ATSDR 1999). In addition, inhalation exposure can also occur in areas of life where formaldehyde is used as a preservative or is present as an impurity in products. Cammalleri et al. (2022) summarized the result of studies from 2004 to 2019 and found formaldehyde concentrations from 10 to 4820 μg/m^3 in 12 studies performed in healthcare and research settings like gross anatomy and pathology laboratories, while the concentration in esthetic and wellness settings like nail, hair, and beauty salons ranged from <LOQ to 2400 μg/m^3 in 11 studies. In this study, the riskiest occupational settings for formaldehyde exposure were the gross anatomy and pathology laboratories, the hairdressing salons and some specific productive settings, such as wooden furniture factories, dairy facilities, and fish hatcheries. For example, the EU Commission has adopted a binding Occupational Exposure Limit (OEL) for formaldehyde (inhalation) of 375 μg/m^3 (0.3 ppm) (as 8 hours time-weighted average) and 749 μg/m^3 (0.6 ppm) as short-term exposure limit (15 min) (ECHA 2019).

3.2.2 Other Aldehydes

Various longer-chain aldehydes—above formaldehyde—with more than one carbon atom, are often quantified as part of the usual VOC measurement. Other aldehydes require a different analytical approach. Typical for the chemical structures of this substance class is the so-called carbonyl group (C=O group), i.e., the presence of a double bond of a carbon with an oxygen atom. Based on the molecular structure, the very extensive substance group of aldehydes can be classified into different groups (Ad-hoc-AG 2009):

- saturated unbranched or branched aliphatic aldehydes such as n-hexanal,
- saturated cyclic aldehydes such as cyclohexylmethanol,
- aromatic aldehydes such as phenylmethanal (benzaldehyde) and aromatic-like unsaturated cyclic aldehydes such as 2-furylmethanal (furfural),
- mono- or polyunsaturated aldehydes (alkenals, alkadienals) such as propenal (acrolein), trans-butenal (crotonaldehyde),
- polyvalent aldehydes (alkanediols) such as ethanedial (glyoxal), propanedial (malondialdehyde),
- aldehydes with adjacent functional groups (hydroxy, keto, acid group) such as hydroxyethanal (glycolaldehyde) and 2,3-dihydroxypropanal (glyceraldehyde).

Figure 3.4 shows the chemical structures of some aldehydes important in indoor environments.

Acetaldehyde (ethanal) (CAS No.: 70–07-0) is formed as an intermediate and by-product of alcoholic fermentation and is therefore found in the environment and in many foods. It is also used on a large scale in industry, e.g., in paper and cosmetics production and as an intermediate. Butanal and propanal are used, among other things, in the production of plastics and plasticizers, and in some cases as solvents. Since a number of aldehydes can be formed by oxidation from unsaturated fatty acids, fatty acid-rich woods and wood-based materials, wood pellets, varnishes, alkyd resin paints, oils and adhesives, natural oil-based coating products, and floor coverings such as linoleum represent an important indoor. Since the above-mentioned decomposition processes often take place relatively slowly, emissions may well persist for longer periods and subsequently cause odor nuisance. For example, the decomposition of linoleic acid produces hexanal, while oleic acid produces heptanal, octanal, nonanal, and decanal.

3.2.2.1 Health Aspects

In the case of acetaldehyde, the irritative effects on the eye and on mucous membranes of the respiratory tract, which have been well documented in experimental studies on human volunteers, are in the foreground. In long-term exposure, irritant effects in the upper respiratory tract, especially in the olfactory epithelium of the nose, and adeno- and squamous cell tumors are described in animal experiments. Acetaldehyde has a characteristic pungent odor. The other aldehydes discussed here, in addition to oral exposure as natural constituents of food, also cause sensory trigeminal irritation effects in the upper respiratory tract after inhalation that decrease with increasing chain length. Several review papers on the health effects of aldehydes have been published (O'Brien et al. 2005; Laskar and Younus 2019).

Acetaldehyde is classified according to EC Regulation 1272/2008 in Annex VI Part 1 Table 3 in category 1B (substance that is probably carcinogenic in humans; classification is mainly based on evidence in animals). With regard to mutagenicity, acetaldehyde is classified in category 2 (substance of concern to humans because it may cause heritable mutations in germ cells of humans). Furaldehyde is listed in category 2 (suspected carcinogenic effect in humans).

3.2.2.2 Occurrence in the Air of Residences

In their review, Cometto-Muñiz and Abraham (2015) compile the results of study from apartments and schools and describe medians of 20.0 μg/m^3 for acetaldehyde, 18.0 μg/m^3 for nonanal, and 7.5 μg/m^3 for hexanal.

In 132 apartments in Sweden, a median of 0.5 μg/m^3 was found for hexanal and 5.0 μg/m^3 for nonanal (Langer and Bekö 2013), while in France in 2002–2005, a median of 11.6 μg/m^3 (90th percentile: 25 μg/m^3) was determined for acetaldehyde in the air of 567 apartments (Langer et al. 2016). In 2013–2016, a study of 169 energy-efficient homes in Switzerland revealed medians of 0.5 μg/m^3 (maximum: 5.1 μg/m^3) for acetaldehyde, 6.9 μg/m^3 (41 μg/m^3) for hexanal, 1.0 μg/m^3 (22 μg/m^3) for benzaldehyde (Yang et al. 2020). Representative data on the occurrence of other aldehydes besides formaldehyde in indoor residential air were collected for Germany as part of the Children's Environmental Survey 2003–2006 and the German Environmental Health Survey 2014–2017, respectively (UBA 2008; Birmili et al. 2022). Table 3.4 summarizes the results of the two studies. Within the

Table 3.4 Aldehydes in indoor air in μg/m^3

IUPAC name	Trivial name	Children's Environmental Survey 2003–06 ($N = 555$) (UBA 2008)				German Environmental Health Study 2014–17 ($N = 639$) (Birmili et al. 2022)			
		% ≥ LOQ	Median	95th percentile	Maxi-mum	% ≥ LOQ	Median	95th percentile	Maxi-mum
Methanal	Formaldehyde	100	23.5	47.7	68.9	96	24.9	55.1	113
Ethanal	Acetaldehyde	99	15.5	50.3	863	96	5.5	16.8	50.0
Propanal	Propionaldehyde	100	2.5	6.1	40.9	87	1.5	4.6	23.0
Butanal	Butyraldehyde	99	2.4	8.1	43.2	80	1.3	4.9	29.1
Pentanal	Valeraldehyde	99	3.7	10.6	28.4	88	2.7	9.2	23.1
Hexanal	Hexaldehyde	100	9.8	30.0	81.7	96	10.9	33.3	117
Heptanal	Heptylaldehyde	88	1.3	3.0	9.1	50	<0.5	3.6	8.8
Octanal	Caprylaldehyde	99	1.6	3.6	6.8	63	1.0	4.2	9.5
Nonanal	Pelargonaldehyde	100	7.2	14.7	30.2	81	6.1	21.4	45.1
Decanal	Caprinaldehyde	97	2.5	5.5	9.9	58	1.3	10.9	24.8
Undecanal	Laurinaldehyde	44	<1.0	3.1	13.2	7	<0.6	0.8	2.0
Furfural	2-Furaldehyde	96	0.9	2.8	19.3	26	<0.2	1.0	7.4
Phenylmethanal	Benzaldehyde	98	2.9	6.6	11.9	55	0.6	4.9	12.1
3-Methylbutanal	Isovaleraldehyde	35	<0.5	3.9	15.9	27	<0.4	4.8	19.0
2-Oxopropanal	Methylglyoxal	50	<1.0	17.8	54.1	–	–	–	–

LOQ limit of quantification

summary of measurement results of different research institutes from 2006–2012, acetaldehyde was found in 96% of 249 samples with a median of 22.0 μg/m^3 (95th percentile: 67.8 μg/m^3) and hexanal was found in 96% of 750 samples with a median of 14.0 μg/m^3 (95th percentile: 108 μg/m^3) (Hofmann et al. 2014). In the German Environmental Health Study 2014–2017, acetaldehyde was determined in 639 children's living rooms in 96% of the samples with concentrations of 5.5 μg/m^3 (median) and 16.8 μg/m^3 (95th percentile), respectively, and hexanal in 96% of the samples with 10.9 μg/m^3 (median) and 33.3 μg/m^3 (95th percentile), respectively (Birmili et al. 2022). Alvarez-Vaca et al. (2022) reported medians of 17.7 μg/m^3 for acetaldehyde, 7.2 μg/m^3 for nonanal, and 4.6 μg/m^3 for hexanal in a study in 159 residences in Luxembourg from 2014 to 2019.

In a very large Canadian study, a total of 3857 apartments or houses were sampled in 2009–2011, and a median of 12.4 μg/m^3 (95th percentile: 63.9 μg/m^3) was determined for hexanal (Zhu et al. 2013). In another nationwide study in Canada, 3524 residential interiors were investigated in 2012–2013 (Li et al. 2019). Here, the following aldehydes could be found with the highest medians (95th percentiles): Hexanal 4.9 μg/m^3 (41 μg/m^3), nonanal 4.3 μg/m^3 (31 μg/m^3), and benzaldehyde 2.5 μg/m^3 (8.9 μg/m^3).

The homes of 5017 randomly-selected participants in the Japan Environment and Children's Study (JECS) performed in two sampling periods between 2014 and 2017 showed medians of 12 and 12 μg/m^3 for acetaldehyde (Jung et al. 2021). Liu et al. (2022) summarized field measurements in nearly 17,000 residences in China from 2000 to 2021 and reported a median of 12.9 μg/m^3 (95% CI: 2.3–93 μg/m^3) for acetaldehyde.

3.2.2.3 Occurrence in the Air of Public Facilities

In the AIRMEX study, a median of 7.1 μg/m^3 (range: 1.4–29.1 μg/m^3) for acetaldehyde and of 11.4 μg/m^3 (range: 1.2–160 μg/m^3) for hexaldehyde could be observed (Geiss et al. 2011). Szabados et al. (2021) investigated in the Transnational Adaption Actions for Integrated Indoor Air Quality Management Project (InAirQ) for 5 consecutive days 64 primary schools in Czech Republic, Hungary, Italy, and Slovenia in 2017–2018. They found a median value 5.0 μg/m^3 (range: 1.4–11.0 μg/m^3) for acetaldehyde and 8.7 μg/m^3 (range: 1.6–32.7 μg/m^3) for hexaldehyde.

French nurseries showed median concentrations of 5.1 and 5.6 μg/m^3 (acetaldehyde), 7.6 and 11.7 μg/m^3 (hexaldehyde), and 4.3 and 14.2 μg/m^3 (butyraldehyde) in the cold and warm seasons, respectively (Roda et al. 2011). Except for acetaldehyde, levels were significantly higher in the warm season. Villanueva et al. (2018) studied a total of 18 facilities in rural and 18 in urban areas of Spain in 2013 and found acetaldehyde with medians of 6.7 and 5.1 μg/m^3 in school classrooms and 6.5 and 4.6 μg/m^3 in daycare centers, and hexanal with medians of 15.2 and 10.8 μg/m^3 in classrooms and 19.1 and 14.8 μg/m^3 in daycare centers. In 345 and 409 measurements in 101 and 116 schools, respectively, from 2004–2009 collected in the

Table 3.5 Results of aldehydes in the air of schools and daycare centers in $\mu g/m^3$

Trivial name	Fromme et al. (2008)		Fromme et al. (2016)		Hofmann et al. (2014)		Geiss et al. (2011)		Szabados et al. (2021)		Sivanantham et al. (2021)	
	Median	95th percentile	Median	95th percentile	Median	95th percentile	Median	95th percentile	Median	95th percentile	Mean	95th percentile
Formaldehyde	14.1	30.4	–	–	44.5	170	14.1	31.5	8.0	17.2	21.2	–
Acetaldehyde	7.6	19.3	–	–	15.0	76.4	7.2	18.8	5.0	8.9	5.7	–
Propionaldehyde	1.0	3.8	–	–	5.0	37.2	2.3	9.1	1.4	3.1	–	–
Crotonaldehyde	<0.5	0.9	–	–	0.5	2.0	–	–	–	–	–	–
Butyraldehyde	0.5	1.9	<1.0	1.4	2.0	17.1	–	–	–	–	–	–
Benzaldehyde	<0.5	5.7	–	–	5.0	60.0	–	–	0.5	1.1	–	–
Valeraldehyde	0.6	1.4	<1.0	2.5	4.0	28.8	–	–	–	–	–	–
Hexaldehyde	6.1	14	2.2	7.9	10.0	81.0	11.4	39.6	8.7	15.9	13.3	–
Heptylaldehyde	–	–	0.5	2.3	2.0	10.0	–	–	–	–	–	–
Caprylaldehyde	2.9	12	2.3	5.1	3.0	18.4	–	–	–	–	–	–
Pelargonaldehyde	8.3	22	7.1	13.4	7.0	28.0	–	–	–	–	–	–
Caprinaldehyde	4.2	15	2.2	4.8	2.0	11.0	–	–	–	–	–	–
	Germany 2004/2005		Germany 2011/12		Germany 2006–12		Different European cities 2003–08		Czech Republic, Hungary, Italy, Poland, and Slovenia 2017/18		France 2013–17	

German exposure database MEGA, 95th percentiles of 60 μg/m^3 for acetaldehyde and 36 μg/m^3 for hexaldehyde were obtained (Neumann et al. 2012).

For aldehydes—except formaldehyde—measurement results are compiled in Table 3.5.

Here, an overall comparable exposure level is shown, with acetaldehyde, hexaldehyde, and nonanal being the substances with the highest concentrations.

Liu et al. (2022) summarized field measurements in nearly 120 residences in China from 2000 to 2021 and reported a median of 7.6 μg/m^3 (95% CI: 5.7–11 μg/m^3) for acetaldehyde.

3.3 Glycol Ethers

Due to their chemical–physical properties, glycols and their derivatives, glycol ethers and glycol ether acetates are used as solvents, as starting materials and as intermediates in a wide variety of industrial processes and in a wide variety of household products, cosmetics, and items for everyday use. They are chemicals that are consumed in large quantities. For example, in the early 2000s, the quantities of ethylene glycol ethers and their acetates produced annually in the USA were about 800,000 metric tons and in Europe about 245,000 metric tons, while the quantities for propylene glycol ethers and their acetates were about 255,000 metric tons and about 190,000 metric tons, respectively (Mangelsdorf and Fromme 2021). Glycols are alcohols with two hydroxy groups (so-called diols) whose alcohol function is usually bonded to adjacent carbon atoms. Glycol ether is a group designation of ethers derived from different glycols in which the hydrogen atoms of one or both hydroxyl groups are replaced (e.g., by alkyl radicals). Figure 3.5 shows an example of the basic chemical structure of glycol ethers and glycol ether acetates.

3.3.1 Health Aspects

The glycol ethers are very well absorbed by the organism after oral uptake, inhalation or dermal exposure and are rapidly metabolized. In terms of effects, it is assumed that generally the alkoxy acids, especially the methoxy- and ethoxyacetic acids, are the critical metabolites in the human organism. They appear to be

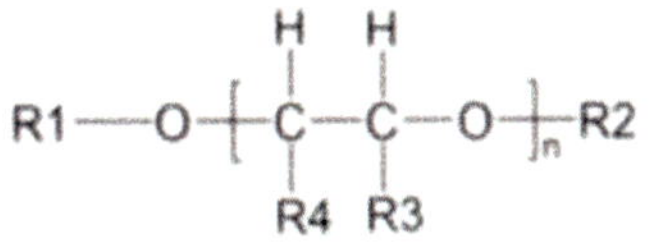

Fig. 3.5 Chemical structure of glycol ethers, glycol esters, and glycol diether (R1 and R2: e.g., other glycol or methyl, ethyl, or acetate group; R3 and R4: H or methyl group)

responsible for the majority of toxic effects. Although the overall toxicity of glycol ethers is rather moderate, their effects should not be underestimated, since the high dermal absorption can lead to considerable exposure on contact with the skin. The effects on the testis and the hematopoietic system described in animal experiments are to be seen as central critical effects (Mangelsdorf et al. 2016; HC 2018; Mangelsdorf and Fromme 2021). They are weak airway irritants and do not expect to induce sensitization (Clausen et al. 2020). Results of epidemiological studies indicate that some glycol ethers may also show hematotoxic or reproductive toxic effects in humans at high exposure levels.

3.3.2 *Occurrence in the Air of Residences*

In the scientific literature, few results of indoor air measurements have been published so far. Generally, only a limited number of buildings and often only individual glycol ethers were investigated.

Representative comparative values for indoor air pollution were collected from 555 dwellings as part of the Children's Environmental Survey in Germany in 2003–2006 (UBA 2008). For sampling, passive samplers were used to determine weekly mean values under real conditions of room use. Sampling was carried out in the room where children (3–14 years) generally spent the longest time during a day. Another German study is a compilation of results from different private research institutes collected in 2006–2012 (Hofmann et al. 2014). In addition, results from the German Environmental Health Study (GerES V) conducted in 2014–2017 from 154 apartments, the majority of which were children's rooms, are available (Bethke et al. 2021). The results of the above-mentioned studies and another Swedish study are summarized in Table 3.6. They show that in Germany, in descending order, especially EGBE, 2PG1ME, EGPhE, and DEGBE can be found above the determination limit of the measurement methods. The 95th percentiles in the aforementioned studies were <1.0–37 μg/m^3 (EGBE), 8.4–23 μg/m^3 (2PG1ME), 3.7–4.6 μg/m^3 (EGPhE), and < 1.0–12 μg/m^3 (DEGBE). Other rarely analyzed glycol ethers such as PGEE, TPGBE, DEGDME, TEGBE, DEGDEE, EGDME, and EGDEE were not found above the limit of quantification of 1.0 μg/m^3 in the study by Hofmann et al. (2014). In the German Environmental Health Study 2014–2017, compared to the 2003–2006 study, there is evidence of a decrease in both the concentration level and the number of results above the limit of quantification for ethylene glycol ethers. In an earlier study, in which a total of 79 rooms in 39 private dwellings in Schleswig-Holstein were investigated in 2000–2001, comparable concentration levels of a maximum of 55 μg/m^3 (EGBE), 32 μg/m^3 (DEGBE), and 8.4 μg/m^3 (2PG1ME) were found (Hippelein 2003).

Geiss et al. (2011) were able to determine EGBE and 2PG1BE in 2003–2008 in 11 European cities with a 95th percentile of 2.7 μg/m^3 in 50 dwellings and of <0.3 μg/m^3 in 8 dwellings, respectively. Langer et al. (2016) found EGBE in 83% of measurements (LOQ: 0.4 μg/m^3) and 2PG1ME in 85% of samples (LOQ: 0.5 μg/

Table 3.6 Glycol ethers in the air in Europe residences in µg/m³

Substance	*N*	% ≥ LOQ	Median	95th percentile	Maximum	Year	Country
EGME							
UBA (2008)	555	5	<1.0	1.2	14,1	2003–06	Germany
Hofmann et al. (2014)	647	0.9			–	2006–12	Germany
EGEE							
UBA (2008)	555	10	<1.0	1.5	7,1	2003–06	Germany
Hofmann et al. (2014)	666	11	<1.0	<2.0	–	2006–12	Germany
Bethke et al. (2021)	154	0			<1,0	2014–17	Germany
EGBE							
UBA (2008)	555	61	1.4	10.3	117,0	2003–06	Germany
Choi et al. (2010)	390	7	2.4	–	24,4		Sweden
Hofmann et al. (2014)	671	67	2.0	37.0	–	2006–12	Germany
Bethke et al. (2021)	154	6			<1,0	2014–17	Germany
DEGME							
UBA (2008)	62	0				2003–06	Germany
Choi et al. (2010)	390	1.5	4.5	–	15,1		Sweden
Hofmann et al. (2014)	600	0.8			–	2006–12	Germany
DEGEE							
Choi et al. (2010)	390	4.1	5.7	–	36,9		Sweden
Hofmann et al. (2014)	640	21	<1.0	8.0	–	2006–12	Germany
DEGBE							
UBA (2008)	555	33	<1.0	6.0	35,5	2003–06	Germany
Choi et al. (2010)	390	18	2.9	–	30,3		Sweden
Hofmann et al. (2014)	662	26	<2.0	12.4	–	2006–12	Germany
Bethke et al. (2021)	154	2			<1,0	2014–17	Germany
EGPhE							
UBA (2008)	555	36	<1.0	3.7	11,5	2003–06	Germany
Hofmann et al. (2014)	668	42	<2.0	4.6	–	2006–12	Germany
Bethke et al. (2021)	154	26	<1.0	7.4	12,0	2014–17	Germany

(continued)

Table 3.6 (continued)

Substance	*N*	% ≥ LOQ	Median	95th percentile	Maximum	Year	Country
2PG1ME							
UBA (2008)	555	64	1.5	8.4	86,1	2003–06	Germany
Choi et al. (2010)	390	22	3.5	–	15,6		Sweden
Hofmann et al. (2014)	670	69	2.0	22.6	–	2006–12	Germany
Bethke et al. (2021)	154	45	<1.0	12.0	91,0	2014–17	Germany
2PG1BE							
UBA (2008)	555	47	<1.0	12.8	126,0	2003–06	Germany
Choi et al. (2010)	390	16.6	4.0	–	53,0		Sweden
Hofmann et al. (2014)	601	40	<2.0	11.0	–	2006–12	Germany
Bethke et al. (2021)	154	50	<1.0	41.6	81,0	2014–17	Germany
2PG1PhE							
UBA (2008)	555	2	<1.0	<1.0	3,1	2003–06	Germany
Hofmann et al. (2014)	426	1.4				2006–12	Germany
Bethke et al. (2021)	154	0			<1,0	2014–17	Germany
DPGBE							
UBA (2008)	555	18	<1.0	3.2	35,0	2003–06	Germany
Hofmann et al. (2014)	661	24	<1.0	5.0	–	2006–12	Germany
Bethke et al. (2021)	154	2		<1.0	12,0	2014–17	Germany

LOQ limit of quantification

m³) in 567 dwellings in France with a median level of 1.6 μg/m³ (90th percentile: 5.7 μg/m³) and 0.9 μg/m³ (90th percentile: 10 μg/m³), respectively. In Canada, EGBE was observed in nearly all rooms with a median of 3.3 μg/m³ (95th percentile: 19.0 μg/m³) in a nationwide survey of 3857 indoor rooms in 2009–2011 (Zhu et al. 2013). In contrast, EGEEA, EGMEA, DEGME, and DEGEE could only be measured in a few samples above the analytical limit of quantification with concentrations between 0.19 and 0.51 μg/m³ (medians).

Table 3.7 Glycol ethers in the air of school classrooms in Germany in µg/m³

Substance	*N*	% ≥ LOQ	Median	95th percentile	Maximum	Year
EGME						
LasD (2009)+	285	31	<1.0	4.0	59	2005–07
Fromme et al. (2016)*	63	2.0			1.1	2011–12
Hofmann et al. (2014)	487	0	<1.0	2.5	–	2006–12
EGEE						
LasD (2009)+	285	3.0			88	2005–07
Fromme et al. (2016)*	63	10		1.2	5.1	2011–12
Hofmann et al. (2014)	492	13		2.0	–	2006–12
EGBE						
LasD (2009)+	285	81	3.0	53	190	2005–07
Fromme et al. (2016)*	63	86	2.3	19.6	159	2011–12
Fromme et al. (2008)	165	86	6.5	65.6	190	2004–05
Hofmann et al. (2014)	492	68.5	2.0	49.4	–	2006–12
DEGME						
LasD (2009)+	285	2.0			6.0	2005–07
Fromme et al. (2016)*	63	5.0			7.2	2011–12
Fromme et al. (2008)	165	4.2			7.0	2004–05
Hofmann et al. (2014)	334	1.8	<1.0	2.0	–	2006–12
DEGEE						
LasD (2009)+	285	14	<2.0	16	232	2005–07
Fromme et al. (2016)*	63	65	1.7	55.5	432	2011–12
Fromme et al. (2008)	165	47	<0.2	30.0	110	2004–05
Hofmann et al. (2014)	459	33	<1.0	22.2	–	2006–12
DEGBE						
LasD (2009)+	285	65	1.0	27	200	2005–07
Fromme et al. (2016)*	63	75	2.2	18.5	87	2011–12
Fromme et al. (2008)	165	80	5.4	35.8	240	2004–05
Hofmann et al. (2014)	495	47	<2.0	27.6	–	2006–12
EGPhE						
Fromme et al. (2016)*	63	89	2.8	13.3	97	2011–12
Hofmann et al. (2014)	493	50	<2.0	38.4	–	2006–12
2PG1ME						
LasD (2009)+	285	11	<2.0	12	120	2005–07
Fromme et al. (2016)*	63	89	5.2	38.0	211	2011–12
Fromme et al. (2008)	165	33	<0.2	13.6	75	2004–05
Hofmann et al. (2014)	491	82	2.0	31.0	–	2006–12
2PG1BE						
Fromme et al. (2016)*	63	41	<1.0	8.1	29.7	2011–12
Hofmann et al. (2014)	336	37	<1.0	8.0	–	2006–12
2PG1PhE						
Hofmann et al. (2014)	134	3.7	<1.0	<1.0	–	2006–12
DPGBE						
Hofmann et al. (2014)	488	3.0	<1.0	4.0	–	2006–12

LOQ limit of quantification; * daycare centers; + schools and daycare centers

3.3.3 *Occurrence in the Air of Public Facilities*

Table 3.7 summarizes the results from recent German studies in the indoor air of school classrooms and common rooms of daycare centers. Essentially, EGBE, 2PG1ME, DEGBE, DEGEE, and 2PG1BE could be determined in the samples in descending order. The 95th percentiles were 20–65 μg/m^3 (EGBE), 12–38 μg/m^3 (2PG1ME), 19–36 μg/m^3 (DEGBE), 16–56 μg/m^3 (DEGEE), and 8.0–8.1 μg/m^3 (2PG1BE). It should be noted in principle that sampling conditions and analytical methods were heterogeneous. In the very large data set compiled by Hofmann et al. (2014), other glycol ethers such as PGEE, TPGBE, DEGDME, TEGBE, DEGDEE, EGDME, and EGDEE could not be quantified above the limit of quantification of 1.0 μg/m^3. Comparable 95th percentiles of 43 μg/m^3 (EGBE), 32 μg/m^3 (2PG1ME), and 44 μg/m^3 (DEGBE) could also be found in the investigation of 41 school containers in Schleswig-Holstein (LasD 2011).

Geiss et al. (2011) were able to measure EGBE at the 95th percentile of 8.3 μg/m^3 in 74 schools and 2PG1BE at 18.6 μg/m^3 in 21 schools in 11 European cities in 2003–2008. In 10 newly constructed energy-efficient schools in France, medians of 0.6 μg/m^3 (EGBE) and 1.3 μg/m^3 (2PG1ME) could be calculated, while EGBEA and 2PG1MEA were both below the quantification limit of 0.1–0.6 μg/m^3 (Verriele et al. 2016).

Hoang et al. (2017) detected EGBE in all samples with a median of 2.9 μg/m^3 (95th percentile approximately 64 μg/m^3) in indoor air during their study of 34 California child-care centers in 2010–2011.

3.3.4 *Factors Influencing Indoor Air Concentrations*

A factor influencing the level of indoor air content seems to be a recent renovation. The results of a study in the vicinity of Zurich in 2000 (Scherer and Maly 2000), in which 22 freshly painted interiors were measured, also substantiate the influence of renovation work on indoor air content. Since a second sampling was also carried out in the investigated objects after 2–13 weeks, a half-life for the decrease in concentration of approx. 30–50 days could be calculated. A corresponding influence of construction measures could also be confirmed in a study in office buildings (Oppl 1999). While the 95th percentiles of the 2PG1ME and EGPhE content in 41 buildings without construction measures in the past 3 months were 34 and 31 μg/m^3, concentrations of 180 and 190 μg/m^3 were found in the 21 buildings in which construction measures were carried out in the past 3 months.

Various test chamber investigations also gave indications of the emission characteristics of glycol ethers. In particular, the emission behavior from wood or wooden objects (Jensen et al. 2001), from (water-soluble) paints and varnishes (Clausen et al. 1991; Gehrig et al. 1993; Wolkoff 1998; Knudsen et al. 1999), from indoor air improvers (Singer et al. 2006a; Uhde and Schulz 2015), from

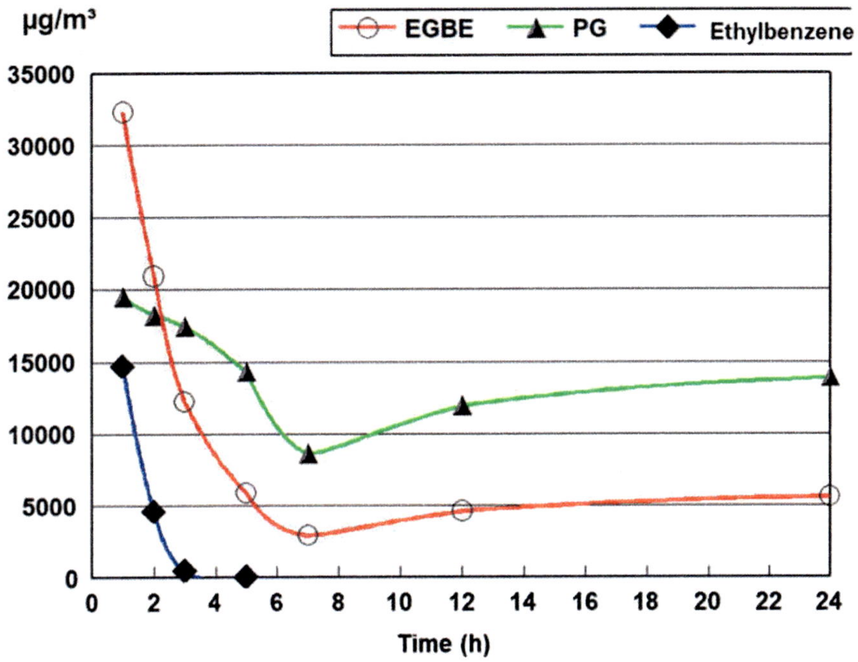

Fig. 3.6 Decay behavior (mean values) of indoor air pollution after manual processing of different varnishes on wood (EGBE: ethylene glycol monobutyl ether, PG: propylene glycol) (after Woitowitz and Knecht 1992)

carpeting/adhesives (Knudsen et al. 1999), and cleaning agents (Gibson et al. 1991; Singer et al. 2006b; Koontz et al. 2006; Rossignol et al. 2013; Gerster et al. 2014) was investigated. Figure 3.6 shows an example of the decay behavior for EGBE, propylene glycol and for a "classical" solvent (ethylbenzene) in the room air of a test chamber. It can be seen that glycol ethers, in contrast to ethyl benzene, can be detected in the room air over a significantly longer period of time.

Bello et al. (2013) tested the effect of bathroom cleaning with a glass cleaner (25–30% EGBE) and an all-purpose cleaner (35–45%) on indoor air levels. During cleaning activities, a mean EGBE concentration of 91 mg/m^3 and a peak of 210 mg/m^3 were reported, depending on room size and ventilation conditions. In another study, cleaning with products containing EGBE (7.0%), 2PG1BE (3.5%), or EGPE (5.0%) was performed in 5–8 office rooms prior to the start of work (Fromme et al. 2013). Following cleaning, mean concentrations of 7.5 mg/m^3 (EGBE), 3.0 mg/m^3 (2PG1BE), and 3.3 mg/m^3 (EGPE) could be found during the working day. On days without cleaning activity, the levels were below the limit of quantification of 0.1 mg/m^3. The people working in the rooms also showed increased internal exposure at the end of working hours.

The emission behavior of the individual glycol ethers seems to depend not only on the room air temperature and room air humidity but also, and in particular, on the nature of the room surfaces. It has been observed on various occasions that, in addition to primary emission, re-emission can occur, e.g., from carpets and furnishings as well as other adsorptive room surfaces. These processes may be largely responsible for the very slow decline in indoor air pollution (Neretnieks et al. 1993; Chang et al. 1998a).

3.4 Isothiazolinones

Isothiazolinones are heterocyclic compounds consisting of a five-membered ring with one sulfur and one nitrogen atom integrated into the ring structure. Of prominent technical importance are 2-methylisothiazol-3-one (MIT) (CAS No. 2682-20-4), 5-chloro-2-methylisothiazol-3-one (CMIT) (CAS No. 26172–55-4), 2-octylisothiazol-3-one (OIT) (CAS No. 26530–20-1), and benzisothiazol-3-one (BIT) (CAS No. 2634-33-5). Some chemical structures are summarized in Fig. 3.7. Isothiazolinones are used in a wide variety of industrial processes, but also in numerous consumer-related products, e.g., as preservatives to prevent microbial contamination in paints, cosmetics, hygiene and consumer products, as well as in detergents, cleaners, and personal care products (Garcia-Hidalgo et al. 2017).

Name	Abbreviation	Structur
Methylisothiazolinone	MIT	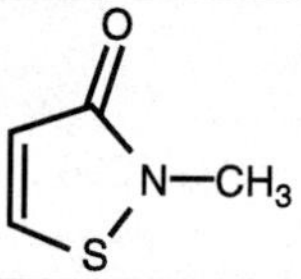
Chloromethylisothiazolinone	CMIT	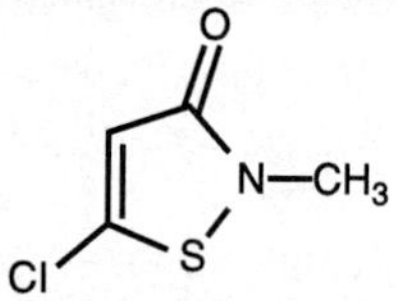
Benzisothiazolinone	BIT	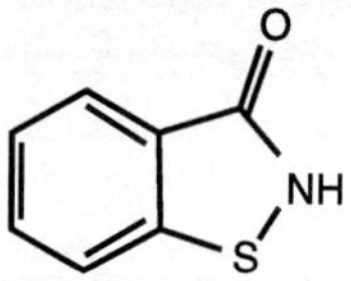
Octylisothiazolinone	OIT	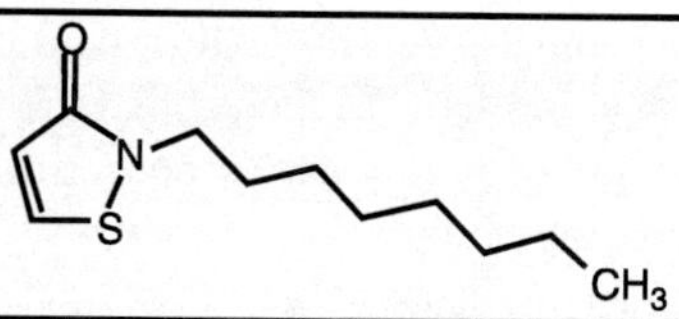

Fig. 3.7 Chemical structure of some common isothiazolinones

3.4.1 *Health Aspects*

Animal experimental studies showed decreased body weight, feed intake, and weight gain after dermal and oral exposure, as well as partial hematological and clinical chemistry abnormalities. The main effect after inhalation exposure represents the irritant effect in the upper respiratory tract. There are several cases in the scientific literature that isothiazolinones can lead to contact dermatitis via the aerogenic route after indoor painting. Extensive reviews have been published so far (Aerts et al. 2017; ECHA 2018a; b; Herman et al. 2019; Silva et al. 2020).

3.4.2 *Occurrence in Air*

Alvarez-Vaca et al. (2022) reported a median (90th percentile) of 1.0 μg/m^3 (4.0 μg/m^3) for MIT in a study in 159 residences in Luxembourg performed from 2014 to 2019.

In an older study, when a wall paint with a high content of a CMIT/MIT mixture in a 3:1 ratio (brand name: Kathon®) was applied, indoor air levels were observed to drop down from about 70 μg/m^3 (CMIT) and about 10 μg/m^3 (MIT) immediately upon application to about 2 μg/m^3 each after 26 days (Bohn et al. 2000). In another study, wall paint was applied in 42 indoor rooms and CMIT concentrations of about 2 μg/m^3 were measured after application, decreasing to 25% after 7 days and to 0.5 μg/m^3 after about 2 weeks (Reinhard et al. 2001). Even after painting with a paint containing MIT and BIT in one room, 45 μg/m^3 of MIT was still present one week after painting (start of the study) (Lundov et al. 2014). Between the eighth and 16th day after painting, room air levels varied from 1 to 5 μg/m^3.

Binder et al. (2001) investigated the time course of concentrations of isothiazolinones after the application of paint containing MIT and CMIT in two rooms following renovation work. In room 1, CMIT levels dropped from about 1.5 μg/m^3 on the first day to about 0.4 μg/m^3 after 4 days and then to about 0.2 μg/m^3 after 17 days. Initially, 0.2 μg/m^3 MIT was present in this room, which dropped to 0.05–0.1 μg/m^3 after 3 days and remained in this range throughout the rest of the observation period. Room 2 showed a slower decrease from an initial 1.9 μg/m^3 (sum of CMIT and MIT) to 1.03 μg/m^3 after 9 days. Nagorka et al. (2015) investigated the emission behavior of wall paints in 2 test rooms. In one room, a maximum OIT of 1.4 μg/m^3 was reached 12 hours after application, which only dropped into the range of the limit of quantification after more than 275 days. In the second room, there was a maximum of 29 μg MIT/m^3 after 4 days and a slow reduction to half after 3 weeks. The OIT maximum of 1.2 μg/m^3 was observed in this experiment only after 12 days, while concentrations in the range of 0.5–1.0 μg/m^2 could still be detected after several hundred days.

In an experimental study, a mixture of CMIT and MIT was nebulized into a test room using a humidifier (Park et al. 2020). When the mixture was used as intended as a preservative and an adjusted air-exchange rate of 0.3 and 1.0 per hour, 0.4 μg/m^3 and 0.3 μg/m^3 of MIT and 0.95 μg/m^3 and 1.0 μg/m^3 of CMIT were detected, respectively. The indoor air concentrations increased linearly with the emission rate of the devices, but were not dependent on the air-exchange rate.

3.5 Methyl Methacrylates

Methacrylates play an important role in the production of high-molecular-weight plastics by reacting as small but reactive molecules in the presence of a reaction accelerator (e.g., N,N-dimethyl-p-toluidine) to form a polymeric plastic (ECHA 2018c). For example, methyl methacrylate (MMA, CAS-No.: 80-62-6) is used in dentistry, in orthopedics for artificial skeletal parts, and as bone cement for filling bones and fixing implants. In cosmetics, MMA is used extensively for nail modeling. Technical uses of MMA are primarily in the manufacture of acrylics, in adhesives and sealants, machine rinse aids and detergents, automotive care products, paints and coatings, and in fragrances and room sprays (ECHA 2018c). MMA is an intense-smelling liquid that is highly volatile and is produced in the EU in quantities of up to approximately 1,000,000 metric tons (ECHA 2018c). The chemical structure of MMA is shown in Fig. 3.8. In addition to MMA, ethyl methacrylate (EMA, CAS-No.: 97-63-2), hydroxyethyl methacrylate (HEMA, CAS-No.: 868-77-9), triethylene glycol dimethacrylate (TEGDMA), and bisphenol glycidyl methacrylate (BisGMA) are of particular technical importance.

3.5.1 *Health Aspects*

MMA irritates the mucous membranes of the nose and eyes upon exposure and has a high sensitizing potential. When inhaled, lesions in the area of the nose and pulmonary changes up to pulmonary edema have been described in animal experiments. Effects on the lungs and central nervous system have also been reported in highly exposed workplaces. MMA is hydrolyzed by carboxylesterases to methacrylic acid in both laboratory animals and humans, e.g., in the nasal mucosa, which is responsible for local toxicity (DFG 2006). Contact dermatitis, asthmatic complaints, and allergies have been described in employees handling MMA, e.g., in nail studios, dental businesses, and dental clinics. In the context of nail modeling,

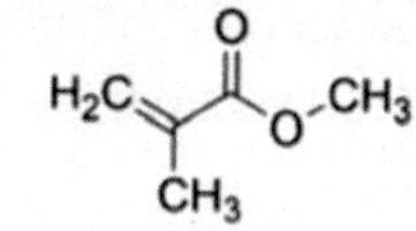

Fig. 3.8 Chemical structure of methyl methacrylate

nail bed or nail fold inflammations, irritative dermatitis, bacterial infections, and mycoses of the nail bed have been observed (Baran 2002; Heymann 2007). Extensive summaries of toxicological and epidemiological knowledge are available (US EPA 1998; WHO 1998; DFG 2006; ECHA 2018c).

3.5.2 Occurrence in Air

So far, only limited results are available on the occurrence of MMA in the air of apartments, schools, kindergartens, and offices. For example, Eis et al. (2005) were able to determine MMA in 39 out of 53 German apartments in 1999–2003 with the 95th percentile being only 1.8 μg/m^3. Based on 3619 measurements in 2006–2012 in German indoor spaces with different uses, concentrations above the limit of quantification of 1 μg/m^3 (95th percentile: 1.5 μg/m^3) were found in only 12% of the samples (Hofmann et al. 2014). In 701 homes, 1776 offices, and 496 classrooms, 95th percentiles of 2.0 μg/m^3, 1.4 μg/m^3, and 1.0 μg/m^3, respectively, were found in this study, while the medians were all below 1.0 μg/m^3 (limit of quantification).

Zhu et al. (2005) found MMA in Canada in 2002–2003 in only 11 of 75 rooms with a mean value of 0.5 μg/m^3 (range: <0.01–1.12 μg/m^3).

In individual cases, however, significantly higher concentrations can occur. For example, MMA concentrations of 3300–13,000 μg/m^3 were measured in the course of improperly executed remediation work in a house in Berlin, Germany, that contained apartments and a daycare center for children (Walt et al. 2015). After about one year, the level dropped to 2–14 μg/m^3 following appropriate remediation measures.

Another known source of exposure for the population is nail studios. For example, in a study in Munich in 2012, MMA concentrations of 2100–15,700 μg/m^3 (mean: 4400 μg/m^3) were measured in 6 nail studios during opening hours in 13 measurements (Nitschke et al. 2013). N,N-dimethyl-p-toluidine was detected in this study in the range < 1–33 μg/m^3 (mean: 10.0 μg/m^3). MMA levels between 499 and 5400 μg/m^3 were also found in the 2008–2009 study of 3 nail salons in California, or a geometric mean of 1430 μg/m^3 in another 77 nail salons in 2013–2016 (Quach et al. 2011, 2018). In a study in 17 Michigan, USA, nail salons MMA and EMA were detected in 85% and 18% of the samples with medians of 550 μg/m^3 (<2.5–34,200 μg/m^3) and 0.5 μg/m^3 (<0,5–1359 μg/m^3) using stationary samplers in 2016/17 (Zhong et al. 2019). In personal air the contractions of MMA were significantly higher (median: 970 μg/m^3). MMA were found above LOD in 10 out of 20 nail studios in Greater New York in 2021 with a lower mean of 5.8 μg/m^3 (Han et al. 2022). In this study sampling period was approximately 72 h, including non-business hours. In 22 Norwegian nail salons, indoor MMA levels of 42–333 μg/m^3 and EMA concentrations of 315 11,270 μg/m^3 were measured (Gjölstad et al. 2006). Also in 3 Finnish stores, majority small but very well ventilated rooms, the mean personal concentrations of MMA were 43 μg/m^3 during nail modeling and 13 μg/m^3 at the end of the work and of EMA 108 μg/m^3 and 61 μg/m^3, respectively (Henriks-Eckerman and Korva 2012).

In 4 dental practices, MMA concentrations between 60 and 400 μg/m^3 were measured during the placement of composite fillings (Marquardt et al. 2009). The levels of HEMA and TEGDMA in this situation varied from 7 to 45 μg/m^3 and 19 to 45 μg/m^3, respectively. In 21 measurements in Swedish dental clinics, the median 8-h values were 0.3 μg/m^3 (< 0.1–13 μg/m^3) for MMA and 2.9 μg/m^3 (1–15 μg/m^3) for HEMA (Hagberg et al. 2005). Significantly higher levels of exposure can be detected in the air of dental technology plants, where concentrations of 100–30,200 μg/m^3 (Marquardt et al. 2008) or 5000–78,000 μg/m^3 (Liljelind et al. 2005) have been described.

Another special exposure situation can be 3D printers, where a thermoplastic material is melted and applied layer by layer to a movable carrier plate. When polylactide (PLA) is used as the plastic material, the mean concentration increased from 1.2 μg/m^3 to 19 μg MMA/m^3 during printing in a smaller, poorly ventilated office room, while only a small increase from 0.1 to 1.7 μg/m^3 was observed in a large, well-ventilated room (Steinle 2016).

3.6 Emerging PFAS

The term per- and polyfluorinated hydrocarbons (PFAS) covers about 8000 organic compounds in which the hydrogen atoms on the carbon skeleton have been replaced completely or partially by fluorine atoms. In addition to the perfluorinated substances discussed in Sect. 4.11, there are various other substances or groups of substances that are used technically on a large scale (Buck et al. 2021), particularly, the fluorotelomer alcohols (FTOH), the fluorotelomer acrylates (FTAC), the perfluorooctane sulfonamides (FOSA), the perfluorooctane sulfonamide ethanols (FOSE), and the perfluoroalkyl phosphates (PAP). The most important substances are shown in Table 3.8 and some chemical structures are given in Fig. 3.9. On the one hand, these are classic substitute products that have been developed specifically, for example, and substances that have been used for a long time and can be detected in the environment, such as sulfonamide ethanols and fluorotelomer alcohols (Jahnke et al. 2007). They can be found as residual contents in consumer products (e.g., carpets, upholstery, fibers, paper fibers). In contrast to PFOS and PFOA, they are uncharged substances that are poorly soluble in water due to the perfluoroalkyl residues and are also comparatively volatile. However, the nonpolar perfluorinated compounds prove to be less persistent and can be degraded via atmospheric and biological processes (Ellis et al. 2004). Thus, in the environment and in the organism, the degradation of perfluorinated sulfonamide ethanols via sulfonamide can lead to the persistent PFOS, while fluorotelomers but also PAP are converted to PFOA (D'eon and Mabury 2011a, b). Here, degradation occurs after oxidation of the telomeral alcohol function via short-lived fluorotelomer aldehydes (FTAL) to saturated and unsaturated fluorotelomer carboxylic acids (FTCA and FTUCA,

Table 3.8 Abbreviations, names, and CAS numbers of emerging PFAS

Acronym	Substance name	CAS No.
6:2 FTAC	2-Perfluorohexylethyl acrylate	17,527–29-6
8:2 FTAC	2-Perfluorooctyl ethyl acrylate	27,905–45-9
10:2 FTAC	2-Perfluorodecyl ethyl acrylate	17,741–60-5
4:2 FTOH	2-Perfluorobutylethanol	647–42-7
6:2 FTOH	2-Perfluorohexylethanol	647–42-7
8:2 FTOH	2-Perfluorooctylethanol	678–39-7
10:2 FTOH	2-Perfluorodecylethanol	865–86-1
MeFOSA	Methyl perfluorooctane sulfonamide	31,506–32-8
EtFOSA	Ethyl perfluorooctane sulfonamide	4151-50-2
MeFOSE	Methylperfluorooctane sulfonamidoethanol	24,448–09-7
EtFOSE	Ethylperfluorooctane sulfonamidoethanol	1691-99-2
6:2 monoPAP	6:2 Polyfluoroalkyl phosphate monoester	57,678–01-0
8:2 monoPAP	8:2 Polyfluoroalkyl phosphate monoester	57,678–03-2
6:2 diPAP	6:2 Polyfluoroalkyl phosphate diester	57,677–95-9
6:2/8:2 diPAP	6:2/8:2 Polyfluoroalkyl phosphate diester	943,913–15-3
8:2 diPAP	8:2 Polyfluoroalkyl phosphate diester	678–41-1
8:2/10:2 diPAP	8:2/10:2 Polyfluoroalkyl phosphate diester	1,158,182–60-5
10:2 diPAP	10:2 Polyfluoroalkyl phosphate diester	865–86-1

respectively) via a ß-oxidative process to PFOA or abiotically to PFNA or PFOA (Rand and Mabury 2017; Kabadi et al. 2018). Due to the aforementioned degradation mechanisms, the aforementioned precursor substances thus advance to emission-remote sources of the persistent perfluorosurfactants PFOS and PFOA. Against this background, in addition to the toxicological properties inherent in these substances, these transformation mechanisms must always be taken into account.

Fluorotelomer alcohols are various chemicals which, in addition to the fluorinated carbon chain, have carbon-bonded hydrogen atoms and an hydroxyl-group. They are named by indicating the number of fluorinated carbon atoms to those bearing a hydrogen atom, separated by a colon (e.g., 8:2 FTOH). Worldwide production is estimated at 6000–12000 tons. Approximately 80% of the total amount produced is used in fluoropolymer production. They are used, for example, in the impregnation of textiles and can be detected as residues on appropriately finished textiles. FTOH and FTAC accounted for approximately 60% of the global fluorotelomer market in 2013. FTAC are used, for example, in the production of methacrylate polymers. In addition, fluorotelomer alcohol is known to be used in firefighting foam.

Polyfluoroalkyl phosphoric acid esters (PAP) are hydrophobic and, in addition to the phosphate group, have mono-, di-, or tri-substituted fluorinated side chains with a chain length typically between C_8 and C_{12}. In the industrial product, di-PAP in particular are present, while mono and tri PAPs appear only as by products. They have been used since the 1960s in many applications, mainly in paper products and

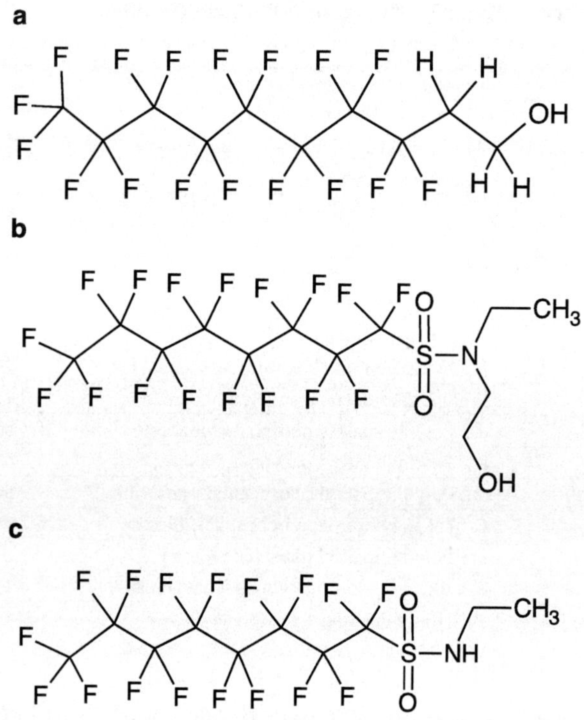

Fig. 3.9 Chemical structure of (**a**) 8:2 fluorotelomer alcohol (8:2 FTOH), (**b**) N-ethyl-perfluorooctanesulfonamidoethanol (NEtFOSE), (**c**) N-ethyl-perfluorooctanesulfonamide (NEtFOSA)

paperboards including packaging in consumer products, detergents, and paints, due to their surfactant properties.

Fluorotelomer sulfonates such as the 6:2 FTSA are used as substitutes for PFOS or their salts in quenching foams and in electroplating in hard and decorative chrome plating processes. For a long time, in the USA since the 1960s, 6:2-FTSA was used in AFFF (Aqueous Film-Forming Foam) firefighting foams and as a wetting agent in many commercial products.

3.6.1 Health Aspects

Effect data from animal studies after oral administration are available for FTOH. They show that major target organs are the liver and kidney (Serex et al. 2014; Rice et al. 2020). This can also be partly attributed to the fact that, for example, 8:2 FTOH can be converted to PFOA in the organism. There is no evidence of mutagenicity or genotoxicity (EFSA 2014). For the other FTOH, only limited results from animal studies are available so far (Rice et al. 2020). Findings from epidemiological studies in humans have not yet been published.

Toxicological knowledge on PAP is very limited to date (Taxvig et al. 2014). In vitro studies showed hormonal activity and interference with steroid hormone production.

Toxicological data are also limited for FTSA. In an animal study with mice, an increase in liver weight, signs of hypertrophy of liver cells, and an increase in inflammation markers in the liver were observed after 28 days of oral administration of 6:2 FTSA (Sheng et al. 2017).

3.6.2 Occurrence in the Air of Residences

The results of the European studies published in the scientific literature to date are summarized in Table 3.9. Studies of apartments are available from Norway, Finland, Spain, England, Sweden, and Germany. The wide range of individual results for FTOH and FOSE/A is striking, indicating differences in use and sources of varying emission strength. For FTOH, the highest concentrations are reported for the 8:2 FTOH and 10:2 FTOH with mean values of 4250–29,400 pg/m^3 and of 1240–12,300 pg/m^3, respectively. The highest mean values for FTOH were found in 61 dwellings in Oslo, Norway, and the lowest in 57 dwellings in Finland. FTAC levels are at least an order of magnitude lower in European studies, but the database is very limited. Considerable variation in individual results can also be observed for FOSE/FOSA, with the highest mean values reported for MeFOSE ranging from 90–2050 pg/m^3.

More comprehensive studies are available from 59 homes from Canada in 2002/03 and 2007/08, where gas and particulate phases were collected in parallel using active or passive sampling on PU foam and glass fiber filters, respectively (Shoeib et al. 2005; Makey et al. 2017). Although different sampling techniques were used, the studies found concentration levels, which are comparable to that in Europe. For MeFOSE and EtFOSE, the geometric mean values range between 380 and 1490 pg/m^3 and 50 and 744 pg/m^3, respectively, with concentrations showing a decreasing trend over time. For 8:2 FTOH and 10:2 FTOH, geometric mean values in 59 Canadian residences in 2007/08 were 2400 and 960 pg/m^3, respectively (Makey et al. 2017).

Yao et al. (2018) collected 81 samples in homes, hotels, textile stores, and cinemas in Tianjin, China, in 2015. Median values were < LOQ for 6:2 FTOH, 2600 pg/m^3 for 8:2 FTOH, and 679 pg/m^3 for 10:2 FTOH in the 22 homes. MeFOSE and EtFOSE could only be determined in 5% of the samples. In Japan, a total of 84 indoor rooms in Kyoto and Osaka were investigated in 2008 and geometric mean values of 260 pg/m^3 for 6:2 FTOH, 5160 pg/m^3 for 8:2 FTOH, and 920 pg/m^3 for 10:2 FTOH were determined (Liu et al. 2013).

Fraser et al. (2012) reported results from 31 offices in Boston, USA in 2009. They found geometric mean values of 1320 pg/m^3 (6:2 FTOH), 9920 pg/m^3 (8:2 FTOH),

Table 3.9 Mean concentrations (range) of FTAC/FTOH and FOSE/FOSA in residential air in Europe in pg/m^3

Reference	Substance	*N*	Concentration	Country, Year
Langer et al. (2010)	6:2 FTAC	2	<LOQ and 200	Germany, 2009/10
Fromme et al. (2015)		13	<54	Germany, 2014
Winkens et al. (2017)		57	<9.6	Finland, 2014/15
Langer et al. (2010)	8:2 FTAC	2	200 and 1500	Germany, 2009/10
Fromme et al. (2015)		13	422 (58–1040)	Germany, 2014
Langer et al. (2010)	10:2 FTAC	2	700 and 1100	Germany, 2009/10
Fromme et al. (2015)		13	205 (27–560)	Germany, 2014
Huber et al. (2011)	4:2 FTOH	6	21* (<16–34.0)	Norway, 2007/08
Haug et al. (2011)		41	4.8 (< LOQ-38)	Norway, 2008
Fromme et al. (2015)		13	<181	Germany, 2014
Huber et al. (2011)	6:2 FTOH	6	42* (16.0–332)	Norway, 2007/08
Haug et al. (2011)		41	1492 (< LOQ-9414)	Norway, 2008
Ericson Jogsten et al. (2012)		10	– (3–47)	Spain, 2009
Padilla-Sánchez et al. (2017)		61	8550 (604–101,000)	Norway, 2013/14
Fromme et al. (2015)		13	2058 (289–5424)	Germany, 2014
Winkens et al. (2017)		57	1810 (<290–8060)	Finland, 2014/15
Huber et al. (2011)	8:2 FTOH	6	10005* (3373–11,109)	Norway, 2007/08
Haug et al. (2011)		41	6438 (< LOQ-25323)	Norway, 2008
Ericson Jogsten et al. (2012)		10	– (7.5–170)	Spain, 2009
Padilla-Sánchez et al. (2017)		61	29,400 (1220–446,000)	Norway, 2013/14
Fromme et al. (2015)		13	9007 (4702–21,698)	Germany, 2014
Winkens et al. (2017)		57	4250 (1290–13,500)	Finland, 2014/15
Sha et al. (2018)		9	6700 (890–14,000)	Sweden, 2016
Huber et al. (2011)	10:2 FTOH	6	3405* (1231–5901)	Norway, 2007/08
Haug et al. (2011)		41	4088 (< LOQ-28898)	Norway, 2008
Ericson Jogsten et al. (2012)		10	– (<0.6–47)	Spain, 2009
Padilla-Sánchez et al. (2017)		61	12,300 (697–255,000)	Norway, 2013/14
Fromme et al. (2015)		13	3296 (759–10,306)	Germany, 2014
Winkens et al. (2017)		57	1240 (210–8950)	Finland, 2014/15
Sha et al. (2018)		9	1600 (250–3400)	Sweden, 2016
Goosey and Harrad (2012)	MeFOSE	20	950 (77–3100)	England, 2008/09
Huber et al. (2011)		6	846* (222–1288)	Norway, 2007/08
Haug et al. (2011)		41	346 (< LOQ-1433)	Norway, 2008
Ericson Jogsten et al. (2012)		10	– (<0.5–6.1)	Spain, 2009
Padilla-Sánchez et al. (2017)		61	2050 (<71–38,800)	Norway, 2013/14
Fromme et al. (2015)		13	552 (27–3796)	Germany, 2014
Winkens et al. (2017)		57	90 (<8.2–394)	Finland, 2014/15
Goosey and Harrad (2012)	EtFOSE	20	600 (80–1900)	England, 2008/09
Huber et al. (2011)		6	156* (65–333)	Norway, 2007/08
Haug et al. (2011)		41	97 (< LOQ-338)	Norway, 2008
Ericson Jogsten et al. (2012)		10	– (<0.5–6.1)	Spain, 2009
Padilla-Sánchez et al. (2017)		61	776 (<50–13,200)	Norway, 2013/14

(continued)

Table 3.9 (continued)

Reference	Substance	*N*	Concentration	Country, Year
Fromme et al. (2015)		13	69 (27–155)	Germany, 2014
Winkens et al. (2017)		57	24.1 (<8.1–132)	Finland, 2014/15
Goosey and Harrad (2012)	MeFOSA	20	<2.5 (<2.5–19)	England, 2008/09
Huber et al. (2011)		6	27* (9–224)	Norway, 2007/08
Haug et al. (2011)		41	14 (<LOQ-70)	Norway, 2008
Ericson Jogsten et al. (2012)		10	– (<1.2–14)	Spain, 2009
Padilla-Sánchez et al. (2017)		61	1350 (<35–78,300)	Norway, 2013/14
Fromme et al. (2015)		13	<54 (<54–166)	Germany, 2014
Winkens et al. (2017)		57	<7.9 (<7.9–20)	Finland, 2014/15
Goosey and Harrad (2012)	EtFOSA	20	<120 (14–350)	England, 2008/09
Huber et al. (2011)		6	29* (7–104)	Norway, 2007/08
Haug et al. (2011)		41	9.6 (<LOQ-95)	Norway, 2008
Ericson Jogsten et al. (2012)		10	– (<0.5–6.1)	Spain, 2009
Padilla-Sánchez et al. (2017)		61	69 (<35–1350)	Norway, 2013/14
Fromme et al. (2015)		13	<54 (<54–119)	Germany, 2014
Winkens et al. (2017)		57	<7.9 (<7.9–340)	Finland, 2014/15

LOQ limit of quantification; *: median; +: geometric mean value

and 2850 pg/m^3 (10:2 FTOH). In another study, different PFAS were measured in two office rooms (Jahnke et al. 2007). Here, the measured concentrations were significantly lower than the concentrations determined in residential rooms. The authors blame the lack of typical sources for this. An experiment showed that when a paraglider was placed in the office room, the content of individual substances increased by a factor of 1.3 to 81. This is confirmed by a study by Schlummer et al. (2013), who measured very high levels of up to 286,000 pg/m^3 of 8:2 FTOH in sales rooms of outdoor clothing, indicating outdoor textiles to be a relevant source of FTOH in indoor workplace environments. The same group estimated concentrations of 510 pg/m^3 for 6:2 FTOH, 8430 pg/m^3 for 8:2 FTOH, and 3370 pg/m^3 for 10:2 FTOH in the indoor air of a nine months old car.

Overall, it must be considered that significantly higher levels of volatile perfluorinated compounds can be expected in indoor air compared to outdoor air. For example, in Canada, indoor air was observed to contain a factor of 18 higher levels of MeFOSE and a factor of 8 higher levels of N-EtFOSE compared to outdoor air (Shoeib et al. 2005).

Only Yao et al. (2018) presented PAP results of 81 samples collected in the indoor air of homes, hotels, textile stores, and cinemas in Tianjin, China, in 2015. The 6:2 diPAP and 8:2 diPAP were found in 77% and 14% of 22 homes, respectively, with medians (maximum) of 1.56 μg/m^3 (6.33 pg/m^3) and 2.14 pg/m^3 (3.84 pg/m^3).

3.6.3 *Occurrence in the Air of Public Facilities*

In the indoor air of schools, only study results from Germany and the US are available to date. In Germany, a total of 14 elementary schools were investigated in 2014 (Fromme et al. 2015). While 4:2 FTOH, MeFOSA and EtFOSE could not be quantified in any room, 6:2 FTAC and MeFOSE could only be determined in two and three classrooms, respectively. The mean concentration of the other substances was 4554 pg/m^3 (8:2 FTAC), 1800 pg/m^3 (10:2 FTAC), 7853 pg/m^3 (6:2 FTOH), 5717 pg/m^3 (8:2 FTOH), and 2066 pg/m^3 (10:2 FTOH). For EtFOSA, a mean level of 274 pg/m^3 was observed. Compared to residential interiors, a comparable exposure level was found for FTOH, while FTAC concentrations were higher in schools. In the US, Morales-McDevitt et al. (2021) sampled air of 9 kindergarten in California in 2018-2020 and reported higher medians of 70,000 pg/m^3 for 6:2 FTOH, 20,000 pg/m^3 8:2 FTOH, and 5,800 pg/m^3 for 10:2 FTOH. 8:2 FTAC and 10:2 FTAC, and EtFOSA and MeFOSE could not be quantified in any sample.

3.6.4 *Occurrence in Dust of Residences*

Xu et al. (2013) reported mean levels of 19.4 ng/g (6:2 FTOH), 29.5 ng/g (8:2 FTOH), and 17.5 ng/g (10:2 FTOH) in dust from 31 apartments in Munich, Germany. The 4:2 FTOH could not be quantified in any sample. In a Finnish study, mean levels of 40 ng/g (6:2 FTOH), 60 ng/g (8:2 FTOH), 20 ng/g (10:2 FTOH), and 30 ng/g (EtFOSE) were observed in 65 apartments in 2014/2015 (Winkens et al. 2018). In 12 Czech homes, mean values in 2013 were 9 ng/g (MeFOSE) and 2 ng/g (EtFOSE) (Karásková et al. 2016). FTOH were not measured in this study.

In the winter of 2002/2003, levels of EtFOSE and MeFOSE of 1–75,400 ng/g (geometric mean: 140 ng/g) and 3–8800 ng/g (geometric mean: 110 ng/g), respectively, were measured in 66 homes in Canada (Shoeib et al. 2005). Geometric mean values of 80 ng/g (MeFOSE) and 20 ng/g (EtFOSE) were also reported in 150 dust samples from 2007/2008 (Makey et al. 2017). For 8:2 FTOH and 10:2 FTOH, the geometric means in this study were 100 ng/g and 50 ng/g, respectively. In addition, Hall et al. (2020) determined lower levels in 184 Canadian homes in 2014–2016 with medians of <13 ng/g for MeFOSE and < 15 ng/g for EtFOSE, while significantly higher fluorotelomer levels were observed at 569 ng/g for 6:2 FTOH and 1435 ng/g for 8:2 FTOH than in the other studies. Zheng et al. (2020) studied 20 dust samples from 7 kindergartens in the USA and found median values for 6:2 FTOH of 130 ng/g, for 8:2 FTOH 20 ng/g, and for 10:2 FTOH 40 ng/g.

Yao et al. (2018) collected dust in 18 homes in Tianjin, China, in 2015. 8:2 FTOH and 10:2 FTOH were found in only 22% and 6% of samples, respectively, with medians of <0,60 and < 0,32 ng/g. Eriksson and Kärrman (2015) collected dust from vacuum cleaner bags (sieved to 150 μm) from individual households in

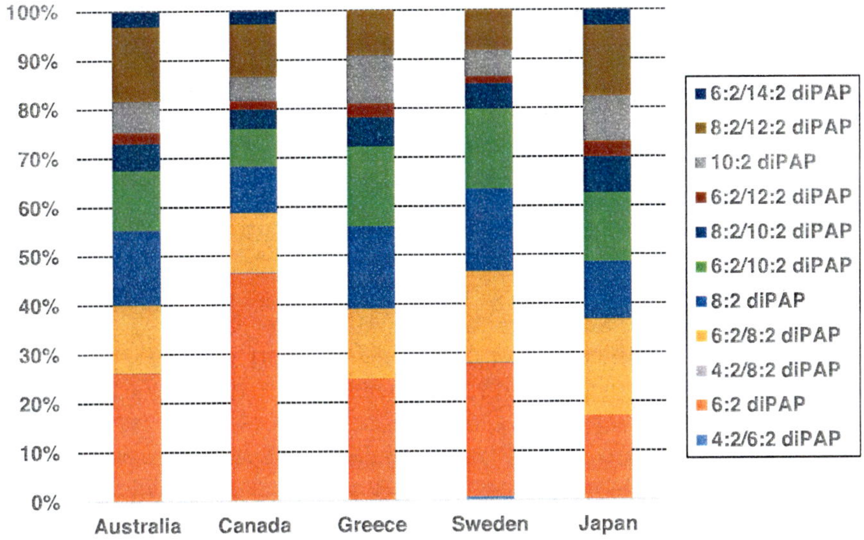

Fig. 3.10 Distribution of diPAP in house dust from different countries using medians (only if 50% of the values were above LOQ) (modified after Eriksson and Kärrman 2015)

Canada, the Faroe Islands, Sweden, Spain, Greece, Nepal, Japan, and Australia between 2009 and 2014. MeFOSE was detected in 33% of the samples, EtFOSE in 8%, MeFOSA and EtFOSA in 3%. They reported medians from for sum of FOSE/A < 0.1 ng/g in all countries, except for Canada (13 ng/g).

Two Canadian studies collected dust samples of 102 residences in 2007/2008 and 184 residences in 2014–2016 (De Silva et al. 2012; Hall et al. 2021). The first study reported medians of 460 ng/g for 6:2 diPAP, 614 ng/g for 6:2/8:2 diPAP, 535 ng/g for 8:2 diPAP, and 149 ng/g for 10:2 diPAP, whilst the second found geometric means for 8:2 diPAP and 10:2 diPAP of 530 and 170 ng/g, respectively. Weiss et al. (2021) obtained 46 dust samples in Sweden in 2013/2014 and reported medians of 140 ng/g for 6:2 diPAP and 100 ng/g for 8:2 diPAP, while 6:2 monoPAP and 8:2 monoPAP had medians of 31 ng/g and 22 ng/g. Eriksson and Kärrman (2015) collected dust from vacuum cleaner bags (sieved to 150 μm) from individual households in Canada, the Faroe Islands, Sweden, Spain, Greece, Nepal, Japan, and Australia between 2009 and 2014. Median levels of quantified and semiquantified monoPAPs ranged from 3.7 ng/g in Nepal to 1023 ng/g in Japan. The median levels of diPAP ranged between 3.6 ng/g in Nepal and 692 ng/g in Japan, with significantly higher levels in Australia, Canada, the Faroe Islands, and Japan (343 ng/g, 352 ng/g, 450 ng/g, and 692 ng/g, respectively) than in Sweden, Greece, Spain, and Nepal (62 ng/g, 25 ng/g, 14 ng/g, 3.6 ng/g). 6:2 diPAP, 6:2/8:2 diPAP, and 8:2 diPAP were found with the highest medians of 1.2–164 ng/g (Nepal, Canada), 0.71–137 ng/g (Nepal, Faroe Island), and 0.72–81 ng/g (Nepal, Japan). Using a simulated diPAP profile from the historical fluorotelomer production by converting the fluorotelomers to corresponding diPAP, similar distribution was observed to the study results (see Fig. 3.10).

3.6.5 Occurrence in Dust of Public Facilities

In the indoor environment of 20 samples, medians of 130 ng/g for 4:2 FTOH, 20 ng/g for 6:2 FTOH, and 40 ng/g for 10:2 FTOH, 11 ng/g for MeFOSE, and 15 ng/g for EtFOSE were measured in a study from seven daycare centers in the USA (Zheng et al. 2020), while in 20 Swedish preschools medians (95th percentiles) of 4.1 ng/g (399 ng/g) for 6:2 FTOH, 18 ng/g (142 ng/g) for 8:2 FTOH, and 12 ng/g (75 ng/g) for 10:2 FTOH were reported in 2018 (Giovanoulis et al. 2019). The medians (95th percentiles) were 151 ng/g (2728 ng/g) for 6:2 monoPAP, 1143 ng/g (42,281 ng/g) for 6:2 diPAP, and 36 ng/g (377 ng/g) for 8:2 diPAP. Additionally, Wu et al. (2020) collected paired 29 carpet and 28 dust samples from 18 California childcare centers in 2018 and found medians of 88 ng/g (22–571 ng/g) for 6:2 FTOH, 32 ng/g (14–297 ng/g) for 8:2 FTOH, 28 ng/g (<4.9–356 ng/g) for 10:2 FTOH, 3.0 ng/g (<3.2–123 ng/g) for MeFOSE, and 4.1 ng/g (<3.5–98 ng/g) for EtFOSE. Significant associations of PFCA with FTOH and FTSA as well as PFOS with MeFOSE and EtFOSE in dust were observed in this study, consistent with the fact that FTOH and FTSA can degrade to PFCA and MeFOSE and EtFOSE to PFOS.

3.7 Oximes (Butanone Oxime)

Oximes are a class of substances that are produced on a large scale and used in a variety of industrial applications. They contain the functional group C=N-OH (Fig. 3.11), to which different residues are attached. Two groups are distinguished, the so-called ketoximes, which are derivatives of ketones, and aldoximes, which are derivatives of aldehydes.

Oximes are used as anti-skin agents in solvent-based surface treatments with alkyd resin or oil-containing binders, such as alkyd enamel paints and oil alkyd paints, and in wood preservation and wood care products (HC 2014). The content in the products can be up to several mass percent (Dörr et al. 2017). Neutral curing oxime-based silicone sealants containing between 4–6% oxime silanes as crosslinkers also release oximes during the curing process (Dörr et al. 2017). In the past, 2-butanone oxime (methyl ethyl ketone oxime, MEKO, CAS-No.: 96–29-

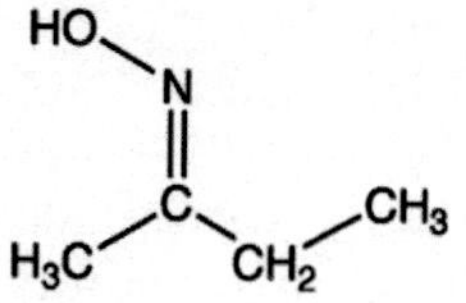

Fig. 3.11 Chemical structure of butanone oxime

7) and sometimes acetone oxime (2-propanone oxime, dimethyl ketone oxime, DMKO, CAS-No.: 127–06-0) were used in particular. Due to the increasing discussion on the health effects of the aforementioned substances, they are increasingly replaced by related oximes such as 2-pentanone oxime (methy propyl ketone oxime, MPKO, CAS-No.: 623–40-5), 4-methyl-2-pentanone oxime (methyl isobutyl ketone oxime, MIBKO, CAS-No.: 105–44-2). Zhao et al. (2022) evaluated the VOC emissions from 30 adhesives that are commonly available on the Chinese market. Butanone oxime was the concentrated compound found in 20 silicone adhesives, responsible for around 70% of the total VOC emissions.

Due to the risks mentioned above, the use of alternative products should be investigated indoors for oxime-based silicone sealants and varnishes.

3.7.1 Health Aspects

Hematological effects such as methemoglobin formation, degenerative changes in the olfactory epithelium, and effects on the liver and spleen (e.g., organ weight) have been reported in animal experiments with long-term exposure (HC 2010). Effects on reproduction and developmental disorders, if any, are observed only at very high concentrations. 2-Butanone oxime did not show mutagenic or genotoxic effects; however, the reactive metabolites could have genotoxic effects. In animal experiments, dose-dependent effects on the liver were found, especially for male animals, with liver adenomas and carcinomas observed. In the European hazardous substances legislation, 2-butanone oxime and acetone oxime are classified as category 1B (substances likely to be carcinogenic in humans) (ECHA 2020; 2018e).

3.7.2 Occurrence in the Air

To date, there are only a few studies on the occurrence of oximes in indoor environments, which are limited to 2-butanone oxime. The emission of 2-butanone oxime could be observed practically immediately when three different alkyd paints were applied in a test chamber with a volume of 0.053 m^3 (Chang et al. 1998b). Approximately 90% was emitted within 10 hours of paint application. For one test house, peak concentrations of 4–20 mg/m^3 were modeled based on the emission behavior during painting with three alkyd paints at an air-exchange rate of 0.5/hour.

Significant variations in the emission profile were observed for the different methods of silicone caulk application (Klewicz et al. 2020). From a neutral silicone sealant, there are three main stages of emission with differences in the emissions of 2-butanone oxime between the first and last stages of emission ranged from factors of 100–1000.

In Germany, results have been published by private research laboratories in 2006–2012 (Hofmann et al. 2014). In 60 dwellings, 2-butanone oxime could only

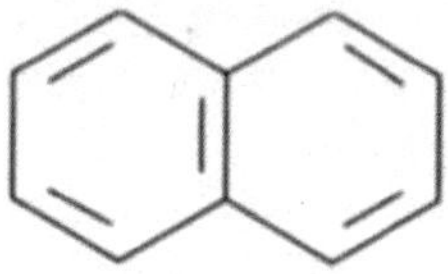

Fig. 3.12 Chemical structure of naphthalene

be found in 8.3% of the samples with a 95th percentile of 1.0 μg/m³ and in 38 classrooms in schools with 0.5 μg/m³ (in 7.9% of the samples).

3.8 Naphthalene

Naphthalene (CAS No.: 91-20-3) was released to ambient air from the combustion of wood and fossil fuels and smaller amounts are introduced to water as the result of discharges from coal-tar production and distillation processes (ATSDR 2005). Not only tobacco smoke and other combustion products, but also tar-containing building products (in moisture barriers, roofing felt, adhesives or paints containing tar or carboline), heated natural insulation materials (e.g., cork), and off-gassing of naphthalene-containing moth repellents are still the source of contamination in indoor spaces. Again, NAP was used as an intermediate in the manufacture of especially phthalic anhydride and azo dyes. NAP may still be used as a household insecticide in different parts of the world, in particular in tropical zones (WHO 2010). The chemical structure of NAP was given in Fig. 3.12.

3.8.1 *Health Aspects*

Various reports described an association between NAP exposure and hemolytic anemia in humans following acute or occupational exposure (ATSDR 2005). They described a relationship between an inherited deficiency in the enzyme glucose 6-phosphate dehydrogenase (G6PD) and susceptibility to NAP-related hemolysis. Newborn infants also appear to be susceptible to naphthalene-induced hemolysis presumably due to a decreased ability to conjugate and excrete naphthalene metabolites (ATSDR 2005).

The IARC classified naphthalene to be possibly carcinogenic to humans (Group 2B) with sufficient evidence in experimental animals and inadequate evidence in humans (IARC 2002). At the present time the mechanism whereby naphthalene produces benign respiratory tract tumors at high concentrations is not fully understood, however, based on the many negative results obtained in genotoxicity tests, a genotoxic mechanism appears unlikely and therefore an inhalation unit risk was not derived by the US EPA. Because nasal lesions were seen at clearly lower concentrations than tumor development, WHO stated that the use of a LOAEL/NOAEL as a threshold, combined with safety factors, is considered to be an appropriate approach for setting indoor air guidelines to minimize also the carcinogenic risks (WHO 2010). NAP is classified as a suspected carcinogen (carc. 2) in European hazardous substances legislation.

There are extensive reviews by international organizations on the hazards and environmental and health risks of NAP (e.g., ATSDR 2005; IARC 2010; WHO 2010; ECHA 2018d; Yost et al. 2021).

3.8.2 *Occurence in the Air of Residences*

Naphthalene is the most volatile member of the PAH. Therefore, NAP is usually included in VOC sampling and analysis and it is discussed here.

Indoor air pollution with NAP appears to have been declining in recent decades. In 1985/1986, for example, a median of 2.2 μg/m^3 (95th percentile: 4.9 μg/m^3) was found in a representative study in 479 apartments in Germany (Krause et al. 1991). In two studies in the early 2000s, median concentrations decreased to only 0.3 μg/m^3 (95th percentile: 3.1 μg/m^3) and 0.5 μg/m^3 (90th percentile: 1.6 μg/m^3) (Schlink et al. 2004; Hippelein 2004). In a comprehensive compilation of measurement results from Germany from 2006 to 2012, NAP could still be detected in 48% of the samples collected in 712 homes with a median of 0.5 μg/m^3 and the 95th percentile was 2.40 μg/m^3 (Hofmann et al. 2014). In another German study 35 residences in four German states were investigated in 2019/2020 and a median (95th percentile) of 0.20 μg/m^3 (0.53 μg/m^3) was found (Fromme et al. 2023). Noticeable in all studies is the skewness of the distribution. In a study performed in Sweden in 2015/16, air samples from 51 rooms associated with building-related health effects or symptoms and 34 control rooms were collected and means of 0.055 and 0.051 μg/m^3 were measured (Veenaas et al. 2020).

Comparable results are obtained from studies of living spaces in North America, where medians of 0.3 μg/m^3 to 1.1 μg/m^3 were found in the early 2000s (compilation in Jia and Batterman 2010). Batterman et al. (2012) described the distribution of vapor-phase NAP concentrations in 288 suburban and urban homes in four communities in southeast Michigan, USA in 2004–2010. Indoors, medians varied in this study from 0.7–1.0 μg/m^3 (95th percentiles: 6.6–37.6 μg/m^3) in the four monitoring sites. Concentrations were extremely skewed and 14% of homes exceeded 3 μg/m^3 (maximum: 201 μg/m^3). Important sources include naphthalene's use as a pest repellent and deodorant, migration from attached garages and, to lesser extents, cigarette smoke and vehicle emissions. Using passive air samplers with polyurethane foam subsequently lower mean levels of 9.6 ng/m^3 were reported in Fairbanks, Alaska in winter 2008/09 (Gouin et al. 2010). In a first population-based national indoor air survey of 3218 houses and 546 apartments in Canada in 2009–2011, a median of 0.72 μg/m^3 (95th percentile: 7.74 μg/m^3) were reported (Zhu et al. 2013). In a second major survey campaign, carried out over a 24-month period in 2012 and 2013, samples in 3524 Canadian residential homes were collected as part of the

Canadian Health Measures Survey (CHMS) (Li et al. 2019). In this study NAP was detected in 98% of the samples with a median of 0.63 μg/m^3 (90th percentile: 3.25 μg/m^3).

In Japan, 68 children's bedrooms were investigated in 2017 and concentrations from 0.032 to 15.0 μg/m^3 (median: 0.22 μg/m^3; 95th percentile: 4.30 μg/m^3) were reported (Yoshida et al. 2021). Oshima et al. (2022) investigated a total of 28 homes at four time points during the year in a 2017/18 nationwide survey in Japan and reported medians of 0.35 μg/m^3 in Summer, 0.34 μg/m^3 in fall and winter, and 0.27 μg/m^3 in spring. The median indoor to outdoor concentration ratio ranged from 1.2 (spring) to 1.9 (summer).

3.8.3 Occurence in the Air of Public Facilities

In a comprehensive compilation of measurement results performed by different private laboratories in Germany from 2006 to 2012 (Hofmann et al. 2014) NAP could still be detected in 33% of the samples collected in 493 schools with a median of 0.5 μg/m^3 and the 95th percentile was 7.20 μg/m^3. In another German study 27 schools in four German states were investigated in 2019/2020 and a median (95th percentile) of 0.30 μg/m^3 (2.8 μg/m^3) was found (Fromme et al. 2023). Data from 115 schools in 54 European cities from 23 countries were collected in the SINPHONIE study in 2011/12 with a median of 0.02 μg/m^3 (90th percentile: 8.11 μg/m^3) (Baloch et al. 2020). In Lithuania, mean NAP concentrations between 0.012 and 0.026 μg/m^3 (range: 0.0001–0.12 μg/m^3) were reported in samples collected in 5 primary schools in 2011/12 (Krugly et al. 2014). A median of 0.014 μg/m^3 (0.014–0.063 μg/m^3) was found in 152 gas-phase samples collected in two preschools in Portugal in 2015 (Oliveira et al. 2017).

St-Jean et al. (2012) investigated 21 daycare centers in Montreal, Canada in 2008 and found mean NAP concentrations of 0.90 μg/m^3 (range: 0.3–3.1 μg/m^3). In the USA, 25 school building samples were collected in 2013–2015 (Rago et al. 2021). NAP were detected in 23% of the samples with a median of 0.26 μg/m^3 (range: <LOQ-5.98 μg/m^3; 90th percentile: 0.60 μg/m^3). In 144 classrooms in 37 elementary schools in four states in the USA, visited from 2015 to 2016, a median of 0.1 μg/m^3 (maximum: 2.9 μg/m^3) was observed in 40% of the samples (Zhong et al. 2017). Again, a median of 0.05 μg/m^3 (maximum: 0.29 μg/m^3) was reported by Bayati et al. (2021) for three classrooms in Missouri in 2016.

In Singapore, 104 daycare centers were examined and a median of 1.33 μg/m^3 (90th percentile: 5.45 μg/m^3) was found in the gas phase (Zuraimi and Tham 2008).

Table 3.10 Selection of very volatile organic compounds (VVOC) ordered by the number of carbon atoms in the molecule (after Salthammer 2016; Schieweck et al. 2018)

Substance	CAS No.	Apartments (Hofmann et al. 2014)		Schools (Hofmann et al. 2014)		Apartments (Sysoltseva et al. 2023)		Schools (Sysoltseva et al. 2023)		Apartments (Weisel et al. 2008)	
		N (% > LOQ)	Median/ 95th P.	*N* (% > LOQ)	Median/ 95th P.	*N* (*N* > LOQ)	Median/ 95th P.	*N* (*N* > LOQ)	Median/ 95th P.	*N* (% > LOQ)	Median/ 95th P.
C1											
Formaldehyde	50–00-0	979 (95)	38/125	260 (99)	45/170	35 (34)	17.6/ 44.0	22	16.8/ 35.0		
Formic acid	64–18-6					35 (34)	15.0/ 20.2	23 (18)	15.8/ 23.7		
Chloromethane	74–87-3									100 (81)	1.4/ 2.2.1
Dichloromethane	75–09-2					35 (1)	<1.0/ 7.5*	23 (2)	<1.0/ 2.5*		
Trichloromethane	67–66-3	323 (7)	<1.0/ 1.0	109 (4)	<1.0/ <1.0						
Carbon disulfide	75–15-0					35 (3)	<3.0/ 1.0*	23 (1)	<3.0/ 3.0*	100 (3)	<1.6/ 4.4*
C2											
Ethanol	64–17-5					35 (35)	431/ 1257	23 (23)	170/690		
Acetaldehyde	75–07-0	249 (96)	22/68	113 (93)	15/76	35 (33)	14.7/ 72.5	23 (18)	9.9/22.5		
1,1-Dichloroethene	75–35-4										
Trichloroethene	79–01-6					35 (0)	<1.0	23 (0)	<1.0	100 (8)	<2.7/ 2.74

(continued)

Table 3.10 (continued)

Substance	CAS No.	Apartments (Hofmann et al. 2014)		Schools (Hofmann et al. 2014)		Apartments (Sysoltseva et al. 2023)		Schools (Sysoltseva et al. 2023)		Apartments (Weisel et al. 2008)	
		N (% > LOQ)	Median/ 95th P.	*N* (% > LOQ)	Median/ 95th P.	*N* (*N* > LOQ)	Median/ 95th P.	*N* (*N* > LOQ)	Median/ 95th P.	*N* (% > LOQ)	Median/ 95th P.
Acetic acid	64–19-7	348 (91)	34/190	402 (83)	26/177	35 (34)	71.8/ 148	23 (18)	85.3/ 140		
Vinyl chloride	75–01-4					35 (0)	<7.0	(0)	<7.0		
C3											
1-Propanol	71–23-8	232 (35)	<1.0/23	218 (49)	1.0/2.1	35 (7)	<6.0/ 49.5*	23 (11)	<6.0/ 302		
2-Propanol	67–63-0	316 (72)	19/124	238 (87)	8.0/85	35 (30)	38.8/ 235	23 (19)	28.5/ 156		
Propanal	123–38-6	237 (83)	4.0/12	113 (84)	5.0/37	35 (14)	<1.0/ 9.1	23 (7)	1.8/3.5		
Acrolein	107–02-8	208 (0.5)	<1.0/ 2.5	112 (0)		35 (25)	5.0/7.5	23 (22)	3.0/7.0		
Acetone	67–64-1	150 (81)	45/312	48 (89)	48/150	35 (35)	27.8/ 214	23	35.1/ 78.6	100 (94)	34.5 (190)
Propylene bichloride	78–87-5					35 (0)	<2.0	23 (0)	<2.0		
Dimethyl carbonate	616–38-6					35 (0)	<4.0	23 (0)	<4.0		
2-Chloropropylene	577–98-2					35 (0)	<4.0	23 (0)	<4.0		
Methyl acetate	79–20-9	454 (74)	1.0/11	327 (68)	1.0/11	35 (22)	3.0/10.5	23 (13)	2.0/5.8		
2-Chloropropane	75–29-6					35 (0)	<4.0	23 (0)	<4.0		

C4											
1,3-Butadiene	106–99-0					35 (0)	<3.0	23 (0)	<3.0	100 (7)	<1.1/1.3
n-Butanal	123–72-8	680 (71)	3.0/17	354 (73)	2.0/17	35 (5)	<1.0/25.0	23 (4)	<1.0/4.0*		
2-Methylpropanal	78–84-2										
Methacrolein	78–85-3					35 (18)	2.0/3.7	23 (3)	<1.0/1.0*		
Crotonaldehyde	4170-30-3					35 (0)	<6.0	23 (0)	<6.0		
2-Butanone	78–93-3							23 (1)		100 (84)	3.5/21.4
Methyl vinyl ketone	78–94-4					35 (1)	<1.0/2.0*	23 (0)	<1.0		
Ethyl acetate	141–78-6					35 (32)	9.3/77.9	23 (14)	3.0/11.7		
Furan	110–00-9					35 (0)	<2.0	23 (1)	<2.0/41*		
Vinyl acetate	108–05-4	269 (12)	1.0/2.0	96 (4.2)	<1.0/1.0	35 (0)	<8.0	23 (0)	<8.0		
C5											
n-Pentane	109–66-0					35 (28)	4.5/43.5	23 (9)	<1.0/5.8		
1-Pentene	109–67-1					35 (0)	<2.0	23 (0)	<2.0		
2-Pentano	6032-29-7					35 (0)	<8.0	23 (0)	<8.0		
Isopentane	78–78-4					35 (33)	14.5/205	23 (12)	5.0/9.5		
Isoprene	78–79-5					35 (35)	13.5/35.1	23 (22)	8.8/25.5		

(continued)

Table 3.10 (continued)

Substance	CAS No.	Apartments (Hofmann et al. 2014)		Schools (Hofmann et al. 2014)		Apartments (Sysoltseva et al. 2023)		Schools (Sysoltseva et al. 2023)		Apartments (Weisel et al. 2008)	
		N (% > LOQ)	Median/ 95th P.	*N* (% > LOQ)	Median/ 95th P.	*N* (*N* > LOQ)	Median/ 95th P.	*N* (*N* > LOQ)	Median/ 95th P.	*N* (% > LOQ)	Median/ 95th P.
Methyl propyl ketone	107–87-9	101 (65)	<1.0/ 2.4	–	–	35 (0)	<3.0	23 (0)	<3.0		
Cyclopentane	287–92-3	181 (34)	<1.0/ 9.0	211 (43)	<1.0/ 5.5	35 (3)	<2.0/ 5.0*	23 (1)	<2.0/ 2.0*		
2-Methylpentane	96–14-0	401 (73)	1.0/14	103 (81)	1.0/14	35 (13)	<1.0/ 49.5*	23 (6)	<1.0/ 3.0*		
3-Methylpentane	96–14-0	403 (64)	1.0/6.7	102 (71)	1.0/6.0	35 (7)	<2.0/ 40.5	23 (1)	<2.0/ 2.0		
Furfural	98–01-1	314 (67)	2.0/7.0	248 (42)	<1.0/ 3.5						
C6											
Methylcylopentane	96–37-7	706 (37)	<1.0/ 4.0	498 (30)	<1.0/ 3.0	35 (4)	<2.0/ 59.5*	23 (0)	<2.0		

LOQ limit of quantification; * Maximum

3.9 Very Volatile Organic Compounds (VVOC)

The so-called very volatile organic compounds (VVOC) are of increasing health significance. First, the World Health Organization published a definition of VVOC based on the boiling point, although not clearly delimited (see Table 3.1) (WHO 1989). In this context, Salthammer et al. (2016) proposes, in distinction to the definition of VOC in ISO 16000-6, to consider all substances as VVOC that have a boiling point below 69 °C or have less than 6 carbon atoms in the molecule. In addition to this basic problem of differentiating VVOC from VOC, the analytical recording of this heterogeneous group represents a considerable difficulty. In contrast to VOC, there is no standardized, validated measurement method and sampling technique. A problem description and first development steps toward a valid sampling and measurement method have been published by Schieweck et al. (2018). Again, Even et al. (2021) highlight the points that still need to be explored toward the standardization of a suitable procedure: investigations on sorbent combinations, the suitability of chromatography columns, and the development and use of gaseous standards.

In a review, Salthammer et al. (2016) classified 79 substances between C_1 and C_6 into the group of VVOC. Based on this work, those VVOC that were determined more frequently in indoor air are compiled in Table 3.10. Some of these substances can already be detected by the classical gas chromatographic VOC method or the liquid chromatographic method for the determination of aldehydes. For aldehydes, and especially for formaldehyde, many measurement results are available so far. This class of substances is therefore discussed in a separate chapter (see Sect. 3.2). In Table 3.10, results on the detected VVOC are summarized for some studies. For example Hofmann et al. (2014) compiled on the basis of measurement results from private research laboratories in Germany from 2006 to 2012. In addition, Weisel et al. (2008) reviewed VVOC in 100 homes in the USA in 2003–2006, among other studies, and compared them to larger North American studies whose results were published before 2003. In another review, results of a total of 267 organic substances and metals from 77 studies were evaluated and compared with health values (Logue et al. 2011). Within the German LUPE project, air samples were collected in 23 classrooms and 35 apartments and 31 VVOC were quantified in 2019/20 (Sysoltseva et al. 2023). For the sum of VVOC between 69 and 2714 μg/m^3 in the apartments and from 16 to 2056 μg/m^3 in classrooms were found. Ethanol was the most abundant substance in this study with median concentrations of 431 μg/m^3 (apartments) and 170 μg/m^3 (classrooms). Moreover, isopropanol, 2-propenal, methyl acetate, ethyl acetate, and isoprene were found in nearly all of the indoor air samples of both sites.

In addition, information are available for some single substances from other studies. In 168 German classrooms studied in 2004–2005, the median for acetone was 30 μg/m^3 and the 95th percentile was 136 μg/m^3 (Fromme et al. 2008).

In 395 measurements in 113 schools in 2004–2009 collected in the German exposure database, the 95th percentile for butanone was 21 μg/m^3 (Neumann et al.

2012). In a study from the USA, the indoor air of 126 houses was examined and 1,2-DCE above the method detection limit of 0.39 μg/m^3 was determined in 13% of the cases. The 95th percentile in this study was 1 μg/m^3 and the maximum was 8 μg/m^3 (Chin et al. 2014). In contrast, in another study of 100 homes in New Jersey, 1,2-DCE was detected in only one sample (3.5 μg/m^3) (Weisel et al. 2008).

During the processing of phenolic resin insulation boards ("Resol" rigid foam boards) in the window and door reveals in daycare centers, 2-chloropropane was used for foaming the insulation. As a result, up to 1700 μg/m^3 could be detected in the indoor air (Nix 2012). In this case, the insulation panels had also been mistakenly installed in the interior. If, on the other hand, insulation was only present in the outdoor area, the concentrations in 4 other objects only ranged between <1 and 51 μg/m^3.

References

Ad-hoc-AG (2009) Indoor air guideline values for saturated acyclic aliphatic C4–C11 aldehydes. Bundesgesundheitsbl – Gesundheitsforsch – Gesundheitsschutz 52:650–659. [in German]

Aerts O, Goossens A, Lambert J, Lepoittevin JP (2017) Contact allergy caused by isothiazolinone derivatives: an overview of non-cosmetic and unusual cosmetic sources. Eur J Dermatol 27: 115–122

AIR (2016a) Indoor air guide value for formaldehyde. Communication from the Committee on Indoor Guide Values. Bundesgesundheitsbl 59:1040–1044. [in German]

AIR (2016b) On the question of an asthma-inducing or worsening potential of formaldehyde in indoor air in children. Bundesgesundheitsbl 59:1028–1039. [in German]

Allegra A, Spatari G, Mattioli S, Curti S, Innao V, Ettari R, Allegra AG, Giorgianni C, Gangemi S, Musolino C (2019) Formaldehyde exposure and acute myeloid leukemia: a review of the literature. Medicina (Kaunas) 55:638

Alvarez-Vaca D, Duca RC, Borras-Santos A, Hardy E, Creta M, Eicher C, Wurth L, Vergison A, Van Nieuwenhuyse A (2022) Surveillance of indoor air concentration of vlatile organic compounds in Luxembourgish households. Int J Environ Res Public Health 19:5467

Andersson K, Bakke JV, Björseth O, Bornehag CG, Clausen G, Hongslo JK, Kjellman M, Kjergaard S, Molhave L, Skerfving S, Sundell J (1997) TVOC and health in non-industrial indoor environments. Indoor Air 7:78–91

ATSDR (1999) Toxicological profile for formaldehyde. shttps://www.atsdr.cdc.gov/toxprofiles/tp111.pdf

ATSDR (2005) Toxicological profile for naphthalene, 1-methylnaphthalene, and 2-methylnaphthalene. https://www.atsdr.cdc.gov/toxprofiles/tp67.pdf

Baloch RM, Maesano CN, Christoffersen J, Banerjee S, Gabriel M, Csobod É, de Oliveira FE, Annesi-Maesano I (2020) Indoor air pollution, physical and comfort parameters related to schoolchildren's health: data from the European SINPHONIE study. Sci Total Environ 739: 139870

Baran R (2002) Nail beauty therapy: an attractive enhancement or a potential hazard? Cosmet Dermatol 1:24–29

Batterman S, Chin JY, Jia C, Godwin C, Parker E, Robins T, Max P, Lewis T (2012) Sources, concentrations, and risks of naphthalene in indoor and outdoor air. Indoor Air 22:266–278

Bayati M, Vu DC, Vo PH, Rogers E, Park J, Ho TL, Davis AN, Gulseven Z, Carlo G, Palermo F, McElroy JA, Nagel SC, Lin CH (2021) Health risk assessment of volatile organic compounds at daycare facilities. Indoor Air 31:977–988

Bello A, Quinn MM, Milton DK, Perry MJ (2013) Determinants of exposure to 2-butoxyethanol from cleaning tasks: a quasi-experimental study. Ann Occup Hyg 57:125–135

Bernardini L, Barbosa E, Charão MF, Brucker N (2020) Formaldehyde toxicity reports from in vitro and in vivo studies: a review and updated data. Drug Chem Toxicol 20:1–13

Bethke R et al (2021) Indoor volatile organic compounds in the German Environmental Survey for Children and Adolescents (GerES, 2014–2017). Manuscript in preparation

Binder M, Obenland H, Maraun W (2001) Indoor air and house dust exposure to isothiazolones from wall paints. In: Environment, Building & Health, Ed. Arbeitsgemeinschaft ökologischer Forschunsinstitute (AGÖF), Springe-Eldagsen. [in German]

Birmili W, Daniels A, Bethke R, Schechner N, Brasse G, Conrad A, Kolossa-Gehring M, Debiak M, Hurraß J, Uhde E, Omelan A, Salthammer T (2022) Formaldehyde, aliphatic aldehydes (C2-C11), furfural, and benzaldehyde in the residential indoor air of children and adolescents during the German Environmental Survey 2014-2017 (GerES V). Indoor Air 32:e12927

Bohn S, Niederer M, Brehm K, Bircher AJ (2000) Airborne contact dermatitis from methylchloroisothiazolinone in wall paint. Abolition of symptoms by chemical allergen inactivation. Contact Dermatitis 42:196–201

Buck RC, Korzeniowski SH, Laganis E, Adamsky F (2021) Identification and classification of commercially relevant per- and poly-fluoroalkyl substances (PFAS). Integr Environ Assess Manag 17:1045–1055

Cammalleri V, Pocino RN, Marotta D, Protano C, Sinibaldi F, Simonazzi S, Petyx M, Iavicoli S, Vitali M (2022) Occupational scenarios and exposure assessment to formaldehyde: a systematic review. Indoor Air 32: e12949

Caron-Beaudoin É, Whyte KP, Bouchard MF, Chevrier J, Haddad S, Copes R, Frohlich KL, Dokkie D, Juul S, Bouchard M, Verner MA (2022) Volatile organic compounds (VOCs) in indoor air and tap water samples in residences of pregnant women living in an area of unconventional natural gas operations: findings from the EXPERIVA study. Sci Total Environ 805:150242

Catalani S, Donato F, Madeo E, Apostoli P, De Palma G, Pira E, Mundt KA, Boffetta P (2019) Occupational exposure to formaldehyde and risk of non hodgkin lymphoma: a meta-analysis. BMC Cancer 19:1245

Chang JCS, Sparks LE, Guo Z (1998a) Evaluation of sink effects on VOCs from a latex paint. J Air Waste Manage Assoc 48:953–958

Chang JCS, Guo Z, Sparks LE (1998b) Exposure and emission evaluations of methyl ethyl ketoxime (MEKO) in alkyd paints. Indoor Air 8:295–300

ChemVerbotsV (2017) Ordinance on prohibitions and restrictions on the marketing and placing on the market of certain substances, mixtures and products under the Chemicals Act. 20th January 2017. BGBl. I: 94. [in German]

Chen X, Li F, Liu C, Yang J, Zhang J, Peng C (2017) Monitoring, human health risk assessment and optimized management for typical pollutants in indoor air from random families of university staff, Wuhan City, China. Sustainability 9:1115

Chin J-Y, Godwin C, Parker E, Robins T, Lewis T, Harbin P, Batterman S (2014) Levels and sources of volatile organic compounds in homes of children with asthma. Indoor Air 24:403–415

Choi H, Schmidbauer N, Spengler J, Bornehag CG (2010) Sources of propylene glycol and glycol ethers in air at home. Int J Environ Res Public Health 7:4213–4237

Clausen PA, Wolkoff P, Holst E, Nielsen PA (1991) Long-term emission of volatile organic compounds from waterborne paints – methods of comparison. Indoor Air 4:562–576

Clausen PA, Frederiksen M, Sejbæk CS, Sørli JB, Hougaard KS, Frydendall KB, Carøe TK, Flachs EM, Meyer HW, Schlünssen V, Wolkoff P (2020) Chemicals inhaled from spray cleaning and disinfection products and their respiratory effects. A comprehensive review. Int J Hyg Environ Health 229:113592

Cometto-Muñiz JE, Abraham MH (2015) Compilation and analysis of types and concentrations of airborne chemicals measured in various indoor and outdoor human environments. Chemosphere 127:70–86

Csobod E, Annesi-Maesano I, Carrer P, Kephalopoulos S, Madureira J, Rudnai P, de Oliveira Fernandes E (2014) SINPHONIE. Schools indoor pollution and Health. Final Report, EUR 26738. https://publications.jrc.ec.europa.eu/repository/handle/JRC91160

D'eon JC, Mabury SA (2011a) Exploring indirect sources of human exposure to perfluoroalkyl carboxylates (PFCAs): evaluating uptake, elimination, and biotransformation of polyfluoroalkyl phosphate esters (PAPs) in the rat. Environ Health Perspect 119:344–350

D'eon JC, Mabury SA (2011b) Is indirect exposure a significant contributor to the burden of perfluorinated acids observed in humans? Environ Sci Technol 45:7974–7984

De Silva AO, Allard CN, Spencer C, Webster GM, Shoeib M (2012) Phosphorus-containing fluorinated organics: polyfluoroalkyl phosphoric acid diesters (diPAPs), perfluorophosphonates (PFPAs), and perfluorophosphinates (PFPIAs) in residential indoor dust. Environ Sci Technol 46:12575–12582

DFG (Senate Commission for the Investigation of Health Hazards of Chemical Compounds in the Work Area) (2006) Methylmethacrylate. In: MAK Value Documentation in German language, methylmethacrylate addendum 2006. https://onlinelibrary.wiley.com/doi/epdf/10.1002/3527600418.mb8062d0041 [in German]

Dörr R, Reinecke T, Kersting K, Rühl R (2017) Methylethyl ketone oxime (MEKO) in lacquers, parquet sealants and silicone jointing compounds. Gefahrst Reinh Luft 77:63–68. [in German]

Dugheri S, Massi D, Mucci N, Marrubini G, Cappelli G, Speltini A, Bonferoni MC, Arcangeli G (2021) Exposure to airborne formaldehyde: sampling and analytical methods—a review. Trends Environ Anal Chem 29:00116

ECHA (European Chemicals Agency) (2012) Opinion of the Committee for Risk Assessment on the proposal fort he harmonised classification and labelling at EU level. Formaldehyde. https://echa.europa.eu/documents/10162/b8dfa022-9544-72e8-dcaa-7491dff3c0d5

ECHA (2018a) Opinion proposing harmonised classification and labelling at EU level of 2-methyl-1,2-benzothiazol-3(2H)-one (MBIT). Committee for Risk Assessment. https://echa.europa.eu/documents/10162/708c2c4e-7b66-51f5-14c2–1090e8203227

ECHA (2018b) Opinion proposing harmonised classification and labelling at EU level of octhilinone (ISO); 2-octyl-2H-isothiazol-3-one (OIT). Committee for Risk Assessment. https://echa.europa.eu/documents/10162/4f2ebbc7-d814-f967-12bb-223c17b9aeec

ECHA (2018c) Substance evaluation conclusion and evaluation report for methyl methacrylate. Report prepared by ANSES, EC No 201-297-1. https://echa.europa.eu/documents/10162/c819add5-3d93-4482-3b70-72cb30debea7

ECHA (2018d) Note on reference dose-response relationship for the carcinogenicity of pitch, coal tar, high temperature and on PBT and vPvB properties. https://echa.europa.eu/documents/10162/13637/ctpht_rac_note_en.pdf/a184ee42-0642-7454-2d18-63324688e13d

ECHA (2018e) Committee for Risk Assessment RAC. Annex 1 Background document to the Opinion proposing harmonised classification and labelling at EU level of butanone oxime; ethyl methyl ketoxime; ethyl methyl ketone oxime. https://echa.europa.eu/documents/10162/fa6a43c5-6b2a-4c20-023b-801846dd19e3

ECHA (2019) Worker exposure to formaldehyde and formaldehyde releasers. https://echa.europa.eu/documents/10162/13641/investigationreport_formaldehyde_workers-exposure_final_en.pdf/ac457a0c-378d-4eae-c602-c7cd59abc4c5

ECHA (2020) CLH report. Proposal for harmonised classification and labelling. Based on Regulation (EC) No 1272/2008 (CLP Regulation), Annex VI, Part 2. Acetone oxime. Helsinki, Finland. https://echa.europa.eu/documents/10162/2d0014c7-15e5-675f-bbe2-d869aaf00c1b

EFSA (2014) External report: Bulla S, Burnett K, Vassaux K, Ashdown L, Brown T, Rushtond L. Extensive literature search and provision of summaries of studies related to the oral toxicity of perfluoroalkylated substances (PFASs), their precursors and potential replacements in experimental animals and humans. EFSA supporting publication 2014: EN-572. https://efsa.onlinelibrary.wiley.com/doi/abs/10.2903/sp.efsa.2014.EN-572

Eis D, Helm D, Laußmann D, Mühlinghaus T, Dietel A, Jordan L, Birkner N, Thierfelder W, Traenckner-Probst I, Worm M (2005) Berlin study on environment-related diseases. https://www.apug.de/archiv/pdf/Berichtsband_Berliner-Studie.pdf. [in German]

Ellis DA, Martin JW, De Silva AO, Mabury SA, Hurley MD, Sulbaek Andersen MP, Wallington TJ (2004) Degradation of fluorotelomer alcohols: a likely atmospheric source of perfluorinated carboxylic acids. Environ Sci Technol 38:3316–3321

Ericson Jogsten I, Nadal M, van Bavel B, Lindström G, Domingo JL (2012) Per- and polyfluorinated compounds (PFCs) in house dust and indoor air in Catalonia, Spain: implications for human exposure. Environ Int 39:172–180

Eriksson U, Kärrman A (2015) World-wide indoor exposure to polyfluoroalkyl phosphate esters (Paps) and other PFASs in household dust. Environ Sci Technol 49:14503–14511

Even M, Juritsch E, Richter M (2021) Measurement of very volatile organic compounds (VVOCs) in indoor air by sorbent-based active sampling: identifying the gaps towards standardisation. Trends in Anal Chem 140:116265

Fang L, Liu N, Liu W, Mo J, Zhao Z, Kan H, Deng F, Huang C, Zhao B, Zeng X, Sun Y, Qian H, Sun C, Guo J, Zheng X, Zhang Y (2022) Indoor formaldehyde levels in residences, schools, and offices in China in the past 30 years: a systematic review. Indoor Air 32:e13141

Fraser AJ, Webster TF, Watkins DJ, Nelson JW, Stapleton HM, Calafat AM, Kato K, Shoeib M, Vieira VM, McClean MD (2012) Polyfluorinated compounds in serum linked to indoor air in office environments. Environ Sci Technol 46:1209–1215

Fromme H, Heitmann D, Dietrich S, Schierl R, Körner W, Kiranoglu M, Zapf A, Twardella D (2008) Air quality in schools – classroom levels of carbon dioxide (CO_2), volatile organic compounds (VOC), aldehydes, endotoxins and cat allergen. Gesundheitswesen 70:88–97. [in German]

Fromme H, Nitschke L, Boehmer S, Kiranoglu M, Göen T (2013) Exposure of German residents to ethylene and propylene glycol ethers in general and after cleaning scenarios. Chemosphere 90: 2714–2721

Fromme H, Dreyer A, Dietrich S, Fembacher L, Lahrz T, Völkel W (2015) Neutral polyfluorinated compounds in indoor air in Germany – The LUPE 4 study. Chemosphere 139:572–578

Fromme H, Lahrz T, Burkhardt R, Fembacher L, Kraft M, Sievering S, Kadler D, Dietrich S (2016) Air quality in daycare centers – exposure to carbon dioxide (CO_2), volatile organic compounds (VOCs), and fine dusts – LUPE 3. Gefahrst Reinh Luft 76:55–61. [in German]

Fromme H, Achten C, Gerull F, Gessner A, Papavlassopoulos H, Kraft M, Röhl C, Burghardt R, Sysoltseva M (2023) Polycyclic aromatic hydrocarbons including dibenzopyrenes in indoor air samples from schools and residences in Germany. Submitted.

Garcia-Hidalgo E, Sottas V, von Goetz N, Hauri U, Bogdal C, Hungerbühler K (2017) Occurrence and concentrations of isothiazolinones in detergents and cosmetics in Switzerland. Contact Dermatitis 76:96–106

Gehrig R, Hill M, Zellweger C, Hofer P (1993) VOC-emissions from wall paints – a test chamber study. Proc Indoor Air 2:431–436

Geiss O, Giannopoulos G, Tirendi S, Barrero-Moreno J, Larsen BR, Kotzias D (2011) The AIRMEX study – VOC measurements in public buildings and schools/kindergartens in eleven European cities: statistical analysis of the data. Atmos Environ 45:3676–3684

Gerster FM, Vernez D, Wild PP, Hopf NB (2014) Hazardous substances in frequently used professional cleaning products. Int J Occup Environ Health 20:46–60

Gibson WB, Keller PR, Foltz DJ, Harvey GJ (1991) Diethylene glycol mono butyl ether concentration in room air from application of cleaner formulations to hard surfaces. J Expo Anal Environ Epidemiol 1:369–383

Giovanoulis G, Nguyen MA, Arwidsson M, Langer S, Vestergren R, Lagerqvist A (2019) Reduction of hazardous chemicals in Swedish preschool dust through article substitution actions. Environ Int 130:104921

Gjölstad M, Thorud S, Molander P (2006) Occupational exposure to airborne solvents during nail sculpturing. J Environ Monit 8:537–542
Gminski R, Marutzky R, Kevekordes S, Fuhrmann F, Bürger W, Hauschke D, Ebner W, Mersch-Sundermann V (2011a) Chemosensory irritations and pulmonary effects of acute exposure to emissions from oriented strand boards. Hum Exp Toxicol 30:1204–1221
Gminski R, Marutzky R, Kevekordes S, Fuhrmann F, Bürger W, Hauschke D, Ebner W, Mersch-Sundermann V (2011b) Sensory irritations and pulmonary effects in human volunteers following short-term exposure to pinewood emissions. J Wood Sci 57:436–445
Goosey E, Harrad S (2012) Perfluoroalkyl substances in UK indoor and outdoor air: spatial and seasonal variation, and implications for human exposure. Environ Int 45:86–90
Gouin T, Wilkinson D, Hummel S, Meyer B, Culley A (2010) Polycyclic aromatic hydrocarbons in air and snow from Fairbanks, Alaska. Atmos Pollut Res 1:9–15
Hagberg S, Ljungkvist G, Andreasson H, Karlsson S, Barregård L (2005) Exposure to volatile methacrylates in dental personnel. J Occup Environ Hyg 2:302–306
Halios CH, Landeg-Cox C, Lowther SD, Middleton A, Marczylo T, Dimitroulopoulou S (2022) Chemicals in European residences – Part I: A review of emissions, concentrations and health effects of volatile organic compounds (VOCs). Sci Total Environ 839:156201
Hall SM, Patton S, Petreas M, Zhang S, Phillips AL, Hoffman K, Stapleton HM (2020) Per- and polyfluoroalkyl substances in dust collected from residential homes and fire stations in North America. Environ Sci Technol 54:14558–14567
Han I, Seo JY, Barr DB, Panuwet P, Yakimavets V, D'Souza PE, An-Han H, Afshar M, Chao Y-Y (2022) Evaluating indoor air phthalates and volatile organic compounds in nail salons in the Greater New York City Area: a pilot study. Int J Environ Res Public Health 19:12411
Hartwig A, Arand M, Epe B, Guth S, Jahnke G, Lampen A, Martus H-J, Monien B, Rietjens ICM, Schmitz-Spanke S, Schriever-Schwemmer G, Steinberg P, Eisenbrand G (2020) Mode of action-based risk assessment of genotoxic carcinogens. Arch Toxicol 94:1787–1877
Haug LS, Huber S, Schlabach M, Becher G, Thomsen C (2011) Investigation on per- and polyfluorinated compounds in paired samples of house dust and indoor air from Norwegian homes. Environ Sci Technol 45:7991–7998
HC (2010) Screening assessment for the challenge 2-butanone, oxime (butanone oxime). Ottawa, Canada. https://www.ec.gc.ca/ese-ees/32AD1FD8-68E2–4782-AFB3-A198A23AF330/batch7_96–29-7_en.pdf
HC (2014) Code of practice for 2-butanone, oxime (butanone oxime) associated with the interior application of consumer alkyd paint and coating products. Ottawa, Canada. https://www.canada.ca/content/dam/hc-sc/migration/hc-sc/ewh-semt/alt_formats/pdf/pubs/contaminants/butanone-oxime/butanone-oxime-eng.pdf
HC (2018) Screening assessment ethylene glycol ethers group. https://www.canada.ca/content/dam/eccc/documents/pdf/pded/eges/201808311-Ethylene-Glycol-Ethers-FSAR-EN.pdf
Heeley-Hill AC, Grange SK, Ward MW, Lewis AC, Owen N, Jordan C, Hodgson G, Adamson G (2021) Frequency of use of household products containing VOCs and indoor atmospheric concentrations in homes. Environ Sci Process Impacts 23:699–713
Heinrich J (2011) Influence of indoor factors in dwellings on the development of childhood asthma. Int J Hyg Environ Health 214:1–25
Henriks-Eckerman M-L, Korva M (2012) Exposure to airborne methacrylates in nail salons. J Occup Environ Hyg 9:D146–D150
Herman A, Aerts O, de Montjoye L, Tromme I, Goossens A, Baeck M (2019) Isothiazolinone derivatives and allergic contact dermatitis: a review and update. J Eur Acad Dermatol Venereol 33:267–276
Heymann WR (2007) Nail cosmetics: potential hazards. J Am Acad Dermatol 57:1069–1070
Hippelein M (2003) Collection and discussion of indoor TVOC concentration reference data. Umweltmed Forsch Praxis 8:87–98. [in German]
Hippelein M (2004) Background concentrations of individual and total volatile organic compounds in residential indoor air of Schleswig-Holstein, Germany. J Environ Monit 6:745–752

Hoang T, Castorina R, Gaspar F, Maddalena R, Jenkins PL, Zhang Q, McKone TE, Benfenati E, Shi AY, Bradman A (2017) VOC exposures in California early childhood education environments. Indoor Air 27:609–621

Hofmann H, Erdmann G, Müller A (2014) Conflicting goals of energy-efficient construction and good indoor air quality – data collection for volatile organic compounds in indoor air of residential and office buildings (solutions). Annex to the final report. [in German] http://www.agoef.de/fileadmin/user_upload/dokumente/forschung/AGOEF-Abschlussericht_VOCDB_II_Anhang-nicht-barrierefrei.pdf

Huber S, Haug LS, Schlabach M (2011) Per- and polyfluorinated compounds in house dust and indoor air from northern Norway – a pilot study. Chemosphere 84:1686–1693

IARC (2002) Some traditional herbal medicines, some mycotoxins, naphthalene and styrene. IARC Monographs on the evaluation of carcinogenic risk of chemicals to humans, vol. 82. Lyon, France

IARC (2006) Formaldehyde, 2-butoxyethanol and 1-tert-butoxypropan-2-ol. IARC Monographs on the evaluation of carcinogenic risks to humans, Vol. 88. Lyon, France. https://publications.iarc.fr/Book-And-Report-Series/Iarc-Monographs-On-The-Identification-Of-Carcinogenic-Hazards-To-Humans/Formaldehyde-2-Butoxyethanol-And-1--Em-Tert-Em--Butoxypropan-2-ol-2006

IARC (2010) Some non-heterocyclic polycyclic aromatic hydrocarbons and some related exposures. IARC Monogr Eval Carcinog Risks Hum 92:1–853. http://monographs.iarc.fr/ENG/Monographs/vol92/mono92-6.pdf

Jahnke A, Ahrens L, Ebinghaus R, Temme C (2007) Urban versus remote air concentrations of fluorotelomer alcohols and other perfluorinated alkyl substances in Germany. Environ Sci Technol 41:745–752

Jensen LK, Larsen A, Molhave L, Hansen MK, Knudsen B (2001) Health evaluation of volatile organic compounds (VOC) emissions from wood and wood-based materials. Arch Environ Health 56:419–432

Jia C, Batterman S (2010) A critical review of naphthalene sources and exposures relevant to indoor and outdoor air. Int J Res Public Health 7:2903–2939

Jung CR, Nishihama Y, Nakayama SF, Tamura K, Isobe T, Michikawa T, Iwai-Shimada M, Kobayashi Y, Sekiyama M, Taniguchi Y, Yamazaki S (2021) Indoor air quality of 5,000 households and its determinants. Part B: Volatile organic compounds and inorganic gaseous pollutants in the Japan Environment and Children's study. Environ Res 197:111135

Kabadi SV, Fisher J, Aungst J, Rice P (2018) Internal exposure-based pharmacokinetic evaluation of potential for biopersistence of 6:2 fluorotelomer alcohol (FTOH) and its metabolites. Food Chem Toxicol 112:375–382

Kanchongkittiphon W, Mendell MJ, Gaffin JM, Wang G, Phipatanakul W (2015) Indoor environmental exposures and exacerbation of asthma: an update to the 2000 review by the Institute of Medicine. Environ Health Perspect 123:6–20

Karásková P, Venier M, Melymuk L, Beanová J, Vojta Š, Prokes R, Diamond ML, Klánová J (2016) Perfluorinated alkyl substances (PFASs) in household dust in Central Europe and North America. Environ Int 94:315–324

Kjærgaard SK, Mølhave L, Pedersen OF (1991) Human reactions to a mixture of indoor air volatile organic compounds. Atmos Environ 25A:1417–1426

Klewicz K, Marć M, Zabiegała B (2020) Emission profile of butan-2-one oxime from commercially available neutral silicone sealant. Microchem J 156:104982

Knudsen HN, Kjaer UD, Nielsen PA, Wolkoff P (1999) Sensory and chemical characterization of VOC emissions from building products: impact of concentration and air velocity. Atmos Environ 33:1217–1230

Koontz M, Price P, Hamilton J, Daggett D, Sielken R, Bretzlaff R, Tyler T (2006) Modeling aggregate exposure to glycol ethers from use of commercial floor products. Int J Toxicol 25:95–107

Koren HS, Graham DE, Devlin RB (1992) Exposure of humans to a volatile organic mixture. III. Inflammatory response. Arch Environ Health 47:39–44

Kotzias D, Geis O, Tirendi S, Barrero-Moreno J, Reina V, Gotti A, Cimino-Reale G, Casati B, Marafante E, Sarigiannis D (2009) Exposure to multiple air contaminants in public buildings, schools and kindergartens. The European Air Monitoring and Exposure Assessment (AIRMEX) Study. Fresenius Environ Bull 18:670–681

Krause C, Chutsch M, Henke M, Kliem C, Leiske M, Schulz C, Schwarz E (1991) German Environmental Survey, Vol. IIIa: Indoor living space: trace element levels in house dust. WaBoLu-Hefte Nr. 2. Berlin, Germany. [in German]

Krugly E, Martuzevicius D, Sidaraviciute R, Ciuzas D, Prasauskas T, Kauneliene V, Stasiulaitiene I, Kliucininkas L (2014) Characterization of particulate and vapor phase polycyclic aromatic hydrocarbons in indoor and outdoor air of primary schools. Atmos Environ 82: 298–306

Kwak K, Paek D, Park JT (2020) Occupational exposure to formaldehyde and risk of lung cancer: a systematic review and meta-analysis. Am J Ind Med 63:312–327

La Torre G, Vitello T, Cocchiara RA, Della Rocca C (2023) Relationship between formaldehyde exposure, respiratory irritant effects and cancers: a review of reviews. Public Health 218:186–196

Langer S, Bekö G (2013) Indoor air quality in the Swedish housing stock and its dependence on building characteristics. Build Environ 69:44–54

Langer V, Dreyer A, Ebinghaus R (2010) Polyfluorinated compounds in residential and nonresidential indoor air. Environ Sci Technol 44:8075–8081

Langer S, Bekö G, Bloom E, Widheden A, Ekberg L (2015) Indoor air quality in passive and conventional new houses in Sweden. Build Environ 93:92e100

Langer S, Ramalho O, Derbez M, Riberon J, Kirchner S, Mandin C (2016) Indoor environmental quality in French dwellings and building characteristics. Atmos Environ 128:82–91

LasD (State Office for Social Services) (2009) Indoor air investigations in public buildings in Schleswig-Holstein Part 1: Background values for schools and kindergartens (School and Kindergarten Study 2005/2007)

LasD (State Office for Social Services) (2011) Indoor air investigations in public buildings in Schleswig-Holstein. Part 2: Container buildings for schools and kindergartens. [in German]

Laskar AA, Younus H (2019) Aldehyde toxicity and metabolism: the role of aldehyde dehydrogenases in detoxification, drug resistance and carcinogenesis. Drug Metab Rev 51:42–64

Li Y, Cakmak S, Zhu J (2019) Profiles and monthly variations of selected volatile organic compounds in indoor air in Canadian homes: results of Canadian national indoor air survey 2012–2013. Environ Int 126:134–144

Liljelind IE, Eriksson KA, Nilsson LO, Jonsson IBM, Burstrom YI (2005) A method for measuring the potential dermal exposure to methyl methacrylate during two different dental technical work tasks. J Environ Monit 7:519–523

Liu W, Takahashi S, Sakuramachi Y, Harada KH, Koizumi A (2013) Polyfluorinated telomers in indoor air of Japanese houses. Chemosphere 90:1672–1677

Liu N, Bu Z, Liu W, Kan H, Zhao Z, Deng F, Huang C, Zhao B, Zeng X, Sun Y, Qian H, Mo J, Sun C, Guo J, Zheng X, Weschler LB, Zhang Y (2022) Indoor exposure levels and risk assessment of volatile organic compounds in residences, schools, and offices in China from 2000 to 2021: a systematic review. Indoor Air 32:e13091

Logue JM, McKone TE, Sherman MH, Singer BC (2011) Hazard assessment of chemical air contaminants measured in residences. Indoor Air 21:92–109

Lundov MD, Kolarik B, Bossi R, Gunnarsen L, Johansen JD (2014) Emission of isothiazolinones from water-based paints. Environ Sci Technol 48:6989–6994

Makey CM, Webster TF, Martin JW, Shoeib M, Harner T, Dix-Cooper L, Webster GM (2017) Airborne precursors predict maternal serum perfluoroalkyl acid concentrations. Environ Sci Technol 51:7667–7675

Mangelsdorf I, Fromme H (2021) VI–4 Organic substances – Glycol ether. In: Wichmann H-E, Fromme H (Hrgs) Handbook of Environmental Health. Ecomed-Verlag, Landsberg. [in German]

Mangelsdorf I, Kleppe SN, Heinzow B, Sagunski H (2016) Indoor air guide values for glycol ethers and glycol esters – a category approach. Int J Hyg Environ Health 219:419–436

Mannan M, Al-Ghamdi SG (2021) Indoor air quality in buildings: a comprehensive review on the factors influencing air pollution in residential and commercial structure. Int J Environ Res Public Health 18:3276

Marquardt W, Seiss M, Spahl W, Hickel R, Reichl FX (2008) Determination of volatile methacrylates in dental laboratories. Quintessence J Dental Technol 6:20–28

Marquardt W, Seiss M, Hickel R, Reichl FX (2009) Volatile methacrylates in dental practices. J Adhes Dent 11:101–107

Maung TZ, Bishop JE, Holt E, Turner AM, Pfrang C (2022) Indoor air pollution and the health of vulnerable groups: a systematic review focused on particulate matter (PM), volatile organic compounds (VOCs) and their effects on children and people with pre-Eeisting lung disease. Int J Environ Res Public Health 19:8752

Mølhave L (2003) Organic compounds as indicators of air pollution. Indoor Air 13(Suppl 6):12–19

Mølhave L, Bach B, Pedersen OF (1986) Human reactions to low concentrations of volatile organic compounds. Environ Int 12:167–175

Mølhave L, Clausen G, Berglund B, De Ceaurriz J, Kettrup A, Lindvall T, Maron M, Pickering AC, Risse U, Rothweiler H, Seifert B, Younes M (1997) Total volatile organic compounds (TVOC) in indoor air quality investigations. Indoor Air 7:225–240

Morales-McDevitt ME, Becanova J, Blum A, Bruton TA, Vojta S, Woodward M, Lohmann R (2021) The air that we breathe: Neutral and volatile PFAS in indoor air. Environ Sci Technol Lett 8:897–902

Nagorka R, Gleue C, Scheller C, Moriske H-J, Straff W (2015) Isothiazolone emissions from building products. Indoor Air 25:68–78

Neretnieks I, Christiansson J, Romero L, Dagerholt L, Yu J-W (1993) Modelling of emission and re-emission of volatile organic compounds from building materials with in-door air applications. Indoor Air 3:2–11

Neumann HD, Buxtrup M, von Hahn N, Koppisch D, Breuer D, Hahn J-U (2012) Proposal for the derivation of indoor workplace reference values in schools. Gefahrst Reinh Luft 72:291–297. [in German]

Nielsen GD, Larsen ST, Wolkoff P (2017) Reevaluation of the WHO (2010) formaldehyde indoor air quality guideline for cancer risk assessment. Arch Toxicol 91:35–61

Nitschke L, Dietrich S, Fromme H (2013) Pilot project on solvents in the air of nail studios. Report of the Bavarian Health and Food Safety Authority. [in German]

Nix N (2012) 2-Chloropropane in indoor air as a result of energy building renovation as a result of energy building renovation. 19th WaBoLu-Innenraumtage, Berlin. [in German] https://www.nuernberg.de/imperia/md/sun/dokumente/umweltanalytik/chlorpropan_praesentation_wabolu.pdf

Nurmatov UB, Tagiyeva N, Semple S, Devereux G, Sheikh A (2015) Volatile organic compounds and risk of asthma and allergy: a systematic review. Eur Respir Rev 24:92–101

O'Brien PJ, Diraki AG, Shangari N (2005) Aldehyde sources, metabolism, molecular toxicity mechanisms, and possible effects on human health. Crit Rev Toxicol 35:609–662

Oliveira M, Slezakova K, Madureira J, de Oliveira FE, Delerue-Matos C, Morais S, do Carmo Pereira M. (2017) Polycyclic aromatic hydrocarbons in primary school environments: levels and potential risks. Sci Total Environ 575:1156–1167

Oppl R (1999) Volatile organic compounds (VOCs) in indoor air of office buildings. Study by MILJÖ Chemie, Hamburg. http://www.rolandbraun.de/downloads/VOC-Studie%20MILJOE-CHEMIE%201999.pdf. [in German]

Oshima N; Tahara M, Sakai S, Karashi Y (2022) A nationwide survey on indoor air concentration of benzene and naphthalene in general residential housings. Indoor Environ 25:177–184

Ostendorp G, Riemer D, Harmel K, Heinzow B (2009) Current background values for VOC pollution in schools and kindergartens in Schleswig-Holstein. Umweltmed Forsch Prax 14: 135–152. [in German]

Paciência I, Madureira J, Rufo J, Moreira A, Fernandes EO (2016) A systematic review of evidence and implications of spatial and seasonal variations of volatile organic compounds (VOC) in indoor human environments. J Toxicol Environ Health B Crit Rev 19:47–64

Padilla-Sánchez JA, Papadopoulou E, Poothong S, Haug LS (2017) Investigation of the best approach for assessing human exposure to poly- and perfluoroalkyl substances through indoor air. Environ Sci Technol 51:12836–12843

Park SK, Seol H-S, Park H-J, Kim Y-S, Ryu S-H, Kim J, Kim S, Lee J-H, Kwon J-H (2020) Experimental determination of indoor air concentration of 5-chloro-2-methylisothiazol-3(2H)-one/2-methylisothiazol-3(2H)-one (CMIT/MIT) emitted by the use of humidifier disinfectant. Environ Anal Health Toxicol 35:e2020008

Pei J, Yin Y, Liu J, Dai X (2020) An eight-city study of volatile organic compounds in Chinese residences: compounds, concentrations, and characteristics. Sci Total Environ 698:134137

Pibiri E, Omelan A, Uhde E, Salthammer T (2020) Effect of surface covering on the release of formaldehyde, acetaldehyde, formic acid and acetic acid from particleboard. Build Environ 178: 106947

Quach T, Gunier R, Tran A, Von Behren J, Doan-Billings PA, Nguyen KD, Okahara L, Lui BY, Nguyen M, Huynh J, Reynolds P (2011) Characterizing workplace exposures in Vietnamese women working in California nail salons. Am J Public Health 101(Suppl 1):S271–S276

Quach T, Von Behren J, Nelson DO, Nguyen TN, Tsoh J, Le M, Fu L, Beckman S, Reynolds P (2018) Evaluating an owner-to-worker training intervention in California nail salons using personal air monitoring. Am J Ind Med 61:831–841

Rago R, Rezendes A, Peters J, Chatterton K, Kammari A (2021) Indoor air background levels of volatile organic compounds and air-phase petroleum hydrocarbons in office buildings and schools. Groundwat Monit Remed 41:27–47

Rand AA, Mabury SA (2017) Is there a human health risk associated with indirect exposure to perfluoroalkyl carboxylates (PFCAs)? Toxicology 375:28–36

Raw GJ, Coward SKD, Brown VM, Crump DC (2004) Exposure to air pollutants in English homes. J Expo Anal Environ Epidemiol 14:S85–S94

Reinhard E, Waeber R, Niederer M, Maurer T, Maly P, Scherer S (2001) Preservation of products with MCI/MI in Switzerland. Contact Dermatitis 45:257–264

Rice PA, Aungst J, Cooper J, Bandele O, Kabadi SV (2020) Comparative analysis of the toxicological databases for 6:2 fluorotelomer alcohol (6:2 FTOH) and perfluorohexanoic acid (PFHxA). Food Chem Toxicol 138:111210

Roda C, Barral S, Ravelomanantsoa H, Dusseaux M, Tribout M, LeMoullec Y, Momasa I (2011) Assessment of indoor environment in Paris child daycare centers. Environ Res 111:1010–1017

Rösch C, Kohajda T, Röder S, von Bergen M, Schlink U (2014) Relationship between sources and patterns of VOCs in indoor air. Atmos Pollut Res 5:129–137

Rossignol S, Rio C, Ustache A, Fable S, Nicolle J, Même A, D'Anna B, Nicolas M, Leoz E, Chiappini L (2013) The use of a housecleaning product in an indoor environment leadingto oxygenated polar compounds and SOA formation: gas and particulate phase chemical characterization. Atmos Environ 75:196–205

Rovira J, Roig N, Nadal M, Schuhmacher M, Domingo JL (2016) Human health risks of formaldehyde indoor levels: an issue of concern. J Environ Sci Health Part A 51:357–363

Rufo JC, Madureira J, Oliveira Fernandes E, Moreira A (2016) Volatile organic compounds in asthma diagnosis: a systematic review and meta-analysis. Allergy 71:175–188

Salthammer T (2019a) Formaldehyde sources, formaldehyde concentrations and air exchange rates in European housings. Build Environ 150:219–232

Salthammer T (2019b) Data on formaldehyde sources, formaldehyde concentrations and air exchange rates in European housings. Data Brief 22:400–435

Salthammer T (2022) TVOC – revisited. Environ Int 167:107440

Salthammer T, Uhde E, Schripp T, Schieweck A, Morawska L, Mazaheri M, Clifford C, He C, Buonanno G, Querol X, Viana M, Kumar P (2016) Children's well-being at schools: impact of climatic conditions and air pollution. Environ Int 94:196–210

Salthammer. (2016) Very volatile organic compounds: an understudied class of indoor air pollutants. Indoor Air 26:25–38

Sarigiannis DA, Karakitsios SP, Gotti A, Liakos IL, Katsoyiannis A (2011) Exposure to major volatile organic compounds and carbonyls in European indoor environments and associated health risk. Environ Int 37:743–765

Scherer S, Maly P (2000) Occurrence of glycol and isothiazolinone compounds in the indoor air of freshly painted rooms. Report by Ökoscience Lufthygiene AG on behalf of the Federal Office of Public Health, Zürich, Schweiz. [in German]

Schieweck A, Gunschera J, Varol D, Salthammer T (2018) Analytical procedure for the determination of very volatile organic compounds (C_3–C_6) in indoor air. Anal Bioanal Chem 410:3171–3183

Schlink U, Rehwagen M, Damm M, Richter M, Borte M, Herbarth O (2004) Seasonal cycle of indoor-VOCs: comparison of apartments and cities. Atmos Environ 38:1181–1190

Schlummer M, Gruber L, Fiedler D, Kizlauskas M, Müller J (2013) Detection of fluorotelomer alcohols in indoor environments and their relevance for human exposure. Environ Int 57–58:42–49

Schroeter JD, Campbell J, Kimbell JS, Conolly RB, Clewell HJ, Andersen ME (2014) Effects of endogenous formaldehyde in nasal tissues on inhaled formaldehyde dosimetry predictions in the rat, monkey, and human nasal passages. Toxicol Sci 138:412–424

Seifert B (1999) Guideline values for indoor air. The assessment of indoor air quality using the sum of volatile organic compounds (TVOC value). Bundesgesundheitsbl 42:270–278. [in German]

Serex T, Anand S, Munley S, Donner EM, Frame SR, Buck RC, Loveless SE (2014) Toxicological evaluation of 6:2 fluorotelomer alcohol. Toxicology 319:1–9

Sha B, Dahlberg A-K, Wiberg K, Lutz A (2018) Fluortelomer alcohols (FTOHs), brominated flame retardants (BFRs), organophosphorus flame retardants (PFRs) and cyclic volatile methylsiloxanes (cVMSs) in indoor air from occupational and home environments. Environ Pollut 241:319–330

Sheng N, Zhou X, Zheng F, Pan Y, Guo X, Guo Y, Sun Y, Dai J (2017) Comparative hepatotoxicity of 6:2 fluorotelomer carboxylic acid and 6:2 fluorotelomer sulfonic acid, two fluorinated alternatives to long-chain perfluoroalkyl acids, on adult male mice. Arch Toxicol 91:2909–2919

Shoeib M, Harner T, Wilford BH, Jones KC, Zhu J (2005) Perfluorinated sulfonamides in indoor and outdoor air and indoor dust: occurrence, partitioning, and human exposure. Environ Sci Technol 39:6599–6606

Shrubsole C, Dimitroulopoulou S, Foxall K, Gadeberg B, Doutsi A (2019) IAQ guidelines for selected volatile organic compounds (VOCs) in the UK. Build Environ 165:106382

Silva V, Silva C, Soares P, Garrido EM, Borges F, Garrido J (2020) Isothiazolinone biocides: chemistry, biological, and toxicity profiles. Molecules 25:991

Singer BC, Coleman BK, Destaillats H, Hodgson AT, Lunden MM, Weschler CJ, Nazaroff WW (2006a) Indoor secondary pollutants from cleaning products and air freshner use in the presence of ozone. Atmos Environ 40:6696–6710

Singer BC, Destaillats H, Hodgson AT, Nazaroff WW (2006b) Cleaning products and air fresheners: emissions and resulting concentrations of glycol ethers and terpenoids. Indoor Air 16: 179–191

Sivanantham S, Dassonville C, Grégoire A, Malingre L, Ramalho O, Mandin C (2021) Coexposure to indoor pollutants in French schools and associations with building characteristics. Energ Buildings 252:111424

Steinle P (2016) Characterization of emissions from a desktop 3D printer and indoor air measurements in office settings. J Occup Environ Hyg 13:121–132

St-Jean M, St-Amand A, Gilbert NL, Soto JC, Guay M, Davis K, Gyorkos TW (2012) Indoor air quality in Montréal area day-care centres, Canada. Environ Res 118:1–7

Sun HZ, Yu P, Lan C, Wan MWL, Hickman S, Murulitharan J, Shen H, Yuan L, Guo Y, Archibald AT (2022a) Cohort-based long-term ozone exposure-associated mortality risks with adjusted metrics: a systematic review and meta-analysis. Innovation (Camb) 20:100246

Sun J, Wang J, Yang J, Shi X, Li S, Cheng J, Chen S, Sun K, Wu Y (2022b) Association between maternal exposure to indoor air pollution and offspring congenital heart disease: a case-control study in East China. BMC Public Health 22:767

Sysoltseva M, Schieweck A, Röhl C, Gerull F, Burghardt R, Gessner A, Papavlassopoulos H, Schober W, Völkel W, Fromme H (2023) Very volatile organic compounds (VVOCs) and aldehydes in indoor air samples of classrooms and apartments in Germany (LUPE10 study). Submitted

Szabados M, Csákó Z, Kotlík B, Kazmarová H, Kozajda A, Jutraz A, Kukec A, Otorepec P, Dongiovanni A, Di Maggio A, Fraire S, Szigeti T (2021) Indoor air quality and the associated health risk in primary school buildings in Central Europe – The InAirQ study. Indoor Air 31: 989–1003

Tagiyeva N, Sheikh A (2014) Domestic exposure to volatile organic compounds in relation to asthma and allergy in children and adults. Expert Rev Clin Immunol 10:1611–1639

Taxvig C, Rosenmai RK, Vinggaard AM (2014) Polyfluorinated alkyl phosphate ester surfactants – current knowledge and knowledge gap. Basic Clin Pharmacol Toxicol 115:41–44

UBA (German Environment Agency) (2008) Comparative values for volatile organic compounds (VOCs and aldehydes) in the indoor air of households in Germany. Bundesgesundheitsbl 51: 109–112. [in German]

Uhde E, Salthammer T (2009) Impact of reaction products from building materials and furnishings on indoor air quality – A review of recent advances in indoor chemistry. Atmos Environ 41: 3111–3128

Uhde E, Schulz N (2015) Impact of room fragrance products on indoor air quality. Atmos Environ 106:492–502

US EPA (1998) Toxicological review of methyl methacrylate (CAS No. 80-62-6). U.-S. Environmental Protection Agency, Washington, DC. https://cfpub.epa.gov/ncea/iris/iris_documents/documents/toxreviews/1000tr.pdf

Vardoulakis S, Giagloglou E, Steinle S, Davis A, Sleeuwenhoek A, Galea KS, Dixon K, Crawford JO (2020) Indoor exposure to selected air pollutants in the home environment: a systematic review. Int J Environ Res Public Health 17:8972

Veenaas C, Ripszam M, Glas B, Liljelind I, Claeson AS, Haglund P (2020) Differences in chemical composition of indoor air in rooms associated/not associated with building related symptoms. Sci Total Environ 720:137444

Verriele M, Schoemaecker C, Hanoune B, Leclerc N, Germain S, Gaudion V, Locoge N (2016) The MERMAID study: indoor and outdoor average pollutant concentrations in 10 low-energy school buildings in France. Indoor Air 26:702–713

Villanueva F, Tapia A, Lara S, Amo-Salas M (2018) Indoor and outdoor air concentrations of volatile organic compounds and NO_2 in schools of urban, industrial and rural areas in Central-Southern Spain. Sci Total Environ 622–623:222–235

Wallenius K, Hovi H, Remes J, Mahiout S, Liukkonen T (2022) Volatile organic compounds in Finnish office environments in 2010–2019 and their relevance to adverse health effects. Int J Environ Res Public Health 19:4411

Wallner P, Munoz U, Tappler P, Wanka A, Kundi M, Shelton JF, Hutter H-P (2015) Indoor environmental quality in mechanically ventilated, energy-efficient buildings vs. conventional buildings. Int J Environ Res Public Health 12:14132–14147

Walt M, Matter K, Hackenschmidt S (2015) Methyl methacrylate (MMA) contamination of daycare center and apartments, report. Zentrum für Arbeits- und Umweltmedizin e.V. (ZAUM Berlin). https://zaumberlin.de/wordpress_zaum/wp-content/uploads/2015/09/MMA_kont_wohn-kita_stick28-Bl-neu.pdf

Weisel CP, Alimokhtari S, Sanders PF (2008) Indoor air VOC concentrations in suburban and rural New Jersey. Environ Sci Technol 42:8231–8238

Weiss JM, Jones B, Koekkoek J, Bignert A, Lamoree MH (2021) Per- and polyfluoroalkyl substances (PFASs) in Swedish household dust and exposure of pet cats. Environ Sci Pollut Res Int 28:39001–39013

Weschler CJ (2009) Changes in indoor pollutants since the 1950s. Atmos Environ 43:153–169

WHO (1989) Indoor air quality: organic pollutants. EURO Reports and Studies, Vol. 111. Copenhagen, Denmark

WHO (1998) Methyl methacrylate. Concise International Chemical Assessment Document 4. WHO Press, Geneva, Switzerland. http://www.inchem.org/documents/cicads/cicads/cicad04.htm

WHO (2010) Guidelines for indoor air quality: selected pollutants. World Health Organization, Regional Office for Europe, Copenhagen (http://www.euro.who.int/en/what-we-do/health-topics/environmental-health/air-quality/publications/2010/who-guidelines-for-indoor-air-quality-selected-pollutants)

WHO (2021) Literature review on chemical pollutants in indoor air in public settings for children and overview of their health effects with a focus on schools, kindergartens and day-care centres. Supplementary publication to the screening tool for assessment of health risks from combined exposure to multiple chemicals in indoor air in public settings for children. WHO Europe. https://apps.who.int/iris/handle/10665/341467

Wickliffe JK, Stock TH, Howard JL, Frahm E, Simon-Friedt BR, Montgomery K, Wilson MJ, Lichtveld MY, Harville E (2020) Increased long-term health risks attributable to select volatile organic compounds in residential indoor air in southeast Louisiana. Sci Rep 10(1):21649

Winkens K, Koponen J, Schuster J, Shoeib M, Vestergren R, Berger U, Karvonen AM, Pekkanen J, Kiviranta H, Cousins IT (2017) Perfluoroalkyl acids and their precursors in indoor air sampled in children's bedrooms. Environ Pollut 222:423–432

Winkens K, Giovanoulis G, Koponen J, Vestergren R, Berger U, Karvonen AM, Pekkanen J, Kiviranta H, Cousins IT (2018) Perfluoroalkyl acids and their precursors in floor dust of children's bedrooms – implications for indoor exposure. Environ Int 119:493–502

Woitowitz HJ, Knecht U (1992) Emissions of organic compounds during the craft processing of varnishes and paints. Research report No. 10408152. Umweltbundesamt, Berlin. [in German]

Wolkoff P (1998) Impact of air velocity, temperature, humidity, and air on long-term VOC emissions from building products. Atmos Environ 14:2659–2668

Wu Y, Romanak K, Bruton T, Blum A, Venier M (2020) Per- and polyfluoroalkyl substances in paired dust and carpets from childcare centers. Chemosphere 251:126771

Xu Z, Fiedler S, Pfister G, Henkelmann B, Mosch C, Völkel W, Fromme H, Schramm K-W (2013) Human exposure to fluorotelomer alcohols, perfluorooctane sulfonate and perfluorooctanoate via house dust in Bavaria, Germany. Sci Total Environ 443:485–490

Yang S, Perret V, Jörin CH, Niculita-Hirzel H, Pernot JG, Licina D (2020) Volatile organic compounds in 169 energy-efficient dwellings in Switzerland. Indoor Air 30:481–491

Yang Y, Liu B, Hua J, Yang T, Dai Q, Wu J, Feng Y, Hopke PK (2022) Global review of source apportionment of volatile organic compounds based on highly time-resolved data from 2015 to 2021. Environ Int 165:107330

Yao Y, Zhao Y, Sun H, Chang S, Zhu L, Alder AC, Kannan K (2018) Per- and polyfluoroalkyl substances (PFASs) in indoor air and dust from homes and various microenvironments in China: implications for human exposure. Environ Sci Technol 52:3156–3166

Yoon C, Lee K, Park D (2011) Indoor air quality differences between urban and rural preschools in Korea. Environ Sci Pollut Res Int 18:333–345

Yoshida T, Mimura M, Sakon N (2021) Estimating household exposure to moth repellents p-dichlorobenzene and naphthalene and the relative contribution of inhalation pathway in a sample of Japanese children. Sci Total Environ 783:146988

Yost EE, Galizia A, Kapraun DF, Persad AS, Vulimiri SV, Angrish M, Lee JS, Druwe IL (2021) Health effects of naphthalene exposure: a systematic evidence map and analysis of potential considerations for dose-response evaluation. Environ Health Perspect 129:76002

You B, Zhou W, Li J, Li Z, Sun Y (2022) A review of indoor gaseous organic compounds and human chemical exposure: insights from real-time measurements. Environ Int 170:107611

Zhao Z, Pei Y, Zhao P, Wu C, Qu C, Li W, Zhao Y, Liu J (2022) Characterizing key volatile pollutants emitted from adhesives by chemical compositions, odor contributions and health risks. Molecules 27:1125

Zheng G, Boor BE, Schreder E, Salamova A (2020) Indoor exposure to per- and polyfluoroalkyl substances (PFAS) in the childcare environment. Environ Pollut 258:113714

Zhong L, Su FC, Batterman S (2017) Volatile organic compounds (VOCs) in conventional and high performance school buildings in the U.S. Int J Environ Res Public Health 14:100

Zhong L, Batterman S, Milando CW (2019) VOC sources and exposures in nail salons: a pilot study in Michigan, USA. Int Arch Occup Environ Health 92:141–153

Zhu J, Newhook R, Marro L, Chan CC (2005) Selected volatile organic compounds in residential air in the city of Ottawa, Canada. Environ Sci Technol 39:3964–3971

Zhu J, Wong SL, Cakmak S (2013) Nationally representative levels of selected volatile organic compounds in Canadian residential indoor air: population-based survey. Environ Sci Technol 47:13276–13283

Zuraimi MS, Tham KW (2008) Effects of child care center ventilation strategies on volatile organic compounds of indoor and outdoor origins. Environ Sci Technol 42:2054–2059

Chapter 4
Semi-volatile Organic Compounds (SVOC)

Contents

Abstract In this chapter, the large group of semi-volatile organic compounds (SVOC) like polycyclic aromatic hydrocarbons, plasticizers (e.g., phthalates, DINCH, adipates), flame retardants (brominated and organophosphate flame retardants), per- and polyfluorinated substances, polychlorinated biphenyls, dioxins, chlorinated paraffins, pyrethroids, bisphenols, parabens, organotin, synthetic musks, perchlorate, organic UV light protection filter, and diisocyanates are presented individually in detail. In addition to a brief description of the chemical–physical characteristics and the use of the substance, an overview of main health

H. Fromme, *Indoor Air Quality*, https://doi.org/10.1007/978-3-031-40078-0_4

risks and, in particular, references to further scientific literature are given. In the foreground, however, are extensive presentations of the occurrence of the substances in indoor air and in house dust.

Keywords SVOC · phthalates · PBDE · dioxins · PAH · PCB · UV-Filter · parabens · pesticides

This heterogeneous group contains a large number of individual substances or substance groups. Often, only limited data on the exposure of room users in indoor spaces, especially in public indoor spaces, are available so far.

4.1 Polycyclic Aromatic Hydrocarbons (PAH)

Polycyclic aromatic hydrocarbons (PAH) are several hundred organic substances composed of two or more aromatic rings (see examples in Fig. 4.1). The rings can be linear, angular, or arranged in clusters, and the hydrogen atoms can be substituted by various polar as well as nonpolar groups. PAH have become a ubiquitous contaminant in the course of industrialization.

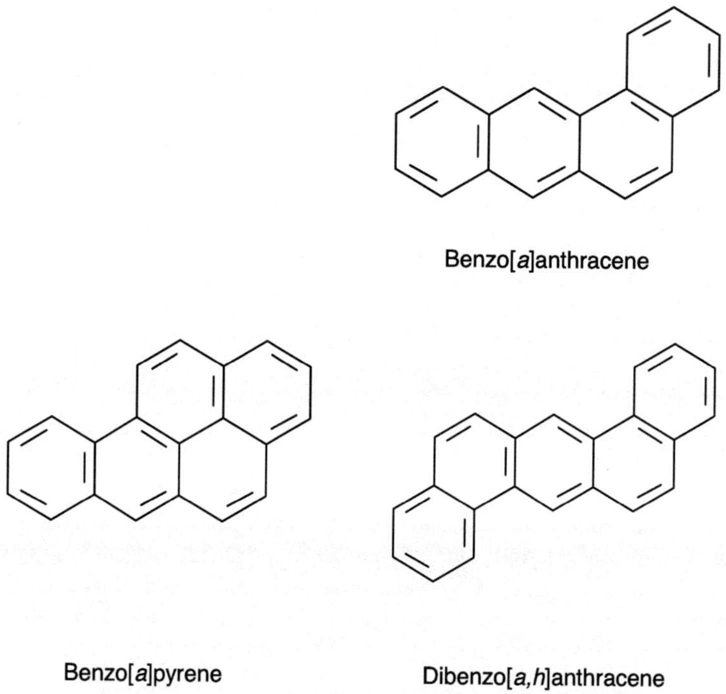

Fig. 4.1 Chemical structure of some polycyclic aromatic hydrocarbons (PAH)

PAH are predominantly formed by anthropogenic activities in the course of incomplete combustion of organic material under oxygen deficiency conditions (so-called pyrolysis). They also occur in liquid and solid products formed by heating organic material under oxygen deficiency conditions, such as tar, petroleum distillates, used motor oils, and soot. PAH may also be present in foodstuffs, particularly if their manufacture or preparation involves a pyrolytic process or if contamination by PAH-containing impurities is present. PAH are also formed during the coalification process, as occurs in petroleum formation or coal formation. The more completely a fuel containing carbon and hydrogen is burned with oxygen to form carbon dioxide and water, the smaller the amount of PAH formed as a by-product.

The following major sources must be considered:

- Industrial processes: coking plants, aluminum smelters (especially Söderberg process). Petroleum refining, production of tar oil, asphalt and bitumen, tire production, power plants, waste incineration.
- Domestic area: heating with coal, wood and oil stoves, open fireplaces, tobacco smoke.
- Road transport: internal combustion engines, especially diesel engine vehicles.
- Agricultural activities: deliberately caused forest, steppe, bush and bog fires.
- Natural processes: unintentional forest, steppe, bush and moor fires, volcanic eruptions.

PAH are always present in the environment as complex mixtures of substances, whereby, depending on the source, very different mixing ratios of the individual substances occur. Due to its volatility naphthalene was discussed in the chapter "volatile organic compounds."

4.1.1 Health Aspects

The health significance of this substance class lies essentially in the carcinogenic effects, which are well documented for individual PAH and mixtures for the inhalation pathway and after application to the skin (Danish-EPA 2013a). For example, tumors of the respiratory tract, bladder, and skin have been observed in epidemiological studies in workplaces and animal studies. Since PAH are always present as a complex mixture in the environment and in workplaces, the carcinogenic effect observed in the workplace studies, for example, is also always proportionally due to the presence of other PAHs besides B[*a*]P. Various studies suggest that PAH with 3–4 rings have lower carcinogenic potency and those with 5–6 rings higher (DECOS 2006; IARC 2010). One explanation is structural differences of the PAH adducts formed, which lead to various effect intensities. A meta-analysis of occupational cohort studies found a 20% increase in relative risk per 100 μg/m^3 years cumulative B[*a*]P exposure (IARC 2012). Based on epidemiological data from studies in coke-oven workers, a unit risk for B[*a*]P as an indicator of PAH in air was estimated to be 8.7×10^{-5} per ng/m^3 by WHO (WHO 2010). The corresponding

Table 4.1 Basic information on some important PAH

Substance	CAS-No.	Abbreviation
Acenaphthene	83–32-9	Ace
Acenaphthylene	208–96-8	Acy
Anthracene	120–12-7	Ant
Benzo[*a*]anthracene	56–55-3	BaAnt
Benzo[*a*]pyrene	50–32-8	BaP
Benzo[*b*]fluoranthene	205–99-2	BbFlu
Benzo[*j*]fluoranthene	205–82-3	BjFlu
Benzo[*k*]fluoranthene	207–08-9	BkFlu
Benzo[*e*]pyrene	192–97-2	BePyr
Benzo[*ghi*]perylene	191–24-2	BghiPer
Benzo[*c*]fluorene	205–12-9	BcFl
Chrysene	218–01-9	Chr
Coronen	191–07-1	Cor
Dibenzo[*a,c*]anthracene	215–58-7	DBacAnt
Dibenzo[*a,h*]anthracene	53–70-3	DBahAnt
Dibenzo[*a,e*]pyrene	192–65-4	DBaePyr
Dibenzo[*a,i*]pyrene	189–55-9	DBahPyr
Dibenzo[*a,h*]pyrene	189–55-9	DBaiPyr
Dibenzo[*a,l*]pyrene	191–30-0	DBalPyr
Fluoranthene	206-44-0	Flu
Fluorene	86-73-7	Fl
Indeno[1,2,3-*cd*]pyrene	193–39-5	IcdPyr
Naphthalene	91–20-3	Nap
Phenanthrene	85–01-8	Phe
Pyrene	129–00-0	Pyr

concentrations for lifetime exposure to B[*a*]P producing excess lifetime cancer risks of 1/10000, 1/100000, and 1/1000000 are approximately 1.2 ng/m^3, 0.12 ng/m^3, and 0.012 ng/m^3, respectively.

Often, the 16 PAH classified as high priority substances by the US Environmental Protection Agency are analyzed in environmental media due to their potential toxicity in humans and other organisms (see Table 4.1). Usually B[*a*]P were used as an indicator substance for the entire group. This is not without controversy, since other PAH, e.g. the dibenzopyrenes, probably have a significantly higher carcinogenic potency than B[*a*]P. However, they have so far been determined in indopor spaces only very rarely.

In the European Union B[*a*]P has been classified as carcinogenic for humans (category carc. 1B), genotoxic (muta. 1B), and toxic to reproduction (repr. 1B). IARC has classified B[*a*]P in group 1 (carcinogenic to humans), DB[*ah*]Ant, DB[al] Pyr, and cyclopenta[cd]pyrene in group 2A (probably carcinogenic to humans), and some other in group 2B (possibly carcinogenic to humans).

There are extensive reviews by international organizations on the hazards and environmental and health risks of PAH and especially B[*a*]P (e.g., EFSA 2008; IARC 2010; 2012; WHO 2010; HC 2016; US EPA 2017a; ECHA 2018a).

4.1.2 Occurrence of PAH in the Air of Residences

Wang et al. (2021a) summarized the results of PAH in indoor air from 131 studies with 6766 samples collected in different countries worldwide and observed a decreasing trend of the concentrations. This can be explained by the implementation of technologies to reduce emissions. Table 4.2 summarizes the results of 7 individual PAH in indoor air reported from several studies worldwide. Moreover, depending on the study, significant differences in concentrations can be found in indoor living spaces. Comparable to outdoor air, an annual cycle can also be found in indoor air with higher levels in winter (heating period) and spring compared to summer. Levels of most indoor particle-bound PAH were heavily influenced by outdoor air (e.g., Ma et al. 2021). In a study in China, the indoor air size distribution of particle-bound B[*a*]P showed that the concentrations found in the 0.32–1.0 μm fraction predominated the total amount (Hu et al. 2018). These associations are given in Fig. 4.2.

In Europe, the lowest concentrations were observed with 0.03 ng B[*a*]P/m^3 (ΣPAH_{15}: 7.2 ng/m^3) in 6 apartments in Paris, the highest in the period from December to February with 7.4 ng/m^3 B[*a*]P (ΣPAH_8: 48.0 ng/m^3) in Krakow, Poland. A median of 0.05 ng B[*a*]P/m^3 (ΣPAH_8: approx. 4.5 ng/m^3) was determined in 111 measurements in 20 properties throughout Germany in 1997–2003 (Köhler et al. 2004). Within the representative German Environmental Health Study 2014–2017, median B[*a*]P levels of 0.11 ng/m^3 and a 95th percentile of 0.79 ng/m^3 (maximum: 1.61 ng/m^3) were determined in $PM_{2.5}$ during weekly measurements (Neumann et al. 2020). Fromme et al. (2023) reported the results of 56 PAH analyzed in PM_{10} of 35 residences in Germany in 2019/20. A median of 0.026 ng/m^3 for B[*a*]P (95th percentile: 0.228 ng/m^2) and of 0.495 ng/m^3 (95th percentile: 2.64 ng/m^3) for priority US EPA PAH (except naphthalene and acenaphthylene) were observed.

Substantially lower B[*a*]P concentrations in $PM_{2.5}$ were found in Hong Kong (median: 0.10 ng/m^3) residences compared to another urban city (Guangzhou) (median: 0.60 ng/m^3) in South China in 2010 (Wang et al. 2013e). Coal combustion and vehicle emissions were considered as the most important PAH sources in this study.

In the absence of significant indoor sources such as smoking or open fires, indoor levels are highly dependent on the respective outdoor air levels. In these cases, the ratio of indoor air concentration to outdoor air concentrations (I/A) for B[*a*]P and PAH is often in the range of 0.7–1.0. In contrast, significantly higher I/A ratios have been reported in those indoor spaces where open fireplaces or stoves were still active as sources of exposure (Liu et al. 2001; Ohura et al. 2004; Mannino and Orecchio 2008; Orecchio 2011; Wang et al. 2013e; Gustafson et al. 2008; Lv et al. 2009;

Table 4.2 Median concentrations of PAH in the air of residences in ng/m^3

References	*N*	Period	BaAnt	Chr	BbFlu	BkFlu	BaP	IcdPyr	DBahAnt	Country, time
Europe										
Fromme et al. (2004a)	57	Heat	0.28	0.32	0.36	0.12	0.25	0.36	0.09	Germany, 1997/98
Fromme et al. (2004a)	58	nHeat	0.04	0.19	0.10	0.05	0.09	0.24	0.07	Germany, 2000
Gustafson et al. (2008)	10	Heat	<0.15	0.15	0.11	0.08	0.12	0.10	–	Sweden, 2003
Halsall et al. (2008)	5		0.55	0.39	0.03	0.02	0.02	–	0.12	UK, 2003 *
Mandin et al. (2016)	285		0.04	0.08	0.31	0.06	0.14	0.18	0.02	France, 2003–05
Choi et al. (2008)	21	nHeat	0.56	0.62	0.99	0.36	0.46	0.75	0.14	Poland, 2007 +
	21	Heat	7.3	5.8	10.8	3.3	7.35	6.7	1.3	Poland, 2007 +
Castro et al. (2011)	1		0.30	0.27	1.33b	0.26	0.61	0.61	2.3	Portugal, 2009 *
Delgado-Saborit et al. (2011)	81		0.06	0.14	0.18	0.11	0.09	0.11	0.03	UK +
Romagnoli et al. (2014)	3	nHeat	0.10	–	0.09		0.08	0.05	0.01	Italy, 2011/12
	6	Heat	0.17	–	2.14		0.81	1.17	0.16	Italy, 2011/12
Moreau-Guigon et al. (2016)	6	nHeat	0.01	0.02	0.02	<0.5	<0.4	<0.6	<0.5	France, 2011/12 *
	6	Heat	1.3	1.30	0.02	<0.5	0.03	<0.6	<0.5	France, 2011/12 *
Romagnoli et al. (2016)	1		0.07	0.25	0.47	0.19	0.37	0.43	0.11	Italy, 2013 *
Neumann et al. (2020)	77		0.14	0.09	0.23	0.11	0.11	0.26	0.05	Germany, 2014–17
Madruga et al. (2019)			0.007	0.015	0.052[a]	0.014	0.016	0.025	0.009	Spain, 2017–18 *
Fromme et al. (2023)	35	Heat	0.010	0.027	0.077	0.032	0.026	0.068	0.006	Germany, 2019/20
Asia										
Wang et al. (2013e)			0.060	0.100	0.210		0.100	0.180	0.020	China, 2010[b]

Wang et al. (2013e)			0.270	0.520	1.310		0.600	1.150	0.190	China, 2010[c]
Chen et al. (2022a)	26		0.05	0.21	0.17	0.12	0.05	0.06	0.01	China, 2014–16[b]*
Zhang et al. (2019)	13	nHeat	0.750	0.690	0.930	0.970	0.980	1.000	0.990	China, 2017
Ma et al. (2021)	15	nHeat	0.018	0.059	0.325	0.025	0.074	0.137	0.019	China, 2017
North America/Australia										
Li et al. (2017)	9		<0.04–0.89	<0.08–1.7	0.04–0.38	–	<0.04–0.15	<0.03–0.11	<0.03–0.11	Canada, 2014[d]
Wang et al. 2019)	28		0.046		<0.007		<0.01	<0.006	<0.014	Australia, 2015[e]

* mean value; + geometric mean; *Heat* heating period, *nHeat* outside the heating period

[a]including BjF

[b] in Hong Kong

[c] in Guangzhou

[d] ranges

[e] houses and offices

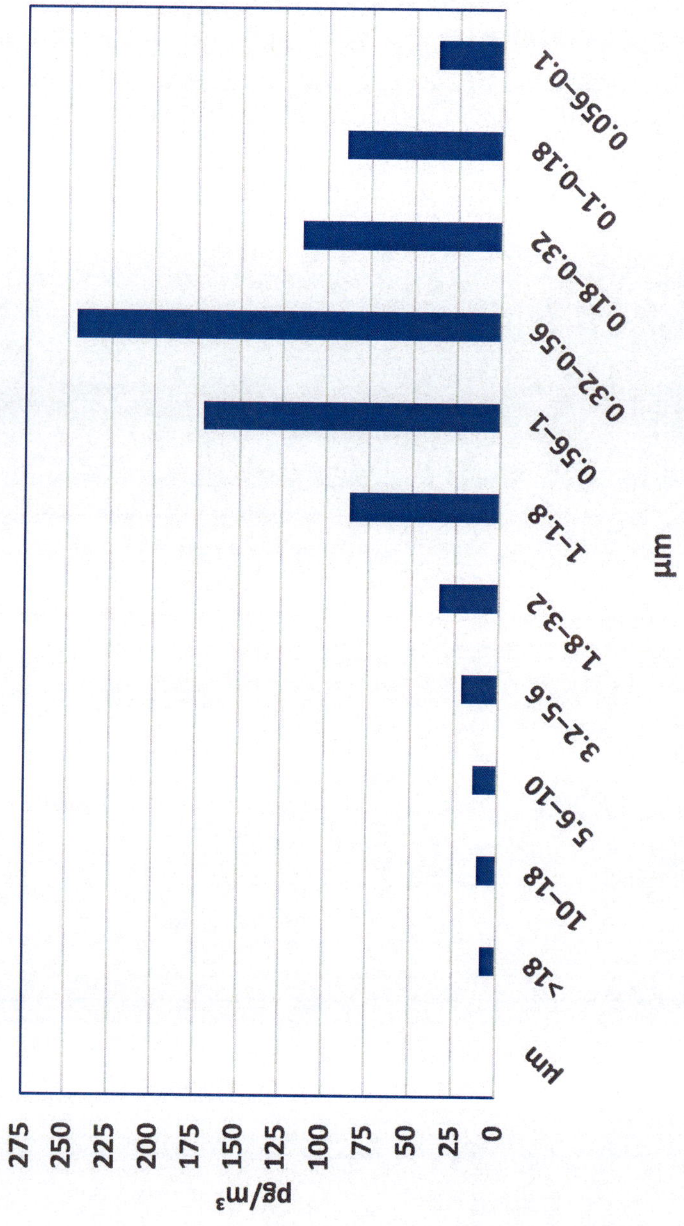

Fig. 4.2 Mean concentrations of B[*a*]P in different size fractions (sum of six duplicates) (according to Hu et al. 2018)

Delgado-Saborit et al. 2011; de Gennaro et al. 2016; Lin et al. 2016; Lui et al. 2017). In a Chinese study I/A, personal/A, and personal/I ratios of 0.92, 0.48, and 0.73, respectively, were found suggesting that personal PAH exposures were mainly attributable to ambient PAH (Chen et al. 2022a).

Indoor cooking activities can also be associated with elevated PAH concentrations (Zhu and Wang 2003; Zhu et al. 2009; Gautam et al. 2013; Yao et al. 2015a; Chen et al. 2016a). For example, ΣPAH_{16} concentrations ranging from 430 to 36,000 ng/m^3 could be measured in kitchens of private residences in China. High B[*a*]P levels of 150–440 ng/m^3 (ΣPAH_{12}: from 10,000–21,000 ng/m^3) could also be observed in commercial kitchens (Zhu and Wang 2003). In particular, frying was associated with high indoor air levels in these studies. The highest concentrations to date have been measured in Indian homes with cooking activities with levels from 2000 to 12,300 ng B[*a*]P/m^3 (Gautam et al. 2013). As well in China, B[*a*]P was not detected in the sample with exclusive LPG/electric use, but exceeded 40 ng/m^3 in the household samples using all three stoves and only the traditional stoves (Lai et al. 2020).

Another source observed in many studies is indoor smoking of tobacco products (Fromme et al. 2004a; Ren et al. 2006; Mannino and Orecchio 2008; Bolte et al. 2008; Aquilina et al. 2010; Castro et al. 2011; Hoh et al. 2012; Whitehead et al. 2013a; Muñoz et al. 2016; Cattaneo et al. 2016; Adesina et al. 2021; Zhang et al. 2022a). For example, Fromme et al. (2004a) found higher concentrations of ΣPAK_{12} in homes of smokers compared to non-smokers, from 5.9 ng/m^3 to 2.4 ng/m^3 in winter and 3.8 ng/m^3 to 2.1 ng/m^3 in summer, respectively. There was no smoking in the rooms during sampling. Castro et al. (2011) studied two apartments in Portugal identical in their baseline data, one where there was no smoking and one where about 6–7 cigarettes were smoked per day. The results are shown in Fig. 4.3. The

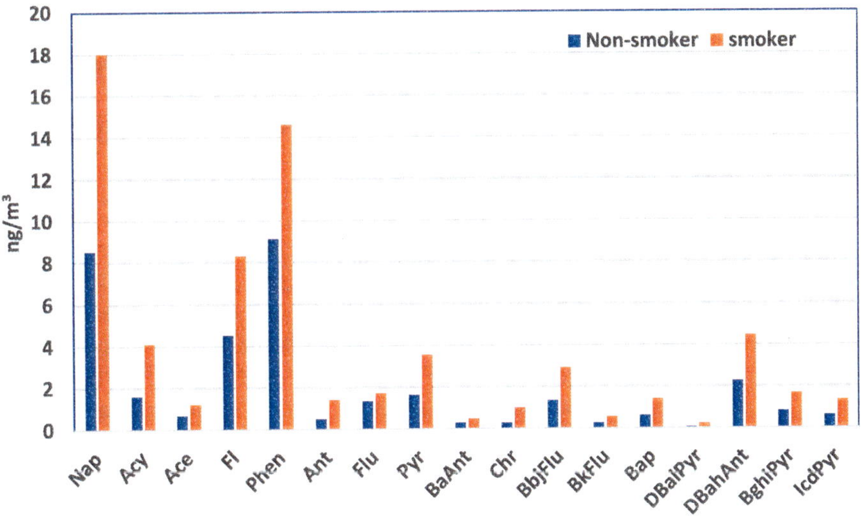

Fig. 4.3 PAH in a smoker's and a non-smoker's apartment (after Castro et al. 2011)

concentrations in the non-smoking apartment reach only 27–80% of those in the smoking apartment (ΣPAH_{17}: 52%), depending on the individual substance. High PAH concentrations were also observed in hospitality venues in Bavaria, Germany, before the smoking ban (Bolte et al. 2008). Here, the mean level for ΣPAK_{16} was 242 ng/m^3 in 11 restaurants, 350 ng/m^3 in 7 bars, and 407 ng/m^3 in 10 discotheques. For B[*a*]P, mean levels in the aforementioned establishments were 1.7 ng/m^3, 3.8 ng/m^3, and 4.8 ng/m^3.

Another exposure possibility is smoking a shisha (water pipe). For example, in an indoor room where a shisha was smoked for 4 hours, concentrations of 270 ng/m^3 were found for ΣPAK_{16} and 4.9 ng/m^3 for the 7 PAHs classified as probable carcinogens by the US EPA (B2-PAH) (Fromme et al. 2009b). Naphthalene alone accounted for 110 ng/m^3. Markowicz et al. (2014) investigated only the particle-bound PAH in a 21 m^3 test chamber and were able to find a comparable level for B2-PAH when smoking a shisha with an average of about 7.5 ng/m^3. Also, in 46 facilities in 3 countries (Russia, Chile, Egypt) where shisha was smoked, a median level of 136 ng/m^3 (1.3–1963 ng/m^3) was found for the sum of particle-bound PAH using a photoelectric sensor (Moon et al. 2015).

Concentrations between 200 and 240 ng/m^3 (ΣPAH_{12}) (Lung et al. 2003), or a mean of 348 ng/m^3 (ΣPAH_{15}) (Li and Ro 2000), were found indoors when incense or incense candles were burned. In rooms where moxibustion was performed, the mean ΣPAH_{16} was as high as 12,900 ng/m^3 (Mo et al. 2014). Indoor burning of wax candles results in ΣPAK_{18} concentrations from 7 to 267 ng/m^3 (B[*a*]P: 0.1–7.5 ng/m^3) (Orecchio 2011).

4.1.3 Occurrence in the Air of Public Facilities

A recent review summarized the results in school microenvironments (Oliveira et al. 2019). Again, Table 4.3 gives an overview of different study results.

In Germany, median B[*a*]P levels of 0.97 ng/m^3 (ΣPAK_{12}: 4.4 ng/m^3) were measured in 74 daycare centers and 0.75 ng/m^3 (ΣPAH_{12}: 4.5 ng/m^3) in 40 schools (Fromme et al. 2005a). In Portugal, Oliveira et al. (2017) reported mean B[*a*]P-levels (ΣPAH_{16}-levels) from 0.04 to 3.5 ng/m^3 (2.8 to 54 ng/m^3) in 20 classrooms of 10 schools in 2014. Average B[*a*]P-levels in a summer and winter monitoring period in 6 schools in the city of Rome, Italy in 2011 and 2012 were 0.07–0.16 ng/m^3 and 0.30–1.58 ng/m^3, respectively (Romagnoli et al. 2014). In Lithuania, mean B[*a*]P-concentrations between 1.0 and 3.2 ng/m^3 were reported in samples collected in 5 primary schools in 2011/12 (Krugly et al. 2014). Medians of 4.0 ng/m^3 and 3.6 ng/m^3 for B[*a*]P were reported in samples collected during spring season in two kindergartens in Poland in 2010 (Blaszczyk et al. 2017). Fromme et al. (2023) reported the results of 56 PAH analyzed in PM_{10} of 27 classrooms in Germany in 2019/20. A median of 0.034 ng/m^3 for B[a]P (95th percentile: 0.124 ng/m^2) and of 0.839 ng/m^3 (95th percentile: 2.77 ng/m^3) for priority US EPA PAH (except naphthalene and acenaphthylene) were observed.

Table 4.3 Median concentrations of PAH in the air of schools and daycare centers in ng/m^3

References	*N*	Period	BaAnt	Chr	BbFlu	BkFlu	BaP	IcdPyr	DBahAnt	Country, time
Fromme et al. (2005a)	74	nHeat	0.28	0.56	1.23	0.49	0.97	0.76	0.16	Germany, 2000/01
Fromme et al. (2005a)	40	Heat	0.34	0.57	0.92	0.41	0.75	1.53	0.10	Germany, 2002/03
Tang et al. (2019a)	43	nHeat	0.040	0.040	0.246	0.103	0.266	0.961	0.185	Taiwan, 2004/05
Blaszczyk et al. (2017)	2		4.7/2.1	4.7/1.9	4.9/3.7	2.5/1.8	4.0/3.6	4.4/3.7	0.5/0.4	Poland, 2010
Oliveira et al. (2015)	1		0.09	0.32	0.79	0.18	0.32	0.58	1.5	Portugal, 2011
Krugly et al. (2014)	5	Heat	–	–	–	–	1.0–3.2	–	–	Lithuania, 2011/12
Romagnoli et al. (2014)	12	Heat	0.39	–	2.36		1.13	1.17	0.16	Italy, 2011/12
Van Drooge et al. (2020)	24		0.06	0.18	0.21		0.03	0.07	–	Spain, 2012–13
Hu et al. (2018)	9	Heat	0.081	0.161	0.392		0.186	0.262	0.129	China, 2014/15[a]
Hu et al. (2018)	9	Heat	0.081	0.161	0.392		0.186	0.262	0.129	China, 2014/15[a]
Oliveira et al. (2017)	2	nHeat	0.03	0.08	0.16	0.04	0.05	0.12	0.53	Portugal, 2015
Lim et al. (2021)	5	Heat	0.007	0.016	0.019	0.008	0.014	0.016	0.003	Sweden, 2016/17
Fromme et al. (2023)	27	Heat	0.018	0.065	0.148	0.043	0.034	0.065	0.007*	Germany, 2019/20

[a] Schools, offices, and residences; *Heat* heating period, *nHeat* outside the heating period

In China, ΣPAK_{19} were 50 ng/m^3 and 51 ng/m^3 in two classrooms in 2012 (Wang et al. 2017c).

4.1.4 Occurrence in the Air of Transport Microenvironments

PAH have so far been measured rather rarely. For example, mean B[*a*]P concentrations of 1.0–4.4 ng/m^3 were observed in busses in Paris and Bordeaux, France, in the 1980s (Limasset et al. 1993). In a car driving through the inner city of Berlin, Germany, on 8 days in summer 1995 and on 8 days in winter 1995/1996, an average of 1.0 ng/m^3 and 3.2 ng/m^3, respectively, was found (Fromme et al. 1998). For ΣPAK_{12}, 10.2 ng/m^3 and 28.7 ng/m^3, respectively, were measured during the two measurement periods. Lower B[*a*]P concentrations of 0.14 ng/m^3 (summer) and 0.72 ng/m^3 (winter) could be detected in two cars traveling on different routes through Rome, Italy, in 2011/2012, while in one bus the level was 0.09 ng/m^3 (summer) and 0.76 ng/m^3 (winter) (Romagnoli et al. 2017). For ΣPAH_8, 0.64 ng/m^3 and 4.8 ng/m^3 were observed in the two measurement periods, respectively.

In the USA, B[*a*]P concentrations of 0.3 ng/m^3 were observed in cars and busses in Atlanta using continuously measuring PAH monitors (Greenwald et al. 2014). Significantly higher levels were described for different traffic modes in Beijing in 2011 (Yan et al. 2015). Here, B[*a*]P concentrations ranged from 1.4 to 26.5 ng/m^3 when walking or from 3.1 to 10.4 ng/m^3 when riding a bus. The mean level of ΣPAH_{18} was 77.8 ng/m^3 when walking and 99.6 ng/m^3 when riding in the bus.

In a subway train, mean B[*a*]P levels of 0.7 ng/m^3 and 4.0 ng/m^3 were found on 8 days in summer 1995 and on 8 days in winter 1995/1996, respectively, while driving through Berlin, Germany (31 km) (Fromme et al. 1998). For ΣPAH_{12}, significantly higher levels than in a car were measured during the two measurement periods with 30.2 ng/m^3 and 67.5 ng/m^3, respectively. During 5 measurement runs in subways in Rome, Italy, mean B[*a*]P concentrations of 0.12 ng/m^3 or a ΣPAH_8 content of 1.8 ng/m^3 was determined (Romagnoli et al. 2017). The highest level was measured by Yan et al. (2015) in the Beijing subway in 2011. Here, concentrations ranged from 0.43 to 4.6 ng/m^3 for B[*a*]P and 46.3 ng/m^3 (mean) for ΣPAH_{18}.

4.1.5 Dibenzopyrenes and Benzo[c]fluorene in the Air

In the scientific literature, it has been discussed whether the concentrations of the most important carcinogenic PAHs can be converted into TEQ (toxicity equivalents) using the PEF (potency equivalency factors) or TEF (toxicity equivalent factor) concept. The TEF value is used to express the ratio of an individual PAH to the carcinogenic potency of B[*a*]P, with the value for B[*a*]P set to 1 by convention. Different TEF values have been proposed so far and it must be emphasized that there are considerable uncertainties associated with the concept, since it cannot be ruled

Table 4.4 PAH and toxicity equivalent factors (TEF)

Substance	Abbreviation	US-EPA PAH[a]	TEF
Acenaphthene	Ace	X	0.001[b]
Acenaphthylene	Acy	X	0.001[b]
Anthracene	Ant	X	0.01[b]
Benzo[*a*]anthracene	BaAnt	X	0.1[b]
Benzo[*a*]pyrene	BaP	X	1.0[b]
Benzo[*b*]fluoranthene	BbFlu	X	0.1[b]
Benzo[*j*]fluoranthene	BjFlu	X	0.1[c]
Benzo[*k*]fluoranthene	BkFlu	X	0.1[b]
Benzo[*e*]pyrene	BePyr		0.01[c]
Benzo[*ghi*]perylene	BghiPer		0.01[b]
Benzo[*c*]fluorene	BcFl		20[d]
Chrysene	Chr	X	0.01[b]
Coronene	Cor		0.001[c]
Dibenzo[*a,c*]anthracene	DBacAnt		0.1[c]
Dibenzo[*a,h*]anthracene	DBahAnt	X	1.0[c]
Dibenzo[*a,e*]pyrene	DBaePyr		1.0[e]
Dibenzo[*a,i*]pyrene	DBahPyr		10[e]
Dibenzo[*a,h*]pyrenee	DBaiPyr		10[e]
Dibenzo[*a,l*]pyrene	DBalPyr		10[e]
Fluoranthene	Flu	X	0.001[b]
Fluorene	Fl	X	0.001[b]
Indeno[1,2,3-*cd*]pyrene	IcdPyr	X	0.1[b]
Naphthalene	Nap	X	0.001[b]
Phenanthrene	Phe	X	0.001[b]
Pyrene	Pyr	X	0.001[b]

[a] priority PAH according to US EPA (U.S. Environmental Protection Agency)
[b]Nisbet and LaGoy (1992)
[c] Malcolm and Dobsen (1994)
[d] US EPA (2010)
[e] Collins et al. (1998)

out that different PAH mixtures differ in, for example, bioavailability, metabolism, and antagonistic effects.

In the above concept, some dibenzopyrenes have obtained very high TEF values compared to B[*a*]P. In addition to dibenzo[*ae*]pyrene (DB[*ae*]Pyr), which like B[*a*]P received a factor of 1, dibenzo[*al*]pyrene (DB[al]Pyr), dibenzo[a*i*]pyrene (DB[*ai*] Pyr) and dibenzo[*ah*]pyrene (DB[*ah*]Pyr) were each assigned a factor of 10, and benzo[*c*]fluorene (B[*c*]Fl) a factor of 20. Therefore, for these substances, it is particularly important to consider TEQ in addition to concentrations. In the following, the PEF/TEF derived by Collins et al. (1998) for the California Environmental Protection Agency and that of Nisbet and LaGoy (1992) are used for TEQ calculation (see Table 4.4).

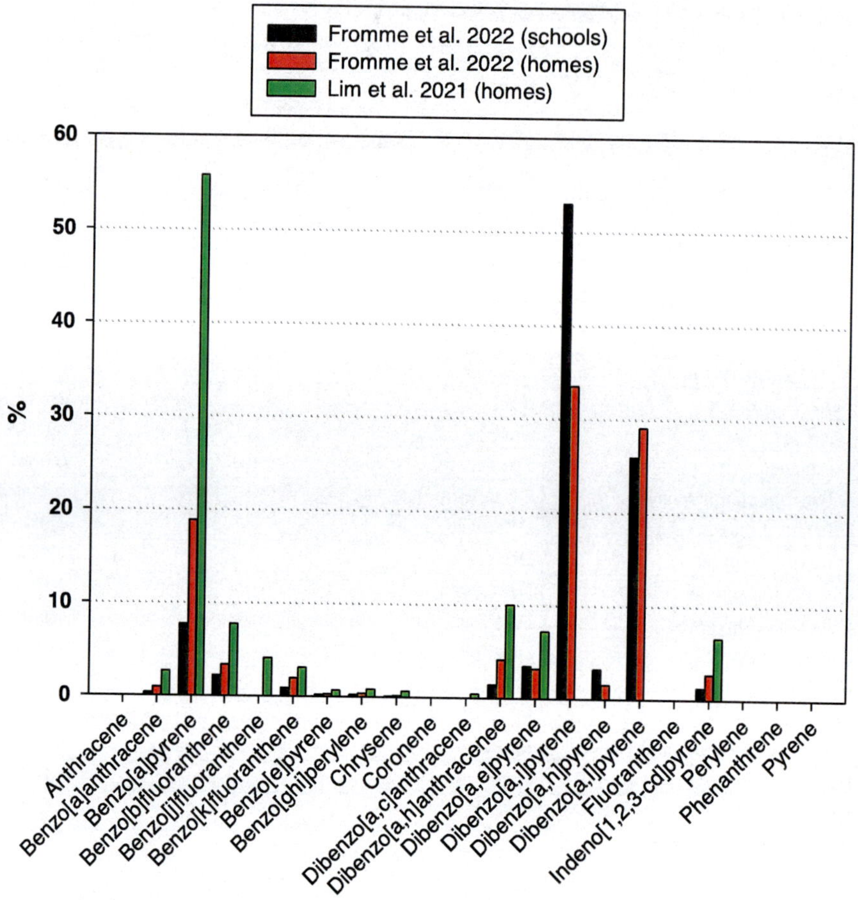

Fig. 4.4 Share of individual PAH on total TEQ in indoor air concentrations (using medians)

In a first study Castro et al. (2011) studied a portuguese house whre smoking occurred and a non-smoking house in 2009 and found DB[*al*]Pyr of 232 pg/m³ and 69 pg/m³, respectively, with DB[*al*]Pyr found only in the particle but not in the gas phase. Moreover, Lim et al. (2021) quantified one dibenzopyrene in the indoor air with a PM_{10}-bound mean of 1.77 pg/m³ for DB[*ae*]Pyr (range: <LOQ – 7.77 pg/m³) in 5 Swedish preschools in 2016/17. A second study was performed in 16 kitchen and 13 bedrooms in Shandong province, an area where bricked biomass stove with a chimney was usually employed in the kitchen for cooking (Zhang et al. 2019). The mean concentrations in kitchen and bedrooms were 6300 pg/m³ and 1600 pg/m³ for DB[*al*]Pyr and 1700 pg/m³ and 100 pg/m³ for DB[*ae*]Pyr, respectively. Just as for B[*a*]P (20,600 and 7000 pg/m³), very high concentrations were observed for dibenzopyrenes in indoor air compared to the other studies. Since the concentrations in the outside air are significantly lower with 200 pg/m³ for DB[*al*]P and 100 pg/m³

for DB[*ae*]P, combustion processes in the interior are responsible for the unusual contamination. In another study PM_{10} was sampled in 27 school classrooms and 35 residences in four German states in 2019/2020 (Fromme et al. 2023). Mean concentrations of 8.3 pg/m^3 and 7.4 pg/m^3 for DB[*al*]Pyr, 16.4 pg/m^3 and 11.9 pg/m^3 for DB[*ae*]Pyr, <LOQ and 5.7 pg/m^3 for DB[*ai*]Pyr, and 6.4 pg/m^3 and 4.2 pg/m^3 for B[*c*]Fl were observed for classrooms and residences, respectively, while DBahP was not found above LOQ in any sample. Significant correlations were observed between the sum of dibenzopyrenes and BaP ($r = 0.743$; $p < 0.001$) as well as the sum of priority US EPA PAH (except acenaphthylene and naphthalene) ($r = 0.802$; $p < 0.001$). While the mass concentrations and their percentage contribution to the concentrations of 20 PAH measured in the German study were low for the dibenzopyrenes and B[*c*]Fl, they have a share of 74% (classrooms) and 67% (residences) to the total BaPeq concentrations of the 20 PAH. If B[*a*]P is also taken into account, the share of total BaPeq concentrations increases to 87% in schools and 86% in the residences. The span of the individual PAH on the total TEQ amount for the Swedish and German study is given in Fig. 4.4. No other measurement results on dibenzopyrenes in indoor air are available so far.

Some other studies reported dibenzopyrene levels in outdoor air have been conducted in Sweden, Italy, Czech Republic, Colombia, and Cyprus so far (Bergvall and Westerholm 2007; Menichini and Merli 2012; Polachova et al. 2020; Mueller et al. 2019; Iakovides et al. 2021). Overall, the dibenzopyrene concentrations were in a comparable concentration range compared to our indoor air values. Only in the Czech Republic higher concentrations were reported for this group of PAH (Polachova et al. 2020). For B[*c*]Fl, Polachova et al. (2020) reported concentration from 2–521 pg/m^3 and 1–441 pg/m^3 in two Czech cities in the course of a year and Mueller et al. (2019) levels from 3.6–35 pg/m^3 at six sampling sites in Medellin, Colombia. In the urban area of Nicosia, Cyprus, a mean of 21.4 pg/m^3 B[*c*]Fl was reported in 62 samples (Iakovides et al. 2021).

4.1.6 Occurrence of Other PAH Classes in the Air

Chen et al. (2017) collected $PM_{2.5}$ in 84 houses in China in 2011 (non-heating period) reported concentrations of nitro-PAH and oxy-PAH of 2.2 ng/m^3 and 230 ng/m^3 for kitchen and 0.73 ng/m^3 and 120 ng/m^3 for living room. As part of the simultaneously conducted personal measurements it was found that the concentrations of nitro-PAH and oxy-PAH were significantly higher for smoking subjects than for non-smoking subjects.

In Beijing, China, in 2015 11 parent PAH and 10 nitro-PAH were analyzed in a school classroom and the concentrations of both classes were consistently higher in the heating period compared to the non-heating period with medians of 223 ng/m^3 vs. 8.75 ng/m^3 for parent PAH and 3.61 ng/m^3 vs. 0.25 ng/m^3 for nitro-PAH (Zhang et al. 2020b). In both periods all PAH were positively correlated with their outdoor concentrations. While means of 0.086 and 0.068 ng/m^3 for

9-nitroanthracene, 0.043 and 0.017 ng/m^3 for 2-nitrofluorene, and 0.050 and 0.005 ng/m^3 for 1-nitropyrene were found in two districts in indoor $PM_{2.5}$, respectively, in winter 2020, in South China (Gao et al. 2022), the nitro-PAH concentrations did not show a significant decrease from winter to summer.

Strandberg et al. (2022) measured particle-bound PAH ($PM_{2.5}$) including parent-, alkylated-, nitro-, and oxy-PAH over a week in 15 occupied homes in southern Sweden during wintertime. Median of the 16 priority US EPA PAH, alkylated-, nitro-, and oxy-PAH concentrations were 0.40 ng/m^3, 0.012 ng/m^3, 0.0027 ng/m^3, and 0.019 ng/m^3, respectively. The low concentrations of nitro- and oxy-compared to parent PAH, compared to other published studies, could possibly be due to low atmospheric photo formation of nitro- and oxy-PAHs because of the cold climate.

4.1.7 Occurrence in Dust of Residences

The results of PAH measurements in dust of residences available to date are summarized in Table 4.5.

In 61 German residences in 2000 ΣPAH_{18} levels ranged from 2400–21,000 ng/g (median: 6400 ng/g) (Fromme et al. 2004a). In 2010, 45 apartments from several German cities an average concentration of ΣPAH_{15} of 2000 ng/g was found (Anders et al. 2012). In Greece, median levels of 441 ng/g for ΣPAH_{16} (range: 26–1394 ng/g) were reported in 11 household dust samples (Christopoulou et al. 2012).

In 51 Canadian homes in 2002–2003 ΣPAH_{13} ranged from 1500 to 325,000 ng/g (median: 9530 ng/g) (Maertens et al. 2008). In the USA, Hoh et al. (2012) described relatively low ΣPAH_{15} concentrations in 2005–2007 in Southern California, significantly higher in smoker homes than non-smoker (by concentration: 990 ng/g vs 756 ng/g). Additionally, lower results were reported with levels of 10–190 ng/g for ΣPAH_{12} in 204 samples from Northern California homes sampled in 2010 (Whitehead et al. 2013a). In a second Canadian study, collected dust in 79 bedrooms in 2014/15, again high levels with a median 2050 ng/g for ΣPAH_{12} (range: 15 to 117,000 ng/g) was observed (Wan et al. 2022). The differences also reflect the breadth of influence on levels in house dust, which can range from the type of sampling and sampling location to the duration and strength of potential sources. In this case, perhaps the preferred indoor heating method could be an explanation.

The ΣPAH_{16} was 1140–3402 ng/g (median: 1972 ng/g) in the indoor dust of houses with smoking in Kuwait significantly higher compared to non-smoking houses with 507–1857 ng/g for ΣPAH_{16} (median: 993 ng/g) (Al-Harbi et al. 2021).

In China, a median of 4540 ng/g for ΣPAH_{16} (range: 1630–29,200 ng/g) was observed in 23 house dust samples from different cities in South China (Kang et al. 2011). Significantly higher median level of 21,000 ng/g for ΣPAH_{15} (range: 8450–121,000 ng/g) was observed in coal-burning households than in homes without residential coal combustion with a median of 6010 ng/g (range: 4–10,600 ng/g) in China in 2011/12 (Wang et al. 2013a). In another Chinese study similar levels were reported with concentrations ranging from 33,900 to 196,400 ng/g for ΣPAH_{16} compared to 21,800 to 329,600 ng/g during summer and winter, respectively, in 31 dust

Table 4.5 Median PAH concentrations in dust of residences in ng/g

References	*N*	Sieve fraction	BaAnt	Chr	BbFlu	BkFlu	BaP	IcdPyr	DBahAnt	Country, time
Europe										
Fromme et al. (2004a)	61	Fine dust	290	550	540	370	270	330	50	Germany, 2000
Tatur et al. (2009)	48	<150 µm	540	160	5280	170	180	180	140	Poland, 2003/04
Langer et al. (2010)	500	n.g.	17	–	–	–	9	–	–	Denmark, 2008
Christopoulou et al. (2012)	11	<62 µm	17	23	23	20	7,8	14	5.2	Greece, 2010
Anders et al. (2012)	53	<63 µm	50	100	120	70	80	80	10	D, UK, USA, 2010
Jakovljevic et al. (2022)	66	<500 µm	19	24	36	13	17	21	4.6	Croatia, 2020/21
North America										
Maertens et al. (2008)	51	< 150 µm	696	1190	1660	532	803	911	185	Canada, 2002/03
Hoh et al. (2012)	43	n.g.	19	46	145		41	37	7.1	USA, 2005–07[a]
Whitehead et al. (2013a)	204	< 150 µm	29	110	61	41	34	35	10	USA, 2010
Kim et al. (2021)	29	<150 µm	<100	<100	<100	–	<250	<500		USA, 2015–18
Asia, Australia										
Kang et al. (2011)	23	<100 µm	130	290	380		240	160	40	China
Wang et al. (2013a)	50	100 µm	450	690	730		250	480	230	China, 2011/12[b]
Wang et al. (2013a)	20	100 µm	1460	3260	2700		2080	490	470	China, 2011/12[c]
Wan et al. (2022)	79	n.g.	162	211	305	< 5.3	221	82	< 1.8	Canada, 2014/15
Wang et al. (2019)	28	sieved	150		63		30	38	21	Australia, 2015
Al-Harbi et al. (2021)	59	< 2 mm	56	127	48	46	50	51	48	Kuwait, 2018/19

* mean value; *n.g.* not given
[a] non-smoker homes
[b] without residential coal burning
[c] with residential coal burning

samples from dormitories, offices, kitchen, and living rooms (Wang et al. 2017a). Median concentration of 660 ng/g for ΣPAH_{13} was reported in 28 dust samples collected from residential houses and offices in Australia in 2015 (Wang et al. 2019).

Another source of dust contamination was the use of building products containing PAH. Especially adhesives containing tar and pitch under wood parquet and end-grain flooring, asphalt flooring (mastic asphalt and asphalt flooring panels), bituminized sealing and roofing membranes and bituminized chipboard, bitumen cork boards, "torfoleum boards" (peat board bonded with bitumen), "tar cork" (tar-bonded granulated cork boards and pipe shells), bitumen grouts, and coal tar oils used as wood preservatives (carbolineum). Use frequently occurred until the late 1960s in Germany, but in some cases continued into the 1990s. The use can, e.g. depending on the condition of the parquet floor, the condition and the PAH content of the underlying adhesive, lead to increased PAH concentrations in house dust. In Berlin, Germany, for example, B[*a*]P concentrations of <100–1,241,000 ng/g (median: 1250 ng/g) were found in 596 house dust samples (in the presence of strip parquet) (ITOX 1999). Also in Frankfurt, Germany, B[*a*]P levels in glue and house dust were measured in 2800 apartments with parquet floors in the former US housing area (Heudorf 1999). In vacuumed samples, this study showed a median of 5000 ng/g and a maximum value of 90,000 ng/g for B[*a*]P. In the USA, Mahler et al. (2010) reported median concentration in dust from parking lots coal tar-based (CT) sealcoats of 4,760,000 ng/g for ΣPAH_{16}, while parking lots with other pavement surface types (asphalt-based sealcoat, unsealed asphalt, concrete) of only 9000 ng/g. In dust samples from apartments with an attached CT parking lot the median was 129 ng/g compared to 5100 µg/g in dust from apartments with parking lots with other pavement surface types.

4.1.8 *Occurrence in Dust of Public Facilities*

The levels of PAH in dust collected from schools and daycare centers are summarized in Table 4.6. Overall, in recent studies the median concentration ranged from around 9 to 177 ng/g. The lowest values were found in six Spanish high schools in the area of Barcelona with a range of 5–21 ng/g (Velázquez-Gómez et al. 2019), the highest in a nationwide survey in 540 classrooms in France in 2013–2017 (Sivanantham et al. 2021).

4.2 Organochlorine Pesticides

Organochlorine pesticides are chlorinated hydrocarbons that have been used as biocides for many decades. The chemical structure of some important compounds is given in Fig. 4.5. They include above all hexachlorobenzene (HCB), hexachlorocyclohexane (HCH), and dichlorodiphenyltrichloroethane (DDT). These are very persistent substances that accumulate in the environment and in the

Table 4.6 Median PAH concentrations in dust of schools and daycare centers in ng/g

References	*N*	Sieve fraction	BaAnt	Chr	BbFlu	BkFlu	BaP	IcdPyr	DBahAnt	Country, time
Langer et al. (2010)	151	n.g.	15	–	–	–	11	–	–	Denmark, 2008
Raffy et al. (2017)	90	<100 μm	–	–	–	–	< 66	–	–	France, 2009/10
Lim et al. (2021)	5	n.g.	44*	–	78*	29*	44*	31*	8.3*	Sweden, 2016/17
Wei et al. (2021)	564	<100 μm	115	–	20	39	78	70	17	France, 2013–17
Sivanantham et al. (2021)	602	n.g.	234*		417*	89*	177*	146*	30*	France, 2013–17
Besis et al. (2023)	20	<1000 μm	83	134	101	69	57	56	26	Greece, 2018
Velázquez-Gómez et al. (2019)	6	<125 μm	16	32	18	23	9	8	29	Spain
Alghamdi et al. (2020)[a]	40	<1000 μm	155	188	248	133	164	149	179	Saudi-Arabia, 2019
Dvoršćak et al. (2022)	10	<500 μm	21	37	36	15	19	23	4.6	Croatia, 2020

* mean value; *n.g.*: not given

[a]air conditioner filter dust

Fig. 4.5 Chemical structure of (**a**) DDT, (**b**) hexachlorobenzene (HCB), (**c**) γ-hexachlorocyclohexane (γ-HCH)

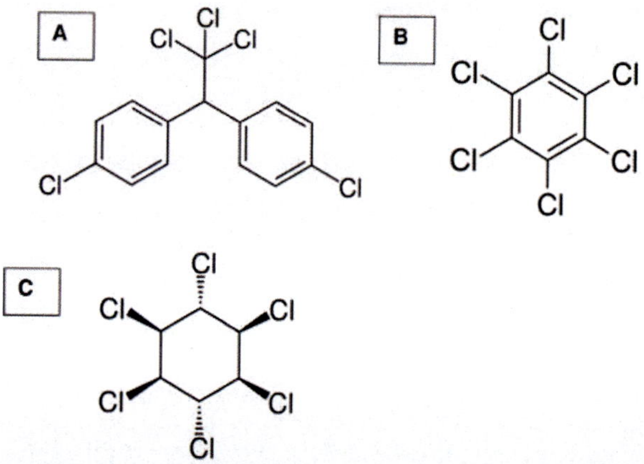

food chain. The main source of human exposure, even if it is clearly declining, is high-fat animal food. Since 2001 at the latest, they have no longer been approved as plant protection products in the EU.

HCH has been used extensively as an insecticide since the 1940s. Its use has since been banned; only lindane (γ-HCH) is still permitted for topical use in scabies and pediculosis. HCH isomers are volatile, relatively persistent substances in the atmosphere and are expected to be capable of long-range transport (ATSDR 2023). The technical mixture contains approximately 60–70% α-HCH, 5–12% β-HCH, 10–15% γ-HCH, 6–10% δ-HCH, and 3–4% ε-HCH (ATSDR 2023).

Technical DDT consists of 65–80% p,p'-DDT and smaller proportions of o,p'-DDT, p,p'-DDE (dichlorodiphenyldichloroethane) and p,p'-DDD (dichlorodiphenyldichloroethane), the latter two also being metabolites of DDT in organisms. DDT has been used extensively as an insecticide, beginning in about 1938. A total of approximately 2.8 million tons of DDT have been produced worldwide, with about 50% produced in the US (UNEP 2015). In Europe, the use of DDT has been banned for decades; only in developing countries, DDT may still be used to control vector-borne diseases such as malaria (ATSDR 2019).

HCB is a white, crystalline solid that is practically insoluble in water and was used as a fungicide and in pyrotechnic materials since the 1940s until it was banned. It has not been produced for decades, but can still occur as an impurity in the production of chlorinated solvents and pesticides.

Due to the Stockholm Convention, HCB, β-HCH, and γ-HCH (lindane) are listed in annex A. Therefore, parties must take measures to eliminate the production and use of these chemicals. DDT is listed in annex B and parties must take measures to restrict the production and use in light of any applicable acceptable purposes and/or specific exemptions.

4.2.1 *Health Aspects*

Extensive substance dossiers from scientific organizations are available on the health effects. For HCB (IARC 2001; EFSA 2006a; ATSDR 2013), HCH (EFSA 2005a; IARC 2015a; ATSDR 2023), and DDT (EFSA 2006b; IARC 2015a; ATSDR 2019).

In terms of acute exposure, γ-HCH has the greatest toxic potential of all HCH isomers, whereas β-HCH is particularly relevant for long-term exposure. A major explanation is the longer half-life in humans of β-HCH of approximately 7 years compared to γ-HCH of 18–163 hours. Effects on the nervous system have been described at very high exposure in animal experiments and in the context of poisoning. Effects on the liver, immune system, reproduction, and hematological changes have been observed in experimental animals at long-term exposure. Studies on carcinogenic potential are inconclusive, partly because there is often mixed exposure with other organochlorine compounds. Lindane has been classified as a human carcinogen (Group 1) by the WHO's International Agency for Research on Cancer (IARC) on the basis of epidemiological findings on non-Hodgkin's lymphoma.

p,p′-DDT and o,p′-DDT are metabolized in the body to p,p′-DDD and o,p′-DDD, while p,p′-DDT is also degraded to DDE. DDE is more lipophilic and is excreted more poorly than the parent compound. Elimination half-lives are 5 years for p, p′-DDT and 10 years for DDE. The main target organs are the liver and nervous system and effects on reproduction. The immune system and fetal development have also been described as target organs. DDT has been classified by IARC as a probable carcinogen (Group 2A) on the basis of the results obtained in animal experiments.

HCB has multiple effects in the organism, with central targets being the liver, ovary, and central nervous system. HCB has been classified by IARC as possibly carcinogenic (Group 2B) based on animal experimental results.

4.2.2 *Occurrence in the Air*

Table 4.7 summarizes the distribution parameters of Σ-HCH (α-HCH, β-HCH, and γ-HCH), DDX (Σ-DDD, Σ-DDE, and Σ-DDT), and HCB.

In European studies, medians for Σ-HCH range from 50 to 746 pg/m^3, for DDX from 22 to 363 pg/m^3, and for HCB from 67 to 711 pg/m^3. Some studies have measured only lindane in indoor air. For example, in 2002, lindane levels of <200–72,200 pg/m^3 (median: 3600 pg/m^3) were measured in 19 French residences (Bouvier et al. 2006). Lindane could also be determined in the gas phase in 30 French homes in 2010–2011 with concentrations ranging from <600–66,000 pg/m^3 (median: 1200 pg/m^3) (Blanchard et al. 2014). The authors do not provide an explanation for the very high values compared to the other European studies. In 602 classrooms from 308 randomly-selected French schools mean concentrations of α-HCH, γ-HCH, DDT, and DDE of 900 pg/m^3, 4200 pg/m^3, 130 pg/m^3, and 70 pg/m^3, respectively, were reported in 2013–2017 (Sivanantham et al. 2021).

Table 4.7 Concentration of organochlorine pesticides in the air (pg/m^3) and in dust (ng/g)

References	Quantity	Σ-HCH		DDX		HCB		Country, time
		Median	Range	Median	Range	Median	Range	
Indoor air								
Bohlin et al. (2008)	5	50	3.3–64	240	63–260	67	48–100	Sweden, 2006
	17	260	47–1300	66	11–340	92	35–890	Great Britain, 2006
De la Torre et al. (2020)	10	536	240–2307	109	24.7–1645	135	<LOQ - 318	Belgium, 2010/11
	10	162	48–292	239	17.0–1236	78	<LOQ - 2531	Italy, 2010/11
	10	375	10.9–6416	22	<LOQ - 397	105	20.3–642	Portugal, 2010/11
	10	693	258–10,657	185	47.5–723	268	153–1475	Spain, 2010/11
Moreau-Guigon et al. (2016)	6	–	–	–	–	110	2–260	France, 2011/12
Audy et al. (2018)	23	274	64–2180	100	41.4–1550	363	160–767	Canada, 2013
	20	746	206–7820	363	48–5650	711	294–1530	Czech Republic, 2013
Demirtepe et al. (2019)	60	235	35–2519	181	17–4214	373	178–1193	Slovakia, 2015
Casey et al. (2022)	12	–	–	260[a]	n.g.-1240[a]	210	50–680	US
	602							France, 2013–17
House dust								
Bräuner et al. (2011)	10	–	–	101	17–19,756	<1.0	<1.0–3.0	Denmark, 2004
Dirtu et al. (2012)	47	170	10–1090	1125	40–56,475	1.3	0.2–105	Romania, 2010
Rasmussen et al. (2022)	1025	–	–	<6.3[b]	n.g.-672[b]	<0.6	n.g.-800	Canada, 2007–10
Audy et al. (2018)	23	7.2	4.0–42	46	24–376	<LOQ	<LOQ	Canada, 2013
	20	52	12–427	42	5.1–2040	<LOQ	<LOQ	Czech Republic, 2013
Demirtepe et al. (2019)	60	6.2	0.3–48	43	7.8–617	6.7	2.3–43	Slovakia, 2015
Wei et al. (2021)	564	32[c]	5.6–340[c,d]	10[b]	<LOQ-82[b,d]	–	–	France, 2013–17

n.g.: not given, *LOQ* limit of quantification, *DDX* sum of o,p'-DDT, p,p'-DDT, o,p'-DDD, p,p'-DDD, o,p'-DDE, and p,p'-DDE
[a] only p,p' DDE
[b] only p,p' DDT
[c] only y-HCH
[d] fifth and 95th percentile

In contrast, concentrations were lower in North Carolina, USA (Morgan et al. 2014). Here, lindane was measured in 129 homes in the range < 90–18,500 pg/m^3 (median < 90 pg/m^3), while concentrations in 13 daycare centers ranged from <90 to 8970 pg/m^3 (median: <90 pg/m^3). In this study, p,p′-DDT was found in 37% of the samples with ranges from <90 to 90200 pg/m^3 (median: <90 pg/m^3). Moreover, Lv et al. (2023) studied a total of 28 homes in southern China in 2019 and reported medians of 19.8 pg/m^3 (10.3–226 pg/m^3) for p,p′-DDE and 11.4 pg/m^3 (1.8–440 pg/m^3) for p,p′-DDT. As part of the Venda Health Examination of Mothers, Babies and their Environment (VHEMBE) study in South Africa, a total of 49 wristbands of children were analyzed and medians of 42.6 pg/m^3 (<4.5–4150 pg/m^3) for p,p′-DDD and 27.3 pg/m^3 (<2.5–7630 pg/m^3) for p,p′-DDT were found (DeLay et al. 2022).

For Σ-HCH, higher concentrations were found in the Czech Republic compared to Canada, probably due to the older age of the houses, and an approximately ten-fold higher content indoors than in outdoor air (Audy et al. 2018). In this study, γ-HCH accounted for 84% of total levels. A study in four European countries by De la Torre et al. (2020) also found γ-HCH to account for 91% of Σ-HCH.

Audy et al. (2018) described p,p'-DDE as the dominant congener, accounting for 61% of total levels in the Czech Republic and 54% in Canada, with indoor air concentrations 4–100 times higher than outdoor air levels. The ratio of p,p'-DDE/p,p'-DDT was above 3 in many samples, indicating older contamination. In contrast, Bohlin et al. (2008) found ratios below 3 in Sweden and England in 2006, indicating a recent use of DDT at that time. The study by De la Torre et al. (2020) in 4 countries could only determine a ratio below 3 (1.7) for Belgium, while it was clearly above 3 in Italy, Portugal, and Spain.

Σ-HCH and DDX correlate significantly with agricultural fields within 1 kilometer of the study site (Demirtepe et al. 2019). For both substance groups, concentrations were higher in older houses (<1977 and <1973, respectively) than in newly constructed ones. While there was a significant correlation between dust and air content for Σ-HCH and DDX in the aforementioned study, this could not be found for HCB.

Figueiredo et al. (2021) compared 58 homes in pesticide spraying sites with 15 controls 500 meter away during non-use periods and application periods. The pesticide concentrations were 5–10 times higher in outdoor air during application periods. Similar concentration differences were observed between exposed homes and controls both during pesticide use and non-use period.

4.2.3 *Occurrence in Dust*

Table 4.7 summarizes the distribution parameters of organochlorine pesticides in dust. For Σ-HCH and DDX the highest medians were described in Romania, while for HCB the medians of all studies were in a comparable low concentration range. Although DDT had been banned since 1980, very high levels of DDX were measured in 47 urban dwellings in eastern Romania compared with other European studies, with p,p′-DDT dominating the total levels by 50–80% (Dirtu et al. 2012). This congener also accounted for 50% of DDX in the Czech Republic (Audy et al. 2018). The β-HCH accounted for the largest proportion of Σ-HCH in the

Czech Republic and in Canada (Audy et al. 2018). Lv et al. (2023) studied a total of 28 homes in southern China in 2019 and reported medians of 5.4 ng/g (<LOQ-653 ng/g) for p,p′-DDD and 26.2 ng/g (<LOQ-3450 ng/g) for p,p′-DDT.

Lindane alone was also measured in 10 Danish dusts, of which 6 were from dwellings (Bräuner et al. 2011). It could only be quantified in one sample with 9 ng/g above the limit of quantification.

4.3 Carbazole

Carbazoles (CZ) are heterocyclic compounds consisting of three rings containing a nitrogen atom. There are also polyhalogenated or polychlorinated carbazoles, in which the degree of chlorination and the position of the chlorine atoms in the molecule differ. The chemical structures of some important carbazoles are summarized in Fig. 4.6. 9H-Carbazole is a constituent of coal tar and petroleum and is formed during incomplete combustion of biomass (IARC 1983; Glarborg et al. 2003; Aydin et al. 2014). Halogenated carbazoles can be formed due to chemical reactions in the environment or are formed as a by-product during various industrial processes, such as the production of indigo dyes, the production of polymers for electrical and electronic products, and the application of specific herbicides (Altarawneh and Dlugogorski 2015; Parette et al. 2015).

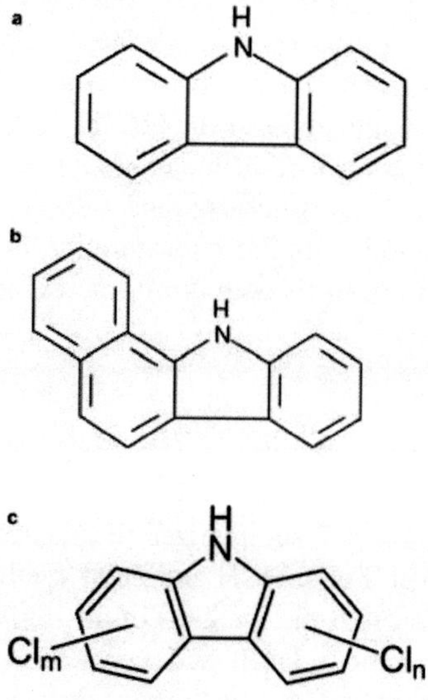

Fig. 4.6 Chemical structure of (**a**) 9H-carbazole, (**b**) 11H-benzo[*a*]carbazole, and (**c**) polychlorinated carbazoles

4.3.1 Health Aspects

Information from studies in humans is not available. There are indications of carcinogenic effects for 9H-carbazole from animal studies, but the IARC has not yet assessed the data situation as sufficient for a classification (IARC 2013). The chlorinated and brominated carbazoles are suspected to be persistent in the environment and to accumulate in organisms (Zhu and Hites 2005). Due to the structural similarity of the chlorinated carbazoles to the halogenated dibenzofurans, they are suspected to have biological effects comparable to those of the halogenated dibenzofurans (Riddell et al. 2015; Mumbo et al. 2015). A recent study demonstrated an endocrine effect for 2,7-dibromo-9H-carbazole (27-BCZ), 3-bromo-9H-carbazole (3-BCZ), and 3,6-dibromo-9H-carbazole (36-BCZ) in multiple assays (Yue et al. 2020).

4.3.2 Occurrence in Indoor Spaces

Exposure data from indoor spaces are hardly available so far. In a German study, 14 dust samples from classrooms of schools, 13 samples from group rooms of daycare centers, and 13 house dust samples from apartments were analyzed for carbazoles and halogenated carbazoles (Fromme et al. 2018). In all indoor rooms, the median level (95th percentile) of the most commonly detected halogenated CZ was 13.3 ng/g (736 ng/g) for 3,6-dichloro-9H-carbazole, 6.2 ng/g (159 ng/g) for 1,3,6-tribromo-9H-carbazole, and 1.2 ng/g (21.1 ng/g) for 2,7-dibromo-9H-carbazole. 9H-carbazole and 11H-benz[a]carbazole had medians and 95th percentiles of 10.3 ng/g and 308 ng/g, respectively, and 3.9 ng/g and 40.2 ng/g, respectively. Again, CZ and 11 chlorinated and brominated CZ were analyzed in indoor dust samples collected from residential houses in rural (n = 51) and urban (n = 55) regions of Hangzhou, China in 2020 (Zhou et al. 2022). While CZ was detected in all dust samples with the concentrations of 0.81–18 ng/g (median 3.7 ng/g), halogenated CZ concentrations were 0.98 ng/g for 3,6-dichloro-9H-carbazole (range: <LOD-1.6 ng/g), 1.2 ng/g for 3,6-dibromo-9H-carbazole (range: <LOD-4.7 ng/g), 0.62 ng/g for 3-monobromo-9H-carbazole (range: <LOD-2.1 ng/g), and 0.41 ng/g for 1,3,6-tribromo-9H-carbazole (range: <LOD-1.0 ng/g).

In addition, the target parameters in the aforementioned German study were also examined in airborne dust (PM_{10}) from 5 homes. In this case, most of the substances could only be found in the detection limit (lower pg/m^3 range). The highest medians were observed for the 9H-carbazole with 7.7 pg/m^3 and for the 2,3,6,7-tetrachloro-9H-carbazole with 6.1 pg/m^3. In indoor air, there are also results available from China that determined 9H-carbazole in $PM_{2.5}$ in the context of using coal as fuel in a 43 m^3 room (Ho et al. 2016). Extremely high concentrations of 7500 to 20,168 ng/m^3 of 9H-carbazole were measured in indoor air during the combustion of 4 different fuels.

4.4 Plasticizers

Plasticizers are substances that are widely added to plastics, rubber, filming coatings, and paints and coatings. They are used to make the products softer, more flexible, and elastic and enable a better manufacturing process and use of the products. Most plasticizers are external plasticizers that are not firmly bonded to the polymer, but are only intended to increase the mobility of the polymer chains. Since they do not form a strong chemical bond, they can be rubbed off or dissolved out of plastics during manufacture, use, and disposal/recycling.

In China and Europe, about 125 and 57 million tons of plastics were produced in 2021 (about 391 million tons worldwide), with about 44% of consumption as packaging materials and about 18% in building equipment (Ceresana 2022).

There are thousands of different substances that could be used as plasticizers, but only about 50–100 of them are also used on a large scale. Table 4.8 shows the most commonly used plasticizers at present. It is clear that this is a chemically very heterogeneous group. This is not surprising in view of the wide range of product

Table 4.8 Key representatives of frequently used plasticizers

Substance group	Name	Abbreviation	CAS-No.
Phthalates	Di-(2-ethylhexyl) phthalate	DEHP	117–81-7
	Di-n-butyl phthalate	DnBP	84–74-2
	Diisobutyl phthalate	DiBP	84–69-5
	Benzyl butyl phthalate	BzBP	85–68-7
	Diisononyl phthalate	DiNP	68,515–48-0; 28,553–12-0
	Diisodecyl phthalate	DiDP	68,515–49-1; 26,761–40-0
Adipates	Diethylhexyl adipate	DEHA	103–23-1
	Di-n-butyl adipate	DnBA	105–99-7
	Diisobutyl adipate	DiBA	141–04-8
	Diisononyl adipate	DiNA	33,703–08-1
	Diisodecyl adipate	DiDA	27,178–16-1
Trimellitate	Trioctyl trimellitate	TOTM	3319-31-1
Terephthalates	Diethylhexyl terephthalate	DEHT	6422-86-2
Hexanoicacid esters	Cyclohexanedicarboxylicaciddiisononyl ester	DINCH	166,412–78-8
Organophosphates	Tri-n-butyl phosphate	TnBP	126–73-8
	Triisobutyl phosphate	TiBP	126–71-6
	Tris(2-chloroethyl) phosphate	TCEP	115–96-8
	Tris(2-chloroisopropyl) phosphate	TCPP	13,674–84-5
	Tri(2-butoxyethyl) phosphate	TBEP	78–51-3
	Triphenyl phosphate	TPhP	115–86-6
	Tris(1,3-dichloro-2-propyl) phosphate	TDCPP	13,674–87-8
	Tricresyl phosphate	TCP	1330-78-5
Citrates	Acetyl tributyl citrate	ATBC	77–90-7
	Butyryl trihexyl citrate	BTHC	82,469–79-2

properties that are to be achieved by the addition of plasticizers. According to manufacturers, approximately 7.7 million tons of plasticizers are consumed worldwide each year, including 1.3 million tons per year in Europe alone (Ceresana 2022). It is estimated that in 2021, the phthalates group accounted for about 80% of use, compared to about 88% in 2005 (IHS 2020). The background to this is the discussions about health risks, in particular reproductive toxicity, which has led to the increased use of other plasticizers as substitutes. Worldwide, however, di (2-ethylhexyl) phthalate (DEHP) with 36% and diisononyl phthalate (DINP) and diisodecyl phthalate (DIDP) with together 31% are still the most widely used plasticizers (Ceresana 2022). Approximately 80–90% of phthalates are used in products made from polyvinyl chloride (PVC) plastic, especially flexible PVC, in which the average plasticizer content can be around 30–35% (IHS 2022).

4.4.1 Phthalate Plasticizers

Phthalates are the diesters of ortho-phthalic acid (1,2-benzenedicarboxylic acid). They do not occur in nature, but are anthropogenic chemical substances that have been used on a large scale for over 50 years. Their chemical structure shows an aromatic ring to which two side chains with different numbers of groups—essentially alkyl groups—are attached (see Fig. 4.7). Plasticizers are added to various materials from plastics, coatings, and adhesives in order to increase the flexibility, molding capacity or elastic properties of products or to enable lower hardness. Di (2-ethylhexyl) phthalate (DEHP) and di-n-butyl phthalate (DnBP), the best-known members of the phthalate group, are used as plasticizers in about 80–90% of cases, especially in the production of flexible PVC and other plastics. Other applications include use as dielectrics in capacitors, defoamers in paper production, emulsifiers for cosmetics, excipients in pharmaceuticals, textile auxiliaries, coating systems, concrete additives, in adhesives, paints and coatings, and sealants. A review on the occurrence in indoor environments can be found in Lucattini et al. (2018).

4.4.1.1 Health Aspects

After inhalation or oral intake, phthalates are well absorbed and rapidly degraded to the respective primary (monoesters) or secondary metabolites and excreted in the urine. Therefore, human biomonitoring of phthalate metabolites in urine is well

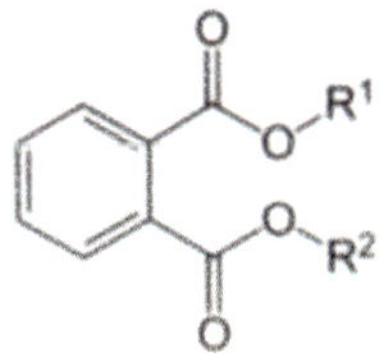

Fig. 4.7 General structural formula of phthalates (R1 and R2 are variable residues)

suited for estimating exposure and back-calculating the total amount of phthalates ingested. In animal experiments, phthalates have a particularly toxic effect on the liver and kidneys in long-term studies. The most significant health impact is the effect on reproduction by influencing Sertoli and Leydig cells as well as testosterone production, which has been demonstrated with some medium-chain representatives of this substance class, e.g. DEHP. From epidemiological studies there is evidence of effects on sex development and in the thyroid gland. Summaries of the various effects can be found in the following literature: HC (2015), Skakkebaek et al. (2016), Rowdhwal and Chen (2018), Radke et al. (2018, 2019, 2020), and Eales et al. (2022).

4.4.1.2 Occurrence in the Air of Residences

The statistical parameters of current studies are summarized in Table 4.9. In Europe, medians varied between 28 and 2850 ng/m^3 for DEHP and 86–1083 ng/m^3 for DnBP, the two most common phthalates in indoor air. The lowest concentrations were detected by Mandin et al. (2016) in PM_{10} from 285 French homes. In this study DnBP median was only <0.8 ng/m^3. A study of 59 German apartments in 2000–2001 found the highest median indoor air levels of 1083 ng/m^3 for DnBP (maximum: 5586 ng/m^3) (Fromme et al. 2004b) and the highest DEHP median of 2150 ng/m^3 (260–11,000 ng/m^3) were measured in 50 apartments in Munich, Germany and the surrounding area in 2005 (Fromme et al. 2010).

Comparable concentration levels were reported in the USA and Asia with medians from 45–68 ng/m^3 and 56–2437 ng/m^3 for DEHP, respectively, and 90–140 ng/m^3 and 200–1939 ng/m^3 for DnBP. For example, Rudel et al. (2010) describe medians of 68 ng/m^3 for DEHP and 140 ng/m^3 for DnBP in 50 Californian houses. Only in 10 newly renovated houses in China significantly higher medians could be found for DEHP at 1322 ng/m^3 and for DnBP at 1066 ng/m^3 (Pei et al. 2013).

Due to their physicochemical characteristics, the individual phthalates are distributed differently in the air between the gas and particle phases. While DEHP and DiNP, and to some extent DnBP, are predominantly found in the particulate fraction, DEP is predominantly found in the gas phase (Fromme et al. 2004b, 2013a; Blanchard et al. 2014; Tran and Kannan 2015; Huang et al. 2021). In light of this, the distribution in air and house dust is also different. Figure 4.8 shows the proportions of each phthalate in the total concentration in air and dust. Some studies have found a statistically significant relationship between air and house dust content, especially for DiBP, DnBP, and BzBP (Bergh et al. 2011a; Fromme et al. 2013a; Dodson et al. 2015; Gaspar et al. 2014).

Mandin et al. (2016) reported significant associations between BzBP or DEHP and the season of sampling, as well as significantly higher concentrations of DEP, DnBP, DiBP, DEHP, and DiNP in smokers' apartments compared to non-smokers' apartments.

Table 4.9 Phthalates in the air of residences in ng/m^3

	N	DEP		DnBP		DiBP		BzBP		DEHP		DiNP		Country
		Median	95th percentile	Median	95th percentile	Median	95th percentile	Median	95th percentile	Median	95th percentile	Median	95th percentile	
Fromme et al. (2004b)	59	643	1860	1083	2453	–	–	18	75	156	390	–	–	Germany, 2000/01
Rudel et al. (2010)	43	330	1200	140	550	130	370	7	38	68	130	–	–	USA
Bergh et al. (2011a)	10	1300	–	850	–	270	–	21	–	200	–	–	–	Sweden
Bergh et al. (2011b)	169	210	–	190	–	230	–	9.0	–	220	–	–	–	Sweden
Mandin et al (2016)	285	<0,8#	9.8#	<0.8#	25#	0.9#	23#	1.6#	32#	46#	389#	7.9#	50#	France, 2003–05
Fromme et al. (2010)	50	320	770	590	1850	660	1330	10	40	2150	9650	–	–	Germany, 2005
Kanazawa et al. (2010)	41	61	–	200	–	75	–	<3	–	147	–	<64	–	Japan, 2006/07
Luongo and Östman (2016)	62	200	–	207	–	266	–	9	–	232	–	–	–	Sweden, 2008
Dallongeville et al. (2016)	150	182	376*	86	176*	354	1350*	1.8	7.9*	28	70*	11	32*	France, 2012/13
Pei et al. (2013)	10	2290+	–	1939+	–	–	–	3975+	–	2437+	–	–	–	China, 2011/12
Wang et al. (2014)	28[a]	–	–	555#		720#	–	–	–	560#				China, 2012/13
Tran and Kannan (2015)	60	432	–	90	–	73	–	3.8	–	45	–	–	–	USA, 2014
Chen et al. (2018a)	134+	5.7	–	376	–	–	–	25	–	56	–	–	–	China, 2014–16
Chen et al. (2022b)	26	2.2	9.3	21	71	–	–	13	316	410	2007	–	–	China, 2014–16
Huang et al. (2021)	100	101	–	932	–	500	–	0.17	–	174				China, 2018

* 90th percentile; +: mean; #: only particle-phase
DEP diethyl phthalate; *DnBP* di-n-butyl phthalate, *DiBP* di-iso-butyl phthalate, *BzBP* butyl benzyl phthalate, *DEHP* Di(2-ethylhexyl) phthalate, *DiNP* Di-isononyl phthalate
[a] residences and offices

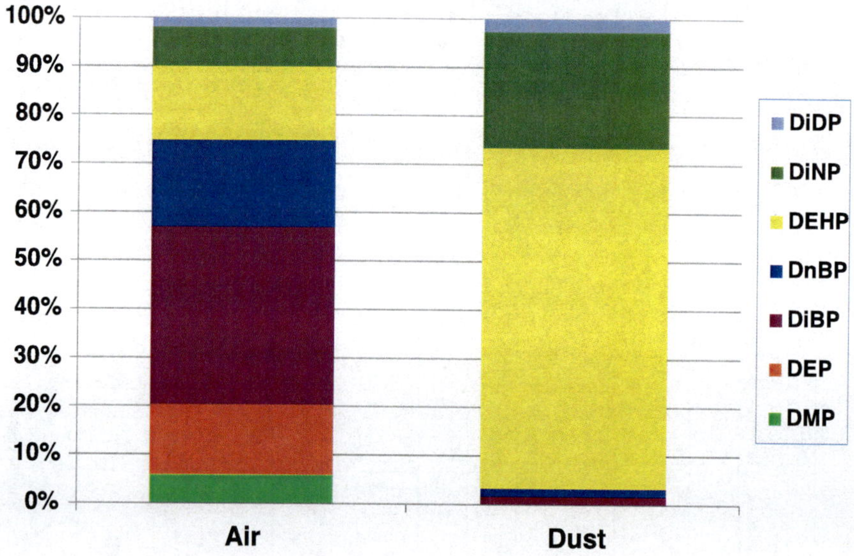

Fig. 4.8 Proportion of individual phthalates in air and dust of daycare centers on the total concentrations using medians (modified from Fromme et al. 2013b)

Little research has been done to date on the relationship between indoor air content and urinary metabolite excretion. Adibi et al. (2003) described for the first time a significant relationship between DEP, DnBP, and BzBP in indoor air and their respective metabolites in urine. This relationship was also confirmed in German kindergarten children for DnBP, DiBP, and DEHP, but not for the other phthalates (Fromme et al. 2013a). Again, Hammel et al. (2019) reported an association of DEHP and DiNP with their corresponding urinary metabolites.

4.4.1.3 Occurrence in the Air of Public Facilities

To date, only a few results have been published on the occurrence of phthalates in these indoor environments. Overall, the studies showed a concentration level comparable to that of apartments and houses (see Table 4.10). In Europe, results from three countries showed medians from 108–458 ng/m^3 for DEHP and 227–1188 ng/m^3 for DnBP. In the daycare centers sampled in 2011/2012, DEHP levels in common areas ranged from 25 to 889 ng/m^3 and DnBP levels ranged from 49 to 1276 ng/m^3 (Fromme et al. 2013b). In the USA, ranges of <LOQ – 2710 ng/m^3 for DEHP and 50–2650 ng/m^3 for DnBP were reported in 40 early childhood education facilities in California (Gaspar et al. 2014) as well as 5.9–92.2 ng/m^3 for DEHP and 21.1–98.7 ng/m^3 for DnBP in six schools in Albany, New York (Tran and Kannan 2015).

Table 4.10 Phthalates in the air of schools and daycare centers in ng/m^3

	N	DEP		DnBP		DiBP		BzBP		DEHP		DiNP		Country
		Median	95th percentile	Median	95th percentile	Median	95th percentile	Median	95th percentile	Median	95th percentile	Median	95th percentile	
Fromme et al. (2004b)	74	353	767	1188	7376	–	–	<10	26	458	1510	–	–	Germany, 2000/01
Bergh et al. (2011a)	10	870	–	600	–	190	–	21	–	240	–	–	–	Sweden
Raffy et al. (2017)	30–62	221	515	228	744	–	–	19	–	108	417	35	214	France, 2009/10
Gaspar et al. (2014)	40	210	1200	520	2340	100	530	10	170	10	340	–	–	USA, 2010/11
Fromme et al. (2013b)	63	183	383	227	884	468	916	<10	<10	194	765	102	341	Germany, 2011/12
Wang et al. (2017b)	2	80, 56	–	255/309	–	–	–	43/44		319/338	–	–	–	China, 2012
Sivananthan et al. (2021)	602+	371	–	241	–	1373	–	26	–	93	–	17	–	France, 2013–17
Tran and Kannan (2015)	6	211	–	66	–	59	–	11	–	10	–	–	–	USA, 2014

+: mean; DEP: diethyl phthalate; DnBP: di-n-butyl phthalate; DiBP: di-iso-butyl phthalate; BzBP: butyl benzyl phthalate; DEHP: Di(2-ethylhexyl) phthalate; DiNP: Di-isononyl phthalate

4.4.1.4 Occurrence in the Air of Transport Microenvironments

Phthalates are widely used in the fabrication of indoor material in cars, busses, and other transportation systems. To date, only a few results have been published in the scientific literature on the concentrations found in indoor spaces of motor vehicles. In one study, 6 brand-new passenger cars were examined on a test rig (Wensing and Bauhof 1999). The mean content was 600 ng/m^3 for DnBP and 9600 ng/m^3 for DEHP. However, DEHP could only be detected at all in two vehicles. In the interior air of a one-year-old truck, a level of plasticizers (with the fan switched on) of 8000 ng/m^3 DEHP and 300 ng/m^3 DnBP was found in summer 2001 (Leimbrock 2001). In a study of 23 cars in 2007, concentrations of <150–1400 ng/m^3 (DEP), <150–1630 ng/m^3 (DnBP), and <150–3656 ng/m^3 (DEHP) were measured inside the car after a collection period of 16 hours (Geiss et al. 2009). Chi et al. (2017) collected air samples from 105 busses, 40 subways, 30 taxis, and 60 private cars in China. They reported mean gas-phase and particle-phase concentrations of 1302 and 275 ng/m^3 for DEP, for 987 and 894 ng/m^3 DnBP, and 1324 and 6935 ng/m^3 for DEHP, respectively, in private cars. The concentrations of particle-phase phthalates followed the trend, private car > busses > taxi > subway. In the subway corresponding mean gas-phase and particle-phase concentrations were 857 and 1528 ng/m^3 for DnBP and 3183 and 3118 ng/m^3 for DEHP. In another study, Xia et al. (2018) investigated airborne phthalate levels in 10 cars and reported means of 3700 ng/m^3 for DnBP and 6300 ng/m^3 for DEHP, respectively.

4.4.1.5 Occurrence in Dust

A compilation of results in dust of residences can be found in Table 4.11. What is striking in all studies is the large variability of the individual results in house dust, which can vary over several orders of magnitude.

In Europe, they showed median levels of 62–1542 μg/g for DEHP, 7.1–130 μg/g for DnBP, and 2.7–219 μg/g for BzBP. The highest DEHP and BzBP concentrations from 231–1763 μg/g and 5.0–816 μg/g, respectively, were found in 30 residences in Berlin, Germany (Fromme et al. 2004b) and the lowest in 18 houses in Belgium (DEHP range: 9–479 μg/g) and in 9 homes in the Netherlands (BzBP range: 0.7–18 μg/g) (Christia et al. 2019). For DnBP, median levels ranged from 3.2 μg/g (range: 1.2–146 μg/g) in the Netherlands to 130 μg/g (range: 17–260 μg/g) in Sweden.

In North America, no substantially different exposure situation could be determined compared to Europe. For example, the lowest medians were reported by Shin et al. (2020) in 38 dust samples collected in Northern California with a median of 39 μg/g for DEHP and a median of 7.2 μg/g for BzBP.

Again, the results from Asian studies showed similar concentration ranges compared to European or American investigations. Nevertheless, the medians for BzBP were lower, ranging from only 0.12 μg/g in 72 residences in seven geographical regions in China (Li et al. 2021) to 2.9 μg/g in 215 urban houses in Nanjing, China

Table 4.11 Phthalates in dust from residences in µg/g

	N	DEP		DnBP		DiBP		BzBP		DEHP		DiNP		Country, year
		Median	95th p.	Median	95th p.	Median	95th p.	Median	95th p.	Median	95th p.	Median	95th p.	
Europe														
Fromme et al. (2004b)	30	159	632	30	141	–	–	219	816	1542	1763	–	–	Germany, 2000/01
Bergh et al. (2011a)	10	3.7	–	130	–	4	–	17	–	680	–	–	–	Sweden
Langer et al. (2010)	497	1.7	200	15	74	27	140	3.7	50	210	980	–	–	Denmark
Abb et al. (2009)	30	–	–	87	–	–	–	15	–	604	–	129	–	Germany
Fromme et al. (2010)	50	2.6	96	17	90	18	104	28	215	670	2000	155	625	Germany, 2005
Preece et al. (2021)	455	2.0[a]	–	22[a]	–	8.2[a]	–	29[a]	–	150[a]	–	–	–	Sweden, 2007–10
Luongo and Östman (2016)	62	14	–	103	–	104	–	16	–	449	–	106	–	Sweden, 2008
Blanchard et al. (2014)	30	2.8	–	12	–	19	–	9	–	289	–	130	–	France, 2010/11
Kademoglcu et al. (2018)	10	1.8	–	10	–	–	–	–	–	225	–	–	–	Norway, 2013/14
Daniels et al. (2021)	646	<0.7	5.3	7.6	44.7	8.7	89.5	3.0	88.5	153	770	193	1815	Germany, 2014–17
Demirtepe et al. (2021)	55	1.42	–	24	–	–	–	2.9	–	319	–	–	–	Slovakia, 2015
Christia et al. (2019)	6	0.64	–	7.1	–	7.0	–	3.9	–	114	–	72	–	Ireland, 2017
Christia et al. (2019)	18	1.4	–	22	–	9.4	–	8.5	–	62	–	26	–	Belgium, 2017
Christia et al. (2019)	9	0.23	–	3.2	–	4.7	–	2.7	–	111	–	34	–	Netherlands, 2017

(continued)

Table 4.11 (continued)

	N	DEP		DnBP		DiBP		BzBP		DEHP		DiNP		Country, year
		Median	95th p.	Median	95th p.	Median	95th p.	Median	95th p.	Median	95th p.	Median	95th p.	
Nagorka et al. (2022)	645	<0.45–0.7	–	7.2	–	7.9	–	2.9	–	147	–	212	–	Germany, 2015–17
America														
Philippat et al. (2015)	145	1.0	6.7	10	37	–	–	13	200	187	656	–	–	USA
Dodson et al. (2015)	50	2.1	6.3	11	35	–	–	19	220	140	460	–	–	USA, 2006
Kubwabo et al. (2013)	126	2.0	–	16.8	–	5.2	–	42.3	–	462	–	111	–	Canada, 2007–10
Hammel et al. (2019)	188	1.9	12	9.6	73	4.4	34	14	133	119	484	79	788	USA, 2014–16
Wan et al. (2022)	79	1.28	4.34[b]	13	37[b]	12	29[b]	88	853[b]	257	625[b]	561	2890[b]	Canada, 2014/15
Shin et al. (2020)	38	1.0	4.2	5.0	16	3.5	26	9.2	135	39	78			USA, 2015/16
Subedi et al. (2017)	11	0.5	–	3.5	–	–	–	51		73	–	–	–	USA, 2016
Asia														
Bamai et al. (2014)	128	<0.5	–	17	–	3.1	–	2.0	–	1110	–	139	–	Japan, 2009/10
Zhang et al. (2013)	215	0.2	3.1	24	176	–	–	2.9	9.7	183	1750	–	–	China, 2011
Wang et al. (2014)	28[c]	–	–	135	–	234	–	–	–	582	–	–	–	China, 2012/13
He et al. (2016)	6	0.47	–	75	–	–	–	0.52	–	698	–	–	–	China, 2014/15
Huang et al. (2021)	100	0.16	–	75	–	11	–	–	–	355	–	–	–	China, 2018
Li et al. (2021)	72	0.16	–	24	–	8.8	–	0.12	–	51	–	–	–	China, 2019

[a] geometric mean
[b] 90th percetile
[c] residences and offices

(Zhang et al. 2013). Moreover, Liu et al. (2022) integrated the national exposure level for indoor dust from literature through systematic review in China and found that DnBP, DiBP, and DEHP were the most abundant phthalates in residences with medians of national dust phase concentration of 96.3, 94.1 μg/g, 634.2 μg/g.

Study results in the house dust of schools and daycare centers have, in contrast to residential interiors, only been published to a limited extent so far (Table 4.12). While dust exposure to most phthalates is low with relatively uniform in level compared to residences, high levels of DEHP were reported. For example, a 95th percentile of 5830 μg/g was observed in 90 classrooms in Brittany, France (Raffy et al. 2017). Comparable results were also reported from 63 German daycare centers with levels from 99–10,086 μg/g studied in 2011 (Fromme et al. 2013b). In this study, the medians for DiNP was 302 μg/g (30–7091 μg/g) and for DiDP 34 μg/g (5.0–571 μg/g). In 602 classrooms from 308 randomly-selected French schools mean values of 106 μg/g for DnBP, 320 μg/g for DiBP, 1900 μg/g for DEHP, and 2115 μg/g for DiNP were reported in 2013–2017 (Sivanantham et al. 2021). Moreover, Liu et al. (2022) integrated the national exposure levels from 26 chinese studies between 2000 and 2017 and found medians of 92.7 μg/g for DnBP, 39.6 μg/g for DiBP, and 1395 μg/g for DEHP.

A comparable exposure situation is also described in two studies in the USA and in two investigations in Asia. For example, Subedi et al. (2017) found levels of 44–190 mg/kg (DEHP), 2.4–18.3 mg/kg (DnBP), and 4.6–138 mg/kg (DiBP) in 12 dust samples from 11 childcare facilities across the USA in 2016. In four schools in three cities in South China, DEHP and DnBP ranged from 372–952 μg/g and 42–74 μg/g, respectively (Kang et al. 2012), while in 50 Korean nursery schools high levels (95th percentile: 9030 μg/g) were reported (Kim et al. 2013). Additionally, high levels of DiNP were found in both studies from Asia.

To date, there are only a few studies that have investigated the significance of specific sources in more detail. For example, in some studies, it was observed that DEHP from PVC flooring materials can migrate into the indoor environment (Xu et al. 2009, 2010; Liang and Xu 2014; Jeon et al. 2016; Larsson et al. 2017; Preece et al. 2021; Li et al. 2021). In others that PVC can also be a source of BzBP (Bergh et al. 2011b; Shu et al. 2019; Preece et al. 2021) and DiNP (Wan et al. 2022). In a Japanese study conducted in 128 dwellings the number of PVC interior materials was significantly positively correlated with the levels of DEHP and DiNP in multi-surface dust (Bamai et al. 2014). In an experimental study the migration of DEHP from flooring materials increased when the flooring was heated (50 °C) (Jeon et al. 2016). Dust levels of DnBP, BzBP, and DEHP were also significantly influenced by polishing wax for solid-wood floors in 215 urban houses in Nanjing, China (Zhang et al. 2013). Moreover, a significant association of DEP with fragrance products was reported by Wan et al. (2022).

Table 4.12 Phthalates in dust from schools and daycare centers in μg/g

	N	DEP		DnBP		DiBP		BzBP		DEHP		DiNP		Country, year
		Median	95th p.	Median	95th p.	Median	95th p.	Median	95th p.	Median	95th p.	Median	95th p.	
Bergh et al. (2011a)	10	4.2	–	150	–	3.0	–	31	–	1600	–	–	–	Sweden
Langer et al. (2010)	151	2.2	7.8	38	160	23	58	17	150	500	1900	–	–	Denmark
Raffy et al. (2017)	30	2.9	6.6	38	< 53	< 53	< 53	105	468	1430	5830	1030	4100	France, 2009/10
Gaspar et al. (2014)	39	1.4	4.5	14	80	9.3	40	47	1046	172	498	–	–	USA, 2010/11
Fromme et al. (2013a)	63	1.4	7.4	21	95	20	174	6	93	888	7616	302	2955	Germany, 2011/12
Wang et al. (2017b)	45	–	–	31	–	95	–	–	–	202	–	–	–	China, 2010/11
Kim et al. (2013)	50	<4.9	–	52	146	–	–	50	–	3030	9030	946	3200	Korea, 2012
Kang et al. (2012)	4	1.69	–	52	–	57	–	10	–	819	–	–	–	China, 2012
Larsson et al. (2017)	100	0.46	130	21	140	6.4	46	8.7	110	290	1900	380	3400	Sweden, 2015
Lee and Choi (2020)	11	33	39[a]	70	317[a]	–	–	–	<4.0	2190	7160[a]	3870	19,000[a]	Korea
Qu et al. (2021)	69	–	–	33	414[a]	21	475[a]	0.4	8.0[a]	186	2260[a]	–	–	China, 2017
Besis et al. (2023)	20	2.5	–	84	–	22	–	118	–	536	–	–	–	Greece, 2018

[a] maximum

4.4.2 Nonphthalate Plasticizers

Due in particular to the reproductive toxicity of some phthalates, alternative plasticizers have increasingly been produced and marketed on a large scale. Particularly interesting are substitutes that can be used in technically equivalent ways and also have more favorable toxicological properties.

The main groups are briefly described below:

- *Cyclohexanedicarboxylic acid diisononyl ester (DINCH).* DINCH is a complex mixture of diisononyl esters of 1,2-cyclohexanedicarboxylic acid ($C_{26}H_{48}O_4$, CAS-No.: 166412–78-8 [mixture of isomers]), which is produced on an industrial scale by BASF AG, in particular under the trade name Hexamoll DINCH®. It was originally intended primarily for sensitive applications in the medical field, in toys and food contact materials. This plasticizer shows improved UV and temperature resistance compared to phthalates. Since its market launch in 2002, production volume has risen sharply.
- *Adipates.* Adipates are esters of the alcohol 2-ethylhexanol and adipic acid, which have long been produced in large quantities. They have a higher volatility than phthalates and are better able to migrate from plastics. The main representatives are dibutyl adipate (DBA) (CAS-No.: 105–99-7), diisobutyl adipate (DiBA) (CAS-No.: 141–04-8), di(2-ethylhexyl) adipate (DEHA, CAS-No.: 103–23-1), and diisononyl adipate (DiNA, CAS-No.: 33703–08-1). DEHA has become one of the main substitutes for DEHP as an alternative plasticizer in medical applications such as transfusion bags and tubing and in food packaging and cosmetics. DEHA is also used in a variety of applications in lubricants, sealants, and adhesives.
- *Terephthalates.* These are the esters of terephthalic acid, another isomeric form of ortho-phthalates. Di(2-ethylhexyl) terephthalate (DEHT, CAS-No.: 6422-86-2), for example, has better resistance to low temperatures and lower volatility than DEHP. However, terephthalates are less resistant and have limited suitability for the production of durable flexible PVC articles. Typical applications are toys, children's articles, and drinking and storage containers.
- *Trimellitates.* They are produced by esterifying C_7-C_{10} alcohols with trimellitic anhydride. Main representatives are tri-2-ethylhexyl trimellitate (TOTM or TEHTM, CAS-No.: 3319-31-1) and tricapryl trimellitate (TCTM, CAS No.: 27251–75-8). The structure of TOTM is similar to that of di(2-ethylhexyl) phthalate, with the exception that TOTM contains a third functionality on the aromatic ring (CPSC 2018d). Due to their low volatility and high temperature resistance, these plasticizers are used in the automotive industry and for insulation or sheathing of power cables. However, the medical sector is also a major area of application.
- *Citrates.* Main representative of thios group is acetyl tri-n-butyl citrate (ATBC, CAS-No.: 77–90-7). Citrate ester plasticizers are produced from the starting material citric acid and are used in numerous areas, e.g. in vinyl resins and films for toys, children's articles, cosmetics, medical materials and in the food

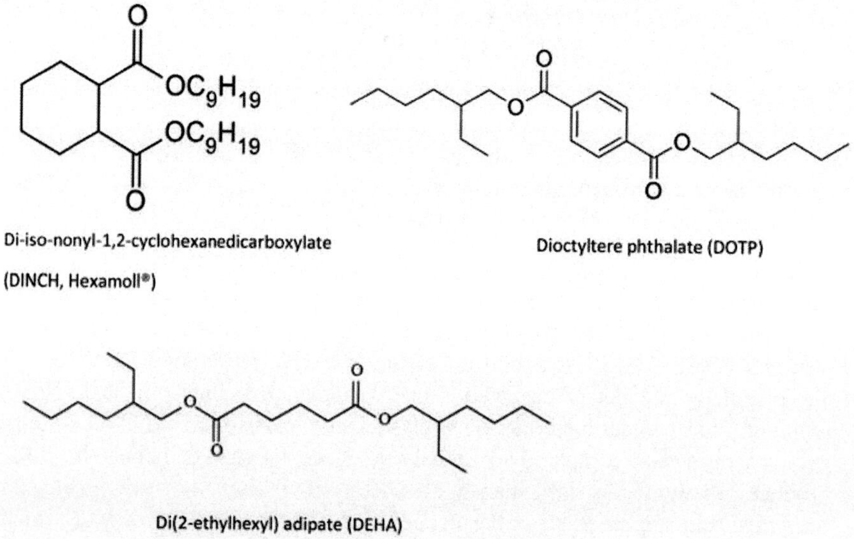

Fig. 4.9 Chemical structure of commonly used plasticizers (non-phthalates)

sector. However, compared to ortho-phthalates, they have lower durability and show higher volatility loss and stronger migration behavior, which limits their use.

- *Benzoates.* Monobenzoates or dibenzoates like isodecyl benzoate (IDB, Cas No.: 131298–44-7) and dipropylene glycol dibenzoate (DPGDB, Cas No.: 27138–31-4) are high solvating plasticizers for PVC and they are used mostly in flooring, wall coverings, automotive—specifically in radiators of the cars—adhesive and sealants, inks and waxes, film and sheet products (EP 2023).
- *Vegetable oil derivatives (bio-plasticizers).* Plasticizers from this group are epoxidized vegetable oils with a generally low volatility. One example is epoxidized soybean oil (ESBO). Methyl oleate (MO, CAS No.: 112–62-9) and glycerol monooleate (GMO, CAS No.: 25496–72-4) are methyl ester of the oleic acid. Overall, they are used to improve heat resistance in the manufacture of PVC because of the epoxy group in the molecule and also in lid gaskets, rubber articles, epoxy resins, paints and coatings.
- *Organophosphates.* This class of substances is discussed in Sect. 4.9.

Some chemical structures of plasticizers are shown in Fig. 4.9.

4.4.2.1 Health Aspects

Results from studies on humans are not yet available. All plasticizers are characterized by low acute toxicity in animal experiments. However, the toxicological data on long-term effects are limited for most substances. DINCH showed mild effects on the liver, urinary tract, and thyroid gland when ingested chronically, but not on the reproductive system. Long-term studies showed effects for DEHA essentially on the liver. It caused liver tumors in mice, but not rats, due to peroxisome proliferation and is probably of little relevance to humans. Studies are available for DEHT showing low systemic toxicity and the absence of reproductive and carcinogenic effects. Trimellitates have very low systemic toxicity and act as peroxisome proliferators in the liver. With repeated administration, decreased birth weight and adaptive effects in the liver are described for ATBC. Extensive compilations of effects can be found in various scientific reviews: DEHA: HC (2011a), CPSC (2018a); DiNA: CPSC (2019a); DBA, DiBA: CPSC (2019b); DINCH: Bhat et al. (2014), ECHA 2016a, Giovanoulis et al. (2018), CPSC (2018b); DEHT: ECHA (2016b), CPSC (2018c); TOTM: CPSC (2018d); ATBC: CPSC (2018e); all plasticizers: SCENIHR (2015), Bui et al. (2016).

4.4.2.2 Occurrence in the Indoor Air

Measurements in indoor air are hardly available for these plasticizer groups so far. First, in the USA, DEHA was measured in 50 homes in all samples above the limit of quantification with a median of 27 ng/m^3 (6.6–76 ng/m^3) (Rudel et al. 2010). In a Japanese study in which six houses were investigated in 2012, levels of 5–27 ng/m^3 (DEHT), 40–590 ng/m^3 (DEHA), 3–2800 ng/m^3 (DnBA), and 27–920 ng/m^3 (DiBA) could be measured (Takeuchi et al. 2014). From Germany, results are available from 63 daycare centers investigated in 2011/2012 (Fromme et al. 2016b). Here, medians of 20 ng/m^3 (5–226 ng/m^3) were observed for DEHT, 34 ng/m^3 (5–206 ng/m^3) for DEHA, 91 ng/m^3 (<2–134 ng/m^3) for DnBA, 35 ng/m^3 (<2–89 ng/m^3) for DiBA, and 45 ng/m^3 (<2–22 ng/m^3) for ATBC. In 15 bedrooms medians of <0.5 ng/m^3 (<0.5–5.6 ng/m^3) for DINCH, 2.2 ng/m^3 (<0.7–36 ng/m^3) for DEHT, 3.7 ng/m^3 (0.8–20 ng/m^3) for DEHA, and 1.7 ng/m^3 (0.6–15 ng/m^3) for ATBC were reported in South China in 2017 (Tang et al. 2020).

4.4.2.3 Occurrence in Dust of Residences

The results in dust samples from residences are summarized in Table 4.13. Overall, the data situation is still limited and the number of samples examined is often small.

Measurements in homes show increasing concentrations of DINCH in house dust over time. For example, an increase in DINCH concentrations, as well as in the number of results above the limit of quantification, was evident in samples from Germany from 2001/2002 to 2009 after the market launch in 2002 (Nagorka et al. 2011a, b). The increase in concentrations in samples from 2009 onward was particularly marked. A similar temporal trend was also seen for DEHT in this study. In the German Environmental Health Study 2014–2017, significantly higher concentrations of DINCH and DEHT were again found in house dust (sieved to 63 μm) in 646 children's living rooms (Daniels et al. 2021). DEHS could be

Table 4.13 Non-phthalate plasticizers in dust of residences and daycare centers in μg/g

References	*N*	DINCH		DEHT		DEHA		TOTM		ATBC		Country, Year
		Median	95th percentile	Median	95th percentile	Median	95th percentile	Median	95th percentile	Median	95th percentile	
Residences												
Nagorka et al. (2011a, b)	50	–	–	<0.8	<0.8	–	–	–	–	–	–	Germany, 1997–99
	274	<0.3	<0.3	<0.8	<0.8	<0.6	2.5	1.5	6.6	–	–	Germany, 2001/02
	593	<0.3	2.5	<0.8	4.6	0.5	3.5	2.0	6.5	–	–	Germany, 2003/06
	36	2.2	76	1.4	24	1.6	7.0	2.0	5.3	2.5	26	Germany, 2009
Preece et al. (2021)	482	37	630+	–	–	–	–	–	–	–	–	Sweden, 2007–10
Giovanoulis et al. (2018)	60	33	162	–	–	–	–	–	–	–	–	Norway, 2013–14
Christia et al. (2019)	18	74	1051+	23	101+	5.4	272+	5.2	130+	1.9	101+	Belgium, 2017
	9	9.5	19+	37	764+	2.4	17+	1.9	46+	2.3	21+	Netherlands, 2017
	6	33	111+	164	247+	2.9	4.7+	2.6	3.2+	7.5	17+	Ireland, 2017
Daniels et al. (2021)	646	17.3	1757	82.9	832	<1.0	6.8	<5.0	8.7	9.9	79.7	Germany, 2014–17
Demirtepe et al. (2021)	55	<40	13,200+	71	4713+	7.5	274+	–	–	13.4	307+	Slovakia, 2015
Shin et al. (2020)	38	–	–	–	–	<5.0	17.5	–	–	35.7	27.5	USA, 2015/16
Tang et al. (2020)	15	3.0	12+	<2.9	48	<0.4	1.5	1.8	101+	2.7	37+	China, 2017

Tan et al. (2023)	36	2.5	–	–	–	0.16	–	5.8	–	3.1	–	North China, 2017–22
	41	11.5	–	–	–	1.4	–	29.7	–	7.2	–	South China, 2017–22
	21	0.8		–	–	0.05	–	0.17	–	0.16	–	Vietnam, 2017–22
	42	1.0	–	–	–	0.5	–	0.12	–	2.43	–	Australia, 2017–22
	17	0.3	–	–	–	0.17	–	0.27	–	1.17	–	US, 2017–22
Daycare centers												
Fromme et al. (2016b)	63	302	1622	40	256	49	307	<7	<7	24	407	Germany, 2011/12
Larsson et al. (2017)	100	49	1300	86	1500	9.7	170	–	–	6.2	82	Sweden, 2015
Christia et al. (2019)	3/2*	22/16	52/23	83/57	237/72	4.8/2.2	70/2.4	1.5/1.0	3.1/1.0	32/6.7	89/9.3	Sweden, 2017

+ maximum values; * results for measuring period in winter/measuring period in spring quantified in this study only in 9% of the samples with a 95th percentile of 5.4 µg/g. In a 2017 study of homes from Belgium ($n = 18$), the Netherlands ($n = 9$), and Ireland ($n = 6$), higher levels were found in dust for DINCH and DEHT, but not for the other plasticizers, compared with the aforementioned German study (Christia et al. 2019). Significant differences were also observed between the 3 countries.

High DEHA levels between 0.9 and 391 µg/g (median: 6.0 µg/g) were observed in 120 residences in the USA in 1999–2001 (Rudel et al. 2003). Additionally, Subedi et al. (2017) found high levels for DEHA and especially ATBC in North American dust samples from 11 homes in 2016. Here, concentrations of DEHA ranged from <LOQ – 225 µg/g and of ATBC from <LOQ – 2180 µg/g.

In 15 bedrooms medians of 2.96 µg/g (1.1–11.6 µg/g) for DINCH, <2.9 (<2.9–47.7 µg/g) for DEHT, <0.4 (<0.4–1.52 µg/g) for DEHA, and 2.7 µg/g (1.3–37.3 µg/g) for ATBC were reported in South China in 2017 (Tang et al. 2020). Tao et al. (2022) analyzed 34 emerging plasticizers in 49 dust samples from South China in 2017–2019 and found medians of 61.6 µg/g for glycerol monooleate (GMO), 16.4 µg/g for methyl oleate (MO), 8.9 µg/g for ATBC, 1.6 µg/g for dipropylene glycol dibenzoate (DPGDB), and of 0.75 µg/g for tributyl citrate.

In a recent study, Tan et al. (2023) investigated 45 nonphthalate plasticizers along with the major phthalates in 157 dust samples from China, Vietnam, Australia, and the USA in 2017–2022. The sum of all nonphthalates were on a median basis 252 µg/g in South China, 199 µg/g in North China, 116 µg/g in the USA, 115 µg/g in Australia, and only 17.8 µg/g in Vietnam. The mean ratios of Σnonphthalates to Σphthalates ranged from 0.19 in Vietnam to 0.72 in Australia. In Fig. 4.10, concentrations of the major nonphthalates are summed for the five study regions. On the one hand, the total concentrations of the individual substances reflect the large differences between the regions and, on the other hand, tricapryl trimellitate (TCTM) appears to be used mainly in southern China and oleate increasingly in China and Australia. Even though the concentrations of the phthalates in dust are still significantly higher than those of the nonphthalates, the latter seem to be increasingly used and found in indoor spaces.

4.4.2.4 Occurrence in Dust of Public Facilities

The results in dust samples from schools or daycare centers are summarized in Table 4.13. In the 2011/2012 study of 63 daycare centers, DINCH was quantified above the limit of quantification in all dust samples examined, with a median of 302 µg/g and a 95th percentile of 1622 µg/g (Fromme et al. 2016b). DEHT and DEHA were also detectable in nearly all of the dusts examined in this study, but the median levels of 40 µg/g for DEHT and 49 µg/g for DEHA are significantly lower. Compared to two Swedish studies in schools, the concentrations in Germany were higher for DINCH and DEHA, but lower for DEHT. This indicates a different spectrum of plasticizer use in these countries. A study of 100 Swedish daycare

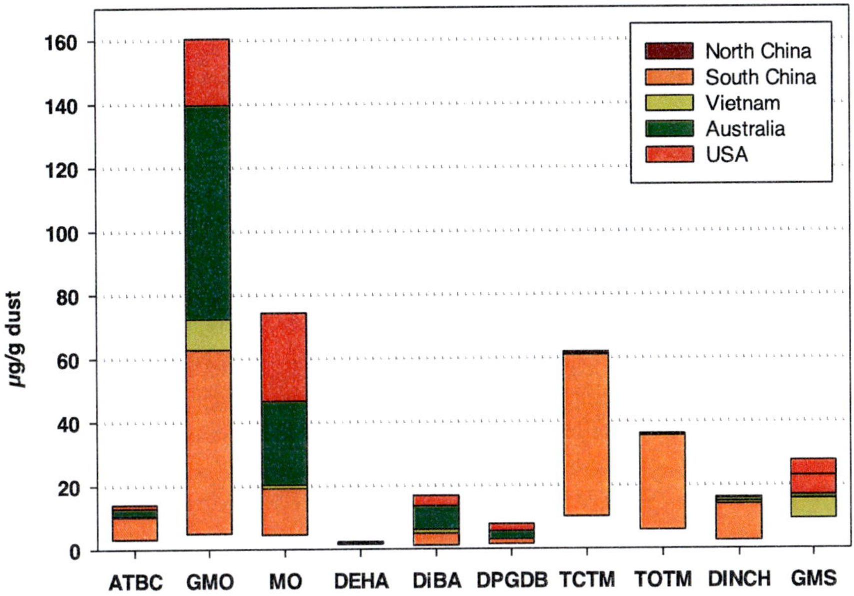

Fig. 4.10 Concentrations of nonphthalate plasticizers in house dust from five study regions using medians (modified after Tan et al. 2023)

centers also showed that concentrations in dust can vary widely, with levels tending to be lower than in Germany, except for DEHT (Larsson et al. 2017).

A high level was observed in 2016 for DEHA and especially ATBC in North American dust samples from 11 daycare centers (Subedi et al. 2017). Here, DEHA concentrations ranged from 15.2 to 275 µg/g and ATBC at 45.6 to 542 µg/g.

4.4.3 Alkanesulfonic Acid Phenyl Ester (Mesamoll®)

The C_{10}-C_{21} alkanesulfonic acid phenyl esters (ASE) are a mixture of various alkylsulfonic acid esters of phenol (CAS No.: 91082–17-6), which are used on the market under the product name Mesamoll® or Mesamoll II®. Mesamoll consists of 75–85% of a mixture of secondary alkanesulfonic acid phenyl esters and additionally contains 15–25% of secondary alkanedisulfonic acid diphenyl esters as well as 2–3% of non-sulfonated alkanes. The alkyl chains are predominantly unbranched, with chain lengths mainly in the C_{13}-C_{17} range. The chemical structures are shown as an example in Fig. 4.11. It is a liquid that is poorly soluble in water and has low volatility (Maag et al. 2010). Alkanesulfonic acid phenyl esters have been used for over 60 years as phthalate-free plasticizers for a variety of polymers, including PVC, polyurethane foam, and rubber. Currently, production volumes are around

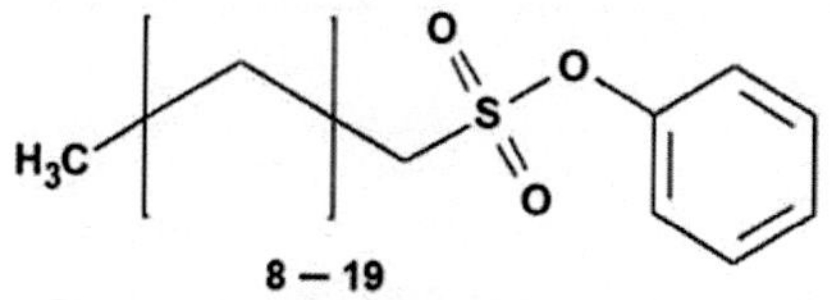

Fig. 4.11 Chemical structure of alkanesulfonic acid phenyl ester (ASE)

10,000–100,000 metric tons per year and the use is particularly in the manufacture of medical devices, children's toys, and food contact materials.

4.4.3.1 Health Aspects

Results from studies in humans are not available; animal experiments have described effects on the liver and kidney and delayed development (EFSA 2009; Maag et al. 2010).

4.4.3.2 Occurrence in Indoor Spaces

For indoor air, no study results have been published so far. For house dust, a study from Germany is available, in which dust from 25 daycare centers and 25 apartments from 3 German states was investigated (Fromme et al. 2017). Tetra-, penta-, and hexadecylphenyl esters were detected overall in 80–91% of the dust samples, while heptadecylphenyl esters were detected in only 60% (apartments) and 27% (daycare centers) of the samples. The median (95th percentile) sum of tetra- to heptadecylphenyl esters was 19.6 μg/g (216 μg/g) in daycare centers and 7.6 μg/g (171 μg/g) in homes. On a median basis, the tetra- and pentadecylphenyl esters were responsible for approximately 70% of the total.

4.5 Dechloranes

Dechloranes are polychlorinated flame retardants. They were originally developed as substitutes for Mirex, which was already banned in the 1970s. Of particular technical importance is Dechlorane Plus (DDC-CO) (CAS-No.: 13560–89-9), which is a commercially available product consisting of two stereoisomers, *syn*-DDC-CO (CAS-No.: 135821–03-3) and *anti-DDC-CO* (CAS-No.: 135821–74-8). The technical DDC-CO is a mixture of two stereoisomers composed of around 25% *syn*-DDC-CO and 75% *anti*-DDC-CO (ECHA 2021). The chemical structure is shown in Fig. 4.12. It was first produced by OxyChem (Niagara Falls, USA) and later also in China and used as an additive flame retardant mainly in cable sheathing, plastic flooring material and less frequently in textiles. The annual global production volume is estimated to be about 5000 tons (Wang et al. 2016a). DDC-CO has a

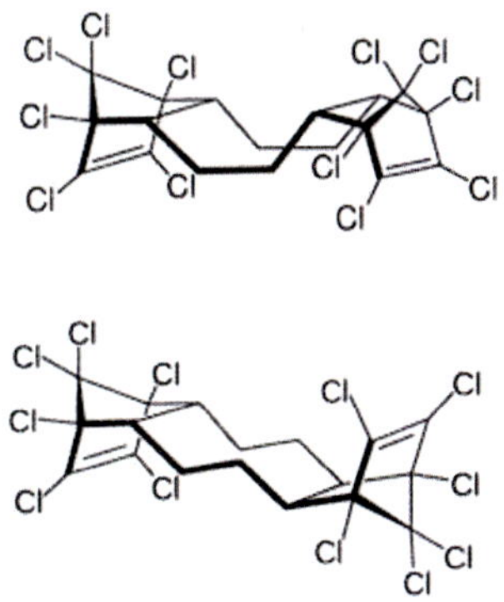

Fig. 4.12 Chemical structure of the stereoisomers *syn*-DDC-CO (top) and *anti*-DDC-CO (bottom) of Dechlorane Plus

long residence time in the environment and is persistent in water, soil, and sediment and is also considered to be persistent in air under natural conditions as a result of its binding to air particles and limited photodegradation (ECHA 2021). It is now found in many environmental media as well as biological matrices. In addition, the dechloranes Dec-602 (DDC-DBF, CAS-No.: 31107–44-5), Dec-603 (DDC-Ant, CAS-No.: 13560–92-4), and Dec-604 (CAS-No: 34571–16-9) are also used technically. However, no production or use data are available for these substances.

4.5.1 Health Aspects

Data on health effects are very limited and do not currently allow a conclusive risk assessment. Acute and chronic toxicity is estimated to be low, with the liver as the major target organ (ECHA 2018c; ECCC/HC 2019a). DDC-CO is not carcinogenic, mutagenic, or toxic to reproduction using experimental animals (ECHA 2021). Epidemiology studies indicate some possible associations between serum DP levels and effects on thyroid- and sex-hormones, but a co-exposure to other flame retardants must be considered (ECHA 2021). In the EU, Dechloran Plus is classified as persistent in the environment and bioaccumulative.

4.5.2 Occurrence in the Air

Results are available from different European countries and show relatively comparable concentration levels. For example, in 2010/2011, Melymuk et al. (2016a) found median DDC-CO concentrations of <LOQ (<LOQ-450 pg/m^3) and 54 pg/m^3 (13–29,000 pg/m^3) in 20 Czech homes in winter and 17 in summer, respectively. Median concentrations were a factor of 6 higher indoors than outdoors and a factor of 1.7 higher in summer than in winter. In 2015, a median of 0.36 pg DDC-CO/m^3 (<LOQ-6.4 pg/m^3) was found in 60 indoor spaces in Slovakia (Demirtepe et al.

2019) and < 2.0 pg/m^3 (<2.0–4.6 pg/m^3) for syn-DDC-CO and < 1.2 pg/m^3 (<1.2–20 pg/m^3) for anti-DDC-CO in 20 indoor spaces in England (Tao et al. 2016). In 47 Norwegian homes, syn-DDC-CO and anti-DDC-CO were detected in only 2% and 4% of samples, respectively, with mean values of 0.18 pg/m^3 (*syn*-DDC-CO) and 0.28 pg/m^3 (*anti-DDC-CO*), whereas in 6 school classrooms, they were not detected in any sample (Cequier et al. 2014). De la Torre et al. (2020) determined a median of 0.30 pg sum DDC-CO/m^3 (<0.1–7.8 pg/m^3) in 40 homes in 2010/2011 in Spain, Belgium, Italy, and Portugal. The highest meddian concentratians were found in Belgium with 0.48 pg/m^3 and the lowest in Spain with < 0.2 pg/m^3. Again, in 10 homes in Spain in 2020 syn- and anti-DDC-CO as well as Dec-604 could only be quantified only in sone sample above LOQ with mean levels of 2.1 pg/m^3, 1.4 pg/m^3, and 10.2 pg/m^3, respectively, whilst DDC-DBF and DDC-Ant were not detected in any sample above the LOQ of 41 and 2.0 pg/m^3 (Esplugas et al. 2022).

Concentrations in $PM_{2.5}$ ranging from 2.91 to 42.6 pg/m^3 (mean: 12.3 pg/m^3; syn-DDC-CO 10.4 pg/m^3 and anti-DDC-CO 1.97 pg/m^3) were measured in 19 Spanish offices and other workplaces (without DDC-CO handling) (Reche et al. 2019). DDC-DBF and DDC-Ant showed mean levels of 0.44 and 0.72 pg/m^3, respectively, with Dec603 detectable in the air of only two rooms. DDC-CO indoor air concentrations were significantly higher than outdoor air in this study and correlated with the presence of new electronic devices.

In Hong Kong, five kindergartens were surveyed and DDC-CO levels in $PM_{2.5}$ of 46.1–99.2 pg/m^3 (*syn*-DDC-CO: 10.0–14.6 pg/m^3, *anti*-DDC-CO: 37.3–84.5 pg/m^3) were measured (Deng et al. 2016).

In contrast, very high mean concentrations of 2500–390,000 pg/m^3 and of 670–53,000 pg/m^3 were found in 4 Swedish recycling plants in 2008 and 2009 (Rosenberg et al. 2011).

4.5.3 Occurrence in Dust

Table 4.14 gives an overview of the measurements in house dust. While median concentrations of <2.0 ng/g (syn-DDC-CO) and 3.0 ng/g (anti-DDC-CO) were found in Canada in 2011, they were even higher in previous Canadian studies. More recent studies from China showed higher levels in house dust compared with the North American studies. It was even higher, with a median of 541 ng/g, only in homes located in areas where electronic waste is recycled (Wang et al. 2011). In Chinese kindergartens, median concentrations of anti-DDC-CO ranged from 0.94 ng/g to 6.6 ng/g and of syn-DDC-CO from 0.32 ng/g to 2.2 ng/g, respectively (Deng et al. 2016). Here, there was no correlation between the concentrations in dust and air. Basically, the content of *anti-DDC-CO* in house dust is always higher than that of syn-DDC-CO.

In 2014, Sun et al. (2018) found levels of <0.05–0.64 ng/g (DDC-DBF), <0.05–0.2 ng/g (DDC-Ant), and < 0.11–2.0 ng/g (Dec604) in 67 samples from different indoor environments in China. In Latvia in 2016–2018 dust samples

Table 4.14 Dechloranes in dust in ng/g

References	*N*	∑-DDC-CO		*syn*-DDC-CO	*anti*-DDC-CO	Country, year
		Median	Minimum-Maximum	Median	Median	
Residences						
Zhu et al. (2007)	69	14	2.3–5683	–	–	Canada, 2002/03
Dodson et al. (2012)	16	10	3.0–47	3.0	7.5	USA, 2006
Zhu et al. (2007)	7	22	14–61	–	–	Canada, 2007
Shoeib et al. (2012)	116	7.3	<0.7–354	2.8	4.0	Canada, 2007/08
Fan et al. (2016)	351	–	<1.2–1570	2.9	5.6	Canada, 2007–10
Rasmussen et al. (2022)	1025	13.2	n.g.-2500	4.4	8.3	Canada, 2007–10
Wang et al. (2011)	27	14	2.8–70	3.0	12.4	China, urban, 2008/09
Wang et al. (2011)	20	4.0	<LOQ-27	1.0	2.5	China, rural, 2008
Li et al. (2015)	81[a]	4.0	0.35–1000	1.0	2.8	China, 2010
Dodson et al. (2012)	16	4.5	<2.0–15	<2.0	3.0	USA, 2011
Sun et al. (2018)	20	34	16–264	–	–	China, urban, 2011
Deng et al. (2016)	2	–	3.6 + 7.2	0.70–1.7	2.9–5.4	China, urban, 2014
Deng et al. (2016)	4[b]	–	1.3–8.7	0.32–2.2	0.94–6.6	China, urban, 2024
Tang et al. (2019b)	51	24.5	<LOQ-1050	5.3	19.4	China, urban
Lee et al. (2020)	111	8.5	0.30–530	1.9	6.5	Korea, 2015
Pasecnaja et al. (2021)	34	–	–	3.5	8.0	Latvia, 2016–18
Esplugas et al. (2022)	4	–	–	<0.0003	6.6[c]	Spain, 2020
Offices						
Reche et al. (2019)	7	1060[c]	168–5118	169	892	Spain, 2011
Sun et al. (2018)	20	83	18–305	–	–	China, 2011
Lee et al. (2020)	4	35	3.7–100	12	23	Korea, 2015

LOQ limit of quatification, *ng* not given
[a] apartments
[b] kindergartens
[c] mean value

collected in 34 households contained medians of 12.9 ng/g (DDC-DBF), 0.77 ng/g (DDC-Ant), and 0.88 ng/g (Dec-604) (Pasecnaja et al. 2021).

4.6 Polybrominated Diphenyl Ethers (PBDE)

Polybrominated diphenyl ethers (PBDE) have been widely used as additive flame retardants in plastics, electronic devices such as televisions and computers, as well as in construction materials and textiles for about 40 years (Fromme et al. 2019c). Depending on the degree of bromination and the position of the bromine atoms in the molecule, there are theoretically 209 individual substances (so-called congeners) of the PBDE, which are divided into 10 homolog groups ranging from mono- to decabromodiphenyl ethers (see Table 4.15). The general chemical structure of PBDEs is shown in Fig. 4.13.

4.6.1 *Health Aspects*

The available toxicological literature on the mode of action of the individual congeners is still limited; important endpoints for toxic effects include the liver, nervous system, immune system, thyroid hormone system, and reproductive organs (summaries in: EFSA 2012a; Lyche et al. 2015; Fromme et al. 2016c; ATSDR 2017; Wu et al. 2020a). Meanwhile, the production and use of octa-, penta-, and deca-BDE

Table 4.15 List of PBDE homolog groups

Homologue group	Acronym	No. of congeners	CAS No.
Monobromodiphenyl ether	MonoBDE	1–3	101–55-3
Dibromodiphenyl ether	DiBDE	4–15	2050-47-7
Tribromodiphenyl ether	TriBDE	16–39	49,690–94-0
Tetrabromodiphenyl ether	TetraBDE	40–81	40,088–47-9
Pentabromodiphenyl ether	PentaBDE	82–127	32,534–81-9
Hexabromodiphenyl ether	HexaBDE	128–169	36,483–60-0
Heptabromodiphenyl ether	HeptaBDE	170–193	68,928–80-3
Octabromodiphenyl ether	OctaBDE	194–205	32,536–52-0
Nonabromodiphenyl ether	NonaBDE	206–208	63,936–56-1
Decabromodiphenyl ether	DecaBDE	209	1163-19-5

Modified according after EFSA (2012a)

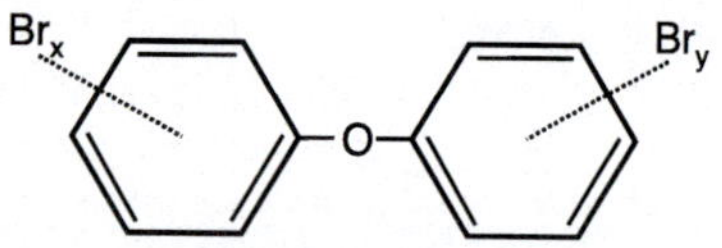

Fig. 4.13 Chemical structure of polybrominated diphenylethers (PBDE)

has been limited or banned for products containing more than 10 mg/kg (0.001%) of each PBDE in the EU. To date, commercial octa-, penta-, and deca-BDE listed in the Annex A of the Stockholm Convention. While the USA has no restrictions on PBDE at the federal (nationwide) level, thirteen states have implemented restrictions which include some limitation on the use and presence of PBDE in certain commercial products (Sharkey et al. 2020).

4.6.2 *Occurrence in the Air of Residences*

A review of indoor occurrence can be found in paper published by Lucattini et al. (2018), Klinčić et al. (2020), and Hoang et al. (2021). The results of the measurements of European studies in indoor air are summarized in Fig. 4.14. Due to the very different application quantities, structural characteristics and analytical methods, a certain range of results can be seen. In Europe, the lowest medians were reported for Σ_6-PBDE in 2013 and 2009–2011, at 4.6 pg/m^3 and 7.6 pg/m^3 in 60 Norwegian and 6 French dwellings, respectively (Tay et al. 2017; Moreau-Guigon et al. 2016). In contrast, the highest Σ_6-PBDE concentration of 211 pg/m^3 was measured in 43 Danish indoor dwellings in 2007 (Vorkamp et al. 2011). In the study by Karlsson et al. (2007), the content in 5 dwellings was still somewhat higher, but at that time the determination limits were relatively high and almost only BDE 28 and BDE 47 were detected. In 34 German apartments, the median for Σ_7-PBDE was 20 pg/m^3 (5.2–439 pg/m^3) (Fromme et al. 2009a). On the basis of median values, BDE 47 and BDE 99 are responsible for approximately 18–84% and 17–66% of the total Σ_6-PBDE concentration, respectively, in the European studies. In 10 Swedish

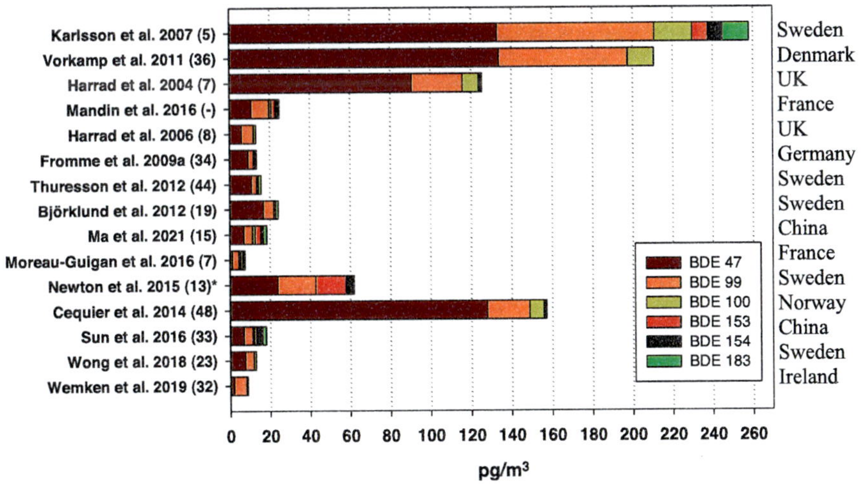

Fig. 4.14 Polybrominated diphenylethers in indoor air using median values (*: geometric means, values in brackets are number of residences)

offices, a median of 13 pg/m^3 $\sum_{17}$-PBDE (<16–87 pg/m^3) was measured in 2016/2017 (Tao et al. 2019). In Europe, the results of more recent studies indicate a decreasing trend in indoor air pollution.

Widely varying results are also reported for BDE 209 levels across Europe, ranging from a median of 1.6 pg/m^3 in 23 Swedish homes in 2014/2015 (Wong et al. 2018) to 290 pg/m^3 in 10 Swedish homes (Thuresson et al. 2012). A median of 10 pg/m^3 BDE 209 (0.9–438 pg/m^3) was observed in 34 German homes (Fromme et al. 2009a). In Ireland, a very high median of 410 pg/m^3 (<7.5–5500 pg/m^2) was reported in 28 residences in 2016/17, while in Sweden in 2014/15 low median concentration of 1.6 pg/m^3 (<2.2–8.6 pg/m^3) was found (Wemken et al. 2019; Wong et al. 2018). Again, all measurement results were below the limit of quantification of 1.2 pg/m^3 in 10 Swedish offices in 2016–2017 (Tao et al. 2019).

Compared to European studies, higher levels of Σ_6-PBDE were found in indoor air in 10 homes in Australia with a median of 634 pg/m^3 (Toms et al. 2009). Very high medians between 79 and 1936 pg/m^3 (Σ_6-PBDE) were also reported in five studies from the USA (Allen et al. 2007; Johnson-Restrepo and Kannan 2009; Batterman et al. 2009; Bennett et al. 2015; Venier et al. 2016). In these studies, BDE 47 and BDE 99 accounted for 65–78% and 18–27% of the total concentrations of Σ_6-PBDE, respectively. One explanation can be seen in the higher use levels of these flame retardants. In contrast, and comparable to the situation in Europe, a Canadian study found a median of only 100 pg/m^3 for Σ_{10}-PBDE in 74 homes in 2002/2003 (Wilford et al. 2004) and 59 ng/m^3 in 23 homes in 2013 (Venier et al. 2016).

In North America, median concentrations of 95 pg/m^3 BDE 209 were reported in 20 bedrooms in Boston in 2006 as well as 257 pg/m^3 in 20 homes in Bloomington in 2013 (Allen et al. 2007; Venier et al. 2016) and 49 ng/m^3 in 23 homes in Canada in 2013 (Venier et al. 2016).

On the contrary, Bannan et al. (2021) observed very low levels in Saudi children rooms in Jeddah with medians of <1.0 pg/m^3 for BDE 47, 2 pg/m^3 for BDE 99, and 75 pg/m^3 for BDE 209. Air samples collected in 33 homes in China 2013 showed medians of 7.0 pg/m^3 for BDE 47 and 4.5 pg/m^3 for BDE 99 (Sun et al. 2016). Two studies from China collected air samples from 8 homes in 2014 in Shanghai and 15 homes in 2017 (Xu et al. 2016; Ma et al. 2021). Medians of BDE 47 were 7.1 pg/m^3 and 1.5 pg/m^3 as well as 4.4 pg/m^3 and 1.2 pg/m^3 for BDE 99, respectively. Medians of BDE 209 in these two studies were 165 pg/m^3 and 37 pg/m^3.

The medians of the congeners BDE 47, BDE 99, and BDE 100 were approximately a factor of 12, 5, and 6 higher in indoor air compared to outdoor air, while BDE 153 and BDE 154 occurred at somewhat lower concentrations indoors than outdoors (Harrad et al. 2004). In Canada, indoor-to-outdoor ratios as high as 75:1 and 11:1 were observed for BDE 47 and BDE 99, respectively (Wilford et al. 2004). In China, the median indoor and outdoor levels were comparable, while for BDE 209 the outdoor median was higher (Ma et al. 2021).

Statistically significant correlations between indoor air and house dust levels have been observed for low-brominated BDEs (e.g., pentaBDE) by several research groups (Wilford et al. 2005; Allen et al. 2008, Fromme et al. 2009a; Tao et al.

2019), while others have not been able to confirm such an association (Vorkamp et al. 2011; Imm et al. 2009). For BDE 209, most studies did not find any association in this regard (Allen et al. 2008; Wilford et al. 2005; Vorkamp et al. 2011). In addition, positive correlations were observed between indoor air concentration and the number of electronic devices, the number of polyurethane mattresses, the number of synthetic carpets, and the size of the room surface, but not with the age of the buildings and the floor area of the room (De Wit et al. 2012).

4.6.3 *Occurrence in the Air of Public Facilities*

In 2006, 10 daycare centers in Sweden were studied and medians (ranges) of 110 pg/m^3 (4.3–380 pg/m^3) for BDE 47, 26 pg/m^3 (3.1–390 pg/m^3) for BDE 99, and 1100 pg/m^3 (62–2300 pg/m^3) for BDE 209 were obtained (Thuresson et al. 2012), while in 6 Norwegian school classrooms medians of 131 pg/m^3 for BDE 47, 23 pg/m^3 for BDE 99, and < LOQ for BDE 209 were reported (Cequier et al. 2014). In Spain in 2019/20, means of 3.2 pg/m^3 for BDE 28, 64.6 pg/m^3 for BDE 47, and 27.8 pg/m^3 for BDE 99 were observed in 5 kindergarten and preschools (Esplugas et al. 2022).

In another study, 15 Korean schools and 41 daycare centers, each in two seasons, could be investigated (Baek et al. 2010). In winter, a median for BDE 47 of 21 pg/m^3 and in summer of below the limit of quantification were found in schools, while for Σ_7-PBDE 223 pg/m^3 and 11 pg/m^3 were measured, respectively. In the daycare centers, medians of 88 pg/m^3 and 10 pg/m^3 (BDE 47) and of 120 pg/m^3 and 23 pg/m^3 (Σ_7-PBDE) were found in the two seasons. In 30 Korean elementary schools means of 399 pg/m^3 for BDE 47, 279 pg/m^3 for BDE 99, and 208 pg/m^3 were reported in 2010/11 (Lim et al. 2014).

4.6.4 *Occurrence in Dust*

A large number of studies have investigated PBDE in dust, summarized by Klinčić et al. (2020) and Yu et al. (2023). Table 4.16 gives an overview of recent results from different countries. From the table, the large regional variations in the exposure situation become clear.

From European studies, medians of 2.3–42 ng/g are reported for BDE 47 and 6.5–66 ng/g for BDE 99. Among these, for Σ_6-PBDE, the lowest median was found at 19 ng/g in 43 Belgian homes (Ali et al. 2011) and the highest at 122 ng/g in 34 Swedish apartments (Thuresson et al. 2012). The BDE 47 and BDE 99 are responsible for approximately 5–49% and 10–60% of the total Σ_6-PBDE levels, respectively, in the aforementioned studies.

For BDE 47 and BDE 99, the highest median concentrations worldwide are observed between 40 and 2300 ng/g and 95 and 4600 ng/g in North America,

Table 4.16 Median concentrations of PBDE congeners in dust in ng/g

References	*N*	BDE 28	BDE 47	BDE 99	BDE 100	BDE 153	BDE 154	BDE 183	BDE 209	Country, Year
Europe										
Esplugas et al. (2022)	4	0.4+	7.4+	11.1+	1.4+	3.0+	1.0+	8.2+	514+	Spain, 2020
Dvoršćak et al. (2022)	10	<0.1	0.83	2.0	0.95	0.50	0.24	0.96	–	Croatia, 2020
Besis et al. (2023)	20	0.23	1.1	1.2	0.34	0.73	0.34	1.2	199	Greece, 2018
Pasecnaja et al. (2021)	34	2.1	14	17	4.5	4.6	3.9	11	1510	Latvia, 2016–18
Wemken et al. (2019)	32	–	7.6	13	–	–	–	1.0	13,000	Ireland, 2016/17
Rantakokko et al. (2019)	40	0.3	12	20	3.3	2.4	1.6	–	411	Finland
Civan and Kara (2016)	40	4.0	10	6.5	13	14	–	21	138	Turkey, 2015/16
Kuang et al. (2016)	11	1.0	13	33	3.2	1.9	0.7	4.2	1500	UK, 2015
Kademoglou et al. (2017)	10	2.6	2.3	–	23	8.9	9.6	3.6	161	Norway, 2013–14
Kademoglou et al. (2017)	10	2.1	11	–	3.2	8.2	4.1	5.8	3351	UK, 2013–14
Tay et al. (2017)	60	0.5	22	27	0.2	2.5	1.8	3.6	940	Norway, 2013
Newton et al. (2015)	27	–	18	38	7.6	8.7	4.7	–	130	Sweden, 2012
Fromme et al. (2014a)	20	0.1	5.7	9.2	1.6	2.1	1.1	9.3	950	Germany, 2010
Sahlström et al. (2015)	27	0.99	21	17	2.6	1.9	–	–	310	Sweden, 2009/10
Cunha et al. (2010)	10	–	17	7.0	2.0	3.0	3.0	4.0	206	Portugal, 2008
Björklund et al. (2012)	18	0.8	38	25	5.5	6.0	2.9	3.0	520	Sweden, 2008
Kalachova et al. (2012)	25	<0.1	8.9	11.6	<0.1	<0.3	<0.3	3.9	375	Czech Republic, 2008
D'Hollander et al. (2010)	43	0.4	8.1	8.9	1.1	2.2	0.9	1.4	313	Belgium, 2008
Ali et al. (2011)	43	–	8	9	–	–	–	2	317	Belgium, 2008
Vorkamp et al. (2011)	42	0.2	16.9	13.6	2.5	2.5	1.3	4.0	432	Denmark, 2007
Harrad et al. (2008b)	30	<LOQ	10	20	3.4	5.0	2.8	4.2	8100	UK, 2006/07
Thuresson et al. (2012)	10	1.3	42	52	–	6.6	–	12	320	Sweden, houses, 2006

	34	0.8	37	66	–	7.8	–	11	1100	Sweden, apartments, 2006
Harrad et al. (2008a)	28/16[a]	0.53	13	23	4.2	5.2	3.3	13	2800	UK, 2006
Fromme et al. (2009a)	34	0.25	9.1	12.5	2.5	2.7	1.6	4.3	312	Germany, 2005
America										
Young et al. (2021)	47[b]	2.8	64	120	27	11	5.3	19	800	USA, 2019
Hammel et al. (2017)	137	–	202	310	44	34	27	–	537	USA, 2014–16
Schreder and La Guardia (2014)	20	–	587	921	187	91.3	61.3	12.2	1140	USA, 2011–12
Dodson et al. (2012)	16	14	1000	1100	240	150	110	18	1200	USA, 2011
Whitehead et al. (2013b)	203	20	1300	2100	330	290	150	17	2500	USA, 2010
Imm et al. (2009)	38*	–	520	614	120	73	51	–	1398	USA, 2008
Johnson et al. (2010)	50	11.7	390	427	99.9	55.9	51.3	17.4	1482	USA, 2002–8
Johnson-Restrepo and Kannan (2009)	12	<0.1	40	95	16.1	25.9	9.6	<21	903	USA, 2007–08
Shoeib et al. (2012)	116	4.5	280	350	67	42	25	14	1300	Canada, 2007–08
Batterman et al. (2009)	12	26	2000	4600	1200	110	190	220	190	USA, 2006–07
Harrad et al. (2008a)	28/16[a]	4.1	140	330	65	43	39	9.0	560	Canada, 2006
Harrad et al. (2008a)	20/17[c]	14	410	820	160	110	89	16	1300	USA, 2006
Allen et al. (2008)	20*	6.4	338	536	77	47	35	15.1	1811	USA, 2006
Dodson et al. (2012)	16	26	2300	2300	520	250	240	28	1400	USA, 2006
Asia										
Al-Harbi et al. (2021)	[d]	1.0	9.2	9.9	0.3	0.69	0.9	0.1	304	Kuwait, 2018/19
Bannan et al. (2021)	30	<0.2	2.0	<0.2	<0.2	2.0	<0.2	<0.2	3150	Saudi Arabia, 2019
Lee et al. (2020)	111	–	12	4.2	–	–	–	1.1	640	Korea, 2015
Zheng et al. (2017)	9	–	<LOQ	3.6	<LOQ	<LOQ	<LOQ	<LOQ	351	China, 2015

(continued)

Table 4.16 (continued)

References	*N*	BDE 28	BDE 47	BDE 99	BDE 100	BDE 153	BDE 154	BDE 183	BDE 209	Country, Year
Xu et al. (2016)	8	0.80	3.9	4.0	0.76	1.3	0.79	5.1	499	China, 2014
Zhang et al. (2022b)	9*	0.80	2.3	1.8	2.4	2.3	1.9	4.7	870	China, 2014
Sun et al. (2016)	10	0.36	4.0	3.6	0.55	0.76	0.48	1.5	228	China, 2013
Wang et al. (2015a)	216	0.32	8.4	1.7	4.1	4.6	0.6	4.0	63,000	China, 2011

* geometric mean; + mean; *LOQ* limit of quantification

[a] 28 samples analyzed for tri- to hexaBDE and 16 samples analyzed for tri-decaBDE

[b] six office suites, 23 common rooms, and 18 classrooms

[c] 20 samples analyzed for tri- to hexaBDE and 17 samples for tri- to decaBDE

[d] non-smoking homes

respectively. Here, the lowest medians with 187 ng/g (Σ_6-PBDE) were reported in a study of 12 homes from New York State (Johnson-Restrepo and Kannan 2009), and the highest with 8320 ng/g (Σ_6-PBDE) from 12 homes in Michigan (Batterman et al. 2009). BDE 47 and BDE 99 are responsible for approximately 22–41% and 41–51% of the total Σ_6-PBDE levels, respectively. Significant differences in exposure levels can also be demonstrated within the USA. The explanation for the very high levels in the USA compared to Europe lies in the considerably larger application quantities of these flame retardants.

In Asia, concentrations of PBDE in dust were very low compared to America. For example, in Korea in 111 houses medians of 12 ng/g and 4.2 ng/g were reported for BDE 47 and BDE 99, respectively (Lee et al. 2020).

While many study results are available on the occurrence of PBDE in house dust of apartments, less data regarding the concentrations in schools and daycare centers are available. For example, a total of 8 Austrian all-day schools were studied in 2006–2007, in which medians (95th percentiles) of 60 ng/g (572 ng/g) for BDE 47, 72 ng/g (508 ng/g) for BDE 99, and 15 ng/g (92 ng/g) for BDE 99 were determined (Hohenblum et al. 2008). The most comprehensive study to date was conducted in 2007/2008 in 43 schools and nurseries in England (Harrad et al. 2010). Here, medians (95th percentiles) of 26 ng/g (85 ng/g) for BDE 47, 36 ng/g (140 ng/g) for BDE 99, and 6.6 ng/g (26 ng/g) for BDE 99 were reported; lower than in the Austrian study. In 100 preschools in Sweden, medians (95th percentiles) of 7.7 ng/g (110 ng/g) for BDE 47, 9.2 ng/g (120 ng/g) for BDE 99, and 1.6 ng/g (20 ng/g) for BDE 99 were observed in 2015 (Larsson et al. 2018). In Korea, Lee et al. (2020) reported medians of 14 ng/g fore BDE 47 and 2.5 ng/g for BDE 99 in four daycare centers in 2015. In 602 classrooms from 308 randomly-selected French schools mean values of 27.6 ng/g for BDE 47, 27.1 ng/g for BDE 99, and 8.6 ng/g for BDE 153 were found in 2013–2017 (Sivanantham et al. 2021). Zhang et al. (2022b) reported a geometric means of 1.7 ng/g for BDE 47 and 1.6 ng/g for BDE 99 in 10 daycare centers in China in 2014.

Concentrations of BDE 209 in house dust in homes can also vary over several orders of magnitude. a high median level has been found particularly in Ireland, the UK, the USA, China, Korea, and Singapore. For example, Wemken et al. (2019) observed a median of 13,000 ng/g in 32 homes in Ireland and Whitehead et al. (2013b) reported a median of 2500 ng/g in 203 homes in northern California. In Germany, lower levels were found in 20 homes in 2011 with a median of 950 ng/g (10.0–3748 ng/g) (Fromme et al. 2014a). While in North America, BDE 209 only accounts for about 2–83% (median: 47%) of total PBDE levels in house dust, in European studies this proportion is about 74–99%. In Europe, BDE 209 is used very intensively in fire protection, especially in England, which explains the high content for this congener. In Asia, the median levels of BDE 209 are comparable to the concentrations found in Europe.

Median (95th percentile) levels of 780 ng/g (2240 ng/g) were found in dust from daycare centers in 8 Austrian schools and 5000 ng/g (24,000 ng/g) in 43 English schools and daycare centers (Hohenblum et al. 2008; Harrad et al. 2010). In 100 preschools in Sweden, a median of 270 ng/g and a 95th percentile of 1600 ng/g were reported for BDE 209 in 2015 (Larsson et al. 2018). Zhang et al. (2022b) reported a geometric mean of 570 ng/g (range: 220–5500 ng/g) in 10 daycare centers in China in 2014.

4.7 Emerging Brominated Flame Retardants

In addition to polybrominated diphenyl ethers, the following other flame retardants are of particular technical importance: 1,2,5,6,9,10-hexabromocyclododecane (HBCDD), tetrabromobisphenol A (TBBPA), 1,2-bis(2,4,6-tribromophenoxy)ethane (BTBPE), 2-ethyl-1-hexyl-2,3,4,5-tetrabromobenzoate (EH-TBB), bis(2-ethyl-1-hexyl)tetrabromophthalate (BEH-TEBP), decabromodiphenylethane (DBDPE), and tetrabromoethylcyclohexane (DBE-DBCH). The chemical structures of the main brominated "non-PBDE" flame retardants are shown in Fig. 4.15. The nomenclature was chosen following Bergman et al. (2012).

HBCDD (CAS-No.: 3194-55-6, 25,637–99-4) was mainly used in polystyrene insulation materials for buildings and to a lesser extent in upholstery textiles and packaging plastics. As of 2016, approximately 12,000 metric tons of HBCDD were used annually in the EU. HBCDD has been included in the candidate list in Annex XIV of REACH and has also been a substance to be eliminated under Annex A of the Stockholm Convention since 2013. Based on the regulations of persistent organic pollutants, no products containing more than 100 mg HBCDD/kg may be manufactured or placed on the market in the EU since March 22, 2016, with a temporary exemption for expanded polystyrene insulation materials.

TBBPA (CAS-No.: 79-94-7) is used as a reactive and additive flame retardant in the production of epoxy and polycarbonate resins, acrylonitrile-butadiene-styrene resins, polystyrene and printed circuit boards. There is no producer in the EU. Globally, >120,000 tons are produced per year and in the EU about 12,000 tons are consumed annually.

BTBPE (CAS-No.: 37853-59-1) is an additive flame retardant that has been widely produced since the 1970s, particularly as a substitute for octaBDE, and used in polymers and foams, probably including toys.

EH-TBB (CAS-No.: 183658-27-7) is an additive flame retardant used in the manufacture of polyurethane foams, originally as a replacement product for PentaBDE. It often occurs in combination with BEH-TEBP in the products Firemaster 550 and Firemaster BZ-54.

BEH-TEBP (CAS-No.: 26040-51-7), a brominated phthalate, is an additive flame retardant specifically designed to replace pentaBDE in many applications, such as polyurethane foam, PVC, neoprene rubber, and cable sheathing (EFSA 2012b). It is used in the products Firemaster 550 and Firemaster BZ-54, to which it is added at 15% and 30%, respectively, and is responsible for the effects together with EH-TBB.

There is no manufacturer for **DBDPE (CAS-No.: 84852–53-9)** within the EU, however, with an increasing tendency, approx. 2000 tons per year are consumed in

CAS-Nr.	Substanzname	Abkürzung	Formel	Struktur
79-94-7	Tetrabromobisphenol A	TBBPA	$C_{15}H_{12}Br_4O_2$	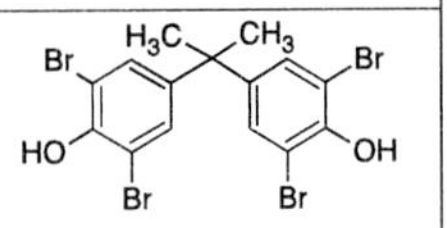
3194-55-6 25637-99-4	Hexabromocyclododecan (α, β, γ)	HBCDD	$C_{12}H_{18}Br_6$	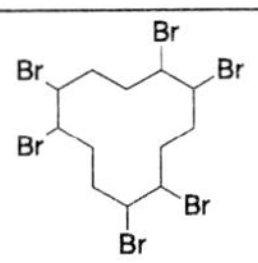
37853-59-1	1,2-Bis(2,4,6-tribromophenoxy)ethan	BTBPE	$C_{14}H_8Br_6O_2$	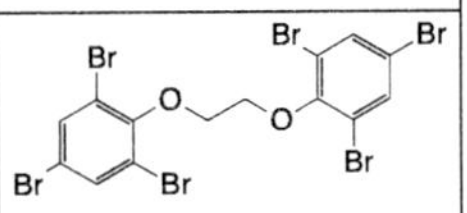
183658-27-7	2-Ethyl-1-hexyl-2,3,4,5-tetrabromobenzoat	EH-TBB	$C_{15}H_{18}Br_4O_2$	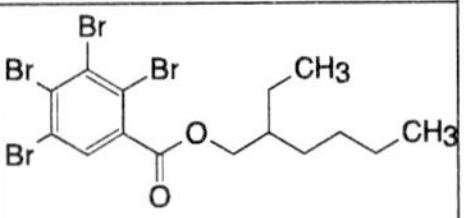
26040-51-7	Bis(2-ethyl-1-hexyl)-tetrabromophthalat	BEH-TEBP (TBPH)	$C_{24}H_{34}Br_4O_4$	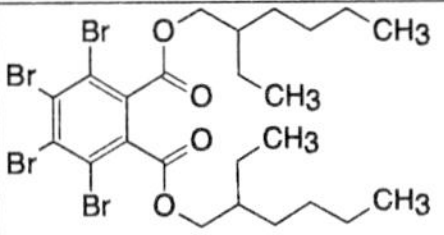
84852-53-9	Decabromodiphenylethane [1,2-Bis(pentabromo-phenyl)-ethan]	DBDPE	$C_{14}H_4Br_{10}$	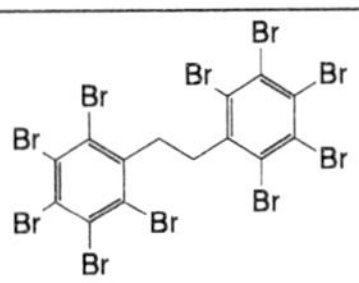
3322-93-8	Tetrabromoethylcyclohexan (α, β, γ, δ) [4-(1,2-Dibromoethyl)-1,2-dibromocyclohexan]	DBE-DBCH (TBECH)	$C_8H_{12}Br_4$	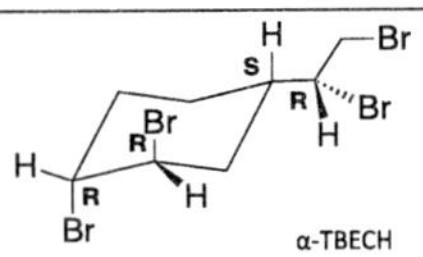 α-TBECH

Fig. 4.15 Chemical structure and characteristics of emerging brominated flame retardants

the EU. It is an additive flame retardant used in various polymers and textiles. It has an application spectrum comparable to DecaBDE.

DBE-DBCH (CAS-No.: 3322-93-8) is an additive flame retardant that is used especially in OSB boards, electrical cables, rigid and flexible foams for insulation, furnishings and in extruded polystyrene and polyurethane. It is a mixture of mainly α- and β-diasteroisomers, although other isomers may occur.

4.7.1 Health Aspects

Information on toxicological and epidemiological knowledge is available in a large number of review papers. This includes **HBCDD** (HC 2011b; NICNAS 2012; US EPA 2020a; EFSA 2021), **TBBPA** (US EPA 2020b; EFSA 2011), **BEH-TEBP** (US EPA 2015), and numerous other significant flame retardants (EFSA 2012b; Xiong et al. 2019).

After oral intake, the brominated flame retardants, except BEH-TEBP and DBDPE, are well absorbed from the gastrointestinal tract into the organism, while dermal absorption is consistently low for all. If any information is available at all, half-lives are weeks to months for HBCDD and hours for TBBPA and BTBPE. In animal experiments, effects on liver, thyroid, and neurodevelopment are most prominent for HBCDD, while effects on metabolic and hormonal processes have been described for TBBPA, with inconsistent data. DBE-DBCH shows androgenic effects in vitro. For many substances (except HBCDD and TBBPA), toxicological data are limited. TBBPA was clasassified as probanly carcinogenic to humnan (group 2A) by the cancer agency of the WHO (IARC 2018). Acute toxicity is estimated to be low for substances in this class.

4.7.2 Occurrence in the Air

Study results have been compiled in review (Malliari and Kalantzi 2017; Zuiderveen et al. 2020). Abdallah et al. (2008b) were able to find median levels of Σ-HBCDD of 180 pg/m^3 (67–1300 pg/m^3) in 33 homes in central England in 2006–2007. The γ-HBCDD were responsible for 67% of the total content. De Wit et al. (2012) reported low median Σ-HBCDD levels of 2.0 pg/m^3 (<1.6–33 pg/m^3) and < LOQ (<1.6–15 pg/m^3) in 10 Swedish houses and 44 apartments in 2006. A negative correlation was found with the age of the objects, but no association with the presence of electronic/electrical equipment, upholstered furniture, and polyurethane foam mattresses. In another Swedish study, a median of <1.3 pg/m^3 for Σ-HBCDD (<1.3–19 pg/m^3) was found in 13 homes in 2012 (Newton et al. 2015). Wong et al. (2018) also observed a median of 1.8 pg Σ-HBCDD/m^3 in 23 Swedish homes. Indoor air measurements in 60 Slovakian apartments were in a similar concentration range with 5.4 pg/m^3 (median, Σ-HBCDD) (Demirtepe et al. 2019). In contrast, Tao et al. (2016) were able to determine a median of 110 pg/m^3 (19–1500 pg/m^3) in 15 homes in Birmingham. In Ireland, a median of 20 pg/m^3 (0.9–2500 pg/m^3) could be observed in 32 residential interiors in 2016–2017 (Wemken et al. 2019). While in Ireland and central England, the γ-isomer accounted for 66 and 67% of Σ-HBCDD, respectively (Wemken et al. 2019; Abdallah et al. 2008b), in the Birmingham study it accounted for approximately 94% (Tao et al. 2016). In a Japanese study, concentrations of 160 pg/m^3 and 12 pg/m^3 (Σ-HBCDD) were found in two houses (Takigami et al. 2009).

TBBPA has been measured at concentrations of 9–22 pg/m^3 in indoor air from 5 houses in England (Abdallah et al. 2008b) and 8 and 20 pg/m^3 in two Japanese houses (Takigamin et al. 2009). Higher levels of TBBPA ranging from 10 to 70 pg/m^3 were observed in 4 samples from rooms with a large number of computers (Sjödin et al. 2001).

BTBPE was measured in 5 houses in an older Swedish study and could not be found above the limit of quantification of 12 pg/m^3 (Karlsson et al. 2007). Tao et al. (2016) found a median of 5 pg/m^3 (<1.0–50 pg/m^3) in 15 houses in Birmingham. A median of 0.23 pg/m^3 (<LOQ-1.6 pg/m^3) in summer and < LOQ (<LOQ-1.2 pg/m^3) in winter was measured in 17 and 20 houses, respectively, in the Czech Republic in 2010/2011 (Melymuk et al. 2016a).

In 47 Norwegian homes and 6 classrooms in 2012, BEH-TEBP was determined in 19% of samples with a median of <2.9 pg/m^3 (<2.9–24.2 pg/m^3) and in 33% of samples <2.9 pg/m^3 (<2.9–6.3 pg/m^3), respectively (Cequier et al. 2014). In contrast, a median of <35 pg/m^3 (<35–150 pg/m^3) was found in 13 homes in a 2012 Swedish study (Newton et al. 2015). Similarly, a median of only 1.8 pg/m^3 (<LOQ-84 pg/m^3) was measured in 60 Slovakian homes in 2015 (Demirtepe et al. 2019). Comparable are the result in Saudi Arabia in 2019 with a median of 2.0 pg/m^3 (range: <2.0–15 pg/m^3) (Bannan et al. 2021).

In the aforementioned Norwegian homes and classrooms, DBDPE was determined in 47% of samples in 2012 with a median of <14 pg/m^3 (<14–963 pg/m^3) and 8.3 pg/m^3 (<14–21 pg/m^3) in 50% of samples, respectively (Cequier et al. 2014). Tao et al. (2016) found a median of <10 pg/m^3 (<10–87 pg/m^3) in 15 houses in Birmingham. In Ireland, a median of 48 pg/m^3 (15–7000 pg/m^3) could be observed in 32 residential interiors in 2016–2017 (Wemken et al. 2019).

DBE-DBCH was determined in all of the samples in the 2012 Norwegian study with a median of 78 pg/m^3 (maximum: 4120 pg/m^3) and 47 pg/m^3 (maximum: 399 pg/m^3), respectively (Cequier et al. 2014). Newton et al. (2015) found comparable concentrations in 13 Swedish houses in 2012 with a median of 55 pg/m^3 (7.0–130 pg/m^3). In contrast, only a median of 3.9 pg/m^3 (0.8–122 pg/m^3) was measured in 60 Slovakian homes in 2015 (Demirtepe et al. 2019).

4.7.3 *Occurrence in Dust*

Study results on emerging flame retardant content in dust are presented in Table 4.17. In addition, study results are also compiled in reviews (Malliari and Kalantzi 2017; Zuiderveen et al. 2020).

Results from 20 dust samples collected in Munich, Germany, found a median of 345 ng/g dust (range: 53–4051 ng/g) for Σ-HBCDD (Fromme et al. 2014a). In the other European studies, median levels for Σ-HBCDD ranged from 9 to 730 ng/g. In contrast, very high concentrations of 1300 ng/g (median, Σ-HBCDD) were found in a study conducted in 2006–2007 in 45 indoor homes in England (Abdallah et al. 2008b). In one house, this resulted in a maximum of 140,000 ng/g. This high level

Table 4.17 Median content of emerging flame retardants in dust in ng/g

References	*N*	α-HBCDD	β-HBCDD	γ-HBCDD	Σ-HBCDD	TBBPA	BTBPE	EH-TBB	BEH-TEBP	DBDPE	Country, Year
Esplugas et al. (2022)	4	2455*	295*	7830*	–	–	–	15.8*	634*	233*	Spain, 2020
Bannan et al. (2021)	30	–	–	–	–	–	8.0	10	15	–	Saudi Arabia, 2019
Besis et al. (2023)	20	–	–	–	–	–	8.3	–	2.9	76	Greece, 2018
Pasecnaja et al. (2021)	34	–	–	–	–	–	8.0	<LOQ	–	114	Latvia, 2016–18
Wemken et al. (2019)	32	200	100	200	490	–	–	–	–	4200	Ireland, 2016/17
Lee et al. (2020)	111	–	–	–	–	–	9.7	–	–	64	Korea, 2015
Rantakokko et al. (2019)	40	–	–	–	–	–	1.4	5.5	106	119	Finland
Demirtepe et al. (2019)	60	–	–	–	153	–	3.7	6.8	70	–	Slovakia, 2015
Kuang et al. (2016)	11	280	67	110	–	–	4.5	12	75	120	UK, 2015
Lankova et al. (2015)	18	50	16	38	–	–	<1.0	–	–	–	Czech Republic, 2013
Cequier et al. (2014)	48	–	–	–	–	–	3.8	2.5	79	147	Norway, 2012
Newton et al. (2015)	27	–	–	–	100	–	19	78	93	37	Sweden, 2012
Brommer et al. (2012)	6	–	–	–	–	–	–	–	–	40	Germany, 2010/11
Fromme et al. (2014a)	20	180	35	114	345	28	<10	<3.0	343	146	Germany, 2010
Sahlström et al. (2015)	27	56	18	37	110	–	6.3	2.6	61	150	Sweden, 2009/10
Abb et al. (2011)	26	–	–	166	–	48	–	–	–	–	Germany/USA
Björklund et al. (2012)	19	–	–	–	86	–	–	–	–	–	Sweden, 2008[a]

Björklund et al. (2012)	18	–	–	–	9	–	–	–	–	–	Sweden, 2008[b]
Geens et al. (2009)	18	–	–	–	–	10	–	–	–	–	Belgium, 2008
D'Hollander et al. (2010)	43	–	–	–	130	–	–	–	–	–	Belgium, 2008
Ali et al. (2011)	39	–	–	–	–	78	2.0	1.0	13	153	Belgium, 2008
Kalachova et al. (2012)	25	26	7.1	61	93	–	<2.0	–	–	141	Czech Republic, 2008
Roosens et al. (2009)	16	69	14	31	114	–	–	–	–	–	Belgium, 2007
Abdallah et al. (2008b)	45	380	93	670	1300	62[c]	–	–	–	–	UK, 2006/07
Abdallah et al. (2008a)	31	170	66	440	730	–	–	–	–	–	UK, 2006
Sahlström et al. (2012)	6	93	25	39	152	–	<8.0	53	430	1700	Sweden, 2006

* mean; *LOQ* limit of quantification
[a] vacuum cleaner dust
[b] self-vacuumed by investigators
[c] $n = 35$

was confirmed by a second English study in 31 houses (median: 730 ng Σ-HBCDD/g) (Abdallah et al. 2008a). The lowest median to date is reported to be 9.0 ng Σ-HBCDD/g dust from Sweden (Björklund et al. 2012). In this case, the dust was collected by the study participants themselves. In contrast, the level in the vacuum cleaner bag from the same indoor locations was 86 ng/g, and no correlation was found between the results of the two sampling techniques. Depending on the product used or the manufacturing conditions, a different pattern of diastereomers was found. The γ-HBCDD still has the highest share of the total content in older studies from England and the Czech Republic with 52–66%, while in the other studies from Europe it only has a share of 15–34% of the total HBCDD. In these, the α-HBCDD dominates. Significant associations were described between concentrations in dust and floor area of rooms, number of foam mattresses and number of pillows in indoor rooms (De Wit et al. 2012).

In recent North American studies, the median level for the Σ-HBCDD ranges from 160 to 300 ng/g (Dodson et al. 2012; Shoeib et al. 2012; Schreder and La Guardia 2014; La Guardia and Hale 2015).

The concentrations of Σ-HBCDD in 150 dust samples collected from 23 provinces and cities across China in 2018 ranged from 6.65 to 1335 ng/g, with an average of 236 ng/g (Li et al. 2022). Higher levels were observed in samples from eastern coastal and economically developed regions.

TBBPA has been found with medians of 10–78 ng/g in European studies and 57–260 ng/g in North America. In some German study, a median of 28 ng/g (2.9–233 ng/g) was found (Fromme et al. 2014a). In a study in 12 countries, medians between 1.6 and 140 ng/g were observed, with the lowest value occurring in Vietnam and the highest in Japan (Wang et al. 2015c). In Korea, the median was 79 ng/g dust in 46 homes in 2009 (Barghi et al. 2017).

For BTBPE, rather low median concentrations between <1.0–42 ng/g are described in Europe, whereas much higher medians (12–140 ng/g) were found in North America.

The concentrations of BEH-TEBP in dust vary over a wide range. While Ali et al. (2011) found a very low median of 13 ng/g in Belgium, it was much higher in Germany at 343 ng/g (Fromme et al. 2014a) and in Sweden at 430 ng/g (Sahlström et al. 2012). In the German study, the range was between 25 and 2275 ng/g. In North America, the median level was between 99 and 260 ng/g.

DBDPE was found in most studies between 40 and 153 ng/g (medians). Only individual studies from Sweden, China, Ireland, and the USA describe higher concentrations (Sahlström et al. 2012; Wang et al. 2010; Wemken et al. 2019; La Guardia and Hale 2015).

DBE-DBCH has previously been studied in 48 Norwegian, 27 Swedish, and 40 Finnish residential samples (Cequier et al. 2014; Newton et al. 2015; Rantakokko et al. 2019). Medians (range) were 1.72 ng/g (<0.2–172 ng/g), 1.2 ng/g (<0.07–3.8 ng/g), and < 0.6 ng/g (<0.06–2.4 ng/g).

4.8 Bromophenols

Brominated phenols are technically produced by bromination of phenol, using 2,4-dibromophenol (2,4-DBPh, CAS-No.: 615–58-7), 2,6-dibromophenol (2,6-DBPh, CAS-No.: 608–33-3), 2,4,6-tribromophenol (2,4,6-TBPh, CAS-No.: 118–79-6), and pentabromophenol (PBPh, CAS-No.: 608–71-9) are of industrial importance in terms of quantity. The chemical structure of some important bromophenols is shown in Fig. 4.16.

2,4-DBPh, 2,4,6-TBPh, and PBPh are used as reactive flame retardants in epoxy resins, phenolic resins, polyester resins, polyolefins, and vinyl-aromatic polymers (EFSA 2012a; Koch and Sures 2018). 2,4,6-TBPh is also used as a fungicide, e.g. for the treatment of wood is a common precursor for other brominated flame retardants (Michałowicz et al. 2022). 2,4- and 2,6-DBPh and 2,4,6-TBPh are also naturally produced in the marine environment (EFSA 2012a).

A recent review on environmental occurrence, toxic effects, and transformation of man-made bromophenols was published by Michałowicz et al. (2022).

Fig. 4.16 Chemical structure of bromophenols (**a**): 2,4-dibromophenol; (**b**) 2,4,6-tribromophenol; (**c**) pentabromophenol)

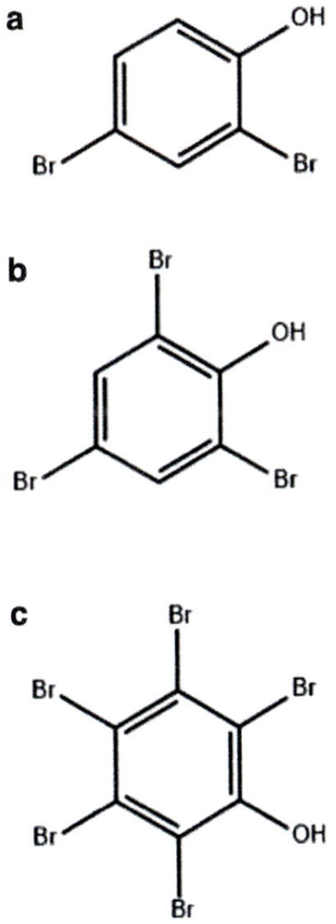

4.8.1 *Health Aspects*

Data on the toxicity of bromophenols are extremely limited and relate predominantly to the 2,4,6-TBPh. In a few short-term oral toxicity studies in rats, liver and kidneys were the main targets, while in a repeated dose oral toxicity study, which was combined with a reproduction/developmental toxicity screening test, reduced neonatal viability and reduced neonatal body weights were observed (EFSA 2012a). No studies on health effects in humans due to exposure to brominated phenols or their derivatives are available. There is also evidence that 2,4,6-TBPh leads to disruption of thyroid homeostasis (Koch and Sures 2018).

4.8.2 *Occurrence in Air and Dust*

Using a quartz-fiber filter, Saito et al. (2007) found levels from <2000 to 6800 pg/m^3 for 2,4,6-TBPh in 18 Japanese homes in 2002 and between <1100 and 1700 pg/m^3 for PBP in 14 offices. In 2006, Takigami et al. (2009) found 2,4,6-TBPh concentrations of 690 and 430 pg/m^3 in living rooms and 220 and 280 pg/m^3 in bedrooms in two rooms of two modern Japanese houses after active collection on quartz filter and polyurethane foam. In Sweden, mean values of 120 pg/m^3 (range: 30–290 pg/m^3) for 2,4,6-TBPh were measured in 9 houses during passive sampling in 2016 (Sha et al. 2018).

In the aforementioned study by Takigami et al. (2009), concentrations of 30 and 15 ng/g for 2,4,6-TBPh were reported in house dust from the two houses. In another Japanese study, dust from 19 houses and 14 offices was examined in 2005, and concentrations of 16–130 ng/g (median: 34 ng/g) and 27–620 ng/g (median: 90 ng/g) were found for 2,4,6-TBPh, respectively (Suzuki et al. 2008). In Italy, in 2019, levels of 26.4 and 51.7 ng/g were measured in two apartments and 16.7 ng/g of 2,4,6-TBPh was measured in an office (Simonetti et al. 2020).

4.9 Organophosphate Flame Retardants

Organophosphorus flame retardants (OPFR) have been widely used for a long time as flame retardants for plastic products, textiles such as upholstered furniture, building materials, as plasticizers, stabilizers, wetting agents, antifoams, and lubricants. In the context of flame retardancy, the group of organophosphate flame retardants, which typically consist of aromatic and aliphatic esters of phosphoric acid, currently accounts for about 20% of the substances used. Increasingly, they are also being used as substitutes for brominated flame retardants. The most important representatives are tri-n-butyl phosphate (TnBP) and tri-(2-butoxyethyl) phosphate (TBEP) as representatives of the aliphatic, halogen-free structures,

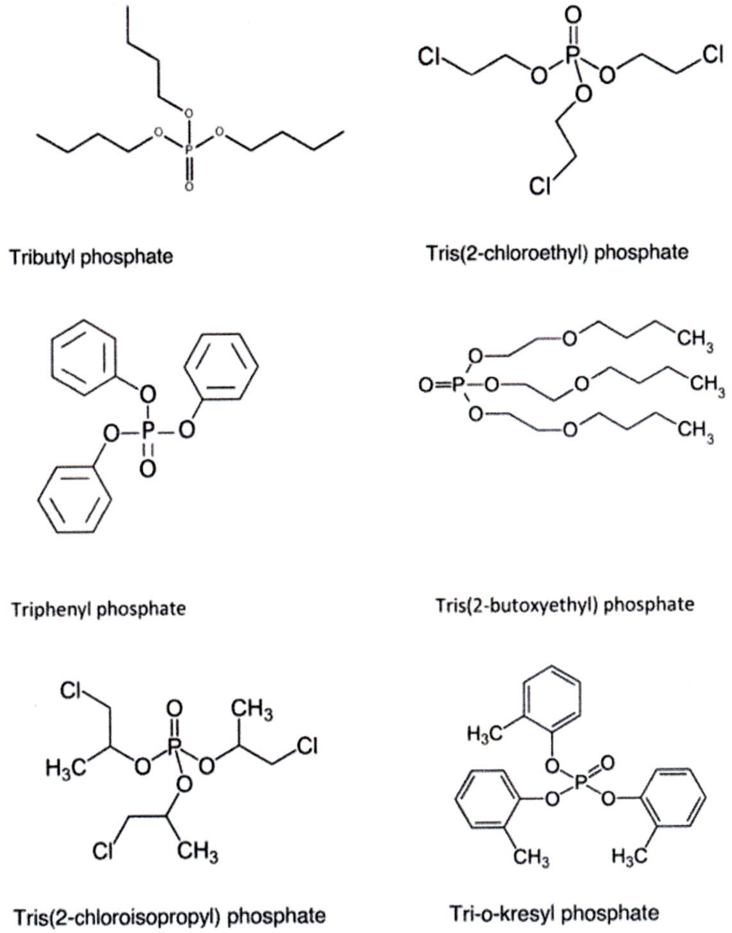

Fig. 4.17 Chemical structure of the most important organophosphate flame retardants

tri-(2-chloroethyl) phosphate (TCEP) and tri-(2-chloroisopropyl) phosphate (TCPP) as representatives of the chlorine-containing structures, and triphenyl phosphate (TPhP) and tricresyl phosphate (TCP) as representatives of the aromatic OPFR. Figure 4.17 shows the chemical structures of some substances and Table 4.18 gives an overview of names, abbreviations, and CAS No. Reviews of their occurrence in indoor environments have been published (Wei et al. 2015; Lucattini et al. 2018; Chupeau et al. 2020; Yao et al. 2021; Dou and Wang 2023). Moreover, organophosphate diesters are biotic or abiotic degradation products of triesters or used in industrial processes and can be detected in a variety of environmental matrices (Liu et al. 2021a).

Table 4.18 Name, abbreviation, formula, and CAS number of important organophosphate flame retardants

Name	Abbriviation	Formula	CAS No.
Trimethyl phosphate	TMP	$C_3H_9O_4P$	512–56-1
Triethyl phosphate	TEP	$C_6H_{15}O_4P$	78–40-0
Tripropyl phosphate	TPrP	$C_9H_{21}O_4P$	513–08-06
Tri-iso-butyl phosphate	TiBP	$C_{12}H_{27}O_4P$	126–71-6
Tri-n-butyl phosphate	TnBP	$C_{12}H_{27}O_4P$	126–73-8
Tri(2-chloroethyl) phosphate	TCEP	$C_6H_{12}Cl_3O_4P$	115–96-8
Tri(2-chloroisopropyl) phosphate	TCPP	$C_9H_{18}Cl_3O_4P$	13,674–84-5
Tri(1,3-dichloropropyl) phosphate	TDCPP	$C_9H_{15}Cl_6O_4P$	13,674–87-8
Tri(2-butoxyethyl) phosphate	TBEP	$C_{18}H_{39}O_7P$	78–51-3
Tri-phenyl phosphate	TPhP	$C_{18}H_{15}O_4P$	115–86-6
Trimethylphenyl phosphate	TMPP	$C_{21}H_{21}O_4P$	563–04-2/1330-78-5
Diphenyl(2-ethylhexyl) phosphate	DPEHP	$C_{20}H_{27}O_4P$	1241-94-7
Tris(2-ethylhexyl) phosphate	TEHP	$C_{24}H_{51}O_4P$	78–42-2

4.9.1 *Health Aspects*

In addition to known neurotoxic effects in the context of acute (accidental) poisoning, neuropsychological effects are also currently being discussed after chronic exposure to organophosphates. Occupational studies are available for TCP and TPhP, which showed only minor health effects. In animal experiments, changes in the liver, and for TnBP in the bladder, have been observed after repeated administration of TCEP and TBEP. TPhP and TCPP are considered to have low toxicity. For OPFR like TPHP and TDCPP endocrine-disrupting effects were discussed in the scientific literature. Some evidence of carcinogenic potential exists for TCEP, TnBP, and TCP. Different reviews on health effects are available (ATSDR 2012; Wei et al. 2015; ECHA 2018b; Chupeau et al. 2020; Yao et al. 2021).

4.9.2 *Occurrence of Triesters in the Air*

Only very few results on indoor air measurements have been published so far (see Fig. 4.18). In most studies TCPP was the substance that appeared with the highest concentrations in homes, while in investigations performed in schools and daycare centers TBEP could be quantified in higher levels.

In Europe, median concentrations of 5.1 ng/m^3 (TnBP), 2.3 ng/m^3 (TCEP), 42 ng/m^3 (TCPP), and < 0.04 ng/m^3 (TCP) were found in 47 Norwegian homes in 2012 (Cequier et al. 2014), while lower median concentrations of 0.2 ng/m^3 (TnBP) and 0.26 ng/m^3 (TCEP) were observed in 10 Swedish homes in 2016 (Sha et al. 2018).

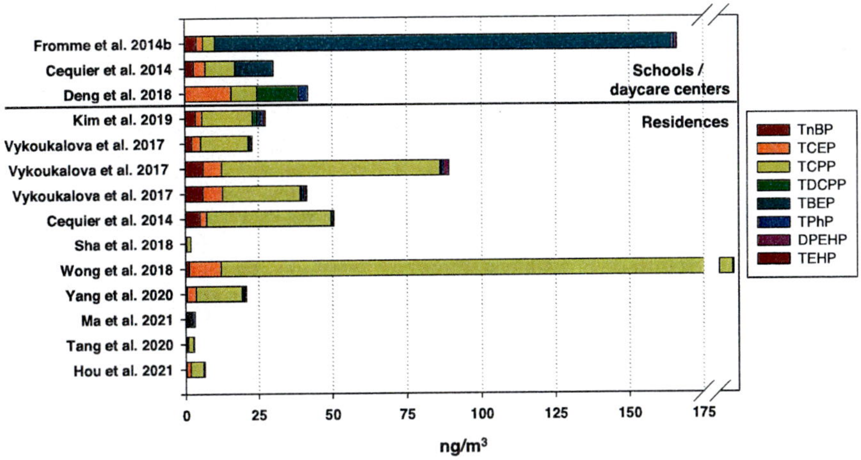

Fig. 4.18 Organophosphate flame retardants in indoor air

Wong et al. (2018) reported medians of 1.3 ng/m^3 for TnBP, 10.9 ng/m^3 for TCEP, and a very high median of 330 ng/m^3 for TCPP during a one-year sampling period in 23 indoor air samples from Stockholm, Sweden in 2014/15. Vykoukalová et al. (2017) found median concentrations of 3.0 ng/m^3 (TCEP), 2.3 ng/m^3 (TnBP), and 16.4 ng/m^3 (TCPP) in 20 homes in the Czech Republic in 2013. Similarly, in 2016/2017, both the profile and occurrence were evident in 10 Swedish offices with medians of 0.3 ng/m^3 (TCEP), 2.2 ng/m^3 (TnBP), and 31 ng/m^3 (TCPP) (Tao et al. 2019).

Similar levels were also found in 54 homes in New York State in 2018, with medians of 2.1 ng/m^3 for TnBP, 1.7 ng/m^3 for TCEP, 15 ng/m^3 for TCPP, and 1.23 ng/m^3 for TBEP (Kim et al. 2019). In 20 residences in the USA and 23 in Canada in 2013, medians of 6.1 and 6.2 ng/m^3 for TnBP, 6.8 and 6.4 ng/m^3 for TCEP, and 26.2 and 73.6 ng/m^3 were reported by Vykoukalová et al. (2017).

Medians of 0.57 and 0.37 ng/m^3 for TnBP ng/m^3, 0.36 and 0.37 ng/m^3 for TCEP, and 0.01 and 0.51 ng/m^3 for TCPP were found in samples of two studies from 15 bedrooms and 15 homes in heavily polluted cities in South China in 2017 (Tang et al. 2020; Ma et al. 2021). For $\sum_{11}$OPFR medians (range) of 4.6 ng/m^3 (1.9–14 ng/m^3) (Tang et al. 2020) and 4.9 ng/m^3 (2.8–15.8 ng/m^3) (Ma et al. 2021) were reported. On the contrary, Yang et al. (2020) observed higher medians of 2.93 ng/m^3 for TCEP and 15.9 ng/m^3 for TCPP (median $\sum_{11}$OPFR: 26.9 ng/m^3; range: 10.4–456 ng/m^3), but not for the other OPFR, in 19 air samples collected in two cities in North Chinas in 2018. In Koreas, median concentrations of 3.8 ng/m^3 for TCEP, 3.8 ng/m^3 for TnBP, 17.3 ng/m^3 for TBEP, and 3.6 ng/m^3 for TPHP in 60 households in 2018–2020 (Lu et al. 2022).

There have been limited studies in the scientific literature measuring OPFR in indoor air from public facilities. For example, Carlsson et al. (1997) reported results from three Swedish schools and one kindergarten with medians of 104 and 144 ng/m^3 (TCEP), 26 and 7.6 ng/m^3 (TnBP), and 2.4 and 5.9 ng/m^3 (TBEP). In another Swedish study in 10 kindergartens, the medians were 25 ng/m^3 (TCEP), 18 ng/m^3 (TnBP), and 84 ng/m^3 (TBEP) (Bergh et al. 2011b). Comparable concentrations were also observed in 6 schools in Norway in 2012 (Cequier et al. 2014). In addition, results from 60 kindergartens in the German daycare centers in 2011/2012, medians (95th percentiles) of <2.0 ng/m^3 (7.5 ng/m^3) (TCEP), 2.2 ng/m^3 (7.4 ng/m^3) (TnBP), and 49 ng/m^3 (833 ng/m^3) (TBEP) were seen (Fromme et al. 2014b). More recently, Deng et al. (2018) observed medians 16 ng/m^3 for TCEP, 8.7 ng/m^3 for TCPP, and 14 ng/m^3 for TDCPP in the air of nine kindergarten and two primary schools in Hong Kong in 2015/16.

In a study collected 30 $PM_{2.5}$ samples, means of 9.1 ng/m^3 for TCEP, 11.0 ng/m^3 for TCPP, and 14.6 ng/m^3 for TDCPP were reported in 11 platforms of subway stations in Barcelona, Spain between 2013 and 2016 (Olivero-Verbel et al. 2021).

Figure 4.19 shows the percentage distribution of each single OPFR in air and dust samples. It can be deduced that TCPP was the dominating substance in indoor air while differences between the USA (Kim et al. 2019) and China (Hou et al. 2021b; Tang et al. 2020) are obvious. Country-specific use pattern of individual OPFR in building materials and consumer products might be responsible for the observable differences.

In South China, the indoor concentrations of TnBP and TBEP were 4.3 and 11 times higher, respectively, than those measured in outdoor air (Ma et al. 2021). Again, Wong et al. (2018) found two orders of magnitude higher levels in indoor air compared to outdoor air in Sweden in 2014/15. Additionally, they observed a high temporal variability of OPFR during the one-year sampling period. A statistically

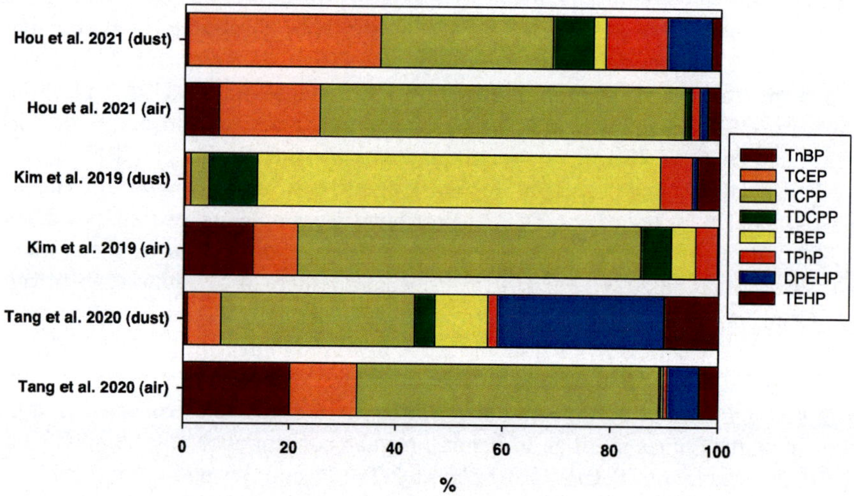

Fig. 4.19 Contribution of individual organophosphate flame retardants on the total amount in simultaneously sampled indoor air and dust (China: Hou et al. 2021b; Tang et al. 2020; U.S.: Kim et al. 2019)

significant association was observed between some OPFR levels in indoor air compared to dust (Cequier et al. 2014; Vykoukalová et al. 2017; Tang et al. 2020).

4.9.3 Occurrence of Triesters in Dust of Residences

A summary of recent measurement results in dust of residences can be found in Table 4.19. Overall, comparison of all study results indicates a significant variation in concentrations and patterns due to different national fire safety regulations and therefore different usage of OPFR, as well as various sampling techniques. For example, Vykoukalová et al. (2017) reported significantly higher dust concentrations in the USA than in Canada, and Czech Republic. Again, Li et al. (2019) analyzed 20 organophosphate triesters in 341 house dust samples collected from 12 countries during the period 2010–2014. ΣOPFR concentrations in indoor dust from more industrialized countries (e.g., South Korea: median, 31.3 μg/g; Japan: 29.8 μg/g; and the USA: 26.5 μg/g) were one or two orders of magnitude higher than those from less industrialized countries (e.g., Saudi Arabia: 5.3 μg/g, Vietnam: 1.2 μg/g, China: 1.1 μg/g, Colombia: 0.37 μg/g, and India: 0.28 μg/g).

In residential interiors, European studies describe very variable results with medians of the most important OPFR from 2.0 to 13.4 μg/g for TBEP, 0.9 to 11 μg/g for TCPP, 0.4 to 10 μg/g for TDCPP, and 0.2 to 4.0 μg/g for TCEP.

In the USA, studies found median concentrations between 0.06 and 0.20 μg/g for TnBP, 0.29–1.44 μg/g for TCEP, and 0.92–3.70 for TDCPP. TBEP showed the highest median concentrations of all OPFR.

So far, studies have been published from China between 2013/14 and 2018, as well as one study each from Japan in 2009/10 and Australia in 2018. Overall, the levels are comparable to the results of European and North American studies. Only in China significantly lower medians of only 0.14–0.6 μg/g are described for TBEP. In Japan, a significant association was observed between TBEP as well as TCPP and their corresponding metabolites in urine samples collected from 128 participants in 2009/10 (Bastiaensen et al. 2019), while Hoffman et al. (2015) found no significant correlation between dust levels of TDPP and TPHP and their metabolites in urine of 11 subjects in the USA in 2012.

Additionally, dust samples of 45 residences in Beijing, China were collected from floor surfaces, furniture, upholstery, desks, and/or windowsills using a soft brush and analyzed for new tri- and diester (Wang et al. 2022). Median level—semiquantitative—were 4.1 ng/g for tridecyl phosphate, 35.1 ng/g for tridodecyl phosphate, 29.3 ng/g dioctyl butoxyethoxyethyl phosphate, 13.6 ng/g for dioctyl (oxo)butoxypropyl phosphate, <LOQ for dioctyl hydroxyethoxyethoxyethyl phosphate, 3.3 ng/g for didodecyl hydroxyethoxyethyl phosphate, 2.1 ng/g tetradecyl dodecyl hydroxyethoxyethyl phosphate, while a median of 5.8 ng/g could be quantified for bis(2-butoxyethyl) hydroxyethyl phosphate.

Table 4.19 Medians of organophosphate flame retardants in dust of residences in µg/g

References	*N*	TiBP	TnBP	TCEP	TCPP	TDCPP	TBEP	TPhP	DPEHP	TEHP	Country, year
Europe											
Esplugas et al. (2022)	4	–	0.025*	0.25*	0.32*	0.32*	<LOQ	–	–	0.25*	Spain, 2020
Dubocq et al. (2021)	6	–	0.006*	0.006*	2.9*	0,60*	–	–	–	–	Sweden, 2019
Daniels et al. (2021)	643	–	<0.9	<4.0	2.4)	–	6.5	–	<1.6)	<0.8	Germany, 2014–17
Rantakokko et al. (2019)	40	<0.05	0.04	0.66	1.9	0.43	11.1	0.77	–	0.20	Finland
Demirtepe et al. (2019)		–	0.42	0.46	0.73	2.1	2.5	1.1	–	–	Slovakia, 2015
Zhou et al. (2017)	15	0.38	0.25	1.1	4.2	<0.02	4.3	1.2)	–	<0.01	Germany, 2015
Vykoukalová et al. (2017)	20	–	0.05	0.16	1.86	0.18 (	–	0.81	0.84	0.15	Czech Rep., 2013
Cequier et al. (2014)	48	0.06	–	0.41	2.7	0.50	13.4	0.98	0.62	–	Norway, 2012
Shoeib et al. (2019)	39	–	–	0.2	0.9	0.7	<LOQ	0.30	0.3	<LOQ	Turkey, 2012
Brommer and Harrad (2015)	32	–	<0.03	0.81	–	0.71)	–	3.3	1.6	–	UK, 2011/12
Van den Eede et al. (2011)	33	3.0	0.13	0.23	1.4	0.36	2.0	0.5	–	–	Belgium
Luongo and Östman (2016)	62	5.3	5.6	4.0	11	2.0	11	4.3	2.7	<21	Sweden, 2008
Bergh et al. (2011a)	10	1.1	0.3	2.1	1.6	10	4.0	1.2	0.5	<LOQ	Sweden
Asia/Australia											
Lu et al. (2022)	60	0.2	0.2	0.2	0.99	0.99	20	0.9	0.99	0.2	Korea, 2018–20
Chen et al. (2021)	10	<0.3	<0.31	<0.27	–	–	31.5	–	<0.34	<0.31	Taiwan
Huang et al. (2020)	43	–	0.50	1.5	6.9	1.28	8.27	2.12	0.60	0.59	Australia, 2018
Yang et al. (2020)	10	–	0.02	0.14	0.51	0.11	0.14	0.31	0.23	0.17	China, 2018
Tang et al. (2020)	15	<0.03	<0.02	0.09	0.52	54	0.14	0.10	–	0.15	China, 2017
Tan et al. (2019)	30	–	0.04	0.42	0.70	1.36	0.41	0.61	0.48	0.71	China, 2015/16
Zheng et al. (2017)	9	–	<LOQ	0.11	0.25	0.33	<LOQ	0.28	0.18	019	China, 2015
Lee et al. (2020)	111	–	–	0.45	0.44	–	0.60	0.54	0.64	0.16	Korea, 2015
He et al. (2015)	25	–	0.14	1.93	1.22	0.15	0.20	1.09	0.31	0.19	China, 2013/14, rural
He et al. (2015)	11	–	0.08	3.78	0.75	0.13	0.32	0.15	0.36	0.14	China, 2013/14, urban
Bastiaensen et al. (2019)	128	–	<LOQ	<LOQ	0.7	2.8	26.1	0.9	–	<LOQ	Japan, 2009/10

North America											
Kim et al. (2019)	8	0.06		0.29	0.9	2.5	20.5	1.6	0.3	1.1	USA, 2018
Tan et al. (2019)	17	–	0.20	0.32	5.6	3.7	20	1.6	0.91	0.67	USA, 2015/16
Vykoukalová et al. (2017)	20	–	0.11	1.44	2.79	3.68	–	3.04	0.89	1.36	USA, 2013
Vykoukalová et al. (2017)	23	–	0.06	0.81	1.47	0.92	–	2.35	0.75	0.10	Canada, 2013

* mean; *LOQ* limit of quantitation

4.9.4 Occurrence of Triesters in Dust of Public Facilities

Some measurement results on OPFR in dust are also available for schools and daycare centers (see Table 4.20).

Overall, all OPFR show a wide range of results, while the medians in Europe, China, and North America are often at comparable concentration levels. Only for TCPP were higher concentrations observed in dust in some European studies, while high medians for TBEP were reported in Japan (Mizouchi et al. 2015) and European studies (e.g., Bergh et al. 2011a; Fromme et al. 2014b).

4.9.5 Occurrence of Diesters in Air and Dust

Organophosphate diesters are well-known degradation products of their related triesters. Nevertheless, ten diesters were produced in approximately 170 factories worldwide, mainly in China, with an annual production volume of around 17,050 tons (Liu et al. 2021a). They were generally used as flame retardants and plasticizers. Liu et al. (2021a) give an overview of the diester sources, chemical properties, environmental and human occurrence, and adverse effects, while Hou et al. (2021a) reviewed the literature on contamination in indoor air and dust. The most information from indoor spaces is available for the following substances: di-n-butyl phosphate (DnBP; CAS-No.: 107–66-4), bis(2-ethylhexyl) phosphate (BEHP; CAS-No.: 298–07-7), Bis(2-butoxyethyl) phosphate (BBOEP; CAS-No.: 14260–97-0), bis (2-chloroethyl) phosphate (BCEP; CAS-No.: 3040-56-0), and diphenyl phosphate (DPhP; CAS-No.: 838–85-7).

In 62 house dust samples collected from 19 buildings in Sweden during the year 2008 diphenyl-2-ethylhexyl phosphate was found with a median of 2.7 μg/g (<LOQ - 20 μg/g) in 84% of the samples, while in indoor air only 3% of the samples were above LOD with a maximum: of 3.9 ng/m^3 (Luongo and Östman 2016). Again, Bjornsdotter et al. (2018) analyzed TPhP and DPhP in indoor dust samples collected in Spain and in the Netherlands ($n = 80$) and reported ranges from 0.17 to 142.5 μg/g and 0.11 to 79.7 μg/g, respectively. DPhP concentrations were strongly correlated to the TPhP levels.

Diesters were measured in dust from 30 Chinese and 17 North American houses in addition to triesters (Tan et al. 2019). The sum of diesters in China and the USA corresponds on average to approximately 23% and 21% to the total concentrations of triesters, respectively. The medians in this study were 6.5 μg/g for DPhP, 1.5 μg/g for BCEP, and 0.9 μg/g for BEHP. They concluded that the measured concentrations of diesters mostly reflect their original abundances in house dust and are not caused by degradation processes during sample treatment. Lower levels were reported from the following Chinese and Australian studies. Here, Hu et al. (2020) collected dust from 2015 to 2018 in China and found medians (ranges) of 0.12 μg/g (0.03–0.29 μg/g) for DnBP, 0.20 μg/g (0.03–0.86 μg/g) for DPhP, 0.01 μg/g (<LOQ - 0.94 μg/g)

Table 4.20 Medians of organophosphate flame retardants in dust of schools and daycare centers in μg/g

References	*N*	TiBP	TnBP	TCEP	TCPP	TDCPP	TBEP	TPhP	DPEHP	TEHP	Country, year
Europe											
Besis et al (2023)	20	–	0.04	0.03	0.55	0.49	1.82	0.25	–	0.04	Greece, 2018
Brommer and Harrad (2015)	28	–	0.12	0.86	16	0.51	–	4.1	29	–	UK, 2015
Zhou et al. (2017)	5	0.42	0.24	1.1	4.0	<0.02	60	1.2	–	<0.01	Germany, 2015
Larsson et al. (2018)	100	–	–	<1.2	<2.6	1.9	79	1.1	–	–	Sweden, 2015
Cequier et al. (2014)	6	–	0.04	1.2	2.0	1.5	87.2	1.5	2.3	–	Norway, 2012
Fromme et al. (2014b)	63	<0.3	<0.3	0.4	2.7	–	225	0.5	1.1	0.5	Germany, 2011/12
Bergh et al. (2011a)	10	0.7	1.2	30	3.1	9.1	1600	1.9	0.8	0.1	Sweden
Asia											
Deng et al. (2018)	11	–	–	0,15	0.06	0.59		0.86	–	–	Hong Kong, 2015/16
Lee et al. (2020)	4	–	–	0.47	0.47	–	0.95	0.94	2.0	0.16	Korea, 2015
Wu et al. (2016)	9	<0.01	<0.01	2.0	0.22	0.41	0.63	0.56	–	–	China, 2012/13
Mizouchi et al. (2015)	18	<0.01	<0.01	0.50	0.63	0.74	270	2.0	0.40	<0.005	Japan, 2009/10
North America											
Young et al. (2021)	18	0.04	0.05	0.29	2.52	1.05	16.3	1.04	0.18	0.04	USA, 2019

* mean; *LOQ* limit of quantitation

for BBOEP, and 0.14 µg/g (<LOQ - 0.35 µg/g) for BCEP. Additionally, medians (ranges) of 0.24 µg/g (0.014–3.10 µg/g) for DnBP, 3.41 µg/g (0.66–316 µg/g) for DPhP, 1.21 µg/g (0.27–3.98 µg/g) for BEHP, and 0.17 µg/g (0.01–6.30 µg/g) for BBOEP were observed in dust samples collected from 43 homes in South Australia in 2018 (Huang et al. 2020). Their sum concentration was approximately 13% of the total triester concentration.

4.9.6 *Emerging Organophosphates in Indoor Spaces*

Due to the toxicological properties of the classic OPFR, especially the chlorinated compounds, replacement products are being intensively sought. Very often, these substances are grouped together under the term emerging OPFR (eOPFR). The most important representatives are listed in Table 4.21. Liu et al. (2020) summarized, for example, that chlorinated OPFR, such as TCEP and TCIPP, were replaced by resorcinol bis(diphenyl phosphate) (RBDP) in some consumer products and cresyl diphenyl phosphate (CDPP), isodecyl diphenyl phosphate (iDDPHP), and tetrakis-(2-chloroethyl)dichloroisopentyl diphosphate (V6) have been documented as potential TPHP replacements.

To date, there are only very limited results of eOPFR in indoor air. In a 2017 study in 15 bedrooms in southern China, indoor air was sampled as well as dust on floors (Tang et al. 2020). Low medians (range) of 0.03 ng/m^3 (<0.007–0.45 ng/m^3) for RBDP, 0.04 ng/m^3 (<0.14–0.38 ng/m^3) for BPA-DP, and 0.05 ng/m^3 (<0.11–0.99 ng/m^3) for iDDPP were reported, while V6 was not found above the limit of quantification (0.02 ng/m^3). eOPFRs accounted for about 13% of the total content of four eOPFR and 11 OPFR in the air samples and about 20% in the dust samples. No significant correlation was observed between the levels in dust and air.

Table 4.22 summarized the results of eOPFR in dust. In these studies, the highest medians were found to be <28–1225 ng/g for iDDPP, from 56 to 739 ng/g for CDPP, from 30 to 530 ng/g for BPA-DP, and from <1.8 to 362 ng/g for RBDP. Particularly striking are the considerable differences between countries, both in terms of concentration levels but also in terms of distribution patterns. In particular, in an Australian study in 43 urban and suburban residences in 2018, where dust was actively collected from the ground with a vacuum cleaner, many eOPFR could be determined at higher concentrations (Huang et al. 2020). Some of these eOPFR, often structurally based on TPHP, are substitutes for other legacy flame retardants (Huang et al. 2020). For example, BPA-DP and RBDP are replacements for TPHP and DecaBDE, RBDP is also applied as a substitute for TCEP and TCPP.

In another study performed in South China in 2018, Liu et al. (2020) were able to detect four eOPFR not only in outdoor air but also in 49 house dust samples. The medians (ranges) were 10.7 µg/g (<0.002–94.5 µg/g) for tris(2,4-di-*tert*-butylphenyl) phosphate (AO 168), 0.22 µg/g (<0.003–1.3 µg/g) for bis(2,4-di-*tert*-butylphenyl) pentaerythritol diphosphate (AO 626), 0.02 µg/g (<0.002–0.3 µg/g) for triisodecyl phosphate, and 0.80 µg/g (0.1–21.6 µg/g) for trisnonylphenol phosphate.

Table 4.21 List of emerging organophosphate flame retardants

Compound	Abbreviation	CAS-No.
Butyl diphenyl phosphate	BDPP	2752-95-6
Isopropyl diphenyl phosphate	iPDPP	60,763–39-5
Dibutyl phenyl phosphate	DBPP	2528-36-1
Octyl diphenyl phosphate	ODPP	115–88-8
Isodecyl diphenyl phosphate	iDDPP	29,761–21-5
Dodecyldiphenyl phosphate	DDDPP	27,460–02-2
Tetradecyl diphenyl phosphate	TDDPP	142,474–86-0
Cresyl diphenyl phosphate	CDPP	26,444–49-5
2-isopropylphenyl diphenyl phosphate	2iPPDPP	64,532–94-1
3-isopropylphenyl diphenyl phosphate	4iPPDPP	55,864–04-5
2,4-diisopropylphenyl diphenyl phosphate	24DiPPDPP	96,107–55-0
2-tert-butylphenyl diphenyl phosphate	2tBPDPP	56,803–37-3
4-tert-butylphenyl diphenyl phosphate	4tBPDPP	981–40-8
Bis(2,4-diisopropylphenyl) phenyl phosphate	B24DiPPPP	2,190,501–29-0
Bis(2-tert-butylphenyl) phenyl phosphate	B2tBPPP	65,652–41-7
Bis(4-tert-butylphenyl) phenyl phosphate	B4tBPPP	115–87-7
Trixylyl phosphate	TXP	25,155–23-1
Tris(2-isopropylphenyl) phosphate	T2iPPP	64,532–95-2
Tris(3-isopropylphenyl) phosphate	T3iPPP	72,668–27-0
Tris(4-isopropylphenyl) phosphate	T4iPPP	2502-15-0
Tris(3,4-dimethylphenyl) phosphate	T34DMPP	3862-11-1
Tris(3,5-dimethylphenyl) phosphate	T35DMPP	25,653–16-1
Tris(4-tert-butylphenyl) phosphate	T4tBPP	78–33-1
Bisphenol A bis(diphenyl phosphate)	BPA-DP	5945-33-5
Resorcinol bis(diphenyl phosphate)	RBDP	57,583–54-7
Tris(2,3-dibromopropyl) phosphate	TDBPP	126–72-7
Tetrakis(2-chloroethyl) dichloroisopentyl diphosphate	V6	38,051–10-4

The concentrations found indicate the use of these substances as flame retardants, although a specific source is not yet known.

4.10 Diisocyanates

Isocyanates are organic chemicals with the chemical formula R-N=C=O, containing one or more functional group (R). Technically and for human exposure, e.g. at specific workplaces are the diisocyanates. In their case, R can be an aliphatic, cycloaliphatic, or aromatic group. Monoisocyanates have only one NCO group in the molecule (e.g., methyl isocyanate) and are mostly used for chemical synthesis and can be formed during the thermal decomposition of plastics (AGS 2009). Di- or triisocyanates carry two or three NCO groups in the molecule, respectively, and thus

Table 4.22 Emerging organophosphate flame retardants in dust in ng/g

Compound	Kubwabo et al. (2021)	Kademoglou et al. (2017)	Kademoglou et al. (2017)	Tan et al. (2018)	Guan et al. (2019), Tan et al. (2019)	Guan et al. (2019), Tan et al. (2019)	Tang et al. (2020)	Huang et al. (2020)	Du et al. (2019)
	Canada, 2007–10	Norway, 2013/14	UK, 2013/14	China, 2015/16	China, 2015/16	USA, 2015/16	China, 2017	Australia, 2018	China, 2019
iDDPP	1225	51.3	401	0.26	164	90	<28	740	37.7
24DiPPDPP					8.7	83.6		370	
B2iPPP	377				5.9	34		147	11.7
B4iPPP					2.5	13.7		33	
2iPPDPP	325				28.6	184		797	39.0
4iPPDPP					9	95.9		284	
CDPP	91			0.08	55.8	362		739	101
T2iPPP				<0.6	7.9	33.7		63.3	9.0
T3iPPP	72				4.3	26.8		146	
T4iPPP					ND	0.8		<0.3	
2tBPDPP	52			80	ND	1.1		<0.2	
4tBPDPP					23.3	59.2		96.4	13.0
B2tBPPP	26				ND	0.9		<0.2	
B4tBPPP					12.5	10.6		47.6	
T34DMPP								19	
T35DMPP				<0.6				16.5	
T4tBPP	<44							6.4	3.7
TDBPP		<0.5	<0.5						
B24DiPPPP	<49				1.6	20		46.2	
BPA-DP	30	35.4	66.8	0.53	405	283	305	193	187
RBDP	14	<1.8	1.9	0.06	55.9	362	76	85.1	131
ODPP	22								

BDPP	<6								
iPDPP	<18								
DDDPP	<48								
TDDPP	<36								
DBPP	<13								
TXP	23	9.1	26.5						
V6	–	4.1	16.6	ND			32	0.14	3.4

ND not detected

can form polymer chains or cross-linked molecules. Diisocyanates are solids or liquids that exhibit high reactivity. Airborne isocyanates occur in different forms, i.e. as vapors, as aerosols, or adsorbed on particulate matter. While monoisocyanates are more volatile, the diisocyanates have a low vapor pressure.

The four technically most important diisocyanates, which are responsible for about 94% of the commercially produced or imported quantities in Europe, are m-tolylidene diisocyanate (TDI, mixture of two isomeres; CAS No.: 26471–62-5), 4,4′-methylenediphenyl diisocyanate (4,4′-MDI; CAS No.: 101–68-8), 4-methyl-m-phenylene diisocyanate (2,4′-TDI; CAS No.: 584–84-9), and the hexamethylene diisocyanate (HDI; CAS No.: 822–06-0) (ECHA 2019). Among them, TDI (as a mixture of 2,4′-TDI and 2,6′-TDI) is responsible for 48%, 4,4′-MDI for 29%, 2,4′-TDI for 12%, and HDI for 4.3% of the total amounts. In the USA, TDI and MDI and their polyisocyanates also account for >90% of the commercial market (ATSDR 2018). Figure 4.20 shows the chemical structures of the four aforementioned substances. Other more commonly used diisocyanates are hydrogenated MDI (H_{12}MDI, CAS No.: 5124-30-1) and isophorone diisocyanate (IPDI, CAS No.: 4098-71-9).

Around five million metric tons MDI and two million tons TDI are manufactured and used worldwide each year in the production of polyurethanes (ACS 2023). Polyurethanes are made by reacting diisocyanates with a range of polyols (an alcohol with more than two reactive hydroxyl groups per molecule) and the properties can be controlled by variation of the amount of blowing agent giving a range of densities and the molecular weight of components such as the polyols which helps to control the flexibility/rigidity of the end material (ISOPA 2020). These polyurethane chemicals are used in a range of continuous or discontinuous processes and often combined with other materials to manufacture polyurethane foams (flexible and rigid

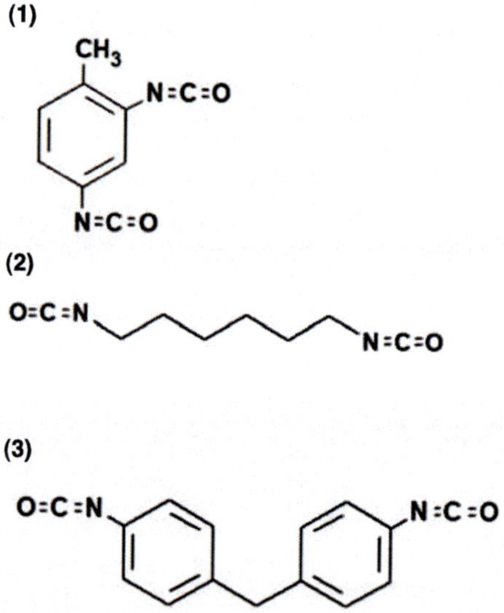

Fig. 4.20 Chemical structure of (1) m-tolylidene diisocyanate (TDI), (2) hexamethylene diisocyanate (HDI), and (3) 4,4′-methylenediphenyl diisocyanate (4,4′-MDI)

foam), assembly foams (e.g. insulation panels), foundry cores (casting), coating materials like paints, lacquers, and varnishes, adhesives and sealants, glues, elastomers, pre-polymers in chemical synthesis, engineering plastics, and polyurethane fibers/composite (ECHA 2019). Liang et al. (2021) provide a high-resolution material flow analysis of polyurethane flows through the US economy and reported that around 3700 kt were consumed in the USA in 2016.

The European Commission has amended Annex XVII to the REACH Regulation with Regulation (EU) 2020/1149, which entered into force on 24 August 2020 and stated: After 24 August 2023 diisocyanates may not be used either as substances on their own, as a constituent in other substances, or in mixtures for industrial and professional use unless (a) the concentration of diisocyanates individually and in combination is less than 0.1% by weight, or (b) the employer or self-employed ensures that industrial or professional users have successfully completed training on the safe use of diisocyanates prior to the use of the substances or mixtures (BAUA 2023).

4.10.1 Health Aspects

Exposure to isocyanates may be inhalation or dermal. Inhaled isocyanates encounter the mucous membranes of the respiratory tract and may react with molecules in the epithelial lining fluid (Schupp and Plehiers 2022). After inhalation and absorption, diisocyanates are primarily bound to albumin in plasma of exposed workers and also to hemoglobin and formed, as reactive molecules, adducts with nucleophilic biological macromolecules as well as glutathione conjugates (ATSDR 2018). After oral uptake they are readily hydrolyzed in the acidic environment of the gastrointestinal tract to the corresponding amines, which may then be metabolized further.

A large number of epidemiology and experimental studies have examined the toxicity of TDI to the respiratory tract. Data from a limited number of acute-duration studies identify respiratory irritation as the primary effect at low concentrations and severe respiratory symptoms and possibly asthma occurring after exposure to high concentrations while long-term occupational exposure leads to bronchial asthma, asthma-like respiratory symptoms, and a decline in pulmonary function (ATSDR 2018). The toxicity of MDI to the respiratory tract has not been as well investigated as TDI, but the effects appear to be similar. The primary effects observed include occupational asthma in sensitized individuals and decreased lung function (ATSDR 2018).

Both animal experiments with various species and humans give proof that diisocyanates can elicit respiratory sensitization (ECHA 2019). The most important clinical manifestation of this respiratory sensitization is occupational asthma. Moreover, there is evidence that also dermal exposure is a risk factor for the induction phase of respiratory sensitization (ECHA 2019). Health effects of non-occupational exposure are still a matter of debate. Krone (2004) summarized reports of allergy-like responses to PU-containing materials like contact dermatitis from the PU in athletic shoes, hypersensitivity reactions to PU medical product, and childhood asthma and wheezing attacks with the use of PU bedding.

The IARC concluded that there is inadequate evidence for the carcinogenicity of TDI in humans and sufficient evidence for the carcinogenicity in experimental animals. The overall conclusion was that TDI is possibly carcinogenic to humans (group 2B), while MDI is not classifiable as to its carcinogenicity to humans (group 3).

4.10.2 Occurrence in the Air

Indoor applications are always of particular health significance in the context of possible exposure of room users if, on the one hand, there is an application in which aerosols can occur (e.g., spray applications) and/or protective measures and sufficient ventilation are not observed.

Bekki et al. (2020) investigated a living room in a newly built single-family house two months after spraying urethane and found 14,000 ng/m^3 isocyanic acid, while the other isocyanates were under the limit of detection. Very high concentration of 3500 ng/m^3 4,4′-MDI and 5800 ng/m^3 phenyl isocyanate were detected in a house where polyurethane foam was sprayed shortly before sampling, indicating an only temporal contamination during the process of spraying polyurethane foam. The large-scale spreading or application of isocyanate-containing foams indoors can lead to exposure in these rooms during the first few hours. For example, in the context of a doctoral thesis, isocyanates were detected in the air of various residential and other rooms in which polyurethane-containing products such as varnishes and adhesives had been applied (quoted from Wolff and Stirn 2000). In this study, individual isocyanates, if above the detection limit, were determined as follows: IPDI 0.6 μg/m^3 (only one value), TDI 0.25–5.2 μg/m^3, MDI 0.08–0.35 μg/m^3.

Bekki et al. (2018) collected indoor dust samples (once or twice a week) in the living rooms of three households using the residences' own vacuum cleaners. Their average concentrations were 11.5 ng/g for methyl isocyanate and 8.4 ng/g for phenyl isocyanate. The authors concluded that isocyanates like 2,4-TDI, 2,6-TDI, and 4,4′-MDI, had concentrations under the detection limit of the analytical method. On the contrary, in 27 dust samples from vacuum cleaners bags collected in Spanish homes, 2,4-TDI was quantified with a high median concentration of 2870 ng/g (1400–5870 ng/g) (Castro et al. 2019).

4.11 Perfluorinated Alkyl Sulfonates and Perfluorinated Alkyl Carboxylic Acids

The term per- and polyfluorinated hydrocarbons (PFAS) covers organic compounds in which all or part of the hydrogen atoms on the carbon skeleton have been replaced by fluorine atoms. Since the polar carbon-fluorine bond is one of the most stable bonds in organic chemistry, these substances exhibit higher thermal and chemical stability than analogous hydrocarbon compounds (ATSDR 2021). Two groups, the perfluorinated alkyl sulfonates (PFSA) and the perfluorinated alkyl carboxylic acids (PFCA), are of particular importance because of their application quantities and physicochemical characteristics (see Table 4.23). Currently, the most intensively discussed substances from the PFSA group are perfluorooctane sulfonate (PFOS) and from the PFCA group perfluorooctanoic acid (PFOA), since both are very persistent in the environment and are excreted in the human organism only with very long half-lives of years. Their chemical structure is shown in Fig. 4.21.

In addition to the single substance PFOS, there are a large number of anthropogenic chemicals, all of which are derivatives of PFOS (PFOS related compounds) and can be converted to PFOS in the environment and partly also in humans. PFOA is a chemical that carries a carboxylic acid group directly on the perfluorinated carbon chain. The term PFOA is also often used as a group name for the actual acid and its salts.

PFAS are substances with an amphiphilic character, which are able to lower the surface tension of a liquid or the interfacial tension between two phases. Due to their thermal and chemical stability, resistance to UV radiation and weathering, and dirt-, paint-, grease-, oil-, and water-repellent properties, PFAS has found application in a wide range of industrial and consumer products. Major applications of PFOS-like substances in the period around 2000 were impregnation of the surfaces of textiles, leather and carpets, surface coating of paper and cardboard products, as surfactants in surface cleaning and surface treatment agents, in paints and varnishes, in fire extinguishing foam, in the photographic industry, in the aircraft industry in hydraulic oils and in the treatment of metal surfaces (ATSDR 2021). This has resulted in environmental contamination. PFOA, on the other hand, has only been used as an intermediate in fluoropolymer production.

PFAS were produced by two technical processes: electrochemical fluorination (ECF) and fluorine telomerization (ATSDR 2021). ECF is an electrolysis of organic substrates in an electric field using a nickel anode in anhydrous hydrofluoric acid. Advantages of this process, despite a yield of only 35–40%, are the good cost–benefit ratio and the fact that the functional groups are retained during the process. However, fluorination produces a number of by-products with shorter chain lengths (alkyl chain degradation) and branched alkyl chains. Therefore, the products formed in ECF are always complex mixtures of substances. In fluorotelomerization, the other important technical process for the production of perfluorinated surfactants, tetrafluoroethene is reacted with iodine and iodine pentafluoride in the first step. The resulting telomer (pentafluoriodoethane) reacts further with tetrafluoroethene several

Table 4.23 Substances from the group of perfluorinated hydrocarbons

Chain length	Acronym	Substance name	CAS number
Perfluorinated alkyl carboxylic acids (PFCA)			
Short-Chain	PFBA	Perfluorobutanoic acid	375–22-4
	PFPA	Perfluoropentanoic acid	2706-90-3
	PFHxA	Perfluorohexanoic acid	307–24-4
	PFHpA	Perfluoroheptanoic acid	375–85-9
Long-chain	PFOA	Perfluorooctanoic acid	335–67-1
	PFNA	Perfluoronanoic acid	375–95-1
	PFDA	Perfluorodecanoic acid	335–76-2
	PFUnDA	Perfluorundecanoic acid	2058-94-8
	PFDoDA	Perfluorododecanoic acid	307–55-1
Perfluorinated alkyl sulfonates (PFSA)			
Short-chain	PFBS	Perfluorobutanesulfonic acid	375–73-5
	PFPS	Perfluoropentanesulfonic acid	2706-91-4
Long-chain	PFHxS	Perfluorohexanesulfonic acid	355–46-4
	PFHpS	Perfluoroheptane sulfonic acid	375–92-8
	PFOS	Perfluorooctane sulfonic acid	1763-23-1
	PFNS	Perfluoronon sulfonic acid	474,511–07-4
	PFDS	Perfluorodecane sulfonic acid	335–77-3
	PFUnDS	Perfluorundecane sulfonic acid	749,786–16-1
	PFDoDS	Perfluorododecane sulfonic acid	79,780–39-5

Fig. 4.21 Chemical structure of (1) perfluorooctane sulfonate (PFOS) and (2) perfluorooctanoic acid (PFOA)

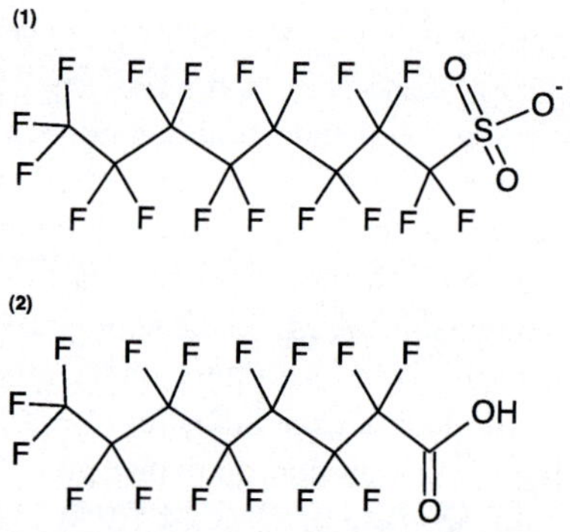

times. The perfluorinated iodoalkanes formed in this way have a linear structure, with the alkyl chain lengths differing by 2 carbon atoms in each case. The addition of ethene forms perfluoroalkyl ethyl iodides, which are intermediates for the preparation of perfluoroalkyl olefins, alcohols, isocyanates, sulfonyl chlorides, and thiols. The reaction is controlled so that mainly substances with alkyl chains of 8 carbon atoms are formed. However, substances with alkyl chains of C_4 to C_{12} are also present in the product (Lehmler 2005).

The production, use, and application of PFOS have been banned in the EU since 2003. For PFOA there is a ban from 2020. The Stockholm Convention listed PFOS, its salts and PFOA-related substances in annex A (parties must take measures to eliminate the production and use), and PFOS and its salts in annex B (parties must take measures to restrict the production and use in light of any applicable acceptable purposes and/or specific exemptions).

4.11.1 Health Aspects

Results from a large number of epidemiological and animal studies, some of which are very extensive, are available for this substance class and have been compiled in various recent reviews (ATSDR 2021; EFSA 2018a, 2020a; Carlson et al. 2022, Steenland et al. 2020, Crawford et al. 2023). In this context, associations of PFAS as a whole or of individual substances and changes in lipid metabolism with higher cholesterol levels, a reduced antibody response after vaccination, a higher risk of preeclampsia and thyroid diseases, and changes in liver enzymes were observed in particular.

4.11.2 Occurrence in the Air

Due to the low volatility of these substances, low concentrations in indoor air are to be expected. In the study by Barber et al. (2007), PFAS were determined in the particulate phase of 4 homes in 2005. The highest content with 17.1 pg/m^3 was determined for PFHxA, followed by PFOA with 4.4 pg/m^3, while for sulfonates only PFDS with 2.6 pg/m^3 was found above the limit of quantification. Geometric mean values of 16.2 pg/m^3 (PFOA), 9.4 pg/m^3 (PFHxA), 4.6 pg/m^3 (PFDA), and 2.5 pg/m^3 (PFNA) could be determined in 57 Finnish apartments in 2014/2015 (Winkens et al. 2017).

A Canadian study conducted in 2007/2008 using passive samplers in 59 homes was able to detect PFOA in 68% and PFNA in 42% of air samples, while PFOS could not be quantified above the quantification limit of 0.02 pg/m^3 (Makey et al. 2017). For PFOA, the geometric mean was 47 pg/m^3 and for PFNA 1.5 pg/m^3.

Yao et al. (2018) collected dust samples in 18 homes in Tianjin, China, in 2015. PFOA and PFOS were found in all samples with medians of 221 and 196 pg/m^3, respectively.

4.11.3 Occurrence in Dust of Residences

Data of the results published so far in the scientific literature are summarized in Table 4.24. The highest concentration can be found in the aforementioned studies for PFCA and here in particular for PFOA, while for PFSA, PFOS determines the total

Table 4.24 Median (ranges) concentrations of PFAS in dust of residences in ng/g

References	*N*	PFHxA	PFOA	PFNA	PFDA	PFUnDA	PFDoDA	PFHxS	PFOS	PFDS	Country, Year
Europe											
Huber et al. (2011)	7	10 (1.6–27)	39 (10–80)	7.0 (3.3–27)	7.5 (2.0–11)	97 (0.9–322)	0.8 (0.2–3.0)	1.4 (0.8–3.1)	9.1 (4.4–24)	1.5 (1.0–14)	Norway, 2007/08
Fromme et al. (2008b)	12	–	11 (2–141)	–	–	–	–	–	16 (3–342)	–	Germany, 2008
Xu et al. (2013)	31	–	39 (6.1–676)	–	–	–	–	–	20 (3.3–1046)	–	Germany, 2008/09
Lankova et al. (2015)	18	– (<1.0–9.7)	2.4 (<1.0–9.1)	– (<1.0–8.9)	– (<1.0–12)	– (<1.0–3.3)	– (<1.0–32)	–	1.5 (<0.5–3.4)	– (<0.5–222)	Czech Republic, 2013
Karásková et al. (2016)	16	3.8 (1.4–69)	2.0 (<1.1–7.8)	<0.7 (<0.7–11)	<0.4 (<0.4–17)	<0.6 (<0.6–4.3)	0.5 (<0.3–13)	2.0 (<0.1–9.3)	10.3 (4.8–118)	0.8 (<0.2–6.6)	Czech Republic, 2013
Eriksson and Kärrman (2015)	7	3.9 (<3.2–26)	13 (<7.0–129)	3.6 (<1.0–18)	5.4 (<1.9–21)	2.9 (0.7–7.9)	3.6 (1.1–8.0)	0.4 (0.3–11)	7.2 (2.8–81)	– (<0.01–0.4)	Greece, 2013/14
Eriksson and Kärrman (2015)	10	3.4 (<3.2–5.5)	8.8 (<7.0–40)	1.7 (<1.0–43)	3.7 (2.1–81)	1.1 (<0.4–33)	2.0 (1.0–60)	0.4 (<0.1–1.0)	5.3 (2.1–7.2)	- (<0.01–0.02)	Spain, 2013/14
Eriksson and Kärrman (2015)	10	7.1 (<3.2–40)	14.4 (<7.0–50)	2.0 (<1.0–8.5)	3.7 (<1.9–35)	0.9 (<0.4–6.1)	3.4 (0.6–17)	0.2 (<0.09–3.6)	2.8 (<0.7–9.7)	0.01 (<0.01–0.4)	Sweden, 2013/14
Eriksson and Kärrman (2015)	10	8.0 (2.6-	15 (<7.0–192)	2.4 (1.1–11)	4.8 (2.1–13)	1.4 (0.7–5.2)	4.1 (1.5–7.8)	0.4 (<0.09–11)	5.8 (2.6–149)	- (<0.01–0.5)	Faroe Islands, 2013/14
Weiss et al. (2021)	46	6.4 (<LOQ-99)	9.0 (<LOQ-650)	3.4 (<LOQ-44)	3.2 (<LOQ-100)	1.0 (<LOQ-22)	<LOQ (<LOQ-110)	<LOQ (<LOQ-8.1)	13 (<LOQ-220)	–	Sweden, 2013/14
Winkens et al. (2018)	63	2.3 (<0.8–55)	5.3 (<2.5–83)	1.1 (<0.7–15)	1.6 (<0.5–67)	0.7 (<0.7–4.6)	0.8 (<0.4–44)	0.7 (<0.5–37)	1.0 (<0.5–11)	0.8 (n.s.-34)	Finland, 2014/15
Gustafsson et al. (2022)	32	39	41	4.5	20	2.3	8.4		13		Sweden, 2015
America											
Strynar and Lindstrom (2008)	112	54 (<10–1250)	142 (<10–1960)	8.0 (<11–263)	6.7 (<9.4–267)	7.6 (<11–588)	7.8 (<11–520)	46 (<13–357,000)	201 (8.9–12,100)	–	USA, 2000/01

Scher et al. (2019)	19	29 (5.4–240)	51 (9.9–970)	26 (<LOQ-450)	13 (<LOQ-370)	7.2 (<LOQ-67)	8.2 (<LOQ-190)	18 (<LOQ-790)	67 (8.4–2000)	–	USA, 2010
Karásková et al. (2016)	15	5.6 (1.7–146)	8.2 (2.1–93)	4.4 (<LOQ-195)	2.4 (0.9–86)	1.1 (n.d.-50)	1.1 (n.d.-61)	1.9 (n.d.-12)	9.1 (3.3–32)	3.2 (0.5–15)	Canada, 2013
Karásková et al. (2016)	14	6.5 (2.5–190)	9.0 (2.9–318)	3.9 (1.1–63)	1.8 (0.4–64)	1.2 (n.d.-13)	0.6 (n.d.-9.0)	8.7 (1.4–84)	2.8 (0.5–9.8)	–	USA, 2013
Eriksson and Kärrman (2015)	10	7.4 (2.7–97)	21 (<7.0–310)	13.5 (2.2–419)	7.1 (2.3–20)	7.8 (1.1–140)	4.2 (1.1–12)	3.8 (0.3–472)	7.3 (1.7–699)	0.4 (0.2–2.7)	Canada, 2013/14
Hall et al. (2020)	184	8.5 (<0.4–1380)	7.9 (0.3–2350)	3.8 (<0.2–208)	6.2 (<0.1–4130)	–	–	2.0 (<0.3–694)	4.4 (<0.2–2810)	–	USA, 2014/16
Asia/Australia											
Eriksson and Kärrman (2015)	10	4.8 (<3.2–84)	13.5 (<7.0–183)	3.0 (<1.0–9.2)	5.2 (<1.9–9.6)	1.9 (<0.4–5.3)	3.9 (1.4–7.8)	4.3 (0.7–607)	9.7 (2.8–319)	0.1 (<0.01–1.3)	Australia, 2013/14
Eriksson and Kärrman (2015)	5	12 (<3.0–119)	26 (<7.0–111)	36 (17–241)	11 (4.3–49)	11 (7.2–70)	5.6 (2.2–8.7)	0.1 (<0.09–1.4)	3.9 (3.0–21)	0.01 (<0.01–0.02)	Japan, 2013/14
Yao et al. (2018)	18	3.4 (<LOQ-41)	49 (7.3–197)	2.4 (1.0–13)	2.2 (<LOQ-11)	1.9 (0.5–4.1)	1.7 (0.6–7.4)	2.9 (<LOQ-18)	9.0 (3.4–30)	–	China, 2015
Wang et al. (2021b)	43	(<LOQ-6.7)	(<LOQ-361)	(<LOQ-3.9)	(<LOQ-4.2)	(<LOQ-2.9)	(<LOQ-2.9)	(<LOQ-2.2)	(<LOQ-31)	–	China, 2017

n.s. not specified, *n.d.* not detected

amount. For PFOA, the medians range from 2.0 to 39 ng/g in Europe, 7.9 to 142 ng/g in America, and 13.5 to 48 ng/g in Asia/Australia. Regarding PFOS, the medians range from 1.0 to 20 ng/g in Europe, 2.8 to 201 ng/g in America, and 3.9 to 9.7 ng/g in Asia/Australia.

In Europe, the highest PFOA level was found in 7 Norwegian homes in 2007/2008 and in 31 German residences in 2008, while the lowest in 63 residences in Finland in 2014/15. In a German study, 12 apartments were investigated and the dust samples were either sieved to 63 μm or not treated at all before analysis (Fromme et al. 2008b). While in the sieved fraction the medians were 11 ng/g for PFOA and 16 ng/g for PFOS, they were significantly lower in the unsieved samples with medians of 7 ng/g (PFOA) and 10 ng/g (PFOS).

A study in Canada in 2002/2003 ($N = 67$) found medians (range) of 20 ng/g for PFOA and 38 ng/g for PFOS, respectively (Kubwabo et al. 2005). In this investigation, lower levels of PFOS and PFOA occurred in older homes compared to newly constructed homes, and the concentrations correlated positively with the percentage of carpet area in the rooms. Median levels observed in Canada ($N = 132$) in 2007/2008 were slightly higher compared to the previous study (PFOA: 30 ng/g; PFOS: 71 ng/g) (Shoeib et al. 2011). Again, significantly lower concentrations were reported in a 2013 Canadian study ($N = 15$) with medians of 8.2 ng/g for PFOA and 9.1 ng/g for PFOS (Karásková et al. 2016). Hall et al. (2020) also reported decreasing concentrations in dust in 184 homes in the USA in 2014–2016, with medians of only 7.9 ng/g for PFOA, 4.4 ng/g for PFOS, and 2.0 ng/g for PFHxS.

In the USA. in 2000/2001 ($N = 112$) very high medians of 142 ng/g for PFOA and 201 ng/g for PFOS were observed (Strynar and Lindstrom 2008), while recently the medians were only 7.9 ng/g for PFOA and 4.4 ng/g for PFOS (Hall et al. 2020). In 2010, testing of 20 homes in a groundwater-contaminated area in the USA, PFOA was measured in dust samples from the entrance area and living space at concentrations of 26 ng/g and 51 ng/g (median), respectively, which were a factor of 35 and 70 lower than concentrations in soil samples from the surrounding area (Scher et al. 2019). The authors consider the results as evidence for the dominance of indoor over outdoor sources.

Yao et al. (2018) used dust samples from 18 homes in Tianjin, China, in 2015. PFOA and PFOS were found in all samples with medians of 49 and 9,0 ng/g, respectively. Eriksson and Kärrman (2015) collected dust from vacuum cleaner bags (sieved to 150 μm) from individual households in Canada, the Faroe Islands, Sweden, Spain, Greece, Nepal, Japan, and Australia between 2009 and 2014. They reported medians from 8.8 to 25.5 ng/g (Spain, Japan) for PFOA and 2.8 to 9.7 ng/g (Sweden, Australia).

Dust samples were collected from 32 vacuum cleaner bags from individual residential homes in Sweden in 2015 and sieved to different size fractions from 3.7 μm to 1000–2000 μm (Gustafsson et al. 2022). The highest concentrations of the 25 PFAS are found in the inhalable particulate fraction which is explained by the increased surface area as the particulate's sizes decrease. The polyfluoroalkyl phosphoric acid diesters (diPAP), followed by PFOA and PFHxA showed the highest concentrations of PFAS analyzed.

4.11.4 *Occurrence in Dust of Public Facilities*

In the indoor environment of 20 daycare center samples, medians of 2.0 ng/g (0.3–5.1 ng/g) for PFOA and 1.2 ng/g (0.2–4.2 ng/g) for PFOS were measured in a study from the USA, with the highest median of 3.2 ng/g (<LOQ - 9.9 ng/g) reported for PFBA (Zheng et al. 2020), while in 20 Swedish preschools medians (95th percentiles) of 12.2 ng/g (49 ng/g) for PFOS, 7.71 ng/g (35 ng/g) for PFOA, and 1.09 ng/g (56 ng/g) for PFNA were reported in 2018 (Giovanoulis et al. 2019). Additionally, Wu et al. (2020b) collected paired 29 carpet and 28 dust samples from 18 California childcare centers in 2018 and found medians of 4.9 ng/g (1.4–27 ng/g) for PFOA, 4.6 ng/g (1.1–39 ng/g) for PFHxA, 4.5 ng/g (<0.4–326 ng/g) for PFBA, and 4.6 ng/g (0.4–44 ng/g) for PFOS. They reported strong associations between PFAS levels in carpet and dust pairs, suggesting that carpets can be a source and a sink for PFAS.

4.12 Polychlorinated Biphenyls (PCB)

PCB consist of two linked aromatic rings and are synthesized by chlorination of biphenyl. Due to the number and position variants of the chlorine atoms on the rings, 209 congeners are theoretically possible. According to Ballschmiter and Zell (1980), the individual mono- to decachlorobiphenyls are designated by sequential numbers. Figure 4.22 shows the general chemical structure of PCB. Of particular importance are the non-ortho- and mono-ortho-substituted PCB, which show similar biological effects as the polychlorinated dibenzodioxins and furans (PCDD/F) and are therefore discussed in the next chapter.

PCB were always produced and used in mixtures under the names Aroclor (USA), Clophen (Germany), Phenochlor (France), Kanechlor (Japan), Pyralene (France), Fenoclor (Italy), and Delor (former Czechoslovakia). For example, for the clophenes produced in Germany in the period between about 1930 and 1983, the degree of chlorination is between 40 and 42% (Clophen A 30) and 60 and 62% (Clophen A 60) (IARC 2015b). See Table 4.25.

Until they were banned, PCB were produced on a large scale due to their physicochemical properties (high heat stability, low flammability, relatively resistant to acids, alkalis, and other chemicals, good thermal conductivity, very low electrical conductivity) and used for decades in a variety of open and closed applications. Major open applications in the past included lubricants, gear oils, screw greases, water-repellent impregnating and flame retardant agents for wood, paper, fabrics and

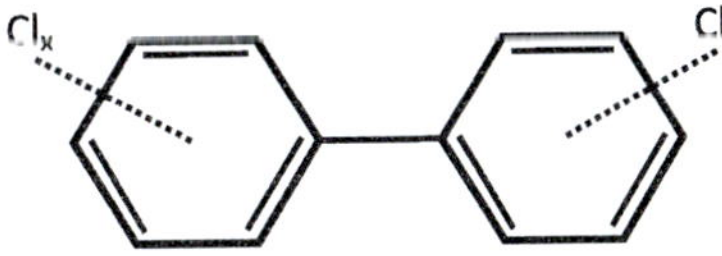

Fig. 4.22 Chemical structure of polychlorinated biphenyls (PCB)

Table 4.25 Share of homologues on two PCB products and range of chlorination (modifiziert nach IARC 2015b)

	Congener No.	Arochlor						
		1221	1232	1016	1242	1248	1254	1260
		Clophen						
		–	–	–	A30	A40	A50	A60
Monochlorobiphenyl	1–3	51	26	1	1	–	–	–
Dichlorobiphenyl	4–15	32	29	20	17	1	–	–
Trichlorobiphenyl	16–39	4	24	57	40	23	–	–
Tetrachlorobiphenyl	40–81	2	15	21	32	50	16	–
Pentachlorobiphenyl	82–127	0.5	0.5	1	10	20	60	12
Hexachlorobiphenyl	128–169	–	–	<0.01	0,5	1	23	46
Heptachlorobiphenyl	170–193	–	–	–	–	–	1	36
Octachlorobiphenyl	194–205	–	–	–	–	–	–	6
Nonachlorobiphenyl	206–208	–	–	–	–	–	–	–
Decachlorobiphenyl	1	–	–	–	–	–	–	–
Chlorination (%)		21	32–33	40–42	40–42	48	52–54	60–62

leather, coating of tracing and carbon copy paper, additives in adhesives, sealants, and joint putties, and as dispersants in printing inks, color pigments, and waxes. In closed systems, they were used in particular in transformers, condensers, heat exchangers, and as hydraulic fluids. Estimated total cumulative worldwide producction was 1 to 1.5 million tons or more (IARC 2015b). The production starting in the 1930s and continued in many countries until the 1970s and 1980s, in Korea up to 2012.

In 1978, their use in open systems was banned, and in 1989, their production, marketing, and use were banned throughout the EU. In the Stockholm Convention PCB are listed in annex A (parties must take measures to eliminate the production and use).

Due to their use in large quantities and their persistence in the environment, PCB are still ubiquitous. Due to their good fat solubility, especially higher chlorinated congeners have a pronounced tendency to bioaccumulate, so that they accumulate in the food chain, are ingested by humans mainly via food (Weitekamp et al. 2021), and can still be detected in human test materials.

4.12.1 Health Aspects

As explained, PCB in indoor environments are always a complex mixture of congeners with different compositions depending on the source. The available findings on health effects are also based almost exclusively on animal experiments and epidemiological studies with mixtures. Major endpoints for the effects of PCB are the liver, the nervous system (e.g., delayed cognitive development in young children), the immune system, the reproductive organs, and the thyroid gland. At high concentrations and dermal exposure, chloracne occurs, as with the other chlorinated compounds. Animal and epidemiological studies have found evidence of an association between PCB exposure and the development of tumors, with tests for genotoxicity yielding predominantly negative results. Thus, mixtures of PCB with higher chlorination levels, as opposed to those with low chlorination levels, resulted in an increased incidence of hepatocellular carcinomas in rats and mice. Various summaries on the toxicological and epidemiological background have been published (EFSA 2005b; IARC 2015b; Faroon and Ruiz 2016).

As part of the Health Effects of PCBs in Indoor Air (HESPAIR) cohort 38,613 residents of two partly PCB-contaminated residential areas in Greater Copenhagen, identified by nationwide registries, were included and diagnoses extracted from the Danish Cancer Registry for the follow-up period of 1970–2018 (Deen et al. 2022). Overall risk of cancer was not associated with PCB_{year} (defined as years in an apartment multiplied by the concentration defined as PCB_{total}), but residents exposed to ≥3000 ng/m^3 PCB × year had higher risk of liver cancer (HR = 2:81; 95%CI: 1.28,6.15) and meningiomas (HR = 3:49; 95%CI: 1.84,6.64), with indications of exposure–response relationships.

On the basis of sufficient evidence in humans for the carcinogenicity of PCB (association for malignant melanoma, non-Hodgkin lymphoma, and cancer of the breast) and sufficient evidence in experimental animals, IARC has classified PCB as carcinogenic to humans (Group 1) (IARC 2015b).

4.12.2 *Occurrence in the Air of Residences*

The six so-called indicator PCB congeners (PCB 28, PCB 52, PCB 101, PCB 138, PCB 153, and PCB 180) are regularly measured in indoor air. From the sum of the concentration of these indicator congeners multiplied by a factor of 5, the total PCB content is approximately determined. In addition, PCB 118 is often determined analytically.

The results of studies that quantified background exposure in buildings for which no active PCB sources are known are summarized in Table 4.26. For example, in 2011, in a French study of 30 dwellings, the medians of the individual congeners were mostly below the limit of quantification; only for some lower-chlorinated PCB were maximum values between 0.3 and 2.2 ng/m^3 observed (Blanchard et al. 2014). Mean concentrations of $\sum PCB_7$ ranging from 0.27 to 0.82 ng/m^3 could be found in 6 other French apartments (Moreau-Guigon et al. 2016). Comparable results were also determined for 8 Italian dwellings ($\sum PCB_6$: 0.2 4.0 ng/m^3) (Menichini et al. 2007), for 17–20 dwellings in the Czech Republic (median $\sum PCB_7$: 0.06–0.08 ng/m^3) (Melymuk et al. 2016a), for 5 Swedish and 17 English houses (median $\sum PCB_7$: 0.16 and 0.33 ng/m^3, respectively) (Bohlin et al. 2008). A similar exposure situation

Table 4.26 Median concentrations of polychlorinated biphenyls (PCB) in the air (without suspected contamination) in ng/m^3

References	*N*	PCB 28	PCB 52	PCB 101	PCB 118	PCB 138	PCB 153	PCB 180	Total PCB[a]	Country, time
Europe										
Blanchard et al. (2014)	30	<0.25	<0.25	<0.25	<0.25	<0.25	<0.25	<0.25		France, 2010–11
Frederiksen et al. (2012)	21	1.3	2.1	0.1	<LOQ	<LOQ	<LOQ	<LOQ	17.8	Denmark, 2011
Moreau-Guigon et al. (2016) [b,c]	6	0.12	0.35	0.10	0.02	0.05	0.07	0.10	4.1	France, 2011–12
Sakhi et al. (2019)[b]	44	0.11	0.14	0.06	0.006	0.009	0.02	0.001	1.7	Norway, 2012
Melymuk et al. (2016b)	1	0.06	0.07	0.03	0.005	0.01	0.02	0.07	1.3	Czech Republic, 2013
Audy et al. (2018)	28	0.11	0.12	0.08	0.02	0.04	0.07	0.01	2.3	Czech Republic, 2013
Andersen et al. (2020)	20	3	6	1	<0.4	<0.4	<0.4	<0.4		Denmark, 2017/18
De la Torre et al. (2020)	40	0.089	0.101	0.053	0.014	0.010	0.018	0.002	1.5	Belgium, Italy, Portugal, Spain, 2020
North America										
Fitzgerald et al. (2011)[b]	176	0.78	–	–	0.23	0.18	0.28	0.11		USA, 2000–02
Zhang et al. (2011)	20	0.005	–	–	–	<0.001	0,003	–		Canada, 2006
Audy et al. (2018)	34	0.17	0.17	0.05	0.013	0.008	0.012	0.002	2.1	Canada, 2013
Asia, Australia										
Wang et al. (2019)	28	0.031	0.011	0.011	0.004	0.0015	0.0018	<0.094	–	Austalia, 2015

[a]Total PCB content: sum of the six indicator congeners multiplied by 5
[b] mean values
[c] heating period: *LOQ* limit of quantification

was also found in 31 English residential interiors (median total PCB: 1.8 ng/m^3; range: 0.5–9.7 ng/m^3) (Harrad et al. 2006) and in 21 Danish apartments ($\sum PCB_6$: <LOQ-50.7 ng/m^3) (Frederiksen et al. 2012). In the study of 44 Norwegian apartments, a median of 0.99 ng/m^3 was found for $\sum PCB_{32}$ (Sakhi et al. 2019). Demirtepe et al. (2019) reported a median $\sum PCB_9$ of 1,09 ng/m^3 (range: 0.28–6.92 ng/m^3) in 60 houses in Slovakia in 2015. In PM_{10} in from 285 French dwellings, medians of 2.3 ng/m^3 (PCB 138), 2.4 ng/m^3 (PCB 153), and 2.1 ng/m^3 (PCB 180) were observed (Mandin et al. 2016).

Total PCB levels ranging from 0.10 to 233 ng/m^3 or means/medians of 0.05–14 ng/m^3 have been reported in the USA, Canada, and Mexico (Bohlin et al. 2008; Fitzgerald et al. 2011; Wilson et al. 2011; Zhang et al. 2011; Ampleman et al. 2015; Audy et al. 2018), while in Australia a median of 0.20 ng/m^3 ($\sum PCB_8$) was found (Wang et al. 2019). In China Wang et al. (2016b) investigated 12 houses near e-waste recycling factories and found concentrations of the $\sum PCB_{57}$ in the gas and particle phase of 34.5 and 8.6 ng/m^3.

Under experimental conditions in both test rooms and homes, there was an association between indoor air temperature or building temperature and PCB indoor air concentrations (Liu et al. 2015; Lyng et al. 2016).

So far, the mono- and dichlorinated PCB congeners have been little studied. Thomas et al. (2012) investigated six schools contained some interior or exterior caulk with PCB levels greater than 10,000 ppm and reported indoor averages of 3.7 ng/m^3 for PCB 31, 3.8 ng/m^3 for PCB 28, 1.5 ng/m^3 for PCB 8, and 0.34 ng/m^3 for PCB 4. Higher levels with median concentrations of 39 ng/m^3 (PCB 31), 26 ng/m^3 (PCB 28), 31 ng/m^3 (PCB 18), 19 ng/m^3 (PCB 20/33), 9.1 ng/m^3 (PCB 8), 1.9 ng/m^3 (PCB 6), and 1.6 ng/m^3 (PCB 4) were found in 35 contaminated office spaces in Germany (Kraft et al. 2018). Median (range) of total PCB level was 479 ng/m^3 (92–2797 ng/m^3) in this study. They described that PCB pattern was determined primarily by trichlorinated biphenyls, while di-and trichlorinated biphenyls represented approximately one-third of the total PCB concentration in the indoor air. Casey et al. (2022) measured 101 congeners in 10 homes contaminated via air from the local General Electric facility and reported a median as sum of all congeners of 20.3 ng/m^3 in the basement. Levels were approximately 0.52 ng/m^3 for PCB 31, 0.64 ng/m^3 for PCB 28, 0.87 ng/m^3 for PCB 18, 0.52 ng/m^3 PCB 8, and 0.98 ng/m^3 for PCB 4/2.

4.12.3 Occurrence in the Air of Public Facilities

Median concentrations of $\sum PCB_7$ of 0.28–3.2 ng/m^3 were found in 6 French daycare centers in 2011/2012 (Moreau-Guigon et al. 2016), whereas a median of 1.14 ng/m^3 for $\sum PCB_{32}$ was determined in 6 classrooms in Norway (Sakhi et al. 2019). In 602 classrooms from 308 randomly-selected French schools mean concentrations of 0.38 ng/m^3 for PCB 28, 0.35 ng/m^3 for PCB 52, 0.21 ng/m^3 for PCB 101, 0.07 ng/m^3

for PCB 118, 0.05 ng/m^3 for PCB 138, 0.05 ng/m^3 for PCB 153, and 0.02 ng/m^3 for PCB 180, respectively, were found in 2013–2017 (Sivanantham et al. 2021).

In the USA, an attempt was made to measure all 209 congeners in the air of 6 schools, if possible (Marek et al. 2017). In this case, the authors describe indoor air levels of ∑PCB of 0.5–194 ng/m^3 and outdoor air levels from 0.03 to 3.0 ng/m^3. In buildings before the PCB ban in 1977, the levels were significantly higher than in newer buildings. In this study, eleven OH-PCB were also analyzed and lelels of sum of OH-PCB ranging from 0.004 to 0.665 ng/m^3 were reported, with 2-OH-PCB 2 and 6-OH-PCB 2 being the most frequently detectable congeners.

4.12.4 Occurrence in the Air of Contaminated Buildings

The occurrence of PCB as contamination in building materials is of particular health significance (Bräuner et al. 2016). For example, especially in the period 1950–1975, buildings of skeletal concrete construction were equipped with polysulfide sealants (so-called Thiokol sealants), from which PCB can outgas. Often, these primary sources also lead to secondary contamination of room surfaces and masonry that require extensive remediation (Brown et al. 2016; Herrick et al. 2016).

In 1990–1996, median total PCB levels of 114 ng/m^3 and a maximum of 7360 ng/m^3 were measured in 308 Berlin schools and 102 day-care centers for children in buildings of skeletal concrete construction with polysulfide sealants suspected of contamination (Fromme et al. 1996). In this study, 3% of the measurements were above 3000 ng/m^3 (intervention value according to the German PCB guideline) and only 30% were below 300 ng/m^3 (precautionary value). Further comparable indoor air results are available from Germany, especially for school buildings with Thiokol sealants (Heudorf et al. 1996; Gabrio et al. 2000; Schwenk et al. 2002; Liebl et al. 2004; Heinzow et al. 2007). Also in Switzerland, 720–13,000 ng/m^3 (total PCBs) were measured in 5 buildings with contaminated sealants, or a median of 410 ng/m^3 (total PCBs) in 160 buildings (Kohler et al. 2002). In this study, 3% of the buildings were above 3000 ng/m^3. In 83 Danish apartments with contaminated sealants in 2011, concentrations as $\sum PCB_6 \times 5$ ranged from 305 to 2781 ng/m^3 (median: 859 ng/m^3) (Frederiksen et al. 2012; Kolarik et al. 2014). Additionally, 53 apartments located in high-rise buildings with indoor PCB containing sealants showed significant elevated total concentrations of PCB in air and dust compared to 20 reference homes in Denmark in 2017/18 (Andersen et al. 2020). The total average concentration in exposed homes was 2330 ng PCB/m^3 in air and 12,000 ng $\sum PCB_{15}$/g in dust, which was 40 and 25 times higher than the averages found in reference homes.

To investigate the human exposure pathways, Andersen and Frederiksen (2021) exposed experimentally three house dust samples (sieved <75 mm) to elevated air concentrations of PCB in a vacated apartment from seven to 40 days and estimated the absorption of PCB into the dust organic matter. Dust exposed directly on the floorboards showed lower concentrations than samples placed on foil, indicating an ongoing sorption to the varnish being a tertiary source contaminated by the air.

4.12.5 *Occurrence in Dust*

A direct comparison of results in dust is hampered by the fact that sampling techniques differed widely, different numbers of congeners were measured, and in some cases high limits of quantification were used.

In a comparative study, Harrad et al. (2009) found median levels in homes as $\sum PCB_6 \times 5$ of 200 ng/g in the USA, 260 ng/g in Canada, 48 ng/g in the United Kingdom, and 42 ng/g in New Zealand. Levels in North America were significantly higher than in Europe and New Zealand in this study. In 19 Belgian homes, a median of 73.5 ng/g (28–153 ng/g) was determined for the sum of congeners PCB 118, 138, 153, 180, and 170, whereas a median of 15 ng/g ($\sum PCB_6$) was observed in 36 elementary school and kindergartens in England (Harrad et al. 2010). Median of 75 ng/g $\sum PCB_7$ was reported in a study in 16 residences in Czech Republic in 2013 (Audy et al. 2018). Additionally, Andersen et al. (2020) found 273 ng/g for $\sum PCB_7$ in 20 Swedish residences in 2017.

Two studies from the USA in 1187 homes in four cities (DellaValle et al. 2013) and in 289 homes in California (Whitehead et al. 2014) report comparable results to those in UK homes. In a Californian study, the PCB concentration in house dust did not correlate with the PCB serum concentration in the blood of the residents of the apartments studied ($n = 48$) (Whitehead et al. 2015).

In contrast, Chinese studies report very different medians, ranging from 1.5 to 180 ng/g (total PCB) (Wang et al. 2013b, 2015b; Meng et al. 2016). Total PCB levels of 22 and 15 ng/g dust were reported in two Japanese houses (Takigami et al. 2009) and very low levels (median: about 7 ng/g for $\sum PCB_6 \times 5$) were reported from Singapore (Tan et al. 2007). The highest total PCB content published to date is from samples of 10 South African houses, at approximately 3700 ng/g (Abafe and Martincigh 2014). In Lahore, Pakistan, a median of 7.7 ng/g for $\sum PCB_6$ (range: 0.9–38 ng/g) was observed in 32 dust samples (Aslam et al. 2021).

The highest PCB amount can be found in the sieve fraction <63 µm, the lowest in the fraction 280–2000 µm (Wang et al. 2013b).

4.13 Polychlorinated Dibenzodioxins/Furans and Dioxin-like PCB

Polychlorinated dibenzo-p-dioxins (PCDD) and polychlorinated dibenzofurans (PCDF) are tricyclic aromatic hydrocarbons in which two benzene rings are linked by one or two oxygen bridges (see Fig. 4.23). Due to the binding possibilities of the chlorine atoms, there are theoretically 75 possible PCDD congeners and 135 PCDF congeners. PCDD/F are not of anthropogenic origin, but are found in trace amounts in chemical products and are formed as by-products in various industrial/chemical processes and in incomplete combustion processes. In particular, the 17 PCDD/F with chlorine substituents in the 2,3,7,8 positions are lipophilic substances and extremely persistent in the environment.

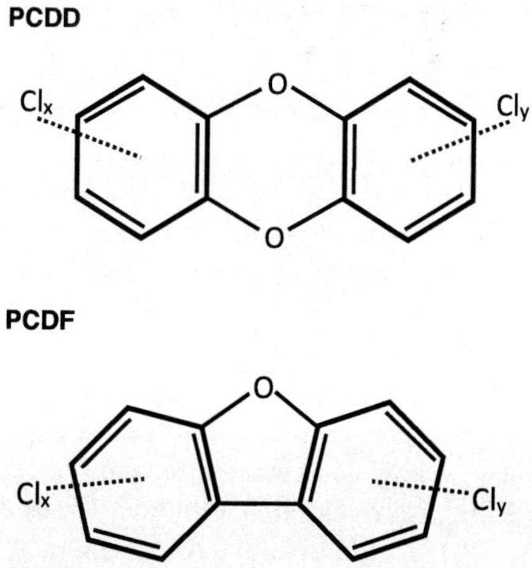

Fig. 4.23 Chemical structure of polychlorinated dibenzo-p-dioxines (PCDD) and dibenzofuranes (PCDF)

In the context of assessment and regulation, the concentration data of PCDD/PCDF and dioxin-like PCB are usually expressed in toxicity equivalents (TEQ). For this purpose, the content in the investigated medium is multiplied by a toxicity equivalency factor (TEF) defined for the individual congener. These TEF were last derived in 2005 by a WHO expert panel based on toxicological data (Van den Berg et al. 2006). A TEF value is used to account for the ratio of this substance to the most toxicologically significant substance, 2,3,7,8-tetrachlorodibenzo-p-dioxin (2,3,7,8-TCDD), for which a value of 1 is set. In the past, other TEF have also been derived, for example by NATO/CCMS (I-TEF, International Toxicity Equivalency Factor) (Kutz et al. 1990) or, in 1998, the older TEFs published by WHO (Van den Berg et al. 1998). For the following measured values, it is therefore important to note which TEF were used in each case. The different TEF are expressed in Table 4.27.

Dioxin-like PCB (dl-PCB) are the non-ortho and mono-ortho substituted PCBs that exhibit similar biological effects to PCDD/F due to their spatial orientation. In contrast to the PCDD/F, the PCB are a class of substances that were produced on a large scale and used in a variety of open and closed applications until they were banned in 1989.

4.13.1 Health Aspects

PCDD/F and dl-PCB are mainly reached humans via animal-derived, high-fat foods (more than 90%) and stored in body fat. The number and position of the chlorine atoms significantly influences the biological activity and toxic effects of the congeners. In recent decades, a large number of studies have been conducted that have characterized the toxic effects in laboratory animals, in workplace exposures, or in the context of accidents (US EPA 2017b). Depending on the dose and chlorination level, reproductive and developmental toxic effects and effects on the immune

Table 4.27 Names, abbreviations, and TEF (toxicity equivalency factors) values assessed from different institutions

Substances	Abbreviations	NATO/CCMS I-TEF	WHO 1998 TEF	WHO 2005 TEF
Chlorinated dibenzo-p-dioxins				
2,3,7,8-Tetrachlorodibenzo-p-dioxin	2,3,7,8-TCDD	1	1	1
1,2,3,7,8-Pentachlorodibenzo-p-dioxin	1,2,3,7,8-PeCDD	0.5	1	1
1,2,3,4,7,8-Hexachlorodibenzo-p-dioxin	1,2,3,4,7,8-HxCDD	0.1	0.1	0.1
1,2,3,6,7,8-Hexachlorodibenzo-p-dioxin	1,2,3,6,7,8-HxCDD	0.1	0.1	0.1
1,2,3,7,8,9-Hexachlorodibenzo-p-dioxin	1,2,3,7,8,9-HxCDD	9.1	0.1	0.1
1,2,3,4,6,7,8-Heptachlorodibenzo-p-dioxin	1,2,3,4,6,7,8-HpCDD	0.01	0.01	0.01
Octachlorodibenzo-p-dioxin	OCDD	0.001	0.0001	0.0003
Chlorinated dibenzofurans				
2,3,7,8-Tetrachlorodibenzofuran	2,3,7,8-TCDF	0.1	0.1	0.1
1,2,3,7,8-Pentachlorodibenzofuran	1,2,3,7,8-PeCDF	0.05	0.05	0.03
2,3,4,7,8-Pentachlorodibenzofuran	2,3,4,7,8-PeCDF	0.5	0.5	0.3
1,2,3,4,7,8-Hexachlorodibenzofuran	1,2,3,4,7,8-HxCDF	0.1	0.1	0.1
1,2,3,6,7,8-Hexachlorodibenzofuran	1,2,3,6,7,8-HxCDF	0.1	0.1	0.1
1,2,3,7,8,9-Hexachlorodibenzofuran	1,2,3,7,8,9-HxCDF	0.1	0.1	0.1
2,3,4,6,7,8-Hexachlorodibenzofuran	2,3,4,6,7,8-HxCDF	0.1	0.1	0.1
1,2,3,4,6,7,8-Heptachlorodibenzofuran	1,2,3,4,6,7,8-HpCDF	0.01	0.01	0.01
1,2,3,4,7,8,9-Heptachlorodibenzofuran	1,2,3,4,7,8,9-HpCDF	0.01	0.01	0.01
Octachlorodibenzofuran	OCDF	0.001	0.0001	0.0003
Non-ortho substituted PCB				
3,3′,4,4′- Tetrachlorobiphenyl	PCB 77		0.0001	0.0001
3,4,4′,5- Tetrachlorobiphenyl	PCB 81		0.0001	0.0003
3,3′,4,4′,5-Pentachlorobiphenyl	PCB 126		0.1	0.1
3,3′,4,4′,5,5′-Hexachlorobiphenyl	PCB 169		0.01	0.03

(continued)

Table 4.27 (continued)

Substances	Abbreviations	NATO/CCMS I-TEF	WHO 1998 TEF	WHO 2005 TEF
Mono-ortho substituted PCB				
2,3,3′,4,4′-Pentachlorobiphenyl	PCB 105		0.0001	0.00003
2,3,4,4′,5-Pentachlorobiphenyl	PCB 114		0.0005	0.00003
2,3′,4,4′,5-Pentachlorobiphenyl	PCB 118		0.0001	0.00003
2′,3,4,4′,5-Pentachlorobiphenyl	PCB 123		0.0001	0.00003
2,3,3′,4,4′,5-Hexachlorobiphenyl	PCB 156		0.0005	0.00003
2,3,3′,4,4′,5′-Hexachlorobiphenyl	PCB 157		0.0005	0.00003
2,3′,4,4′,5,5′-Hexachlorobiphenyl	PCB 167		0.00001	0.00003
2,3,3′,4,4′,5,5′-Heptachlorobiphenyl	PCB 189		0.0001	0.00003

system have been observed in animal experiments. In addition, the liver, lipid and carbohydrate metabolism, and the nervous system are known target systems. In dermal exposure, typical chloracne has been described at high concentrations. Epidemiological studies in workers or in connection with accidents (e.g., Seveso accident) indicate an increased incidence of tumors, but there was concurrent co-exposure to other chlorinated compounds. Several reviews have been published on the toxicological and epidemiological knowledge (IARC 1997; EFSA 2015a, 2018b).

The WHO's cancer agency IARC has classified 2,3,7,8-TCDD in Group 1 (carcinogenic to humans). Other polychlorinated dibenzo-p-dioxins and dibenzo-p-furans could not be classified as to their carcinogenicity to humans. DL-PCBs, with a toxicity equivalency factor (TEF) according to WHO, are carcinogenic to humans (Group 1) too (IARC 2015b).

4.13.2 *Occurrence in the Air*

Measurement results of indoor air investigations have been reported only to a limited extent. First in 1986/1987, a median PCDD/F concentration of 0.42 pg I-TEQ/m^3 (0.01–1.74 pg I-TEQ/m^3) was reported in the investigation of 10 Hamburg day-care centers (Von Manikowsky et al. 1998). In 1994/1995, during the investigation of 27 schools, gymnasiums, and daycare centers, where the source was large areas of wood paneling treated with wood preservative glazes, low PCDD/F levels up to a maximum of 0.18 pg I-TEQ/m^3 were found in 22 buildings (Blessing 1997). In the remaining 5 buildings, the concentrations ranged between 0.5 and 1.6 pg I-TEQ/m^3. The homolog distribution pattern showed an emphasis of hepta- and especially octa-PCDD/F, which is typical for wood preservative contamination. A study conducted in 4 buildings with contaminated sealants or ceiling tiles revealed concentrations of 0.041–0.972 pg WHO-TEQ_{1998}/m^3 for the sum of PCDD/F and 0.387–5.85 pg WHO-TEQ_{1998}/m^3 for the sum of dl-PCB and PCDD/F, with PCDD/F accounting for only 5–17% of the total TEQ (Heinzow et al. 2007). A much higher content was found in buildings with ceiling tiles compared to those with sealants. In contrast, there were no significant differences in indoor air levels for ndl-PCBs from the two sources. In both studies, collection was on PU foam and filters. Levels in indoor air from 5 buildings with PCB-treated sound-absorbing ceiling tiles can vary considerably, depending on the time of year, room temperature, and ventilation behavior of occupants (Volland et al. 2006). Depending on conditions at the time of sampling, levels of dl-PCB in these buildings ranged from 3.5 to 13.7 pg WHO-TEQ_{1998}/m^3 under standard conditions. In summer, indoor air concentrations were higher and PCB 118 accounted for the largest proportion of total levels, approximately 60–70%.

Again, biomass burning is a sufficient source of different combustion-related substances like PCDD/F. Samples in 25 kitchen, 26 living rooms, and 34 bedrooms were collected over seven consecutive days in China in 2019 (Wu et al. 2021). The average PCDD/F concentrations in kitchens of 0.545 pg I-TEQ/m^3 were 3.1- and 4.3-fold higher than those in living rooms with 0.122 pg I-TEQ/m^3 and bedrooms with 0.079 pg I-TEQ/m^3, respectively. In ambient air a mean level of 0.041 pg I-TEQ/m^3 was found in this study.

4.13.3 Occurrence in Dust

Measurements of concentrations in dust have often been made against a background of environmental contamination to estimate population exposure in these areas. In the USA, dust in 100 homes from 4 cities was studied in 1998–2000 and a relationship was found between exposure levels and the presence of an industrial source at a median for PCDD/F of 24 pg WHO-TEQ_{2005}/g (3.0–169 pg WHO-TEQ_{2005}/g) (Deziel et al. 2017). Slightly lower results were obtained by Franzblau et al. (2009), who studied dust samples from 764 Michigan homes in 2004–2005. Here, a median of 16.2 pg WHO-TEQ_{2005}/g (1.4–1750 pg WHO-TEQ_{2005}/g) was found for the sum of PCDD/F and dl-PCB. Only in a few outliers a specific external source for the exposure could be identified. Relatively low concentrations were also found in 21 homes in New York State in 2005/2006, with a median of 8.2 pg WHO-TEQ_{2005}/g (0.09–170 pg WHO-TEQ_{2005}/g) (∑PCDD/F and dl-PCB) (Tue et al. 2013).

Dioxin-like PCB were found in ten residential samples from two Japanese cities with mean concentrations of 1.6 and 2.9 pg WHO_{1998}-TEQ/g dust (Saito et al. 2003). In another study performed in 33 Japanese residences medians of 4.8 pg WHO-TEQ_{2005}/g (0.69–51 pg WHO-TEQ_{2005}/g) for PCDD and 4.6 pg

WHO-TEQ$_{2005}$/g (0.57–82 pg WHO-TEQ$_{2005}$/g) for PCDF were found (Suzuki et al. 2010).

4.14 Polybrominated Dibenzodioxins/Furans (PBDD/F)

Theoretically, depending on the number and position of bromine atoms in the molecule, 75 polybrominated dibenzodioxins (PBDD) and 135 dibenzofurans (PBDF) can occur. Compared to the chlorinated analogues, they have higher molecular weight, lower volatility and water solubility, and lower environmental persistence (Birnbaum et al. 2003; Piskorska-Pliszczyńska and Maszewski 2014; Yang et al. 2021).

PBDD/F occur, for example, as impurities in brominated flame retardants and in the course of their degradation and are formed by combustion processes, open fires and fires, (Altarawneh and Dlugogorski 2013; Yang et al. 2021). It must be assumed that PBDD/F are also formed from PBDE and other brominated substances during thermal processes (UNEP 2010). In terms of quantity, this appears to be possible in the lower percentage range. PBDD/F have been detected in a wide variety of environmental media and foodstuffs (Venkatesan and Halden 2014; Rose et al. 2015; Fernandes and Falandysz 2021).

4.14.1 Health Aspects

Overall, the toxicological data on this substance class is very limited and findings on effects in humans have hardly been published so far (Van den Berg et al. 2013; Piskorska-Pliszczyńska and Maszewski 2014). Only Zober et al. (1992) found no adverse effects on the immune system when they studied exposed workers in the plastics processing industry. Studies in rats and mice showed effects for 2,3,7,8-TBDD in the same dose range as 2,3,7,8-TCDD, suggesting comparable toxic potency (Van den Berg et al. 2013).

A joint expert group of the World Health Organization (WHO) and the United Nations Environment Programme (UNEP) concluded that, based on limited data, PBDD/Fs have effects mainly on the liver, thymus, and weight development that are at least qualitatively comparable to those of chlorinated dibenzodioxins and dibenzofurans (van den Berg et al. 2013). Therefore, the working group provisionally recommends using the TEF established for the chlorinated compounds in the context of an assessment.

4.14.2 *Occurrence in the Indoor Environment*

There is little information available about PBDD/F in indoor spaces. Schacht et al. (1995) analyzed different environmental media and reported 320 pg/g for sum of PBDD/F (without OBDD/F) in one vacuum cleaner bag from a residence and 3639 pg/g in dust obtained from a computer room. Suzuki et al. (2010) collected dust from 19 households and 14 offices In residences, the median concentrations of PBDD and PBDF were 36 pg/g (range: <2.5–140 pg/g) and 2900 pg/g (range: 1100–12,000 pg/g, respectively, while in office dust medians of 82 pg/g (23–2300 pg/g) and 5200 pg/g (1600–29,000 pg/g) were found. In a second study sulfuric-acid-treated extracts dust obtained from the vacuum cleaner bags of 21 houses from New York State in 2005/06 were analyzed with DR-CALUX assay (Tue et al. 2013). PBDD could be quantified only in some samples with a maximum of 230 pg/g, while PBDF were found between 330 and 150,000 pg/g. The authors stated that the most plausible sources of indoor PBDD/F are widely used PBDE. Compared to the median concentrations reported in the before mentioned Japanese study, the median in the US study was slightly lower for PBDD/F (2100 vs. 2900 pg/g).

Gou et al. (2016) reported results in six urban and four rural schools in Taiwan in 2014–2015. They found ΣPBDD/F medians of 4950 pg/g (339 pg WHO-TEQ_{2005}/g) and 6390 pg/g (57.7 pg WHO-TEQ_{2005}/g) in urban normal classrooms and computer classrooms, respectively, and 1460 pg/g (11.8 pg WHO-TEQ_{2005}/g) and 3670 pg/g (35.9 pg WHO-TEQ_{2005}/g) in rural normal classrooms and computer classrooms. PBDF accounted for at least 90% of the total PBDD/F in all classrooms. Indoor mean ΣPBDD/F levels in $PM_{2.5}$ and dust of in six gyms were investigated in Taiwan in 2018–2019 and concentrations of 0.202 pg/m^3 (0.0065 pg WHO-TEQ_{2005}/g) and 7090 pg/g (37.8 pg WHO-TEQ_{2005}/g), respectively, were reported (Su et al. 2022).

4.15 Polychlorinated Naphthalenes (PCN)

PCN are a class of substances comprising a total of 75 congeners of varying degrees of chlorination (Table 4.28). The basic chemical structure is shown in Fig. 4.24. The physicochemical characteristics can be very different. For example, the more highly chlorinated tetra- to octa-CN are lipophilic, while volatility decreases with increasing degree of chlorination (UNEP 2013). Congener mixtures are always present in commercial products, which may be liquids or waxes. PCN have been used as coolants in transformers, hydraulic oil, or as plasticizers. Different mixtures ranging from predominantly mono-CN (Halowax 1031) to octa-CN (Halowax 1051) were found under the trade name Halowax. Although no longer used for decades, exposure via the environment, e.g. via thermal processes in waste incineration and through old waste deposits, is still possible today (Domingo 2004; Liu et al. 2014). PCN are listed under Annex C of the Stockholm Convention. All parties

Table 4.28 List of polychlorinated naphthalene (PCN) homolog groups

Congener	Acronym	Formula	No. of single congeners
Monochloronaphthalenes	MonoCN	$C_{10}H_7Cl$	2
Dichloronaphthalenes	DiCN	$C_{10}H_6Cl_2$	10
Trichloronaphthalenes	TriCN	$C_{10}H_5Cl_3$	14
Tetrachloronaphthalenes	TetraCN	$C_{10}H_4Cl_4$	22
Pentachloronaphthalenes	PentaCN	$C_{10}H_3Cl_5$	14
Hexachloronaphthalenes	HexaCN	$C_{10}H_2Cl_6$	10
Heptachloronaphthalenes	HeptaCN	$C_{10}HCl_7$	2
Octachloronaphthalene	OctaCN	$C_{10}Cl_8$	1

Fig. 4.24 Chemical structure of polychlorinated naphthalenes (PCN) (n = 1–8)

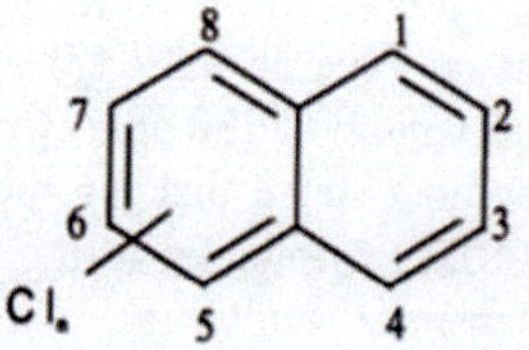

must take measures to reduce the unintentional releases of the chemicals with the goal of continuing minimization and, where feasible, ultimate elimination.

4.15.1 Health Aspects

Environmental fate and toxicity are comparable to polychlorinated biphenyls and are based on interaction with the cytosolic Ah receptor. At very high doses, hyperkeratosis of the skin was observed in experimental animals. Overall, toxicity data are limited. The focus is on liver toxicity, on the one hand, and a dioxin-like effect, on the other hand, due to the comparable mechanism of action of PCN (WHO 2001; Falandysz et al. 2014; Fernandes et al. 2022).

4.15.2 Occurrence in the Air

Within the scope of standard VOC analysis, 1-chloronaphthalene, 2-chloronaphthalene, and 1,4-dichloronaphthalene could be found in a few hundred German residences in only 0.3–1.3% above the limit of quantification of 1 µg/m^3, while 1,5-dichloro naphthalene could not be determined in any sample (Hofmann et al. 2014). In comparison, geometric mean values of 31 and 11 µg/m^3 (monochloronaphthalene), 13 and 8 µg/m^3 (dichloronaphthalene), and 2 and 2 µg/m^3 (trichloronaphthalene) were measured in 4 homes and 5 offices with tar-impregnated wood in 2014–2019 (Kozicki and Niesłochowski 2020).

High-resolution mass spectrometry was used to analyze a wider range of PCN in air in two studies. For example, in China in 2013, concentrations of $\sum PCN_{16}$ of 276 and 828 pg/m^3 were found in two indoor rooms, one workplace and one electronic equipment recycling room, while in outdoor air the levels were 43.3 and 118 pg/m^3 (Li et al. 2014). The total content is dominated by CN 66/67 and CN 1. Since the profile in the air differs significantly from that in the technical product, it can be assumed that secondary emissions now determine the concentration levels. In another study from Pakistan, 125 indoor air samples were analyzed from 5 regions where recycling of electrical appliances was carried out on a large scale (Waheed et al. 2020). In this case, mean concentrations of $\sum PCN_{39}$ of 51,843 pg/m^3 (8645–124,999 pg/m^3) could be found, with tetra-CN at 39%, tri-CN at 25%, and penta-CN at 22% determining the total content. CN 35, CN 51, CN 39, and CN 54 were the individual congeners with the highest mean content.

4.15.3 Occurrence in Dust

Examination of 19 dust samples (sieved to <1 mm) from Japan measured $\sum PCN_{39}$ of 4.8 ng/g (1.7–26.0 ng/g) (Suzuki et al. 2010). In a study from Pakistan, 250 dust samples (sieved to <0.5 mm) were analyzed from 5 regions where electrical appliances were recycled (Waheed et al. 2020). Mean concentrations of 2350 ng/g (319–7814 ng/g) were found for $\sum PCN_{39}$, with total concentration dominated substantially by CN 36/45, CN 62, CN 16, and CN 46.

4.16 Synthetic Musk Compounds

For a long time, natural musk was extracted from the exocrine glands of the male musk ox and the Siberian musk deer native to East Asia and used as fragrances in soaps and perfumes. Then, at the beginning of the twentieth century, di- and trinitro-compounds were synthesized, which were similar in olfactory impression to natural substances, but were less expensive and could be produced on a larger scale. In the 1950s, polycyclic compounds were additionally synthesized (David 2020). Overall, the synthetic musk compounds can be divided into the three groups of aromatic nitro compounds, polycyclic musk compounds, and macrocyclic musk compounds. The nitro compounds, such as musk xylene (MX) and musk ketone (MK), have been steadily replaced in recent decades by the polycyclic compounds, which appear to be more toxicologically favorable. Among the polycyclic musk compounds, HHCB (galaxolides; 1,3,4,6,7,8-hexahydro-4,6,6,7,8,8-hexamethyl- cyclopenta-[g]-2-benzopyran) and AHTN (tonalides; 7-acetyl-1,1,3,4,4,6-hexamethyl-1,2,3,4-tetrahydronaphthalene) are of particular technical importance. In addition, the following polycyclic compounds are often determined: ADBI (celestolide; 4-acetyl-1,1-dimethyl-6-tert-butylindane), AHMI (phantolides; 6-acetyl- 1,1,2,3,3,5-

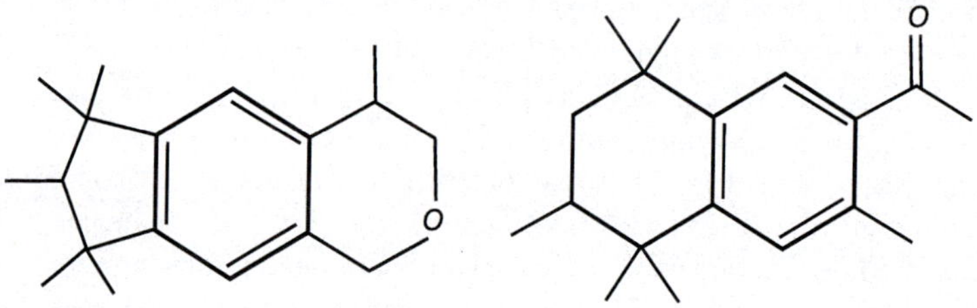

Fig. 4.25 Chemical structure of polycyclic musk compounds. Left HHCB (galaxolides) and right AHTN (tonalides)

hexamethylindane), ATII (traseolides; 5-acetyl-1,1,2,6-tetramethyl-3-isopropylindane), and DPMI (cashmerane; 1,2,3,5,6,7-hexahydro-1,1,2,3,3-pentamethyl-4H-inden-4-one).

Musk compounds are used in Europe in quantities of several thousand tons per year as fragrances in cosmetics and detergents. Worldwide, polycyclic musk compounds account for about 65% of production. The chemical structures are shown in Fig. 4.25.

4.16.1 *Health Aspects*

This group of substances has low acute toxicity and only moderate developmental toxic effects even with repeated administration, but only at doses at which toxic effects already occurred in the dams. They are not genotoxic or, in some cases, doubtfully carcinogenic and show low endocrine potency. An overview of toxicity can be found in Taylor et al. (2014). In addition, comprehensive risk reports from the European Chemicals Agency (ECHA) are available for the individual substances on the ECHA homepage.

4.16.2 *Occurrence in the Air*

In indoor air, only a limited number of studies are available so far. For example, in a French study of 29 indoor spaces in 2011/2011, a median level of 61.5 ng/m^3 (15.2–687 ng/m^3) was determined for HHCB and 14.8 ng/m^3 (4.2–180 ng/m^3) for AHTN in the gas phase (Blanchard et al. 2014), while the target parameters could not be determined in the particulate phase.

In a private house situated in Rome, Italy, in 2021 concentrations of 2.4 ng/m^3 and 1.3 ng/m^3 were reported for musk xylene and musk ketone, while HHCB were detected at concentrations of approximately 10 ng/m^3 (Balducci et al. 2022).

In 2006, a study in 20 Canadian homes reported mean concentrations (estimated from a figure) of approximately 4 ng/m^3 (HHCB), 2 ng/m^3 (AHTN), 0.1 ng/m^3

(ADMI), and 0.03 ng/m^3 (AHMI) (Wong et al. 2019). Dodson et al. (2019) analyzed samples from 37 homes in greater Bosten area, the USA in 2013–2015. HHCB and AHTN were found in all samples with medians (ranges) of 93 ng/m^3 (3.5–430 ng/m^3) and 17 ng/m^3 (1.3–87 ng/m^3), while MK and MX were observed in only 24% and 27% of the samples with ranges from <0.17 to 9.9 ng/m^3 and < 0.17 to 31 ng/m^3, respectively.

In 2000/2001, a total of 75 daycare centers in Berlin were investigated during opening hours and both the gas and particulate phases were collected (Fromme et al. 2004b). The medians were 101 ng/m^3 (maximum: 299 ng/m^3) for HHCB, 44 ng/m^3 (maximum: 107 ng/m^3) for AHTN, and 20 ng/m^3 (maximum: 77 ng/m^3) for AHMI. Musk ketone (MK) could only be found in one sample at 12 ng/m^3, and ATII in 4 samples up to a maximum of 17 ng/m^3. Sofuoglu et al. (2010) investigated a school classroom in Turkey for musk compounds in 2008. Sampling was conducted over 8 hours on each of 10 different days in the gas and particulate ($PM_{2.5}$) phases, and the compounds were found predominantly in the gas phase. The mean concentrations in the gas phase were 9.9 ng/m^3 for MX, 0.12 ng/m^3 for MK, 267 ng/m^3 for HHCB, 59.6 ng/m^3 for AHTN, 1.5 ng/m^3 for ADB), 58.7 ng/m^3 for ATII, and 84.5 ng/m^3 for DPMI. In another study, sampling 62 classrooms from 30 schools in 2009/2010 in France resulted in medians of <50 ng/m^3 for HHCB and AHTN (Raffy et al. 2017).

In all studies, HHCB and AHTN are the substances with the highest concentrations of polycyclic compounds in indoor air. In the Turkish study, a higher level was observed than in the German and French studies. In the Canadian indoor air, a significantly lower level of contamination was determined. It should be noted here that passive sampling with polyurethane foams was carried out for 30 days in Canada, while active sampling was used in the other studies.

4.16.3 Occurrence in Dust

Table 4.29 shows the results of measurements in dust. The wide range of individual results in dust is striking. In the case of nitromusks, musk xylene (MX) in particular dominates with medians between 45 and 300 ng/g, while musk moskene was found only in some samples up to 15 ng/g. Among the polycyclic compounds, HHCB and AHTN are mainly found with medians of 340–992 ng/g and 79–900 ng/g, respectively.

So far, results from schools have only been described from two studies. In the first musk compounds were investigated in a total of 9 Austrian schools between 2006 and 2007 (Hohenblum et al. 2008). While the nitromusk compounds could only be quantified above the limit of quantification in one sample, the other musk compounds showed overall low levels. The second study performed 2009/10 in French schools reported concentrations in the same order of magnitude (Raffy et al. 2017). In comparison with data from apartments, which are also shown in Table 4.29, a similar exposure level is obtained.

Table 4.29 Medians (ranges) of musk compounds in dust in ng/g

References	*N*	MX	MK	HHCB	AHTN	ADBI	ATII	AHMI	DPMI	Country, time
Residences										
Fromme et al. (2004b)	30	–	300 (<500–47,000)	700 (<500–11,400)	900 (<500–3100)	<500	<500	<500	–	Germany, 2000/01
Regueiro et al. (2007)	8	69 (<0,1–692)	65 (<3,3–2303)	–	–	–	–	–	–	Spain
Kubwabo et al. (2012)	5	41 (<39–2130)	45 (<42–559)	992 (36–31,100)	405 (91–2360)	<63 (<63–1800)	<42 (<42–127)	<71 (<71–3930)	all <124	Canada
Rasmussen et al. (2022)	1025	46 (<12–6360)	49 (<13–2200)	949 (n.g.-35,000)	511 (n.g.-24,000)	<20 (<20–2550)	<13 (<13–411)	<22 (<22–10,000)	all <39	Canada, 2007–10
Lu et al. (2011)	56	12 (<LOQ-56)	14 (<LOQ-203)	38 (1.9–577)	17 (0.8–114)	–	–	–	–	China, 2009
Mercier et al. (2014)	7	–	–	- (296–4180)	- (127–1170)	–	–	–	–	France, 2009/12
Blanchard et al. (2014)	29	–	–	980	410	–	–	–	–	France, 2011
Nakata et al. (2015)	10	<LOQ	93 (<LOQ-390)	340 (84–1600)	79 (<LOQ-220)	–	–	–	–	Japan, 2012/14
Castro et al. (20190	27	–	–	750	520	–	–	–	–	Spain
Kim et al. (2021)	29	–	–	308 (<50-5669[b])	76 (<100-640[b])	–	–	–	–	USA, 2015–18
Schools										
Hohenblum et al. (2008)	9	–	–	740	640	–	–	–	–	Austria

Raffy et al. (2017)	30	–	–	965 (304–2191[a])	337 (<66–950[a])	–	–	–	–	France, 2009/10

LOQ limit of quantitation, *n.g.* not given, *MX* musk xylene, *MK* musk ketone, *HHCB* galaxolides, *AHTN* tonalides, *ADBI* celestolides, *ATII* traesolides, *AHMI* phantolides, *DPMI* cashmeran

[a] fifth and 95th percentile

[b] 95th percentile

4.17 Pyrethroids/Pyrethrum

Today, indoor pesticide applications are dominated by pyrethroids or pyrethrum, which, in addition to their use in plant and stock protection, are applied as pharmaceuticals and veterinary drugs, for wood protection, as hygiene and disinfestation agents, and as material and textile protection agents.

Pyrethrum is one of the oldest known natural insecticides and has been used indoors in Central Europe since the beginning of the nineteenth century. It is extracted and purified from the dried flower heads of various agriculturally grown chrysanthemum species (e.g., Chrysanthemum cinerariaefolium (Tanacetum cinerariifolium)). Pyrethrum is an extract of 6 insecticidally active esters of Chrysanthemum or pyrethric acid with the cyclic ketoalcohols pyrethrolone, cinerolone, and jasmolone (ATSDR 2003). The main constituents of the pyrethrum extract are the pyrethrins with approx. 40% followed by the cinerins with 10% and the jasmolins with 5%.

Pyrethroids (see Fig. 4.26) are organic compounds which, starting from the structure of natural pyrethrum, have been produced on a large scale by modification and introduction of various functional groups, especially since the 1980s. Depending on their properties (duration of action, stability, volatility), a distinction is made between short-term and long-term pyrethroids. The substances are present as mixtures of enantiomers. In general, a distinction can be made between type I pyrethroids (without α-cyano substitution (e.g., allethrin, permethrin, tetramethrin) and type II pyrethroids with α-cyano substitution (e.g., cyfluthrin, deltamethrin, fenvalerate).

Pyrethroids are commonly used indoors to control pests and nuisances. They are applied both by professional pest controllers and by the occupants themselves in the form of household products containing insecticides (insect sprays, gels, strips, dusters, electric vaporizers). They are also used as moth and beetle repellents for the treatment of home textiles. In order to avoid misapplication and any associated

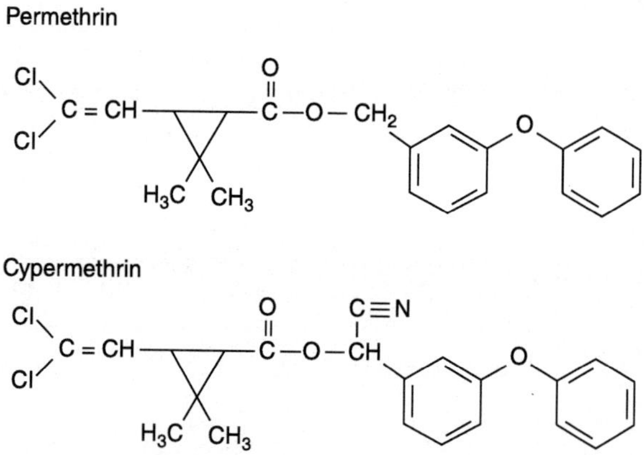

Fig. 4.26 Chemical structure of permethrin (type I pyrethroid) and cypermethrin (type II pyrethroid)

health effects it is essential to use only registered products and use it in accordance with the manufacturer's instructions for use.

4.17.1 Health Aspects

Despite their strong insecticidal activity and pronounced fish toxicity, they are only moderately toxic to mammals after oral ingestion (Chrustek et al. 2018). However, there are considerable differences in the potency of the individual pyrethroids. The mechanism of action is assumed to be an influence on the sodium permeability of the nerve membrane, leading to permanent opening of the sodium channels and subsequently longer depolarization (Soderlund 2012). The effects of type I pyrethroids are typically characterized by the so-called T-syndrome, which is usually characterized by tremor, hyperexcitability, ataxia, and convulsions. In contrast, type II pyrethroids lead to long-lasting nerve impulses and in some cases reduce the resting membrane potential to such an extent that no action potential is formed. In this case, the typical poisoning pattern is the so-called CS syndrome, in which choreatic-athetotic mixed hyperkinesis as well as salivation and tonic-clonic seizures are in the foreground (Zhu et al. 2020a).

4.17.2 Occurrence in the Air

Leng et al. (2005) determined background concentrations in 18 German indoor environments with medians below the limit of quantification of 1–3 ng/m^3 and 95th percentiles of 23.0 ng/m^3 (permethrin), 7.3 ng/m^3 (cyfluthrin), and 5.5 ng/m^3 (cypermethrin). Mandin et al. (2016) studied a total of 285 French homes in 2003–2005 and determined a median level of permethrin in PM_{10} of 0.06 ng/m^3 (95th percentile: 1.8 ng/m^3). In Spain, López et al. (2021) found geometric means of 10.4 ng/m^3 for cypermethrin, 9.7 ng/m^3 for lambda-cyhalothrin, 7.5 for permethrin, and 3.8 ng/m^3 for pyrimethanil in 41 residences in 2015.

Levels of <0.09–40.9 ng/m^3 and < 0.09–34.4 ng/m^3 for trans- and cis-permethrin, respectively, and < 0.97–183 ng/m^3 for cyfluthrin were observed in 129 homes in North Carolina in 2001 (Morgan et al. 2014). The 95th percentiles in this study were 7.62 ng/m^3 (trans-permethrin), 7.9 ng/m^3 (cis-permethrin), and < 0.97 ng/m^3 (cyfluthrin). A total of 102 homes were surveyed in New York between 2001 and 2004, with 61% of residents reporting pesticide application during the collection period (Whyatt et al. 2007). Concentrations of trans- and cis-permethrin ranged from <0.1 to 164 ng/m^3 and < 0.4 to 125 ng/m^3, respectively, while levels of piperonyl butoxide were < 0.2–608 ng/m^3. Medians were always below the limit of quantitation of 0.1–0.4 ng/m^3. Even in 20 Boston homes where pest control operations were more frequent, only lower concentrations of <LOQ-3.8 ng/m^3 for allethrin, <LOQ-9.2 ng/m^3 for cyhalothrin, <LOQ-5.5 ng/

m^3 for cypermethrin, <LOQ-3.0 ng/m^3 for permethrin, and < LOQ-0.4 ng/m^3 for tefluthrin could be found in 2011 (Lu et al. 2013).

In Japan, 85 bedrooms of children were investigated in 2017, and only transfluthrin were detected in 92% of the samples (median: 12 ng/m^3, while the other 19 pyrethroids were detected only in 0–21% (Yoshida et al. 2021). Resmethrin, Tetramethrin, Cyfluthrin, and Deltamethrin were observed with maximum levels of 0.82 ng/m^3, 2.5 ng/m^3, 1.6 ng/m^3, and 1.2 ng/m^3.

Permethrin and deltamethrin were often detected in low concentrations in 28 residences and offices in two Australian cities in 2015 with medians of 0.084 ng/m^3 and 0.001 ng/m^3, respectively (Wang et al. 2019).

Levels of <0.09–2.76 ng/m^3 and < 0.09–3.05 ng/m^3 for trans- and cis-permethrin, respectively, and < 0.97–1.74 ng/m^3 for cyfluthrin were found in 13 daycare centers in North Carolina in 2001 (Morgan et al. 2014). The 95th percentiles in this study were 2.14 ng/m^3 (trans-permethrin), 2.45 ng/m^3 (cis-permethrin), and < 0.97 ng/m^3 (cyfluthrin). Comparable results were also described by the same research group in 16 daycare centers in Ohio (Morgan et al. 2007).

4.17.3 Occurrence in Dust

Table 4.30 summarizes the results of dust measurements. The content of pyrethroids in vacuum cleaner bag dust was measured in 1998 as part of the representative environmental survey in Germany (Becker et al. 2004). Only permethrin and piperonyl butoxide could be quantitatively determined in a larger number of samples, whereas this was not possible for cyfluthrin, cypermethrin, deltamethrin, d-phenothrin, and λ-cyhalothrin. Permethrin levels here, with a median of 0.17 μg/g, were lower than in a study conducted in Hamburg, Germany, at about the same time (median 5.7 μg/g) (Kersten and Reich 2003). In this study, the medians of the other pyrethroids were below 0.1 μg/g (limit of quantification) and the 95th percentiles were 0.3 μg/g for cyfluthrin, 0.2 μg/g for cypermethrin, and 0.4 μg/g for deltamethrin. In another German study examining 18 homes, medians (95th percentiles) were found to be 1.2 μg/g (5.9 μg/g) for permethrin, <3 μg/g (35 μg/g) for cyfluthrin, <3 μg/g (3 μg/g) for cypermethrin, and < 0.5 μg/g (2.3 μg/g) for deltamethrin (Leng et al. 2005).

In a pilot study dust from 52 rural homes in Taiwan was collected and cypermethrin and tetramethrin were the most abundant pyrethroids with medians of 0.37 μg/g and 0.38 μg/g, suggesting that indoor use of such pesticide products was the major source (Hung et al. 2018). On the contrary, permethrin and deltamethrin were often detected in 28 residences and offices in two Australian cities in 2015 (Wang et al. 2019).

Results from two studies in homes and daycare centers in North Carolina and Ohio have previously been published from the USA, which found no significant

Table 4.30 Median (95th percentile) concentrations of pyrethroids in dust in µg/g

References	*N*	trans-permethrin	cis-permethrin	Cyfluthrin	Allethrin	Tetramethrin	Cypermethrin	Deltamethrin	Cyhalothrin	Country, Time
Residences										
Becker et al. (2002)	738	0.17 (14.5)		–	–	–	–	–	–	Germany, 1998
Kersten and Reich (2003)	65	5.7 (110)		<0.1 (0.3)	–	<0.1 (0.2)	<0.1 (0.2)	<0.1 (0.4)	<0.1 (–)	Germany, 1998–2000
Uhl et al. (2004)	22	0.86 (35)		<0.76 (<0.76)	–	<0.28 (<0.28)	<0.51 (<0.51)	<0.45 (<0.45)	<0.34 (<0.34)	Austria, 2003
Deziel et al. (2013)	21	1.00		<0.02	–	–	0.34	–	–	USA, 2003–05
Leng et al. (2005)	18	1.2 (5.9)		0.8 (34.9)	–	–	<0.5 (3.0) -	<0.5 (2.3) -	–	Germany
Morgan et al. (2014)	129	0.63 (19.4)	0.84 (21.1)	<0.02 (1.7)	–	–	–	–	–	USA, North Carolina, 2000/01
Morgan et al. (20070	120	0.34 (9.2)	0.47 (7.6)	–	–	–	–	–	–	USA, Ohio, 2001
Regueiro et al. (2007)	8	2.9 (–)		<0.003	<0.004	<0.007 (–)	<0.003 (–)	<0.003 (–)	<0.001 (–)	Spain
Glorennec et al. (2017)	245	0.77 (26.3)		<0.003 (0.07)	–	<0.003 (1.4)	0.14 (3.1)	<0.003 (0.07)	–	France, 2009–12
Wang et al. (2019)	28	5.1		0.069	–	–	–	0.082	–	Australia, 2015
Hung et al. (2018)	52	<LOQ (0.35)		–	–	0.38 (280)	0.37 (106)	<LOQ	–	Taiwan, 2016
Kim et al. (2021)	29	1.37 (2.7)		–	–	–	<1.0 (15.0)	<0.5 (1.8)	<0.25 (–)	USA, 2015–18

(continued)

Table 4.30 (continued)

References	*N*	trans-permethrin	cis-permethrin	Cyfluthrin	Allethrin	Tetramethrin	Cypermethrin	Deltamethrin	Cyhalothrin	Country, Time
Daycare centers										
Morgan et al. (2014)	13	0.86 (20.9)	0.81 (19.7)	<0.02/ 1.8/1.8	–	–	–	–	–	USA, North Carolina, 2000/01
Morgan et al. (2007)	23	0.54 (3.4)	1.9 (3.8)	–	–	–	–	–	–	USA, 2001

LOQ limit of quantitation

differences for permethrin between indoor environments, but did find differences between study areas (Morgan et al. 2007, 2014).

Piperonyl butoxide (PBO) has been analyzed in dust at levels ranging from <0.1 to 5.1 μg/g (95th percentile: 3.1 μg/g) (Kersten and Reich 2003) and in the 1998 German Environmental Survey at concentrations ranging from <0.1–200 μg/g (95th percentile: 3.7 μg/g). Walker et al. (1999) also describe PBO concentrations between <0.1 and 270 μg/g in Lower Saxony, Germany. In the USA, Deziel et al. (2013) reported a median of 0.28 μg/g in 21 residences in California in 2003–2005.

4.17.4 Occurrence Indoors Following Applications

Significantly higher concentrations are found in indoor air during and following pyrethroid applications. For example, one day after a cyfluthrin application in 11 homes (spot treatment), median dust levels increased from 0.25 μg/g to 33.8 μg/g and air concentrations increased from <3 ng/m^3 to 4.9 ng/m^3 (Leng et al. 2005). Median cyfluthrin levels were also 4.9 μg/g after 4–6 months and only 1.5 μg/g after 10–12 months. After permethrin application in 4 homes, a median increase from 4.3 μg/g to 70 μg/g in house dust and from <1 ng/m^3 to 18.1 ng/m^3 in air was observed on the first day. As time progressed, dust levels dropped above 28.5 μg/g after 4–6 months and to 14.2 μg/g after 10–12 months. The level in the air was 8.9 ng/m^3 after 4–6 months and dropped to 4.9 ng/m^3 after 10–12 months. It was also observed in another experimental study that the concentration of cypermethrin and cyfluthrin in dust had decreased to 13% and 16%, respectively, after 112 days (Nakagawa et al. 2017). On the contrary, 89% of the initial permethrin concentration persisted on aluminum foil in a test house for 112 days (Nakagawa et al. 2020). After a spray application of permethrin, 9 trials with smooth flooring (low dust loading) and 15 trials with textile flooring (high dust loading) were determined in indoor air in a test room (Zhou et al. 2019). Simulating movement of a young child using a mobile robot resulted in mean personal permethrin concentrations of 143 ng/m^3 (smooth flooring) and 61 ng/m^3 (textile flooring), respectively, which were approximately a factor of 2 higher than in the stationary measurements conducted at the same time.

In a model room (40 m^3), 5 insecticidal aerosol sprays were applied once for 10–20 seconds (Berger-Preiss et al. 2009). During application, mean concentrations in the alveolar dust fraction of 68–587 μg/m^3 (tetramethrin), 121 μg/m^3 (cyfluthrin), 855 μg/m^3 (chlorpyrifos), and 345 μg/m^3 (pyrethrins) were measured. When an electric vaporizer was used against mosquitoes, indoor air levels were from 5 to 12 μg/m^3 for allethrin and 0.5 to 2.0 μg/m^3 for PBO. Vesin et al. (2013) confirmed these results when testing two electric vaporizers. Here, a maximum concentration of 4.9–8.5 μg/m^3 was measured. In another study, 0.19–0.34 μg/m^3 of dimefluthrin was measured in air during the application of a liquid vaporizer, 15.1–24.3 μg/m^3 of allethrin was measured with a vaporizer platelet, and 0.5–0.8 μg/m^3 of dimefluthrin was measured with a so-called mosquito coil (Li et al. 2016).

Also, during model execution of pest control actions in aircraft, levels between 11 and 65 μg/m^3 (pyrethrins) and 200–485 μg/m^3 (PBO) were found during application and 40 minutes later (Berger-Preiss et al. 2004). In other indoor biocide applications, levels of 36–150 μg/m^3 (permethrin) (Berger-Preiss et al. 1997) or 55–300 μg/m^3 (pyrethrum, tetramethrin, cyfluthrin) (Class 1991) were found during the control action.

Oudejans et al. (2020) provided information about decontamination approaches for residential or commercial buildings to reduce occupant exposure following pesticide misuse. They concluded that decontamination practices should be tailored to the specific pesticide involved to maximize its efficacy and minimize the risk of transferring contamination within and from the site. In a first approximation, application of household bleach or a hydrogen peroxide-based decontaminant offered the highest efficacy reducing pyrethroids.

4.18 Bisphenols

Bisphenol A (BPA) (CAS-No.: 80–05-7) has been manufactured on a commercial scale for decades and is used extensively in polycarbonates and epoxy resins. These are used in plastic products such as CDs, faucets and plastic parts in cars, household items, plastic baby bottles, food, and beverage packaging. It is also used as an additive in thermal paper and PVC. Adhesives and epoxy-coated inner surfaces of food and beverage cans may also contain BPA. The global production of BPA is around eight million tons per year, with China being one of the main producers (Vasiljevic and Harner 2021). Bisphenol S (BPS, CAS-No.: 80–09-1), bisphenol F (BPF, CAS-No.: 620–92-8), and bisphenol AF (BPAF, CAS-No.: 1487-61-1) are sometimes used as substitutes for BPA or in special applications. The chemical structures of the most important bisphenols are summarized in Fig. 4.27.

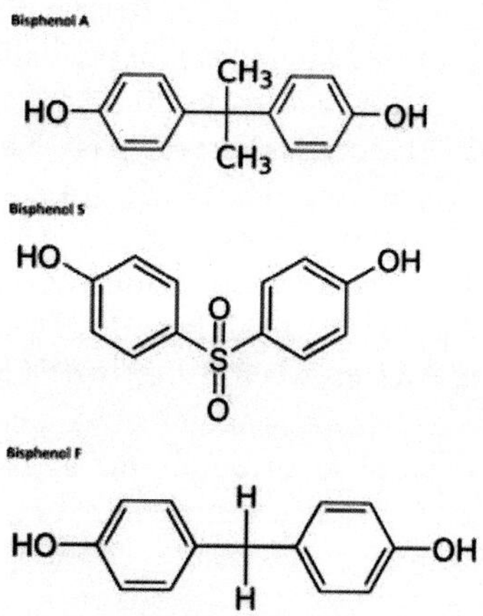

Fig. 4.27 Chemical structure of bisphenols

4.18.1 Health Aspects

BPA has a low acute toxicity and in long-term animal studies mainly affects the liver and kidney, while the development of cancer is not considered likely. In vitro and in vivo test systems showed hormone-like effects, including estrogenic, anti-androgenic, and anti-thyroidal effects. Effects on the mammary gland and reproductive system of test organisms are considered likely (EFSA 2017). The significance of the aforementioned effects for humans has not yet been fully clarified and is the subject of controversial debate.

The European Chemicals Agency added BPA to the Annex XIV candidate list in 2018 due to its endocrine disrupting effects.

In the scientific literature, there are several comprehensive reviews on the occurrence in environmental media and health effects of bisphenol A (Chen et al. 2016b; EFSA 2015b; Giulivo et al. 2016; Qiu et al. 2019; Catenza et al. 2021) or its analogs such as BPS (Chen et al. 2016b; Wu et al. 2018; Qiu et al. 2019; Pelch et al. 2019; EFSA 2020b; Catenza et al. 2021), BPF (Usman et al. 2019; Pelch et al. 2019), and BPAF (Escrivá et al. 2019; Pelch et al. 2019).

4.18.2 Occurrence in the Air

Results for Europe are available from a French study in which a total of 6 apartments and 6 daycare centers were investigated in 2011/2012 (Moreau-Guigon et al. 2016). The mean BPA content was 0.39 ng/m^3 in the apartments during the heating period and 0.68 ng/m^3 during periods without heating activities, and 0.34 ng/m^3 and 1.1 ng/m^3 in the daycare centers. The indoor air content was significantly higher than the values determined in parallel in the outdoor air. In another French study a median of 0.64 ng/m^3 was found in airborne particles in 30 residences in 2011 (Blanchard et al. 2014). Sangiorgi et al. (2013) reported average levels of BPA in $PM_{2.5}$ of 0.23 ng/m^3 in warm season and 0.38 in cold season in four offices in Milan, Italy.

In addition, median concentrations for BPA of 0.59 ng/m^3 (0.15–4.65 ng/m^3), for BPS of 0.07 ng/m^3 (<LOQ-0.29 ng/m^3), and for BPF 0.78 ng/m^3 (<LOQ-84.1 ng/m^3) were found in 26 homes in the USA in 2014 (Xue et al. 2016), while in Hong Kong a median of 0.64 ng/m^3 (maximum: 1.0 ng/m^3) was observed in $PM_{2.5}$ samples from nine kindergartens and two primary schools in 2015–2016 (Deng et al. 2018).

4.18.3 Occurrence in Dust

Overall, relatively low concentrations in dust has been found in European indoor spaces so far. In principle, it should be taken in mind that sample pretreatment varies;

in some cases, the samples were sieved, in others only dust from specific room surfaces was taken and analyzed.

For example, in 2003, in an Austrian study, BPA concentrations of 120–8800 ng/g (median: 2500 ng/g) could be found in 21 dust samples from apartments (Uhl et al. 2004). Comparable medians were measured in 2008 at 1461 ng/g (535–9729 ng/g) in 18 Belgian apartments and at 1300 ng/g (<90–15,000 ng/g) in 100 preschools in Stockholm (Geens et al. 2009; Larsson et al. 2017). Similar results were reported in 16 bedrooms and 16 living rooms in Spain in 2017/18, with medians of 1817 ng/g (140–10,848 ng/g) and 2533 ng/g (<LOQ-13846 ng/g), respectively (Duenas-Mas et al. 2019). A somewhat lower level of 27–4440 ng/g (median: 150 ng/g) was observed in 28 indoor environments in Greece in 2014 (Wang et al. 2015c). In Germany Völkel et al. (2008) reported a median of 553 ng/g (range: 117–1486 ng/g) in 12 dust samples from vacuum cleaner bags (sieved to 63 μm).

Comparable levels to the European studies were observed in 38 and 22 indoor spaces in the USA, with BPA medians of 1600 and 1500 ng/g, respectively (Liao et al. 2012; Wang et al. 2015c). A median of 3816 ng/g was also found in 2014–2018 when 186 homes were studied in North Carolina, USA (Levasseur et al. 2021). In 44 households vacuum dust collected 2007–2010 in Canada median value of 3100 ng/g was observed, while in fresh dust samples median was 2700 ng/g (Kubwabo et al. 2016).

In a nationwide survey in China in 2017–2019 a BPA median of 214 ng/g (range: <LOQ - 37,000 ng/g) was found in dust of 289 houses (Zhu et al. 2020b), while in Hong Kong a median of 74 ng/g (range: 10–200 ng/g) was observed in samples from nine kindergartens and two primary schools in 2015–2016 (Deng et al. 2018). In 47 houses in southern China in 2016-2019 a median of 1780 ng/g (260 - 37,000 ng/g) was reported (Shi et al. 2021). Moreover, 33 and 27 indoor dust samples from homes in rural and urban areas in China in 2018 contained 1100 and 1700 ng/g (median) of BPS, respectively, with significant higher amounts in urban dust (Liu et al. 2021b).

After BPA, BPS can be detected, in lower concentrations in some studies, with medians of 860 ng/g in Greece (Wang et al. 2015c), 240 ng/g in Sweden (Larsson et al. 2017), <2.0 as well as 630 ng/g in the USA (Wang et al. 2015c; Liao et al. 2012), 28 ng/g in Canada (Rasmussen et al. 2022), and 5.64 ng/g in China (Zhu et al. 2020b). In this Chinese study also other bisphenol analogs were quantified with median of 32.9 ng/g for BPF (all other with medians <LOQ), while Wang et al. (2015c) reported medians for BPF in 12 countries <1.0 ng/g. Similar results were reported in 16 bedrooms and 16 living rooms in Spain in 2017/18, with BPS medians of 135 ng/g (<LOQ-711 ng/g) and 192 ng/g (<LOQ-483 ng/g), respectively (Duenas-Mas et al. 2019). In this study the medians for BPF were 48 ng/g and 68 ng/g.

In a Chinese study of 40 students and their homes, a statistically significant correlation was found between BPA house dust levels and urinary concentrations, with BPA accounting for 79% of the bisphenol content in house dust (Zhang et al. 2020a). Concentrations (medians) in this case were 2980 ng/g for BPA, 220 ng/g for BPS, and 150 ng/g for BPF. Moreover, 33 and 27 indoor dust samples from homes in rural and urban areas in China in 2018 contained medians of 460 ng/g BPS and < 100 ng/g BPF as well as 560 ng/g BPS and 220 ng/g BPF, respectively, with significant higher amounts in urban dust (Liu et al. 2021b).

4.19 Parabens

The parabens are alkyl and iso-alkyl esters as well as benzyl esters of p-hydroxybenzoic acid and their sodium salts. The chemical structures of the frequently used parabens are summarized in Fig. 4.28. Parabens are antibacterial preservatives. The methyl and ethyl esters have been used since about the 1920s, especially as preservatives in food, cosmetic products, detergents, and pharmaceuticals to prevent microbial contamination, while the propyl and butyl esters are used only in cosmetics. Currently, methyl and propyl parabens are among the most commonly used substances from this class.

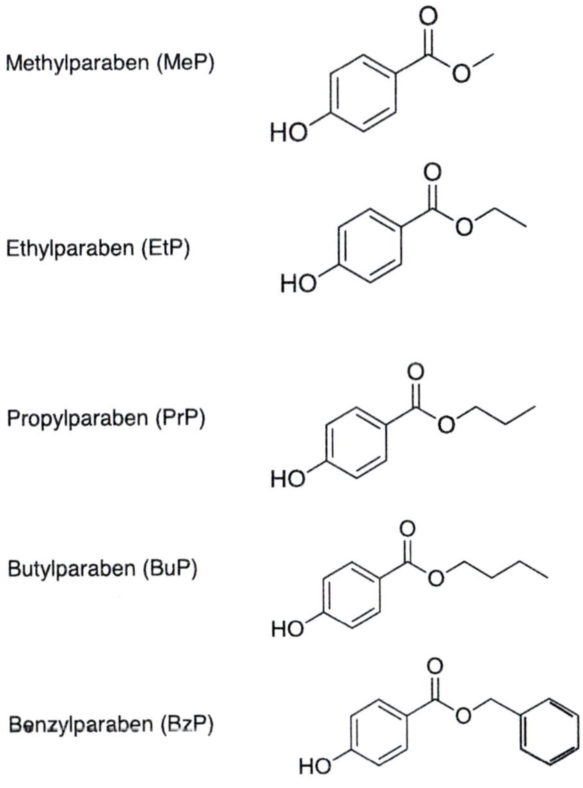

Fig. 4.28 Chemical structure of parabenes

4.19.1 Health Aspects

Parabens have a low acute toxicity, are not genotoxic, and show no carcinogenic effects in animal experiments. Both in vitro and in vivo studies show endocrine activity, which can be explained mainly due to the binding capacity to the estrogen receptor (SCCS 2013). However, it is orders of magnitude lower than those of natural 17β-estradiol. In animal experiments, parabens showed effects on the male reproductive system in particular (alteration of sperm count and testosterone levels), with effects increasing with increasing chain length and branching. Several reviews on the occurrence in environmental media and health effects are available (Danish-EPA 2013b; Błędzka et al. 2014; Giulivo et al. 2016; Matwiejczuk et al. 2020; Petric et al. 2021).

Methylparaben (MeP) and ethylparaben (EtP) have been assessed as safe for humans in a concentration of up to 0.4% in cosmetic products (BfR 2011; SCCS 2013). The use of butylparaben (BuP) and propylparaben (PrP) should be limited to 0.19% overall due to insufficient dermal uptake and exposure data (BfR 2011; SCCS 2013). In 2020, the European Chemicals Agency classified butyl paraben as a substance of concern due to its endocrine activity and included it in the Annex XIV candidate list (ECHA 2020). The aim is to abandon its use. Isopropyl, isobutyl, pentyl, benzyl, and phenyl paraben are no longer permitted in cosmetics in the EU.

4.19.2 Occurrence in the Air

In 2011/2012, when 6 French apartments were studied, mean concentrations of 0.22 and 0.81 ng/m^3 for MeP, 0.01 and 0.03 ng/m^3 for PrP, 0.07 and 0.22 ng/m^3 for EtP, and 0.06 and 0.23 ng/m^3 for BuP (Moreau-Guigon et al. 2016) were detected as a sum in the gas and particle phases, respectively, during periods when there was no heating and during the heating season. The total content is almost exclusively determined by the concentrations in the gas phase. In a living room and a barbershop in Isfahan, Iran, levels of 0.55 and 0.63 ng/m^3 für MeP, 0.37 and 0.65 ng/m^3 for EtP, 0.83 and 1.07 ng/m^3 for PrP, and 0.04 and 0.21 ng/m^3 for BuP were reported (Hajizadeh et al. 2021).

4.19.3 Occurrence in Dust

In Spain, medians of 455 ng/g were measured for MeP (53–970 ng/g), 415 ng/g for PrP (16–1950 ng/g), 58 ng/g for EtP (8–370 ng/g), and 43 ng/g for BuP (4–210 ng/g) (Canosa et al. 2007).

In the USA, two studies in 118 and 80 homes, respectively, measured median levels of 978 and 1920 ng/g for MeP, <200 and 180 ng/g for EtP, and < 300 and

80 ng/g for BuP (Rudel et al. 2003; Chen et al. 2018b). Similarly, in 2014–2018, a median of 1874 ng/g for MeP, 100 ng/g for EtP, and 18 ng/g for BuP was found in a survey of 186 homes in North Carolina (Levasseur et al. 2021). Concentrations in dust from 63 Canadian indoor environments were comparable, with medians of 1080 ng/g for MeP, 463 ng/g for PrP, 25 ng/g for EtP, and 59 ng/g for BuP (Fan et al. 2010). In another nationwide study in Canada in 2007–2010 medians of 1480 ng/g for MeP, 681 ng/g for PrP, 46 ng/g for EtP, and 68 ng/g for BuP were observed in 1025 homes (Rasmussen et al. 2022).

Concentrations from 30 to 340 ng/g for MeP, 13 to 233 ng/g for PrP, <1.8 to 61 ng/g for EtP, and < 1.9 to 52 ng/g for BuP were measured in surveys of 4 homes in Taiwan (Chung et al. 2019). In a nationwide survey in China in 2017–2019 medians of 184 ng/g (MeP), 11.0 ng/g (EtP), 53.4 ng/g (PrP), and 1.36 ng/g (BuP) were found in dust of 289 houses (Zhu et al. 2020b). Between 2006 and 2012, 158 dust samples were collected from 13 cities in four countries, including the USA ($n = 40$), China ($n = 55$), Korea ($n = 41$), and Japan ($n = 22$) (Wang et al. 2012). MeP and PrP were the predominant compounds with medians ranged from 320 ng/g (China) to 1470 (Japan) and from 182 ng/g (China) to 800 ng/g (Korea). In 47 houses in southern China in 2026–2019 concentration ranged from 20 to 474 ng/g for MeP, 8 to 1360 ng/g for PrP, and 4 to 8380 ng/g for EtP (Shi et al. 2021).

4.20 Perchlorates/Chlorates

Perchlorates are the salts of perchloric acid ($HClO_4$), with sodium, ammonium, magnesium, and lithium perchlorates being of particular technical importance. Some perchlorates occur naturally in the soil, but they are mainly of anthropogenic origin. They are used, particularly ammonium perchlorate (CAS No.: 77–90-9), as an oxidizer in explosives and propellants, but also in fireworks, matches, in analytical chemistry as an oxidizer, and in various industrial processes such as paper finishing (Charnley 2008). In addition, perchlorate can also be formed in trace amounts in the atmosphere due to natural processes and deposited in the soil compartment via precipitation.

Chlorates are salts of chloric acid ($HClO_3$) containing the anion ClO_{3-}, especially sodium and potassium chlorate were used in the past for weed control. Nowadays, no applications of chlorate-containing plant protection products or biocidal products are permitted in the EU (BfR 2014). However, possible applications may exist in the field of flammables and explosives, for bleaching paper or for tanning leather. In addition, chlorate is formed as a by-product when chlorine-containing substances are used for cleaning or disinfection, e.g. in the context of drinking and surface water disinfection.

4.20.1 Health Aspects

It has long been known that perchlorate inhibits iodide uptake into the thyroid gland. It is a competitive inhibitor of iodide transport that blocks the sodium-iodide symporter without itself entering the thyroid via the iodide transporter (Lisco et al. 2020). Animal study showed blockade of iodine uptake into the thyroid with subsequent decrease in thyroid hormone levels and increase in serum TSH levels. Reactively, there is hyperplasia and hypertrophy of the organ with signs of hypothyroidism. Thyroid tumors have also been observed in rats, but are not considered significant in humans due to the mechanism. Overall, perchlorate was absorbed after inhalation or oral exposure, not metabolized in the body and excreated in urine with a half-life of around 10 hours from the thyroid (Niziński et al. 2020).

This active principle of perchlorates is also used therapeutically. For example, sodium perchlorate (Irenat®) is used in the prophylaxis of iodine-induced reaction in the context of diagnostics with iodine-containing X-ray contrast media. Compilations of the spectrum of effects can be found in NRC (2005), ATSDR (2008), Pleus and Corey (2018), and Niziński et al. (2020).

Single doses of chlorates in the context of accidental ingestion and poisoning showed hemolysis, methemoglobin formation, and renal damage at higher doses (Steffen and Wetzel 1993). Subchronic and chronic exposure toxicity studies in rats observed an effect on animal body weight and, comparable to perchlorate, hypertrophies, and hyperplasias of thyroid cells and inhibition of iodide uptake into the thyroid gland. However, these effects were not observed in a study with mice. Sodium chlorate may increase the incidence of thyroid neoplasms in animal experiments, but only in females. The spectrum of effects is summarized in the review by Smith et al. (2012).

4.20.2 Occurrence in the Air

In a Chinese study in 2013, the indoor air of a university building in a large city and the outdoor air were analyzed in parallel (Yao et al. 2015b). The mean levels of 67 $PM_{2.5}$ samples in indoor air were 3.54 ng/m^3 (0.14–125 ng/m^3) for perchlorate and 0.94 ng/m^3 (0.10–1.80 ng/m^3) for chlorate. In outdoor air, a mean content of 4.18 ng/m^3 (perchlorate) and 2.82 ng/m^3 (chlorate) was determined for comparison. A significant correlation was found between the indoor air content and the concentrations in the outdoor air. Moreover, Zheng et al. (2023) analyzed perchlorate in $PM_{2.5}$ samples from four stationary outdoor sites in China in 2020–2021 and reported mean values from 0.08–0.09 ng/m^3 (range: <LOQ-1.14 ng/m^3). Further measurement results in indoor air have not yet been published in the scientific literature.

4.20.3 *Occurrence in Dust*

In dust from vacuum cleaner bags sieved to 150 µm prior to analysis, medians for perchlorate of 0.16 mg/kg (0.03–1.83 mg/kg) and 0.37 mg/kg (0.08–4.67 mg/kg) were found in 23 Romanian samples in 2012 and 30 Greek samples in 2014, respectively (Wan et al. 2015). In 38 houses in Malta and Gozo, perchlorate could be found in 76% of indoor spaces in 2011–2013 with a median of 7.8 mg/kg (0.79–53 mg/kg) (Vella et al. 2015). The authors assume that the concentrations were largely determined by sources in the outdoor air (e.g., fireworks at festivals).

Comparable levels were found in the USA in 2014 with perchlorate concentrations from 0.02 to 1.18 mg/kg (median: 0.41 mg/kg) (Wan et al. 2015). This study group also reported a comparable exposure pattern in 9 other countries from Asia and South America, with the highest median of 4.25 mg/kg in China. Another Chinese study, which collected 51 samples in 2014 in two cities reported concentrations from 0.11–119 mg/kg, with higher means in one city (30.1 mg/kg) compared to the second (9.5 mg/kg) (Gan et al. 2015). Li et al. (2018) collected a total of 97 indoor and outdoor samples in summer 2015 showed a median of 11.4 mg/kg (range: <LOQ-753 mg/kg) for perchlorate. The highest concentrations published to date were observed in 8 homes and 8 classrooms in 2012/2013 in China with medians for perchlorate of 73.2 mg/kg and 212 mg/kg, respectively (Zhang et al. 2015).

4.21 Chlorinated Paraffins

Chlorinated paraffins (CP) are polychlorinated, saturated, mostly unbranched hydrocarbons formed by chlorination from alkanes. In this process, complex mixtures of theoretically 7000 to 900,000 individual substances of different chain lengths and degrees of chlorination are formed. CP are dense, viscous oils that appear either colorless or slightly yellow, with the exception of LCCP with high chlorine contents, which are solids (South et al. 2022). In general, the analytical techniques are able to separate and detect isomeric groups, but there is no ideal technique that can be used for analysis of CP and all available methods have limitations and are a compromise in some aspect (EFSA 2020c). A distinction is made between very short-chain chlorinated paraffins (vSCCP; $<C_9$), short-chain CP (SCCP; C_{10-13}; CAS-No.: 85535–84-8), medium-chain CP (MCCP; C_{14-17}; CAS-No.: 85535–85-9), and long-chain CP (LCCP; $>C_{18-30}$; CAS-No.: 85535–86-0). A typical structure of a short-chain chlorinated paraffins is shown in Fig. 4.29. Depending on the application, the average degree of chlorination can range from 40 to 70%. CP have been produced on a large scale since about the 1930s. A global annual production volume of about 920,000 tons was estimated for the 2000s (Coelhan and Hilger 2014). Commercial names for CP are for example: Arubren CP®, Cereclor®, Cloparin®, Chlorowax®, Essechlor®, FLX-0012®, 597 Hordaflex®, Paroil®, Ribeclor®, Uniclor®, and Witaclor® (EFSA 2020c). The letters and figures used in the

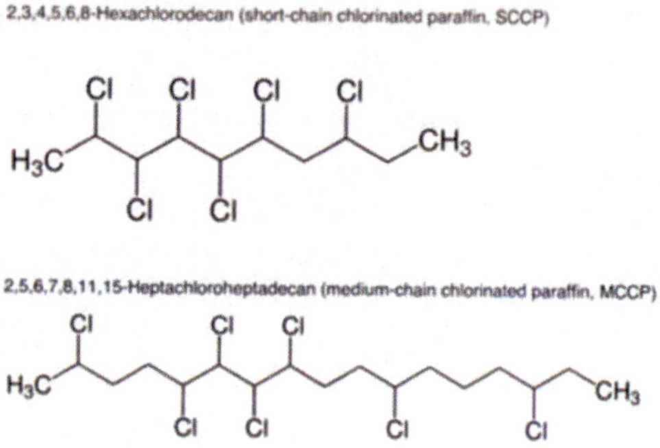

Fig. 4.29 Chemical structure of two chlorinated paraffins

commercial names differ from country to country and describe both the average chain length and the chlorine content. CP have a wide range of applications as flame retardants and plasticizers in a variety of products, as additives in paints, rubber, adhesives, and sealants. In the EU, the use of SCCP has been banned since 2012 under the EU POP Regulation and classified as persistent organic pollutants by the Stockholm Convention in 2017.

Short-chain CP have a high persistence in the environment and accumulate in the food chain. Also against this background, the use and marketing of SCCP above 1% by weight has been banned in the EU since 2012, and since 2017 SCCP have been subject to the bans of the Stockholm Convention.

4.21.1 Health Aspects

CP have low acute toxicity and in long-term animal experiments with SCCP and MCCP mixtures, the liver, kidney, and thyroid (alteration of thyroid hormones) were major target organs (Glüge et al. 2018; EFSA 2020c). In addition, developmental toxic effects have been found for SCCP. For LCCP, effects on the liver have been described with repeated administration. SCCP increase the incidence of liver and thyroid tumors in rats and mice and renal tumors in rats, whereas for LCCP only tumors in mice were observed. No studies were available on MCCP for these endpoints. Short-chain CPs have been classified by IARC in Group 2B, as a possible human carcinogen.

4.21.2 Occurrence in the Air

In Sweden, CP were measured for the first time in indoor air of 44 dwellings in 2006/2007 (Fridén et al. 2011). It resulted in a median of 69 ng/m^3 for the sum of SCCP and MCCP at that time (range < 5.0–210 ng/m^3). In the 2012 study of 44 homes and 6 classrooms in Norway, higher medians for SCCP of 128 ng/m^3 (36–289 ng/m^3)

were found in homes than in schools at 43 ng/m^3 (23–150 ng/m^3) (Sakhi et al. 2019). In this study, a positive correlation of the level with the number of TV sets in the home was found. In a third study from Oslo, Norway, in 2013–2014 medians of 9.6 ng/m^3 for ∑SCCP, 1.2 ng/m^3 for ∑MCCP, and 0.02 ng/m^3 for ∑LCCP were observed in 61 residences using a stationary sampler (Yuan et al. 2021). The estimated inhalation exposure to total CP was around 5 times higher when the personal 24 h air results were used rather than the corresponding stationary indoor air results, indicating that exposure situations other than living rooms contributed significantly to the overall personal exposure.

Gao et al. (2018) studied a total of 39 homes and workplaces in China in 2014/2015 and found a median of 72 ng/m^3 (9.8–966 ng/m^3) for ∑SCCP and 3.5 ng/m^3 (<LOQ-613 ng/m^3) for ∑MCCP. Comparable results were obtained in 2016 in a study in Beijing with ∑SCCP of 61.6 ng/m^3 (38.3–87.7 ng/m^3) and for ∑MCCP of 6.9 ng/m^3 (3.2–9.6 ng/m^3) (Huang et al. 2017). CP were found in higher concentrations in the PM_{10} fraction compared to the $PM_{2.5}$ and $PM_{1.0}$ fractions, and levels were higher in indoor air than in outdoor air. Concentrations of ∑SCCP and ∑MCCP in indoor $PM_{2.5}$ ranged from 2.9 to 45.4 ng/m^3 (geometric mean: 12.9 ng/m^3) and from 5.4 to 182 ng/m^3 (geometric mean: 29.9 ng/m^3) in 24 homes in China in 2017 (Zhuo et al. 2019). ∑SCCP and ∑MCCP indoors were approximately twice as high as in the outdoor environment.

In an Australian study, 34.5–121 ng/m^3 (median 51.9 ng/m^3) was observed for ∑SCCP in residential air (n = 10) (Gillett et al. 2010).

4.21.3 Occurrence in Dust

In a first German study, dust from 65 apartments was investigated in the period 1998–2000 (Kersten and Reich 2003). Median levels of ∑SCCP were 26 µg/g (95th percentile: 180 µg/g) and ∑MCCP were 36 µg/g (95th percentile: 150 µg/g). In 11 apartments from Germany, the authors reported levels for ∑SCCP between <14.4 and 27 µg/g (median: 6 µg/g) and for ∑MCCP between 8 and 892 µg/g (median: 176 µg/g) (Hilger et al. 2013). Around the same time, SCCP concentrations from <1.0 to 12.8 µg/g were measured in 11 Belgian homes and < 0.12 to 13 µg/g in 10 pooled dust samples from the UK (Al Bitar 2004; Santillo et al. 2003). Levels between 3.2 and 18 µg/g (sum of SCCPs and MCCPs) were found in 6 dust samples from Stockholm in 2006/2007 (Fridén et al. 2011). In another study from Oslo, Norway, in 2013–2014 medians of 5.8 µg/g for ∑SCCP, 21 µg/g for ∑MCCP, and 8.1 µg/g for ∑LCCP were observed in 61 residences using a stationary sampler (Yuan et al. 2021). In 50 Belgian dust samples collected in 2019 overall median concentrations of 6.1 µg/g (range; 0.61 to 120 µg/g) for SCCP, 45 µg/g (range; 4.5 to 520 µg/g) for MCCP, and 4.5 µg/g (range; 0.3 to 50 µg/g) for LCCP (McGrath et al. 2022). Concentrations were significantly higher in the winter samples than summer for each of the three groups. On a median basis ΣSCCP, ΣMCCP, and ΣLCCP accounted for 8%, 80%, and 7% in winter and for 12%, 79%, and 27% in summer.

Concentrations comparable to Europe were found in 2007–2010 in a representative study of vacuum cleaner bag dust in 48 Canadian homes with ∑SCCP ranging from 4.0–57 μg/g (median: 6.2 μg/g) and for ∑MCCP at 5.9–901 μg/g (median: 19 μg/g) (Shang et al. 2019). Levels in the 23 parallel actively collected samples correlated with those in the dust. While the medians in the freshly collected dust were higher for ∑SCCP (8.1 μg/g vs. 6.7 μg/g), no difference was found for ∑MCCP.

In South Africa, in addition to SCCPs and MCCPs, LCCPs were quantified in 20 homes (Brits et al. 2020). Medians were 14 μg/g (5.1–214 μg/g) for ∑SCCP, 46 μg/g (13–200 μg/g) for ∑MCCP, and 13 μg/g (4.3–47 μg/g) for ∑LCCP.

Gao et al. (2018) described the highest concentrations in 2014/2015 in 115 dust samples from Chinese indoor environments with a median for ∑SCCP of 99 μg/g (5.4–1022 μg/g) and for ∑MCCP of 90 μg/g (2.1–725 μg/g). In descending order, median levels were in homes (201 μg/g), followed by dormitories (113 μg/g) and offices (84 μg/g). In contrast, lower levels were described in 4 Chinese indoor dust samples in 2013: ∑SCCP of 4.2–35.6 μg/g and ∑MCCP of 6.7–57.0 μg/g (Wu et al. 2019). Moreover, Wu et al. (2023) collected 53 dust samples in China in 2013 and reported medians of 47 μg/g (4.2–304 μg/g) for SCCP, 72 μg/g (6.7–495 μg/g) for MCCP, and 27 μg/g (3.7–331 μg/g) for LCCP. vSCCP could only be detected in some samples with a median of 0.05 μg/g (maximum: 0.47 μg/g).

Average concentrations of 13 (1.0–42) for ∑SCCP, 110 (8.8–380) for ∑MCCP, and 3.0 (<0.0014–15) for ∑LCCP in 27 dust samples from residences collected in Australia in 2015 (He et al. 2019).

4.22 Organotin Compounds

Organotin compounds (OTC) were first synthesized by Frankland and Löwig in 1852. They are derivatives of tetravalent tin characterized by the presence of at least one covalent carbon-tin bond. Their general chemical structure is shown in Fig. 4.30, where the substituent R is an alkyl or aryl group, Sn is the central tin atom, and X is organic or inorganic radicals. The number and ratio of organic substituents and

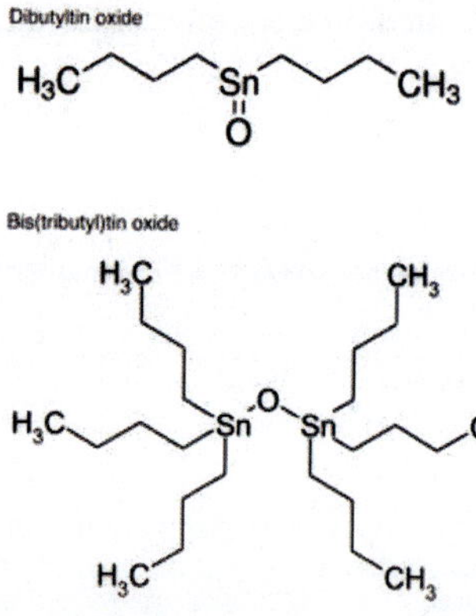

Fig. 4.30 Chemical structure of two organotin compounds

inorganic groups attached to tin has a significant influence on the chemical-physical properties and biological effects and also determines to a large extent the various applications. In the organotin compounds that are of technical importance, the corresponding methyl, butyl, octyl, and phenyl groups occur essentially as R. The anions are usually chlorides. Chlorides, fluorides, oxides, hydroxides, carboxylates, or thiolates are usually present as anions. Of particular importance and determined by measurement are monobutyltin (MBT), dibutyltin (DBT), tributyltin (TBT), monooctyltin (MOT), dioctyltin (DOT), and triphenyltin (TPT).

Of the various OTC, it is primarily the mono-, di-, and tri-substituted compounds that are used. In particular, the mono- and di-substituted ones are widely used as heat and light stabilizers in the production of polyvinyl chloride (PVC). In addition, they are used industrially as catalysts in the manufacturing process of polyurethane foams, silicone sealants and in the coating of glass surfaces. In addition, OTC are also used in pesticides, in surface disinfectants and as protective agents for wood, paper, and textiles. Tributyltin oxide (TBTO), in particular, was added to paints for marine coatings as an antifouling agent until it was banned and had thus led to pollution of the aquatic ecosystem.

4.22.1 *Health Aspects*

OTC have high toxicity, e.g. via endocrine effects, to marine organisms and accumulate in the food chain (Yi et al. 2012). Several reviews also summarize effects on terrestrial organisms (EFSA 2004; RPA 2005; SCHER 2006). In animal experiments, effects on reproduction and, consistently for DOT, DBT, TBT, and TPT, on the immune system have been described, with DBT and DOT appearing to be the most potent members of the OTC group in terms of effects. Regarding the underlying mechanisms, a review of the current literature indicates that OTC can directly damage endocrine glands, as well as interfere with neurohormonal control of endocrine function (i.e., in the hypothalamic-pituitary axis), altering hormone synthesis and/or bioavailability or activity of hormone receptors in the target cells (Marques et al. 2018). As a result they induce biochemical and morphological changes in gonads and abnormal steroidogenesis.

4.22.2 *Occurrence in the Indoor Environment*

No results of organotin compounds in indoor air have been published to date.

While very many measurement results are available, e.g. in the aquatic ecosystem (especially on TBT), there is only limited information on indoor exposure. Table 4.31 summarizes the results available in the scientific literature on dust studies. In the individual studies, the wide range of results is striking. This is largely due to the inhomogeneity of house dust and to the different methods of sample collection and sample preparation (e.g., sieving for different dust fractions). The OTC detected

Table 4.31 Medians (ranges) of organotin compounds in dust in µg/g

	N	MBT	DBT	TBT	MOT	DOT	TPhT	Country, time
Haumann and Tumulla (2002)	33	0.5 (n.s.-24)	3.4 (n.s.-2.2)	0.5 (n.s.-15)	–	–	–	Germany
Kersten and Reich (2003)	50	1.4 (n.s.-18)	0.2 (n.s.-5.6)	0.03 (<0.001–0.2)	0.20 (n.s.-2.8)	0.01 (<0.001–0.2)	0.004 (<0.001–0.02)	Germany, 1998–2000
Fromme et al. (2005b)	28	0.05 (<0.01–1.5)	0.03 (<0.01–5.6)	0.01 (<0.01–0.08)	0.01 (<0.01–0.04)	<0.01 (<0.01–0.36)	<0.01 (<0.01–0.36)	Germany, 2000
Santillo et al. (2003)	10*	1.4 (0.81–2.8)	0.52 (0.16–1.3)	0.05 (0.02–0.76)	0.35 (0.08–1.3)	0.06 (0.02–0.55)	<0.001	UK, 2002
Uhl et al. (2004)	24	1.1 (0.09–20)	0.55 (<0.05–24)	<0.05 (<0.05–0.26)	–	–	<0.05	Austria, 2003
Fromme et al. (2005c)	20	0.08 (0.02–0.50)	0.09 (0.02–2.5)	0.02 (0.001–0.11)	0.03 (0.007–0.12)	0.03 (0.001–0.22)	–	Germany, 2005
Kannan et al. (2010)	24	2.5 (0.32–11)	0.32 (0.01–12)	0.02 (<0.002–0.3)	0.73 (0.04–7.4)	0.16 (<0.02–6.4)	<0.002	USA, 2005/06

MBT monobutyltin, *DBZ* dibutyltin, *TBT* tributyltin, *MOT* monooctyltin, *DOT* dioctyltin, *TPhT* triphenyltin
* Pool samples from 100 households; *n.s.* not specified

most frequently and with the highest contents in all investigations were MBT and DBT. For example, Fromme et al. (2005c) found a median of 0.31 µg/g (range: 0.08–3.1 µg/g) for the sum of six organotin compounds in the 63 µm dust fraction, but only a median of 0.14 µg/g (range: 0.03–0.84 µg/g) in the 2 mm fraction. For the sum of OZV in the 63 µm fraction, a correlation with PVC flooring was found for apartments. Thus, the median in 6 apartments with PVC flooring was 1.4 µg/g, in 14 apartments with other floor coverings only 0.24 µg/g. In the period from 1998 to 2000, Kersten and Reich (2003) studied 50 Hamburg apartments and found 95th percentiles of 8.7 µg/g (MBT) and 1.4 µg/g (DBT), respectively. Also in the study of Haumann and Tumula (2002), higher levels for MBT and DBT were observed in 33 dust samples from apartments with PVC flooring. Very high levels between 1.3 and 2222 mg/kg have been determined in a study in 33 apartments in southern Germany (Hagenau 2001). Here, the samples were taken by means of a vacuum cleaner with a special filter system, which may explain the level of contamination.

Studies from the UK and the USA tended to show higher results compared to the aforementioned studies (Santillo et al. 2003, Kannan et al. 2010).

4.23 Volatile Methylsiloxanes

Siloxanes (or silicones) include both aliphatic and cyclic oxygenated silicon compounds with the basic structure Si-O-Si, to which various molecular groups such as methyl, ethyl, or phenyl groups are attached.

Of particular importance are the volatile cyclic methylsiloxanes. Typical representatives of this group are D3 (hexamethylcyclotrisiloxan, CAS-No.: 541-05-9), D4 (octamethylcyclotetrasiloxane, CAS-No.: 556–67-2), D5 (decamethylcyclopentasiloxane, CAS-No.: 541–02-6), and D6 (dodecamethylcyclohexasiloxane, CAS-No.: 540–97-6) (see structural formula in Fig. 4.31). They are listed by the OECD as "High Production Volume Chemicals" (>1000 t/year in at least one OECD member country). The global capacity of siloxanes production reached around 10,000 mio. tons per year in 2017 (Tran et al. 2019). Due to

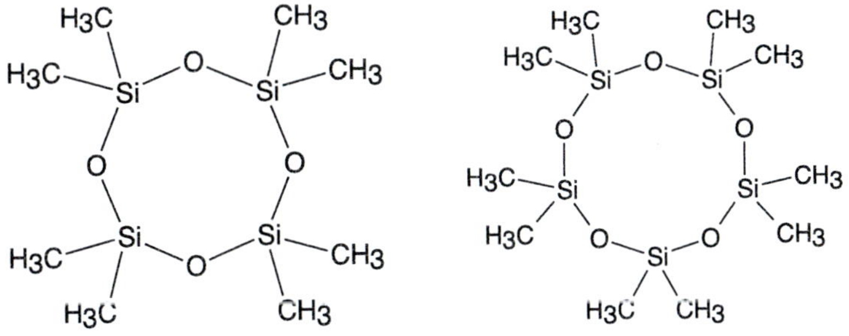

Fig. 4.31 Chemical structure of octamethylcyclotetrasiloxane (D4) (left) and decamethylcyclopentasiloxane (D5) (right)

their large production volume and increasing use, resulting in increased human exposure and increased input into the environment, siloxanes are increasingly discussed critically (Wang et al. 2013c). The focus is particularly on ecotoxicological criteria such as environmental persistence and bioaccumulation. Cyclic siloxanes are primarily used as raw materials in the production of higher molecular weight polydimethylcyclosiloxanes and copolymers. These, in turn, are used in industry as well as in a wide variety of consumer products, in pharmaceutical formulations, as antifoaming agents in foods, and also in medical devices (e.g., breast implants, catheters, infusion sets) (CES 2016). Residues of volatile siloxanes could be detected in numerous higher molecular weight polymers. Due to their volatility, low surface tension, transparency and hydrophobicity, and odorlessness, siloxanes are used in cosmetics and personal care products such as shampoos, conditioners, antiperspirants, soaps, moisturizers, and body lotions, as well as in decorative cosmetics (makeup, mascara, lipsticks).

Linear siloxanes like hexamethyldisiloxane (L2, CAS-No.: 107-46-0), octamethyltrisiloxane (L3, CAS-No.: 107-51-7), decamethyltetrasiloxane (L4, CAS-No.: 141-62-8), and dodecamethylpentasiloxane (L5, CAS-No.: 141-63-9) were continuously numbered according to the number of silicon atoms per molecule. L2, L4, and L5 are primarily used as intermediates, solvents, skin conditioning agents, surface active agents, polymers and functional fluids in products available to consumers such as cosmetics, natural health products, electronics, medical devices, adhesives and sealants, as well as in industrial applications such as paints and coatings (ECCC/HC 2019b).

4.23.1 Health Aspects

Results of epidemiological studies have not yet been published in the scientific literature. However, numerous animal studies and some experimental studies on test subjects are available (Dan-EPA 2014). The results show that siloxanes have very low acute toxicity and are not mutagenic or genotoxic. In addition, none of the siloxanes were found to have an irritant effect on the eyes or skin (SCCS 2010, 2017). In studies with repeated-dose exposure to relatively high doses, effects on the liver, lungs, and reproductive toxicity were observed in experimental animals (Dan-EPA 2014; ECCC/HC 2019b).

Inhalation uptake and, to a lesser extent, dermal uptake after application of products close to the body must be seen as the main route of intake for humans. From a health point of view, the use of D4 is undesirable, while the use of D5 in cosmetics and cosmetic products is considered safe (SCCS 2015, 2017). Only hair sprays from aerosol cans and sunscreen sprays containing D5 are associated with unnecessarily high indoor air pollution and should be avoided. D4 is classified as a category 2 (probable reproductive toxicant) in the EU. D4, D5, and D6 are classified by ECHA as very persistent and very bioaccumulative (vPvB), whereas D4 is also persistent, bioaccumulative, and toxic (PBT).

4.23.2 Occurrence in the Air of Residences

Results from studies are summarized in Table 4.32. In general, the levels of siloxanes in indoor air appear to be highly variable and strongly dependent on the individual use pattern of the room in question.

In a field study in Scandinavia, the indoor air of a total of 400 Swedish children's rooms was investigated (NILU 2007). While D3 could only be detected in one room at 7.3 μg/m^3, concentrations of 0.6–51 μg/m^3, 0.5–79 μg/m^3 and 0.6–164 μg/m^3 were found for D4, D5, and D6, respectively. Other study results from Italy and England in 2011 (Pieri et al. 2013) and Sweden in 2016 (Sha et al. 2018) gave mean values of 0.2–34 μg/m^3 (D4), 1.6–84 μg/m^3 (D5), and 4.4–45 μg/m^3 (D6). In 79 rooms of 39 apartments or houses in Germany, samples were taken in 2000/2001 in two rooms each that had not been aired for 8 hours before sampling (Hippelein 2004). Concentrations ranged from <0.2 to 688 μg/m^3 for D4 and < 0.2 to 1086 μg/m^3 for D5, with medians below the limit of quantification for D4 and at 3.9 μg/m^3 for D5, respectively. An extensive data set comes from a German project by private research institutes (Hofmann et al. 2014). The medians for the 4 cyclic siloxanes were all in the range at 1.0–2.5 μg/m^3, with the highest median for D5 (5.0 μg/m^3). In dwellings where renovation had taken place shortly before sampling or up to 6 weeks before, median levels of D4 and D5 can be found that are higher by a factor of 2–3. With a longer time interval to the renovation, however, this effect was no longer observed. As part of the German Environmental Health Study (GerES V), which was conducted in 2014–2017, representative samples in 639 homes were collected (Bethke et al. 2021). In this study, the medians of D3 (<1.0 μg/m^3) and D5 (14.0 μg/m^3) are in the range described from the other studies, while lower concentrations were measured for D4 (<1.0 μg/m^3).

In the USA, considerable lower mean levels of 0.057 for D4, 0.456 for D5, and 0.075 μg/m^3 for D6 were found in 20 homes, while L8 was the linear siloxane with highest mean value (0.99 μg/m^3) (Tran et al. 2015).

In China, 60 residential houses were investigated in 2012 and no linear compound, but D4 (67% > LOQ) and D5 (83% > LOQ) with ΣvCMS ranging from 0.045 to 0.139 μg/m^3 (Xu et al. 2015). In an additional Chinese study in 2014, average air levels in 40 homes during redecoration and 10 residences without such activities were 1.5 and 0.58 μg/m^3 for D4, 6.67 and 4.99 μg/m^3 for D5, and 1.7 and 0.65 μg/m^3 for D6, respectively (Meng and Wu 2015). The levels found in indoor air samples (both gas-phase and particulate phase) in 30 residences in China in 2015 are comparable to concentrations observed in Europe with means of 2.4 μg/m^3 for D4, 6.9 μg/m^3 for D5, and 1.5 μg/m^3 for D6 (Guo et al. 2021). In this study linear siloxanes (L5–L12) could not be detected in the air. In another Chinese study performed in 2017 indoor gas-phase concentrations ranged from 0.33–2.1 μg/m^3 (D4), 0.95–18 μg/m^3 (D5), and 0.29–2.6 μg/m^3 (D6) (Guo et al. 2019).

From the results of the previous studies, it can be concluded that D5 is most commonly found indoors. The nature of the distribution of the measured values indicates that specific siloxane sources must be present in residential interiors to

Table 4.32 Siloxanes in the air of residences in μg/m^3

	n	% > LOQ	Median	95. percentile	Maximum	Country
D3						
Schleibinger et al. (2001)	53	n.s.	<1.0	5.0	14	Germany
Eis et al. (2005)	39	44	<1.0	5.0	6.0	Germany
Conrad and Brenske (2010)	586	44	<1.0	9.0	44	Germany
Pieri et al. (2013)	5	100	3.6*	–	8.2	Italy
Pieri et al. (2013)	5	100	160*	–	270	UK
Hofmann et al. (2014)	571	74	2.5	25.5	n.s.	Germany
Tran et al. (2015)	20	100	0.025*	–	n.s.	U.S.
Bethke et al. (2021)	639	30	<1.0	6.8	36.0	Germany
D4						
Schleibinger et al. (2001)	173	n.s.	<1.0	15.0	374	Germany
Hippelein (2004)	22	28	<0.2	3.2[a]	n.s.	Germany
Eis et al. (2005)	39	56	<1.0	17.0	17	Germany
Conrad and Brenske (2010)	586	75	2.5	11.3	187	Germany
Pieri et al. (2013)	5	100	8.2*	–	22	Italy
Pieri et al. (2013)	5	100	43*	–	80	UK
Hofmann et al. (2014)	696	68	1.0	12.6	n.s.	Germany
Tran et al. (2015)	20	100	0.057*	–	0.21	U.S.
Bethke et al. (2021)	639	19	<1.0	2.1	200	Germany
Guo et al. (2021)	30		2.4*	–	–	China
D5						
Schleibinger et al. (2001)	174	n.s.	<1.0	27.0	168	Germany
Scholz (2000)	602	n.s.	<1.0	24.0	n.s.	Germany
Hippelein (2004)	71	90	3.9	101[a]	n.s.	Germany
Eis et al. (2005)	40	82	3.0	324	369	Germany
Conrad and Brenske (2010)	586	89	10.9	133	432	Germany
Pieri et al. (2013)	5	100	38*	–	79	Italy
Pieri et al. (2013)	5	100	84*	–	160	UK
Hofmann et al. (2014)	648	85	5.0	86.0	n.s.	Germany
Tran et al. (2015)	20	100	0.45*	–	1.8	U.S.
Bethke et al. (2021)	639	91	14.0	152	800	Germany
Guo et al. (2021)	30		6.9*	–	–	China
D6						
Pieri et al. (2013)	5	100	45*	–	180	Italy
Pieri et al. (2013)	5	100	12*	–	32	UK
Hofmann et al. (2014)	104	18	<1.5	23.7	n.s.	Germany
Tran et al. (2015)	20	100	0.075*	–	0.31	U.S.
Guo et al. (2021)	30		1.5*	–	–	China

n.s. not specified, *LOQ* limit of quantification; * mean
[a] 90. perzentile

explain the large range of concentration levels. For example, Tang et al. (2015) reported the results of occupant-related VOC emissions from engineering students in a classroom in Chicago, USA. They reported that D5 contributes to around 30% of the total indoor VOC mass concentration. Molinier et al. (2022) performed four multiweek indoor air measurement campaigns and found that the D5 emissions are mainly attributable to personal care product use, while D3 and D4 are emitted from oven and personal care product use, and L4 and L5 from personal care product use. As part of an experimental work, it was observed whether siloxanes from 14 silicone baking molds are given off to indoor air when baking cakes (Fromme et al. 2019). The highest median indoor air concentrations were found immediately after baking with molds with a high siloxane content. They were 301 μg/m^3 for D7, 212 μg/m^3 for D6, 130 μg/m^3 for D8, 78 μg/m^3 for D5, and 10 μg/m^3 for D4. When the baking pans were used more frequently, the observed indoor air concentrations also decreased.

4.23.3 *Occurrence in the Air of Public Facilities*

The siloxane levels measured in the indoor air of public facilities are summarized in Table 4.33.

From 2004 to 2005, the indoor air quality of 92 classrooms (winter) and 76 classrooms (summer) of schools in southern Germany was investigated (Fromme et al. 2008a). A concentration range from 0.7 to 300 μg/m^3 for D5 and 0.4 to 53 μg/m^3 for D6 was found. The content in indoor air tends to show higher concentrations in the winter measurements, whereby a lower air circulation and a restricted air exchange due to less frequent ventilation of the rooms in the cold season can presumably be assumed as the cause. In this study, a significantly lower content was found in the school rooms of the first to fourth grade compared to the seventh to ninth grade. While the median content in elementary school classrooms was 10.5 μg/m^3 (D5) and 3.4 μg/m^3 (D6), it was 51 μg/m^3 (D5) and 7.3 μg/m^3 (D6) in classrooms attended by older students. This strongly suggests the increased use of personal care products as a source of siloxane. In an investigation of 63 daycare centers for children from three German states (LUPE 3 program) conducted in 2011/2012 in the group rooms, higher levels could be detected especially for D5 with 57 μg/m^3 (95th percentile) (Fromme et al. 2016a).

Indoor air samples collected in 2010–2011 from 34 early childhood education facilities serving California children reported medians of 1.5 (D3), 0.9 (D4), and 51 μg/m^3 (D5) (Hoang et al. 2017).

Comparable levels found in indoor air samples (both gas-phase and particulate phase) in 22 classrooms of six kindergartens and 18 classrooms of five primary schools in China in 2015 (Guo et al. 2021). For the linear siloxanes (L5–L12) levels were often under the LOD or with averages up to 0.09 μg/m^3.

Table 4.33 Siloxane content in the air in schools and daycare centers in μg/m³

References	Number	% > LOQ	Median	95th percentile	Maximum	Country
D3						
Neumann et al. (2012)	68	67	n.s.	48.0	n.s.	Germany
Hofmann et al. (2014)	211	67	2.5	25.0	35	Germany
Tran et al. (2015)	6	100	0.016*	–	0.025	USA
Hoang et al. (2017)	34	50	1.5	8.0	9.3	USA
Fromme et al. (2016a)	63	8	<1.5	2.0	5.3	Germany
D4						
Neumann et al. (2012)	68	43	n.s.	32.0	n.s.	Germany
Hofmann et al. (2014)	496	65	1.0	14.3	n.s.	Germany
Tran et al. (2015)	6	100	0.076*	–	0.025	USA
Hoang et al. (2017)	34	91	0.9			USA
Fromme et al. (2016a)	63	19	<1.0	2.6	5.2	Germany
Guo et al. (2021)[a]	22		4.9*	–	–	China
Guo et al. (2021)[b]	18		2.1*	–	–	China
D5						
Fromme et al. (2008a)	165	100	20.0	168	300	Germany
Neumann et al. (2012)	118	46	n.s.	23.0	n.s.	Germany
Hofmann et al. (2014)	405	79	2.0	19.0	n.s.	Germany
Tran et al. (2015)	6	100	0.65*	–	1.77	USA
Hoang et al. (2017)	34	100	51.4			USA
Fromme et al. (2016a)	63	98	10.6	56.5	132	Germany
Guo et al. (2021)[a]	22		12.2*	–	–	China
Guo et al. (2021)[b]	18		5.4*	–	–	China
D6						
Fromme et al. (2008a)	164	100	5.4	23.9	53	Germany
Hofmann et al. (2014)	58	36	<0.5	4.5	n.s.	Germany
Tran et al. (2015)	6	100	0.11*	–	0.20	USA
Guo et al. (2021)[a]	22		4.4*	–	–	China
Guo et al. (2021)[b]	18		1.0*	–	–	China

n.s. not specified, *LOQ* limit of quantification; *: arithmetic mean

[a] kindergarten

[b] primary school

4.23.4 *Occurrence in the Office Air*

Pieri et al. (2013) collected samples in 5 offices in Italy and 4 offices in the UK in 2011 and found average levels of 3.3 and 6.6 μg/m^3 (D3), 2.2 and 9.8 μg/m^3 (D4), and 7.5 and 54 μg/m^3 (D5), respectively. D6 was found only in the UK with an arithmetic mean of 4.6 μg/m^3. In Sweden, means of 0.28 μg/m^3 for D4, 1.1 μg/m^3 for D5, and 0.32 μg/m^3 for D6 in 2016 (Sha et al. 2018). Significant lower levels were observed in rooms with forced-ventilation systems. In addition, results from the aforementioned German AGÖF research project are available in which 27.7 μg/m^3 (D3), 11.5 μg/m^3 (D4), 31.0 μg/m^3 (D5), and 18.0 μg/m^3 (D6) were determined as the 95th percentile (Hofmann et al. 2014). 95th percentile values of 51.5 μg/m^3 (D3), 33.6 μg/m^3 (D4), and 120 μg/m^3 (D5) were observed in the study of office spaces in 257 to 347 establishments sampled between 2006 and 2010 (von Hahn et al. 2011).

In various states of the USA, measurements were carried out in a total of 70 office buildings occupied by very different numbers of people (Shields et al. 1996). A direct relationship was found between the occupancy of the respective room and the measured concentrations of volatile siloxanes. The authors consider this to indicate that cyclic siloxanes are primarily due to the use of personal care products by people who are in the room. In contrast, other factors, such as different ventilation systems or floor coverings, seem to play a minor role. In 4 university offices in Chicago, concentrations ranged from 0.07–0.50 μg/m^3 for D4, 2.3–53.0 μg/m^3 for D5, and 0.03–0.16 μg/m^3 (Yucuis et al. 2013). Moreover, in 7 offices in the USA in 2014 means of 42.7 μg/m^3 (D3), 35 μg/m^3 (D4), 795 μg/m^3 (D5), and 262 μg/m^3 (D6) were reported (Tran et al. 2015).

4.23.5 *Occurrence in Dust*

Dust has hardly been studied so far, also against the background of the physical characteristics of siloxanes.

A comprehensive study examined siloxanes in dust samples from 12 countries (Tran et al. 2015). The highest levels were observed at 6.7 ng/g (D3), 66 ng/g (D4), 1200 ng/g (D5), and 83 ng/g (D6) in Greece. The lowest medians were found at 11 ng/g (D3), <3.0 ng/g (D4), 1.7 ng/g (D5), and 10 ng/g (D6) in Pakistan. Total cyclic siloxanes accounted for a major proportion of total siloxanes concentrations in dust collected from Colombia (90%), India (80%), and Saudi Arabia (70%), whereas total linear siloxanes predominated in samples collected from Japan (89%), Kuwait (85%), and South Korea (78%). The composition profiles of 15 siloxanes in dust using median values are given in Fig. 4.32.

In 32 dust samples collected in 2009, medians of 5.1 ng/g (D4), 14.8 ng/g (D5), and 11.0 ng/g (D6) were observed in China (Lu et al. 2010). In 60 residential Chinese houses investigated in 2012, ΣvCMS (D4–D6) and sum of L8 to L12 were 23.3–73.5 ng/g (detection frequency: 25–83%) and 18.0–156 ng/g (detection

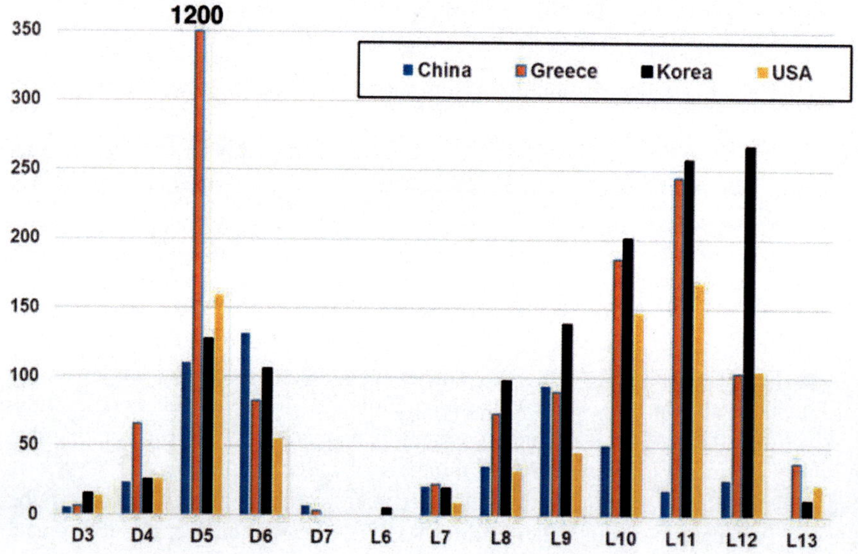

Fig. 4.32 Median concentration of cyclic and linear siloxanes in dust in ng/g (according to Tran et al. 2015)

frequency: 40–92%), respectively (Xu et al. 2015). In another Chinese study, the medians were considerably higher at 74 ng/g (D4), 83 ng/g (D5), and 90 ng/g (D6) (Meng and Wu 2015). In 30 urban households in China high mean values of 84 ng/g for D4, 203 ng/g for D5, and 79 ng/g for D6 were reported, while the sum of L4–L16 showed a mean value of 2084 ng/g (Liu et al. 2018). The linear compounds L 10 (mean: 387 ng/g) and L11 (mean: 463 ng/g) have the highest share on the total amounts. In Vietnam, medians of 103 ng/g for D3, 230 ng/g for D4, 483 ng/g for D5, and 185 ng/g for D6 were reported in dust collected from 17 homes in 2019/20 (Hoang et al. 2023).

Comparable levels found in indoor air samples (both gas-phase and particulate phase) in 22 rooms of six kindergartens, 18 classrooms of five primary schools, and 50 residences in China in 2018 (Guo et al. 2021). In kindergarten the average levels were 160 ng/g (D3), 380 ng/g (D4), and 240 ng/g (D5), while in primary schools means were 49 ng/g (D3), 120 ng/g (D4), and 71 ng/g (D5). In 30 residences levels of 79 ng/g (D3), 210 ng/g (D4), and 100 ng/g (D5) were reported. Moreover, Hoang et al. (2023) collected dust from 11 classrooms in Vietnam in 2019/2020 and reported medians of 33.3 ng/g for D3. 97.9 ng/g for D4, 255 ng/g for D5, and 163 ng/g for D6.

Table 4.34 Common organic UV light protection filters

Abbreviation	Substance	CAS No.
4-MBC	3-(4-Methylbenzylidene) camphor	36,861–47-9
HMS	Homosalate (homomethyl salicylate)	118–56-9
EHS	Octisalate (2-ethylhexyl salicylate)	118–60-5
OCT	Octocrylene	6197–30-4
Benzophenone		
BP-3	2-Hydroxy-4-methoxybenzophenone	131–57-7
4-OH-BP	4-Hydroxybenzophenone	1137-42-4
2,4-2OH-BP	2,4-Dihydroxybenzophenone	131–56-6
22′-2OH-4-MeOBP	2,2′-Dihydroxy-4-methoxybenzophenone	131–53-3
22′44′-4OH-BP	2,2′,4,4′-Tetrahydroxybenzophenone	131–55-5
Benzothiazoles		
BTH	1,3-Benzothiazole	95–16-9
2-OH-BTH	2-Hydroxybenzothiazole	934–34-9
2-NH_2-BTH	2-Aminobenzothiazole	136–95-8
2-Me-S-BTH	2-methylthiobenzothiazole	615–22-5
Benzotriazoles		
1-H-BTR	1H- Benzotriazole	95–14-7
1-OH-BTR	1- Hydroxybenzotriazole	2592-95-2
5-Cl-BTR	5-Chloro-1-hydrogenbenzotriazole	94–97-3
5,6-2Me-1-HBTR	5,6-dimethyl-1-hydrogenbenzotriazole	4184-79-6
Cinnamic acid esters		
IAMC	Isoamyl-4-methoxycinnamate	71,617–10-2
EHMC	2-ethylhexyl-4-methoxycinnamate	5466-77-3

4.24 Organic UV Light Protection Filter

Due to increasing human exposure to the sun, the so-called UV filters are increasingly used in sunscreen products, hair and facial care products, and lip and lip balms. They are also used as stabilizers in many industrial products such as textiles and rubber to increase their resistance to light. For sun protection, both organic and inorganic substances such as nanoscale zinc oxide or titanium dioxide are used in humans—in addition to textiles. Organic UV light protection filters absorb light quanta in a specific wavelength range and convert the energy into heat. Often, several UV filters must be combined in products to achieve a sufficient effect in the entire wavelength range from 290 to 400 nm. These are very different classes of substances, some of which have other applications besides use as UV filters, e.g. as corrosion inhibitors and fungicides (Table 4.34). Figure 4.33 shows the chemical structure of some important representatives of the organic UV filters.

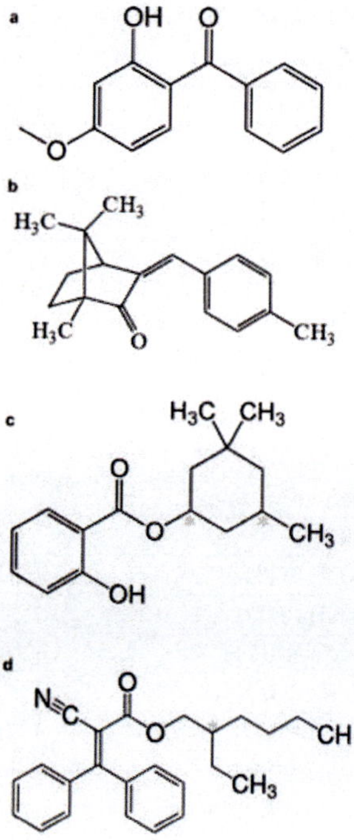

Fig. 4.33 Chemical structure of (**a**) 4-methylbenzylidene-camphor (4-MBC), (**b**) homosalate (HMS), (**c**) octocrylene (OC), and (**d**) benzophenone (BP-3)

4.24.1 Health Aspects

The data available on the health effects of UV filters are limited. In vitro studies have described estrogenic and anti-androgenic effects for various UV filters. Sensitizing and mucosal irritant effects and effects on the kidney were found for 1,3-benzothiazole (1,3-BTH) (Liao et al. 2018). Homosalates causes mild irritant effects to the eye in sunscreens. It has very low systemic toxicity and there is no evidence of sensitizing potential (SCCP 2007). Octocrylene (OCT) and 3-(4-methylbenzylidene) camphor (4-MBC) show effects on the thyroid gland in animal experiments, while there is no evidence of carcinogenic potential or effects on fertility/development (Berardesca et al. 2019). There is evidence that OCT may be a contact allergen (Manová et al. 2014). 1H-benzotriazole (1-H-BTR) has low irritant and sensitizing potency and animal experiments show neoplastic lesions in the nervous system.

4.24.2 Occurrence in the Air

For benzothiazoles (BTH) and benzotriazoles (BTR) results from only different indoor environments have been published so far in 2014 (Wan et al. 2016, Xue et al. 2017). In the gas and particulate phase, a median level of 41.9 ng/m^3 for 1,3-BTH (4.7–1233 ng/m^3), 2.5 ng/m^3 for 2-OH-BTH, 0.08 ng/m^3 for 2-NH_2-BTH, and 0.53 ng/m^3 for 2-Me-S-BTH were found in 26 indoor residential spaces in the USA in 2014 (Wan et al. 2016). For the sum of benzothiazoles in this study, the median in the particulate fraction was 6.2 ng/m^3 and in the gas phase 32.7 ng/m^3. Significantly higher concentrations were found in 7 car interiors with 146 ng/m^3 of 1,3-BTH (median). A median for 1H-BTR of 2.41 ng/m^3 (0.46–14.5 ng/m^3) was reported by the same research group in the same 26 homes (Xue et al. 2017). For two BTR, tolyltriazole (TTC; isomeric mixture of 4-Me-1H-BTR and 5-Me-1H-BTR) and xylyltriazole (XTR; 5,6-diMe-1-H-BTR), medians were 1.0 ng/m^3 and 0.65 ng/m^3, respectively. 1-H-BTR was present in 65% of the gas phase, while XTR was found in 94% of the particulate phase.

4.24.3 Occurrence in Dust

Table 4.35 summarizes the measurement results of the studies published in the scientific literature to date. Overall, the results show a considerable range of variation over several orders of magnitude. First, Negreira et al. (2009) studied dust from 10 homes in Spain and found high median levels of OCT and 2-ethylhexyl-4-methoxycinnamate (EHMC) at 4441 ng/g and 1865 ng/g, respectively. Another Spanish study found comparable results for EHMC but even higher concentrations of OCT (17,700 ng/g) (Castro et al. 2019). Median levels of 1,3-BTH and 2-OH-BTH are also significantly higher compared with data from the USA, at <0.5 ng/g and 1140 ng/g, respectively (Wang et al. 2013d). In two Chinese studies in 110 and 203 indoor rooms, respectively, median concentrations of 4-MBC were higher and OCT lower than in Spain (Ao et al. 2017, 2018). In 158 dust samples collected from homes and offices in 2006–2012, median concentrations of ∑BTR in 4 countries could be found to be 87.1 ng/g in Korea, 36.2 ng/g in the USA, 33.7 ng/g in Japan, and 19.3 ng/g in China (Wang et al. 2013d). For the ∑BTH, this study found 2000 ng/g in Korea, 1290 ng/g in the USA, 857 ng/g in China, and 605 ng/g in Japan.

For benzophenone-3 (BP-3), the median concentration varied in the USA, China, Korea, and Japan between 40 and 296 ng/g, significantly than in Spain. In a recent study, dust were collected in 12 houses in the USA in 2013 and in 20 Canadian houses in 2015 (Wu et al. 2020c). They reported relatively high BP-3 medians of 832 ng/g (119–8180 ng/g) and 313 ng/g (6.1–3430 ng/g).

Table 4.35 Substances used as photoprotective UV filters in dust of residences in ng/g (medians, maxima in parentheses)

Abbreviation	Negreira et al. (2009)	Wang et al. (2013d)*	Wang et al. (2013d)*	Wang et al. (2013d)*	Wang et al. (2013d)*	Ao et al. (2017)	Ao et al. (2018)	Castro et al. (2019)
	Spain	USA	China	Japan	Korea	China	China	Spain
4-MBC	<20 (1900)	–	–	–	–	161 (1444)	118 (1889)	–
HMS	–	–	–	–	–	238 (2732)	83 (2731)	–
EHS	114 (2747)	–	–	–	–	–	–	–
OCT	4441 (41000)	–	–	–	–	1170 (55059)	326 (55059)	17,700 (–)
Benzophenone								
BP-3	–	243 (1190)	40 (285)	74 (3280)	75 (1690)	296 (5209)	241 (5261)	1960 (–)
4-OH-BP	–	16 (689)	14 (280)	8.7 (29)	13 (53)	–	–	–
2,4-2OH-BP	–	102 (36400)	7.4 (776)	10 (537)	26 (600)	–	–	–
22′-2OH-4-MeOBP	–	8.0 (95)	2,7 (19)	2.4 (37)	2.1 (77)	–	–	–
22′44′-4OH-BP	–	50 (98)	3.0 (219)	5.5 (85)	20 (8810)	–	–	–
Benzothiazoles								
BTH	–	<0.5 (546)	360 (2780)	<0.5 (43)	5.9 (392)	–	–	2960 (–)
2-OH-BTH	–	1140 (13600)	472 (4020)	552 (5550)	1760 (9060)	–	–	3610 (–)
2-NH_2-BTH	–	<0.5 (12)	<0.5 (6.2)	<0.5 (<0.5)	<0.5 (13)	–	–	–
2-Me-S-BTH	–	66 (1340)	32 (264)	47 (938)	50 (772)	–	–	–
Benzotriazoles								
1-H-BTR	–	<0.5 (125)	6.1 (1940)	<0.5 (508)	16 (72)	–	–	–
1-OH-BTR	–	<0.5 (<0.5)	2.3 (42)	<0.5 (19)	<0.5 (10)	–	–	–
5-Cl-BTR	–	4.6 (24)	1.3 (132)	10 (41)	12 (54)	–	–	–
5,6-2Me-1-HBTR	–	<0.5 (2.8)	<0.5 (18)	<0.5 (17)	<0.5 (15)	–	–	–

Cinnamic acid esters								
IAMC	<5 (1290)	–	–	–	–	–	–	–
EHMC	1865 (15000)	–	–	–	–	–	–	1540 (–)

References

Abafe OA, Martincigh BS (2014) Polybrominated diphenyl ethers and polychlorinated biphenyls in indoor dust in Durban, South Africa. Indoor Air 25:547–556

Abb M, Heinrich T, Sorkau E, Lorenz W (2009) Phthalates in house dust. Environ Int 35:965–970

Abb M, Stahl B, Lorenz W (2011) Analysis of brominated flame retardants in house dust. Chemosphere 85:1657–1663

Abdallah MAE, Harrad S, Ibarra C, Robson M, Covaci A (2008a) Hexabromocyclododecanes in indoor dust from Canada, the United Kingdom and the United States. Environ Sci Technol 42: 459–464

Abdallah MAE, Harrad S, Covaci A (2008b) Hexabromocyclododecanes and tetrabromobisphenol – A in indoor air and dust in Birmingham, UK: implications for human exposure. Environ Sci Technol 42:6855–6861

ACS (American Chemistry Council) (2023) Applications & benefits. https://www.americanchemistry.com/industry-groups/center-for-the-polyurethanes-industry-cpi/applications-benefits

Adesina OA, Nwogu AS, Sonibare JA (2021) Indoor levels of polycyclic aromatic hydrocarbons (PAHs) from environment tobacco smoke of public bars. Ecotoxicol Environ Saf 208:111604

Adibi JJ, Perera FP, Jedrychowski W, Camann DE, Barr D, Jacek R, Whyatt RM (2003) Prenatal exposure to phthalates among women in New York City and Krakow, Poland. Environ Health Perspect 111:1719–1722

Al Bitar F (2004) Hazardous chemicals in Belgian house dust. Greenpeace, Bruxelles. https://greenpeace.org.br/toxicos/pdf/hazardous_chemicals_belgium.pdf

Alghamdi MA, Hassan SK, Alzahrani NA, Al Sharif MY, Khoder MI (2020) Classroom dust-bound polycyclic aromatic hydrocarbons in Jeddah primary schools, Saudi Arabia: level, characteristics and health risk assessment. Int J Environ Res Public Health 17:2779

Al-Harbi M, Al-Enzi E, Al-Mutairi H, Whalen JK (2021) Human health risks from brominated flame retardants and polycyclic aromatic hydrocarbons in indoor dust. Chemosphere 282: 131005

Ali N, Harrad S, Goosey E, Neels H, Covaci A (2011) "Novel" brominated flame retardants in Belgian and UK indoor dust: implications for human exposure. Chemosphere 83:1360–1365

Allen JG, McClean MD, Stapleton HM, Nelson JW, Webster TF (2007) Personal exposure to polybrominated diphenyl ethers (PBDEs) in residential indoor air. Environ Sci Technol 41: 4574–4579

Allen JG, McClean MD, Stapleton HM, Webster TF (2008) Critical factors in assessing exposure to PBDEs via house dust. Environ Int 34:1085–1091

Altarawneh M, Dlugogorski BZ (2013) A mechanistic and kinetic study on the formation of PBDD/Fs from PBDEs. Environ Sci Technol 47:5118–5127

Altarawneh M, Dlugogorski BZ (2015) Formation and chlorination of carbazole, phenoxazine, and phenazine. Environ Sci Technol 49:2215–2221

Ampleman MD, Martinez A, DeWall J, Rawn DF, Hornbuckle KC, Thorne PS (2015) Inhalation and dietary exposure to PCBs in urban and rural cohorts via congener-specific measurements. Environ Sci Technol 49:1156–1164

Anders N, Abb M, Sorkau E, Kubinec R, Lorenz W (2012) Analysis and occurrence of polycyclic aromatic hydrocarbons in household dust. Fresenius Environ Bull 21:372–379

Andersen HV, Frederiksen M (2021) Sorption of PCB from air to settled house dust in a contaminated indoor environment. Chemosphere 266:129139

Andersen HV, Gunnarsen L, Knudsen LE, Frederiksen M (2020) PCB in air, dust and surface wipes in 73 Danish homes. Int J Hyg Environ Health 229:113429

Ao J, Yuan T, Ma Y, Gao L, Ni N, Li D (2017) Identification, characteristics and human exposure assessments of triclosan, bisphenol-A, and four commonly used organic UV filters in indoor dust collected from Shanghai, China. Chemosphere 184:575–583

Ao J, Yuan T, Gu Y, Ma Y, Shen Z, Tian Y, Shi R, Zhou W, Zhang J (2018) Organic UV filters in indoor dust and human urine: a study of characteristics, sources, associations and human exposure. Sci Total Environ 640–641:1157–1164

Aquilina NJ, Delgado-Saborit JM, Meddings C, Baker S, Harrison RM, Jacob 3rd P, Wilson M, Yu L, Duan M, Benowitz NL (2010) Environmental and biological monitoring of exposures to PAHs and ETS in the general population. Environ Int 36:763–771

Aslam I, Baqar M, Qadir A, Mumtaz M, Li J, Zhang G (2021) Polychlorinated biphenyls in indoor dust from urban dwellings of Lahore, Pakistan: congener profile, toxicity equivalency, and human health implications. Indoor Air 31:1417–1426

ATSDR (2003) Toxicological profile for pyrethrins and pyrethroids. https://www.atsdr.cdc.gov/ToxProfiles/tp155-p.pdf

ATSDR (2008) Toxicological profile for perchlorates. Atlanta, GA, USA. https://www.atsdr.cdc.gov/toxprofiles/tp.asp?id=895&tid=181

ATSDR (2012) Toxicological profile for phosphate ester flame retardants. https://www.atsdr.cdc.gov/toxprofiles/tp.asp?id=1119&tid=239

ATSDR (2013) Toxicological profile for hexachlorobenzene. (Draft for Public Comment) Atlanta, GA, USA. https://www.atsdr.cdc.gov/ToxProfiles/tp.asp?id=627&tid=115

ATSDR (2017) Toxicological profile for polybrominated diphenyl ethers (PBDEs). https://www.atsdr.cdc.gov/ToxProfiles/tp.asp?id=901&tid=183

ATSDR (2018) Toxicological profile for toluene diisocyanate and methylenediphenyl diisocyanate. https://www.atsdr.cdc.gov/ToxProfiles/tp206-p.pdf

ATSDR (2019) Toxicological profile for DDT, DDE, DDD (Draft for Public Comment). Atlanta, GA, USA. https://www.atsdr.cdc.gov/ToxProfiles/tp.asp?id=81&tid=20

ATSDR (2021) Toxicological profile for perfluoroalkyls. Released May 2021. https://www.atsdr.cdc.gov/toxprofiles/tp200.pdf

ATSDR (2023) Toxicological profile for hexachlorocyclohexane. Atlanta, GA, USA. Online: https://www.atsdr.cdc.gov/Toxprofiles/tp43.pdf

Audy O, Melymuk L, Venier M, Vojta S, Becanova J, Romanak K, Vykoukalova M, Prokes R, Kukucka P, Diamond ML, Klanova J (2018) PCBs and organochlorine pesticides in indoor environments – a comparison of indoor contamination in Canada and Czech Republic. Chemosphere 206:622–631

Aydin YM, Kara M, Dumanoglu Y, Odabasi M, Elbir T (2014) Source apportionment of polycyclic aromatic hydrocarbons (PAHs) and polychlorinated biphenyls (PCBs) in ambient air of an industrial region in Turkey. Atmos Environ 97:271–285

Baek SY, Fang M, Chang Y-S (2010) Concentrations of PBDEs in indoor air of daycare facilities and elementary schools. Proceedings DIOXIN 2009 conference

Balducci C, Cerasa M, Avino P, Ceci P, Bacaloni A, Garofalo M (2022) Analytical determination of allergenic fragrances in indoor air. Separations 9:99

Ballschmiter K, Zell M (1980) Analysis of polychlorinated biphenyls (PCB) by glass capillary gas chromatography. Fresen Z Anal Chem 302:20–31

Bamai YA, Araki A, Kawai T, Tsuboi T, Saito I, Yoshioka E, Kanazawa A, Tajima S, Shi C, Tamakoshi A, Kishi R (2014) Associations of phthalate concentrations in floor dust and multi-surface dust with the interior materials in Japanese dwellings. Sci Total Environ 468–469:147–157

Bannan D, Ali N, Alhakamy NA, Alfaleh MA, Alharbi WS, Rashid MI, Rajeh N, Malarvannan G (2021) Brominated flame retardants in children's room: concentration, composition, and health risk assessment. Int J Environ Res Public Health 18:6421

Barber JL, Berger U, Chaemfa C, Huber S, Jahnke A, Temme C, Jones KC (2007) Analysis of per- and polyfluorinated alkyl substances in air samples from Northwest Europe. J Environ Monit 9: 530–541

Barghi M, Shin ES, Kim JC, Choi SD, Chang YS (2017) Human exposure to HBCD and TBBPA via indoor dust in Korea: estimation of external exposure and body burden. Sci Total Environ 593–594:779–786

Bastiaensen M, Ait Bamai Y, Araki A, Van den Eede N, Kawai T, Tsuboi T, Kishi R, Covaci A (2019) Biomonitoring of organophosphate flame retardants and plasticizers in children: associations with house dust and housing characteristics in Japan. Environ Res 172:543–551

Batterman SA, Chernyak S, Jia C, Godwin C, Charles C (2009) Concentrations and emissions of polybrominated diphenyl ethers from U.S. houses and garages. Environ Sci Technol 43:2693–2700

BAUA (German Federal Institute for Occupational Safety and Health) (2023) Regulation on the restriction of diisocyanates. https://www.baua.de/EN/Topics/Work-design/Hazardous-substances/REACH-assessment-unit/Diisocyanates.html

Becker K, Seiwert M, Kaus S, Krause C, Schulz C, Seifert B (2002) German Environmental Survey 1998 (GerES III): pesticides and other pollutants in house dust. In: INDOOR AIR '02. Proceedings of the 9th International Conference on Indoor Air Quality and Climate. Monterey, USA

Becker K, Kaus S, Seifert M, Heidrich F, Rosskamp E, Schulz C, Schlüter C, Seifert B (2004) German Environmental Survey 1998, Report V: House dust. Substance levels in house dust from households of the population in Germany. WaBoLu-Hefte 05/2004. https://www.umweltbundesamt.de/sites/default/files/medien/publikation/short/k2786.pdf [in German]

Bekki K, Uchiyama S, Kunugita N (2018) Analysis of isocyanates in indoor dust. Anal Bioanal Chem 410:4247–4251

Bekki K, Uchiyama S, Kunugita N (2020) A novel analytical method for simple and low-cost detection of isocyanates in ambient air. Anal Bioanal Chem 412:103–111

Bennett DH, Moran RE, Wu XM, Tulve NS, Clifton MS, Colón M, Weathers W, Sjödin A, Jones R, Hertz-Picciotto I (2015) Polybrominated diphenyl ether (PBDE) concentrations and resulting exposure in homes in California: relationships among passive air, surface wipe and dust concentrations, and temporal variability. Indoor Air 25:220–229

Berardesca E, Zuberbier T, Sanchez Viera M, Marinovich M (2019) Review of the safety of octocrylene used as an ultraviolet filter in cosmetics. JEADV 33(Suppl. 7):25–33

van den Berg M, Denison MS, Birnbaum LS, Devito MJ, Fiedler H, Falandysz J, Rose M, Schrenk D, Safe S, Tohyama C, Tritscher A, Tysklind M, Peterson RE (2013) Polybrominated dibenzo-p-dioxins, dibenzofurans, and biphenyls: inclusion in the toxicity equivalency factor concept for dioxin-like compounds. Toxicol Sci 133:197–208

Berger-Preiss E, Preiß A, Sielaff K, Raabe M, Ilgen B, Levsen K (1997) The behaviour of pyrethroids indoors. A model study. Indoor Air 7:248–261

Berger-Preiss E, Koch W, Behnke W, Gerling S, Kock H, Elflein L, Appel KE (2004) In-flight spraying in aircrafts: determination of the exposure scenario. Int J Hyg Environ Health 207:419–430

Berger-Preiss E, Koch W, Gerling S, Kock H, Appel KE (2009) Use of biocidal products (insect sprays and electro-vaporizer) in indoor areas – exposure scenarios and exposure modeling. Int J Hyg Environ Health 212:505–518

Bergh C, Torgrip R, Emenius G, Ostman C (2011a) Organophosphate and phthalate esters in air and settled dust – a multi-location indoor study. Indoor Air 21:67–76

Bergh C, Magnus Åberg K, Svartengren M, Emenius G, Östman C (2011b) Organophosphate and phthalate esters in indoor air: a comparison between multi-storey buildings with high and low prevalence of sick building symptoms. J Environ Monit 13:2001–2009

Bergman Å, Rydén A, Law RJ, de Boer J, Covaci A, Alaee M, Birnbaum L, Petreas M, Rose M, Sakai S, van den Eede N, van der Veen I (2012) A novel abbreviation standard for organobromine, organochlorine and organophosphorus flame retardants and some characteristics of the chemicals. Environ Int 49:57–82

Bergvall C, Westerholm R (2007) Identification and determination of highly carcinogenic dibenzopyrene isomers in air particulate samples from a street canyon, a rooftop, and a subway station in Stockholm. Environ Sci Technol 41:731–737

Besis A, Avgenikou A, Pantelaki I, Serafeim E, Georgiadou E, Voutsa D, Samara C (2023) Hazardous organic pollutants in indoor dust from elementary schools and kindergartens in Greece: implications for children's health. Chemosphere 310:136750

Bethke R et al (2021) Indoor volatile organic compounds in the German Environmental Survey for Children and Adolescents (GerES, 2014–2017). Manuscript in preparation

BfR (2011) Use of parabens in cosmetic products Opinion No. 009/2011. https://www.bfr.bund.de/cm/343/verwendung_von_parabenen_in_kosmetischen_mitteln.pdf [in German]

BfR (German Federal Institute for Risk Assessment) (2014) BfR proposals for the health assessment of chlorate residues in foodstuffs. Opinion No. 028/2014 of the 12th May 2014. https://www.bfr.bund.de/cm/343/vorschlaege-des-bfr-zur-gesundheitlichen-bewertung-von-chloratrueckstaenden-in-lebensmitteln.pdf [in German]

Bhat VS, Durham JL, English JC (2014) Derivation of an oral reference dose (RfD) for the plasticizer, di-(2-propylheptyl) phthalate (Palatinol® 10-P). Regul Toxicol Pharmacol 70:65–74

Birnbaum LS, Staskal DF, Diliberto JJ (2003) Health effects of polybrominated dibenzo-p-dioxins (PBDDs) and dibenzofurans (PBDFs). Environ Int 29:855–860

Björklund JA, Sellström U, de Wit CA, Aune M, Lignell S, Darnerud PO (2012) Comparisons of polybrominated diphenyl ether and hexabromocyclododecane concentrations in dust collected with two sampling methods and matched breast milk samples. Indoor Air 22:279–288

Blanchard O, Glorennec P, Mercier F, Bonvallot N, Chevrier C, Ramalho O, Mandin C, Le Bot B (2014) Semivolatile organic compounds in indoor air and settled dust in 30 French dwellings. Environ Sci Technol 48:3959–3969

Blaszczyk E, Rogula-Kozowska W, Klejnowski K, Fulara I, Mielyska-Švach D (2017) Polycyclic aromatic hydrocarbons bound to outdoor and indoor airborne particles ($PM_{2.5}$) and their mutagenicity and carcinogenicity in Silesian kindergartens, Poland. Air Qual Atmos Health 10:389–400

Błędzka D, Gromadzińska J, Wąsowicz W (2014) Parabens. From environmental studies to human health. Environ Int 67:27–42

Blessing R (1997) Polychlorinated dioxins and furans in public buildings. Gefahrst Reinh Luft 57: 305–309. [in German]

Bohlin P, Jones KC, Tovalin H, Strandberg B (2008) Observations on persistent organic pollutants in indoor and outdoor air using passive polyurethane foam samplers. Atmos Environ 42:7234–7241

Bolte G, Heitmann D, Kiranoglu M, Schierl R, Diemer J, Körner W, Fromme H (2008) Exposure to environmental tobacco smoke in German restaurants, pubs and discotheques. J Expo Sci Environ Epidemiol 18:262–271

Bouvier G, Blanchard O, Momas I, Seta N (2006) Pesticide exposure of non-occupationally exposed subjects compared to some occupational exposure: a French pilot study. Sci Total Environ 366:74–91

Bräuner EV, Mayer P, Gunnarsen L, Vorkamp K, Raaschou-Nielsen O (2011) Occurrence of organochlorine pesticides in indoor dust. J Environ Monit 13:522–526

Bräuner EV, Andersen ZJ, Frederiksen M, Specht IO, Hougaard KS, Ebbehøj N, Bailey J, Giwercman A, Steenland K, Longnecker MP, Bonde JP (2016) Health effects of PCBs in residences and schools (HESPERUS): PCB – health cohort profile. Nat Sci Rep 6:24571

Brits M, de Boer J, Rohwer ER, De Vos J, Weiss JM, Brandsma SH (2020) Short-, medium-, and long-chain chlorinated paraffins in South African indoor dust and cat hair. Chemosphere 238: 124643

Brommer S, Harrad S (2015) Sources and human exposure implications of concentrations of organophosphate flame retardants in dust from UK cars, classrooms, living rooms, and offices. Environ Int 83:202–207

Brommer S, Harrad S, Van den Eede N, Covaci A (2012) Concentrations of organophosphate esters and brominated flame retardants in German indoor dust samples. J Environ Monit 14:2482–2487

Brown KW, Minegishi T, Cummiskey CC, Fragala MA, Hartman R, MacIntosh DL (2016) PCB remediation in schools: a review. Environ Sci Pollut Res 23:1986–1997

Bui TT, Giovanoulis G, Cousins AP, Magnér J, Cousins IT, deWit CA (2016) Human exposure, hazard and risk of alternative plasticizers to phthalate esters. Sci Total Environ 541:451–467

Canosa P, Rodriguez I, Rubi E, Cela R (2007) Determination of parabens and triclosan in indoor dust using matrix solid-phase dispersion and gas chromatography with tandem mass spectrometry. Anal Chem 79:1675–1681

Carlson LM, Angrish M, Shirke AV, Radke EG, Schulz B, Kraft A, Judson R, Patlewicz G, Blain R, Lin C, Vetter N, Lemeris C, Hartman P, Hubbard H, Arzuaga X, Davis A, Dishaw LV, Druwe IL, Hollinger H, Jones R, Kaiser JP, Lizarraga L, Noyes PD, Taylor M, Shapiro AJ, Williams AJ, Thayer KA (2022) Systematic evidence map for over one hundred and fifty per- and polyfluoroalkyl substances (PFAS). Environ Health Perspect 130:56001

Carlsson H, Nilsson U, Becker G, Östman C (1997) Organophosphate ester flame retardants and plasticizers in the indoor environment: analytical methodology and occurrence. Environ Sci Technol 31:2931–2936

Casey A, Bush B, Carpenter DO (2022) PCBs in indoor air and human blood in Pittsfield, Massachusetts. Chemosphere 293:133551

Castro D, Slezakova K, Delerue-Matos C, Alvim-Ferraz MC, Morais S, Pereira MC (2011) Polycyclic aromatic hydrocarbons in gas and particulate phases of indoor environments influenced by tobacco smoke: levels, phase distributions, and health risks. Atmos Environ 45: 1799–1808

Castro G, Rodríguez I, Ramil M, Cela R (2019) Assessment of gas chromatography time-of-flight mass spectrometry for the screening of semi-volatile compounds in indoor dust. Sci Total Environ 688:162–173

Catenza CJ, Farooq A, Shubear NS, Donkor KK (2021) A targeted review on fate, occurrence, risk and health implications of bisphenol analogues. Chemosphere 268:129273

Cattaneo A, Fermo P, Urso P, Perrone MG, Piazzalunga A, Tarlassi J, Carrer P, Cavallo DM (2016) Particulate-bound polycyclic aromatic hydrocarbon sources and determinants in residential homes. Environ Pollut 218:16–25

Cequier E, Ionas AC, Covaci A, Marcé RM, Becher G, Thomsen C (2014) Occurrence of a broad range of legacy and emerging flame retardants in indoor environments in Norway. Environ Sci Technol 48:6827–6835

Ceresana (2022) Plasticizers market report. https://www.ceresana.com/en/market-studies/chemicals/plasticizers/

CES (Centre Européen des Silicones) (2016) The socioeconomic impact of the silicones industry in Europe. London, UK. http://www.cyclosiloxanes.eu/

Charnley G (2008) Perchlorate: overview of risks and regulation. Food Chem Toxicol 46:2307–2315

Chen Y, Shen G, Huang Y, Zhang Y, Han Y, Wang R, Shen H, Su S, Lin N, Zhu D, Pei L, Zheng X, Wu J, Wang X, Liu W, Wong M, Tao S (2016a) Household air pollution and personal exposure risk of polycyclic aromatic hydrocarbons among rural residents in Shanxi, China. Indoor Air 26: 246–258

Chen D, Kannan J, Tan H, Zheng Z, Feng Y-L, Wu Y, Widelka M (2016b) Bisphenol analogues other than BPA: environmental occurrence, human exposure, and toxicity. A review. Environ Sci Technol 50:5438–5453

Chen Y, Du W, Shen G, Zhuo S, Zhu X, Shen H, Huang Y, Su S, Lin N, Pei L, Zheng X, Wu J, Duan Y, Wang X, Liu W, Wong M, Tao S (2017) Household air pollution and personal exposure to nitrated and oxygenated polycyclic aromatics (PAHs) in rural households: influence of household cooking energies. Indoor Air 27:169–178

Chen Y, Lv D, Li X, Zhu T (2018a) $PM_{2.5}$-bound phthalates in indoor and outdoor air in Beijing: seasonal distributions and human exposure via inhalation. Environ Poll 241:369–377

Chen J, Hartmann EM, Kline J, Van Den Wymelenberg K, Halden RU (2018b) Assessment of human exposure to triclocarban, triclosan and five parabens in U.S. indoor dust using dispersive

solid phase extraction followed by liquid chromatography tandem mass spectrometry. J Hazard Mater 360:623–630
Chen CY, Liu YH, Chieh CH, Chang WH (2021) Fast and environment-friendly GC-MS method for eleven organophosphorus flame retardants in indoor air, dust, and skin wipes. Toxics 9:350
Chen XC, Ward TJ, Ho KF, Sarkar C, Webster C (2022a) Characteristics and health risks of personal exposure to particle-bound PAHs for Hong Kong adult residents: from ambient pollution to indoor exposure. Indoor Air 32:e12956
Chen J, Ward TJ, Ho SSH, Ho KF (2022b) Occurrence and risk assessment of personal $PM_{2.5}$-bound phthalates exposure for adults in Hong Kong. Int J Environ Res Public Health 19:13425
Chi C, Xia M, Zhou C, Wang X, Weng M, Shen X (2017) Determination of 15 phthalate esters in air by gas-phase and particle-phase simultaneous sampling. J Environ Sci (China) 55:137–145
Choi H, Perera F, Pac A, Wang L, Flak E, Mroz E, Jacek R, Chai-Onn T, Jedrychowski W, Masters E, Camann D, Spengler J (2008) Estimating individual-level exposure to airborne polycyclic aromatic hydrocarbons throughout the gestational period based on personal, indoor, and outdoor monitoring. Environ Health Perspect 116:1509–1518
Christia C, Poma G, Harrad S, de Wit CA, Sjostrom Y, Leonards P, Lamoree M, Covaci A (2019) Occurrence of legacy and alternative plasticizers in indoor dust from various EU countries and implications for human exposure via dust ingestion and dermal absorption. Environ Res 171: 204–212
Christopoulou OD, Sakkas VA, Albanis TA (2012) Evaluation of matrix solid-phase dispersion extraction for the determination of polycyclic aromatic hydrocarbons in household dust with the aid of experimental design and response surface methodology. J Sep Sci 35:3554–3560
Chrustek A, Hołyńska-Iwan I, Dziembowska I, Bogusiewicz J, Wróblewski M, Cwynar A, Olszewska-Słonina D (2018) Current research on the safety of pyrethroids used as insecticides. Medicina (Kaunas) 54:61
Chung W-H, Lin J-S, Ding WH (2019) Dual-vortex-assisted matrix solid-phase dispersion coupled withisotope-dilution ultrahigh-performance liquid chromatography-highresolution mass spectrometry for the rapid determination of parabensin indoor dust samples. J Chromatogr A 1605: 460367
Chupeau Z, Bonvallot N, Mercier F, Le Bot B, Chevrier C, Glorennec P (2020) Organophosphorus flame retardants: a global review of indoor contamination and human exposure in Europe and epidemiological evidence. Int J Environ Res Public Health 17:6713
Civan MY, Kara UM (2016) Risk assessment of PBDEs and PAHs in house dust in Kocaeli, Turkey: levels and sources. Environ Sci Pollut Res Int 23:23369–23384
Class TJ (1991) Pyrethroids as household insecticides: analysis, indoor exposure and persistence. Fresenius J Anal Chem 340:446–453
Coelhan M, Hilger B (2014) Chlorinated paraffins in indoor dust samples: a review. Curr Organ Chem 18:2209–2217
Collins JF, Brown JP, Alexeeff GV, Salmon AG (1998) Potency equivalency factors for some polycyclic aromatic hydrocarbons and polycyclic aromatic hydrocarbon derivatives. Regul Toxicol Pharmacol 28:45–54
Conrad A, Brenske KR (2010) Evaluation of the data of the Children's Environment Survey of the German Federal Environment Agency. Personal communication, January 2010
CPSC (United States Consumer Product Safety Commission) (2018a) Toxicity review for bis (2-ethylhexyl)adipate (DEHA). https://www.cpsc.gov/s3fs-public/Toxicity%20Review%20of%20DEHA.pdf?TSiSSb20aUy68dV0qk1AllBUrlaPFSaE
CPSC (2018b) Toxicity review for 1,2-cyclohexanedicarboxylic acid, dinonyl ester, branched and linear (DINX). https://www.cpsc.gov/s3fs-public/Toxicity%20Review%20of%20DINX.pdf?n_tDo9yqCvnxEdDVuINE7tQba9lrg_XQ
CPSC (2018c) Toxicity review for di-2-ethylhexylterephthalate (DEHT). https://www.cpsc.gov/s3fs-public/Toxicity%20Review%20of%20DEHT.pdf?FObpuBBqgypVtw7gIEGMFXHN5H7vbeEz

CPSC (2018d) Toxicity review for trioctyltrimellitate (TOTM). https://www.cpsc.gov/s3fs-public/Toxicity%20Review%20of%20TOTM.pdf?Yjo0hEI05eJsEziyutApCzEobdUITWhX
CPSC (2018e) Toxicity review for acetyl tri-n-butylcitrate (ATBC). https://www.cpsc.gov/s3fs-public/Toxicity%20Review%20of%20ATBC.pdf?AsgeTCxYY0_3F.wMkaokmHmiD5LWh.zO
CPSC (2019a) Toxicity review for diisononyladipate (DINA). https://www.cpsc.gov/s3fs-public/ToxicityReviewforDiisononylAdipate062019.pdf?vM1E2MpKwInTRd11A7yyAUAZq8gAn2Xl
CPSC (2019b) Toxicity review for dibutyladipate (DBA) and di-isobutyl adipate (DiBA). https://www.cpsc.gov/s3fs-public/ToxicityReviewforDBAandDiBA062019.pdf?3w6prfRnPImRtgB5PbGvGD62fsV568dD
Crawford L, Halperin SA, Dzierlenga MW, Skidmore B, Linakis MW, Nakagawa S, Longnecker MP (2023) Systematic review and meta-analysis of epidemiologic data on vaccine response in relation to exposure to five principal perfluoroalkyl substances. Environ Int 172:107734
Cunha SC, Kalachova K, Pulkrabova J, Fernandes JO, Oliveira MBPP, Alves A, Hajslova J (2010) Polybrominated diphenyl ethers (PBDEs) contents in house and car dust of Portugal by pressurized liquid extraction (PLE) and gas chromatography–mass spectrometry (GC–MS). Chemosphere 78:1263–1271
D'Hollander W, Roosens L, Covaci A, Cornelis C, Reynders H, Campenhout KV, de Voogt P, Bervoets L (2010) Brominated flame retardants and perfluorinated compounds in indoor dust from homes and offices in Flanders, Belgium. Chemosphere 81:478–487
Dallongeville A, Costet N, Zmirou-Navier D, Le Bot B, Chevrier C, Deguen S, Annesi-Maesano I, Blanchard O (2016) Volatile and semi-volatile organic compounds of respiratory health relevance in French dwellings. Indoor Air 26:426–438
Daniels A et al (2021) Phthalates, substitutes for phthalates, flame retardants and aromatic musk compounds in house dust in the German Environmental Survey for Children and Adolescents (GerES, 2014–2017). Manuscript in preparation
Danish-EPA (Danish Environmental Protection Agency) (2013a) Polyaromatic hydrocarbons (PAH). Evaluation of health hazards and estimation of a quality criterion in soil. Copenhagen, Denmark. https://mst.dk/service/publikationer/publikationsarkiv/2013/dec/polyaromatic-hydrocarbons-pah/
Dan-EPA (2014) Siloxanes (D3, D4, D5, D6, HMDS). Evaluation of health hazards and proposal of a health-based quality criterion for ambient air. Environmental Project No. 1531. https://www2.mst.dk/udgiv/publications/2014/01/978-87-93026-85-8.pdf
Danish-EPA (2013b) Survey of parabens. Environmental Project No. 1474. Copenhagen, Denmark. https://www2.mst.dk/Udgiv/publications/2013/04/978–87–93026-02-5.pdf
David ORP (2020) A chemical history of polycyclic musks. Chemistry 26:7537–7555
De Gennaro G, Dambruoso PR, Di Gilio A, Di Palma V, Marzocca A, Tutino M (2016) Discontinuous and continuous indoor air quality monitoring in homes with fireplaces or wood stoves as heating system. Int J Environ Res Public Health 13:78
De la Torre A, Sanz P, Navarro I, de los Ángeles Martínez M (2020) Investigating the presence of emerging and legacy POPs in European domestic air. Sci Total Environ 746:141348
De Wit CA, Björklund JA, Thuresson K (2012) Tri-decabrominated diphenyl ethers and hexabromocyclododecane in indoor air and dust from Stockholm microenvironments 2: Indoor sources and human exposure. Environ Int 39:141–147
DECOS (Dutch Expert Committee on Occupational Standards – Health Council of the Netherlands) (2006) BaP and PAH from coal-derived sources. No. 2006/01OSH, The Hague, Netherlands. https://www.ser.nl/api/Mfiles/DownloadFirstDocument?Id=73d8a46e-cbcf-4efb-aeb5-506d29c79cc4
Deen L, Hougaard KS, Clark A, Meyer HW, Frederiksen M, Gunnarsen L, Andersen HV, Hougaard T, Petersen KKU, Ebbehøj NE, Bonde JP, Tøttenborg SS (2022) Cancer risk following residential exposure to airborne polychlorinated biphenyls: a Danish register-based cohort study. Environ Health Perspect 130:107003

DeLay K, Lin EZ, Koelmel JP, Bornman R, Obida M, Chevrier J, Godri Pollitt KJ (2022) Personal air pollutant exposure monitoring in South African children in the VHEMBE birth cohort. Environ Int 170:107524

Delgado-Saborit JM, Stark C, Harrison RM (2011) Carcinogenic potential, levels and sources of polycyclic aromatic hydrocarbon mixtures in indoor and outdoor environments and their implications for air quality standards. Environ Int 37:383–392

DellaValle CT, Wheeler DC, Deziel NC, De Roos AJ, Cerhan JR, Cozen W, Severson RK, Flory AR, Locke SJ, Colt JS, Hartge P, Ward MH (2013) Environmental determinants of polychlorinated biphenyl concentrations in residential carpet dust. Environ Sci Technol 47: 10405–10414

Demirtepe H, Melymuk L, Diamond ML, Bajard L, Vojta Š, Prokeš R, Sáňka O, Klánová J, Palkovičová Murínová Ľ, Richterová D, Rašplová V, Trnovec T (2019) Linking past uses of legacy SVOCs with today's indoor levels and human exposure. Environ Int 127:653–663

Demirtepe H, Melymuk L, Codling G, Murínová ĽP, Richterová D, Rašplová V, Trnovec T, Klánová J (2021) Targeted and suspect screening of plasticizers in house dust to assess cumulative human exposure risk. Sci Total Environ 781:146667

Deng WJ, Zheng HL, Tsui AK, Chen XW (2016) Measurement and health risk assessment of $PM_{2.5}$, flame retardants, carbonyls and black carbon in indoor and outdoor air in kindergartens in Hong Kong. Environ Int 96:65–74

Deng WJ, Li N, Wu R, Richard WKS, Wang Z, Ho W (2018) Phosphorus flame retardants and Bisphenol A in indoor dust and $PM_{2.5}$ in kindergartens and primary schools in Hong Kong. Environ Pollut 235:365–371

Deziel NC, Ward MH, Bell EM, Whitehead TP, Gunier RB, Friesen MC, Nuckols JR (2013) Temporal variability of pesticide concentrations in homes and implications for attenuation bias in epidemiologic studies. Environ Health Perspect 121:565–571

Deziel NC, Nuckols JR, Jones RR, Graubard BI, De Roose AJ, Pronk A, Gourley C, Colt JS, Ward MH (2017) Comparison of industrial emissions and carpet dust concentrations of polychlorinated dibenzo-p-dioxins and polychlorinated dibenzofurans in a multi-center U.-S. study. Sci Total Environ 580:1276–1286

Dirtu AC, Ali N, Van den Eede N, Neels H, Covaci A (2012) Country specific comparison for profile of chlorinated, brominated and phosphate organic contaminants in indoor dust. Case study for Eastern Romania, 2010. Environ Int 49:1–8

Dodson RE, Perovich LJ, Covaci A, Van den Eede N, Ionas AC, Dirtu AC, Brody JG, Rudel RA (2012) After the PBDE phase-out: a broad suite of flame retardants in repeat house dust samples from California. Environ Sci Technol 46:13056–13066

Dodson RE, Camann DE, Morello-Frosch R, Brody JG, Rudel RA (2015) Semivolatile organic compounds in homes: strategies for efficient and systematic exposure measurement based on empirical and theoretical factors. Environ Sci Technol 49:113–122

Dodson RE, Bessonneau V, Udesky JO, Nishioka M, McCauley M, Rudel RA (2019) Passive indoor air sampling for consumer product chemicals: a field evaluation study. J Expo Sci Environ Epidemiol 29:95–108

Domingo JL (2004) Polychlorinated naphthalenes in animal aquatic species and human exposure through the diet: a review. J Chromatogr A 1054:327–334

Dou M, Wang L (2023) A review on organophosphate esters: physiochemical properties, applications, and toxicities as well as occurrence and human exposure in dust environment. J Environ Manag 325(Pt B):116601

Du B, Zhang Y, Chen H, Shen M, Zhou W, Zeng L (2019) Development and validation of a liquid chromatography-tandem mass spectrometry method for the simultaneous determination of 17 traditional and emerging aryl organophosphate esters in indoor dust. J Chromatogr A 1603:199–207

Dubocq F, Kärrman A, Gustavsson J, Wang T (2021) Comprehensive chemical characterization of indoor dust by target, suspect screening and nontarget analysis using LC-HRMS and GC-HRMS. Environ Pollut 276:116701

Duenas-Mas MJ, Ballesteros-Gómez A, Rubio S (2019) Emerging bisphenol a replacements (colour developers) in indoor dust from Spain. Emerg Contam 5:168–172

Dvoršćak M, Jakovljević I, Jagić K, Tariba Lovaković B, Klinčić D (2022) Polybrominated diphenyl ethers and polycyclic aromatic hydrocarbons in dust from different indoor environments in Zagreb, Croatia: levels and human exposure assessment. Indoor Air 32:e13145

Eales J, Bethel A, Galloway T, Hopkinson P, Morrissey K, Short RE, Garside R (2022) Human health impacts of exposure to phthalate plasticizers: an overview of reviews. Environ Int 158: 106903

ECCC/HC (Environment and Climate Change Canada, Health Canada) (2019a) Screening assessment. Certain organic flame retardants substance grouping. 1,4:7,10-Dimethanodibenzo[a,e] cyclooctene,1,2,3,4,7,8,9,10,13,13,14,14-dodecachloro-1,4,4a,5,6,6a,7,10,10a,11,12,12a-dodecahydro-dechlorane plus (DP). https://www.canada.ca/en/environment-climate-change/services/evaluating-existing-substances/certain-organic-flame-retardants-grouping-risk-management-approach-dimethanodibenzo-cyclooctene-dodecachloro-dodecahydro-dechlorane-plus-dp.html

ECCC/HC (2019b) Draft screening assessment siloxanes group. https://www.canada.ca/en/environment-climate-change/services/evaluating-existing-substances/draft-screening-assessment-siloxanes-group.html

ECHA (2016a) Analysis of the most appropriate risk management option (RMOA). 1,2-CYCLOHEXANEDICARBOXYLIC ACID, DIISONONYL ESTER (DINCH). https://echa.europa.eu/documents/10162/9f62e492-4028-5700-e5ec-019c396b13bf

ECHA (2016b) Risk management options analysis. Di ethyl hexyl terephthalate (DEHTP). https://echa.europa.eu/documents/10162/ad151e2f-b765-c69c-ef8e-e9d996e50651

ECHA (2018a) Note on reference dose-response relationship for the carcinogenicity of pitch, coal tar, high temperature and on PBT and vPvB properties. https://echa.europa.eu/documents/10162/13637/ctpht_rac_note_en.pdf/a184ee42-0642-7454-2d18-63324688e13d

ECHA (2018b) Screening report – an assessment of wether the use of TCEP, TCPP and TDCP in articles should be restricted. Helsinki, Finland. https://echa.europa.eu/documents/10162/13641/screening_report_tcep_tcpp_td-cp_en.pdf/e0960aa7-f703-499c-24ff-fba627060698

ECHA (2018c) Draft background document for 1,6,7,8,9,14,15,16,17,17,18,18-dodecachloropentacyclo[12.2.1.16,9.02,13.05,10]octadeca-7,15-diene ("Dechlorane Plus"™). https://echa.europa.eu/registry-of-restriction-intentions/-/dislist/details/0b0236e184a168c4

ECHA (2019) Scientific report for evaluation of limit values for diisocyanates at the workplace. https://echa.europa.eu/documents/10162/db15bfdf-eec8-c10a-67c4-f65166c5110a

ECHA (2020) Member State Committee support document for identification of butyl 4-hydroxybenzoate as a substance of very high concern because of its endocrine disrupting properties (article 57(f) – human health) properties. https://echa.europa.eu/documents/10162/9c45b069-0fbf-c247-08c0-d0567c58a43e

ECHA (2021) Dechlorane Plus. Revised draft risk profile. 9 April 2021. https://assets.publishing.service.gov.uk/government/uploads/system/uploads/attachment_data/file/979985/dechlorane-plus-draft-risk-profile.pdf

EFSA (2005a) Opinion of the Scientific Panel on contaminants in the food chain [CONTAM] related to gamma-HCH and other hexachlorocyclohexanes as undesirable substances in animal feed. EFSA J 250:1–39. https://www.efsa.europa.eu/de/efsajournal/pub/250

EFSA (2005b) Opinion of the Scientific Panel on contaminants in the food chain [CONTAM] related to the presence of non dioxin-like polychlorinated biphenyls (PCB) in feed and food. EFSA J 284:1–137. https://efsa.onlinelibrary.wiley.com/doi/abs/10.2903/j.efsa.2005.284

EFSA (2006a) Opinion of the Scientific Panel on contaminants in the food chain [CONTAM] related to Hexachlorobenzene as undesirable substance in animal feed. EFSA J 402:1–49. https://www.efsa.europa.eu/en/efsajournal/pub/402

EFSA (2006b) Opinion of the Scientific Panel on contaminants in the food chain [CONTAM] related to DDT as an undesirable substance in animal feed. EFSA J 433:1–69. https://www.efsa.europa.eu/en/efsajournal/pub/433

EFSA (2008) Polycyclic aromatic hydrocarbons in food. Scientific opinion of the Panel on Contaminants in the Food Chain. The EFSA J 724:1–114

EFSA (2009) Scientific Opinion on the safety evaluation of the substance, alkyl (C_{10}–C_{21}) sulphonic acid, esters with phenol, CAS No. 91082-17-6, for use in food contact materials. EFSA J 7:1398

EFSA (2011) Scientific opinion on tetrabromobisphenol A (TBBPA) and its derivatives in food. EFSA J 9(12):2477. https://efsa.onlinelibrary.wiley.com/doi/epdf/10.2903/j.efsa.2011.2477

EFSA (2012a) Scientific opinion on brominated flame retardants (BFRs) in food: brominated phenols and their derivatives. EFSA Panel on Contaminants in the Food Chain (CONTAM). EFSA J 10:2634

EFSA (2012b) Scientific opinion on emerging and novel brominated flame retardants (BFRs) in Food. EFSA Panel on Contaminants in the Food Chain (CONTAM). EFSA J 10:2908

EFSA (2015a) Scientific statement on the health-based guidance value for dioxins and dioxin-like PCBs. EFSA J 13:4124. https://www.efsa.europa.eu/de/efsajournal/pub/4124

EFSA (2015b) Scientific opinion on the risks to public health related to the presence of bisphenol A (BPA) in foodstuffs. EFSA J 13:3978. https://www.efsa.europa.eu/en/efsajournal/pub/3978

EFSA (2017) Bisphenol A (BPA) hazard assessment protocol. EFSA supporting publication 2017: EN-1354. https://www.efsa.europa.eu/en/supporting/pub/en-1354

EFSA (2018a) Risk to human health related to the presence of perfuorooctane sulfonic acid and perfuorooctanoic acid in food. EFSA J 16:5194. https://www.efsa.europa.eu/en/efsajournal/pub/5194

EFSA (2018b) Risk for animal and human health related to the presence of dioxins and dioxin-like PCBs in feed and food. EFSA J 16:5333. http://www.efsa.europa.eu/en/efsajournal/pub/5333

EFSA (2020a) Risk to human health related to the presence of perfluoroalkyl substances in food. EFSA J 18:6223. https://www.efsa.europa.eu/en/efsajournal/pub/6223

EFSA (2020b) Assessment of new information on Bisphenol S (BPS) submitted in response to the Decision. Technical report. EFSA Supporting publication 2020: EN-1844. https://www.efsa.europa.eu/en/supporting/pub/en-1844

EFSA (2020c) Risk assessment of chlorinated paraffins in feed and food. EFSA J 18:5991. https://www.efsa.europa.eu/en/efsajournal/pub/5991

EFSA (2021) Update of the risk assessment of hexabromocyclododecanes (HBCDDs) in food. EFSA J 19(3):6421. https://efsa.onlinelibrary.wiley.com/doi/epdf/10.2903/j.efsa.2021.6421

EFSA (European Food Safety Authority) (2004) Opinion of the Scientific Panel on Contaminants in the Food Chain on a request from the Commission to assess the health risks to consumers associated with exposure to organotins in foodstuffs. EFSA J 102:1–119. https://efsa.onlinelibrary.wiley.com/doi/pdf/10.2903/j.efsa.2004.102

Eis D, Helm D, Laußmann D, Mühlinghaus T, Dietel A, Jordan L, Birkner N, Thierfelder W, Traenckner-Probst I, Worm M (2005) Berlin study on environment-related diseases. https://www.apug.de/archiv/pdf/Berichtsband_Berliner-Studie.pdf. [in German]

Eriksson U, Kärrman A (2015) World-wide indoor exposure to polyfluoroalkyl phosphate esters (Paps) and other PFASs in household dust. Environ Sci Technol 49:14503–14511

Escrivá L, Hanberg A, Zilliacus J, Beronius A (2019) Assessment of the endocrine disrupting properties of Bisphenol AF according to the EU criteria and ECHA/EFSA guidance. EFSA J 17: e170914

Esplugas R, Rovira J, Mari M, Fernández-Arribas J, Eljarrat E, Domingo JL, Schuhmacher M (2022) Emerging and legacy flame retardants in indoor air and dust samples of Tarragona Province (Catalonia, Spain). Sci Total Environ 806(Pt 1):150494

Falandysz J, Fernandes A, Gregoraszczuk E, Rose M (2014) The toxicological effects of halogenated naphthalenes: a review of aryl hydrocarbon receptor-mediated (dioxin-like) relative potency factors. J Environ Sci Health, Part C 32:239–272

Fan X, Kubwabo C, Rasmussen P, Jones-Otazo H (2010) Simultaneous quantitation of parabens, triclosan, and methyl triclosan in indoor house dust using solid phase extraction and gas chromatography-mass spectrometry. J Environ Monit 12:1891–1897

Fan X, Kubwabo C, Rasmussen PE, Wu F (2016) Non-PBDE halogenated flame retardants in Canadian indoor house dust: sampling, analysis, and occurrence. Environ Sci Pollut Res Int 23: 7998–8007

Faroon O, Ruiz P (2016) Polychlorinated biphenyls: new evidence from the last decade. Toxicol Ind Health 32:1825–1847

Fernandes AR, Falandysz J (2021) Polybrominated dibenzo-p-dioxins and furans (PBDD/Fs): contamination in food, humans and dietary exposure. Sci Total Environ 761:143191

Fernandes AR, Kilanowicz A, Stragierowicz J, Klimczak M, Falandysz J (2022) The toxicological profile of polychlorinated naphthalenes (PCNs). Sci Total Environ 837:155764

Figueiredo DM, Duyzer J, Huss A, Krop EJM, Gerritsen-Ebben MG, Gooijer Y, Vermeulen RCH (2021) Spatio-temporal variation of outdoor and indoor pesticide air concentrations in homes near agricultural fields. Atmos Environ 262:118612

Fitzgerald EF, Shrestha S, Palmer PM, Wilson LR, Belanger EE, Gomez MI, Cayo MR, Hwang SA (2011) Polychlorinated biphenyls (PCBs) in indoor air and in serum among older residents of upper Hudson River communities. Chemosphere 85:225–231

Franzblau A, Zwica L, Knutson K, Chen Q, Lee SY, Hong B, Adriaems P, Demond A, Garabrant D, Gillespie B, Lepkowski J, Luksemburg W, Maier M, Towey T (2009) An investigation of homes with high concentrations of PCDDs, PCDFs, and/or dioxin-like PCBs in house dust. J Occup Environ Hyg 6:188–199

Frederiksen M, Meyer HW, Ebbehøj NE, Gunnarsen L (2012) Polychlorinated biphenyls (PCBs) in indoor air originating from sealants in contaminated and uncontaminated apartments within the same housing estate. Chemosphere 89:473–479

Fridén UE, McLachlan MS, Berger U (2011) Chlorinated paraffins in indoor air and dust: concentrations, congener patterns, and human exposure. Environ Int 37:1169–1174

Fromme H, Baldauf AM, Klautke O, Piloty M, Bohrer L (1996) Polychlorinated biphenyls (PCBs) in joint sealants of buildings – inventory for Berlin and new indoor sources. Gesundheitswesen 58:666–672. [in German]

Fromme H, Oddoy A, Lahrz T, Krause M, Piloty M (1998) Polycyclic aromatic hydrocarbons (PAH) and diesel engine emission (elemental carbon) inside a car and a subway train. Sci Total Environ 217:165–173

Fromme H, Lahrz T, Piloty M, Gebhardt H, Oddoy A, Rüden H (2004a) Polycyclic aromatic hydrocarbons inside and outside of apartments in an urban area. Sci Total Environ 326:143–149

Fromme H, Lahrz T, Piloty M, Gebhart H, Oddoy A, Rüden H (2004b) Occurrence of phthalates and musk fragrances in indoor air and dust from apartments and kindergartens in Berlin (Germany). Indoor Air 14:188–195

Fromme H, Lahrz T, Piloty M, Oddoy A (2005a) Polycyclic aromatic hydrocarbons in the indoor air of apartments, kindergarten and schools. Umweltmed Forsch Prax 10:35–41. [in German]

Fromme H, Mattulat A, Lahrz T, Rüden H (2005b) Occurrence of organotin compounds in house dust in Berlin (Germany). Chemosphere 58:1377–1383

Fromme H, Jahr D, Roscher E (2005c) Organotin compounds in house dust of Bavarian apartments. Poster presented on the 155. Scientific ÖGD Congress, Bamberg, 14.–16.04.2005. [in German]

Fromme H, Heitmann D, Dietrich S, Schierl R, Körner W, Kiranoglu M, Zapf A, Twardella D (2008a) Air quality in schools – classroom levels of carbon dioxide (CO_2), volatile organic compounds (VOC), aldehydes, endotoxins and cat allergen. Gesundheitswesen 70:88–97. [in German]

Fromme H, Nitschke L, Kiranoglu M, Albrecht M, Völkel W (2008b) Perfluorinated substances in house dust in Bavaria, Germany. Organohalogen Comp 70:1048–1050

Fromme H, Körner W, Shahin N, Wanner A, Albrecht M, Boehmer S, Parlar H, Mayer R, Liebl B, Bolte G (2009a) Human exposure to polybrominated diphenyl ethers (PBDE), as evidenced by data from duplicate diet study, indoor air, house dust, and biomonitoring in Germany. Environ Int 35:1125–1135

Fromme H, Dietrich S, Heitmann D, Dressel H, Diemer J, Schulz T, Jörres RA, Berlin K, Völkel W (2009b) Indoor air contamination during a waterpipe (narghile) smoking session. Food Chem Toxicol 47:1636–1641

Fromme H, Körner W, Gruber L, Heitmann D, Schlummer M, Völkel W, Bolte G (2010) Exposure of the population to phthalates. Gefahrst Reinh Luft 70:1–5. [in German]

Fromme H, Gruber L, Schuster R, Schlummer M, Kiranoglu M, Bolte G, Völkel W (2013a) Phthalate and di-(2-ethylhexyl) adipate (DEHA) intake by German infants based on the results of a duplicate diet study and biomonitoring data (INES 2). Food Chem Toxicol 53:272–280

Fromme H, Lahrz T, Kraft M, Fembacher L, Dietrich S, Sievering S, Burghardt R, Schuster R, Bolte G, Völkel W (2013b) Phthalates in German daycare centers: occurrence in air and dust and the excretion of their metabolites by children (LUPE 3). Environ Int 61:64–72

Fromme H, Hilger B, Kopp E, Miserok M, Völkel W (2014a) Polybrominated diphenyl ethers (PBDEs), hexabromocyclododecane (HBCD) and "novel" brominated flame retardants in house dust in Germany. Environ Int 64:61–68

Fromme H, Lahrz T, Kraft M, Fembacher L, Mach C, Dietrich S, Burkardt R, Völkel W, Göen T (2014b) Organophosphate flame retardants and plasticizers in the air and dust in German daycare centers and human biomonitoring in visiting children (LUPE 3). Environ Int 71:158–163

Fromme H, Lahrz T, Burkhardt R, Fembacher L, Kraft M, Sievering S, Kadler D, Dietrich S (2016a) Air quality in daycare centers – exposure to carbon dioxide (CO_2), volatile organic compounds (VOCs), and fine dusts – LUPE 3. Gefahrst Reinh Luft 76:55–61. [in German]

Fromme H, Schütze A, Lahrz T, Kraft M, Fembacher L, Siewering S, Burkardt R, Dietrich S, Koch HM, Völkel W (2016b) Non-phthalate plasticizers in German daycare centers and human biomonitoring of DINCH metabolites in children attending the centers (LUPE 3). Int J Hyg Environ Health 219:33–39

Fromme H, Becher G, Hilger B, Völkel W (2016c) Brominated flame retardants – exposure and risk assessment for the general population. Int J Hyg Environ Health 219:1–23

Fromme H, Schwarzbauer J, Lahrz T, Kraft M, Fembacher L (2017) Alkylsulfonic acid phenylesters (ASEs, Mesamoll®) in dust samples ofGerman residences and daycare centers (LUPE 3). Int J Hyg Environ Health 220:440–444

Fromme H, Mi W, Lahrz T, Kraft M, Aschenbrenner B, Bruessow B, Ebinghaus R, Xie Z, Fembacher L (2018) Occurrence of carbazoles in dust and air samples from different locations in Germany. Sci Total Environ 610–611:412–418

Fromme H, Witte M, Fembacher L, Gruber L, Hagl T, Smolic S, Fiedler D, Sysoltseva M, Schober W (2019) Siloxane in baking moulds, emission to indoor air and migration to food during baking with an electric oven. Environ Int 126:145–152

Fromme H, Achten C, Gerull F, Gessner A, Papavlassopoulos H, Kraft M, Röhl C, Burghardt R, Sysoltseva M (2023) Polycyclic aromatic hydrocarbons including dibenzopyrenes in indoor air samples from schools and residences in Germany. Submitted.

Gabrio T, Piechotowski I, Wallenhorst T, Klett M, Cott L, Friebel P, Link B, Schwenk M (2000) PCB-blood levels in teachers working in PCB-contaminated schools. Chemosphere 40:1055–1062

Gan Z, Pi L, Li Y, Hu W, Su S, Qin X, Ding S, Sun H (2015) Occurrence and exposure evaluation of perchlorate in indoor dust and diverse food from Chengdu, China. Sci Total Environ 536: 288–294

Gao W, Cao D, Wang Y, Wu J, Wang Y, Wang Y, Jiang G (2018) External exposure to short- and medium-chain chlorinated paraffins for the general population in Beijing, China. Environ Sci Technol 52:32–39

Gao P, Deng F, Chen W-S, Zhong Y-J, Cai X-L, Ma W-M, Hu J, Feng S-R (2022) Health risk assessment of inhalation exposure to airborne particle-bound nitrated polycyclic aromatic hydrocarbons in urban and suburban areas of South China. Int J Environ Res Public Health 19:15536

Gaspar FW, Castorina R, Maddalena LR, Nishioka MG, McKone TE, Bradman A (2014) Phthalate exposure and risk assessment in California child care facilities. Environ Sci Technol 48:7593–7601

Gautam SK, Suresh R, Sharma VP, Sehgal M (2013) Indoor air quality in the rural India. Manag Environ Qual Int J 24:244–255

Geens T, Roosens L, Neels H, Covaci A (2009) Assessment of human exposure to bisphenol-A, triclosan and tetrabromobisphenol – a through indoor dust intake in Belgium. Chemosphere 76: 755–760

Geiss O, Tirendi S, Barrero-Moreno J, Kotzias D (2009) Investigation of volatile organic compounds and phthalates present in the cabin air of used private cars. Environ Int 35:1188–1195

Gillett RW, Keywood MD, Galbally IE (2010) Indoor project – Persistent organic pollutants and metals. Report prepared for Commonwealth of Australia, Canberra

Giovanoulis G, Bui T, Xu F, Papadopoulou E, Padilla-Sanchez JA, Covaci A, Haug LS, Cousins AP, Magnér J, Cousins IT, de Wit CA (2018) Multi-pathway human exposure assessment of phthalate esters and DINCH. Environ Int 112:115–126

Giovanoulis G, Nguyen MA, Arwidsson M, Langer S, Vestergren R, Lagerqvist A (2019) Reduction of hazardous chemicals in Swedish preschool dust through article substitution actions. Environ Int 130:104921

Giulivo M, Lopez de Alda M, Capri E, Barceló D (2016) Human exposure to endocrine disrupting compounds: their role in reproductive systems, metabolic syndrome and breast cancer. A review. Environ Res 151:251–264

Glarborg P, Jensen AD, Johnsson JE (2003) Fuel nitrogen conversion in solid fuel fired systems. Prog Energy Combust Sci 29:89–113

Glorennec P, Serrano T, Fravallo M, Warembourg C, Monfort C, Cordier S, Viel JF, Le Gléau F, Le Bot B, Chevrier C (2017) Determinants of children's exposure to pyrethroid insecticides in western France. Environ Int 104:76–82

Glüge J, Schinkel L, Hungerbühler K, Cariou R, Bogdal C (2018) Environmental risks of medium-chain chlorinated paraffins (MCCPs): a review. Environ Sci Technol 52:6743–6760

Gou YY, Que DE, Chuang CY, Chao HR, Shy CG, Hsu YC, Lin CW, Chuang KP, Tsai CC, Tayo LL (2016) Dust levels of polybrominated diphenyl ethers (PBDEs) and polybrominated dibenzo-p-dioxins/furans (PBDD/Fs) in the Taiwanese elementary school classrooms: assessment of the risk to school-age children. Sci Total Environ 572:734–741

Greenwald R, Bergin MH, Yip F, Boehmer T, Kewada P, Shafer MM, Schauer JJ, Sarnat JA (2014) On-roadway in-cabin exposure to particulate matter: measurement results using both continuous and time-integrated sampling approaches. Aerosol Sci Technol 48:664–675

Guan Q, Tan H, Yang L, Liu X, Fiedler H, Li X, Chen D (2019) Isopropylated and tert-butylated triarylphosphate isomers in house dust from South China and Midwestern United States. Sci Total Environ 686:1113–1119

Guo J, Zhou Y, Cui J, Zhang B, Zhang J (2019) Assessment of volatile methylsiloxanes in environmental matrices and human plasma. Sci Total Environ 668:1175–1182

Guo J, Zhou Y, Wang Y, Zhang B, Zhang J (2021) Assessment of internal exposure to methylsiloxanes in children and associated non-dietary exposure risk. Environ Int 154:106672

Gustafson P, Östman C, Sällsten G (2008) Indoor levels of polycyclic aromatic hydrocarbons in homes with or without wood burning for heating. Environ Sci Technol 42:5074–5080

Gustafsson Å, Bergman Å, Weiss JM (2022) Estimated daily intake of per- and polyfluoroalkyl substances related to different particle size fractions of house dust. Chemosphere 303:135061

Hagenau W (2001) Organotin compounds in PVC – floor coverings and in house dust. Diploma thesis, University of Applied Sciences Weihenstephan. [in German]

Hajizadeh Y, Pourzamani H, Ebrahimpour K, Chavoshani A, Rahimi B (2021) Monitoring of paraben compounds in indoor and outdoor air of a populated city. Atmos Pollut Res 12:43–49

Hall SM, Patton S, Petreas M, Zhang S, Phillips AL, Hoffman K, Stapleton HM (2020) Per- and polyfluoroalkyl substances in dust collected from residential homes and fire stations in North America. Environ Sci Technol 54:14558–14567

Halsall CJ, Maher BA, Karloukovski VV, Shah P, Watkins SJ (2008) A novel approach to investigating indoor/outdoor pollution links: combined magnetic and PAH measurements. Atmos Environ 42:8902–8909

Hammel SC, Hoffman K, Lorenzo AM, Chen A, Phillips AL, Butt CM, Sosa JA, Webster TF, Stapleton HM (2017) Associations between flame retardant applications in furniture foam, house dust levels, and residents' serum levels. Environ Int 107:181–189

Hammel SC, Levasseur JL, Hoffman K, Phillips AL, Lorenzo AM, Calafat AM, Webster TF, Stapleton HM (2019) Children's exposure to phthalates and non-phthalate plasticizers in the home: the TESIE study. Environ Int 132:105061

Harrad S, Wijesekera R, Hunter S, Halliwell C, Baker R (2004) Preliminary assessment of U.-K. human dietary and inhalation exposure to polybrominated diphenyl ethers. Environ Sci Technol 38:2345–2350

Harrad S, Hazrati S, Ibarra C (2006) Concentrations of polychlorinated biphenyls in indoor air and polybrominated diphenyl ethers in indoor air and dust in Birmingham, United Kingdom: implications for human exposure. Environ Sci Technol 40:4633–4638

Harrad S, Ibarra C, Diamond M, Melymuk L, Robson M, Douwes J, Roosens L, Dirtu AC, Covaci A (2008a) Polybrominated diphenyl ethers in domestic indoor dust from Canada, New Zealand, United Kingdom and United States. Environ Int 34:232–238

Harrad S, Ibarra C, Abdallah MAE, Boon R, Neels H, Covaci A (2008b) Concentrations of brominated flame retardants in dust from United Kingdom cars, homes, and offices: causes of variability and implications for human exposure. Environ Int 34:1170–1175

Harrad S, Ibarra C, Robson M, Melymuk L, Zhang X, Diamond M, Douwes J (2009) Polychlorinated biphenyls in domestic dust from Canada, New Zealand, United Kingdom and United States: implications for human exposure. Chemospher 76:232–238

Harrad S, Goosey E, Desborough J, Abdallah MA, Roosens L, Covaci A (2010) Dust from U.K. primary school classrooms and daycare centers: the significance of dust as a pathway of exposure of young U.K. children to brominated flame retardants and polychlorinated biphenyls. Environ Sci Technol 44:4198–4202

Haumann T, Thumulla J (2002) Semi volatile organochemicals in indoor environment – chlorinated phosphorus and organotin compounds in material and house dust. Proc Indoor Air 2002:865–870

HC (Health Canada) (2011a) Screening assessment for the challenge hexanedioic acid, bis (2-ethylhexyl) ester (DEHA). Ottawa, Canada. https://www.ec.gc.ca/ese-ees/39958D25-E900-488F-9190-F776660D4A0A/Batch%2011_103-23-1_EN.pdf

HC (2011b) Screening assessment report on hexabromocyclododecane. http://www.ec.gc.ca/ese-ees/default.asp?lang=En&n=7882C148-1

HC (2015) Proposed approach for cumulative risk assessment of certain phthalates under the chemicals management plan. http://www.ec.gc.ca/ese-ees/default.asp?lang=En&n=723C9007-1

HC (2016) Guidelines for Canadian Drinking Water Quality: Guideline Technical Document – Benzo[a]pyrene. Ottawa, Ontario, Canada. https://www.canada.ca/content/dam/phac-aspc/documents/services/publications/healthy-living/guidelines-canadian-drinking-water-quality-guideline-technical-document-benzo-alpha-pyrene/pub-eng.pdf

He CT, Zheng J, Qiao L, Chen SJ, Yang JZ, Yuan JG, Yang ZY, Mai BX (2015) Occurrence of organophosphorus flame retardants in indoor dust in multiple microenvironments of southern China and implications for human exposure. Chemosphere 133:47–52

He R, Li Y, Xiang P, Li C, Zhou C, Zhang S, Cui X, Ma LQ (2016) Organophosphorus flame retardants and phthalate esters in indoor dust from different microenvironments: Bioaccessibility and risk assessment. Chemosphere 150:528–535

He C, Brandsma SH, Jiang H, O'Brien JW, van Mourik LM, Banks AP, Wang X, Tha PK, Mueller JF (2019) Chlorinated paraffins in indoor dust from Australia: levels, congener patterns and preliminary assessment of human exposure. Sci Total Environ 682:318–323

Heinzow B, Mohr S, Ostendorp G, Kerst M, Körner W (2007) PCB and dioxin-like PCB in indoor air of public buildings contaminated with different PCB sources – deriving toxicity equivalent concentrations from standard PCB congeners. Chemosphere 67:1746–1753

Herrick RF, Stewart JH, Allen JG (2016) Review of PCBs in US schools: a brief history, an estimate of the number of impacted schools, and an approach for evaluating indoor air samples. Environ Sci Pollut Res 23:1975–1985

Heudorf U (1999) Indoor contamination with polycyclic aromatic hydrocarbons from PAH-containing parquet adhesives. Status report on the assessment and handling of this "new" contaminant in the indoor environment. Gesundheitswesen 61:567–572. [in German]

Heudorf U, Salzmann N, Angerer J, Wittsiepe J (1996) Biomonitoring for PCDD/Fs and PCBs in the case of highly elevated indoor air pollution levels. Umweltmed Forsch Prax 1:6–12. [in German]

Hilger B, Fromme H, Völkel W, Coelhan M (2013) Occurrence of chlorinated paraffins in house dust samples from Bavaria, Germany. Environ Pollut 175:16–21

Hippelein M (2004) Background concentrations of individual and total volatile organic compounds in residential indoor air of Schleswig-Holstein, Germany. J Environ Monit 6:745–752

Ho K-F, Chang C-C, Tian L, Chan C-S, Bandowe BAM, Lui K-H, Lee K-Y, Chuang K-J, Liu C-Y, Ning Z, Chuang H-C (2016) Effects of polycyclic aromatic compounds in fine particulate matter generated from household coal combustion on response to EGFR mutations in vitro. Environ Pollut 218:1262–1269

Hoang T, Castorina R, Gaspar F, Maddalena R, Jenkins PL, Zhang Q, McKone TE, Benfenati E, Shi AY, Bradman A (2017) VOC exposures in California early childhood education environments. Indoor Air 27:609–621

Hoang AQ, Tran TM, Tu MB, Takahashi S (2021) Polybrominated diphenyl ethers in indoor and outdoor dust from Southeast Asia: an updated review on contamination status, human exposure, and future perspectives. Environ Pollut 272:116012

Hoang AQ, Trinh HT, Nguyen HMN, Nguyen TQ, Nguyen TX, Duc TV, Nguyen TT, Do TQ, Minh TB, Tran TM (2023) Assessment of cyclic volatile methyl siloxanes (CVMSs) in indoor dust from different micro-environments in northern and central Vietnam. Environ Geochem Health 45:1711–1722

Hoffman K, Garantziotis S, Birnbaum LS, Stapleton HM (2015) Monitoring indoor exposure to organophosphate flame retardants: hand wipes and house dust. Environ Health Perspect 123:160–165

Hofmann H, Erdmann G, Müller A (2014) Conflicting goals of energy-efficient construction and good indoor air quality – data collection for volatile organic compounds in indoor air of residential and office buildings (solutions). Annex to the final report. [in German] http://www.agoef.de/fileadmin/user_upload/dokumente/forschung/AGOEF-Abschlussericht_VOCDB_II_Anhang-nicht-barrierefrei.pdf

Hoh E, Hunt RN, Quintana PJE, Zakarian JM, Chatfield DA, Wittry BC, Rodriguez E, Matt GE (2012) Environmental tobacco smoke as a source of polycyclic aromatic hydrocarbons in settled household dust. Environ Sci Technol 46:4174–4183

Hohenblum P, Kundi M, Gundacker C, Hutter H-P, Jansson M, Moosmann L, Scharf S, Tappler P, Uhl M (2008) LUKI - Air and children. Influence of indoor air on the health of children in all-day schools. Report of the Austrian Environment Protection Agency. https://www.umweltbundesamt.at/fileadmin/site/publikationen/rep0181.pdf

Hou M, Shi Y, Na G, Cai Y (2021a) A review of organophosphate esters in indoor dust, air, hand wipes and silicone wristbands: implications for human exposure. Environ Int 146:106261

Hou M, Fang J, Shi Y, Tang S, Dong H, Liu Y, Deng F, Giesy JP, Godri Pollitt KJ, Cai Y, Shi X (2021b) Exposure to organophosphate esters in elderly people: relationships of OPE body burdens with indoor air and dust concentrations and food consumption. Environ Int 157:106803

Hu YJ, Bao LJ, Huang CL, Li SM, Liu P, Zeng EY (2018) Assessment of airborne polycyclic aromatic hydrocarbons in a megacity of South China: spatiotemporal variability, indoor-outdoor interplay and potential human health risk. Environ Pollut 238:431–439

Hu Q, Xu L, Liu Y, Zeng X, Yu Z (2020) Co-occurrence and distribution of organophosphate tri- and di-esters in indoor dust from different indoor environments in Guangzhou and their potential human health risk. Environ Pollut 262:114311

Huang H, Gao L, Xia D, Qiao L, Wang R, Su G, Liu W, Liu G, Zheng M (2017) Characterization of short- and medium-chain chlorinated paraffins in outdoor/indoor $PM_{10}/PM_{2.5}/PM_{1.0}$ in Beijing, China. Environ Pollut 225:674–680

Huang Y, Tan H, Li L, Yang L, Sun F, Li J, Gong X, Chen D (2020) A broad range of organophosphate tri- and di-esters in house dust from Adelaide, South Australia: concentrations, compositions, and human exposure risks. Environ Int 142:105872

Huang C, Zhang YJ, Liu LY, Wang F, Guo Y (2021) Exposure to phthalates and correlations with phthalates in dust and air in South China homes. Sci Total Environ 782:146806

Huber S, Haug LS, Schlabach M (2011) Per- and polyfluorinated compounds in house dust and indoor air from northern Norway – a pilot study. Chemosphere 84:1686–1693

Hung CC, Huang FJ, Yang YQ, Hsieh CJ, Tseng CC, Yiin LM (2018) Pesticides in indoor and outdoor residential dust: a pilot study in a rural county of Taiwan. Environ Sci Pollut Res Int 25: 23349–23356

Iakovides M, Iakovides G, Stephanou EG (2021) Atmospheric particle-bound polycyclic aromatic hydrocarbons, n-alkanes, hopanes, steranes and trace metals: $PM_{2.5}$ source identification, individual and cumulative multi-pathway lifetime cancer risk assessment in the urban environment. Sci Total Environ 752:141834

IARC (1983) Polynuclear aromatic compounds, part 1, Chemical, environmental and experimental data. IARC Monographs on the evaluation of the carcinogenic risk of chemicals to humans, Vol. 32, Lyon, France

IARC (1997) Polychlorinated dibenzo-para-dioxins and polychlorinated dibenzofurans. IARC Monographs on the evaluation of carcinogenic risks to humans, Vol. 69. https://publications.iarc.fr/Book-And-Report-Series/Iarc-Monographs-On-The-Identification-Of-Carcinogenic-Hazards-To-Humans/Polychlorinated-Dibenzo%2D%2DEm-Para-Em%2D%2DDioxins-And-Polychlorinated-Dibenzofurans-1997

IARC (2001) Hexachlorobenzene. In Some thyrotropic agents. IARC Monographs on the evaluation of carcinogenic risk of chemicals to humans, Vol. 79. Lyon, France. http://www.inchem.org/documents/iarc/vol79/79–13.html

IARC (2010) Some non-heterocyclic polycyclic aromatic hydrocarbons and some related exposures. IARC Monogr Eval Carcinog Risks Hum 92:1–853. http://monographs.iarc.fr/ENG/Monographs/vol92/mono92-6.pdf

IARC (2012) Benzo[a]pyrene. IARC monographs – Vol 100F. Lyon, France. https://publications.iarc.fr/Book-And-Report-Series/Iarc-Monographs-On-The-Identification-Of-Carcinogenic-Hazards-To-Humans/Chemical-Agents-And-Related-Occupations-2012

IARC (2013) Re-evaluation of some organic chemicals, hydrazine and hydrogen peroxide. IARC monographs on the evaluation of carcinogenic risks to humans. Vol. 103, Lyon, France

IARC (2015a) DDT, Lindane, and 2,4-D. IARC Working Group. Monographs Volume 113: Lyon, France. https://publications.iarc.fr/Book-And-Report-Series/Iarc-Monographs-On-The-Identification-Of-Carcinogenic-Hazards-To-Humans/DDT-Lindane-And-2-4-D-2016

IARC (2015b) Polychlorinated biphenyls and polybrominated biphenyls. IARC Monographs on the evaluation of carcinogenic risks to humans, Vol. 107. https://publications.iarc.fr/Book-And-Report-Series/Iarc-Monographs-On-The-Identification-Of-Carcinogenic-Hazards-To-Humans/Polychlorinated-Biphenyls-And-Polybrominated-Biphenyls-2015

IARC (2018) Some industrial chemicals. Tetrabromobisphenol A. IARC monographs on the evaluation of carcinogenic risks to humans, vol 115. https://publications.iarc.fr/563

IHS (2020) Plasticizers. Chemical Economics Handbook. Online: https://ihsmarkit.com/products/plasticizers-chemical-economics-handbook.html

IHS (2022) Plasticizers. Chemical Economics Handbook. https://ihsmarkit.com/products/plasticizers-chemical-economics-handbook.html

Imm P, Knobeloch L, Buelow C, Anderson HA (2009) Household exposures to polybrominated diphenyl ethers (PBDEs) in a Wisconsin cohort. Environ Health Perspect 117:1890–1895

ISOPA (European Diisocyanate and Polyol Producers Association) (2020) Polyurethanes – Benefits of Polyurethanes. https://www.isopa.org/wp-content/uploads/2022/09/isopa-factsheet_benefits-of-polyurethanes_v05_a.pdf

ITOX (Institute for Environmental Analysis and Human Toxicology) (1999) Contributions to environmental and health protection. Berlin, Germany. [in German]

Jakovljevic I, Dvoršcak M, Jagic K, Klincic D (2022) Polycyclic aromatic hydrocarbons in indoor dust in Croatia: levels, sources, and human health risks. Int J Environ Res Public Health 19: 11848

Jeon S, Kim KT, Choi K (2016) Migration of DEHP and DINP into dust from PVC flooring products at different surface temperature. Sci Total Environ 547:441–446

Johnson PI, Stapleton HM, Sjodin A, Meeker JD (2010) Relationships between polybrominated diphenyl ether concentrations in house dust and serum. Environ Sci Technol 44:5627–5632

Johnson-Restrepo B, Kannan K (2009) An assessment of sources and pathways of human exposure to polybrominated diphenyl ethers in the United States. Chemosphere 76:542–548

Kademoglou K, Xu F, Padilla-Sanchez JA, Haug LS, Covaci A, Collins CD (2017) Legacy and alternative flame retardants in Norwegian and UK indoor environment: implications of human exposure via dust ingestion. Environ Int 102:48–56

Kademoglou K, Giovanoulis G, Palm-Cousins A, Padilla-Sanchez JA, Magnér J, De Wit CA, Collins CD (2018) In vitro inhalation bioaccessibility of phthalate esters and alternative plasticizers present in indoor dust Using Artificial Lung Fluids. Environ Sci Technol Lett 5: 329–334

Kalachova K, Hradkova P, Lankova D, Hajslova J, Pulkrabova J (2012) Occurrence of brominated flame retardants in household and car dust from the Czech Republic. Sci Total Environ 441: 182–193

Kanazawa A, Saito I, Araki A, Takeda M, Ma M, Saijo Y, Kishi R (2010) Association between indoor exposure to semi-volatile organic compounds and building-related symptoms among the occupants of residential dwellings. Indoor Air 20:72–84

Kang Y, Cheung KC, Wong MH (2011) Mutagenicity, genotoxicity and carcinogenic risk assessment of indoor dust from three major cities around the Pearl River Delta. Environ Int 37:637–643

Kang Y, Man YB, Cheung KC, Wong MH (2012) Risk assessment of human exposure to bioaccessible phthalate esters via indoor dust around the Pearl River Delta. Environ Sci Technol 46:8422–8430

Kannan K, Takahashi S, Fujiwara N, Mizukawa H, Tanabe S (2010) Organotin compounds, including butyltins and octyltins, in house dust from Albany, New York, USA. Arch Environ Contam Toxicol 58:901–907

Karásková P, Venier M, Melymuk L, Beanová J, Vojta Š, Prokes R, Diamond ML, Klánová J (2016) Perfluorinated alkyl substances (PFASs) in household dust in Central Europe and North America. Environ Int 94:315–324

Karlsson M, Julander A, van Bavel B, Hardell L (2007) Levels of brominated flame retardants in blood in relation to levels in household air and dust. Environ Int 33:62–69

Kersten W, Reich T (2003) Semi volatile organic environmental chemicals in Hamburg house dusts. Gefahrst – Reinh Luft 63:85–91. [in German]

Kim W, Choi I, Jung Y, Lee J, Min S, Yoon C (2013) Phthalate levels in nursery schools and related factors. Environ Sci Technol 47:12459–12468

Kim U-J, Wang Y, Li W, Kannan K (2019) Occurrence of and human exposure to organophosphate flame retardants/plasticizers in indoor air and dust from various microenvironments in the United States. Environ Int 125:342–349

Kim K, Shin HM, Wong L, Young TM, Bennett DH (2021) Temporal variability of indoor dust concentrations of semivolatile organic compounds. Indoor Air 31:693–701

Klinčić D, Dvoršćak M, Jagić K, Mendaš G, Herceg RS (2020) Levels and distribution of polybrominated diphenyl ethers in humans and environmental compartments: a comprehensive review of the last five years of research. Environ Sci Pollut Res Int 27:5744–5758

Koch C, Sures B (2018) Environmental concentrations and toxicology of 2,4,6-tribromophenol (TBP). Environ Pollut 233:706–713

Kohler M, Zennegg M, Waeber R (2002) Coplanar polychlorinated biphenyls (PCB) in indoor air. Environ Sci Technol 36:4735–4740

Köhler M, Weis N, Zorn C (2004) Indoor airborne PAH contamination – occurrence, sources and assessment 7. AGÖF-Fachkongress, 156–165. [in German]

Kolarik B, Frederiksen M, Meyer HW, Ebbehøj NE, Gunnarsen LB (2014) Investigation of the importance of tertiary contamination, temperature and human behaviour on PCB concentrations in indoor air. Indoor Built Environ 25:229–241

Kozicki M, Niesłochowski A (2020) Materials contamination and indoor air pollution caused by tar products and fungicidal impregnations: Intervention research in 2014–2019. Sensors 20:4099

Kraft M, Sievering S, Grün L, Rauchfuss K (2018) Mono-, di-, and trichlorinated biphenyls (PCB 1-PCB 39) in the indoor air of office rooms and their relevance on human blood burden. Indoor Air 28:441–449

Krugly E, Martuzevicius D, Sidaraviciute R, Ciuzas D, Prasauskas T, Kauneliene V, Stasiulaitiene I, Kliucininkas L (2014) Characterization of particulate and vapor phase polycyclic aromatic hydrocarbons in indoor and outdoor air of primary schools. Atmos Environ 82: 298–306

Kuang J, Ma Y, Harrad S (2016) Concentrations of "legacy" and novel brominated flame retardants in matched samples of UK kitchen and living room/bedroom dust. Chemosphere 149:224–230

Kubwabo C, Stewart B, Zhu JP, Marro L (2005) Occurrence of perfluorosulfonates and other perfluorochemicals in dust from selected homes in the City of Ottawa, Canada. J Environ Monit 7:1074–1078

Kubwabo C, Fan X, Rasmussen PE, Wu F (2012) Determination of synthetic musk compounds in indoor house dust by gas chromatography–ion trap mass spectrometry. Anal Bioanal Chem 404: 467–477

Kubwabo C, Rasmussen PE, Fan X, Kosarac I, Wu F, Zidek A, Kuchta SL (2013) Analysis of selected phthalates in Canadian indoor dust collected using household vacuum and standardized sampling techniques. Indoor Air 23:506–514

Kubwabo C, Rasmussen PE, Fan X, Kosarac I, Grenier G, Coleman K (2016) Simultaneous quantification of bisphenol A, alkylphenols and alkylphenol ethoxylates in indoor dust by gas chromatography-tandem mass spectrometry and a comparison between two sampling techniques. Anal Methods 8:4093

Kubwabo C, Fan X, Prasad Katuri G, Habibagahi A, Rasmussen PE (2021) Occurrence of aryl and alkyl-aryl phosphates in Canadian house dust. Emerg Contam 7:149–159

Kutz FW, Barnes DG, Bottimore DP, Greim H, Bretthauer EW (1990) The international toxicity equivalency factor (I-TEF) method of risk assessment for complex mixtures of dioxins and related compounds. Chemosphere 20:751–757

La Guardia MJ, Hale RC (2015) Halogenated flame-retardant concentrations in settled dust, respirable and inhalable particulates and polyurethane foam at gymnastic training facilities and residences. Environ Int 79:106–114

Lai AM, Clark S, Carter E, Shan M, Ni K, Yang X, Baumgartner J, Schauer JJ (2020) Impacts of stove/fuel use and outdoor air pollution on chemical composition of household particulate matter. Indoor Air 30:294–305

Langer S, Weschler CJ, Fischer A, Bekö G, Toftum J, Clausen G (2010) Phthalate and PAH concentrations in dust collected from Danish homes and daycare centers. Atmos Environ 44: 2294–2301

Lankova D, Svarcova A, Kalachova K, Lacina O, Pulkrabova J, Hajslova J (2015) Multi-analyte method for the analysis of various organohalogen compounds in house dust. Anal Chim Acta 854:61–69

Larsson K, Lindh CH, Jönsson BAG, Giovanoulis G, Bibi M, Bottai M, Bergström A, Berglund M (2017) Phthalates, non-phthalate plasticizers and bisphenols in Swedish preschool dust in relation to children's exposure. Environ Int 102:114–124

Larsson K, de Wit CA, Sellström U, Sahlström L, Lindh CH, Berglund M (2018) Brominated flame retardants and organophosphate esters in preschool dust and children's hand wipes. Environ Sci Technol 52:4878–4888

Lee Y-S, Choi I (2020) Phthalates contamination in indoor dust in elementary schools in Seoul: a pilot study. J Environ Health Sci 46:548–554

Lee HK, Kang H, Lee S, Kim S, Choi K, Moon HB (2020) Human exposure to legacy and emerging flame retardants in indoor dust: a multiple-exposure assessment of PBDEs. Sci Total Environ 719:137386

Lehmler H-J (2005) Synthesis of environmentally relevant fluorinated surfactants – a review. Chemosphere 58:1471–1496

Leimbrock W (2001) Investigation of hazardous substance exposure in the interior air of a truck. Report of the Saxon State Institute for Occupational Safety and Health. [in German]

Leng G, Berger-Preiß E, Levsen K, Ranft U, Sugiri D, Hadnagy W, Idel H (2005) Pyrethroids used indoors – ambient monitoring of pyrethroids following a pest control operation. Int J Hyg Environ Health 208:193–199

Levasseur JL, Hammel SC, Hoffman K, Phillips AL, Zhang S, Ye X, Calafat AM, Webster TF, Stapleton HM (2021) Young children's exposure to phenols in the home: associations between house dust, hand wipes, silicone wristbands, and urinary biomarkers. Environ Int 147:106317

Li C, Ro Y (2000) Indoor characteristics of polycyclic aromatic hydrocarbons in the urban atmosphere of Taipei. Atmos Environ 34:611–620

Li F, Jin J, Sun X, Wang X, Li Y, Shah MM, Chen J (2014) Gas chromatography–triple quadrupole mass spectrometry for the determination of atmospheric polychlorinated naphthalenes. J Hazard Mater 280:111–117

Li WL, Qi H, Ma WL, Liu LY, Zhang Z, Zhu NZ, Mohammed MO, Li YF (2015) Occurrence, behavior and human health risk assessment of dechlorane plus and related compounds in indoor dust of China. Chemosphere 134:166–171

Li H, Lydy MJ, You J (2016) Pyrethroids in indoor air during application of various mosquito repellents: occurrence, dissipation and potential exposure risk. Chemosphere 144:2427–2435

Li Y, Xian Q, Li L (2017) Development of a short path thermal desorption-gas chromatography/mass spectrometry method for the determination of polycyclic aromatic hydrocarbons in indoor air. J Chromatogr A 1497:127–134

Li Y, Liao R, Gan Z, Qu B, Wang R, Chen M, Ding S, Su S (2018) Seasonal variation and exposure risks of perchlorate in soil, indoor dust, and outdoor dust in China. Arch Environ Contam Toxicol 75:367–376

Li W, Wang Y, Asimakopoulos AG, Covaci A, Gevao B, Johnson-Restrepo B, Kumosani TA, Malarvannan G, Moon HB, Nakata H, Sinha RK, Tran TM, Kannan K (2019) Organophosphate esters in indoor dust from 12 countries: concentrations, composition profiles, and human exposure. Environ Int 133(Pt A):105178

Li X, Zhang W, Lv J, Liu W, Sun S, Guo XJ (2021) Distribution, source apportionment, and health risk assessment of phthalate esters in indoor dust samples across China. Environ Sci Eur 33:19

Li H, Liu Y, Lan Y, Zhao Y, Lu A, Li C, Lei R, Xue J, Liu W (2022) nationwide survey of 20 legacy brominated flame retardants in indoor dust from China: continuing occurrence, national distribution, and implication for human exposure. Environ Sci Pollut Res Int 29:58828–58842

Liang Y, Xu Y (2014) Emission of phthalates and phthalate alternatives from vinyl flooring and crib mattress covers: the influence of temperature. Environ Sci Technol 48:14228–14237

Liang C, Gracida-Alvarez UR, Gallant ET, Gillis PA, Marques YA, Abramo GP, Hawkins TR, Dunn JB (2021) Material flows of polyurethane in the United States. Environ Sci Technol 55: 14215–14224

Liao C, Liu F, Guo Y, Moon HB, Nakata H, Wu Q, Kannan K (2012) Occurrence of eight bisphenol analogues in indoor dust from the United States and several Asian countries: implications for human exposure. Environ Sci Technol 46:9138–9145

Liao C, Kim U-J, Kannan K (2018) A review of environmental occurrence, fate, exposure, and toxicity of benzothiazoles. Environ Sci Technol 52:5007–5026

Liebl B, Schettgen T, Kerscher G, Broding HC, Otto A, Angerer J, Drexler H (2004) Evidence for increased internal exposure to lower chlorinated polychlorinated biphenyls (PCB) in pupils attending a contaminated school. Int J Hyg Environ Health 207:315–324

Lim YW, Kim HH, Lee CS, Shin DC, Chang YS, Yang JY (2014) Exposure assessment and health risk of poly-brominated diphenyl ether (PBDE) flame retardants in the indoor environment of elementary school students in Korea. Sci Total Environ 470–471:1376–1389

Lim H, Sadiktsis I, de Oliveira Galvão MF, Westerholm R, Dreij K (2021) Polycyclic aromatic compounds in particulate matter and indoor dust at preschools in Stockholm, Sweden: occurrence, sources and genotoxic potential in vitro. Sci Total Environ 755(Pt 1):142709

Limasset J-C, Diebold F, Hubert G (1993) Exposition des conducteurs de bus urbains aux pollutants de la circulation automobile. Sci Total Environ 134:39–49

Lin N, Chen Y, Du W, Shen G, Zhu X, Huang T, Wang X, Cheng H, Liu J, Xue C, Liu G, Zeng EY, Xing B, Tao S (2016) Inhalation exposure and risk of polycyclic aromatic hydrocarbons (PAHs) among the rural population adopting wood gasifier stoves compared to different fuel-stove users. Atmos Environ 147:485–491

Lisco G, De Tullio A, Giagulli VA, De Pergola G, Triggiani V (2020) Interference on iodine uptake and human thyroid function by perchlorate-contaminated water and food. Nutrients 12:1669

Liu Y, Zhu L, Shen X (2001) Polycyclic aromatic hydrocarbons (PAHs) in indoor and outdoor air of Hangzhou, China. Environ Sci Technol 35:840–844

Liu G, Cai Z, Zheng M (2014) Sources of unintentionally produced polychlorinated naphthalenes. Chemosphere 94:1–12

Liu X, Guo Z, Krebs KA, Stinson RA, Nardin JA, Pope RH, Roache NF (2015) Chamber study of PCB emissions from caulking materials and light ballasts. Chemosphere 137:115–121

Liu N, Xu L, Cai Y (2018) Methyl siloxanes in barbershops and residence indoor dust and the implication for human exposures. Sci Total Environ 618:1324–1330

Liu X, Chen D, Yu Y, Zeng X, Li L, Xie Q, Yang M, Wu Q, Dong G (2020) Novel organophosphate esters in airborne particulate matters: Occurrences, precursors, and selected transformation products. Environ Sci Technol 54:13771–13777

Liu Y, Gong S, Ye L, Li J, Liu C, Chen D, Fang M, Letcher RJ, Su G (2021a) Organophosphate (OP) diesters and a review of sources, chemical properties, environmental occurrence, adverse effects, and future directions. Environ Int 155:106691

Liu D, Wu P, Zhao N, Nie S, Cui J, Zhao M, Jin H (2021b) Differences of bisphenol analogue concentrations in indoor dust between rural and urban areas. Chemosphere 276:130016

Liu W, Sun Y, Liu N, Hou J, Huo X, Zhao Y, Zhang Y, Deng F, Kan H, Zhao Z, Huang C, Zhao B, Zeng X, Qian H, Zheng X, Liu W, Mo J, Sun C, Su C, Zou Z, Li H, Guo J, Bu Z (2022) Indoor exposure to phthalates and its burden of disease in China. Indoor Air 32:e13030

López A, Yusa V, Villoldo E, Corpas-Burgos F, Coscolla C (2021) Indoor air pesticide in dwellings of breastfeeding mothers of the Valencian Region (Spain): levels, exposure and risk assessment. Atmos Environ 248:118231

Lu Y, Yuan T, Yun SH, Wang W, Wu Q, Kannan K (2010) Occurrence of cyclic and linear siloxanes in indoor dust from China, and implications for human exposures. Environ Sci Technol 44:6081–6087

Lu Y, Yuan T, Yun SH, Wang W, Kannan K (2011) Occurrence of synthetic musks in indoor dust from China and implications for human exposure. Arch Environ Contam Toxicol 60:182–189

Lu C, Adamkiewicz G, Attfield KR, Kapp M, Spengler JD, Tao L, Xie SH (2013) Household pesticide contamination from indoor pest control applications in urban low-income public housing dwellings: a community-based participatory research. Environ Sci Technol 47:2018–2025

Lu Q-Q, Jung C-C, Liu Y-H, Chang W-H (2022) Occurrences, season, and source characteristics of organophosphorus flame retardants (OPFRs) in residences in Taiwan: implications for young children's exposure and risk assessment using the probabilistic approach. Res Square May 27th 2022

Lucattini L, Poma G, Covaci A, de Boer J, Lamoree MH, Leonards PEG (2018) A review of semi-volatile organic compounds (SVOCs) in the indoor environment: occurrence in consumer products, indoor air and dust. Chemosphere 201:466–482

Lui KH, Bandowe BAM, Tian L, Chan C-S, Cao J-J, Ning Z, Lee SC, Ho KF (2017) Cancer risk from polycyclic aromatic compounds in fine particulate matter generated from household coal combustion in Xuanwei, China. Chemosphere 169:660–668

Lung SC, Kao MC, Hu SC (2003) Contribution of incense burning to indoor PM_{10} and particle-bound polycyclic aromatic hydrocarbons under two ventilation conditions. Indoor Air 13:194–199

Luongo G, Östman C (2016) Organophosphate and phthalate esters in settled dust from apartment buildings in Stockholm. Indoor Air 26:414–425

Lv J, Xu R, Wu G, Zhang Q, Li Y, Wang P, Liao C, Liu J, Jiang G, Wei F (2009) Indoor and outdoor air pollution of polycyclic aromatic hydrocarbons (PAHs) in Xuanwei and Fuyuan, China. J Environ Monit 11:1368–1374

Lv YZ, Luo XJ, Li QQ, Yang Y, Zeng YH, Mai BX (2023) A new insight into the emission source of DDT in indoor environment from rural area of South China and comprehensive human health exposure assessment. Environ Sci Pollut Res Int 30:35189–35199

Lyche JL, Rosseland C, Berge G, Polder A (2015) Human health risk associated with brominated flame-retardants (BFRs). Environ Int 74:170–180

Lyng NL, Clausen PA, Lundsgaard C, Andersen HV (2016) Modelling the impact of room temperature on concentrations of polychlorinated biphenyls (PCBs) in indoor air. Chemosphere 144:2127–2133

Ma S, Yue C, Tang J, Lin M, Zhuo M, Yang Y, Li G, An T (2021) Occurrence and distribution of typical semi-volatile organic chemicals (SVOCs) in paired indoor and outdoor atmospheric fine particle samples from cities in southern China. Environ Pollut 269:116123

Maag J, Lassen C, Brandt UK, Kjølholt J, Molander L, Mikkelsen SH (2010) Identification and assessment of alternatives to selected phthalates. Report for the Danish Environmental Protection Agency, Environmental Project No. 13412010. https://www2.mst.dk/udgiv/publications/2010/978-87-92708-00-7/pdf/978-87-92708-01-4.pdf

Madruga DG, Ubeda RM, Terroba JM, Dos Santos SG, García-Cambero JP (2019) Particle-associated polycyclic aromatic hydrocarbons in a representative urban location (indoor-outdoor) from South Europe: assessment of potential sources and cancer risk to humans. Indoor Air 29: 817–827

Maertens RM, Yang X, Zhu J, Gagne R, Douglas GR, White PA (2008) Mutagenic and carcinogenic hazards of settled house dust I: polycyclic aromatic hydrocarbon content and excess lifetime cancer risk from preschool exposure. Environ Sci Technol 42:1747–1753

Mahler BJ, Van Metre PC, Wilson JT, Musgrove M, Burbank TL, Ennis TE, Bashara TJ (2010) Coal-tar-based parking lot sealcoat: an unrecognized source of PAH to settled house dust. Environ Sci Technol 44:894–900

Makey CM, Webster TF, Martin JW, Shoeib M, Harner T, Dix-Cooper L, Webster GM (2017) Airborne precursors predict maternal serum perfluoroalkyl acid concentrations. Environ Sci Technol 51:7667–7675

Malcolm HM, Dobsen S (1994) The calculation of an environmental assessment level (EAL) for atmospheric PAH using relative potencies. Department of the Environment, Report No DoE/HMIP/RR/94/041, ITE Project TO7061i1 Report to the Her Majesty's Inspectorate of Pollution, London

Malliari E, Kalantzi O (2017) Children's exposure to brominated flame retardants in indoor environments – a review. Environ Int 108:146–169

Mandin C, Mercier F, Ramalho O, Lucas J-P, Gilles E, Blanchard O, Bonvallot N, Glorennec P, Le Bot B (2016) Semi-volatile organic compounds in the particulate phase in dwellings: a nationwide survey in France. Atmos Environ 136:82–94
Mannino MR, Orecchio S (2008) Polycyclic aromatic hydrocarbons (PAHs) in indoor dust matter of Palermo (Italy) area: extraction, GC-MS analysis, distribution and sources. Atmos Environ 42:1801–1817
Manová E, von Goetz N, Hungerbühler K (2014) Ultraviolet filter contact and photocontact allergy: consumer exposure and risk assessment for octocrylene from personal care products and sunscreens. Br J Dermatol 171:368–1374
Marek RF, Thorne PS, Herkert NJ, Awad AM, Hornbuckle KC (2017) Airborne PCBs and OH-PCBs inside and outside urban and rural U.S. schools. Environ Sci Technol 18:7853–7860
Markowicz P, Löndahl J, Wierzbicka A, Suleiman R, Shihadeh A, Larsson L (2014) A study on particles and some microbial markers in waterpipe tobacco smoke. Sci Total Environ 499:107–113
Marques VB, Faria RA, Dos Santos L (2018) Overview of the pathophysiological implications of organotins on the endocrine system. Front Endocrinol (Lausanne) 16:101
Matwiejczuk N, Galicka A, Brzóska MM (2020) Review of the safety of application of cosmetic products containing parabens. J Appl Toxicol 40:176–210
McGrath TJ, Christia C, Poma G, Covaci A (2022) Seasonal variation of short-, medium- and long-chain chlorinated paraffin distribution in Belgian indoor dust. Environ Int 170:107616
Melymuk L, Bohlin-Nizzetto P, Kukuka P, Vojta Š, Kalina J, Cupr P, Klánová J (2016a) Seasonality and indoor/outdoor relationships of flame retardants and PCBs in residential air. Environ Pollut 218:392–401
Melymuk L, Bohlin-Nizzetto P, Vojta S, Kratka M, Kukucka P, Audy O, Pribylova P, Klanova J (2016b) Distribution of legacy and emerging semivolatile organic compounds in five indoor matrices in a residential environment. Chemosphere 153:179–186
Meng F, Wu H (2015) Indoor air pollution by methylsiloxane in household and automobile settings. PLoS One 17:e0135509
Meng G, Nie Z, Feng Y, Wu X, Yin Y, Wang Y (2016) Typical halogenated persistent organic pollutants in indoor dust and the associations with childhood asthma in Shanghai, China. Environ Pollut 211:389–398
Menichini E, Merli (2012). Dibenzopyrenes, other PAHs with molecular weight 302, and selected carcinogenic PAHs seldom determined: identification and one-year quantification in urban air. Int J Environ Anal Chem 92: 1609–1625
Menichini E, Iacovella N, Monfredini F, Turrio-Baldassarri L (2007) Atmospheric pollution by PAHs, PCDD/Fs and PCBs simultaneously collected at a regional background site in central Italy and at an urban site in Rome. Chemosphere 69:422–434
Mercier F, Gilles E, Saramito G, Glorennec P, Le Bot B (2014) A multi-residue method for the simultaneous analysis in indoor dustof several classes of semi-volatile organic compounds by pressurizedliquid extraction and gas chromatography/tandem mass spectrometry. J Chromatogr A 1336:101–111
Michałowicz J, Włuka A, Bukowska B (2022) A review on environmental occurrence, toxic effects and transformation of man-made bromophenols. Sci Total Environ 10:152289
Mizouchi S, Ichiba M, Takigami H, Kajiwara N, Takamuku T, Miyajima T, Kodama H, Someya T, Ueno D (2015) Exposure assessment of organophosphorus and organobromine flame retardants via indoor dust from elementary schools and domestic houses. Chemosphere 123:17–25
Mo F, Chi C, Guo M, Chu X, Li Y, Shen X (2014) Characteristics of selected indoor air pollutants from moxibustion. J Hazard Mater 270:53–60
Molinier B, Arata C, Katz EF, Lunderberg DM, Liu Y, Misztal PK, Nazaroff WW, Goldstein AH (2022) Volatile methyl siloxanes and other organosilicon compounds in residential air. Environ Sci Technol 56:15427–15436
Moon KA, Magid H, Torrey C et al (2015) Secondhand smoke in waterpipe tobacco venues in Istanbul, Moscow, and Cairo. Environ Res 142:568–574

Moreau-Guigon E, Alliot F, Gasperi J, Blanchard M, Teil M-J, Mandin C, Chevreuil M (2016) Seasonal fate and gas/particle partitioning of semi-volatile organic compounds in indoor and outdoor air. Atmos Environ 47:423–433

Morgan MK, Sheldon LS, Croghan CW, Jones PA, Chuang JC, Wilson NK (2007) An observational study of 127 preschool children at their homes and day care centers in Ohio: environmental pathways to cis- and trans-permethrin exposure. Environ Res 104:266–274

Morgan MK, Wilson NK, Chuang JC (2014) Exposures of 129 preschool children to organochlorines, organophosphates, pyrethroids, and acid herbicides at their homes and daycares in North Carolina. Int J Environ Res Public Health 11:3743–3764

Mueller A, Ulrich N, Hollmann J, Zapata Sanchez CE, Rolle-Kampczyk UE, von Bergen M (2019) Characterization of amultianalyte GC-MS/MS procedure for detecting and quantifying polycyclic aromatic hydrocarbons (PAHs) and PAH derivatives from air particulate matter for an improved risk assessment. Environ Pollut 255:112967

Mumbo J, Henkelmann B, Abdelaziz A, Pfister G, Nguyen N, Schroll R, Munch JC, Schramm KW (2015) Persistence and dioxin-like toxicity of carbazole and chlorocarbazoles in soil. Environ Sci Pollut Res Int 22:1344–1356

Muñoz C, Droppelmann A, Erazo M, Aceituno P, Orellana C, Parro J, Mesias S, Marchetti N, Navas-Acien A, Iglesias V (2016) Occupational exposure to polycyclic aromatic hydrocarbons: a cross-sectional study in bars and restaurants in Santiago, Chile. Am J Ind Med 59:887–896

Nagorka R, Conrad A, Scheller C, Sussenbach B, Moriske HJ (2011a) Diisononyl 1,2-cyclohexanedicarboxylic acid (DINCH) and di(2-ethylhexyl) terephthalate (DEHT) in indoor dust samples: concentration and analytical problems. Int J Hyg Environ Health 214: 28–37

Nagorka R, Conrad A, Scheller C, Sussenbach B, Moriske H.J. 2011b. Plasticizers and flame retardants in household dust – part 2: non-phthalates and flame retardants. Gefahrst Reinh Luft 71: 286–292. [in German]

Nagorka R, Birmili W, Schulze J, Koschorreck J (2022) Diverging trends of plasticizers (phthalates and non-phthalates) in indoor and freshwater environments—why? Environ Sci Eur 34:46

Nakagawa LE, Costa AR, Polatto R, Nascimento CMD, Papini S (2017) Pyrethroid concentrations and persistence following indoor application. Environ Toxicol Chem 36:2895–2898

Nakagawa LE, do Nascimento CM, Costa AR, Polatto R, Papini S (2020) Persistence of indoor permethrin and estimation of dermal and non-dietary exposure. J Expo Sci Environ Epidemiol 30:547–553

Nakata H, Hinosaka M, Yanagimoto H (2015) Macrocyclic-, polycyclic-, and nitromusks in cosmetics, household commodities and indoor dusts collected from Japan: Implications for their human exposure. Ecotoxicol Environ Saf 111:248–255

Negreira N, Rodriguez I, Rubi E, Cela R (2009) Determination of selected UV filters in indoor dust by matrix solid-phase dispersion and gas chromatography-tandem mass spectrometry. J Chromatogr A 1216:5895–5902

Neumann HD, Buxtrup M, von Hahn N, Koppisch D, Breuer D, Hahn J-U (2012) Proposal for the derivation of indoor workplace reference values in schools. Gefahrst Reinh Luft 72:291–297. [in German]

Neumann K, Birmili W, Daniels A, Pietsch A, Kura J, Freudenthal G, Siemers U (2020) Indoor/outdoor relationships for polycyclic aromatic hydrocarbons in $PM_{2.5}$ in 80 homes during the German Environmental Survey, 2014–2017. Proceedings AEROSOLS 2020

Newton S, Sellström U, de Wit CA (2015) Emerging flame retardants, PBDEs, and HBCDDs in indoor and outdoor media in Stockholm, Sweden. Environ Sci Technol 49:2912–2920

NICNAS (National Industrial Chemicals Notification and Assessment Scheme, Australian Department of Health and Ageing) (2012) Hexabromocyclododecane. Priority Existing Chemical Assessment Report No. 34. https://www.nicnas.gov.au/__data/assets/word_doc/0009/34848/PEC34-HBCD.docx)

NILU (Norwegian Institute for Air Research) (2007) Siloxanes in the environment of the Inner Oslofjord. http://www.sft.no/publikasjoner/2269/ta2269.pdf

Nisbet ICT, LaGoy PK (1992) Toxic equivalency factors (TEFs) for polycyclic aromatic hydrocarbons (PAHs). Regul Toxicol Pharmacol 1:290–300

Niziński P, Błażewicz A, Kończyk J, Michalski R (2020) Perchlorate – properties, toxicity and human health effects: an updated review. Rev Environ Health 36:199–222

NRC (National Research Council) (2005) NTP technical report on the toxicology and carcinogenesis studies of sodium chlorate (CAS no. 7775-09-9) in F344/N rats and B6C3F1 mice (drinking water studies). National Toxicology Program, NTP TR 517

Ohura T, Amagai T, Fusaya M, Matsushita H (2004) Polycyclic aromatic hydrocarbons in indoor and outdoor environments and factors affecting their concentrations. Environ Sci Technol 38: 77–83

Oliveira M, Slezakova K, Delerue-Matos C, Pereira MC, Morais S (2015) Polycyclic aromatic hydrocarbons: levels and phase distributions in preschool microenvironment. Indoor Air 25: 557–568

Oliveira M, Slezakova K, Madureira J, de Oliveira FE, Delerue-Matos C, Morais S, do Carmo Pereira M. (2017) Polycyclic aromatic hydrocarbons in primary school environments: levels and potential risks. Sci Total Environ 575:1156–1167

Oliveira M, Slezakova K, Delerue-Matos C, Pereira MC, Morais S (2019) Children environmental exposure to particulate matter and polycyclic aromatic hydrocarbons and biomonitoring in school environments: a review on indoor and outdoor exposure levels, major sources and health impacts. Environ Int 124:180–204

Olivero-Verbel R, Moreno T, Fernández-Arribas J, Reche C, Minguillón MC, Martins V, Querol X, Johnson-Restrepo B, Eljarrat E (2021) Organophosphate esters in airborne particles from subway stations. Sci Total Environ 769:145105

Orecchio S (2011) Polycyclic aromatic hydrocarbons (PAHs) in indoor emission from decorative candles. Atmos Environ 45:1888–1895

Oudejans L, Mysz A, Gibb Snyder E, Wyrzykowska-Ceradini B, Nardin J, Tabor D, Starr J, Stout 2nd D, Lemieux P (2020) Remediating indoor pesticide contamination from improper pest control treatments: persistence and decontamination studies. J Hazard Mater 397:122743

Parette R, McCrindle R, McMahon KS, Pena-Abaurrea M, Reiner E, Chittim B, Riddell N, Voss G, Dorman FL, Pearson WN (2015) Halogenated indigo dyes: a likely source of 1,3,6,8-tetrabromocarbazole and some other halogenated carbazoles in the environment. Chemosphere 127:18–26

Pasecnaja E, Perkons I, Bartkevics V, Zacs D (2021) Legacy and alternative brominated, chlorinated, and organophosphorus flame retardants in indoor dust-levels, composition profiles, and human exposure in Latvia. Environ Sci Pollut Res Int 28:25493–25502

Pei XQ, Song M, Guo M, Moa FF, Shen XY (2013) Concentration and risk assessment of phthalates present in indoor air from newly decorated apartments. Atmos Environ 68:17–23

Pelch K, Wignall JA, Goldstone AE, Ross PK, Blain RB, Shapiro AJ, Holmgren SD, Hsieh JH, Svoboda D, Auerbach SS, Parham FM, Masten SA, Walker V, Rooney A, Thayer KA (2019) A scoping review of the health and toxicological activity of bisphenol A (BPA) structural analogues and functional alternatives. Toxicology 424:152235

Petric Z, Ružić J, Žuntar I (2021) The controversies of parabens – an overview nowadays. Acta Pharma 71:17–32

Philippat C, Bennett DH, Krakowiak P, Rose M, Hwang HM, Hertz-Picciotto I (2015) Phthalate concentrations in house dust in relation to autism spectrum disorder and developmental delay in the Childhood Autism Risks from Genetics and the Environment (CHARGE) study. Environ Health 26(14):56

Pieri F, Katsoyiannis A, Martellini T, Hughes D, Jones KC, Cincinelli A (2013) Occurrence of linear and cyclic volatile methyl siloxanes in indoor air samples (UK and Italy) and their isotopic characterization. Environ Int 59:363–371

Piskorska-Pliszczyńska J, Maszewski J (2014) Brominated dioxins: little-known new health hazards – a review. Bull Vet Inst Pulawy 58:327–335

Pleus RC, Corey LM (2018) Environmental exposure to perchlorate: a review of toxicology and human health. Toxicol Appl Pharmacol 358:102–109

Polachova A, Gramblicka T, Parizek O, Sram RJ, Stupak M, Hajslova J, Pulkrabova J (2020) Estimation of human exposure to polycyclic aromatic hydrocarbons (PAHs) based on the dietary and outdoor atmospheric monitoring in the Czech Republic. Environ Res 182:108977

Preece AS, Shu H, Knutz M, Krais AM, Bekö G, Bornehag CG (2021) Indoor phthalate exposure and contributions to total intake among pregnant women in the SELMA study. Indoor Air 31: 1495–1508.u

Qiu W, Zhan H, Hu J, Zhang T, Xu H, Wong M, Xu B, Zheng C (2019) The occurrence, potential toxicity, and toxicity mechanism of bisphenol S, a substitute of bisphenol A: a critical review of recent progress. Ecotoxicol Environ Saf 173:192–202

Qu M, Wang L, Liu F, Zhao Y, Shi X, Li S (2021) Characteristics of dust-phase phthalates in dormitory, classroom, and home and non-dietary exposure in Beijing, China. Environ Sci Pollut Res Int 28:38159–38172

Radke EG, Braun JM, Meeker JD, Cooper GS (2018) Phthalate exposure and male reproductive outcomes: a systematic review of the human epidemiological evidence. Environ Int 121(Pt 1): 764–793

Radke EG, Galizia A, Thayer KA, Cooper GS (2019) Phthalate exposure and metabolic effects: a systematic review of the human epidemiological evidence. Environ Int 132:104768

Radke EG, Braun JM, Nachman RM, Cooper GS (2020) Phthalate exposure and neurodevelopment: a systematic review and meta-analysis of human epidemiological evidence. Environ Int 137:105408

Raffy G, Mercier F, Blanchard O, Derbez M, Dassonville C, Bonvallot N, Glorennec P, Le Bot B (2017) Semi-volatile organic compounds in the air and dust of 30 French schools: a pilot study. Indoor Air 27:114–127

Rantakokko P, Kumar E, Braber J, Huang T, Kiviranta H, Cequier E, Thomsen C (2019) Concentrations of brominated and phosphorous flame retardants in Finnish house dust and insights into children's exposure. Chemosphere 223:99–107

Rasmussen PE, Kubwabo C, Gardner HD, Levesque C, Beauchemin S (2022) Relationships between house characteristics and exposures to metal(loid)s and synthetic organic contaminants evaluated using settled indoor dust. Int J Environ Res Public Health 19:10329

Reche C, Viana M, Querol X, Corcellas C, Barceló D, Eljarrat E (2019) Particle-phase concentrations and sources of legacy and novel flame retardants in outdoor and indoor environments across Spain. Sci Total Environ 649:1541–1552

Regueiro J, Llompart M, Garcia-Jares C, Cela R (2007) Development of a high-throughput method for the determination of organochlorinated compounds, nitromusks and pyrethroid insecticides in indoor dust. J Chromatogr A 1174:112–124

Ren Y, Cheng T, Chen J (2006) Polycyclic aromatic hydrocarbons in dust from computers: one possible indoor source of human exposure. Atmos Environ 40:6956–6965

Riddell N, Jin UH, Safe S, Cheng Y, Chittim B, Konstantinov A, Parette R, Pena-Abaurrea M, Reiner EJ, Poirier D, Stefanac T, McAlees AJ, McCrindle R (2015) Characterization and biological potency of mono- to tetra-halogenated carbazoles. Environ Sci Technol 49:10658–10666

Romagnoli P, Balducci C, Perilli M, Gherardi M, Gordiani A, Gariazzo C, Gatto MP, Cecinato A (2014) Indoor PAHs at schools, homes and offices in Rome, Italy. Atmos Environ 92:51–59

Romagnoli P, Balducci C, Perilli M, Vichi F, Imperiali A, Cecinato A (2016) Indoor air quality at life and work environments in Rome, Italy. Environ Sci Pollut Res Int 23:3503–3516

Romagnoli P, Balducci C, Cecinato A, L'Episcopo N, Gariazzo C, Gatto MP, Gordiani A, Gherardi M (2017) Fine particulate-bound polycyclic aromatic hydrocarbons in vehicles in Rome, Italy. Environ Sci Pollut Res Int 24:3493–3505

Roosens L, Abdallah MAE, Harrad S, Neels H, Covaci A (2009) Exposure to Hexabromocyclododecanes (HBCDDs) via dust ingestion, but not diet, correlates with concentrations in human serum: preliminary results. Environ Health Perspect 117:1707–1712

Rose M, Fernandes A, Mortimer D, Baskaran C (2015) Contamination of fish in UK fresh water systems: risk assessment for human consumption. Chemosphere 122:183–189

Rosenberg C, Hämeilä M, Tornaeus J, Säkkinen K, Puttonen K, Korpi A, Kiilunen M, Linnainmaa M, Hesso A (2011) Exposure to flame retardants in electronics recycling sites. Ann Occup Hyg 55:658–665

Rowdhwal SSS, Chen J (2018) Toxic effects of di-2-ethylhexyl phthalate: an overview. Biomed Res Int 22:1750368

RPA (Risk & Policy Analysts Limited) (2005) Risk assessment studies on targeted consumer applications of certain organotin compounds. Report prepared for the European Commission. https://ec.europa.eu/docsroom/documents/13041/attachments/1/translations/en/renditions/native

Rudel RA, Camann DE, Spengler JD, Korn LR, Brody JG (2003) Phthalates, alkylphenols, pesticides, polybrominated diphenyl ethers, and other endocrinedisrupting compounds in indoor air and dust. Environ Sci Technol 37:4543–4553

Rudel RA, Dodson RE, Perovich LJ, Morello-Frosch R, Camann DE, Zuniga MM, Yau AY, Just AC, Brody JG (2010) Semivolatile endocrine-disrupting compounds in paired indoor and outdoor air in two northern California communities. Environ Sci Technol 44:6583–6590

Sahlström L, Sellström U, de Wit CA (2012) Clean-up method for determination of established and emerging brominated flame retardants in dust. Anal Bioanal Chem 404:459–466

Sahlström L, Sellström U, de Wit CA, Lignell S, Darnerud PO (2015) Estimated intake of brominated flame retardants via diet and dust compared to internal concentrations in a Swedish mother-toddler cohort. Int J Hyg Environ Health 218:422–432

Saito K, Takekuma M, Ogawa M, Kobayashi S, Sugawara Y, Ishizuka M, Nakazawa H, Matsuki Y (2003) Extraction and cleanup methods of dioxins in house dust from two cities in Japan using accelerated solvent extraction and a disposable multi-layer silica-gel cartridge. Chemosphere 53:137–142

Saito I, Onuki A, Seto H (2007) Indoor organophosphate and polybrominated flame retardants in Tokyo. Indoor Air 17:28–36

Sakhi AK, Cequier E, Becher R, Bølling AK, Borgen AR, Schlabach M, Schmidbauer N, Becher G, Schwarze P, Thomsen C (2019) Concentrations of selected chemicals in indoor air from Norwegian homes and schools. Sci Total Environ 674:1–8

Sangiorgi G, Ferrero L, Ferrini BS, Lo Porto C, Perrone MG, Zangrando Z, Gambaro A, Lazzati Z, Bolzacchini E (2013) Indoor airborne particle sources and semi-volatile partitioning effect of outdoor fine PM in offices. Atmos Environ 65:205–214

Santillo D, Labunska I, Davidson H, Johnston P, Strutt M, Knowles O (2003) Consuming chemicals. Hazardous chemicals in house dust as an indicator of chemical exposure in the home. https://www.greenpeace.to/publications/housedust_uk_2003.pdf

SCCP (Scientific Committee on Consumer Products) (2007) Opinion on homosalate. https://ec.europa.eu/health/ph_risk/committees/04_sccp/docs/sccp_o_097.pdf

SCCS (Scientific Committee on Consumer Safety) (2010) Cyclomethicone, octamethylcyclotetrasiloxane (cyclotetrasiloxane, D4) and decamethylcyclopentasiloxane (cyclopentasiloxane, D5). SCCS/1241/10

SCCS (2013) Opinion on parabens. https://ec.europa.eu/health/scientific_committees/consumer_safety/docs/sccs_o_132.pdf

SCCS (2015) Opinion on decamethylcyclopentasiloxane (cyclopentasiloxane, D5) in cosmetic products. SCCS/1549/15

SCCS (2017) Opinion of the Scientific Committee on Consumer Safety (SCCS) – Final version of the opinion on decamethylcyclopentasiloxane (cyclopentasiloxane, D5) in cosmetic products. Regul Toxicol Pharmacol 83:117–118

SCENIHR (Scientific Committee on Emerging and Newly-Identified Health Risks) (2015) Opinion on The safety of medical devices containing DEHP plasticized PVC or other plasticizers on neonates and other groups possibly at risk (2015 update). https://ec.europa.eu/health/scientific_committees/emerging/docs/scenihr_o_047.pdf

Schacht U, Gras B, Sievers S (1995) Determination of polybrominated and polychlorinated dibenzodioxins and -furans in various environmentally relevant materials. Organohalogen Comp 22:325–334. [in German]

SCHER (Scientific Committee on Health and Environmental Risks) (2006) Revised assessment of the risks to health and the environment associated with the use of organotin compounds TBT, DBT, DOT and TPT. https://ec.europa.eu/health/ph_risk/committees/04_scher/docs/scher_o_047.pdf

Scher DP, Kelly JE, Huset CA, Barry KM, Yingling VL (2019) Does soil track-in contribute to house dust concentrations of perfluoroalkyl acids (PFAAs) in areas affected by soil or water contamination? J Expo Sci Environ Epidemiol 29:218–226

Schleibinger H, Hott U, Marchl D, Braun P, Plieninger P, Rüden H (2001) VOC-Konzentrationen in Innenräumen des Großraums Berlin im Zeitraum 1988-1999. Gefahrstoffe–Reinhaltung der Luft 1/2: 26–38. [in German]

Scholz H (2000) Communication to Marchl, D. (B.A.U.C.H.) on the results of investigations of indoor air in the period from 1995 to 2000 by GfU. (citied after Schleibinger et al. 2001). [in German]

Schreder ED, La Guardia MJ (2014) Flame retardant transfers from U.S. households dust and laundry wastewater to the aquatic environment. Environ Sci Technol 48:11575–11583

Schupp T, Plehiers PM (2022) Absorption, distribution, metabolism, and excretion of methylene diphenyl diisocyanate and toluene diisocyanate: many similarities and few differences. Toxicol Ind Health 38:500–528

Schwenk M, Gabrio T, Päpke O, Wallenhorst T (2002) Human biomonitoring of polychlorinated biphenyls and polychlorinated dibenzodioxins and dibenzofuranes in teachers working in a PCB-contaminated school. Chemosphere 47:229–233

Sha B, Dahlberg A-K, Wiberg K, Lutz A (2018) Fluortelomer alcohols (FTOHs), brominated flame retardants (BFRs), organophosphorus flame retardants (PFRs) and cyclic volatile methylsiloxanes (cVMSs) in indoor air from occupational and home environments. Environ Pollut 241:319–330

Shang H, Fan X, Kubwabo C, Rasmussen PE (2019) Short-chain and medium-chain chlorinated paraffins in Canadian house dust and NIST SRM 2585. Environ Sci Pollut Res Int 26:7453–7462

Sharkey M, Harrad S, Abou-Elwafa Abdallah M, Drage DS, Berresheim H (2020) Phasing-out of legacy brominated flame retardants: the UNEP Stockholm Convention and other legislative action worldwide. Environ Int 144:106041

Shi Y, Liu X, Xie Q, Pan XF, Mei Z (2021) Plastic additives and personal care products in south China house dust and exposure in child-mother pairs. Environ Pollut 281:116347

Shields HC, Fleischer DM, Weschler CJ (1996) Comparisons among VOCs measured in three types of U.S. commercial buildings with different occupant densities. Indoor Air 6:2–17

Shin HM, Moschet C, Young TM, Bennett DH (2020) Measured concentrations of consumer product chemicals in California house dust: Implications for sources, exposure, and toxicity potential. Indoor Air 30:60–75

Shoeib M, Harner T, Webster GM, Lee SC (2011) Indoor sources of poly- and perfluorinated compounds (PFCS) in Vancouver, Canada: implications for human exposure. Environ Sci Technol 45:7999–8005

Shoeib M, Harner T, Webster GM, Sverko E, Cheng Y (2012) Legacy and current-use flame retardants in house dust from Vancouver, Canada. Environ Pollut 169:175–182

Shoeib T, Webster GM, Hassan Y, Tepe S, Yalcin M, Turgut C, Kurt-Karakus PB, Jantunen L (2019) Organophosphate esters in house dust: a comparative study between Canada, Turkey and Egypt. Sci Total Environ 650:193–201

Shu H, Jönsson BAG, Gennings C, Lindh CH, Nånberg E, Bornehag CG (2019) PVC flooring at home and uptake of phthalates in pregnant women. Indoor Air 29:43–54

Simonetti G, Di Filippo P, Riccardi C, Pomata D, Sonego E, Buiarelli F (2020) Occurrence of halogenated pollutants in domestic and occupational indoor dust. Int J Environ Res Public Health 17:3813

Sivanantham S, Dassonville C, Grégoire A, Malingre L, Ramalho O, Mandin C (2021) Coexposure to indoor pollutants in French schools and associations with building characteristics. Energ Buildings 252:111424

Sjödin A, Carlsson H, Thuresson K, Sjölin S, Bergman A, Östman C (2001) Flame retardants in indoor air at an electronics recycling plant and at other work environments. Environ Sci Technol 35:448–454

Skakkebaek NE, Rajpert-De Meyts E, Buck Louis GM, Toppari J, Andersson AM, Eisenberg ML, Jensen TK, Jørgensen N, Swan SH, Sapra KJ, Ziebe S, Priskorn L, Juul A (2016) Male reproductive disorders and fertility trends: Influences of environment and genetic susceptibility. Physiol Rev 96:55–97

Smith DJ, Oliver CE, Taylor JB, Anderson RC (2012) Efficacy, metabolism, and toxic responses to chlorate salts in food and laboratory animals. J Anim Sci 90:4098–4117

Soderlund DM (2012) Molecular mechanisms of pyrethroid insecticide neurotoxicity. Recent Adv Arch Toxicol 86:165–181

Sofuoglu A, Kiymet N, Kavcar P, Sofuoglu SC (2010) Polycyclic and nitro musks in indoor air: a primary school classroom and a women's sport center. Indoor Air 20:515–522

South L, Saini A, Harner T, Niu S, Parnis JM, Mastin J (2022) Medium- and long-chain chlorinated paraffins in air: a review of levels, physicochemical properties, and analytical considerations. Sci Total Environ 843:157094

Steenland K, Fletcher T, Stein CR, Bartell SM, Darrow L, Lopez-Espinosa MJ, Barry Ryan P, Savitz DA (2020) Review: evolution of evidence on PFOA and health following the assessments of the C8 Science Panel. Environ Int 145:106125

Steffen C, Wetzel E (1993) Chlorate poisoning: mechanism of toxicity. Toxicology 84:217–231

Strandberg B, Omelekhina Y, Klein M, Krais AM, Wierzbicka A (2022) Particulate-bound polycyclic aromatic hydrocarbons (PAHs) and their nitro and oxy-derivative compounds collected inside and outside occupied homes in Southern Sweden. Polycyclic Arom Compds 27 Oct 2022

Strynar MJ, Lindstrom AB (2008) Perfluorinated compounds in house dust from Ohio and North Carolina. USA. Environ Sci Technol 42:3751–3756

Su C-K, Lu J-H, Chao H-R, Chang W-H, Tsai M-H, Wang C-L, Lu I-C, Chang Y-T, Chuang H-C, Mansor W-N-W, Hsu Y-C, Tsai Y-I, Ma S-M (2022) Polybrominated dibenzo-p-dioxins/furans (PBDD/Fs) and diphenyl ethers (PBDEs) in the indoor and outdoor of gymnasiums. Aerosol Air Qual Res 22:220264

Subedi B, Sullivan KD, Dhungana B (2017) Phthalate and non-phthalate plasticizers in indoor dust from childcare facilities, salons, and homes across the USA. Environ Pollut 230:701–708

Sun J, Wang Q, Zhuang S, Zhang A (2016) Occurrence of polybrominated diphenyl ethers in indoor air and dust in Hangzhou, China: level, role of electric appliances, and human exposure. Environ Pollut 218:942–949

Sun J, Xu Y, Zhou H, Zhang A, Qi H (2018) Levels, occurrence and human exposure to novel brominated flame retardants (NBFRs) and Dechlorane Plus (DP) in dust from different indoor environments in Hangzhou, China. Sci Total Environ 631–632:1212–1220

Suzuki G, Takigami H, Watanabe M, Takahashi S, Nose K, Asari M, Sakai S (2008) Identification of brominated and chlorinated phenols as potential thyroid-disrupting compounds in indoor dusts. Environ Sci Technol 42:1794–1800

Suzuki G, Someya M, Takahashi S, Tanabe S, Sakai S, Takigami H (2010) Dioxin-like activity in Japanese indoor dusts evaluated by means of in vitro bioassay and instrumental analysis: brominated dibenzofurans are an important contributor. Environ Sci Technol 44:8330–8336

Takeuchi S, Kojima H, Saito I, Jin K, Kobayashi S, Tanaka-Kagawa T, Jinno H (2014) Detection of 34 plasticizers and 25 flame retardants in indoor air from houses in Sapporo, Japan. Sci Total Environ 491:28–33

Takigami H, Suzuki G, Hirai Y, Sakai S (2009) Brominated flame retardants and other polyhalogenated compounds in indoor air and dust from two houses in Japan. Chemosphere 76:270–277

Tan J, Cheng SM, Loganath A, Chong YS, Obbard JP (2007) Selected organochlorine pesticide and polychlorinated biphenyl residues in house dust in Singapore. Chemosphere 68:1675–1682

Tan H, Chen D, Peng C, Liu X, Wu Y, Li X, Du R, Wang B, Guo Y, Zeng EY (2018) Novel and traditional organophosphate esters in house dust from South China: association with hand wipes and exposure estimation. Environ Sci Technol 52:11017–11026

Tan H, Yang L, Yu Y, Guan Q, Liu X, Li L, Chen D (2019) Co-existence of organophosphate di- and tri-Esters in house dust from South China and Midwestern United States: Implications for human exposure. Environ Sci Technol 53:4784–4793

Tan H, Yang L, Liang X, Huang D, Qiao X, Dai Q, Chen D, Cai Z (2023) Nonphthalate plasticizers in house dust from multiple countries: an increasing threat to humans. Environ Sci Technol 57: 3634–3644

Tang X, Misztal PK, Nazaroff WW, Goldstein AH (2015) Siloxanes are the most abundant volatile organic compound emitted from engineering students in a classroom. Environ Sci Technol Lett 2:303–307

Tang CS, Lung SC, Chang TY, Tu HH, Chang LT (2019a) Investigation of microenvironmental exposures to particle-bound polycyclic aromatic hydrocarbons for elementary school children. Int J Environ Res Public Health 16:4390

Tang S, Tan H, Liu X, Chen D (2019b) Legacy and alternative flame retardants in house dust and hand wipes from South China. Sci Total Environ 656:1–8

Tang B, Christia C, Malarvannan G, Liu YE, Luo XJ, Covaci A, Mai BX, Poma G (2020) Legacy and emerging organophosphorus flame retardants and plasticizers in indoor microenvironments from Guangzhou, South China. Environ Int 143:105972

Tao F, Abdallah MA, Harrad S (2016) Emerging and legacy flame retardants in UK indoor air and dust: evidence for replacement of PBDEs by emerging flame retardants? Environ Sci Technol 50:13052–13061

Tao F, Sellström U, de Wit CA (2019) Organohalogenated flame retardants and organophosphate esters in office air and dust from Sweden. Environ Sci Technol 53:2124–2133

Tao L, Tan H, Qiao X, Li L, Yu Y, Xie J, Chen D (2022) Emerging plasticizers in South China house dust and hand wipes: calling for potential concern? Environ Sci Technol 56:12190–12199

Tatur A, Kiciska E, Wasiowska A, Gromadka P (2009) Polycyclic aromatic hydrocarbons in house dust from Warsaw. Ecol Chem Eng A 16:867–874

Tay JH, Sellström U, Papadopoulou E, Padilla-Sánchez JA, Haug LS, de Wit CA (2017) Human exposure to legacy and emerging halogenated flame retardants via inhalation and dust ingestion in a Norwegian Cohort. Environ Sci Technol 51:8176–8184

Taylor KM, Weisskopf M, Shine J (2014) Human exposure to nitro musks and the evaluation of their potential toxicity: an overview. Environ Health 13:14

Thomas K, Xue J, Williams R, Jones P, Whitaker D (2012) Polychlorinated biphenyls (PCBs) in school buildings: sources, environmental levels, and exposures. Report EPA/600/R-12/051. https://www.epa.gov/sites/default/files/2015–08/documents/pcb_epa600r12051_final.pdf

Thuresson K, Björklund AW, de Wit CA (2012) Tri-decabrominated diphenyl ethers and hexabromocyclododecane in indoor air and dust from Stockholm microenvironments 1: levels and profiles. Sci Total Environ 414:713–721

Toms LM, Hearn L, Kennedy K, Harden F, Bartkow M, Temme C, Mueller JF (2009) Concentrations of polybrominated diphenyl ethers (PBDEs) in matched samples of human milk, dust and indoor air. Environ Int 35:864–869

Tran TM, Kannan K (2015) Occurrence of phthalate diesters in particulate and vapor phases in indoor air and implications for human exposure in Albany, New York, USA. Sci Total Environ 511:138–144

Tran TM, Abualnaja KO, Asimakopoulos AG, Covaci A, Gevao B, Johnson-Restrepo B, Kumosani TA, Malarvannan G, Minh TB, Moon HB, Nakata H, Sinha RK, Kannan K (2015) A survey of cyclic and linear siloxanes in indoor dust and their implications for human exposures in twelve countries. Environ Int 78:39–44

Tran TM, Hoang AQ, Le ST, Minh TB, Kannan K (2019) A review of contamination status, emission sources, and human exposure to volatile methyl siloxanes (VMSs) in indoor environments. Sci Total Environ 691:584–594

Tue NM, Suzuki G, Takahashi S, Kannan K, Takigami H, Tanabe S (2013) Dioxin-related compounds in house dust from New York State: occurrence, in vitro toxic evaluation, and implications for indoor exposure. Environ Pollut 181:75–80

Uhl M, Hohenblum P, Scharf S, Trimbacher C (2004) House dust - an indicator of indoor pollution. Report BE-258, Umweltbundesamt Österreich. https://www.umweltbundesamt.at/fileadmin/site/publikationen/BE258.pdf [in German]

UNEP (UN Environment Programme) (2010) Supporting document for technical review of the implications of recycling commercial pentabromodiphenyl ether and commercial octabromodiphenyl ether. Persistent Organic Pollutants Review Committee, Sixth meeting

UNEP (2013) Report of the Persistent Organic Pollutants Review Committee on the work of its ninth meeting. Addendum – Risk management evaluation on chlorinated naphthalenes. Rome, Italy. http://chm.pops.int/TheConvention/POPsReviewCommittee/Meetings/POPRC9/Documents/tabid/3281/Default.aspx

UNEP (2015) Road map for the development of alternatives to DDT. UNEP Chemicals Branch, DTIE, Switzerland

US EPA (2010) Development of a relative potency factor (RPF) approach for polycyclic aromatic hydrocarbon (PAH) Mixtures. In support of summary information on the Integrated Risk Information System (IRIS). EPA/635/R-08/012A

US EPA (2015) High production volume information system (HPVIS). Detailed chemical results: 1,2-Benzenedicarboxylic acid, 3,4,5,6-tetrabromo-, bis(2-ethylhexyl) ester: http://ofmpub.epa.gov/oppthpv/quicksearch.display?pChem=102247)

US EPA (2017a) Toxicological review of benzo[a]pyrene. EPA/635/R-17/003Fa. Washington, DC, USA. https://cfpub.epa.gov/ncea/iris/iris_documents/documents/toxreviews/0136tr.pdf

US EPA (2017b) Health assessment of 2,3,7,8-tetrachlorodibenzo-p-dioxin (TCDD) and related compounds. https://cfpub.epa.gov/ncea/iris_drafts/dioxin/nas-review/index.cfm

US EPA (2020a) Risk evaluation for cyclic aliphatic bromide cluster (HBCD). https://www.epa.gov/node/227433

US EPA (2020b) Final scope of the risk evaluation for 4,4′-(1-Methylethylidene)bis[2, 6-dibromophenol] (TBBPA). https://www.epa.gov/sites/default/files/2020-09/documents/casrn_79-94-7_44-1-methylethylidenebis2_6-dibromophenol_tbbpa_finalscope.pdf

Van den Berg M, Birnbaum L, Bosveld AT, Brunström B, Cook P, Feeley M, Giesy JP, Hanberg A, Hasegawa R, Kennedy SW, Kubiak T, Larsen JC, van Leeuwen FX, Liem AK, Nolt C, Peterson RE, Poellinger L, Safe S, Schrenk D, Tillitt D, Tysklind M, Younes M, Waern F, Zacharewski T (1998) Toxic equivalency factors (TEFs) for PCBs, PCDDs, PCDFs for humans and wildlife. Environ Health Perspect 106:775–792

Van den Berg M, Birnbaum LS, Denison M, De Vito M, Farland W, Feeley M, Fiedler H, Hakansson H, Hanberg A, Haws L, Rose M, Safe S, Schrenk D, Tohyama C, Tritscher A, Tuomisto J, Tysklind M, Walker N, Peterson RE (2006) The 2005 World Health Organization reevaluation of human and Mammalian toxic equivalency factors for dioxins and dioxin-like compounds. Toxicol Sci 93:223–241

Van den Eede N, Dirtu AC, Neels H, Covaci A (2011) Analytical developments and preliminary assessment of human exposure to organophosphate flame retardants from indoor dust. Environ Int 37:454–461

Van Drooge B, Rivas I, Querol X, Sunyer J, Grimalt JO (2020) Organic air quality markers of indoor and outdoor $PM_{2.5}$ aerosols in primary schools from Barcelona. Int J Environ Res Public Health 17:3685

Vasiljevic T, Harner T (2021) Bisphenol A and its analogues in outdoor and indoor air: properties, sources and global levels. Sci Total Environ 789:148013

Velázquez-Gómez M, Hurtado-Fernández E, Lacorte S (2019) Differential occurrence, profiles and uptake of dust contaminants in the Barcelona urban area. Sci Total Environ 648:1354–1370

Vella AJ, Chircop C, Micallef T, Pace C (2015) Perchlorate in dust fall and indoor dust in Malta: an effect of fireworks. Sci Total Environ 521–522:46–51

Venier M, Audy O, Vojta S, Bečanová J, Romanak K, Melymuk L, Krátká M, Kukučka P, Okeme J, Saini A, Diamond ML, Klánová J (2016) Brominated flame retardants in the indoor environment — comparative study of indoor contamination from three countries. Environ Int 94:150–160

Venkatesan AK, Halden RU (2014) Contribution of polybrominated dibenzo-p-dioxins and dibenzofurans (PBDD/Fs) to the toxic equivalency of dioxin-like compounds in archived biosolids from the U.S. EPA's 2001 National Sewage Sludge Survey. Environ Sci Technol 48:10843–10849

Vesin A, Glorennec P, Le Bot B, Worthama H, Bonvallot N, Quivet E (2013) Transfluthrin indoor air concentration and inhalation exposure during application of electric vaporizers. Environ Int 60:1–6

Völkel W, Kiranoglu M, Fromme H (2008) Determination of free and total bisphenol A in human urine to assess daily uptake as a basis for a valid risk assessment. Toxicol Lett 179:155–162

Volland G, Hansen D, Neuwirth A, Zöltzer D, Krause G (2006) Dioxin-like polychlorinated biphenyls (PCBs) and polychlorinated dibenzodioxins/furans (PCDD/Fs) in indoor air and house dust. Gefahrst Reinh Luft 66:99–106. [in German]

Von Hahn N, Van Gelder R, Breuer D, Hahn J-U, Gabriel S, Kleine H (2011) Derivation of indoor workplace reference values. Gefahrst Reinh Luft 71:314–322. [in German]

Von Manikowsky S, Sagunski H, Päpke O, Fabig KR, Neus H, Schümann M, Spannhake K, Karmaus W, Osius N, Zier B (1998) PCDD/PCDF Indoor exposure in day-care centers and PCDD/PCDF blood concentrations of female employees. Environ Health Perspect 106(Suppl 2):707–714

Vorkamp K, Thomsen M, Frederiksen M, Pedersen M, Knudsen LE (2011) Polybrominated diphenyl ethers (PBDEs) in the indoor environment and associations with prenatal exposure. Environ Int 37:1–10

Vykoukalová M, Venier M, Vojta Š, Melymuk L, Bečanová J, Romanak K, Prokeš R, Okeme JO, Saini A, Diamond ML, Klánová J (2017) Organophosphate esters flame retardants in the indoor environment. Environ Int 106:97–104

Waheed S, Khan MU, Sweetman AJ, Jones KC, Moon H-B, Malik RN (2020) Exposure of polychlorinated naphthalenes (PCNs) to Pakistani populations via non-dietary sources from neglected e-waste hubs: a problem of high health concern. Environ Pollut 259:113838

Walker G, Hostrup O, Hoffmann W, Butte W (1999) Biocides in dust. Staub-Reinh Luft 59:33–41. [in German]

Wan Y, Wua Q, Abualnaja KO, Asimakopoulos AG, Covaci A, Gevao B, Johnson-Restrepo B, Kumosani TA, Malarvannan G, Moon H-B, Nakata H, Sinha RA, Minhk TB, Kannan K (2015) Occurrence of perchlorate in indoor dust from the United States and eleven other countries: implications for human exposure. Environ Int 75:166–171

Wan Y, Xue J, Kannan K (2016) Benzothiazoles in indoor air from Albany, New York, USA, and its implications for inhalation exposure. J Hazard Mater 311:37–42

Wan Y, North ML, Navaranjan G, Ellis AK, Siegel JA, Diamond ML (2022) Indoor exposure to phthalates and polycyclic aromatic hydrocarbons (PAHs) to Canadian children: the Kingston allergy birth cohort. J Expo Sci Environ Epidemiol 32:69–81

Wang JY, Ma J, Chen SJ, Tian M, Luo XJ, Mai BX (2010) Brominated flame retardants in house dust from e-waste recycling and urban areas in South China: implications on human exposure. Environ Int 36:535–541

Wang J, Tian M, Chen SJ, Zheng J, Luo XJ, An TC, Mai BX (2011) Dechlorane Plus in house dust from E-waste recycling and urban areas in South China: sources, degradation, and human exposure. Environ Toxicol Chem 30:1965–1972

Wang L, Liao C, Liu F, Wu Q, Guo Y, Moon HB, Nakata H, Kannan K (2012) Occurrence and human exposure of p-hydroxybenzoic acid esters (parabens), bisphenol A diglycidyl ether (BADGE), and their hydrolysis products in indoor dust from the United States and three East Asian countries. Environ Sci Technol 46:11584–11593

Wang W, Wu F, Zheng J, Wong MH (2013a) Risk assessments of PAHs and Hg exposure via settled house dust and street dust, linking with their correlations in human hair. J Hazard Mater 263:627–637

Wang W, Huang MJ, Zheng JS, Cheung KC, Wong MH (2013b) Exposure assessment and distribution of polychlorinated biphenyls (PCBs) contained in indoor and outdoor dusts and the impacts of particle size and bioaccessibility. Sci Total Environ 463–464:1201–1209

Wang DG, Norwood W, Alaee M, Byer JD, Brimble S (2013c) Review of recent advances in research on the toxicity, detection, occurrence and fate of cyclic volatile methyl siloxanes in the environment. Chemosphere 93:711–725

Wang L, Asimakopoulos AG, Moon HB, Nakata H, Kannan K (2013d) Benzotriazole, benzothiazole, and benzophenone compounds in indoor dust from the United States and east Asian countries. Environ Sci Technol 47:4752–4759

Wang W, Huang M-j, Chan C-Y, Cheung KC, Wong MH (2013e) Risk assessment of non-dietary exposure to polycyclic aromatic hydrocarbons (PAHs) via house $PM_{2.5}$, TSP and dust and the implications from human hair. Atmos Environ 73:204–213

Wang X, Tao W, Xu Y, Feng J, Wang F (2014) Indoor phthalate concentration and exposure in residential and office buildings in Xi'an, China. Atmos Environ 87:146–152

Wang BL, Pang ST, Zhang XL, Li XL, Sun YG, Lu XM, Zhang Q (2015a) Levels of polybrominated diphenyl ethers in settled house dust from urban dwellings with resident preschool-aged children in Nanjing, China. Arch Environ Contam Toxicol 68:9–19

Wang BL, Pang ST, Sun JP, Zhang XL, Li XL, Sun YG, Lu XM, Zhang Q (2015b) Levels of polychlorinated biphenyls in settled house dust from urban dwellings in China and their neurodevelopmental effects on preschool-aged children. Sci Total Environ 505:402–408

Wang W, Abualnaja KO, Asimakopoulos AG, Covaci A, Gevao B, Johnson-Restrepo B, Kumosani TA, Malarvannan G, Minh TB, Moon HB, Nakata H, Sinha RK, Kannan K (2015c) A comparative assessment of human exposure to tetrabromobisphenol A and eight bisphenols including bisphenol A via indoor dust ingestion in twelve countries. Environ Int 83:183–191

Wang P, Zhang Q, Zhang H, Wang T, Sun H, Zheng S, Li Y, Liang Y, Jiang G (2016a) Sources and environmental behaviors of Dechlorane Plus and related compounds – a review. Environ Int 88: 206–220

Wang Y, Hu J, Lin W, Wang N, Li C, Luo P, Hashmi MZ, Wang W, Su X, Chen C, Liu Y, Huang R, Shen C (2016b) Health risk assessment of migrant workers' exposure to polychlorinated biphenyls in air and dust in an e-waste recycling area in China: indication for a new wealth gap in environmental rights. Environ Int 87:33–41

Wang Z, Wang S, Nie J, Wang Y, Liu Y (2017a) Assessment of polycyclic aromatic hydrocarbons in indoor dust from varying categories of rooms in Changchun city, northeast China. Environ Geochem Health 39:15–27

Wang J, Guinot B, Dong Z, Li X, Xu H, Xiao S, Ho SSH, Liu S, Cao J (2017b) $PM_{2.5}$-Bound polycyclic aromatic hydrocarbons (PAHs), oxygenated-PAHs and phthalate esters (PAEs) inside and outside middle school classrooms in Xi'an, China: concentration, characteristics and health risk assessment. Aerosol Air Qual Res 17:1811–1824

Wang L, Gong M, Xu Y, Zhang Y (2017c) Phthalates in dust collected from various indoor environments in Beijing, China and resulting non-dietary human exposure. Build Environ 124:315–322

Wang X, Banks APW, He C, Drage DS, Gallen CL, Li Y, Li Q, Thai PK, Mueller JF (2019) Polycyclic aromatic hydrocarbons, polychlorinated biphenyls and legacy and current pesticides in indoor environment in Australia – occurrence, sources and exposure risks. Sci Total Environ 693:133588

Wang M, Jia S, Lee SH, Chow A, Fang M (2021a) Polycyclic aromatic hydrocarbons (PAHs) in indoor environments are still imposing carcinogenic risk. J Hazard Mater 409:124531

Wang B, Yao Y, Wang Y, Chen H, Sun H (2021b) Per- and polyfluoroalkyl substances in outdoor and indoor dust from mainland China: contributions of unknown precursors and implications for human exposure. Environ Sci Technol Mar 26

Wang L, Jia Y, Hu J (2022) Nine alkyl organophosphate triesters newly identified in house dust. Environ Int 165:107333

Wei G-L, Li D-Q, Zhuo M-N, Liao Y-S, Xie Z-X, Guo T-L, Li J-J, Zhang S-Y, Liang Z-Q (2015) Organophosphorus flame retardants and plasticizers: sources, occurrence, toxicity and human exposure. Environ Pollut 196:29–46

Wei W, Dassonville C, Sivanantham S, Gregoire A, Mercier F, Le Bot B, Malingre L, Ramalho O, Derbez M, Mandin C (2021) Semivolatile organic compounds in French schools: partitioning between the gas phase, airborne particles and settled dust. Indoor Air 31:156–169

Weiss JM, Jones B, Koekkoek J, Bignert A, Lamoree MH (2021) Per- and polyfluoroalkyl substances (PFASs) in Swedish household dust and exposure of pet cats. Environ Sci Pollut Res Int 28:39001–39013

Weitekamp CA, Phillips LJ, Carlson LM, DeLuca NM, Cohen Hubal EA, Lehmann GM (2021) A state-of-the-science review of polychlorinated biphenyl exposures at background levels: relative contributions of exposure routes. Sci Total Environ 776:145912

Wemken N, Drage DS, Abdallah MA, Harrad S, Coggins MA (2019) Concentrations of brominated flame retardants in indoor air and dust from Ireland reveal elevated exposure to decabromodiphenyl ethane. Environ Sci Technol 53:9826–9836

Wensing M, Bauhof H (1999) Air contaminants in passenger car interiors. Handbook for Bioclimate and Air Hygiene. [in German]

Whitehead T, Metayer C, Petreas M, Does M, Buffler PA, Rappaport SM (2013a) Polycyclic aromatic hydrocarbons in residential dust: sources of variability. Environ Health Perspect 121: 543–550

Whitehead T, Brown FR, Metayer C, Park J-S, Does M, Petreas MX, Buffler PA, Rappaport SM (2013b) Polybrominated diphenyl ethers in residential dust: sources of variability. Environ Int 57–58:11–24

Whitehead TP, Brown FR, Metayer C, Park JS, Does M, Dhaliwal J, Petreas MX, Buffler PA, Rappaport SM (2014) Polychlorinated biphenyls in residential dust: sources of variability. Environ Sci Technol 48:157–164

Whitehead TP, Crispo Smith S, Park JS, Petreas MX, Rappaport SM, Metayer C (2015) Concentrations of persistent organic pollutants in California children's whole blood and residential dust. Environ Sci Technol 49:9331–9340

WHO (2001) International Programme on Chemical Safety. Chlorinated naphthalenes. Concise international chemical assessment document 34. Geneva, Switzerland

WHO (2010) Guidelines for indoor air quality: selected pollutants. World Health Organization, Regional Office for Europe, Copenhagen (http://www.euro.who.int/en/what-we-do/health-topics/environmental-health/air-quality/publications/2010/who-guidelines-for-indoor-air-quality-selected-pollutants)

Whyatt RM, Garfinkel R, Hoepner LA, Holmes D, Borjas M, Williams MK, Reyes A, Rauh V, Perera FP, Camann DE (2007) Within and between-home variability in indoor air insecticide levels during pregnancy among an inner-city cohort from New York City. Environ Health Perspect 115:383–389

Wilford BH, Harner T, Zhu J, Shoeib M, Jones K (2004) Passive sampling survey of polybrominated diphenyl ether flame retardants in indoor and outdoor air in Ottawa, Canada: implications for sources and exposure. Environ Sci Technol 38:5312–5318

Wilford BH, Shoeib M, Harner T, Zhu J, Jones K (2005) Polybrominated diphenyl ethers in indoor dust in Ottawa, Canada: implications for sources and exposure. Environ Sci Technol 39:7027–7035

Wilson LR, Palmer PM, Belanger EE, Cayo MR, Durocher LA, Hwang SA, Fitzgerald EF (2011) Indoor air polychlorinated biphenyl concentrations in three communities along the Upper Hudson River, New York. Arch Environ Contam Toxicol 61:530–538

Winkens K, Koponen J, Schuster J, Shoeib M, Vestergren R, Berger U, Karvonen AM, Pekkanen J, Kiviranta H, Cousins IT (2017) Perfluoroalkyl acids and their precursors in indoor air sampled in children's bedrooms. Environ Pollut 222:423–432

Winkens K, Giovanoulis G, Koponen J, Vestergren R, Berger U, Karvonen AM, Pekkanen J, Kiviranta H, Cousins IT (2018) Perfluoroalkyl acids and their precursors in floor dust of children's bedrooms – implications for indoor exposure. Environ Int 119:493–502

Wolff T, Stirn H (2000) Indoor air guideline values: diisocyanates. Bundesgesundheitsbl 43:505–512. [in German]

Wong F, de Wit CA, Newton SR (2018) Concentrations and variability of organophosphate esters, halogenated flame retardants, and polybrominated diphenyl ethers in indoor and outdoor air in Stockholm, Sweden. Environ Pollut 240:514–522

Wong F, Robson M, Melymuk L, Shunthirasingham C, Alexandrou N, Shoeib M, Luk E, Helm P, Diamond ML, Hung H (2019) Urban sources of synthetic musk compounds to the environment. Environ Sci Processes Impacts 21:74

Wu M, Yu G, Cao Z, Wu D, Liu K, Deng S, Huang J, Wang B, Wang Y (2016) Characterization and human exposure assessment of organophosphate flame retardants in indoor dust from several microenvironments of Beijing, China. Chemosphere 150:465–471

Wu L-H, Zhang X-M, Wang F, Gao C-J, Chen D, Palumbo JR, Guo Y, Zeng EY (2018) Occurence of bisphenol S in the environment and implications for human exposure: a short review. Sci Total Environ 615:87–98

Wu Y, Gao S, Liu Z, Zhao J, Ji B, Zeng X, Yu Z (2019) The quantification of chlorinated paraffins in environmental samples by ultra-high-performance liquid chromatography coupled with Orbitrap Fusion Tribrid mass spectrometry. J Chromatogr A 1593:102–109

Wu Z, He C, Han W, Song J, Li H, Zhang Y, Jing X, Wu W (2020a) Exposure pathways, levels and toxicity of polybrominated diphenyl ethers in humans: a review. Environ Res 187:109531

Wu Y, Romanak K, Bruton T, Blum A, Venier M (2020b) Per- and polyfluoroalkyl substances in paired dust and carpets from childcare centers. Chemosphere 251:126771

Wu Y, Venier M, Hites RA (2020c) Broad exposure of the North American environment to phenolic and amino antioxidants and to ultraviolet filters. Environ Sci Technol 54:9345–9355

Wu D, Qi J, Li Q, Chen J, Chen Y, Chen J (2021) Extreme exposure levels of PCDD/Fs inhaled from biomass burning activity for cooking in typical rural households. Environ Sci Technol 55: 7299–7306

Wu Y, Gao S, Zeng X, Liang Y, Liu Z, He L, Yuan J, Yu Z (2023) Levels and diverse composition profiles of chlorinated paraffins in indoor dust: possible sources and potential human health related concerns. Environ Geochem Health 45:4631–4642

Xia M, Ouyang X, Wang X, Shen X, Zhan Y (2018) Occupational exposure assessment of phthalate esters in indoor and outdoor microenvironments. J Environ Sci (China) 72:75–88

Xiong P, Yan X, Zhu Q, Qu G, Shi J, Liao C, Jiang G (2019) A review of environmental occurrence, fate, and toxicity of novel brominated flame retardants. Environ Sci Technol 53:13551–13569

Xu Y, Hubal EAC, Clausen PA, Little JC (2009) Predicting residential exposure to phthalate plasticizer emitted from vinyl flooring: a mechanistic analysis. Environ Sci Technol 43:2374–2380

Xu Y, Hubal EAC, Little JC (2010) Predicting residential exposure to phthalate plasticizer emitted from vinyl flooring: sensitivity, uncertainty, and implications for biomonitoring. Environ Health Perspect 118:253–258

Xu Z, Fiedler S, Pfister G, Henkelmann B, Mosch C, Völkel W, Fromme H, Schramm K-W (2013) Human exposure to fluorotelomer alcohols, perfluorooctane sulfonate and perfluorooctanoate via house dust in Bavaria, Germany. Sci Total Environ 443:485–490

Xu L, Shi Y, Liu N, Cai Y (2015) Methyl siloxanes in environmental matrices and human plasma/fat from both general industries and residential areas in China. Sci Total Environ 505:454–463

Xu F, Tang W, Zhang W, Liu L, Lin K (2016) Levels, distributions and correlations of polybrominated diphenyl ethers in air and dust of household and workplace in Shanghai, China: implication for daily human exposure. Environ Sci Pollut Res Int 23:3229–3238

Xue J, Wan Y, Kannan K (2016) Occurrence of bisphenols, bisphenol A diglycidyl ethers (BADGEs), and novolac glycidyl ethers (NOGEs) in indoor air from Albany, New York, USA, and its implications for inhalation exposure. Chemosphere 151:1–8

Xue J, Wan Y, Kannan K (2017) Occurrence of benzotriazoles (BTRs) in indoor air from Albany, New York, USA, and its implications for inhalation exposure. Toxicol Environ Chem 99:402–414

Yan C, Zheng M, Yang Q, Zhang Q, Qiu X, Zhang Y, Fu H, Zhu Y, Zhu T, Li X (2015) Commuter exposure to particulate matter and particlebound PAHs in three transportation modes in Beijing, China. Environ Pollut 204:199–206

Yang G, Liu Y, Li X (2020) Spatiotemporal distribution of ground-level ozone in China at a city level. Sci Rep 29:7229

Yang L, Liu G, Shen J, Wang M, Yang Q, Zheng M (2021) Environmental characteristics and formations of polybrominated dibenzo-p-dioxins and dibenzofurans. Environ Int 152:106450

Yao Z, Li J, Wu B, Hao X, Yin Y, Jiang X (2015a) Characteristics of PAHs from deep-frying and frying cooking fumes. Environ Sci Pollut Res Int 22:16110–16120

Yao L, Yang L, Chen J, Toda K, Wanga X, Zhang J, Yamasaki D, Nakamura Y, Sui X, Zheng L, Wen L, Xu C, Wang W (2015b) Levels, indoor-outdoor relationships and exposure risks of airborne particle-associated perchlorate and chlorate in two urban areas in Eastern Asia. Chemosphere 135:31–37

Yao Y, Zhao Y, Sun H, Chang S, Zhu L, Alder AC, Kannan K (2018) Per- and polyfluoroalkyl substances (PFASs) in indoor air and dust from homes and various microenvironments in China: implications for human exposure. Environ Sci Technol 52:3156–3166

Yao C, Yang H, Li Y (2021) A review on organophosphate flame retardants in the environment: Occurrence, accumulation, metabolism and toxicity. Sci Total Environ 15:148837

Yi AX, Leung KMY, Lam MHW, Lee J-S, Giesy JP (2012) Review of measured concentrations of triphenyltin compounds in marine ecosystems and meta-analysis of their risks to humans and the environment. Chemosphere 89:1015–1025

Yoshida T, Mimura M, Sakon N (2021) Estimating household exposure to pyrethroids and the relative contribution of inhalation pathway in a sample of Japanese children. Environ Sci Pollut Res Int 28:19310–19324

Young AS, Hauser R, James-Todd TM, Coull BA, Zhu H, Kannan K, Specht AJ, Bliss MS, Allen JG (2021) Impact of "healthier" materials interventions on dust concentrations of per- and polyfluoroalkyl substances, polybrominated diphenyl ethers, and organophosphate esters. Environ Int 150:106151

Yu X, Liu B, Yu Y, Li H, Li Q, Cui Y, Ma Y (2023) Polybrominated diphenyl ethers (PBDEs) in household dust: a systematic review on spatio-temporal distribution, sources, and health risk assessment. Chemosphere 314:137641

Yuan B, Tay JH, Padilla-Sánchez JA, Papadopoulou E, Haug LS, de Wit CA (2021) Human exposure to chlorinated paraffins via inhalation and dust ingestion in a Norwegian cohort. Environ Sci Technol 55:1145–1154

Yucuis RA, Stanier CO, Hornbuckle KC (2013) Cyclic siloxanes in air, including identification of high levels in Chicago and distinct diurnal variation. Chemosphere 92:905–910

Yue S, Zhang T, Shen Q, Song Q, Ji C, Chen Y, Mao M, Kong Y, Chen D, Liu J, Sun Z, Zhao M (2020) Assessment of endocrine-disrupting effects of emerging polyhalogenated carbazoles (PHCZs): in vitro, in silico, and in vivo evidence. Environ Int 140:105729

Zhang X, Diamond ML, Robson M, Harrad S (2011) Sources, emissions, and fate of polybrominated diphenyl ethers and polychlorinated biphenyls indoors in Toronto, Canada. Environ Sci Technol 45:3268–3274

Zhang Q, Lu X-M, Zhang X-L, Sun J-G, Zhu D-M, Wang B-L, Zhao R-Z, Zhang Z-D (2013) Levels of phthalate esters in settled house dust from urban dwellings with young children in Nanjing, China. Atmos Environ 69:258–264

Zhang T, Chen X, Wang D, Li R, Ma Y, Mo W, Sun H, Kannan K (2015) Perchlorate in indoor dust and human urine in China: contribution of indoor dust to total daily intake. Environ Sci Technol 49:2443–2450

Zhang J, Liu W, Xu Y, Cai C, Liu Y, Tao S, Liu W (2019) Distribution characteristics of and personal exposure with polycyclic aromatic hydrocarbons and particulate matter in indoor and outdoor air of rural households in Northern China. Environ Pollut 255(Pt 1):113176

Zhang H, Quan Q, Zhang M, Zhang N, Zhang W, Zhan M, Xu W, Lu L, Fan J, Wang Q (2020a) Occurrence of bisphenol A and its alternatives in paired urine and indoor dust from Chinese university students: Implications for human exposure. Chemosphere 247:125987

Zhang L, Morisaki H, Wei Y, Li Z, Yang L, Zhou Q, Zhang X, Xing W, Hu M, Shima M, Toriba A, Hayakawa K, Tang N (2020b) $PM_{2.5}$-bound polycyclic aromatic hydrocarbons and nitro-polycyclic aromatic hydrocarbons inside and outside a primary school classroom in Beijing: concentration, composition, and inhalation cancer risk. Sci Total Environ 705:135840

Zhang L, Li P, Han J, Gao S, Wang X, Yang W, Bai Z, Zhang W, Han B (2022a) Characteristics, source contributions, and source-specific health risks of $PM_{2.5}$-bound polycyclic aromatic hydrocarbons for senior citizens during the heating season in Tianjin, China. Int J Environ Res Public Health 19:4440

Zhang Y, Wu M, Xu M, Hu P, Xu X, Liu X, Cai W, Xia J, Wu D, Xu X, Yu G, Cao Z (2022b) Distribution of flame retardants among indoor dust, airborne particles and vapour phase from Beijing: spatial-temporal variation and human exposure characteristics. Environ Int 170:107557

Zheng X, Qiao L, Covaci A, Sun R, Guo H, Zheng J, Luo X, Xie Q, Mai B (2017) Brominated and phosphate flame retardants (FRs) in indoor dust from different microenvironments: Implications for human exposure via dust ingestion and dermal contact. Chemosphere 184:185–191

Zheng G, Boor BE, Schreder E, Salamova A (2020) Indoor exposure to per- and polyfluoroalkyl substances (PFAS) in the childcare environment. Environ Pollut 258:113714

Zheng Q, Qiu H, Zhu Z, Gong W, Zhang D, Ma J, Chen X, Yang J, Lin Y, Lu S (2023) Perchlorate in fine particulate matter in Shenzhen, China, and implications for human inhalation exposure. Environ Geochem Health 45:2857–2867

Zhou L, Hiltscher M, Püttmann W (2017) Occurrence and human exposure assessment of organophosphate flame retardants in indoor dust from various microenvironments of the Rhine/Main region, Germany. Indoor Air 27:1113–1127

Zhou J, Mainelis G, Weisel CP (2019) Pyrethroid levels in toddlers' breathing zone flowing a simulated indoor pesticide spray. J Expo Sci Environ Epidemiol 29:389–396

Zhou H, Dong X, Zhao N, Zhao M, Jin H (2022) Polyhalogenated carbazoles in indoor dust from Hangzhou, China. Sci Total Environ 7:159971

Zhu L, Hites RA (2005) Identification of brominated carbazoles in sediment cores from Lake Michigan. Environ Sci Technol 39:9446–9451

Zhu L, Wang J (2003) Sources and patterns of polycyclic aromatic hydrocarbons pollution in kitchen air, China. Chemosphere 50:611–618

Zhu J, Feng YL, Shoeib M (2007) Detection of dechlorane plus in residential indoor dust in the city of Ottawa, Canada. Environ Sci Technol 41:7694–7698

Zhu L, Lu H, Chen S, Amagai T (2009) Pollution level, phase distribution and source analysis of polycyclic aromatic hydrocarbons in residential air in Hangzhou, China. J Hazard Mater 162: 1165–1170

Zhu Q, Yang Y, Zhong Y, Lao Z, O'Neill P, Hong D, Zhang K, Zhao S (2020a) Synthesis, insecticidal activity, resistance, photodegradation and toxicity of pyrethroids (a review). Chemosphere 254:126779

Zhu Q, Wang M, Jia J, Hu Y, Wang X, Liao C, Jiang G (2020b) Occurrence, distribution, and human exposure of several endocrine-disrupting chemicals in indoor dust: a nationwide study. Environ Sci Technol 54:11333–11343

Zhuo M, Ma S, Li G, Yu Y, An T (2019) Chlorinated paraffins in the indoor and outdoor atmospheric particles from the Pearl River Delta: characteristics, sources, and human exposure risks. Sci Total Environ 650(Pt 1):1041–1049

Zober MA, Ott MG, Päpke O, Senft K, Germann C (1992) Morbidity study of extruder personnel with potential exposure to brominated dioxins and furans. I. Results of blood monitoring and immunological tests. Br J Ind Med 49:532–544

Zuiderveen EAR, Slootweg JC, de Boer J (2020) Novel brominated flame retardants – a review of their occurrence in indoor air, dust, consumer goods and food. Chemosphere 255:126816

Chapter 5
Particulate Matter (PM) and Fibers

Contents

Abstract In this chapter, the particulate matter (PM), ultrafine particles, engineered nanoparticles, fibers like asbestos and man-made mineral fibers, as well as microplastic are presented individually in detail. In addition to a brief description of the chemical–physical characteristics and the use of the substance, an overview of main health risks and, in particular, references to further scientific literature are given. In the foreground, however, are extensive presentations of the occurrence of the substances in indoor air and in house dust.

Keywords Particles · Nanoparticles · UFP · Asbestos · Microplastic · Fibers

Particulate airborne substances, when dispersed in the air, form colloidal systems with the gases, also known as aerosols. The term includes both "solid" (dust) and "liquid" particles (mist). In addition to natural aerosol sources (e.g., sea salt from the oceans, volcanic eruptions, whirled-up Saharan sand) and biogenic aerosols (e.g., pollen, spores, bacteria, viruses), anthropogenic sources (e.g., combustion processes) play an important role. After their release, the inorganic and organic reaction intermediates quickly form ultrafine particles (nucleation), which in the further course agglomerate into larger units for thermodynamic reasons (coagulation). The particles of the fine dust fraction are primarily formed by conversion processes from gases or in the course of combustion processes. They are typically composed of nitrates, sulfates, ammonium, elemental carbon, a large number of organic

H. Fromme, *Indoor Air Quality*, https://doi.org/10.1007/978-3-031-40078-0_5

Table 5.1 Number of particles and particle surface area at different particle diameters per 10 μg/m^3 (according to Oberdörster et al. 2005)

Diameter (nm)	Number concentration (particles/cm^3)	Surface (μm^2/cm^3)
5000	0.15	12
250	1.2	240
20	2,400,000	3016
5	153,000,000	12,000

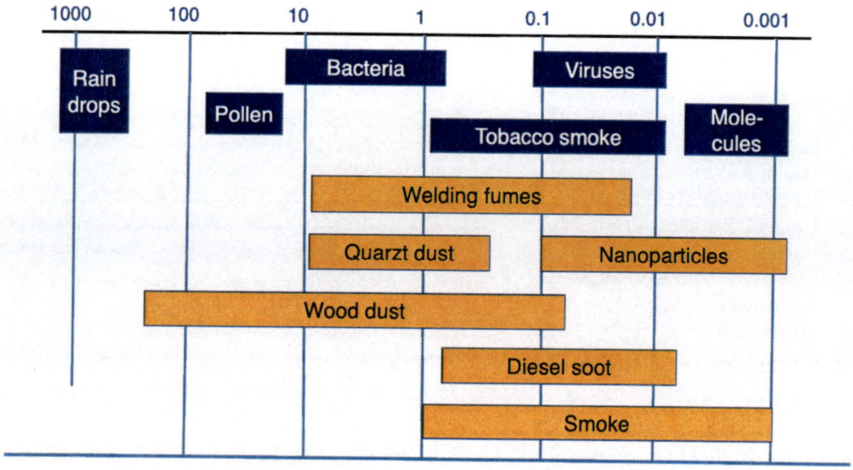

Fig. 5.1 Size classification in μm of different dust and aerosol fractions (after Schober et al. 2012)

compounds and trace elements. The particles in the coarser particulate matter fraction (coarse particles), on the other hand, are largely produced mechanically by the decay of larger solid particles. In urban environments, this fraction typically includes dust stirred up from roads and from industrial processes, as well as biological material such as pollen and bacteria and their fragments. It also includes wind-borne earth crust material from agricultural processes, uncovered soil, gravel roads, and mining activities.

Ultrafine particulate matter includes all particles of natural origin as well as particles released as a by-product or pollutant or waste whose aerodynamic diameter is less than 100 nm. Ultrafine particles (UFP) are mainly generated in road traffic as tire abrasion or during the incomplete combustion of fossil fuels in outdoor air. Due to their small diameter, UFP have an extremely low mass, so that they contribute only about 1% to the total mass of the environmental aerosol. On the other hand, they are present in large numbers. These particular characteristics are shown in Table 5.1.

To give an impression of the size distribution of different dust and aerosol fractions, this is graphically illustrated in Fig. 5.1. It can be seen that cigarette smoke and diesel soot span a similar size range with a center at about 100 nm, while UFP and nanoparticles can have even significantly smaller diameters.

From a historical and regulatory point of view, the particulate mass is in the foreground as a dose parameter for the toxicity of biopersistent dust. This is essentially also related to the relatively simple measurability of this parameter. See further basics on measurement methodology in Sect. 1.5 and important definitions of mass-related particle fractions in Table 5.2. In addition, it is considered increasingly important to include other metrics such as particle number concentration (PNC) and particle surface area as dose parameters. Moreover, the type of material, crystal structure, bioavailability and charge must be taken into account.

5.1 Particle Indoors

Due to their heterogeneity, very complex exposure patterns arise indoors, where important sources have to be considered in addition to inputs from outdoor air. Furthermore, it has been shown that the level of particulate matter occurring indoors can be highly variable spatially, but also temporally. In addition to the prevailing outdoor air conditions in the near vicinity of the indoor space (e.g., location near a busy road or in a rural area) and the current climatic conditions, the structural conditions of the building and the ventilation conditions are important. In addition, indoor activities themselves can contribute significantly to the current exposure situation. Therefore, the following influencing factors should be taken into account when considering and assessing particulate matter in indoor air:

Environmental conditions.

- Outdoor air concentration.
- Meteorology (e.g., wind pressure, wind speed, temperature, humidity).
- Location of the object (e.g., to potential sources).

Building conditions.

- Ventilation conditions (e.g., natural vs. technical ventilation).
- Construction method, "tightness" of the building.

Indoor conditions.

- Smoking tobacco products.
- Direct indoor sources (e.g., open fires, candle fires, cleaning and cooking activities).
- Indirect sources: Deposition and resuspension processes (e.g., air movement, physical activity in the room).
- Hobby and leisure activities.
- Humans or pets as a source (e.g., through clothing, skin, and hair).

Jeong et al. (2021) summarized the percentages of PNC size distribution in seven particle size bins ranging from 2 to 400 nm for six indoor emission sources observed in the experimental studies (see Fig. 5.2). The total particle number concentrations at the start of decay was for the candle burning >510,000 particles/cm^3, for gas stove

Table 5.2 Important definitions (in accordance with DIN EN 1993; EC 2022)

Term	Definition
Common definitions in occupational safety and health	
Inhalable fraction	The mass fraction of all suspended particles that is inhaled through the mouth and nose.
Extrathoracic fraction	The mass fraction of inhaled particles that does not penetrate beyond the larynx.
Thoracic fraction	The mass fraction of inhaled particles that penetrates beyond the larynx (particles up to a maximum of approx. 50 μm*).
Alveolar fraction	The mass fraction of inhaled particles that penetrates into the non-ciliated airways (so-called A-dust) (corresponds approximately to PM_4; particles up to a maximum of approx. 16 μm*).
RPM or RSP	Respirable particulate matter, respirable suspended particulate; corresponds to the alveolar fraction (defined by the American conference of governmental industrial hygienists).
Common definitions in environmental protection and health protection	
TSP	Total suspended particulates; total suspended particulate matter up to a particle diameter of approx. 70 μm*.
PM_{10}	Particulate fraction passing through a size-selective air inlet that has a collection efficiency of 50% for an aerodynamic diameter of 10 μm* (particles up to a maximum of approx. 20 μm*).
PM_4	Corresponds approximately to the alveolar fraction and RPM
$PM_{2.5}$	Particulate fraction passing through a size-selective air inlet, which has a collection efficiency of 50% for an aerodynamic diameter of 2.5 μm* (particles up to a maximum of approx. 5–6 μm*)
Coarse mode particles	Particle fraction larger than 2.5 μm* in diameter, consisting essentially of mechanical abrasion and swirling up of ground dust (also known as $PM_{10-2.5}$)
Fine mode particles	Particulate matter with a diameter smaller than 2.5 μm*, formed primarily from gases or secondarily by nucleation and condensation
Ultrafine particles	Particles with a diameter < 100 nm, formed primarily by nucleation
Nanomaterial	A natural, incidental or manufactured material consisting of solid particles that are present, either on their own or as identifiable constituent particles in aggregates or agglomerates, and where 50% or more of these particles in the number-based size distribution fulfill at least one of the following conditions: (a) One or more external dimensions of the particle are in the size range 1 nm to 100 nm; (b) the particle has an elongated shape, such as a rod, fiber or tube, where two external dimensions are smaller than 1 nm and the other dimension is larger than 100 nm; (c) the particle has a plate-like shape, where one external dimension is smaller than 1 nm and the other dimensions Are larger than 100 nm.

*: given as the aerodynamic diameter. This is the geometric diameter in relation to a sphere with a density of 1 g/cm^3 which has the same sinking speed in still or laminar flowing air

>140,000 particles/cm^3, for clothes dryer 73,000 particles/cm^3, for toast produced 67,000 particles/cm^3, for broiled fish 258,000 particles/cm^3, and 21,000 particles/cm^3 for incense burning.

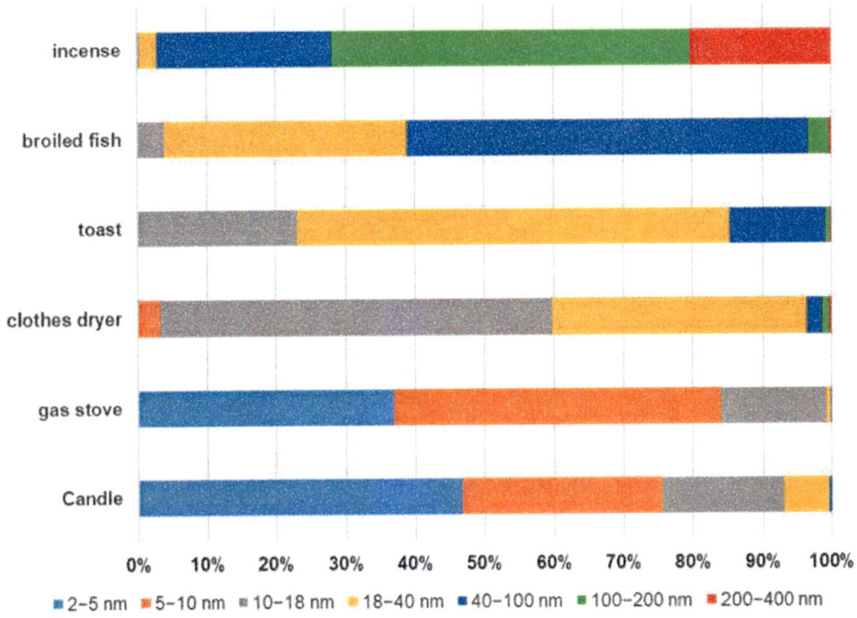

Fig. 5.2 Percentages of PNC size distribution in seven particle size bins for six indoor emission sources observed in experimental studies (according to Jeong et al. 2021)

5.2 Health Aspects

Scientific studies consistently show that short-term, higher exposure to particulate matter can lead to an increase in hospital admissions and increased physician visits, particularly for cardiovascular and respiratory diseases. Mortality for the aforementioned disease groups is also increased. Long-term exposure is equally associated with an increase in respiratory disease and an increase in all-cause and cause-specific mortality from cardiovascular and respiratory disease (e.g., COMEAP 2016; HC 2016; Chen and Hoek 2020; Han et al. 2022; Niu et al. 2023). Older adults, children, and individuals with pre-existing respiratory and cardiovascular conditions represent particularly vulnerable populations. The "Effects of Low-level Air Pollution: a Study in Europe" (ELAPSE) collaboration pools seven cohorts from across Europe and observed a higher risk for lung cancer with higher exposure to $PM_{2.5}$, also if the association were adjusted for other pollutants (Hvidtfeldt et al. 2021). Additionally, associations with $PM_{2.5}$ were observed for cancer (So et al. 2021; Chen et al. 2022), stroke (Wolf et al. 2021), and non-accidental mortality (Stafoggia et al. 2022).

In addition to the papers mentioned above, other scientific reviews are available: Pope et al. (2009), Williams et al. (2012), Beelen et al. (2014), Lippmann (2014), Kaufman et al. (2016), Akintoye et al. (2016), Raaschou-Nielsen et al. (2016), Nhung et al. (2017), Luben et al. (2017), Ritz et al. (2019), and Orellano et al. (2020).

Additionally, the Global Burden of Disease Study (GBD) in 2019 summarized the global, regional, and national data on ambient $PM_{2.5}$-attributable death and disability-adjusted life years (DALYs), and the age-standardized rates of mortality (ASMR) and disability-adjusted life years (ASDR) (Sang et al. 2022). Globally, the ASMR or ASDR of ambient $PM_{2.5}$-attributable disease burden showed an increase in exposure to $PM_{2.5}$ in the low socio-demographic index SDI up to middle SDI but a decrease in the high SDI countries. The authors summarized that 23% of year-2010 deaths attributable to ambient $PM_{2.5}$ could be avoided when aggressive global programs in line with WHO interim guidelines were implemented. Even in the less polluted regions modest improvements in $PM_{2.5}$ would result in surprising large avoided mortality because of the supra-linear concentration-response relationship at lower concentrations.

For ultrafine particles (UFP), aerosols with an aerodynamic diameter of 0.1 μm (100 nm) or less, a relationship between particle number concentration in outdoor air and selected health endpoints is also considered as probable (WHO 2016; Ohlwein et al. 2019; Kwon et al. 2020; Boyes and van Thriel 2020; Schraufnagel 2020; Bouma et al. 2023; Schwarz et al. 2023). But the database is not yet sufficient to derive, e.g., limit or guide values. For example, Ohlwein et al. (2019) concluded that most consistent associations were identified for short-term effects on pulmonary/ systemic inflammation, heart rate variability, and blood pressure and that the evidence suggests adverse short-term associations with inflammatory and cardiovascular changes, which may be at least partly independent of other pollutants.

5.3 Particle Mass

5.3.1 Occurrence in the Air of Residences

A large number of measurements of particle mass concentrations in residential air are available in the scientific literature (Fromme 2019). It should be noted that different sampling times and measurement methods can strongly influence the individual results. In Table 5.3, current results on mass-related measurements in indoor living spaces are compiled. In studies from European countries, median concentrations range between 6.5 and 23 μg/m^3 for $PM_{2.5}$ and between 12 and 39 μg/m^3 foe PM_{10}. Similar results were described in studies performed in the USA.

For China, Zhang et al. (2021) summarized $PM_{2.5}$ results of articles published between 1980 and 2019. They reported that mean indoor $PM_{2.5}$ concentrations decreased from 128 μg/m^3 prior to 2010 to 70.1 μg/m^3 following 2015. The mean indoor $PM_{2.5}$ levels in rural areas (276 μg/m^3) were much higher than that in urban areas (95.4 μg/m^3). As part of the Chinese Indoor Environment and Health Surveillance 2018, a total of 610 residential households were investigated and $PM_{2.5}$ geometric means in living rooms and bedrooms of 40 and 37 μg/m^3 in the summer and 71 and 68 μg/m^3 in the winter (Zhu et al. 2021b). In summer and winter approximately 70% and 90% of indoor $PM_{2.5}$ exceeded the guidance value of

Table 5.3 Particulate matter in the air of residences in μg/m³

References	*N*	$PM_{2.5}$ Median (Min-Max)	PM_{10} Median (Min-Max)	Country, Year
Europe				
Lai et al. (2004)	42	17+ (–)	–	UK, 1998–2000
Fromme et al. (2005)	27[a] 16[a]	30# (12–47) 27# (12–59)	–	Germany, winter, 1997/98 Germany, summer, 2000
Franck et al. (2011)	129	32 (10–115[b])	76 (33–229[b])	Germany; 2001/02
Stranger et al. (2007)	19a	36+ (11–105)	39+ (12–112)	Belgium, 2002/03
Hoek et al. (2008)	37 35 50 30	6.5+ (–) 20+ (–) 14+ (–) 7.5+ (–)	12+ (–) 32+ (–) 22+ (–) 16+ (–)	Finland, 2002–04 Greece, 2002–04 Netherlands, 2002–04 UK, 2002–04
Kirchner et al. (2008)	567	19 (17–21)	32 (28–34)	France, 2003
Wichmann et al. (2010)	29	10 (4.8–19)	–	Sweden, 2003/04
Langer et al. (2016)	186[a]	16 (5.1–164)	26 (6.5–374)	France, 2003–05
Osman et al. (2007)	75	18 (6.0–77[c])	–	UK, 2004/05
Santen et al. (2009)	50	5.6 (0.1–51)	17.3 (1.1–181)	Germany, 2007
Cattaneo et al. (2011)	107	23 (17–37[d])	33 (23–44[d])	Italy, 2007/08
Schembari et al. (2013)	54	23+ (6.7–73)	–	Spain, 2008/09
Vu et al. (2022)	132	9.6 (1.2–437)	17.2 (1.9–644)	UK, 2016–18
Wyss et al. (2016)	22	13+ (0–1765)	–	Norway, 2012
Martins et al. (2020)	4		– (11–38)	Portugal, 2017/18
Madureira et al. (2020)	65	34 (3–2500)	37 (4–2610)	Portugal, 2018/19
Camilleri et al. (2022)	159	4.5 (3.4–7.6 i)	– (–)	Malta, 2018/19
North America				
Héroux et al. (2010)	96	6.4* (1.0–82) 5.5* (0.9–92)	– –	Canada, summer, 2007 Canada, winter, 2007
Jung et al. (2010)	286	14 (0.6–105)[e] 14 (6.3–75)[f]	– –	USA, 2005–10

(continued)

Table 5.3 (continued)

References	*N*	$PM_{2.5}$ Median (Min-Max)	PM_{10} Median (Min-Max)	Country, Year
Tunno et al. (2016)	21	15 (–) 21 (–)	– –	USA, winter, 2011/12 USA, summer, 2011/12
Gould et al. (2018)	48	13 (–)	–	USA, 2013/14
Asia				
Lim et al. (2011)	60	48 (14.4–91)	–	Korea, 2008
Li et al. (2016)	54 53	72 (14–212) 104 (38–368)	–	China, 2013, non-heating China, 2013, heating
Yoda et al. (2017)	55	12 (3.1–33)[g]		Japan, 2014
Nishihama et al. (2021)	4916	13 (<1.3–351)	18 (<2.2–361)	Japan, 2014–16
Barkjohn et al. (2021)	42	34 (–)	– (–)	China, 2017
Zhu et al. (2021b)	610	67 (22–292[h]) 43 (9–171[h])	98 (30–338[h]) 65 (18–219[h])	China, 2018/19, winter China, 2018/19, summer

[#]: PM_4; [+]: arithmetic mean; * geometric mean
[a] Nonsmoking residences only
[b] tenth and 90th percentiles
[c] interquartile range
[d] 25th and 75th percentiles
[e] heating season
[f] nonheating season
[g] weekdays
[h] fifth and 95th percentile

25 μg/m^3, while 60% and 80% exceeded the PM_{10} level of 50 μg/m^3. From 2003 to 2005, PM was measured at fixed-site monitoring stations in 6 Japanese cities and inside and outside of 78 residences during four seasons (Michikawa et al. 2014). $PM_{2.5}$ concentrations ranged from 17.1 to 24.4 μg/m^3 indoors, from 17.1 to 31.2 μg/m^3 outdoors, and from 16.3 to 33.3 μg/m^3 at the fixed sites. The correlations between the fixed-site and indoor measurements were only moderate to low in this study. The Japan Environment and Children's Study (JECS) measured $PM_{2.5}$ and PM_{10} in approximately 5000 households with means of 16.2 and 22.0 μg/m^3, respectively, from 2014 to 2017 (Nishihama et al. 2021).

Several studies have shown that smoking was the most important factor influencing $PM_{2.5}$ levels, and to a somewhat lesser extent cooking, burning incense or mosquito coils (e.g., Özkaynak et al. 1995; Wallace and Howard-Reed 2002; Wallace et al. 2003; Lai et al. 2004; Simoni et al. 2004; Fromme et al. 2005; Breysse et al. 2005; Goel et al. 2021; Ghatge Goel et al. 2021). In the large Japanese study mentioned before, major predictors of indoor $PM_{2.5}$ were outdoor $PM_{2.5}$, indoor smoking, observable smoke, and indoor/outdoor temperature (Nishihama et al. 2021).

5.3.2 *Occurrence in the Air of Schools*

Figure 5.3 summarizes results in school classrooms (see also Fromme 2019). In the majority of the studies, $PM_{2.5}$ levels range on average between about 8 and 30 µg/m^3. In contrast, a study in 27 Belgian schools described relatively high concentrations with a median of 61 µg/m^3 (Stranger et al. 2007). In contrast, PM_{10} levels in schools were highly variable, with medians in the range of about 30–90 µg/m^3. In a Europe-wide survey of 45 schools, PM_{10} levels ranged from 14 to 260 µg/m^3 (median: 106 µg/m^3) (HESE 2006). An investigation of 64 primary school buildings in five Central European countries found median $PM_{2.5}$ levels of 28.4 µg/m^3 (3.2–77.8 µg/m^3) in 2017/18 (Szabados et al. 2021). Again, 34 classrooms were analyzed in Turkey in fall and spring with means of 142 and 84 µg/m^3 for PM_{10} and 73 and 44 µg/m^3 for $PM_{2.5}$ (Babaoglu et al. 2022).

$PM_{2.5}$ medians in fall, winter, and spring of 5.2 µg/m^3 (range: 0.9–24.3 µg/m^3), 4.7 µg/m^3 (1.1–30.1 µg/m^3), and 6.2 µg/m^3 (1.0–28.4 µg/m^3), respectively, were seen in 32 classrooms during school hours in the USA in 2015–2017 (Majd et al. 2019). In another study performed in the USA in 2008–2013 and 2015–2019 in 309 classrooms of 74 inner-city schools during the teaching time, median concentration levels was 5.4 µg/m^3 (range: 0.13–27 µg/m^3) for $PM_{2.5}$ (Matthaios et al. 2022). Classroom exposures were significantly positively associated with outdoor

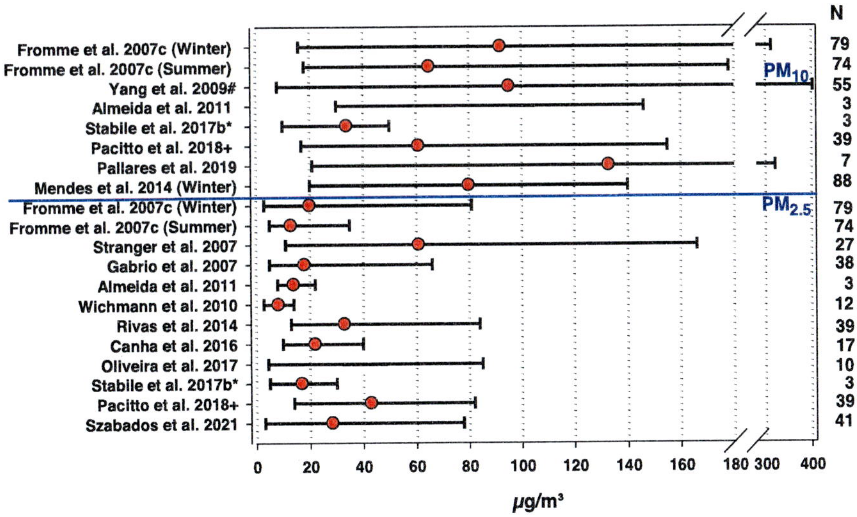

Fig. 5.3 Minimum, median, and maximum PM_{10} and $PM_{2.5}$ concentrations in European schools in µg/m^3 (# geometric mean; * arithmetic mean; + fifth and 95th percentile)

levels and ambient temperature and concentrations during winter period were significantly greater from those measured during fall and spring. Environmental factors were the most important predictors, explaining 57.4% of the classroom, while school originated predictors accounted for 18.6% and classroom-based factors for 24.0% of the variance.

Within the framework of the Chinese Indoor Environment and Health Surveillance 2018 a total of 66 classrooms nationwide in 2018–2019 were investigated and $PM_{2.5}$ and PM_{10} medians of 54 μg/m^3 (7–247 μg/m^3) and 78 μg/m^3 (9–274 μg/m^3) were observed, respectively (Zhu et al. 2021a), while Che et al. (2021) observed means of 35 μg/m^3 for PM_{10} and 19 μg/m^3 for $PM_{2.5}$ in 16 classrooms in Hong Kong. Outdoor concentration is the single largest predictor for indoor concentrations with 74% and 61% of the variations observed for PM_{10} and $PM_{2.5}$ during school hours. Sohn et al. (2009) reported for PM_{10} a mean value of 107 μg/m^3 (8–403 μg/m^3) in 55 classrooms in Korea. Mohd Isa et al. (2022) found medians of 22 μg/m^3 for $PM_{2.5}$ and 37 μg/m^3 for PM_{10} in 16 classrooms in Malaysia in 2018–2019. Using multiple logistic regression models only the concentrations of $PM_{2.5}$ were significantly associated with elevated of fractional exhaled nitric oxide (FeNO) levels in 470 school children attending these classrooms. Cai et al. (2021) investigated 33 Chinese classrooms with natural ventilation and air cleaners (NV) compared to mechanical ventilation with fresh air ventilators (MV) in 2017–2018. The daily occupied-time averaged concentrations of indoor $PM_{2.5}$ in classrooms with MV and NV systems were 34 and 34 μg/m^3 in summer and 26 and 27 μg/m^3 in winter, respectively.

Various studies show for $PM_{2.5}$ a ratio of indoor-to-outdoor air concentration (I/O) in the range of 1 or a strong dependence on outdoor air content (Fromme et al. 2007; Diapouli et al. 2008; Guo et al. 2010; Wichmann et al. 2010; Che et al. 2021). The situation is different when coarser particle fractions are considered. For example, in a German study, it was observed that 90% of the variability in daily medians of PM_{10} was due to differences between schools or days, whereas only 45% of the PM_{10} variance could be explained in this way (Fromme et al. 2007). The I/O ratio for PM_{10} measurements ranged from 1.0 to 36, with the majority ranging from 2 to 5. Indoor sources themselves appear to be of greater importance in this case. In order to estimate the contribution of different sources or outdoor air, several studies also determined the elemental or ionic composition of the dust (Molnár et al. 2007; Diapouli et al. 2008; Fromme et al. 2008). It was found that especially the coarser particles did not originate from the outside air, but had their source in the classroom itself. Investigations of the filters using EDX (energy dispersive X-ray spectroscopy) also indicated that the PM_{10} content was composed in particular of soil particles and other mineral substances, abrasion from building materials, and chalk dust (Fromme et al. 2008). Also another study showed inside significantly more silicate particles (36% of all particles), organic particles (29%, probably from human skin), and Ca-carbonate particles (12%, probably from paper), while in the corresponding outdoor air filters especially Ca-sulfate containing particles (38%) were found.

Specifically, the physical activity of students and the associated stirring up of suspended soil particles appears to be responsible for the high PM_{10} concentrations in classrooms (Fromme et al. 2007, 2008; Guo et al. 2010; Almeida et al. 2011; Öder et al. 2012).

5.3.3 *Occurrence in the Air of Daycare Centers*

Table 5.4 summarizes the results of measurements in the indoor air of daycare centers.

In 73 Berlin kindergartens surveyed in 2000/2001 during childcare hours, relatively high PM_4 levels were found, ranging from 13 to 128 μg/m³ (median: 53 μg/m³) (Fromme et al. 2005). Lower concentrations of $PM_{2.5}$ were observed in 2011/2012 in the LUPE survey (Länderuntersuchungsprogramme) of 63 daycare centers in 3 German states, with a median of 21 μg/m³ (7–60 μg/m³) (Fromme et al. 2016).

Table 5.4 Particulate matter in the air of daycare centers in μg/m³

References	*N*	Median (Min-Max)	Description
$PM_{2.5}$			
Alves et al. (2013)	1	70	Portugal, naturally ventilated
Fromme et al. (2005)	74	53 (13–128)*	Germany, 2000/01, naturally ventilated
Wichmann et al. (2010)	18	5.9 (3.2–9.3)	Sweden, 2003/04, 80% naturally ventilated
Szoboszlai et al. (2011)	1	~12	Hungary, 2009, naturally ventilated
Canha et al. (2016)	7	21+ (~9–45)	France, 2010, 73% naturally ventilated
Fromme et al. (2016)	63	21 (7–60)	Germany, 2011/12, naturally ventilated
Kim et al. (2014)	4	22 (13–31)	Korea, 2012
Branco et al. (2014)	8	23 (4.3–158)	Portugal, 2013
Oh et al. (2014)	10	46 (20–72)	Korea, 2013/14, 50% naturally ventilated
Persson et al. (2018)	4	2.5–9.2+	Portugal, mechanically ventilated
Mainka and Fantke (2022)	4	70 (winter), 75 (spring)+	Poland, 2014/15
PM_{10}			
Szoboszlai et al. (2011)	1	~104+	Hungary, 2009, naturally ventilated
Alves et al. (2013)	1	97	Portugal, naturally ventilated
Carreiro-Martins et al. (2016)	125	130 (90-150a)	Portugal, 2011
Fromme et al. (2016)	63	146 (48–368)	Germany, 2011/12, naturally ventilated
Branco et al. (2014)	3	24 (5–197)	Portugal, 2013
Oh et al. (2014)	10	80 (53–115)	Korea, 2013/14, 50% naturally ventilated

[a] 25th and 75th percentile; *: PM_4, +: arithmetic mean

Overall, median $PM_{2.5}$ and PM_{10} levels in Europe range from about 6–75 μg/m^3 and from 24–146 μg/m^3, respectively. Similar concentrations were observed in Korea. Differences in concentrations can be explained by the absence or presence of ventilation systems, proximity to specific sources (e.g., motorized traffic, industry), and climatic conditions. Higher concentrations were observed in two urban and two rural area kindergartens of the Upper Silesia Region, Poland, in winter and spring 2014/15 typified by the use of fossil fuels for power and heat purposes (Mainka and Fantke 2022). In China, $PM_{2.5}$ was collected in 34 rooms of 10 daycare centers and high levels with a median of 221 μg/m^3 (16–757 μg/m^3) was found (Zhang et al. 2018). In 10 daycare centers in Malysia in 2014 mean concentrations of 69 μg/m^3 (17–170 μg/m^3) and 113 μg/m^3 $PM_{2.5}$ (30–430 μg/m^3) were observed in playrooms and bedroom, respectively (Khamal et al. 2019).

5.3.4 Occurrence in the Office Air

A summary of more recent measurements of particulate matter is given in Table 5.5 and Felgueiras et al. (2022). The mean levels of $PM_{2.5}$ and PM_{10} in non-smoking offices range from about 16–51 μg/m^3 and about 3–33 μg/m^3, respectively. Significantly lower indoor air concentrations are found in mechanically ventilated buildings compared to naturally ventilated rooms (Burton et al. 2000; Quang et al. 2013; Li et al. 2015; Chatoutsidou et al. 2015). Very high levels were reported from 40 Greek offices where smoking was still allowed and where high outdoor air levels were also described by the authors (Gemenetzis et al. 2006). Mean indoor air to outdoor air (I/O) concentrations for $PM_{2.5}$ and PM_{10} varied between 0.5–1.0 and < 0.3–0.5, respectively, in mechanically ventilated rooms, while a French study found an I/O ratio of approximately 1.7 ($PM_{2.5}$) in naturally ventilated rooms (Mosqueron et al. 2002). The level of particles measured in indoor air is thereby, if ventilation systems are present, dependent on the effectiveness of the filter system, but also on the location of the air inlet. In an EU-wide project, $PM_{2.5}$ concentrations ranging from 4.7 to 38 μg/m^3 were observed in 9 office buildings from 6 countries in 2012/2013, with a median of 12 μg/m^3 (Szigeti et al. 2017). Eight green office buildings were compared to six non-green buildings in Singapore in 2014–2016 and means of 29 and 45 μg/m^3 were reported, respectively (Lee et al. 2019). The mean ratings for satisfaction with temperature, humidity, lighting level, air quality, and indoor environment were significant higher in green buildings.

5.3.5 Occurrence in Ground Transport Microenvironments

Exposure in passenger cars is influenced by multiple factors such as outdoor air concentrations, ventilation, driving style, traffic density and flow, roadside development, meteorology, and indoor sources (Lee et al. 2010; de Nazelle et al. 2017). For

Table 5.5 Particulate matter in the office air in µg/m³

Reference	*N*	Median (Min-Max)	Place, year, ventilation type
PM_{10}			
Gemenetzis et al. (2006)	40	103 (25–370)	Greece, naturally ventilated
Liu et al. (2004)	11	63 (14–166)	China, 2002/2003
Mersch-Sundermann et al. (2007)	63	33 (12–94)[a]	Germany, 2006
Horemans & Van Grieken (2010)	10	20 (11–29)	Belgium, 2008, naturally ventilated
Li et al. (2015)	1	27*	China, 2012–13, mechanically ventilated
Chatoutsidou et al. (2015)	1	2.7	Norway, 2014, mechanically ventilated
$PM_{2.5}$			
Mosqueron et al. (2002)	55	26 (5–265)	France, 1999/2000
Gemenetzis et al. (2006)	40	77 (11–250)	Greece, naturally ventilated
Liu et al. (2004)	11	28 (3–103)	China, 2002/2003
Vardavas et al. (2007)	6	51* (39–63)	Greece, 2006, naturally ventilated
Mersch-Sundermann et al. (2007)	63	18 (4–70)[a]	Germany, 2006
Sangiorgi et al. (2013)	4	– (6.2–72)	Italy, 2007–08, naturally ventilated
Horemans and Van Grieken (2010)	10	16 (8.1–24)	Belgium, 2008, naturally ventilated
Quang et al. (2013)	3	~6 (4.4–9.9)	Australia, mechanically ventilated
Jung et al. (2016)	21	29 (6.1–144)	Taiwan, 2011/12
Li et al. (2015)	1	20*	China, 2012–13, mechanically ventilated
Szigeti et al. (2014)	5	– (2.9–23)	Hungary, mechanically ventilated

* arithmetic mean
[a] without activity of office machines

example, $PM_{2.5}$ levels in cars can be reduced by over 75% by modifying ventilation (Ham et al. 2017). Air conditioning and filtration systems should also be taken into account, as they can have a significant influence on the content in vehicle interiors, depending on their quality and degree of separation. Overall, car occupants seem to be exposed to higher levels than cyclists and pedestrians (Adams et al. 2001; Kaur et al. 2005; de Nazelle et al. 2012).

The levels of particulate matter in surface-operated transport vehicles are summarized in Table 5.6. Again, Xu et al. (2018) summarized findings in the literature on air quality inside vehicle cabins. Overall, recent European studies show mean exposure in cars and busses in the range of 7.4–68 µg/m³ for $PM_{2.5}$ and 8.3–50 µg/m³ for PM_{10}. The mean exposure levels in cars and busses in the USA (Riediker et al. 2004; Sabin et al. 2005; Rim et al. 2008) appear to be comparable to those in Europe, while higher levels have been observed in Asian metropolitan areas by comparison (Chan et al. 2002). In Colorado, 45 non-smoking participants were

Table 5.6 Particulate matter in the air of vehicles in µg/m³

Reference	Mean value (Min-Max)	*N*, Type	Country, time
$PM_{2.5}$			
Fromme et al. (1998)	37[a] (18.7–70)* summer 38[a] (27.0–72)* winter	1 car during 2 weeks 1 car during 2 weeks	Germany, 1995/96
Adams et al. (2001)	39 (7.9–97) 38 (15.1–77)	36 busses 42 cars	UK, summer, 1999/2000
Adams et al. (2001)	39 (5.9–87) 34 (6.6–94)	32 busses 12 cars	UK, winter, 1999/2000
Riediker et al. (2004)	23 (7.1–38.7))	9 highway patrol cars	USA, 2001
Sabin et al. (2004)	– (36–60) – (15–36)	31 school busses windows closed 31 school busses windows open	USA, 2002
Chertok et al. (2004)	20,8 [b] (9.1–33) 29.6 [b] (21.4–45)	16 cars, first week 16 cars, second week	Australia, 2002
Kaur et al. (2005)	35 (6.0–65) 38 (15.2–59) 42 (17.9–72)	42 busses 29 cars 22 taxis	UK, 2003
Krausse and Mardaljevic (2005)	– (9.7–26)	133 vans	UK
Sohn et al. (2006)	152 (–) 176 (–)	20 busses 20 taxis	Korea, 2003/04
McNabola et al. (2008)	128 (–) 83 (–)	1 bus over 18 months 1 car over 18 months	Ireland, 2005/06
Rim et al. (2008)	7–20 (–)	6 school busses	USA, 2006
Zuurbier et al. (2011)	39[a] (16–324) 28[a] (11–131) 60[a] (2.7–358) 74[a] (5.3–403)	Diesel bus e-bus Diesel car Gasoline car	Netherlands, 2007/08
Geiss et al. (2010)	27 (0.9–94)	18 cars	Italy, 2009
Ozgen et al. (2016)	3.9 (0.3–27)	Car	Italy, 2010
Assimakopoulos e al. (2018)	11 (7.1–18)	Car	Greece, 2015
Rivas et al. (2017)	13[b] (7.7–27) 7.4[b] (2.0–25)	Bus Car	UK, 2016
Boniardi et al. (2021)	14.1 (1.7–117)	4 cars	Italy, 2020
Murzyn et. al. (2021)	26 (4–330)	45 field trips with a car	France, 2016/17
PM_{10}			
Gulliver and Briggs (2004)	16 (–) 43 (–)	2 autoroutes;	UK, 1999/2000
Zuurbier et al. (2011)	47[a] (23–221) 44[a] (30–107) 50[a] (29–427) 48[a] (27–161)	Diesel bus e-bus Diesel car Gasoline car	Netherlands, 2007/08

(continued)

Table 5.6 (continued)

Reference	Mean value (Min-Max)	*N*, Type	Country, time
Geiss et al. (2010)	49 (0.9–332)	18 cars	Italy, 2009
Ozgen et al. (2016)	13 (0.8–96)	Car	Italy, 2010
Kim et al. (2019)	59 (12–96)[c] winter 60 (7.2–96)[d] winter 22 (3.7–62)[c] summer 18 (3.2–44)[d] summer	10 taxis	Korea, 2013
Assimakopoulos et al. (2018)	58 (32–131)	Car	Greece, 2015
Rivas et al. (2017)	39[b] (18–77) 8.3[b] (2.4–27)	Bus Car	UK, 2016
Moreno et al. (2019)	75[a] (42–106 50[a] (35–75))	Diesel cab e-taxi	Spain, 2017
Boniardi et al. (2021)	19.2 (5.4–137)	4 cars	Italy, 2020
Murzyn et al. (2021)	27 (4–582)	45 field trips with a car	France, 2016/17
Salthammer et al. (2022)	8.7–41.1 (n.a.-101)	5 busses	Germany, 2020/21

* PM_4; *n.a.* not available
[a] median
[b] geometric mean
[c] non-congestion hours
[d] congestion hours

recruited to investigate the role of microenvironment on personal exposures (Koehler et al. 2019). In transit median $PM_{2.5}$ level was 6.9 μg/m^3 (range: 0.7–95 μg/m^3). Although transit typically accounted for less than 7% of participant's time per day, this microenvironment contributed a higher share of the total exposure.

In Korea, indoor air PM_{10} was not significantly different between seasons or during non-congestion hours and congestion hours in ten taxis and showed comparable levels to Europe and the USA (Kim et al. 2019).

Various studies have investigated the exposure of cyclists and pedestrians. For cyclists, these studies have observed mean $PM_{2.5}$ concentrations ranging from 8.5 μg/m^3 in the USA in 2012 (Hankey and Marshall 2015) to 81 μg/m^3 in Ireland in 2005/2006 (McNabola et al. 2008). For pedestrians, exposures were determined between 14 μg/m^3 $PM_{2.5}$ in London in 2016 (Rivas et al. 2016) to 65 μg/m^3 in Ireland (McNabola et al. 2008). Four other studies found mean $PM_{2.5}$ levels of 15 μg/m^3 (Northampton, England), 22 μg/m^3 (Barcelona, Spain), 28 μg/m^3 (London), and 9.1 μg/m^3 for pedestrians walking in the city (Gulliver and Briggs 2004; Kaur et al. 2005; de Nazelle et al. 2012; Boniardi et al. 2021). Important factors influencing the exposure of these road users were traffic density, modal split, and meteorological conditions. For example, in Minneapolis, bicyclists equipped with an online video system found that each passing motor vehicle was associated with high $PM_{2.5}$ concentrations for the bicyclists (Hankey and Marshall 2015). It is also important to consider that it is not only the external exposure situation that needs to be taken into account, but also the dose deposited in the respiratory tract (Cepeda et al. 2017). This is, precisely because of the high respiratory minute volume of cyclists, higher for them by a factor of about 4.3 than for persons in motor vehicles (McNabola et al. 2008; Zuurbier et al. 2010; Int Panis et al. 2010). For example, a study in Brussels estimated a dose to the respiratory tract of 0.8 μg$PM_{2.5}$/km for cyclists, but only 0.1 μg $PM_{2.5}$/km for motorists using the same route (Int Panis et al. 2010).

A special exposure is present when smoking in the interiors of means of transport. In a German study, the concentrations when smoking one cigarette, one e-cigarette, and one IQOS were determined in the air during trips in 7 cars through Munich, with the passenger window open about 2–5 cm (Schober et al. 2019). While mean values of 4–11 μg/m^3 $PM_{2.5}$ were found in the control journeys without smoking activities, the mean concentrations were 64–1988 μg/m^3 when smoking a cigarette and 8–490 μg/m^3 when smoking an e-cigarette. Using IQOS had no effect on indoor air levels. During a car trip in Wellington, New Zealand, a mean $PM_{2.5}$ content of 169 μg/m^3 (peak concentration: 217 μg/m^3) was measured with the side window fully open while smoking a cigarette (Edwards et al. 2006). Subsequently, smoking was performed with the window closed, and the average level during the smoking phase was 2962 μg/m^3, with a peak level of 3645 μg/m^3. The level did not return to the baseline level until 40 minutes after the end of smoking. In a Canadian study, the baseline exposure of about 24 μg/m^3 with closed windows in three cars increased during smoking to a mean $PM_{2.5}$ level between 790 and 4626 μg/m^3 (maximum: 7635 μg/m^3) (Sendzig et al. 2006). In contrast, in a car with wide-open side windows, the average concentration was 49 μg/m^3. Rees and Connolly (2006) determined mean $PM_{2.5}$ values of 271 μg/m^3 (maximum: approx. 500 μg/m^3) in 45 measurements with a continuous measuring device during smoking with closed windows. With open side windows, the mean values were about 50 μg/m^3 (maximum: about 100 μg/m^3). Baseline levels in car interiors in this study were about 10–20 μg/m^3 before the experiment. In urban areas in Texas in 2004, one cigarette was smoked during each trip with different ventilation conditions (Liu and Zhu 2010). Mean $PM_{2.5}$ concentrations when parking with the air conditioner running were 1741 μg/m^3, and when driving with the windows closed were 780–1670 μg/m^3. When driving with the passenger window slightly open, values ranged from 47 to 158 μg/m^3. When not smoking, levels ranged from 2.6–8.4 μg/m^3.

5.3.6 *Occurrence in Underground Transport Microenvironments*

Especially in larger cities, rapid development of subway systems occurs with an increasing number of commuters using the metro for a longer time period (Xu and

Hao 2017). In underground transportation, special exposure conditions exist due to the spatial confinement and the high number of users, which are characterized in particular by the ventilation intensity, frequency of trains, driving conditions, depth of the tunnel, and the year of construction (Wen et al. 2020). Particulate matter originate mainly from mechanical abrasion from the tracks and brakes, building materials, ballast bed, and resuspension of tunnel dust. Due to the complexity of sources there is no distinct composition of particutate matter. Nevertheless, main components sampled in London Underground $PM_{2.5}$ comprised 47% iron oxide, 7% elemental carbon, 11% organic carbon, and 14% metallic and mineral oxides (Smith et al. 2020). In this study, the most abundant mean element concentrations in $PM_{2.5}$ are iron (184,000 ng/m^3), calcium (17,000 ng/m^3), and aluminum (6900 ng/m^3). Compared to the levels of lead (8 ng/m^3), cadmium (0.1 ng/m^3), and arsenic (0.9 ng/m^3) found in London ambient air the levels underground are clearly higher with values of 20 ng/m^3 for lead, 3 ng/m^3 for cadmium, and 13 ng/m^3 for arsenic (Smith et al. 2020).

In this microenvironment, the exposure of commuters is often greater than the values measured in aboveground modes. In vitro studies suggest that underground particulate matter may be more inflammogenic to the airway epithelium than exposure to ambient/urban particulate matter, which is predominantly a result of the metal-rich components (Cooper and Loxham 2019). Overall, Loxham and Nieuwenhuijsen (2019) summarized the results of 27 studies and concluded that from the small number of studies, there is little evidence that the physicochemical characteristics of underground PM is responsible for a significantly increased risk of adverse health effects in underground railway workers or commuters.

In Germany the pollution in the Berlin subway system was measured in 1995/1996 in underground coaches of a central route through the city (Fromme et al. 1998). The mean PM_4 content in this study was 153 µg/m^3 (121–176 µg/m^3) in summer and 141 µg/m^3 (124–169 µg/m^3) in winter. Comparable results were measured in the London Underground in 1996 with mean $PM_{2.5}$ concentrations of 247 µg/m^3 (105–371 µg/m^3) in summer and 157 µg/m^3 (12–263 µg/m^3) in winter (Adams et al. 2001). In another study in three subway lines in London, mean $PM_{2.5}$ (130–200 µg/m^3) was measured in train compartments, while mean $PM_{2.5}$ levels of 270–480 µg/m^3 were determined on the platforms of three subway stations (Seaton et al. 2005). In contrast, significantly lower mean $PM_{2.5}$ levels were found in Boston 2000 (70 µg/m^3), Helsinki 2004 (21 µg/m^3), Los Angeles 2010 (14 µg/m^3), Montreal and Toronto 2010–2013 (35 and 95 µg/m^3), Athens 2012 (17 µg/m^3 for new metro cars and 48 µg/m^3 for old metro cars), Vienna 2016 (~8–50 µg/m^3), London (88 µg/m^3), and in Milano 2020 (15.1 µg/m^3) (Levy et al. 2002; Aarnio et al. 2005; Kam et al. 2011; Van Ryswyk et al. 2017; Barmparesos et al. 2016; Posselt et al. 2019; Smith et al. 2020; Boniardi et al. 2021).

In the underground of Mexico City, the mean values amounted to 61 µg $PM_{2.5}$/m^3 (31–99 µg/m^3) (Gomez-Perales et al. 2004). Measurements in the Seoul metro system in 2005/06 observed mean levels of 148 µg/m^3 (range: 40–257 µg/m^3) for PM_{10} (Sohn et al. 2008) and in Hong Kong in 1999/2000 and in 2014, $PM_{2.5}$ concentrations of 67 µg/m^3 and 10.2 µg/m^3 (26–123 µg/m^3) were found (Chan et al. 2002; Yang et al. 2015). Goel et al. (2015) reported mean $PM_{2.5}$ concentrations of 87 and 76 μg/m^3 in April and May 2014 in Delhi, respectively. During the summer and winter of 2013, five non-smoking field technicians collected PM_{10} and $PM_{2.5}$ samples with means of 59 μg/m^3 (summer) vs. 79 μg/m^3 (winter) and 24 and 50 μg/m^3, respectively, in Seoul in 2013 (Hwang and Lee 2018).

Passi et al. (2021) summarized the results of indoor air studies performed between 1996 and 2018 in underground metro station in a critical review. Additionally, study results of PM levels were given in Table 5.7.

High level of PM_4 of 128–311 μg/m^3 was reported in Berlin underground platforms (Fromme et al. 1998), as well as in London with means of 270–480 μg/m^3 $PM_{2.5}$ (Seaton et al. 2005). Again, Ripanucci et al. (2006) observed a mean of 330 μg/m^3 PM_{10} on two platforms and tunnels in Rome, Italy. The concentrations were around three times higher than outdoors and the sedimented dust consists mainly of iron and silica. Lower level of mean $PM_{2.5}$ levels of 61 μg/m^3 (normal operation) and 93 μg/m^3 (rush hour) was measured on platforms in the Paris metro in 2007 (Raut et al. 2009), while in Barcelona, concentrations were found to be 20–51 μg/m^3 in the warm season and 41–91 μg/m^3 in the colder season (Martins et al. 2015). In Boston, 130 μg/m^3 $PM_{2.5}$ was measured at an underground station (Levy et al. 2002), while the concentrations in Taipei were only 25–40 μg/m^3 on average (Cheng and Yan 2011). PM_{10} and $PM_{2.5}$ concentrations on underground platforms were 2.7 and 2.5 times greater than on ground level platforms in Sydney (Mohsen et al. 2018), while in Italy (Rome) the concentrations were 13 times higher on the metro platform compared to outdoor air (Perrino et al. 2015).

5.4 Ultrafine Particles (UFP)

García-Hernández et al. (2020), Ali et al. 2020, and Marval and Tronville (2022) reviewed the scientific literature regarding sources, personal exposure to UFP in different microenvironments.

Also with regard to ultrafine particles, indoor tobacco smoking is a major source of particles, where concentrations can rise to values of up to 213,000 particles/cm^3 (He et al. 2004; Afshari et al. 2005; Hussein et al. 2006). An increase was also observed when burning candles or oil lamps were used indoors (Fine et al. 1999, Hussein et al. 2006; Wallace & Ott 2011). Significantly higher concentrations occurred when extinguishing candles than during the burning process itself (Hussein et al. 2006).

As an example, the impact of cleaning was investigated in the HOMEChem study, where the living room and kitchen floors were mopped with a bleach solution for 10 min and particle formation and growth up to particles with a diameter of ~30 nm were found within 5 min of the start of mopping (Patel et al. 2020).

In addition, ultrafine particles are released during cooking. In this case, various working groups were able to demonstrate very high peak PNC of 100,000 to 560,000 particles/cm^3 during cooking with electric stoves and especially with gas stoves

Table 5.7 Particulate matter exposure in underground platforms in μg/m³

References	Mean PM_{10} (Min-Max)	Mean $PM_{2.5}$ (Min-Max)	Specifications	Country, time
Fromme et al. (1998)	–	– (128–311)[a]	1 station, Berlin	Germany, 1995
Seaton et al. (2005)	–	270–480	3 stations, London	UK, 2003
Branis (2006)	103	–	1 station, Prague	Czech Republic, 2003–2004
Aarnio et al. (2005)	–	60 (23–193)	1 station, Helsinki	Finland, 2004
Salma et al. (2007)	180 (85–234)	–	1 station, Budapest	Hungary, 2006
Raut et al. (2009)	28, 13[b]	21, 12[b]	1 station, Paris, nighttime	France, 2006
Raut et al. (2009)	200, 111[b]	61, 35[b]	1 station, Paris, normal hours	France, 2006
Raut et al. (2009)	320, 127[b]	93, 39[b]	1 station, Paris, rush hours	France, 2006
Mugica-Álvarez et al. (2012)	90 (77–130)[c]	48 (41–67)[c]	1 station, Mexico City	Mexico, 2007
Sahin et al. (2012)	170 (74–294)	–	6 stations, Istanbul	Turkey, 2007
Kim et al. (2012)	116 (90–141)	66 (43–89)	1 station, Seoul	Korea, 2007–08
Cheng et al. (2008)	32–66 (15–137)	25–44 (12–100)	4 stations, Taipei	Taiwan, 2010
Kam et al. (2011)	78 (14–197)	57 (9–130)	Los Angeles	USA, 2010
Querol et al. (2012)	46, 125	–	New and old station, Barcelona	Spain, 2011
Perrino et al. (2015)	409	–	1 station, Rome	Italy, 2010–12
Gerber et al. (2014)	101 (n.s.-166)	59 (n.s.-85)	1 station, Frankfurt	Germany, 2013
Grydaki et al. (2021)	107 (28–152)	–	1 station, Athens	Greece, 2014
Cartenì et al. (2015)	195 (172–262	52 (45–60)	14 stations, Naples	Italy, 2014
Mohsen et al. (2018)	55	41	Sydney	Australia, 2015
Tu and Olofsson (2021)	68–315 (–)	–	6 stations, Stockholm	Sweden, 2016–20
Mao et al. (2019)	– (75–362)	–	6 stations, Nanjing	China, 2018
Parchami et al. (2022)	168–275[d]	81–115[d]	4 stations, Tehran	Iran, 2018

n.s. not specified
[a] PM_4
[b] week-ends
[c] tenth and 90th percentile
[d] depending on the season

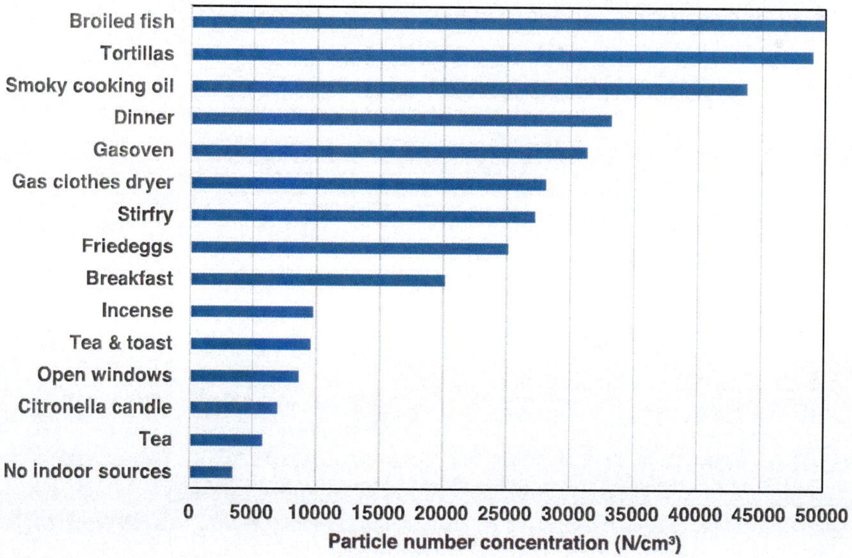

Fig. 5.4 Indoor particle number concentration during different typical indoor activities (according to Wallace 2006)

(Morawska et al. 2003; Dennekamp et al. 2001; He et al. 2004; Afshari et al. 2005; Ogulei et al. 2006; Hussein et al. 2006; Torkmahalleh et al. 2012; Patel et al. 2020). In this context, the wide range of concentrations can be explained by the different cooking activities (e.g., baking, frying, deep-frying and toasting), the energy input, the respective cooking food, the ventilation conditions, and the room geometry. For example, Dennekamp et al. (2001) describe PNC of up to 110,000 and 150,000 particles/cm^3 when using 4 electric and gas cooktops, respectively. Peak values of up to 590,000 particles/cm^3 were reached when frying ham on a gas stove. After a short time, coagulation of the particles in the room air causes a shift toward larger particle diameters (Abt et al. 2000; Dennekamp et al. 2001; Hussein et al. 2006). After the end of the cooking activity, the concentration drops rapidly. Zhao and Zhao (2018) summarized the results on the emission as well as factors influencing the emission from Chinese cooking and found that the cooking method could have a predominant impact on emissions, and oil-based cooking produces air pollutants at much higher levels than water-based cooking. In this review, the average concentrations of UFP were within a range of 119,000 to 6,643,000 particles/cm^3.

Ultrafine and accumulation mode particles were monitored over 37 months in an occupied suburban house under widely varying conditions such as different seasons, high and low temperatures, air-exchange rates, and relative humidities to consider the variability of indoor sources (Wallace 2006). Figure 5.4 gives the concentration differences after subtracting the daily average indoor background levels for each activities. In this house the major indoor source of particles was cooking with gas and indoor sources can produce PNC that are much higher than outdoor peak concentrations and provide a substantial fraction of total exposure.

5.4.1 *Occurrence in the Air of Residences*

In Germany, a median of 7230 particles/cm^3 (range: 700–69,640 particles/cm^3) was found in 2007 when 50 apartments were examined in Hamburg and Berlin (Santen et al. 2009), and a median of 16,000 particles/cm^3 was found in 59 apartments in Leipzig (tenth and 90th percentiles: 4000 and 30,000 particles/cm^3, respectively) (Franck et al. 2011). In the German Environmental Health Study (GerES V), a median of 5470 particles/cm^3, a 90th percentile of 18,300 particles/cm^3, and a maximum of 96,800 particles/cm^3 could be determined in 639 indoor rooms (Birmili 2021). McLaughlin et al. (2005) reported mean PNC values (0.02–1 μm) in 7 Irish dwellings between 4900 and 105,200 particles/cm^3 with a maximum value up to 485,300 particles/cm^3. In another Swedish study, 3 dwellings were investigated, with mean daily values ranging from about 1800 to 8300 particles/cm^3 (indoor) and 2000 to 7000 particles/cm^3 (outdoor) (Matson 2005). The indoor-to-outdoor ratio ranged from 0.7 to 2.5. In a non-smoking apartment in Athens, measurements were taken over 24 hours in the bedroom (no people present during the day) during one week and mean PNC of 14,500 particles/cm^3 (4400–108,000 particles/cm^3) was obtained (Diapouli et al. 2007). The most extensive measurements in Europe were performed in 2002–2004 as part of the RUPIOH study in a combined 152 homes in 4 European cities (Helsinki, Athens, Amsterdam, and Birmingham) (Hoek et al. 2008). In this study, the lowest mean levels were observed in Helsinki with 3000 particles/cm^3 indoors (outdoors: 4500 particles/cm^3) and the highest in Amsterdam with 12,000 particles/cm^3 (outdoors: 26300 particles/cm^3). In Portugal, a median of 7390 particles/cm^3 (1450–390,000 particles/cm^3) was measured in 65 dwellings in 2018/2019 (Madureira et al. 2020). Outdoor air levels were significantly lower compared to indoor air in this study. Again, in seven residences in Sweden average PNC of 6200 particles/cm^3 was found, depending mainly on frequency and type of occupants' activities in studied residences (Omelekhina et al. 2022).

In Australia, 2017 until 2018, 40 non-smoking adults living in 40 residences were recruited and a 24-hour median of 9218 particles/cm^3 (range: 975–35,941 particles/cm^3) were reported (Gilbey et al. 2022).

In a recent study, indoor and outdoor particle number size distributions (diameter: 10–800 nm) were measured in 40 German households under real-use conditions over 500 days, and emission rates were derived for around 800 reported indoor source events (Zhao et al. 2021). Using mean values, penetration from outdoors and open windows is responsible for 26% and 15% of the daily-integrated exposure to UFP (see Fig. 5.5). However, due to the different behavior patterns of residents in the cold and warm seasons, the proportions of indoor and outdoor contributions to total exposure were rather different.

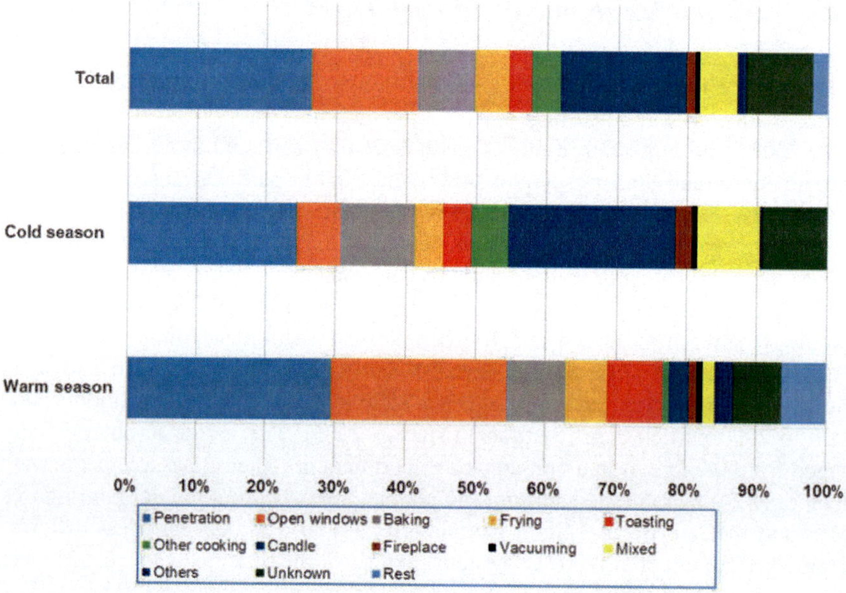

Fig. 5.5 Daily mean source contribution to daily-integrated exposure to indoor PNC (cm^3 h day). (according to Zhao et al. 2021)

5.4.2 *Occurrence in the Air of Public Facilities*

Measurements of PNC have so far only been carried out sporadically in schools.

In 36 German classrooms, PNC of 2600 to 12,100 particles/cm^3 was measured (Fromme et al. 2007). In another German study, levels ranged from 2400 to 75,500 particles/cm^3 (urban) and 1720 to 47,100 particles/cm^3 (rural) (Link et al. 2004). In Greece, mean PNC of 24,000 particles/cm^3 was determined in 7 elementary schools in Athens in 2002/2003 during class time, which correlated well with outdoor air content (32,000 particles/cm^3) (Diapouli et al. 2008). Moreover, in 2013, concentrations of 800–71,000 particles/cm^3 were determined in nine Greek elementary school (Dorizas et al. 2015). Comparable results were also found in 2012/2013 in a Spanish study with mean values of 18,800 particles/cm^3 in winter and 13,700 particles/cm^3 in summer (Rivas et al. 2014), in 2014/2015 in 5 Italian classrooms during the heating season with about 10,400 particles/cm^3 (6000–13,600 particles/cm^3) (Stabile et al. 2017), and about 8000 particles/cm^3 (300–16,500 particles/cm^3) in 65 classrooms in France (Paunescu et al. 2017). Again, Slezakova et al. (2019) reported mean values of 7800 particles/cm^3 (medians of individual classrooms: 1600–17,000 particles/cm^3) in 71 classrooms of 20 schools in Portugal in 2014/2015. A median of 6400 particles/cm^3 (95th percentile: 36400 particles/cm^3) was determined in 151 daycare centers in Sweden and Denmark in 2008 (Clausen et al. 2012), while higher concentrations of 2200–113,000 particles/cm^3 (median: 9300–11,300 particles/cm^3) were observed in three naturally ventilated Portuguese daycare centers in 2013 (Fonseca et al. 2014). If only a size range of 0.3–10 μm were quantified, lower levels from 33 to 168 particles/cm^3 were reported in 11 primary schools in Lithuania in 2017–2018 (Juskiene et al. 2022).

In six classrooms in northern California during normal occupancy and use, concentrations from 5200 to 16,500 particles/cm^3 (on average 10,800 particles/cm^3) were observed in 2008 (Mullen et al. 2011). Gaspar et al. (2018) measured UFP in 40 California child-care facilities in 2010–2012 and reported 1500 to 75,000 particles/cm^3 (median: 14000 particles/cm^3). In Brazil, particles in the range from 10 to 420 nm were collected in two preschools in 2016/17 and means of 244,000 particles/cm^3 and 13,700 particles/cm^3 were reported (Portela et al. 2021).

Among urban high school students in China geometric mean personal exposure was 18,300 particles/cm^3 (Zhou et al. 2020). The maximum proportion to daily UFP exposure resulted by the transport microenvironment (36.7%) followed by the household environment (36.6%). Quantifying school children's personal UFP exposure and exposure intensity against their indoor and outdoor activities during a school day, on average 10,500 particles/cm^3 and 5100 particles/cm^3 were found in Australia and China in 2016 (Mazaheri et al. 2014, 2019).

In school classrooms, where typical sources such as combustion processes or cooking activities are absent, the exposure situation is strongly determined by the outdoor air (Diapouli et al. 2008; Guo et al. 2010; Paunescu et al. 2017).

5.4.3 Occurrence in the Office Air

Ultrafine particles were found in 5 offices of smokers in Sweden and Denmark at concentrations ranging from about 1000 to 13,000 particles/cm^3, even when a ventilation system was present (Matson 2005). In another study in northern Sweden Orru et al. (2021) reported average PNC from 316 to 3288 during the heating season and from 323 to 1241 in the non-heating season. In three mechanically ventilated offices in Australia, PNC ranged from 720–18,000 particles/cm^3 during working hours (Quang et al. 2013). In contrast, very low concentrations of 11–601 particles/cm^3 were found in a Norwegian office (Chatoutsidou et al. 2015).

There are hardly any results on the indoor air to outdoor air (I/O) ratio. Mean I/O of 0.25–0.40 were observed in three Australian offices (Quang et al. 2013), while I/A ratios of 0.5–0.8 were described in 5 offices in Denmark and Norway (Matson 2005).

5.4.4 Occurrence in the Air of Ground Transport Microenvironments

Table 5.8 summarizes published results of studies on the exposure of cars and busses to ultrafine particles. On average/median, PNC results in the range of 9000–60,000

Table 5.8 Ultrafine particles in transport vehicles (in 10^3 particles/cm^3)

Reference	Mean value (Min-Max)	Motor vehicle	Measuring range (µm)	country, time
Dennekamp et al. (2002)	53[a] (–)	11 busses	**0.02 - > 1.0**	**Aberdeen, UK, 2000/1**
Levy et al. (2002)	~32 (12–80) ~39 (11–83)	Bus Car	**0.02- > 1.0**	**Boston, USA, 2000**
Mackay (2004)	44 (9.9–143) 58 (8.0–282)	Bus Car	**0.01- > 1.0**	**Leeds, UK**
Krausse & Mardaljevic (2005)	– (46.1–116)	133 vans	**0.02- > 1.0**	**Leicester, UK**
Kaur et al. (2005)	101 (64.5–159) 100 (36.5–152) 88 (51.5–114)	Bus Car Cab	**0.02- > 1.0**	**London, UK, 2003**
Diapouli et al. (2007)	94 (25.4–217)	Car	**0.01- > 1.0**	**Athens, Greece, 2004**
Eiguren-Fernandez et al. (2005)	24.8 54.7 68.6 246	Small road, AC on Small road, AC off Freeway, AC on Freeway, AC off	**0.005- ~ 2.0**	**Los Angeles, USA**
Rim et al. (2008)	6.1–32 (–)	6 school busses	**0.01–1.0**	**Austin, USA, 2006**
Zuurbier et al. (2011)	37[a] (23–72) 29[a] (16–47) 38[a] (23–63) 37[a] (30–68)	Diesel bus E-bus Diesel car Gasoline car	**0.01- > 1.0**	**Arnhem, Netherlands, 2007/8**
Geiss et al. (2010)	16 (8–30)	18 cars	**0.02–1.0**	**Ispra, Italy, 2009**
De Nazelle et al. (2012)	52 (–) 118 (–)	Bus Car	**0.01- > 1.0**	**Bacelona, Spain, 2009**
Ozgen et al. (2016)	11 (1–71)	Car	**0.02–1.0**	**Milan, Italy, 2010**
Ragettli et al. (2013)	14[a] (–) 32[a] (–)	Bus Car	**0.01–0.30**	**Basel, Switzerland, 2010/11**
Rivas et al. (2017)	8.9[b] (3.9–22) 9.7[b] (3.8–21)	Bus Car	**0.02–1.0**	**London, UK, 2016**
Murzyn et al. (2021)	32 (2–391)	45 field trips with a car	**0.02–1.0**	**France, 2016/17**
Moreno et al. (2019)	32[a] (7.9–88) 11[a] (7.3–13)	Diesel cab E-taxi	**0.03–0.70**	**Barcelona, Spain, 2017**

[a] median
[b] geometric mean

particles/cm^3 were reported. In contrast, a very high content was described by Kaur et al. (2005) in the City of London, which averaged about 88,000–101,000 particles/cm^3, and in a car trip in Barcelona in 2009 with 118,000 particles/cm^3 (De Nazelle et al. 2012). During 499 trips conducted by 50 Parisian taxi drivers a median value of 27,000 particles/cm was observed in 2019/20 (Hachem et al. 2020). Very high mean concentrations of 246,000 particles/cm^3 were reported in New York in 2004 when driving on the freeway with the window open (Eiguren-Fernandez et al. 2005). Striking in this study is the lower level when the air conditioning was on (mean: 69000 particles/cm^3). Driving a plug-in hybrid electric vehicle on highways near Beijing, China, means of 10,800 particles/cm^3 without and 6100 particles/cm^3 with pre-ionization were reported by Wei et al. (2020) in 2019.

Overall, all studies indicate that, in addition to the specific ventilation conditions in the car, the outside air concentration and thus the driving conditions are of particular importance for the exposure inside. Under specific conditions, e.g., a diesel truck driving ahead, short-term peak concentrations of up to 500,000 particles/cm^3 were observed (Abraham et al. 2002; Zhu et al. 2007). In contrast, only 6100–32,000 particles/cm^3 were measured on average and 7300 particles/cm^3 in school busses traveling in rural Texas (Rim et al. 2008; Zhang and Zhu 2010).

The indoor air concentration is particularly dependent on the particle influx from the outside air and the decrease due to deposition on the surfaces of the vehicle cabin and the ventilation systems (Hudda et al. 2011). Various studies were able to show that filters have only a minor effect on the indoor air content and that deposition on the interior surfaces is the main influencing factor (Pui et al. 2008; Hudda et al. 2011). In contrast, Zhu et al. (2007) describe in car trips on two interstates in the USA that PNC decreased significantly when a HEPA filter was in use (3800 vs. 134,000 particles/cm^3 and 2100 vs. 84,000 particles/cm^3). The same working group also describes that, when manufacturers install particulate filters, a 50% reduction is possible for particles 7–40 nm in diameter and 20–30% for particles 40–200 nm (Zhu et al. 2007).

To date, few studies have described the exposure of cyclists and pedestrians to ultrafine particles (Boogaard et al. 2009; Zuurbier et al. 2010; Int Panis e al. 2010, De Nazelle et al. 2012; Ragettli et al. 2013; Hankey and Marshall 2015; Ham et al. 2017). The lowest mean concentrations were described for cyclists at 22000 and 23,000 particles/cm^3 in Sacramento and Switzerland in 2010/2011 (Ham et al. 2017; Ragettli et al. 2013; Patel et al. 2023), and the highest in a study in London in 2003 at 78000 particles/cm^3 (Kaur and Nieuwenhuijsen 2009). For pedestrians, mean concentrations of 19,000 particles/cm^3 were observed in Switzerland and 63,000 particles/cm^3 in London (Ragettli et al. 2013; Kaur and Nieuwenhuijsen 2009). Pedestrian exposure to UFP was measured using personal monitoring along a customized route through an Australian city during different COVID-19-affected traffic flow conditions (Patel et al. 2023). Without any restriction the median exposure of pedestrians was 20,884 particles/cm^3 (range: 1382–384,873 particles/cm^3) while under the most stringent traffic reduction of 82%, median UFP concentrations reduced by 73%. In a Chinese city, pedestrian UFP exposure was measured during rush hours on sidewalks with and without vegetation barrier near an arterial road and concentration was 28.7% lower with weakened fluctuations on vegetation barrier sidewalk than on open sidewalk in average (He et al. 2023).

In Colorado, 45 non-smoking participants were recruited to investigate the role of microenvironment on personal exposures (Koehler et al. 2019). In transit median PNC level was 2700 particles/cm^3. Although transit typically accounted for less than 7% of participant's time per day, this microenvironment contributed of around 55% to the total exposure.

A special exposure is present when smoking takes place in the interiors of means of transport. In the only study available so far from Germany, PNC was determined when smoking one cigarette, one e-cigarette, and one IQOS in the air during journeys of 7 different cars through Munich, with the passenger window open about 2–5 cm (Schober et al. 2019). While mean concentrations of 8400–73,900 particles/cm^3 were found in the control trips without any smoking activities, the mean concentrations when smoking a cigarette were 24,300–236,000 particles/cm^3 and when smoking an e-cigarette were 10,200–74,000 particles/cm^3. Using IQOS had no effect on indoor UFP levels. In urban areas in Texas, one cigarette was smoked during each trip with different ventilation conditions (Liu and Zhu 2010). Mean concentrations when parked with the air conditioner running were 250,000 particles/cm^3, when driving with the windows closed were 170,000–250,000 particles/cm^3, and when driving with the passenger window slightly open were between 22,000 and 29,000 particles/cm^3. When not smoking, the levels ranged from 3000–23,000 particles/cm^3.

5.4.5 *Occurrence in the Air of Underground Transport Microenvironments*

At present, there are hardly any measurements of particle counts for the interiors of subways. An average of 17,000–23,000 particles/cm^3 (>50 nm) was measured in three subway lines in London, while average contents of 14,000–29,000 particles/cm^3 were determined on the platforms of three subway stations (Seaton et al. 2005). Similar results were obtained from measurements in the Boston subway in 2000 (Levy et al. 2002), where mean levels were about 21,000 particles/cm^3, and in Helsinki in 2004, where 27,000 particles/cm^3 was measured (range: 14000–50,000 particles/cm^3) (Aarnio et al. 2005). In Milan, mean PNC of 13,000 particles/cm^3 (range: 5000–22,000 particles/cm^3) was observed in 2010 (Ozgen et al. 2016). A comparable level of mean 19,400 particles/cm^3 (range: 16157–22,395 particles/cm^3) was also found in the Beijing metro in 2011 (Yan et al. 2015). Compared to the results reported by Seaton et al. (2005) in the London Metro, only mean PNC of 6300 particles/cm^3 (range: 2500–12,500 particles/cm^3) was found in 87 trips in London in 2016 (Rivas et al. 2017). In Vienna, Austria, mean levels of around 7000–12,500 particles/cm^3 were reported in 5 subway lines in 2016, while means in two underground and ground level subway stations were 10,000 and 8300 particles/

cm^3, respectively (Posselt et al. 2019). System-representative commuter air pollution exposure data were collected for three Canadian metro systems in 2010–2013, describing mean concentrations of 10,500 to 19,500 particles/cm^3 (95th percentiles: 23700–39,300 particles/cm^3) (Van Ryswyk et al. 2017).

Mean underground platform level ranged from 9500–17,000 particles/cm^3, significantly lower than those in outdoors by approximately 0.3–0.5 times, in 2010 in Taipei, Taiwan (Cheng and Yan 2011).

Lower overall PNC was measured in subways compared to surface transportation (Yan et al. 2015; Rivas et al. 2017), while in Milan this difference was only observed compared to the exposure situation of cyclists and pedestrians, but not car occupants (Ozgen et al. 2016).

5.5 Engineered Nanoparticles (eNP)

In contrast to ultrafine particles (UFP) engineered nanoparticles (eNP) are manufactured through controlled engineering processes (Li et al. 2016). In the context of nanotechnology, novel nanomaterials are created at the atomic, molecular, and macromolecular level through miniaturization or synthetic production. They are characterized by having different physicochemical properties compared to the original material of conventional size (Bundschuh et al. 2018). These new material properties (solubility, reactivity, conductivity) are based on quantum physical principles that only occur at very small particle sizes. Physicochemical characteristics of eNP like particle size, surface area, zeta potential/surface charge, colloidal stability, and core-shell composition and environmental conditions like pH, ionic strength, organic and inorganic colloids, and temperature are the most important parameters which regulated environmental transformation of eNP (Abbas et al. 2020). Therefore, the interactions between nanomaterials and other substances differ significantly from those of macroscopic materials, although they consist of the same chemical components. This is due to the special surface and interface properties as well as the geometric shape of the nanomaterial.

This opens up innovative opportunities to decisively improve applications and production processes. These include above all the cosmetics sector (e.g., zinc oxide and titanium dioxide in sun creams), the food industry, which uses, e.g., silver-doped packaging to improve shelf life, and the textile industry, which is increasingly relying on increased functionality through nanomaterials (Schober et al. 2012). In addition, nanotechnology applications in medicine support the development of new formulations of active pharmaceutical ingredients, and in the automotive industry, chemical industry, and communications technology, for example, in the production of exhaust gas catalysts, surface coatings, or organic light-emitting diodes. Commonly used forms of engineered nanoparticles as raw materials in industry are oxide eNPs of aluminum, manganese, silicon or titanium, non-oxide eNP (e.g., titanium or silicon carbide) and metallic eNP of silver, gold, or copper. In addition, single-walled (SWCNT, single-walled carbon nanotubes) or multi-walled (MWCNT,

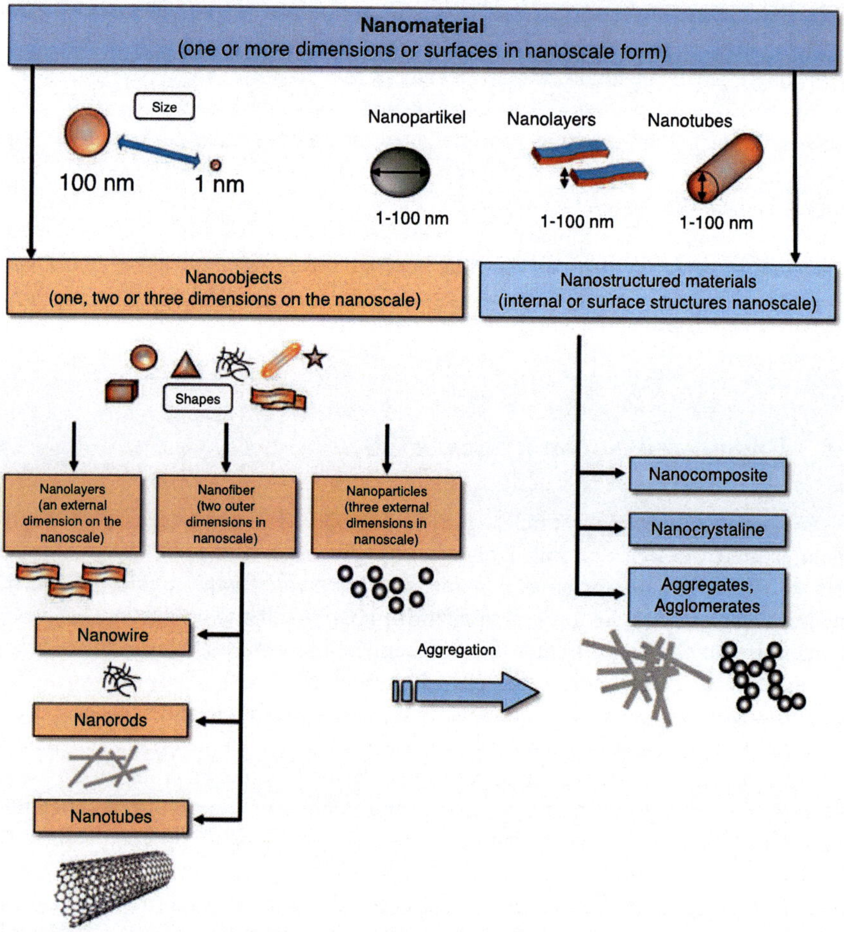

Fig. 5.6 Classification of engineered nanomaterials (Schober and In-Iam 2020)

multi-walled carbon nanotubes) carbon nanotubes, fullerenes, and quantum dots have recently been increasingly used industrially (Schober et al. 2012). A broad overview of, e.g., history, sources, synthesis strategies, their toxicity, risks, and aspects of regulations was presented by Barhoum et al. (2022).

In a first definition by the EU, a nanomaterial is a natural, process-generated or manufactured material that contains particles in an unbound state, as an aggregate or as an agglomerate, and in which at least 50% of the particles in the number size distribution have one or more external dimensions in the range from 1 nm to 100 nm (EU 2011). On this basis, different nanomaterials (nanoobjects) can be distinguished, which are compiled in Fig. 5.6.

As of January 1, 2020, companies that manufacture or import nanoforms are subject to explicit legal requirements under REACH. These notification requirements apply to all new and existing registrations involving nanoforms.

5.5.1 *Health Aspects*

The experimental evidence points to a very heterogeneous risk potential, influenced both by the new properties of the materials and by the diversity of structures, products, and fields of application. The high reactivity of certain nanoparticles, their ability to penetrate body and cell barriers, and their biopersistence must be considered. Valid risk extrapolation of knowledge from environmental aerosol ultrafine particles to eNP is limited because nanomaterials often have different structures and physicochemical properties. Recent accounts of the state of scientific knowledge have been published (Li et al. 2016; Stone et al. 2017; Ganguly et al. 2017; IARC 2017; Laux et al. 2018; Schulte et al. 2019; Boyes and van Thriel 2020). Overall, depending on the aforementioned properties, they indicate local inflammatory responses in the lung and signs of oxidative stress after inhalation. For biopersistent nanofibers such as MWCNT, there is increasing evidence for pulmonary and peritoneal carcinogenicity, similar to asbestos fibers. There is preliminary evidence that some metallic nanoparticles and carbon nanotubes may exacerbate existing asthma and contribute to allergic respiratory disease (Ihrie and Bonner 2018).

IARC classified MWCNT-7 (Mitsui-7, Hodogaya Chemical Co.) multi-walled carbon nanotubes as possibly carcinogenic to humans (Group 2B) and MWCNT other than MWCNT-7 as well as single-walled carbon nanotubes as not classifiable as to their carcinogenicity to humans (Group 3) (IARC 2017).

5.5.2 *Occurrence in the Air*

Despite the partly rapid development of nanotechnology and products manufactured by means of nanotechnology in recent years, little is still known about human exposure to them. Essentially, this knowledge relates to workplaces where nanomaterials are handled (Schober et al. 2012; Winterhalter et al. 2012; Debia et al. 2016; Basinas et al. 2018). Based on product databases, Vance and Marr (2015) conclude that approximately 11% of products on the market have the ability to form aerosols indoors. These are often sprays or cleaning products, with nanoscale silver being the most common.

Studies on exposure to eNP in indoor living spaces are currently not available. Often, due to the blurred distinction to ultrafine particles, UFP of the environmental aerosol in the context of, e.g., combustion processes or indoor pressure activities are also described as nanoparticles. Results on this can be found in particular in Sect. 5.4.

In the scientific literature, results of two studies in parking garages have been published so far. For example, concentrations of fullerenes ranging from 0.08 to 2.27 pg/m^3 were found in the investigation of 8 outdoor air samples from a northern Spanish city in 2015/2016 (Encinas and Gómez-de-Balugera 2018). A garage with 172 parking spaces was also sampled at the same time. Overall, fullerenes ranging from approximately 1 pg/m^3 to 10.5 pg/m^3 were determined in 29% of the 36 samples here. Similar results were obtained in a study that examined a parking garage in Arizona (Benn et al. 2012). In this case, all fullerene concentrations were below the determination limit of 3 pg/m^3. Moreover, Walser et al. (2012) simulated scenarios of equipment failure during gas phase production of nanoparticles in a laboratory and found indoor concentrations of up to 10^6 eNP/m^3 in a 300 m^3 production room only 60 seconds after the accident.

5.6 Man-Made Vitreous Fibers (MMVF)

The generic term man-made mineral fibers (MMMF) or man-made vitreous fibers (MMVF) includes in particular glass fibers as well as stone and slag wool, which are used in large quantities for heat and sound insulation in buildings. Other MMVF, such as ceramic fibers (aluminum silicate fibers), have only special applications due to their temperature resistance and therefore play a secondary role in the exposure of the general population. The MMVF can be divided into different categories, which are compiled following the IARC in Fig. 5.7.

As part of the MMVF production process, inorganic materials are melted at high temperatures and drawn or blown into fibers. Depending on the choice of starting materials, the chemical composition of the fibers can be selectively changed during production, thus influencing the product properties. This option can also be used to

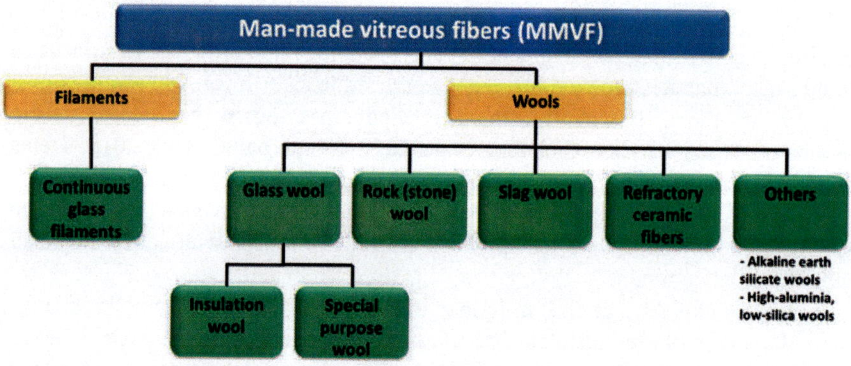

Fig. 5.7 Categories of man-made mineral fibers (MMVF) (according to IARC 2002)

produce fibers with lower biological resistance and thus a better toxic risk profile. Often the products used, such as ceiling tiles, are mixtures of, for example, MMVF, perlite, clay, and starch.

According to IARC (2002), an overall annual production of around nine million tons of MMVF was estimated.

5.6.1 Health Aspects

With regard to the health effects of inorganic fiber dust, in addition to local irritation symptoms on the skin and mucous membranes, inflammatory reactions and the carcinogenic effect after inhalation are in the foreground (ATSDR 2004; Greim et al. 2014). Basically, carcinogenicity is based on three fiber characteristics: sufficiently long, sufficiently thin, and sufficiently persistent. According to the generally accepted WHO definition, fibers are considered particularly critical if they have the following dimensions: longer than 5 µm, thinner than 3 µm, and a length-to-width ratio greater than 3:1 (WHO 2000; AGS 2021). Biosolubility (biopersistence) is a measure of how long the fibers remain in the lungs and is determined in animal experiments. It is assumed that the carcinogenic risk is low with lower biopersistence, i.e. faster removal from the lungs.

In animal studies, long-term inhalation of high doses of MMVF causes chronic inflammation, fibrosis, and cell proliferation of lung alveolar cells and mesothelium. The strength of these effects usually correlates with the biopersistence of the fibers. There is also an irritant effect of coarse fiber dust on the skin and mucous membranes of the eyes and upper respiratory tract.

There are some reviews compiling the health effects (Marsh et al. 2011; Lippmann 2014; Harrison et al. 2015; Suder Egnot et al. 2020). As before Marsh et al. (2011), a recent systematic review by Suder Egnot et al. (2020) of 10 peer-reviewed epidemiological studies published after 2002 found no consistent or credible association between occupational exposure to glass wool and rock/slag wool and respiratory tract cancer.

As of June 2000, a ban on the manufacture and use of biopersistent "old" mineral fibers applies in accordance with Annex IV No. 22 of the German Hazardous Substances Ordinance. In the case of mineral wool installed before 1996, it must be assumed that it is the aforementioned "old" mineral wool and that it can only be broken down with difficulty in the lungs. These are therefore to be assessed as carcinogenic in the sense of the hazardous substances legislation. The mineral wool made of artificial mineral fibers produced since about 1996, which meet the exemption criteria of Annex IV No. 22 of the German Hazardous Substances Ordinance, are referred to as "new" mineral wool. There is no obligation to clean up "old" mineral wool indoors, but during renovation and remediation work on the fibers, the protective measures for employees and other persons must be observed, which are intended, for example, to prevent the release and carryover of fibers. In any case, a specialist company must be commissioned. As long as there are no structural defects in the insulation or structural interventions are carried out on it, exposure of the indoor air to fibers is unlikely.

IARC classified glass fibers such as E-glass and "475" glass fibers as possibly carcinogenic to humans (Group 2B), refractory ceramic fibers as possibly carcinogenic to humans (Group 2B), and insulation glass wool, continuous glass filament, rock (stone) wool and slag wool as not classifiable as to their carcinogenicity to humans (Group 3) (IARC 2002). IARC did not perform an overall evaluation of the newly developed fibers designed to be less biopersistent such as the alkaline earth silicate or high alumina low-silica wools.

5.6.2 Occurrence in the Air

In principle, when measuring dust or fibers indoors, it must be taken into account that the composition of airborne dust is an expression of various sources. For example, it was shown that about 53% of indoor dust consists of inorganic (mineral) particles and about 18% of organic particles, while inorganic fibers (including MMVF) account for only about 1% (Mølhave et al. 2000). In addition, different analytical methods are used for fiber determination, and scanning electron microscopy was used in the majority of the following investigation.

The following results have been described from Europe so far: In 1984, Rindel et al. (1987) examined a total of 25 Danish kindergartens, in which acoustic ceiling tiles and probably other insulating materials were installed, for inhalable MMVF and found 41–110 fibers/m^3. In control rooms, concentrations were essentially below 40–80 fibers/m^3 (limit of quantification). Mean levels of inhalable MMVF of 17–213 fibers/m^3 (overall mean: 82 fibers/m^3) were also found in 105 rooms with acoustic ceiling tiles in public buildings in Denmark (Schneider et al. 1990). In comparison, mean concentrations of 62 fibers/m^3 were found in 10 rooms without acoustic ceiling tiles. Concentrations of other inhalable fibers in this study ranged from 20,000–210,000 fibers/m^3. Compared to natural ventilation, concentrations were lower in the presence of mechanical ventilation. In a German study, a total of 134 air samples were taken in 24 buildings in 1990/1991 and a median of 227 product fibers/m^3 (fibers <3 μm) was found (Bake 2003). In this case, product fibers accounted for approximately 12% of the total inorganic fibers measured. In another German study, contents of 1000–3000 fibers/m^3 (< 3 μm) were measured when 6 offices in a building with acoustic ceiling tiles were examined (Thriene et al. 1996). Very low concentrations of MMVF were found in 20 schools and daycare centers in Schleswig-Holstein (Walker et al. 2012). In this study, the mean concentration of inhalable calcium sulfate fibers was 97 fibers/m^3 (30–248 fibers/m^3) and that of other inorganic fibers was 82 fibers/m^3 (21–186 fibers/m^3).

Very comprehensively, an American study examined 205 air samples in 51 public and private buildings with known MMVF use. Inhalable MMVF in all rooms was below the limit of quantification of 100 fibers/m^3 (Carter et al. 1999). The mean value of all inhalable fibers in this study was 2400 fibers/m^3. Dodgson et al. (1987)

studied 5 houses in which insulation material had been present for a long time. In these houses, all measurements were below the limit of determination of the analytical method of 5000 fibers/m^3. 25 residences and 9 building-interior common areas in New York, USA, were measured to characterize the background levels of residential indoor concentration of synthetic vitreous fiber and on average (upper bound) 80 fibers/m^3 (range: 72–260 fibers/m^3) were found (Tang et al. 2004).

In Canada in 2000–2002, a total of 130 personal and area samples were collected to characterize worker's exposure handling insulation materials (Verma et al. 2004). Mean concentrations from 20,000 fibers/m^3 to 780,000 fibers/m^3 were reported. A Polish study summarized results during different working processes in 2001 and found 150,000 fibers/m^3 during the disassembly and assembly of thermal insulation of mineral wool, 70,000 fibers/m^3 during the disassembly and assembly of thermal insulation of mineral wool mats on the heating pipeline, and 30,000 fibers/m^3 during the installation of thermal insulation of mineral wool plates on fans (Kupczewska-Dobecka et al. 2020). The same working group presented results of measurements in the inhalable fraction of dust at 12 workplaces of mineral wool insulation fitters in 2018. They observed concentrations between 5000 and 35,000 fibers/m^3.

5.7 Asbestos

The term asbestos refers to different types of silicate minerals that occur naturally in the earth's crust. They are divided into the so-called serpentine (chrysotile, or white asbestos), amphibole (crocidolite or blue asbestos), amosite (brown asbestos), tremolite, actinolite, and anthophyllite. In the past, more than 90% of chrysotile was used, while crocidolite and amosite were used in only about 3%. Asbestos is a persistent, chemically inert mineral fiber that can be found in soils, sediments, lofted in air and windblown dust, surface water, groundwater, and biota (US EPA 2020).

The EU Directive 2009/148/EC defines that "asbestos" means the following fibrous silicates:

- asbestos actinolite, CAS-No. 77536–66-4,
- asbestos grunerite (amosite), CAS-No. 12172–73-5,
- asbestos anthophyllite, CAS-No. 77536–67-5,
- chrysotile, CAS-No. 12001–29-5,
- crocidolite, CAS-No. 12001–28-4,
- asbestos tremolite, CAS-No. 77536–68-6.

The following two product forms were mainly used: Asbestos cement is a mixture of cement and approx. 10–15% asbestos with a density of at least 1400 kg/m^3. In this case, the asbestos fibers are firmly bound in the product. They found application on roofs with corrugated sheets, in floor coverings (so-called flex sheets) and in bitumen adhesives. In contrast, sprayed asbestos is a weakly bound asbestos product with an asbestos content of 20% to sometimes over 60%. Release is possible not only during processing, but also during aging of the products and vibration. Spray asbestos was used for heat protection, especially in steel frame buildings and for fire protection (e.g., as asbestos cords, fire dampers, asbestos-containing lightweight panels (Promabest) and asbestos boards).

It must be expected that in the period from about 1965 to about 1985—possibly also until 1990—asbestos was also frequently used in plaster materials, fillers, or tile adhesives in buildings during construction or renovation. In particular, the following products are known to contain asbestos: Colored stone plasters, fire protection plasters, decorative plasters (e.g., rubbing plasters), thermal insulating compounds, floor, wall, and ceiling fillers, thin-bed adhesives, bitumen adhesives for flex panels, tar/bitumen-based coating materials, paints. Under normal use, fiber release into indoor air is not expected. However, fibers may be released by abrasion during mechanical processing of components and surfaces containing asbestos. This is particularly to be expected during dry blasting, milling, brushing, and drilling.

In the 1980s concentrations of 100 to 1000 fibers/m^3 and 1000 to 16,000 fibers/m^3 were observed in ambient air of rural and urban areas in the Netherlands, respectively (DECOS 2010). But near-by busy roads and inside tunnels up to 80,000 fibers/m^3 were measured. Below 2000 fibers/m^3 in urban outdoor air and below 400 fibers/m^3 in rural areas were reported in Germany in 1987 (ECHA 2021).

ECHA (2021) gave an overview of techniques and methods for monitoring of asbestos fibers in air with phase contrast microscopy (PCM), scanning electron microscopy (SEM), and transmission electron microscopy (TEM).

5.7.1 Health Aspects

In the indoor environment inhalation is the most important route of exposure to asbestos. Inhalation at high concentrations and long-term exposure is associated with the development of non-malignant diseases of the lungs and pleura, and malignant diseases arising especially in the lung, larynx, and pleural and peritoneal linings (IARC 2012; Carlin et al. 2015; Baur 2018; US EPA 2020; ECHA 2021). The recent reviews and meta-analyses indicate that for lung cancer the overall data set is compatible either with a more than additive or a multiplicative joint effect between asbestos exposure and smoking (ECHA 2021). Asbestosis, a diffuse pulmonary fibrosis, and mesothelioma are typical asbestos related diseases. Globally, the age-standardized rate of mesothelioma was 0.30 per 100,000 persons in 2020 with the highest rates in Northern Europe (1.4), Australia and New Zealand (1.3), and Western Europe (0.79) (Huang et al. 2023).

However, the carcinogenic mechanism of action of asbestos is not fully understood yet. The first step is considered to be the initiation of an inflammatory reaction. In addition, fibrotic remodeling, known as asbestosis, is evident in the lungs. The shape and size (length, thickness) of the asbestos fibers and their chemical composition, surface structure and solubility in body fluids (biopersistence) determine the cytotoxicity as well as the carcinogenic potential. The individual types of asbestos fibers have different carcinogenic potentials, which are attributed to differences in the element content, e.g., iron.

Different cellular and molecular effects of asbestos after inhalation were discussed (Solbes and Harper 2018):

- Direct mechanical injury.
- Generation of reactive oxygen and nitrogen species.
- Alteration of cellular signal transduction pathways.
- Loss of tumor suppressor genes and activation of oncogenes.
- Chromosomal and DNA alteration.

According to the generally accepted definition of the WHO, fibers are considered particularly critical if they have the following dimensions: longer than 5 μm, thinner than 3 μm and a length-to-width ratio greater than 3:1 (WHO 2000).

In the 1950s and 1960s, there was increasing evidence of the serious health hazards of asbestos, and in 1973 the WHO cancer organization found sufficient evidence for a carcinogenic effect of all 6 types of asbestos in commerce (IARC 1973). IARC subsequently saw sufficient evidence from epidemiological studies and animal experiments for a carcinogenic effect of all forms of asbestos and classified it in Group 1 (IARC 2012).

A corresponding European regulation then came into force in 2004 (EU 2004). Up to this time, however, approximately 3000 products containing asbestos were used in large quantities. Again, in the EU, the manufacture, placing on the market, and use of asbestos fibers and of articles and mixtures containing these fibers are prohibited since 2006 (EU 2006). Commission Regulation 2016/1005 amended Annex XVII of the REACH regulation by specifying that the derogation concerning chrysotile asbestos use in existing electrolysis installations would apply only until 1 July 2025 (EU 2016). In 2022, the EU Commission presented a proposal to revised Directive 2009/148/EC on the protection of workers from the risks related to exposure to asbestos at work with reduced exposure limit value from 0.1 fibers/cm^3 to 0.01 fibers/cm^3 (calculated as an 8-hour time-weighted average). The new value corresponds to the value currently in force in Germany and France.

No national law to ban asbestos existed so far in the USA, but on the basis of the Toxic Substances Control Act (TSCA) US EPA was required to promulgate regulations to prevent or reduce asbestos hazards and banned the use of some but not all asbestos-containing products (US EPA 2020).

Although more than 50 countries worlwide have restricted or banned asbestos, it is still in use and was produced in Russia with 700,000 metric tons, Kazakhstan with 230,000 metric tons, Brazil with 190,000 metric tons, and China with 130,000 metric tons in 2022 (MC 2023). For example in the USA, where production ended 2002, the annual consumption was estimated to be only 260 metric tons for special applications in the chloralkali industry in 2022.

5.7.2 *Occurrence in the Air*

In the context of the asbestos directive, e.g., during building renovations, measurements have frequently been carried out on an ad hoc basis in recent years, but these have generally not been published in the scientific literature. Measurement results from the 1980s and 1990s have been summarized by WHO (2000). For example, in 50 buildings with suspected exposure, an average of 38 inhalable fibers/m^3 was found in 144 measurements, with asbestos fibers detected in only a few indoor locations. Also at that time, significantly higher concentrations were observed in indoor air when simulating work on products and surfaces containing asbestos.

Schneider et al. (1996) published results on 24-h personal exposure over one year of 5 school children from Paris (suburban area), 5 pensioners in Denmark, 5 office workers, and 5 taxi drivers in Germany. They found, when an outlier measurement was not considered, mean values (95th confidence interval) of 44 fibers/m^3 (12–112 fibers/m^3), 66 fibers/m^3 (21–154 fibers /m^3), 10 fibers/m^3 (0.3–58 fibers/m^3), and 47 fibers/m^3 (13–120 fibers/m^3).

A summary reports on 59 Italian schools where asbestos cement products and vinyl ceiling tiles containing asbestos were found by inspection during the period 1992–2002 (Campopiano et al. 2004). No airborne asbestos could be detected during operation in 29 schools (<400 fibers/m^3). However, in one school a maximum of 700 fibers/m^3, in 7 schools a maximum of 1000 fibers/m^3, in 16 schools a maximum of 1600 fibers /m^3, and in 6 schools a maximum of 2200 fibers/m^3. Other distribution parameters are not described. In England, 3 schools with insulation materials containing amosite in contact with space heating were sampled (Burdett et al. 2016). Background concentrations were always below the pooled mean detection limit of 50 fibers/m^3. Even when classrooms were used, no change in levels was observed in 2 schools, while maximum concentrations of 4300 fibers /m^3 were detected in the third school. In a study from Switzerland, an attempt was made to perform a risk assessment in a school exposed to asbestos by measuring indoor air pollution under different scenarios and corresponding modeling (Vernez et al. 2019). In this case, the authors assume a mean annual exposure of 52–320 fibers/m^3.

Older study results in public and commercial indoor environments in the USA are compiled by the Health Effects Institute (HEI 1991). In a first major summary, Corn et al. (1991) conclude that mean values of 230 fibers /m^3 and 120 fibers/m^3 were found in stationary and personal measurements in 71 schools (473 samples). Lee and Van Orden (2008) also summarized measurement results from 752 public buildings and some of residences (a total of 3978 samples) in the USA over the past decades. On average, 120 fibers/m^3 were found, with no fibers detected in 98% of the buildings.

5.8 Microplastics

The topic of microplastics in environmental media has increasingly moved into the scientific and public discussion. The focus has been primarily on the fate, occurrence and effects in the aquatic system, but also in the soil, while occurrence in the air has received little attention to date. The sources of microplastics, i.e. particles with a diameter of <5000 μm, are manifold. They range from natural, organic—often fibrous—particles from everyday objects such as clothing, shoes, tire wear, to such components that are formed due to abrasion or aging processes of plastic components (RIVM 2021). Global annual plastic production alone is estimated at 348 million tons, with 10% of this ending up as waste and 3% being recycled (Verla et al. 2019). Against this background, microplastics must now be assumed to be ubiquitously distributed in all environmental media and also in food (Zhang et al. 2020a; Yee et al. 2021).

To date, there are no standardized or validated methods for the determination of airborne microplastics. For example, total dust or PM_{10} is collected and analytical characterization is then performed using microscopic techniques, fluorescence spectroscopy, Raman spectroscopy, infrared spectroscopy, and pyrolysis-gas chromatography mass chromatography (Enyoh et al. 2019; Beaurepaire et al. 2021). Meanwhile, some extensive reviews are available which summarized not only the analytical methods but also the occurrence, sources, and fate of microplastic in indoor spaces (Salthammer 2022; Zhao et al. 2022; Ageel et al. 2022; Abad Lopez et al. 2023).

5.8.1 *Health Aspects*

To date, limited scientific studies are available on the health effects of microplastics in the case of airborne exposure. Certainly, as far as the particles are inhalable at all due to their size, analogies can be drawn to other particles of the environmental aerosol or of fibers.

In a first study Amato-Lourenço et al. (2021) investigated the presence of microplastic in human parenchymal lung tissue from autopsy using Raman spectroscopy. Overall, polymeric particles (1.6–5.6 μm) and fibers (8.1–16.8 μm) were observed in 13 of 20 tissue samples. In addition, the substances like, e.g., phthalates and flame retardants that are already present in the plastic since manufacture and those that are additionally adsorbed during their stay in the environment must be taken into account in terms of health evaluation. For example, Lim et al. (2021) found no concentration-response relationship on several endpoints in physiological, biochemical, hematological, and respiratory function markers after exposure of rats for 24 days to polystyrene microparticles but an increase in inflammatory proteins in lung tissue. Again, Lu et al. (2021) tested the effect of exposure to green fluorescent spherical polymer particles (diameters between 1–5 μm) after nasal instillation and found respiratory effects in normal and asthmatic mice like inflammatory cell infiltration and increase in cytokine production in bronchoalveolar lavage fluid.

Facciolà et al. (2021) and Ageel et al. (2022) gave an overview of the current knowledge of the health impact. Overall, there is a significant need for further research (Wright and Kelly 2017; Barbosa et al. 2020).

5.8.2 Occurrence in the Air

Results of indoor air investigations are still limited. As a rule, a distinction is made between fibrous particles and irregular fragments, mostly oval or trapezoidal in shape.

In a first study, Dris et al. (2017) investigated microplastic particles in two apartments and an office in Paris, and in parallel in outdoor air. In the before mentioned study, 0.4–59 fibers/m^3 (median: 5.4 fibers/m^3) were observed. The majority had a length between 50 and 450 μm, but there were none above 3200 μm. The fibers consisted of 67% natural material, mainly cellulose, and 33% petrochemical substances, especially polypropylene. No effect on indoor air concentrations was due to the season of sampling, but the content was significantly different between the two dwellings. In the outdoor air, 0.3–1.5 fibers/m^3 (median: 0.9 fibers/m^3) were measured. Comparable results were obtained in a Portuguese study in 6 dwellings, where 6 fibers/m^3 with a length of 17–3669 μm were observed, 8.5% of which were of synthetic origin (Prata et al. 2020). In outdoor air, the content was in a comparable range. Another study reported microplastic concentrations in the range from 0.6 to 1.3 particles/m^3 (median 0.7 particles/m^3 in five houses in Portugal in March and June of 2020 (Xumiao et al. 2021). Mean concentrations of 5 fibers/m^3 were also determined in a study from the UK (Catarino et al. 2018). All fibers had a length below 500 μm. Vianello et al. (2019) found 1.7–16.3 particles/m^3 in an experimentally designed study in 3 Scottish homes. Here, they averaged 4% natural particles (91% proteins, 4% cellulose) and 81% polyester, 5% polyethylene, and 3% nylon. In 20 UK residences a particle deposition of 6204 particles/m^2/day (range: 902–14,551 particles/m^2/day were quantified and polyethylene terephthalate accounted for 62% of the microplastic (Jenner et al. 2021). In Barcelona, Spain, the highest average microparticle concentrations were found in three busses (17.3 particles/m^3) followed by 5.8 particles/m^3 in three subways cars, 4.8 particles/m^3 in six houses, and 4.2 particles/m^3 in three workplaces in a scientific institute (Torres-Agullo et al. 2022). Overall, fibers and fragments frequency decrease with increasing size (see distribution in Fig. 5.8) and in houses and working rooms 27% of the fibers and 44% of the fragments were in the <100 μm range.

Gaston et al. (2020) studied 10 indoor and 11 outdoor air samples in coastal California between January and March 2019. A mean of 3.3 fibers/m^3 was observed indoors and only 0.6 fibers/m^3 was observed outdoors, with lengths of 22–8961 μm (mean: 641 μm) and 25–2061 μm (mean: 616 μm), respectively. For fragments, a mean of 12.6 fragments/m^3 were measured inside and 5.6 fragments/m^3 outside,

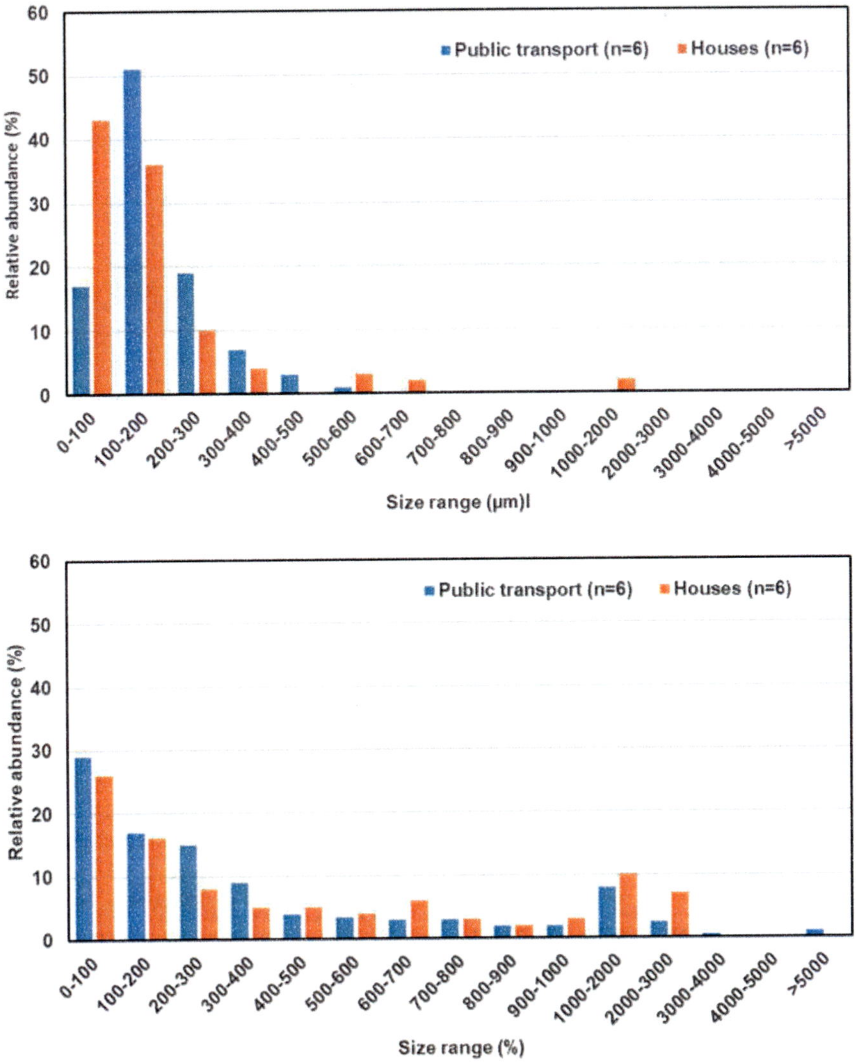

Fig. 5.8 Distribution of fibers (**a**) and fragments (**b**) in different indoor microenvironments (according to Torres-Agullo et al. 2022)

with mean diameters of 58 μm and 100 μm, respectively. In Australia, particles from 20 to 5000 μm were sampled and average concentrations of 2.3 particles/m^3 in childcare sites, 1.2 particles/m^3 in an office, 1.0 particles/m^3 in a school and 0.2 particles/m^3 inside vehicles, comparable to ambient air, were observed (Perera et al. 2023).

In China total suspended particulate were collected in urban and rural areas of a coastal city in July and August of 2019 (Liao et al. 2021). In indoor air a mean value of 1583 particles/m^3 with a size range of 5–4665 μm were reported, while in outdoor air only 189 particles/m^3 were found. Fragments were the dominant shape (indoor: around 90% and 10% fibers). Concentrations in urban areas (mean: 224 particles/m^3) were higher than rural areas (101 particles/m^3). The most common polymers among indoor particles were polyester (28.4%), polyamide (20.5%), and polypropylene (16.3%).

In Australia, O'Brien et al. (2020) reported 1.6 fibers/m^3 (average length: 764 μm) in the indoor air during the mechanical drying of a blue polyester fleece blanket with a common domestic dryer.

5.8.3 Occurrence in Dust

One office, one hallway, one classroom within an university campus, and one single-family house were investigated in the USA in 2019 and deposition rates of 14,400 fibers/m^2/day, 6430 fibers/m^2/day, 62,000 fibers/m^2/day, and 14,400 fibers/m^2/day were reported (Yao et al. 2022). These results are similar to that found by Dris et al. (2017) in Paris (range: 1600 to 11,000 fibers/m^2/day) and Zhang et al. (2020b) in China (average: 9900 MP/m^2/day), while in Australia in 2019, the deposition rate of fibrous MP ranged from 22 to 6169 fibers/m^2/day with an average of 3095 fibers/m^2/day (Soltani et al. 2021). Overall, the variation of deposition rates reflects the impact of different indoor conditions such as the human activities in the room, airflow turbulence, the textile quantity and quality (Yao et al. 2022). In a worldwide study collected 108 dust samples from 29 countries in 2019, an average deposition rate of MP of 1257 fibers/m^2/day in high income, 1268 fibers/m^2/day in middle income, and 3518 fibers/m^2/day in low income countries were observed (Soltani et al. 2022).

Dris et al. (2017) investigated microplastic particles in two apartments in Paris, France, using vacuum cleaner bags and observed oncentrations of fibers in the dust vary between 190 and 670 fibers/mg.

Indoor dust samples were collected from 39 residences in major cities of China in 2017–2018 to investigate the concentration and spatial distribution of typical MP in dust (Liu et al. 2019). The detected polyethylene terephthalate (PET-based) MP concentrations ranged from 1550 to 120,000 mg/kg with a median concentration of 26,800 mg/kg (outdoor 2780 mg/kg) and PC MP from < LOQ to 107 mg/kg (median: 4.6 mg/kg). To identify the relationship between PET MP concentrations and MP abundance, 15 dust samples were selected and the abundance of MP was 17–620 fibers/mg dust. In another Chinese study, very low average concentrations of 1.174 MP/mg were found in 47 residential apartments and 0.896 MP/mg in 50 offices in 2021 (Zhu et al. 2022). Average numbers of MP in 30 indoor dust samples from various locations in two Iranian cities were 90.8 and 80.8 MP/mg with the highest concentrations of 121 MP/mg and 104 MP/mg found in kindergartens (Kashfi et al. 2022). Using micro-Fourier transform infrared (μ-FT-IR) spectrophotometry Lim et al. (2022) investigated dust samples from vacuum cleaner bags of ten houses with mainly wooden floorings in Japan. The samples were composed of cellulose,

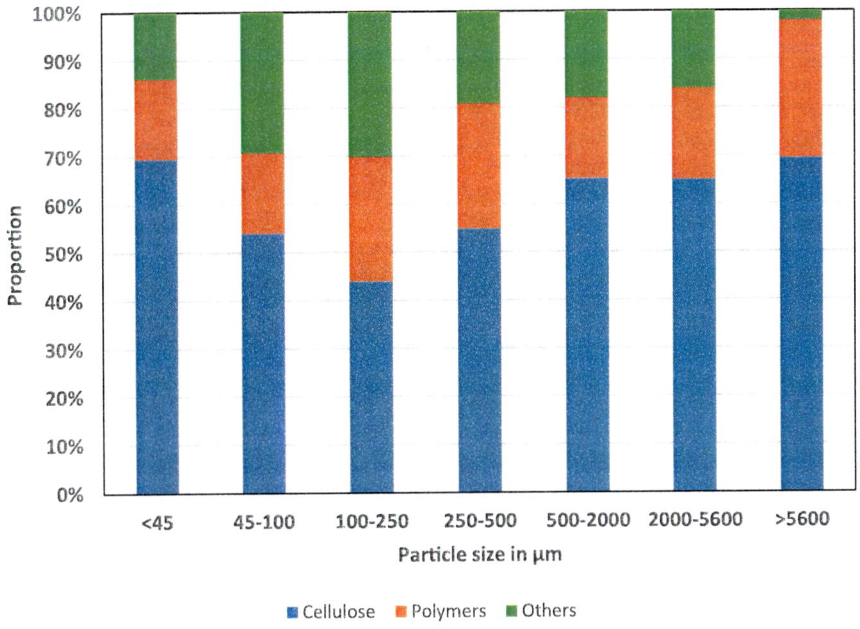

Fig. 5.9 Component proportion in relation to sizes using µ-FT-IR analysis (according to Lim et al. 2022). Polymers containing mainly on polyethylene terephthalate, polyester, polyacrylonitrile

polyethylene terephthalate, polyethylene, alkyd resins, polysulfone, polyvinyl alcohol, and polyamides. Overall, cellulose was the most common in all classification sizes with 55%, polymers with 23%, and silk with 3.1%. A size specific proportion of the components is given in Fig. 5.9.

In a worldwide study 286 dust samples from bedrooms or living rooms were collected and PET- and PC-based MP were quantified in a study in 12 countries in 2010–2014 (Zhang et al. 2020c). PET-based and PC-based MP were detected in all dust samples at concentrations of 38–120,000 mg/kg (median: 5900 mg/kg) and < 0.11–1700 mg/kg (median: 8.8 mg/kg), respectively. The highest PT-based medians were found in South Korea and Japan (25,000 and 23,000 mg/kg), the lowest in India and Colombia (16,000 and 1000 mg/kg), while economic development and human activity are important factors in the distribution of PET- and PC-based MP in indoor dust.

References

Aarnio P, Yli-Tuomi T, Kousa A, Mäakelä T, Hirsikko A, Hämeri K, Räisänen M, Hillamo R, Koskentalo T, Jantunen M (2005) The concentrations and composition of and exposure to fine particles ($PM_{2.5}$) in the Helsinki subway system. Atmos Environ 39:5059–5066

Abad Lopez AP, Trilleras J, Arana VA, Garcia-Alzatec LS, Grande-Tovar CD (2023) Atmospheric microplastics: exposure, toxicity, and detrimental health effects. RSC Adv 13:7468

Abbas Q, Yousaf B, Amina AMU, Munir MAM, El-Naggar A, Rinklebe J, Naushad M (2020) Transformation pathways and fate of engineered nanoparticles (ENPs) in distinct interactive environmental compartments: a review. Environ Int 138:105646

Abraham JL, Siwinski G, Hunt A (2002) Ultrafine particulate exposures in indoor, outdoor, personal and mobile environments: effects of diesel, traffic, pottery kiln, cooking and HEPA filtration on micro-environmental particle number concentration. Ann Occup Hyg 46(suppl 1): 406–411

Abt E, Suh HH, Allen G, Koutrakis P (2000) Characterization of indoor particle sources: a study conducted in the metropolitan Boston area. Environ Health Perspect 108:35–44

Adams HS, Nieuwenhuijsen MJ, Colvile RN, McMullen MAS, Khandelwal P (2001) Fine particle ($PM_{2.5}$) personal exposure levels in transport microenvironments, London, UK. Sci Total Environ 279:29–44

Afshari A, Matson U, Ekberg LE (2005) Characterization of indoor sources of fine and ultrafine particles: a study conducted in a full-scale chamber. Indoor Air 15:141–150

Ageel HK, Harrad S, Abdallah MA (2022) Occurrence, human exposure, and risk of microplastics in the indoor environment. Environ Sci Process Impacts 24:17–31

AGS (2021) TRGS 905. List of carcinogenic, mutagenic or reprotoxic substances. Inorganic fiber dusts (except asbestos). http://www.baua.de/de/Themen-von-A-Z/Gefahrstoffe/TRGS/pdf/905/905-anorganische-fasern.pdf?__blob=publicationFile&v=3 [in German]

Akintoye E, Shi L, Obaitan I, Olusunmade M, Wang Y, Newman JD, Dodson JA (2016) Association between fine particulate matter exposure and subclinical atherosclerosis: a meta-analysis. Eur J Prev Cardiol 23:602–612

Ali MU, Lin S, Yousaf B, Abbas Q, Ahmed M, Munir M, Rashid A, Zheng C, Kuang X, Wong MH (2020) Pollution characteristics, mechanism of toxicity and health effects of the ultrafine particles in the indoor environment: current status and future perspectives. Crit Rev Environ Sci Technol 52:436–473

Almeida SM, Canha N, Silva A, do Carmo Freitas M, Pegas P, Alves C, Evtyugina M, Pio CA (2011) Children exposure to atmospheric particles in indoor of Lisbon primary schools. Atmos Environ 45:7594–7599

Alves C, Nunes T, Silva J, Duarte M (2013) Comfort parameters and particulate matter (PM_{10} and $PM_{2.5}$) in school classrooms and outdoor air. Aerosol Air Quality Res 13:1521–1535

Amato-Lourenço LF, Carvalho-Oliveira R, Júnior GR, Dos Santos GL, Ando RA, Mauad T (2021) Presence of airborne microplastics in human lung tissue. J Hazard Mater 416:126124

Assimakopoulos VD, Bekiari T, Pateraki S, Maggos T, Stamatis P, Nicolopoulou P, Assimakopoulos MN (2018) Assessing personal exposure to PM using data from an integrated indoor-outdoor experiment in Athens-Greece. Sci Total Environ 636:1303–1320

ATSDR (2004) Toxicological profile for synthetic vitreous fibers. http://www.atsdr.cdc.gov/ToxProfiles/tp.asp?id=908&tid=185

Babaoglu UT, Ogutcu H, Erdogdu M, Taskiran F, Gullu G, Oymak S (2022) Assessment of indoor air quality in schools from Anatolia, Turkey. Pollution 8:57–67

Bake D (2003) Asbestos and mineral fibers. In: Morawska L, Salthammmer T (eds) Indoor environment. Airborne particles and settled dust. Wiley-VCH Verlag, Weinheim

Barbosa F, Adeyemi JA, Bocato MZ, Comas A, Campigli A (2020) A critical viewpoint on current issues, limitations, and future research needs on micro- and nanoplastic studies: from the detection to the toxicological assessment. Environ Res 182:109089

Barhoum A, García-Betancourt ML, Jeevanandam J, Hussien EA, Mekkawy SA, Mostafa M, Omran MM, Abdalla MS, Bechelany M (2022) Review on natural, incidental, bioinspired, and engineered nanomaterials: history, definitions, classifications, synthesis, properties, market, toxicities, risks, and regulations. Nano 12:177

Barkjohn KK, Norris C, Cui X, Fang L, Zheng T, Schauer JJ, Li Z, Zhang Y, Black M, Zhang JJ, Bergin MH (2021) Real-time measurements of PM2.5 and ozone to assess the effectiveness of residential indoor air filtration in Shanghai homes. Indoor Air 31:74–87
Barmparesos N, Assimakopoulos VD, Assimakopoulos MN, Tsairidi E (2016) Particulate matter levels and comfort conditions in the trains and platforms of the Athens underground metro. AIMS Environ Sci 3:199–219
Basinas I, Jiménez AS, Galea KS, Tongeren MV, Hurley F (2018) A systematic review of the routes and forms of exposure to engineered nanomaterials. Ann Work Expo Health 62:639–662
Baur X (2018) Asbestos-related disorders in Germany: background, politics, incidence, diagnostics and compensation. Int J Environ Res Public Health 15:143
Beaurepaire M, Dris R, Gasperi J, Tassin B (2021) Microplastics in the atmospheric compartment: a comprehensive review on methods, results on their occurrence and determining factors. Curr Opin Food Sci 42:159–168
Beelen R, Stafoggia M, Raaschou-Nielsen O, Andersen ZJ, Xun WW, Katsouyanni K, Dimakopoulou K, Brunekreef B, Weinmayr G, Hoffmann B, Wolf K, Samoli E, Houthuijs D, Nieuwenhuijsen M, Oudin A, Forsberg B, Olsson D, Salomaa V, Lanki T, Yli-Tuomi T, Oftedal B, Aamodt G, Nafstad P, De Faire U, Pedersen NL, Östenson CG, Fratiglioni L, Penell J, Korek M, Pyko A, Eriksen KT, Tjønneland A, Becker T, Eeftens M, Bots M, Meliefste K, Wang M, Bueno-de-Mesquita B, Sugiri D, Krämer U, Heinrich J, de Hoogh K, Key T, Peters A, Cyrys J, Concin H, Nagel G, Ineichen A, Schaffner E, Probst-Hensch N, Dratva J, Ducret-Stich R, Vilier A, Clavel-Chapelon F, Stempfelet M, Grioni S, Krogh V, Tsai MY, Marcon A, Ricceri F, Sacerdote C, Galassi C, Migliore E, Ranzi A, Cesaroni G, Badaloni C, Forastiere F, Tamayo I, Amiano P, Dorronsoro M, Katsoulis M, Trichopoulou A, Vineis P, Hoek G (2014) Effects of long-term exposure to air pollution on natural-cause mortality: an analysis of 22 European cohorts within the multicentre ESCAPE project. Lancet 383:785–795
Benn T, Herckes P, Westerhoff P (2012) Fullerenes in environmental samples: C60 in atmospheric particulate matter. Comp Anal Chem 59:291–303
Birmili W (2021) The German Environmental Survey 2014–2017 (GerES V): particles in indoor air. Personal communication
Boniardi L, Borghi F, Straccini S, Fanti G, Campagnolo D, Campo L, Olgiati L, Lioi S, Cattaneo A, Spinazzè A, Cavallo DM, Fustinoni S (2021) Commuting by car, public transport, and bike: exposure assessment and estimation of the inhaled dose of multiple airborne pollutants. Atmos Environ 262:118613
Boogaard H, Borgman F, Kamminga J, Hoek G (2009) Exposure to ultrafine and fine particles and noise during cycling and driving in 11 Dutch cities. Atmos Environ 43:4234–4242
Bouma F, Janssen NA, Wesseling J, van Ratingen S, Strak M, Kerckhoffs J, Gehring U, Hendricx W, de Hoogh K, Vermeulen R, Hoek G (2023) Long-term exposure to ultrafine particles and natural and causespecific mortality. Environ Int 175:107960
Boyes WK, van Thriel C (2020) Neurotoxicology of nanomaterials. Chem Res Toxicol 33:1121–1144
Branco PTBS, Alvim-Ferraz MCM, Martins FG, Sousa SIV (2014) Indoor air quality in urban nurseries at Porto city: Particulate matter assessment. Atmo Environ 84:133–143
Branis M (2006) The contribution of ambient sources to particulate pollution in spaces and trains of the Prague underground transport system. Atmos Environ 40:348–356
Breysse PN, Buckley TJ, Williams DA, Beck CM, Jo S-J, Merriman B, Kanchanaraksa S, Swartz LJ, Callahan KA, Butz AM, Rand CS, Diette GB, Krishnan JA, Moseley AM, Curtin-Brosnan J, Durkin NB, Eggleston PA (2005) Indoor exposures to air pollutants and allergens in the homes of asthmatic children in inner-city Baltimore. Environ Res 98:167–176
Bundschuh M, Filser J, Lüderwald S, McKee MS, Metreveli G, Schaumann GE, Schulz R, Wagner S (2018) Nanoparticles in the environment: where do we come from, where do we go to? Environ Sci Eur 30:6

Burdett GJ, Dewberry K, Staff J (2016) Airborne asbestos exposures from warm air heating systems in schools. Ann Occup Hyg 60:27–39
Burton LE, Girman JG, Womble SE (2000) Airborne particulate matter within 100 randomily selected office buildings in the United States (BASE). Proc Healthy Build 1:157–162
Cai C, Sun Z, Weschler LB, Li T, Xu W, Zhang Y (2021) Indoor air quality in schools in Beijing: field tests, problems and recommendations. Build Environ 205:108179
Camilleri R, Vella AJ, Harrison RM, Aquilina NJ (2022) Source apportionment of indoor $PM_{2.5}$ at a residential urban background site in Malta. Atmos Environ 278:119093
Campopiano A, Casciardi S, Fioravanti F, Ramires D (2004) Airborne asbestos levels in school buildings in Italy. J Occup Environ Hyg 1:256–261
Canha N, Mandin C, Ramalho O, Wyart G, Riberon G, Dassonville C, Hänninen O, Almeida SM, Derbez M (2016) Assessment of ventilation and indoor air pollutants in nursery and elementary schools in France. Indoor Air 26:350–365
Carlin DJ, Larson TC, Pfau JC, Gavett SH, Shukla A, Miller A, Hines R (2015) Current research and opportunities to address environmental asbestos exposures. Environ Health Perspect 123: A194–A197
Carreiro-Martins P, Papoila AL, Caires I, Azevedo S, Cano MM, Virella D, Leiria-Pinto P, Teixeira JP, Rosado-Pinto J, Annesi-Maesano I, Neuparth N (2016) Effect of indoor air quality of day care centers in children with different predisposition for asthma. Ped Allergy Immunol 27:299–306
Cartenì A, Cascetta F, Campana S (2015) Underground and ground-level particulate matter concentrations in an Italian metro system. Atmos Environ 101:328–337
Carter CM, Axten CW, Byers CD, Chase GR, Koenig AR, Reynolds JW, Rosinski KD (1999) Indoor airborne fiber levels of MMVF in residential and commercial buildings. Am Ind Hyg Assoc J 60:794–800
Catarino AI, Valeria M, William GS, Richard CT, Theodore BH (2018) Low levels of microplastics (MP) in wild mussels indicate that MP ingestion by humans is minimal compared to exposure via household fibres fallout during a meal. Environ Pollut 237:675–684
Cattaneo A, Peruzzo C, Garramone G, Urso P, Ruggeri R, Carrer P, Cavallo DM (2011) Airborne particulate matter and gaseous air pollutants in residential structures in Lodi Province, Italy. Indoor Air 21:489–500
Cepeda M, Schoufour J, Freak-Poli R, Koolhaas CM, Dhana K, Bramer WM, Franco OH (2017) Level of ambient air pollution according to mode of transport: a systematic review. Lancet Public Health 2:e23–e34
Chan LY, Lau WL, Lee SC, Chan CY (2002) Commuter exposure to particulate matter in public transportation modes in Hong Kong. Atmos Environ 36:3363–3373
Chatoutsidou SE, Ondrácek J, Tesar O, Torseth K, Zdimal V, Lazaridis M (2015) Indoor/outdoor particulate matter number and mass concentration in modern offices. Build Environ 92:462–474
Che W, Li ATY, Frey HC, Tang KTJ, Sun L, Wei P, Hossain MS, Hohenberger TL, Leung KW, Lau AKH (2021) Factors affecting variability in gaseous and particle microenvironmental air pollutant concentrations in Hong Kong primary and secondary schools. Indoor Air 31:170–187
Chen J, Hoek G (2020) Long-term exposure to PM and all-cause and cause-specific mortality: a systematic review and meta-analysis. Environ Int 143:105974
Chen H, Zhou Z, Li Z, Liang S, Zhou J, Zou G, Zhou S (2022) Time trends in the burden of stroke and subtypes attributable to $PM_{2.5}$ in China from 1990 to 2019. Front Public Health 10:1026870
Cheng YH, Yan JW (2011) Comparisions of particulate matter, CO, and CO_2 levels in underground and ground-level stations in the Taipei Mass Rapid Transit System. Atmos Environ 45:4882–4891
Cheng YH, Lin L, Liu CC (2008) Levels of PM_{10} and $PM_{2.5}$ in Taipei rapid transit system. Atmos Environ 42:7242–7249
Chertok M, Voukelatos A, Sheppeard V, Rissel C (2004) Comparison of air pollution exposure for five commuting modes in Sydney – car, train, bus, bicycle and walking. Health Prom J Australia 15:63–67

Clausen G, Høst A, Toftum J, Bekç G, Weschler C, Callesen M, Buhl S, Ladegaard MB, Langer S, Andersen B, Sundell J, Bornehag C-G, Sigsgaard T (2012) Children's health and its association with indoor environments in Danish homes and daycare centres – methods. Indoor Air 22:467–475

COMEAP (2016) Long-term exposure to air pollution and chronic bronchitis. A report by the Committee on the Medical Effects of Air Pollutants. UK. https://assets.publishing.service.gov.uk/government/uploads/system/uploads/attachment_data/file/541745/COMEAP_chronic_bronchitis_report_2016__rev_07-16_.pdf

Cooper DM, Loxham M (2019) Particulate matter and the airway epithelium: the special case of the underground? Eur Respir Rev 28:190066

Corn M, Crump K, Farrar DB, Lee RJ, McFee DR (1991) Airborne concentrations of asbestos in 71 school buildings. Regul Toxicol Pharmacol 13:99–114

De Nazelle A, Fruin S, Westerdahl D, Martinez D, Ripoll A, Kubesch N, Nieuwenhuijsen M (2012) A travel mode comparison of commuters' exposures to air pollutants in Barcelona. Atmos Environ 59:151–159

De Nazelle A, Bode O, Orjuela JP (2017) Comparison of air pollution exposures in active vs. passive travel modes in European cities: a quantitative review. Environ Int 99:151–160

Debia M, Bakhiyi B, Ostiguy C, Verbeek JH, Brouwer DH, Murashov V (2016) A systematic review of reported exposure to engineered nanomaterials. Ann Occup Hyg 60:916–935

DECOS (2010) Asbestos: Risks of environmental and occupational exposure. Publication no. 2010/10E. https://www.healthcouncil.nl/documents/advisory-reports/2010/06/03/asbestos-risks-of-environmental-and-occupational-exposure

Dennekamp M, Howarth S, Dick CAJ, Cherrie W, Donaldson K, Seaton A (2001) Ultrafine particles and nitrogen oxides generated by gas and electric cooking. Occup Environ Med 58: 511–516

Dennekamp M, Mehenni O, Cherrie J, Seaton A (2002) Exposure to ultrafine particles and $PM_{2.5}$ in different microenvironments. Ann Occup Hyg 46(suppl. 1):412–414

Diapouli E, Chaloulakou A, Spyrellis N (2007) Levels of ultrafine particles in different microenvironments – implications to children exposure. Sci Total Environ 388:128–136

Diapouli E, Chaloulakou A, Mihalopoulos N, Spyrellis N (2008) Indoor and outdoor PM mass and number concentrations at schools in the Athens area. Environ Monit Assess 136:13–20

DIN EN (1993) DIN EN 481:1993–09. Workplaces atmospheres; size fraction definitions for measurement of airborne particles

Dodgson J, Harrison GE, Cherrie JW, Sneddon E (1987) Assessment of airborne mineral wool fibres in domestic houses. Research Report TM/87/18, Institute of Occupational Medicine, Edinburgh, UK

Dorizas PV, Assimakopoulos M-N, Helmis C, Santamouris M (2015) An integrated evaluation study of the ventilation rate, the exposure and the indoor air quality in naturally ventilated classrooms in the Mediterranean region during spring. Sci Total Environ 502:557–570

Dris R, Gasperi J, Mirande C, Mandin C, Guerrouache M, Langlois V, Tassin B (2017) A first overview of textile fibers, including microplastics, in indoor and outdoor environments. Environ Pollut 221:453–458

EC (European Commission) (2022) European Commission Recommendations of 10 June 2022 on the definition of nanomaterial. Official J European Union C 229/1. Online: https://eurlex.europa.eu/legal-content/EN/TXT/PDF/?uri=CELEX:32022H0614(01)&from=EN

ECHA (2021) Scientific report for evaluation of limit values for asbestos at the workplace. Helsinki, Finland. https://echa.europa.eu/documents/10162/4605fc92–18a2-ae48-f977-4dffdecfec11

Edwards R, Wilson N, Pierse N (2006) Highly hazardous air quality associated with smoking in cars: New Zealand pilot study. NZMJ 119: No. 1244

Eiguren-Fernandez A, Miguel AH, Zhu YF, Hering SV (2005) In-cabin passenger exposure to ultrafine and nano-particles during daily commute in Los Angeles roads and freeways: evaluation of a HEPA filtration system. Proc Indoor Air Peking 1763–1767

Encinas D, Gómez-de-Balugera Z (2018) Fullerene C 60 in atmospheric aerosol and its relationship to combustion processes. Arch Environ Contam Toxicol 75:616–624

Enyoh CE, Verla AW, Verla EN, Ibe FC, Amaobi CE (2019) Airborne microplastics: a review study on method for analysis, occurrence, movement and risks. Environ Monit Assess 191:668

EU (2006) Regulation (EC) No 1907/2006 of the European Parliament and of the Council of 18 December 2006 concerning the Registration, Evaluation, Authorisation and Restriction of Chemicals (REACH), establishing a European Chemicals Agency. Official Journal of the European Union L 396. https://eur-lex.europa.eu/legal-content/EN/TXT/?uri=CELEX:02006R1907–20140410

EU (2011) Commission recommendation of 18 October 2011 on the definition of nanomaterial. Official Journal of the European Union L 275/38. https://eur-lex.europa.eu/legal-content/EN/TXT/PDF/?uri=CELEX:32011H0696&from=DE

EU (2016) Commission Regulation (EU) 2016/1005 of 22 June 2016 amending Annex XVII to Regulation (EC) No 1907/2006 of the European Parliament and of the Council concerning the Registration, Evaluation, Authorisation and Restriction of Chemicals (REACH) as regards asbestos fibres (chrysotile). Official Journal of the European Union L 165/4. https://eur-lex.europa.eu/legal-content/EN/TXT/?uri=CELEX:32016R1005

EU (European Union) (2004) Directive 2004/37/EC of the European Parliament and of the Council of 29 April 2004 on the protection of workers from the risks related to exposure to carcinogens or mutagens at work (Sixth individual Directive within the meaning of Article 16(1) of Council Directive 89/391/EEC). Official Journal of the European Union L 158

Facciolà A, Visalli G, Pruiti Ciarello M, Di Pietro A (2021) Newly emerging airborne pollutants: current knowledge of health impact of micro and nanoplastics. Int J Environ Res Public Health 18:2997

Felgueiras F, Mourão Z, Moreira A, Gabriel MF (2022) A systematic review of ventilation conditions and airborne particulate matter levels in urban offices. Indoor Air 32:e13148

Fine PM, Cass GR, Simoneit BRT (1999) Characterization of fine particle emissions from burning church candles. Environ Sci Technol 33:2352–2362

Fonseca J, Slezakova K, Morais S, Pereira MC (2014) Assessment of ultrafine particles in Portuguese preschools: levels and exposure doses. Indoor Air 24:618–628

Franck U, Herbart O, Röder S, Schlink U, Borte M, Diez U, Krämer U, Lehmann I (2011) Respiratory effects of indoor particles in young children are size dependent. Sci Total Environ 409:1621–1631

Fromme H (2019) Particulate matter and ultrafine particles in indoor air. Encyclopedia of Environmental Health, 2nd edn. Pages 36–48. https://www.sciencedirect.com/science/article/pii/B9780124095489112436

Fromme H, Oddoy A, Lahrz T, Krause M, Piloty M (1998) Polycyclic aromatic hydrocarbons (PAH) and diesel engine emission (elemental carbon) inside a car and a subway train. Sci Total Environ 217:165–173

Fromme H, Lahrz T, Hainsch A, Oddoy A, Piloty M, Rüden H (2005) Elemental carbon and respirable particulate matter in the indoor air of apartments and nursery schools and outdoor air in Berlin (Germany). Indoor Air 15:335–341

Fromme H, Twardella D, Dietrich S, Heitmann D, Schierl R, Liebl B, Rüden H (2007) Particulate matter in the indoor air of classrooms – exploratory results from Munich and surrounding. Atmos Environ 41:854–866

Fromme H, Diemer J, Dietrich S, Cyrys J, Heinrich J, Lang W, Kiranoglu M, Twardella D (2008) Chemical and morphological properties of particulate matter (PM_{10}, $PM_{2.5}$) from school classrooms and outdoor air. Atmos Environ 42:6597–6605

Fromme H, Lahrz T, Burkhardt R, Fembacher L, Kraft M, Sievering S, Kadler D, Dietrich S (2016) Air quality in daycare centers – exposure to carbon dioxide (CO_2), volatile organic compounds (VOCs), and fine dusts – LUPE 3. Gefahrst Reinh Luft 76:55–61. [in German]

Ganguly K, Ettehadieh D, Upadhyay S, Takenaka S, Adler T, Karg E, Krombach F, Kreyling WG, Schulz H, Schmid O, Stoeger T (2017) Early pulmonary response is critical for extra-pulmonary

carbon nanoparticle mediated effects: comparison of inhalation versus intra-arterial infusion exposures in mice. Part Fibre Toxicol 14:19
García-Hernández C, Ferrero A, Estarlich M, Ballester F (2020) Exposure to ultrafine particles in children until 18 years of age: a systematic review. Indoor Air 30:7–23
Gaspar FW, Maddalena R, Williams J, Castorina R, Wang ZM, Kumagai K, McKone TE, Bradman A (2018) Ultrafine, fine, and black carbon particle concentrations in California child-care facilities. Indoor Air 28:102–111
Gaston E, Woo M, Steele C, Sukumaran S, Anderson S (2020) Microplastics differ between indoor and outdoor air masses: insights from multiple microscopy methodologies. Appl Spectrosc 74: 1079–1098
Geiss O, Barrero-Moreno J, Tirendi S, Kotzias D (2010) Exposure to particulate matter in vehicle cabins of private cars. Aerosol Air Qual Res 10:581–588
Gemenetzis P, Moussas P, Arditsoglou A (2006) Mass concentration and elemental composition of indoor $PM_{2.5}$ and PM_{10} in University rooms in Thessaloniki, northern Greece. Atmos Environ 40:3195–3206
Gerber A, Bohn J, Groneberg A, Schulze J, Bundschuh M (2014) Airborne particulate matter in public transport: a field study at major intersection points in Frankfurt am Main (Germany). J Occup Med Toxicol 9:13–16
Ghatge Goel S, Somwanshi S, Mankar S, Srimuruganandam B, Gupta R (2021) Characteristics of indoor air pollutants and estimation of their exposure dose. Air Qual Atmos Health 14:1033–1047
Gilbey SE, Reid CM, Zhao Y, Soares MJ, Huxley RR, Rumchev KB (2022) Residential indoor exposure to fne and ultrafne particulate air pollution in association with blood pressure and subclinical central haemodynamic markers of cardiovascular risk among healthy adults living in Perth, Western Australia. Air Qual Atmos Health Oct 2022
Goel R, Gani S, Guttikunda SK, Wilson D, Tiwari G (2015) On-road $PM_{2.5}$ pollution exposure in multiple transport microenvironments in Delhi. Atmos Environ 123:129–138
Goel SG, Somwanshi S, Mankar S, Srimuruganandam B, Gupta R (2021) Characteristics of indoor air pollutants and estimation of their exposure dose. Air Qual Atmos Health 14:1033–1047
Gomez-Perales JE, Colvile RN, Nieuwenhuijsen MJ, Fernandez-Bremauntz A, Gutiérrez-Avedoy VJ, Páramo-Figueroa VH, Blanco-Jiménez S, Bueno-López E, Mandujanoc F, Bernabé-Cabanillas R, Ortiz-Segovia E (2004) Commuters' exposure to $PM_{2.5}$, CO, and benzene in public transport in the metropolitan area of Mexico City. Atmos Environ 38:1219–1229
Gould CF, Chillrud SN, Phillips D, Perzanowski MS, Hernández D (2018) Soot and the city: Evaluating the impacts of clean heat policies on indoor/outdoor air quality in New York City apartments. PLoS One 13:e0199783
Greim H, Utell MJ, Maxim LD, Niebo R (2014) Perspectives on refractory ceramic fiber (RCF) carcinogenicity: comparisons with other fibers. Inhal Toxicol 26:789–810
Grydaki N, Colbeck I, Mendes L, Eleftheriadis K, Whitby C (2021) Bioaerosols in the Athens Metro: Metagenetic insights into the PM_{10} microbiome in a naturally ventilated subway station. Environ Int 146:106186
Gulliver J, Briggs DJ (2004) Personal exposure to particulate air pollution in transport microenvironments. Atmos Environ 38:1–8
Guo H, Morawska L, He C, Zhang YL, Ayoko G, Cao M (2010) Characterization of particle number concentrations and $PM_{2.5}$ in a school: influence of outdoor air pollution on indoor air. Environ Sci Pollut Res 17:1268–1278
Hachem M, Saleh N, Bensefa-Colas L, Momas I (2020) Determinants of ultrafine particles, black carbon, nitrogen dioxide, and carbon monoxide concentrations inside vehicles in the Paris area: PUF-TAXI study Indoor Air Dec 22
Ham W, Vijayan A, Schulte N, Herner JD (2017) Commuter exposure to $PM_{2.5}$, BC, and UFP in six common transport microenvironments in Sacramento, California. Atmos Environ 167:335–345

Han C, Xu R, Ye T, Xie Y, Zhao Y, Liu H, Yu W, Zhang Y, Li S, Zhang Z, Ding Y, Han K, Fang, Ji B, Zhai W, Guo Y (2022) Mortality burden due to long-term exposure to ambient $PM_{2.5}$ above the new WHO air quality guideline based on 296 cities in China. Environ Int 166:107331

Hankey S, Marshall JD (2015) On-bicycle exposure to particulate air pollution: particle number, black carbon, $PM_{2.5}$, and particle size. Atmos Environ 122:65–73

Harrison P, Holmes P, Bevan R, Kamps K, Levy L, Greim H (2015) Regulatory risk assessment approaches for synthetic mineral fibres. Regul Toxicol Pharmacol 73:425–441

HC (2016) Human health risk assessment for coarse particulate matter. Ottawa, Canada. https://www.canada.ca/en/health-canada/services/publications/healthy-living/human-health-risk-assessment-coarse-particulate-matter-executive-summary.html

He C, Morawska L, Hitchins J, Gilbert D (2004) Contribution from indoor sources to particle number and mass concentrations in residential houses. Atmos Environ 38:3405–3415

He Y, Lan X, Zhu L (2023) Effect of urban green infrastructure on pedestrian exposure to ultrafine particles: a case study of Guangzhou, China. Urban Climate 49:101453

HEI (Health Effects Institute) (1991) Asbestos in public and commercial buildings: a literature review and synthesis of current knowledge. Health Effects Institute – Asbestos Research. http://pubs.healtheffects.org

Héroux ME, Clark N, Van Ryswyk K, Mallick R, Gilbert NL, Harrison I, Rispler K, Wang D, Anastassopoulos A, Guay M, MacNeill M, Wheeler AJ (2010) Predictors of indoor air concentrations in smoking and non-smoking residences. Int J Environ Res Public Health 7:3080–3099

HESE (Health Effects of School Environment) (2006) Final scientific report prepared for the Health & Consumer Protection Directorate General. Siena, Italy. http://ec.europa.eu/health/ph_projects/2002/pollution/fp_pollution_2002_frep_04.pdf

Hoek G, Kos G, Harrison R, de Hartog J, Meliefste K, ten Brink H, Katsouyanni K, Karakatsani A, Lianou M, Kotronarou A, Kavouras I, Pekkanen J, Vallius M, Kulmala M, Puustinen A, Thomas S, Meddings C, Ayresi J, van Wijnen J, Hameri K (2008) Indoor–outdoor relationships of particle number and mass in four European cities. Atmos Environ 42:156–169

Horemans B, Van Grieken R (2010) Speciation and diurnal variation of thoracic, fine thoracic and sub-micrometer airborne particulate matter at naturally ventilated office environments. Atmos Environ 44:1497–1505

Huang J, Chan SC, Pang WS, Chow SH, Lok V, Zhang L, Lin X, Lucero-Prisno 3rd DE, Xu W, Zheng ZJ, Elcarte E, Withers M, Wong MCS, NCD Global Health Research Group, Association of Pacific Rim Universities (APRU) (2023) Global incidence, risk factors, and temporal trends of mesothelioma: a population-based study. J Thorac Oncol 10:S1556-0864(23)00125-9

Hudda N, Kostenidou E, Sioutas C, Delfino RJ, Fruin SA (2011) Vehicle and driving characteristics that influence in-cabin particle number concentrations. Environ Sci Technol 45:8691–8697

Hussein T, Glytsos T, Ondrácek J, Dohányosová P, Zdimal V, Hämeri K, Lazaridis M, Smolik J, Kulmala M (2006) Particle size characterization and emission rates during indoor activities in a home. Atmos Environ 40:4285–4307

Hvidtfeldt UA, Severi G, Andersen ZJ, Atkinson R, Bauwelinck M, Bellander T, Boutron-Ruault MC, Brandt J, Brunekreef B, Cesaroni G, Chen J, Concin H, Forastiere F, van Gils CH, Gulliver J, Hertel O, Hoek G, Hoffmann B, de Hoogh K, Janssen N, Jöckel KH, Jørgensen JT, Katsouyanni K, Ketzel M, Klompmaker JO, Krog NH, Lang A, Leander K, Liu S, Ljungman PLS, Magnusson PKE, Mehta AJ, Nagel G, Oftedal B, Pershagen G, Peter RS, Peters A, Renzi M, Rizzuto D, Rodopoulou S, Samoli E, Schwarze PE, Sigsgaard T, Simonsen MK, Stafoggia M, Strak M, Vienneau D, Weinmayr G, Wolf K, Raaschou-Nielsen O, Fecht D (2021) Long-term low-level ambient air pollution exposure and risk of lung cancer – a pooled analysis of 7 European cohorts. Environ Int 146:106249

Hwang Y, Lee K (2018) Contribution of microenvironments to personal exposures to PM_{10} and $PM_{2.5}$ in summer and winter. Atmos Environ 175:192–198

IARC (International Agency for Research on Cancer) (1973) Some inorganic and organometallic compounds. IARC Monographs on the evaluation of carcinogenic risks of chemicals to man 2: 1–181. Lyon, France

IARC (2002) Man-made vitreous fibres/IARC working group on the evaluation of carcinogenic risks to humans, Vol. 81. Lyon, France. https://monographs.iarc.who.int/wp-content/uploads/2018/06/mono81.pdf

IARC (2012) Arsenic, metals, fibres, and dusts. IARC Monographs on the evaluation of carcinogenic risks to humans 100 C: 219–309. Lyon, France. https://publications.iarc.fr/Book-And-Report-Series/Iarc-Monographs-On-The-Identification-Of-Carcinogenic-Hazards-To-Humans/Arsenic-Metals-Fibres-And-Dusts-2012

IARC (2017) Some nanomaterials and some fibres. IARC Monographs on the evaluation of carcinogenic risks to humans, Vol. 111. https://publications.iarc.fr/Book-And-Report-Series/Iarc-Monographs-On-The-Identification-Of-Carcinogenic-Hazards-To-Humans/Some-Nanomaterials-And-Some-Fibres-2017

Ihrie MD, Bonner JC (2018) The toxicology of engineered nanomaterials in asthma. Curr Environ Health Rep 5:100–109

Int Panis L, de Geus B, Vandenbulcke G, Willems H, Degraeuwe B, Bleux N, Mishra V, Thomas I, Meeusen R (2010) Exposure to particulate matter in traffic: a comparison of cyclists and car passengers. Atmos Environ 44:2263–2270

Jenner LC, Sadofsky LR, Danopoulos E, Rotchell JM (2021) Household indoor microplastics within the Humber region (United Kingdom): quantification and chemical characterisation of particles present. Atmos Environ 259:118512

Jeong SG, Wallace L, Rim D (2021) Contributions of coagulation, deposition, and ventilation to the removal of airborne nanoparticles in indoor environments. Environ Sci Technol 55:9730–9739

Jung KH, Patel MM, Moors K, Kinney PL, Chillrud SN, Whyatt R, Hoepner L, Garfinkel R, Yan B, Ross J, Camann D, Perera FP, Miller RL (2010) Effects of heating season on residential indoor and outdoor polycyclic aromatic hydrocarbons, black carbon, and particulate matter in an urban birth cohort. Atmos Environ 44:4545–4552

Jung CC, Su HJ, Liang HH (2016) Association between indoor air pollutant exposure and blood pressure and heart rate in subjects according to body mass index. Sci Total Environ 539:271–276

Juskiene I, Prokopciuk N, Franck U, Valiulis A, Valskys V, Mesceriakova V, Kvedariene V, Valiulyte I, Poluzioroviene E, Sauliene I, Valiulis A (2022) Indoor air pollution effects on pediatric asthma are submicron aerosol particle-dependent. Eur J Pediatr 181:2469–2480

Kam W, Cheung K, Daher N, Sioutas C (2011) Particulate matter (PM) concentrations in underground and ground-level rail systems of the Los Angeles Metro. Atmos Environ 45:1506–1516

Kashfi FS, Ramavandi B, Arfaeinia H, Mohammadi A, Saeedi R, DE-la-Torre GE, Dobaradaran S (2022) Occurrence and exposure assessment of microplastics in indoor dusts of buildings with different applications in Bushehr and Shiraz cities. Iran Sci Total Environ 829:154651

Kaufman JD, Adar SD, Barr RG, Budoff M, Burke GL, Curl CL, Daviglus ML, Diez Roux AV, Gassett AJ, Jacobs Jr DR, Kronmal R, Larson TV, Navas-Acien A, Olives C, Sampson PD, Sheppard L, Siscovick DS, Stein JH, Szpiro AA, Watson KE (2016) Association between air pollution and coronary artery calcification within six metropolitan areas in the USA (the Multi-Ethnic Study of Atherosclerosis and Air Pollution): a longitudinal cohort study. Lancet 388: 696–704

Kaur S, Nieuwenhuijsen. (2009) Determinants of personal exposure to $PM_{2.5}$, ultrafine particle counts, and CO in a transport microenvironment. Environ Sci Technol 43:4737–4743

Kaur S, Nieuwenhuijsen M, Colvile R (2005) Personal exposure of street canyon intersection users to $PM_{2.5}$, ultrafine particle counts and carbon monoxide in central London, UK. Atmos Environ 39:3629–3641

Khamal R, Isa ZM, Sutan R, Noraini NMR, Ghazi HF (2019) Indoor particulate matters, microbial count assessments, and wheezing symptoms among toddlers in urban day care centers in the district of Seremban, Malaysia. Ann Glob Health 85:15

Kim M, Sankararao B, Kang O, Kim J, Yoo C (2012) Monitoring and prediction of indoor air quality (IAQ) in subway or metro systems using season dependent models. Energ Buildings 46: 48–55

Kim H-H, Lee G-W, Yang J-Y, Jeon J-M, Lee W-S, Lim J-Y, Lee H-S, Gwak Y-K, Shin D-C, Lim Y-W (2014) Indoor exposure and health risk of polycyclic aromatic hydrocarbons (PAHs) via public facilities $PM_{2.5}$, Korea (II). Asian J Atmos Environ 8-1:35–47

Kim H-H, Park G-Y, Lee J-H (2019) Concentrations of particulate matter, carbon dioxide, VOCs and risk assessment inside Korean taxis and ships. Environ Sci Pollut Res 26:9619–9631

Kirchner S, Derbez M, Duboudin C, Elias P, Gregoire A, Lucas J-P, Pasquier N, Ramalho O, Weiss N (2008) Indoor air quality in French dwellings. Proceedings Indoor Air 2008, Copenhagen, Denmark

Koehler K, Good N, Wilson A, Mölter A, Moore BF, Carpenter T, Peel JL, Volckens J (2019) The Fort Collins commuter study: Variability in personal exposure to air pollutants by microenvironment. Indoor Air 29:231–241

Krausse B, Mardaljevic J (2005) Patterns of driver's exposure to particulate matter. In: Williams K (ed) Spatial planning, urban form and sustainable transport. Ashgate, Aldershot, UK

Kupczewska-Dobecka M, Konieczko K, Czerczak S (2020) Occupational risk resulting from exposure to mineral wool when installing insulation in buildings. Int J Occup Med Environ Health 33:757–769

Kwon H-S, Ryu MH, Carlsten C (2020) Ultrafine particles: unique physicochemical properties relevant to health and disease. Exp Mol Med 52:318–328

Lai HK, Kendall M, Ferrier H, Lindup I, Alm S, Hänninen O, Jantunen M, Mathys P, Colvile R, Ashmore MR, Cullinan P, Nieuwenhuijsen MJ (2004) Personal exposures and microenvironment concentrations of $PM_{2.5}$, VOC, NO_2 and CO in Oxford, UK. Atmos Environ 38:6399–6410

Langer S, Ramalho O, Derbez M, Riberon J, Kirchner S, Mandin C (2016) Indoor environmental quality in French dwellings and building characteristics. Atmos Environ 128:82–91

Laux P, Tentschert J, Riebeling C, Braeuning A, Creutzenberg O, Epp A, Fessard V, Haas KH, Haase A, Hund-Rinke K, Jakubowski N, Kearns P, Lampen A, Rauscher H, Schoonjans R, Störmer A, Thielmann A, Mühle U, Luch A (2018) Nanomaterials: certain aspects of application, risk assessment and risk communication. Arch Toxicol 92:121–141

Lee RJ, Van Orden DR (2008) Airborne asbestos in buildings. Regul Toxicol Pharmacol 50:218–251

Lee K, Sohn H, Putti K (2010) In-vehicle exposures to particulate matter an black carbon. J Air Waste Manage Assoc 60:130–136

Lee JY, Wargocki P, Chan YH, Chen L, Tham KW (2019) Indoor environmental quality, occupant satisfaction, and acute building-related health symptoms in Green Mark-certified compared with non-certified office buildings. Indoor Air 29:112–129

Levy JI, Dumyahn T, Sprengler JD (2002) Particulate matter and polycyclic hydrocarbon concentrations in indoor and outdoor microenvironments in Boston, Massachusetts. J Expo Anal Environ Epidemiol 12:104–114

Li Y, Chen L, Minh Ngoc D, Duan Y-P, Lu Z-B, Wen Z-H, Meng X-Z (2015) Polybrominated diphenyl ethers (PBDEs) in $PM_{2.5}$, PM_{10}, TSP and gas phase in office environment in Shanghai, China: occurrence and human exposure. PLoS One 10:e0119144

Li N, Georas S, Alexis N, Fritz P, Xia T, Williams MA, Horner E, Nel A (2016) A work group report on ultrafine particles (American Academy of Allergy, Asthma & Immunology): why ambient ultrafine and engineered nanoparticles should receive special attention for possible adverse health outcomes in human subjects. J Allergy Clin Immunol 138:386–396

Liao Z, Ji X, Ma Y, Lv B, Huang W, Zhu X, Fang M, Wang Q, Wang X, Dahlgren R, Shang X (2021) Airborne microplastics in indoor and outdoor environments of a coastal city in Eastern China. J Hazard Mater 417:126007

Lim JM, Jeong JH, Lee JH, Moon JH, Chung YS, Kim KH (2011) The analysis of $PM_{2.5}$ and associated elements and their indoor/outdoor pollution status in an urban area. Indoor Air 21: 145–155

Lim D, Jeong J, Song KS, Sung JH, Oh SM, Choi J (2021) Inhalation toxicity of polystyrene micro (nano)plastics using modified OECD TG 412. Chemosphere 262:128330

Lim E, Tanaka H, Ni Y, Bai Y, Ito K (2022) Microplastics/microfibers in settled indoor house dust—exploratory case study for 10 residential houses in the Kanto area of Japan. Jpn Architech Rev 5:682–690

Link B, Gabrio T, Zöllner I, Schwenk M, Siegel D, Schultz E, Scharring S, Borm P (2004) Particulate matter exposure and its health effects in children. Report of the State Health Office Baden-Württemberg

Lippmann M (2014) Toxicological and epidemiological studies on effects of airborne fibers: coherence and public health implications. Crit Rev Toxicol 44:643–695

Liu S, Zhu Y (2010) A case study of exposure to ultrafine particles from secondhand tobacco smoke in an automobile. Indoor Air 20:412–423

Liu Y, Chen R, Shen X, Mao X (2004) Wintertime indoor air levels of PM_{10}, $PM_{2.5}$ and PM_1 at public places and their contributions to TSP. Environ Int 30:189–197

Liu C, Li J, Zhang Y, Wang L, Deng J, Gao Y, Yu L, Zhang J, Sun H (2019) Widespread distribution of PET and PC microplastics in dust in urban China and their estimated human exposure. Environ Int 128:116–124

Loxham M, Nieuwenhuijsen MJ (2019) Health effects of particulate matter air pollution in underground railway systems – a critical review of the evidence. Part Fibre Toxicol 16:12

Lu K, Lai KP, Stoeger T, Ji S, Lin Z, Lin X, Chan TF, Fang JK, Lo M, Gao L, Qiu C, Chen S, Chen G, Li L, Wang L (2021) Detrimental effects of microplastic exposure on normal and asthmatic pulmonary physiology. J Hazard Mater 416:126069

Luben TJ, Nichols JL, Dutton SJ, Kirrane E, Owens EO, Datko-Williams L, Madden M, Sacks JD (2017) A systematic review of cardiovascular emergency department visits, hospital admissions and mortality associated with ambient black carbon. Environ Int 107:154–162

Mackay E (2004) An investigation of the variation in personal exposure to carbon monoxide and particulates on the A660 in Leeds. M.Sc. thesis, University of Leeds

Madureira J, Slezakova K, Costa C, Pereira MC, Teixeira JP (2020) Assessment of indoor air exposure among newborns and their mothers: levels and sources of PM_{10}, $PM_{2.5}$ and ultrafine particles at 65 home environments. Environ Pollut 264:114746

Mainka A, Fantke P (2022) Preschool children health impacts from indoor exposure to $PM_{2.5}$ and metals. Environ Int 160:107062

Majd E, McCormack M, Davis M, Curriero F, Berman J, Connolly F, Leaf P, Rule A, Green T, Clemons-Erby D, Gummerson C, Koehler K (2019) Indoor air quality in inner-city schools and its associations with building characteristics and environmental factors. Environ Res 170:83–91

Mao P, Li J, Xiong L, Wang R, Wang X, Tan Y, Li H (2019) Characterization of urban subway microenvironment exposure – a case of Nanjing in China. Int J Environ Res Public Health 16: 625

Marsh GM, Buchanich JM, Youk AO (2011) Fiber glass exposure and human respiratory system cancer risk: lack of evidence persists since 2001 IARC reevaluation. Regul Toxicol Pharmacol 60:84–92

Martins V, Moreno T, Minguillón MC, Amato F, de Miguel E, Capdevila M, Querol X (2015) Exposure to airborne particulate matter in the subway system. Sci Total Environ 511:711–722

Martins V, Faria T, Diapouli E, Manousakas MI, Eleftheriadis K, Viana M, Almeida SM (2020) Relationship between indoor and outdoor size-fractionated particulate matter in urban microenvironments: levels, chemical composition and sources. Environ Res 183:109203

Marval J, Tronville P (2022) Ultrafine particles: A review about their health effects, presence, generation, and measurement in indoor environments. Build Environ 216:108992

Matson U (2005) Indoor and outdoor concentrations of ultrafine particles in some Scandinavian rural and urban areas. Sci Total Environ 343:169–176

Matthaios VN, Kang CM, Wolfson JM, Greco KF, Gaffin JM, Hauptman M, Cunningham A, Petty CR, Lawrence J, Phipatanakul W, Gold DR, Koutrakis P (2022) Factors influencing classroom exposures to fine particles, black carbon, and nitrogen dioxide in inner-city schools and their implications for indoor air quality. Environ Health Perspect 130:47005

Mazaheri M, Clifford S, Jayaratne R, Megat Mokhtar MA, Fuoco F, Buonanno G, Morawska L (2014) School children's personal exposure to ultrafine particles in the urban environment. Environ Sci Technol 48:113–120

Mazaheri M, Lin W, Clifford S, Yue D, Zhai Y, Xu M, Rizza V, Morawska L (2019) Characteristics of school children's personal exposure to ultrafine particles in Heshan, Pearl River Delta, China – a pilot study. Environ Int 132:105134

MC (Mesothelioma Center) (2023) Mesothelioma & Asbestos Worldwide. https://www.asbestos.com/mesothelioma/worldwide/

McLaughlin J, Hogg C, Guo LY (2005) Ultrafine and coarse mode aerosol measurements in selected dwellings in Ireland. Proc Indoor Air Peking 698–701

McNabola A, Broderick BM, Gill LW (2008) Relative exposure to fine particulate matter and VOCs between transport microenvironments in Dublin: personal exposure and uptake. Atmos Environ 42:6496–6512

Mersch-Sundermann V, Gminski R et al (2007) Pilot study – evaluation of possible relationships between emissions from office machines, in particular from photocopiers and laser printers, and adverse health effects or health damage in exposed office workers. Final Report Federal Institute for Risk Assessment, Berlin. https://www.bfr.bund.de/cm/343/pilotstudie_evaluierung_moeglicher_beziehungen_zwischen_emissionen_aus_bueromaschinen_abschlussbericht.pdf. [in German]

Michikawa T, Nakai S, Nitta H, Tamura K (2014) Validity of using annual mean particulate matter concentrations as measured at fixed site in assessing personal exposure: an exposure assessment study in Japan. Sci Total Environ 466-467:673–680

Mohd Isa KN, Jalaludin J, Mohd Elias S, Mohamed N, Hashim JH, Hashim Z (2022) Evaluation of the relationship between fractional exhaled nitric oxide (FeNO) with indoor PM_{10}, $PM_{2.5}$ and NO_2 in suburban and urban schools. Int J Environ Res Public Health 19:4580

Mohsen M, Ahmed MB, Zhou JL (2018) Particulate matter concentrations and heavy metal contamination levels in the railway transport system of Sydney, Australia. Transport Res Part D 62:112–124

Mølhave L, Schneider T, Kjaergaard SK, Larsen L, Norn S, Jørgensen O (2000) House dust in seven Danish offices. Atmos Environ 34:4767–4779

Molnár P, Bellander T, Sällsten G, Boman J (2007) Indoor and outdoor concentrations of $PM_{2.5}$ trace elements at homes, preschools and schools in Stockholm, Sweden. J Environ Monit 9:348–357

Morawska L, He C, Hitchins J, Mengersen K, Gilbert D (2003) Characteristics of particle number and mass concentrations in residential houses in Brisbane, Australia. Atmos Environ 37:4195–4203

Moreno T, Pacitto A, Fernandez A, Amato F, Marcoa E, Grimalt JO, Buonanno G, Querol X (2019) Vehicle interior air quality conditions when travelling by taxi. Environ Res 172:529–542

Mosqueron L, Momas I, Le Moullec Y (2002) Personal exposure of Paris office workers to nitrogen dioxide and fine particles. Occup Environ Med 59:550–556

Mugica-Álvarez V, Figueroa-Lara J, Romero-Romo M, Sepúlveda-Sánchez J, López-Moreno T (2012) Concentrations and properties of airborne particles in the Mexico City subway system. Atmos Environ 49:284–293

Mullen NA, Bhangar S, Hering SV, Kreisberg NM, Nazaroff WW (2011) Ultrafine particle concentrations and exposures in six elementary school classrooms in northern California. Indoor Air 21:77–87

Murzyn F, Sioutas C, Cavellin LD, Joly F, Baudic A, Mehel A, Cuvelier P, Varea E, Rouland BP (2021) Assessment of air quality in car cabin in and around Paris from on-board measurements and comparison with 2007 data. J Aerosol Sci 158:105822

Nhung NTT, Amini H, Schindler C, Kutlar Joss M, Dien TM, Probst-Hensch N, Perez L, Künzli N (2017) Short-term association between ambient air pollution and pneumonia in children: a systematic review and meta-analysis of time-series and case-crossover studies. Environ Pollut 230:1000–1008

Nishihama Y, Jung CR, Nakayama SF, Tamura K, Isobe T, Michikawa T, Iwai-Shimada M, Kobayashi Y, Sekiyama M, Taniguchi Y, Yamazaki S, Japan Environment and Children's Study Group (2021) Indoor air quality of 5,000 households and its determinants. Part A: Particulate matter ($PM_{2.5}$ and $PM_{10-2.5}$) concentrations in the Japan Environment and Children's Study. Environ Res 198:111196

Niu Z, Duan Z, Yu H, Xue L, Liu F, Yu D, Zhang K, Han D, Wen W, Xiang H, Qin W (2023) Association between long-term exposure to ambient particulate matter and blood pressure, hypertension: an updated systematic review and meta-analysis. Int J Environ Health Res 33: 268–283

O'Brien S, Okoffo ED, O'Brien JW, Ribeiro F, Wang X, Wright SL, Samanipour S, Rauert C, Toapanta TYA, Albarracin R, Thomas KV (2020) Airborne emissions of microplastic fibres from domestic laundry dryers. Sci Total Environ 747:141175

Oberdörster G, Oberdörster E, Oberdörster J (2005) Nanotoxicology: an emerging discipline evolving from studies of ultrafine particles. Environ Health Perspect 113:823–839

Öder S, Weichenmeier I, Dietrich S, Schober W, Pusch G, Jörres RA, Schierl R, Nowak D, Fromme H, Behrendt H, Buters J (2012) Toxicity and elemental composition of particulate matter from outdoor and indoor air of elementary schools in Munich, Germany. Indoor Air 22: 148–158

Ogulei D, Hopke PK, Wallace LA (2006) Analysis of indoor particle size distributions in an occupied townhouse using positive matrix factorization. Indoor Air 16:204–215

Oh H-J, Nam I-S, Yun H, Kim J, Yang J, Sohn J-R (2014) Characterization of indoor air quality and efficiency of air purifier in childcare centers, Korea. Build Environ 82:203–214

Ohlwein S, Kappeler R, Kutlar Joss M, Künzli N, Hoffmann B (2019) Health effects of ultrafine particles: a systematic literature review update of epidemiological evidence. Int J Public Health 64:547–559

Omelekhina Y, Nordquist B, Alce G, Caltenco H, Wallenten P, Borell J, Wierzbicka A (2022) Effect of energy renovation and occupants' activities on airborne particle concentrations in Swedish rental apartments. Sci Total Environ 806(Pt 1):149995

Orellano P, Reynoso J, Quaranta N, Bardach A, Ciapponi A (2020) Short-term exposure to particulate matter (PM_{10} and $PM_{2.5}$), nitrogen dioxide (NO_2), and ozone (O_3) and all-cause and cause-specific mortality: systematic review and meta-analysis. Environ Int 142:105876

Orru H, Hagenbjörk A, Olstrup H (2021) Indoor and outdoor nanoparticle concentrations in an urban background area in Northern Sweden: The NanoOffice Study. Environment 8:75

Osman LM, Douglas JG, Garden C, Reglitz K, Lyon J, Gordon S, Ayres JG (2007) Indoor air quality in homes of patients with chronic obstructive pulmonary disease. Am J Respir Crit Care Med 176:465–472

Ozgen S, Ripamonti G, Malandrini A, Ragettli MS, Lonati G (2016) Particle number and mass exposure concentrations by commuter transport modes in Milan, Italy. AIMS Environ Sci 3: 168–184

Özkaynak H, Xue J, Weker R, Butler D, Koutrakis P, Spengler JD (1995) The Particle TEAM (PTEAM) Study: analysis of the data. Final report. Vol. III, US-Environmental Protection Agency

Parchami M, Motlagh MS, Pardakhti A (2022) Particulate matter concentrations and characterization in urban subway system-case study Tehran, Iran. J Environ Health Sci Eng 20:953–962

Passi A, Shiva Nagendra SM, Maiya MP (2021) Characteristics of indoor air quality in underground metro stations: a critical review. Build Environ 198:107907

Patel S, Sankhyan S, Boedicker EK, DeCarlo PF, Farmer DK, Goldstein AH, Katz EF, Nazaroff WW, Tian Y, Vanhanen J, Vance ME (2020) Indoor particulate matter during HOMEChem: concentrations, size distributions, and exposures. Environ Sci Technol 54:7107–7116

Patel H, Talbot N, Dirks K, Salmond J (2023) The impact of low emission zones on personal exposure to ultrafine particles in the commuter environment. Sci Total Environ 874:162540

Paunescu A-C, Attoui M, Bouallala S, Sunyer J, Momas I (2017) Personal measurement of exposure to black carbon and ultrafine particles in schoolchildren from PARIS cohort (Paris, France). Indoor Air 27:766–779

Perera K, Ziajahromi S, Nash SB, Leusch FDL (2023) Microplastics in Australian indoor air: abundance, characteristics, and implications for human exposure. Sci Total Environ 19:164292

Perrino C, Marcovecchio F, Tofful L, Canepari S (2015) Particulate matter concentration and chemical composition in the metro system of Rome, Italy. Environ Sci Pollut Res Int 22: 9204–9214

Persson J, Wang T, Hagberg S (2018) Indoor air quality of newly built low-energy preschools – Are chemical emissions reduced in houses with eco-labelled building materials? Indoor Built Environ 28:506–519

Pope 3rd CA, Burnett RT, Krewski D, Jerrett M, Shi Y, Calle EE, Thun MJ (2009) Cardiovascular mortality and exposure to airborne fine particulate matter and cigarette smoke: shape of the exposure-response relationship. Circulation 120:941–948

Portela NB, Teixeira EC, Agudelo-Castañeda DM, Civeira MDS, Silva LFO, Vigo A, Kumar P (2021) Indoor-outdoor relationships of airborne nanoparticles, BC and VOCs at rural and urban preschools. Environ Pollut 268(Pt A):115751

Posselt K-P, Neuberger M, Köhler D (2019) Fine and ultrafine particle exposure during commuting by subway in Vienna. Wien Klin Wochenschr 131:374–380

Prata JC, Castro JL, da Costa JP, Duarte AC, Rocha-Santos T, Cerqueira M (2020) The importance of contamination control in airborne fibers and microplastic sampling: experiences from indoor and outdoor air sampling in Aveiro, Portugal. Marine Pollut Bull 159:111522

Pui DY, Qi C, Stanley N, Oberdörster G, Maynard A (2008) Recirculating air filtration significantly reduces exposure to airborne nanoparticles. Environ Health Perspect 116:863–866

Quang TN, He C, Morawska L, Knibbs LD (2013) Influence of ventilation and filtration on indoor particle concentrations in urban office buildings. Atmos Environ 79:41–52

Querol X, Moreno T, Karanasiou A, Reche C, Alastuey A, Viana M, Font O, Gil J, de Miguel E, Capdevila M (2012) Variability of levels and composition of PM_{10} and $PM_{2.5}$ in the Barcelona metro system. Atmos Chem Phys 12:5055–5076

Raaschou-Nielsen O, Beelen R, Wang M et al (2016) Particulate matter air pollution components and risk for lung cancer. Environ Int 87:66–73

Ragettli MS, Corradi E, Braun-Fahrländer C, Schindler C, de Nazelle A, Jerrett M, Ducret-Stich RE, Künzli N, Phuleria HC (2013) Commuter exposure to ultrafine particles in different urban locations, transportation modes and routes. Atmos Environ 77:376–384

Raut JC, Chazette P, Fortain A (2009) Link between aerosol optical, microphysical and chemical measurements in an underground railway station in Paris. Atmos Environ 43:860–868

Rees VW, Connolly GN (2006) Measuring air quality to protect children from second hand smoke in cars. Am J Prev Med 31:363–368

Riediker M, Cascio WE, Griggs TR, Herbst MC, Bromberg PA, Neas L, Williams RW, Devlin RB (2004) Particulate matter exposure in cars is associated with cardiovascular effects in healthy young men. Am J Respir Crit Care Med 169:934–940

Rim D, Siegel J, Spinhirne J, Webb A, McDonald-Buller E (2008) Characteristics of cabin air quality in school buses in Central Texas. Atmos Environ 42:6453–6464

Rindel A, Bach E, Breum NO, Hugod C, Schneider T (1987) Correlating health effect with indoor air quality in kindergartens. Int Arch Occup Environ Health 59:363–373

Ripanucci G, Grana M, Vicentini L, Magrini A, Bergamaschi A (2006) Dust in the underground railway tunnels of an Italian town. J Occup Environ Hyg 3:16–25

Ritz B, Hoffmann B, Peters A (2019) The Effects of fine dust, ozone, and nitrogen dioxide on health. Dtsch Arztebl Int 116:881–886

Rivas I, Viana M, Moreno T, Pandolfi M, Amato F, Reche C, Bouso L, Àlvarez-Pedrerol M, Alastuey A, Sunyer J, Querol X (2014) Child exposure to indoor and outdoor air pollutants in schools in Barcelona, Spain. Environ Int 69:200–212

Rivas I, Donaire-Gonzalez D, Bouso L, Esnaola M, Pandolfi M, de Castro M, Viana M, Alvarez-Pedrerol M, Nieuwhuijsen M, Alastuey A, Sunyer J, Querol X (2016) Spatiotemporally resolved black carbon concentration, schoolchildren's exposure and dose in Barcelona. Indoor Air 26: 391–402

Rivas I, Kumar P, Hagen-Zanker A, de Fatima AM, Slovic AD, Pritchard JP, Geurs KT (2017) Determinants of black carbon, particle mass and number concentrations in London transport microenvironments. Atmos Environ 161:247–262

RIVM (National Institute for Public Health and the Environment) (2021) Microplastics in indoor air. A literature review. RIVM letter report 2021-0059. https://www.rijksoverheid.nl/documenten/rapporten/2021/06/25/microplastics-in-indoor-air-a-literature-review

Sabin LD, Behrentz E, Winer AM, Jeong S, Fitz DR, Pankratz DV, Colome SD, Fruin SA (2004) Characterizing the range of children's air pollutant exposure during school bus commutes. J Expo Anal Environ Epidemiol 15:377–387

Sabin LD, Behrentz E, Winer AM, Jeong S, Fitz DR, Pankratz DV, Colome SD, Fruin SA (2005) Characterizing the range of children's air pollutant exposure during school bus commutes. J Expo Anal Environ Epidemiol 15:377–387

Sahin ÜA, Onat B, Stakeeva B, Ceran T, Karim P (2012) PM_{10} concentrations and the size distribution of Cu and Fe-containing particles in Istanbul's subway system. Transport Res Transport Environ 17:48–53

Salma I, Weidinger T, Maenhaut W (2007) Time-resolved mass concentration, composition and sources of aerosol particles in a metropolitan underground railway station. Atmos Environ 41: 8391–8405

Salthammer T (2022) Microplastics and their additives in the indoor environment. Angew Chem Int Ed Engl 61:e202205713

Salthammer T, Fauck C, Omelan A, Wientzek S, Uhde E (2022) Time and spatially resolved tracking of the air quality in local public transport. Sci Rep 12(1):3262

Sang S, Chu C, Zhang T, Chen H, Yang X (2022) The global burden of disease attributable to ambient fine particulate matter in 204 countries and territories, 1990–2019: a systematic analysis of the Global Burden of Disease Study 2019. Ecotoxicol Environ Saf 238:113588

Sangiorgi G, Ferrero L, Ferrini BS, Lo Porto C, Perrone MG, Zangrando Z, Gambaro A, Lazzati Z, Bolzacchini E (2013) Indoor airborne particle sources and semi-volatile partitioning effect of outdoor fine PM in offices. Atmos Environ 65:205–214

Santen M, Wesselmann M, Fittschen U, Cremer R, Braun P, Lüdecke A, Moriske HJ (2009) Measurements of fine and ultrafine particles in indoor environment of living rooms. Gefahrst Reinh Luft 69:63–70

Schembari A, Triguero-Mas M, de Nazelle A, Dadvand P, Vrijheid M, Cirach M, Martinez D, Figueras F, Querol X, Basagaña X, Eeftens M, Meliefste K, Nieuwenhuijsen MJ (2013) Personal, indoor and outdoor air pollution levels among pregnant women. Atmos Environ 64: 287–295

Schneider T, Nielsen O, Bredsdorff P, Linde P (1990) Dust in buildings with man-made mineral fiber ceiling boards. Scand J Work Environ Health 16:434–439

Schneider T, Burdett G, Martinon L, Brochard P, Guillemin M, Teichert U, Draeger U (1996) Ubiquitous fiber exposure in selected sampling sites in Europe. Scand J Work Environ Health 22:274–284

Schober & In-Iam (2020) NanoWissen Bayern. https://www.nanowissen.bayern.de/nanowissen/nano_was_ist_das/index.htm [in German]

Schober W, Winterhalter R, Kohlhuber M, Fromme H (2012) Nanomaterials in the workplace. Exposure, health risks and prevention measures. Volume 5 of the publication series Occupational Health and Safety and Product Safety of the Bayerisches Landesamt für Gesundheit und Lebensmittelsicherheit. https://www.bestellen.bayern.de/application/applstarter?APPL=eshop&DIR=eshop&ACTIONxSETVAL(artdtl.htm,APGxNODENR:13305,AARTxNR:lgl_all_00009,AARTxNODENR:281720,USERxBODYURL:artdtl.htm,KATALOG:StMUG,AKATxNAME:StMUG,ALLE:x)=X [in German]

Schober W, Fembacher L, Frenzen A, Fromme H (2019) Passive exposure to pollutants from conventional cigarettes and new electronic smoking devices (IQOS, e-cigarette) in passenger cars. Int J Hyg Environ Health 222:486–493

Schraufnagel DE (2020) The health effects of ultrafine particles. Exp Mol Med 2020(52):311–317

Schulte PA, Leso V, Niang M, Iavicoli I (2019) Current state of knowledge on the health effects of engineered nanomaterials in workers: a systematic review of human studies and epidemiological investigations. Scand J Work Environ Health 45:217–238

Schwarz M, Schneider A, Cyrys J, Bastian S, Breitner S, Peters A (2023) Impact of ambient ultrafine particles on cause-specific mortality in three German cities. Am J Respir Crit Care Med 207:1334–1344

Seaton A, Cherrie J, Dennekamp M, Donaldson K, Hurley JF, Tran CL (2005) The London underground: dust and hazards to health. Occup Environ Med 62:355–362

Sendzig T, Fong G, Travers M, Hyland A (2006) The hazard of tobacco smoke pollution in cars: evidence from an air quality monitoring study. 13th World Conference on Tobacco and Health, Washington DC

Simoni M, Scognamiglio A, Carrozzi L, Baldacci S, Angino A, Pistelli F, Di Pede F, Viegi G (2004) Indoor exposures and acute respiratory effects in two general population samples from a rural and an urban area in Italy. J Expo Anal Environ Epidemiol 14(Suppl 1):S144–S152

Slezakova K, de Oliveira FE, do Carmo Pereira M. (2019) Assessment of ultrafine particles in primary schools: emphasis on different indoor microenvironments. Environ Pollut 246:885–895

Smith JD, Barratt BM, Fuller GW, Kelly FJ, Loxham M, Nicolosi E, Priestman M, Tremper AH, Green DC (2020) $PM_{2.5}$ on the London Underground. Environ Int 134:105188

So R, Chen J, Mehta AJ, Liu S, Strak M, Wolf K, Hvidtfeldt UA, Rodopoulou S, Stafoggia M, Klompmaker JO, Samoli E, Raaschou-Nielsen O, Atkinson R, Bauwelinck M, Bellander T, Boutron-Ruault MC, Brandt J, Brunekreef B, Cesaroni G, Concin H, Forastiere F, van Gils CH, Gulliver J, Hertel O, Hoffmann B, de Hoogh K, Janssen N, Lim YH, Westendorp R, Jørgensen JT, Katsouyanni K, Ketzel M, Lager A, Lang A, Ljungman PL, Magnusson PKE, Nagel G, Simonsen MK, Pershagen G, Peter RS, Peters A, Renzi M, Rizzuto D, Sigsgaard T, Vienneau D, Weinmayr G, Severi G, Fecht D, Tjønneland A, Leander K, Hoek G, Andersen ZJ (2021) Long-term exposure to air pollution and liver cancer incidence in six European cohorts. Int J Cancer 149:1887–1897

Sohn JR, Choi DW, Choi JS, WG U. (2006) Assesment of indoor air quality within public transport vehicles operating in specified locations throughout Seoul. Korean J Sanit 21:12–20

Sohn JR, Kim J-C, Kim MY, Son Y-S, Sunwoo Y (2008) Particulate behavior in subway airspace. Asian J Atmos Environ 2-1:54–59

Sohn J, Yang W, Kim J, Son B, Park J (2009) Indoor air quality investigation according to age of the school buildings in Korea. J Environ Manag 90:348–354

Solbes E, Harper RW (2018) Biological responses to asbestos inhalation and pathogenesis of asbestos-related benign and malignant disease. J Investig Med 66:721–727

Soltani NS, Taylor MP, Wilson SP (2021) Quantification and exposure assessment of microplastics in Australian indoor house dust. Environ Pollut 283:117064

Soltani NS, Taylor MP, Wilson SP (2022) International quantification of microplastics in indoor dust: prevalence, exposure and risk assessment. Environ Pollut 312:119957

Stabile L, Dell'Isola M, Russi A, Massimo A, Buonanno G (2017) The effect of natural ventilation strategy on indoor air quality in schools. Sci Total Environ 595:894–902

Stafoggia M, Oftedal B, Chen J, Rodopoulou S, Renzi M, Atkinson RW, Bauwelinck M, Klompmaker JO, Mehta A, Vienneau D, Andersen ZJ, Bellander T, Brandt J, Cesaroni G, de Hoogh K, Fecht D, Gulliver J, Hertel O, Hoffmann B, Hvidtfeldt UA, Jöckel KH, Jørgensen JT, Katsouyanni K, Ketzel M, Kristoffersen DT, Lager A, Leander K, Liu S, Ljungman PLS, Nagel G, Pershagen G, Peters A, Raaschou-Nielsen O, Rizzuto D, Schramm S, Schwarze PE, Severi G, Sigsgaard T, Strak M, van der Schouw YT, Verschuren M, Weinmayr G, Wolf K, Zitt E, Samoli E, Forastiere F, Brunekreef B, Hoek G, Janssen NAH (2022) Long-term exposure

to low ambient air pollution concentrations and mortality among 28 million people: results from seven large European cohorts within the ELAPSE project. Lancet Planet Health 6:e9–e18
Stone V, Miller MR, Clift MJD, Elder A, Mills NL, Møller P, Schins RPF, Vogel U, Kreyling WG, Alstrup Jensen K, Kuhlbusch TAJ, Schwarze PE, Hoet P, Pietroiusti A, De Vizcaya-Ruiz A, Baeza-Squiban A, Teixeira JP, Tran CL, Cassee FR (2017) Nanomaterials versus ambient ultrafine particles: an opportunity to exchange toxicology knowledge. Environ Health Perspect 125:106002
Stranger M, Potgieter-Vermaak SS, Van Grieken R (2007) Comparative overview of indoor air quality in Antwerp, Belgium. Environ Int 33:789–797
Suder Egnot N, Benson SM, Vater MF, Hazan R, Patel O, Marsh GM (2020) Systematic review and meta-analysis of epidemiological literature evaluating the association between exposure to man-made vitreous fibers and respiratory tract cancers. Regul Toxicol Pharmacol 112:104585
Szabados M, Csákó Z, Kotlík B, Kazmarová H, Kozajda A, Jutraz A, Kukec A, Otorepec P, Dongiovanni A, Di Maggio A, Fraire S, Szigeti T (2021) Indoor air quality and the associated health risk in primary school buildings in Central Europe – The InAirQ study. Indoor Air 31: 989–1003
Szigeti T, Kertész Z, Dunster C, Kelly FJ, Záray G, Mihucz VG (2014) Exposure to $PM_{2.5}$ in modern office buildings through elemental characterization and oxidative potential. Atmos Environ 94:44–52
Szigeti T, Dunster C, Cattaneo A, Spinazzè A, Mandin C, Le Ponner E, de Oliveira FE, Ventura G, Saraga DE, Sakellaris IA, de Kluizenaar Y, Cornelissen E, Bartzis JG, Kelly FJ (2017) Spatial and temporal variation of particulate matter characteristics within office buildings - The OFFICAIR study. Sci Total Environ 587–588:59–67
Szoboszlai Z, Furu E, Angyal A, Szikszai Z, Kertész Z (2011) Investigation of indoor aerosols collected at various educational institutions in Debrecen, Hungary. X-Ray Spectrom 40:176–180
Tang KM, Nace Jr CG, Lynes CL, Maddaloni MA, LaPosta D, Callahan KC (2004) Characterization of background concentrations in upper Manhattan, New York apartments for select contaminants identified in World Trade Center dust. Environ Sci Technol 38:6482–6490
Thriene B, Sobottka A, Willer H, Weidhase J (1996) Man-made mineral fibre boards in buildings – health risks caused by quality deficiencies. Toxicol Lett 88:299–303
Torkmahalleh MA, Goldasteh I, Zhao Y, Udochu NM, Rossner A, Hopke PK, Ferro AR (2012) $PM_{2.5}$ and ultrafine particles emitted during heating of commercial cooking oils. Indoor Air 22: 483–491
Torres-Agullo A, Karanasiou A, Moreno T, Lacorte S (2022) Airborne microplastic particle concentrations and characterization in indoor urban microenvironments. Environ Pollut 308: 119707
Tu M, Olofsson U (2021) PM_{10} in underground stations from different types of trains. Trans Res Part D 95:102867
Tunno BJ, Dalton R, Cambal L, Holguin F, Lioy P, Clougherty JE (2016) Indoor source apportionment in urban communities near industrial sites. Atmos Environ 139:30–36
US EPA (2020) Risk evaluation for asbestos. Part I: Chrysotile asbestos. EPA Document EPA-740-R1-8012. https://www.epa.gov/assessing-and-managing-chemicals-under-tsca/final-risk-evaluation-asbestos-part-1-chrysotile
Van Ryswyk K, Anastasopolos AT, Evans G, Sun L, Sabaliauskas K, Kulka R, Wallace L, Weichenthal S (2017) Metro commuter exposures to particulate air pollution and PM2.5-associated elements in three Canadian cities: The Urban Transportation Exposure Study. Environ Sci Technol 51:5713–5720
Vance ME, Marr LC (2015) Exposure to airborne engineered nanoparticles in the indoor environment. Atmos Environ 106:503–509
Vardavas CI, Kondilis B, Travers MJ, Petsetaki E, Tountas Y, Kafatos AG (2007) Environmental tobacco smoke in hospitality venues in Greece. BMC Public Health 7:302

Verla AW, Enyoh CE, Verla EN (2019) Microplastics, an emerging concern: a review of analytical techniques for detecting and quantifying microplastic. Anal Methods Environ Chem J 12:15–32

Verma DK, Sahai D, Kurtz LA, Finkelstein MM (2004) Current man-made mineral fibers (MMMF) exposures among ontario construction workers. J Occup Environ Hyg 1:306–318

Vernez D, Duperrex O, Herrera H, Perret V, Rossi I, Regamey F, Guillemin M (2019) Exposure to amosite-containing ceiling boards in a public school in Switzerland: a case study. Int J Environ Res Public Health 16:5069

Vianello A, Rasmus LJ, Li L, Jes V (2019) Simulating human exposure to indoor airborne microplastics using a breathing thermal manikin. Sci Reports 9:8670

Vu TV, Stewart GB, Kitwiroon N, Lim S, Barratt B, Kelly FJ, Thompson R, Smith RB, Toledano MB, Beevers SD (2022) Assessing the contributions of outdoor and indoor sources to air quality in London homes of the SCAMP cohort. Build Environ 222:109359

Walker G, Ostendorp G, Heinzow B (2012) Man-made mineral fibers and fine dust in schools and kindergartens. Gefahrst Reinh Luft 72:89–93. [in German]

Wallace L (2006) Indoor sources of ultrafine and accumulation mode particles: size distributions, size-resolved concentrations, and source strengths. Aerosol Sci Technol 40:348–360

Wallace L, Howard-Reed C (2002) Continuous monitoring of ultrafine, fine and coarse particles in a residence for 18 months in 1999–2000. J Air Waste Manage Assoc 52:828–844

Wallace L, Ott W (2011) Personal exposure to ultrafine particles. J Expos Sci Environ Epidemiol 21:20–30

Wallace LA, Mitchell H, O'Connor GT, Neas L, Lippmann M, Kattan M, Koeng J, Stout JW, Vaughn BJ, Wallace D, Walter M, Adams K, Liu L-JS (2003) Particle concentrations in inner-city homes of children with asthma: the effect of smoking, cooking, and outdoor pollution. Environ Health Perspect 111:1265–1272

Walser T, Hellweg S, Juraske R, Luechinger NA, Wang J, Fierz M (2012) Exposure to engineered nanoparticles: model and measurements for accident situations in laboratories. Sci Total Environ 420:119–126

Wei D, Nielsen F, Ekberg L, Löfvendahl A, Bernander M, Dalenbäck JO (2020) $PM_{2.5}$ and ultrafine particles in passenger car cabins in Sweden and northern China-the influence of filter age and pre-ionization. Environ Sci Pollut Res Int 27:30815–30830

Wen Y, Leng J, Shen X, Han G, Sun L, Yu F (2020) Environmental and health effects of ventilation in subway stations: a literature review. Int J Environ Res Public Health 17:1084

WHO (2000) Air Quality Guidelines for Europe. WHO Regional Publications, European Series No. 91. https://www.euro.who.int/en/publications/abstracts/air-quality-guidelines-for-europe

WHO (2016) WHO expert consultation: available evidence for the future update of the WHO global air quality guidelines (AQGs). WHO Regional Office for Europe, Kopenhagen

Wichmann J, Lind T, Nilsson MAM, Belland T (2010) $PM_{2.5}$, soot and NO_2 indoor outdoor relationships at homes, pre-schools and schools in Stockholm, Sweden. Atmos Environ 44: 4536–4544

Williams R, Brook R, Bard R, Conner T, Shin H, Burnett R (2012) Impact of personal and ambient-level exposures to nitrogen dioxide and particulate matter on cardiovascular function. Int J Environ Health Res 22:71–91

Winterhalter R, Berlin K, Dietrich S, Schierl R, Fromme H (2012) Measurement of synthetic nanoparticles and ultrafine dust at selected workplaces. Gefahrst Reinh Luft 72:359–366. [in German]

Wolf K, Hoffmann B, Andersen ZJ, Atkinson RW, Bauwelinck M, Bellander T, Brandt J, Brunekreef B, Cesaroni G, Chen J, de Faire U, de Hoogh K, Fecht D, Forastiere F, Gulliver J, Hertel O, Hvidtfeldt UA, Janssen NAH, Jørgensen JT, Katsouyanni K, Ketzel M, Klompmaker JO, Lager A, Liu S, MacDonald CJ, Magnusson PKE, Mehta AJ, Nagel G, Oftedal B, Pedersen NL, Pershagen G, Raaschou-Nielsen O, Renzi M, Rizzuto D, Rodopoulou S, Samoli E, van der Schouw YT, Schramm S, Schwarze P, Sigsgaard T, Sørensen M, Stafoggia M, Strak M, Tjønneland A, Verschuren WMM, Vienneau D, Weinmayr G, Hoek G, Peters A, Ljungman PLS (2021) Long-term exposure to low-level ambient air pollution and incidence of stroke and

coronary heart disease: a pooled analysis of six European cohorts within the ELAPSE project. Lancet Planet Health 5:e620–e632
Wright SL, Kelly FJ (2017) Plastic and human health: a micro issue? Environ Sci Technol 51:6634–6647
Wyss AB, Jones AC, Bølling AK, Kissling GE, Chartier R, Dahlman HJ, Rodes CE, Archer J, Thornburg J, Schwarze PE, London SJ (2016) Particulate matter 2.5 exposure and self-reported use of wood stoves and other indoor combustion sources in urban nonsmoking homes in Norway. PLoS One 11:e0166440
Xu B, Hao J (2017) Air quality inside subway metro indoor environment worldwide: a review. Environ Int 107:33–46
Xu B, Chen X, Xiong J (2018) Air quality inside motor vehicles' cabins: a review. Indoor Built Environ 27:452–465
Xumiao L, Prata JC, Alves JR, Duarte AC, Rocha-Santos T, Cerqueira M (2021) Airborne microplastics and fibers in indoor residential environments in Aveiro, Portugal. Environ Adv 6:100134
Yan C, Zheng M, Yang Q, Zhang Q, Qiu X, Zhang Y, Fu H, Zhu Y, Zhu T, Li X (2015) Commuter exposure to particulate matter and particlebound PAHs in three transportation modes in Beijing, China. Environ Pollut 204:199–206
Yang F, Kaul D, Wong KC, Westerdahl D, Sun L, Ho K-f, Tian L, Brimblecombe P, Ning Z (2015) Heterogeneity of passenger exposure to air pollutants in public transport microenvironments. Atmos Environ 109:42–51
Yao Y, Glamoclija M, Murphy A, Gao Y (2022) Characterization of microplastics in indoor and ambient air in northern New Jersey. Environ Res 207:112142
Yee MS, Hii LW, Looi CK, Lim WM, Wong SF, Kok YY, Tan BK, Wong CY, Leong CO (2021) Impact of microplastics and nanoplastics on human health. Nanomaterials (Basel) 11:496
Yoda Y, Tamura K, Shima M (2017) Airborne endotoxin concentrations in indoor and outdoor particulate matter and their predictors in an urban city. Indoor Air 27:955–964
Zhang Q, Zhu Y (2010) Measurements of ultrafine particles and other vehicular pollutants inside school buses in South Texas. Atmos Environ 44:253–261
Zhang X, Fan Q, Bai X, Li T, Zhao Z, Fan X, Norbäck D (2018) Levels of fractional exhaled nitric oxide in children in relation to air pollution in Chinese day care centres. Int J Tuberc Lung Dis 22:813–819
Zhang Q, Xu EG, Li J, Chen Q, Ma L, Zeng EY, Shi H (2020a) A review of microplastics in table salt, drinking water, and air: direct human exposure. Environ Sci Technol 54:3740–3751
Zhang Q, Zhao Y, Du F, Cai H, Wang G, Shi H (2020b) Microplastic fallout in different indoor environments. Environ Sci Technol 54:6530–6539
Zhang J, Wang L, Kannan K (2020c) Microplastics in house dust from 12 countries and associated human exposure. Environ Int 134:105314
Zhang A, Liu Y, Zhao B, Zhang Y, Kan H, Zhao Z, Deng F, Huang C, Zeng X, Sun Y, Qian H, Liu W, Mo J, Sun C, Zheng X (2021) Indoor $PM_{2.5}$ concentrations in China: a concise review of the literature published in the past 40 years. Build Environ 198:107898
Zhao Y, Zhao B (2018) Emissions of air pollutants from Chinese cooking: a literature review. Build Simul 11:977–995
Zhao J, Birmili W, Hussein T, Wehner B, Wiedensohler A (2021) Particle number emission rates of aerosol sources in 40 German households and their contributions to ultrafine and fine particle exposure. Indoor Air 31:818–831
Zhao X, Zhou Y, Liang C, Song J, Yu S, Liao G, Zou P, Tang KHD, Wu C (2022) Airborne microplastics: occurrence, sources, fate, risks and mitigation. Sci Total Environ 858(Pt 2): 159943
Zhou Y, Shao Y, Yuan Y, Liu J, Zou X, Bai P, Zhan M, Zhang P, Vlaanderen J, Vermeulen R, Downward GS (2020) Personal black carbon and ultrafine particles exposures among high school students in urban China. Environ Pollut 265(Pt A):114825

Zhu Y, Eiguren-Fernandez A, Hinds WC, Miguel AH (2007) In-cabin commuter exposure to ultrafine particles on Los Angeles freeways. Environ Sci Technol 41:2138–2145

Zhu YD, Li X, Fan L, Li L, Wang J, Yang WJ, Wang L, Yao XY, Wang XL (2021a) Indoor air quality in the primary school of China-results from CIEHS 2018 study. Environ Pollut 291: 118094

Zhu YD, Fan L, Wang J, Yang WJ, Li L, Zhang YJ, Yang YY, Li X, Yan X, Yao XY, Wang XL (2021b) Spatiotemporal variation in residential $PM_{2.5}$ and PM_{10} concentrations in China: national on-site survey. Environ Res 202:111731.z

Zhu J, Zhang X, Liao K, Wu P, Jin H (2022) Microplastics in dust from different indoor environments. Sci Total Environ 833:155256

Zuurbier M, Hoek G, Oldenwening M, Lenters V, Meliefste K, van den Hazel P, Brunekreef B (2010) Commuters' exposure to particulate matter air pollution is affected by mode of transport, fuel type, and route. Environ Health Perspect 118:783–789

Zuurbier M, Hoek G, Oldenwening M, Meliefste K, van den Hazel P, Brunekreef B (2011) Respiratory effects of commuters' exposure to air pollution in traffic. Epidemiology 22:219–227

Chapter 6
Bioaerosols

Contents

Abstract The focus of this chapter are bioaerosols in indoor spaces. Bioaerosol refers to all particles of biological origin suspended in the air like plant and animal components, bacteria, fungi, and viruses. In this chapter mold, microbial volatile organic compounds (mVOC), mycotoxins, viruses, inhalation allergens, endotoxins, and viruses are presented individually in detail. In addition to a brief description of the chemical–physical characteristics and the use of the substance, an overview of main health risks and, in particular, references to further scientific literature are given. In the foreground, however, are extensive presentations of the occurrence of the substances in indoor air and in sedimented dust.

Keywords MVOC · Endotoxin · Cat allergens · Mite allergens · Mold · Virus · SARS · Mycotoxins

Bioaerosol refers to all particles of biological origin suspended in the air. In particular, these are plant components (e.g., pollen, fibers, cell assemblies), animal components (e.g., dander, hair, animal epithelia), bacteria (e.g., bacterial spores and cell fragments, metabolic products), and fungi (e.g., fungal spores, cell wall components, metabolic products), as well as viruses. Due to the large number of possible sources of bioaerosols in the environment, their composition varies greatly in terms of quality and quantity. The various sampling techniques and measurement methods are summarized, for example, by Ghosh et al. (2015) and Cox et al. (2020). As diverse as the sources can be the potential adverse health effects of indoor biological

H. Fromme, *Indoor Air Quality*, https://doi.org/10.1007/978-3-031-40078-0_6

contaminants can range from intoxications and infections to allergies, for example. Under normal exposure conditions, humans are well adapted to bioaerosol constituents. However, adverse health effects may have to be expected at high concentrations (e.g., in workplaces) and in individuals with impaired immune responses, allergic individuals, and with pre-damaged respiratory tracts (Walser et al. 2017). Conflicting results regarding health effects may be due to factors such as co-exposures or gene-by-environment interactions, inconsistent exposure measurement approaches, and knowledge gaps about which method of exposure assessment is most pertinent to human health (Wolf and Lai 2020).

6.1 Molds

Mold can be carried indoors directly via the outside air or via soil and dust particles that adhere to shoes, for example. In addition, direct growth inside is possible in the case of moisture damage. In principle, molds can grow on many surfaces of residential interiors, but this is strongly dependent on the nutrient supply on the growth surface, the temperature, and especially the humidity (Du et al. 2021; Rai et al. 2021). In the context of moisture damage, materials such as wallpaper, plasterboard walls, and objects made of paper or leather in particular are colonized by mold relatively quickly (UBA 2019). In the case of prolonged exposure to moisture, there may be moisture penetration of building components and colonization of wood-based materials, plasters, and insulating materials (e.g., mineral wool, polystyrene). The following molds occur mainly with moisture damage in the room: *Acremonium* spp., *Aspergillus penicillioides, Aspergillus restrictus, Aspergillus versicolor, Chaetomium* spp., *Phialophora* spp., *Penicillium chrysogenum, Penicillium brevicompactum, Scopulariopsis brevicaulis, Scopulariopsis fusca, Scopulariopsis brumtii, Scopulariopsis chartarum, Stachybotrys chartarum, Tritirachium (Engyodontium) album,* and *Trichoderma* spp.

Due to the metabolic activity of molds, spores can be actively released into the ambient air in a temporally and spatially variable manner, as well as pieces of mycelium, cell components of the fungi, or substances of secondary metabolism. Mold spores often have a diameter of about 10 μm (with a few exceptions between 2 and 30 μm) and can float in the air for a long time, even over longer distances (UBA 2019). The different sampling techniques and measurement methods are summarized in two recent reviews (Nevalainen et al. 2015; Holme et al. 2020).

6.1.1 Health Aspects

In the context of epidemiological studies, non-specific irritations of the mucous membranes of the upper respiratory tract have been described in connection with mold exposure indoors. Frequently occurring symptoms are burning and tearing of the eye, sneezing, rhinitis, nasal congestion, and dryness of the throat. Reviews of the health effects are available in the scientific literature (WHO 2009; Eduard 2009; Sauni et al. 2013; Walser et al. 2015, 2017; Holme et al. 2020; Fakunle et al. 2021). It is assumed that these health reactions are essentially caused by components of molds (e.g., 1,3-β-D-glucan) and/or other substances produced by molds. In addition, depending on the individual predisposition as well as the sensitizing potential of molds, or their spores, allergic reactions, or the worsening of an existing allergic disease have been described (Walser et al. 2017). In a recent meta-analysis Fakunle et al. (2021) summarized that exposure to a combination of Aspergillus, Penicillium, and Cladosporium species increased the risk of wheeze by 67% that the association of indoor microbial exposure with allergic rhinitis was not significant and that exposure to microbial agents such as endotoxins was protective against asthma, although the estimate was from two studies.

6.1.2 *Occurrence in the Air*

Comprehensive compilations of mold measurement results in the scientific literature up to 2013 are provided by Ghosh et al. (2015) and Nevalainen et al. (2015).

The following literature, which is not listed there, provides additional information on the exposure situation in residences and schools. In Germany, mold concentrations of 5–15,000 colony-forming units/m^3 (CFU/m^3) (median: 105 CFU/m^3) were found in 397 children's bedrooms in 1999/2000 (Jovanovic et al. 2004). In this study, the indoor climatic conditions or the location of the buildings in an urban or rural region had no significant influence on the content. As part of a study in Sweden, Latvia, and Iceland, concentrations between 30 and 20,000 CFU/m^3 (geometric mean: 209 CFU/m^3) were determined in 2001/2002 in a total of 159 indoor rooms with and without visible mold contamination (Sahlberg et al. 2013). Concentrations of 34–6528 CFU/m^3 (median: 250 CFU/m^3) were measured in 68 apartments in Portugal in 2011–2013 (Madureira et al. 2015a). Medians of 38,000 particles/m^3 (~3800 CFU/m^3), 27,000 particles/m^3 (~2700 CFU/m^3), and 69,000 particles/m^3 (~6900 CFU/m^3) were found when investigating 7 dwellings with mold infestation, 3 control rooms, and 10 parallel outdoor air samples in Norway (Afanou et al. 2019). The conversion of the number of spores to CFU is done using the formula 10 spore particles/m^3 = 1 CFU/m^3. This estimation is based on rough assumptions that should be interpreted very cautiously. Only between the outdoor air content and the polluted dwellings was there a significant difference. In a Danish study (n = 127), the lowest and highest median levels in 2010/2011 were found to be 26 CFU/m^3 in winter and 235 CFU/m^3 in summer (Frankel et al. 2012). Thereby, outdoor air temperature and air-exchange rate were positively associated with concentrations. In 48 mold-damaged French dwellings higher total viable fungi from 16.7 to 361,000 CFU/m^3 (mean: 10200 CFU/m^3) of air (Delanoe et al. 2020). The concentration of fungi in 15 Danish homes varied from 20 to 1780 CFU/m^3, with ageometric mean value of 148 CFU/m^3 (Lu et al. 2021).

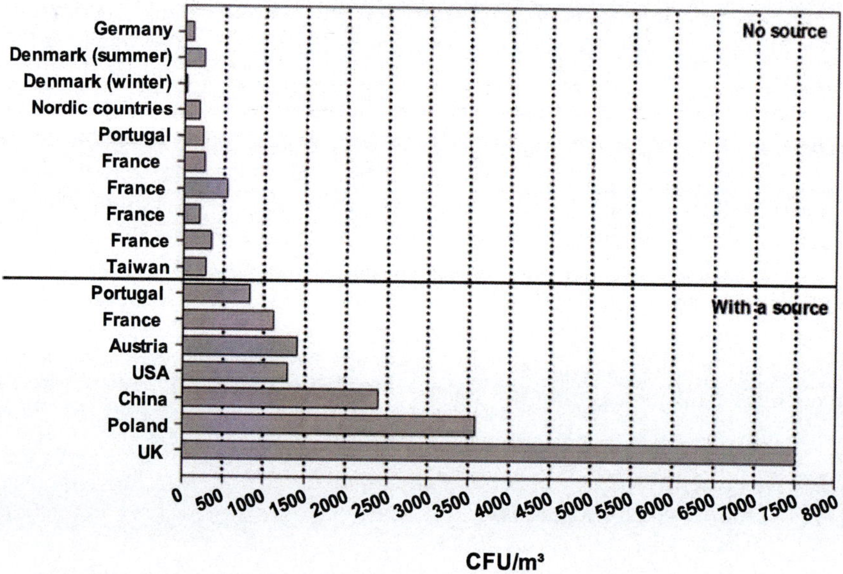

Fig. 6.1 Fungal spore concentrations in indoor spaces with and without a suspicious mold contamination (modified and extended from Al Hallak et al. 2023)

A US study found concentrations from 59 to 2051 CFU/m^3 (geometric mean: 451 CFU/m^3) in 23 homes with mold infestations, with *Cladosporium* and *Penicillium* essentially detected (Ryan and Beaucham 2013). Fan et al. (2021a) collected airborne dust in 50 houses in 12 Chinese cities in 2018 and reported a median of all samples in the bedrooms and living rooms of 350 CFU/m^3 (0-9371 CFU/m^3) and 340 CFU/m^3 (0-8800 CFU/m^3), respectively. THe median concentrations were statistically signicant higher in the bedrooms compared to livingrooms, except one sampling site.

Figure 6.1 gives an overview of the median or average airborne spore concentration in residences without a contamination and in residences that are suspicious of a mold contamination published in some studies.

Concentrations from 63 to 1322 CFU/m^3 (median: 240 CFU/m^3) could be found in 73 classrooms of Portuguese schools in 2011–2013 and 280–711 CFU/m^3 (median: 415 CFU/m^3) in 9 daycare centers (Madureira et al. 2015a, b). No significant difference was found between indoor and outdoor air levels. In another study from Portugal, a median of 460 CFU/m^3 (spring) and 313 CFU/m^3 (winter) was found in 2011/2012 in a total of 52 common rooms in rooms for 0–2-year-old children, and 270 CFU/m^3 (spring) and 510 CFU/m^3 (winter) in rooms for 3–5-year-olds (Mendes et al. 2014).

A Canadian study examined the air in 116 classrooms from 39 schools and found a geometric mean of 323 CFU/m^3, with concentrations twice as high in classrooms with natural ventilation than in classrooms with mechanical ventilation systems (Bartlett et al. 2004).

In Taiwan, median airborne fungal concentration of 314 CFU/m^3 (44–4014 CFU/m^3) was reported in 120 schools in 2010–2012 (Yen et al. 2020a).

6.1.3 *Occurrence in Dust*

In the Netherlands, Verhoeff et al. (1995) found significantly higher geometric mean of 12,880 CFU/g in dust from carpeted floors than in dust from smooth floors with 3530 CFU/g in 60 residences. Higher total concentration of culturable fungi of 2700 CFU/g was quantified in the floor dust of 18 Polish residences (Augustyniuk-Kram & Dmowska 2013). As part of the Chinese Indoor Environment and Health Surveillance 2018 a total of 642 dust samples were collected in 12 Chinese cities and mean concentration of 806 CFU/m^3 (range: 0–9371 CFU/m^3) was reported (Fan et al. 2021b). In some, but not all regions, the concentrations in the warm season were higher than the cold season, and the levels were positively correlated with indoor humidity but negatively correlated with residential temperature and CO_2 concentration. In 2017/18 dust was collected in 37 long-term uninhabited houses and 17 inhabited houses in Japan and approximately an order of magnitude higher fungal concentrations in the uninhabited to the inhabited houses (means: 2.1 × 107 CFU/g dust vs. median 7.7 × 106 CFU/g dust) (Shinohara et al. 2023).

A study in Germany examined 371 classrooms in 111 schools over a measurement period of 6 years and differentiated 105 mold species and genera (Neumann et al. 2018). In dust samples collected from higher areas in classrooms, the median in dust was 452,416 CFU/g. Especially, *Aspergillus fumigatus, Penicillium* spp., and *Cladosporium* spp. were detected in over half of the dust samples. Depending on the mold genus, medians ranged from 7092 CFU/g (*Aspergillus fumigatus*) to 48,916 CFU/g (*Cladosporium* spp.). The seasonal dependence was particularly pronounced for the genus *Cladosporium*, with higher levels in summer.

6.2 Microbial Volatile Organic Compounds (mVOC)

A subgroup of VOC that is not fully covered by standard VOC measurements are the so-called microbial volatile organic compounds (mVOC). These are substances from very different chemical groups such as alcohols, esters, aldehydes, and ketones, which are formed as metabolic products of molds or bacteria (Bos et al. 2013; Mısztal et al. 2018).

The use of mVOC as an indicator for hidden mold occurrence is possible in principle, but is critically discussed (Laußmann et al. 2004; Schuchardt and Strube 2009; Moularat et al. 2011). In the evaluation, it is important to note that the indicator substances can also originate from other sources in the indoor environment or user activities. The most suitable indicators of microbial infestation appear to be 3-methylfuran, dimethyl disulfide, 1-octen-3-ol, 3-octanone, and 3-methyl-1-butanol (UBA 2019), while hexanone, heptanone, 1-butanol, and isobutanol are less specific indicators. Moisture plays an important role in regulating microbial growth and mVOC emissions in indoor air both occurred at lower relative humidity in carpet and floor dust compared to drywall (Haines et al. 2021).

Rering et al. (2020) compared the volatile data derived via three solventless collection techniques and concluded that direct headspace injections were not effective, and the resultant chromatography was poor despite the use of on-column cryofocusing while SPME (solid-phase microextraction) and SPE (solid-phase extraction) detected a similar number of volatiles, but SPE collected a greater abundance of microbial volatiles than SPME.

6.2.1 Health Aspects

Some studies indicate a link between mVOC exposure and health complaints such as mucous membrane irritation and headaches (summary in UBA 2019). Currently, it is assumed that toxic effects at the concentrations found in indoor living spaces are unlikely. However, odor nuisances caused by mVOCs are certainly possible. In the scientific literature, two review papers on health effects have been published so far (Korpi et al. 2009; Thorn and Greenman 2012).

6.2.2 Occurrence in the Air

Table 6.1 summarizes various results of studies conducted in rooms with a mold infestation, moisture damage, or control rooms. In a study conducted by the Unfallkasse Nordrhein-Westfalen, a median sum value for the main indicators (2,3-methyl-1-butanol, 1-octen-3-ol, 3-octanone, 3-methylfuran, dimethyl disulfide) of 655 ng/m^3 was determined (Unfallkasse 2014). According to the above-mentioned mold guideline, a microbial infestation would have been very likely in 50% of the examined rooms. However, there were no further indications of this in the study, so that the authors critically discuss a use of mVOC to determine hidden mold infestation (Neumann 2011). In another German study in 44 rooms with known mold infestation and 44 without evidence of it, significant differences between both room types could be found for most of the analyzed mVOC (Schleibinger et al. 2008). However, when other known influencing factors were taken into account, the associations with mold occurrence were statistically significant only for 2-methyl-1-butanol and 1-octen-3-ol. In a very comprehensive study by Matysik et al. (2009) in 550 apartments in Leipzig, Germany, higher levels of many mVOC were found indoors where Aspergilli or Penicillia were detectable

Table 6.1 Median mVOC concentrations in the air of residences in μg/m^3

References	N	2-methyl-1-butanol	3-methyl-1-butanol	1-octen-3-ol	3-octanone	2-Methylfuran	3-Methylfuran	Dimethyl disulfide	2-Heptanone	2-Hexanone	Country, Year
Moldy											
Schleibinger et al. (2008)	44	<	0.23	<0.1	–	0.65	0.20	<0.1	1.05	0.58	Germany
Matysik et al. (2009)	193	–	1.04	–	–	0.77	0.20	0.05	0.46	0.16	Germany, 2000–06
Neumann (2011)	379+	0.125		0.21	0.145	–	0.015	0.005	–	–	Germany
Sahlberg et al. (2013)	92	0.067	0.26	0.05	0.04	–	0.02	0.03	0.31	0.06	Sweden, Iceland, Estonia
Choi et al. (2017)	202	–	2.42	–	–	< LOQ	–	–	1.95	< LOQ	USA, 2001–02
Araki et al. (2012)	182	–	0.49	< 0.25	< 0.25	–	–	–	< 0.25	0.33	Japan, 2006
Mold-free											
Schleibinger et al. (2008)	40	0.08	0.93	<0.1	–	1.16	0.32	0.10	1.40	0.60	Germany
Matysik et al. (2009)*	201	–	1.03	–	–	1.64	0.41	0.09	0.57	0.22	Germany, 2000–06
Sahlberg et al. (2013)	64	0.08	0.30	0.06	0.04	–	0.03	0.05	0.33	0.05	Sweden, Iceland, Estonia
Choi et al. (2017)	198	–	1.98	–	–	1.45	–	–	2.24	< LOQ	USA, 2001/02
Ryan and Beaucham (2013)	23	–	–	11.8	11.6	6.3	–	–	16.9	–	USA

[a] buildings contaminated with Aspergillus; + classrooms; *LOQ* limit of determination

compared to rooms where the difference in colony-forming units between indoors and outdoors was less than 100 CFU/m^3. Around 150 compounds were identified by screening twelve fungal species (Fiedler et al. 2001). Not a single compound was common to all species, but 1-octen-3-ol, 3-octanone, 2-methyl-1-butanol and 3-methyl-1-butanol occurred in the large majority of them.

As part of the European Community Respiratory Health Survey (ECRHS II), a total of 159 indoor rooms in three Nordic countries were studied and very low MVOC concentrations were measured in both homes with and without mold exposure (Sahlberg et al. 2013). A positive association could be found in this case between 5 mVOC (2-pentanol, 2-hexanone, 2-pentylfuran, 1-octen-3-ol, and 3-methylfuran) and non-specific upper respiratory complaints in terms of sick building syndrome (SBS). Thus, a total of 30% of the 159 participants showed at least one symptom typical of SBS.

Higher mVOC concentrations, the cause of which is not clear, could be found indoors in two North American studies (Ryan and Beaucham 2013; Choi et al. 2017).

6.3 Mycotoxins

The term mycotoxin was derived from the Greek "myko" and the Latin "toxicum," meaning fungus and poison, respectively. To date, more than 300 different mycotoxins are known, which can be assigned to about 25 structural types and have very different chemical structures. The chemical structures of some important mycotoxins are summarized in Fig. 6.2.

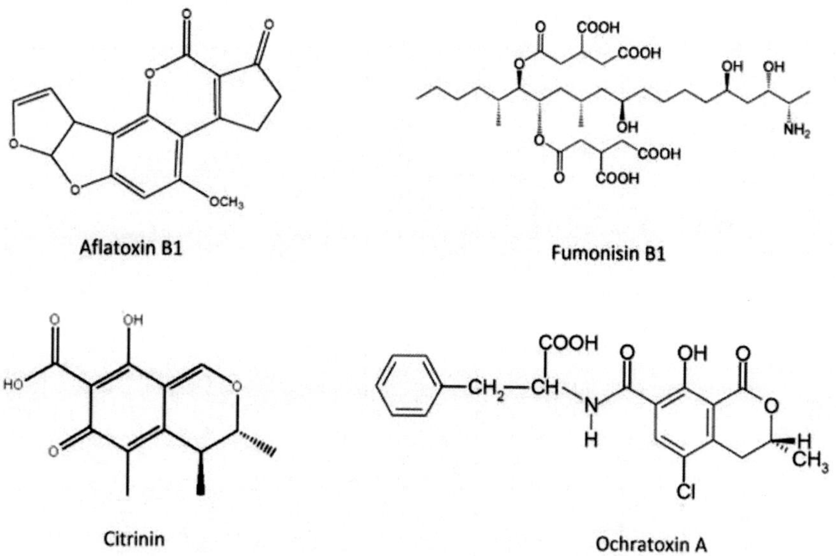

Fig. 6.2 Chemical structure of different mycotoxins

Mycotoxins are natural, secondary metabolic products of molds. Unlike the products of primary metabolism, these substances are not found in all organisms and they are characteristic of their producer. In addition, they are often formed only under certain temperature and humidity conditions, when nutrients are abundant, or at certain stages of development. Often they consist not only of a single substance, but also of a whole family of chemically related compounds. Along with antibiotics, mycotoxins represent the second major group of active substances synthesized by microorganisms. Like antibiotic-producing microorganisms, mycotoxin-producing mold species are distributed worldwide. The function of mycotoxin formation in fungal metabolism is not yet known. Metabolic control functions and an ecological role in the interaction with other organisms are discussed.

Humans are mainly exposed to mycotoxins via moldy food and at specific workplaces, e.g., in the context of sorting recyclable materials or in grain mills (Fromme et al. 2016). However, exposure to mold spores or mycotoxins may also have to be expected in homes and public buildings with mold infestations. In addition to direct inhalation of spores or mold fragments, exposure via contaminated house dust must also be taken into account.

Indoor relevant molds are especially Aspergilli, Penicillia, and Fusarium. The main sources of mycotoxins in this case are *Penicillium chrysogenum, P. brevicom pactum, P. expansum, P. commune,* and *Aspergillus versicolor. Aspergillus ochraceus, A. niger, A. avus,* and *A. fumigatus* occur less frequently. In addition, *Stachybotrys chartarum* and *Trichoderma viride* also occur indoors (Jarvis and Miller 2005).

Table 6.2 shows the most important mycotoxins and their main fungal producers.

6.3.1 *Health Aspects*

The spectrum of health effects of mycotoxins is very broad. It ranges from odor nuisance to non-specific symptoms (e.g., headache and pain in the *limbs,* irritation of the mucous membranes), toxic effects, especially on the liver, kidney, and nervous system, immunosuppressive effects, and mutagenic, carcinogenic, and teratogenic effects (see Table 6.2). Various reviews can be found in the scientific literature (Bennett and Klich 2003; Wu et al. 2014). EFSA provides information on individual mycotoxins on a central website (EFSA 2020).

6.3.2 *Occurrence in the Air*

So far, there are only very limited results of measurements in indoor rooms outside workplaces, mostly in those with previous water damage. Test results are summarized in Table 6.3. For example, Polizzi et al. (2009) analyzed 20 mycotoxins in 99 air samples from 7 buildings with visible mold infestation and found in

Table 6.2 Important mycotoxins and health effects (modified after Fromme et al. 2016 and Degen et al. 2017)

Mycotoxin	Major producer	Effects
Aflatoxin B1/B2, G1/G2, M1	Aspergillus (A.) flavus, A. parasiticus	Liver toxicity, mutagenic, carcinogenic, immunosuppressive
Alternaria toxins	Alternaria spp., A. alternata, A. lycopersici, A. tenuissima	Some substances genotoxic, teratogenic
Citrine	Monascus ruber; Penicillium (P.) citrinum; Aspergillus spp.	Renal toxic, genotoxic
Deoxynivalenol (type B trichothecene)	Fusarium (F.) graminearum, F. culmorum	Immunotoxic, neuroendocrine, emetogenic, gastrointestinal tract lesions
Nivalenol (type B trichothecene)	Fusarium spp.	Immunotoxic, hematotoxic
Fumonisin B1/B2	F. verticillioides, F. proliferatum; A. niger	Toxic to liver and kidneys, carcinogenic
Ochratoxin A	A. ochraceus, A. niger; P. chrysogenum	Nephrotoxicity, genotoxic, carcinogenic
Patulin	Aspergillus spp.; Penicillium spp.; Byssochlamys spp.	Lesions in the gastrointestinal tract, genotoxic
Satratoxin G and H	Stachybotrys chartarum	Effects on the lungs
T-2/HT-2 toxin (type A trichothecene)	F. sporotrichioides, F. langsethiae	Hematotoxic, immunotoxic, cytotoxic
Zearalenone	F. graminearum, F. equiseti, F. culmorum, F. verticillioides	Reprotoxic, estrogenic activity

62 samples mainly roquefortine C (maximum: 4 ng/m^3), chaetoglobosin A (maximum: 3.4 ng/m^3), and sterigmatocystin (maximum: 1.8 ng/m^3), and less frequently also roridin E, OTA, AFB1, and AFB2. In conclusion, the mycotoxins found corresponded to the molds identified in the indoor environments. In a German study, 0.25 ng/m^3 of satratoxin G was detected in the air of a building with water damage and an infestation of *Stachybotrys chartarum* (Gottschalk et al. 2008).

6.3.3 Occurrence in Dust

The results published to date in dust are summarized in Table 6.4. Essentially, the concentrations were in the range of the determination limits of the analytical methods or slightly above. No clear differences could be found in the content of control buildings and those with water damage or visible mold infestation. Only Charpin-Kadouch et al. (2006) reported statistically significantly higher trichothecene concentrations in 15 flooded buildings contaminated with *Stachybotrys chartarum* or *Chaetomium compared to* 9 control buildings. Several studies from the 1990s and two more recent studies are available from Germany. For example, in

Table 6.3 Concentrations of mycotoxins in the air of living spaces in ng/m³ (modified after Fromme et al. 2016)

Reference	N	N > LOQ	Min-Max	County, comments
Aflatoxin B1				
Polizzi et al. (2009)	99 99	5 4	0.002–0.15 0.0003–0.021 AFB2	Belgium; 7 buildings with moisture damage
Satratoxin G				
Gottschalk et al. (2008)	1		0.25	Germany; 1 building with moisture damage
Satratoxin H				
Gottschalk et al. (2008)	1		0.43	Germany; 1 building with moisture damage
Sterigmatocystin				
Polizzi et al. (2009)	99	3	0.003–1.77	Belgium; 7 buildings with moisture damage
Roquefortine C				
Polizzi et al. (2009)	99	1	0.009–4.0	Belgium; 7 buildings with moisture damage
Roridin E				
Polizzi et al. (2009)	99	1	0.003–0.082	Belgium; 7 buildings with moisture damage
Macrocylic trichothecenes				
Brasel et al. (2005)	8	8	0.003–1.4	USA, 8 buildings contaminated with Stachybotrys
Charpin-Kadouch et al. (2006)	15 9		0.62 ppb* 0.29 ppb*	France; apartments with mold France; control apartments
Ochratoxin A				
Polizzi et al. (2009)	99	1	0.012–0.228	Belgium; 7 buildings with moisture damage
Chaetoglobosin A				
Polizzi et al. (2009)	99	3	0.007–3.42	Belgium; 7 buildings with moisture damage

[a] ELISA for the quantitative determination of trichothecenes, including roridin A, E, H, and L-2, satrotoxin G and H, isosatratoxin F, verrucarin A and I, and verrucarol; *LOQ* limit of quantification

200 dust samples analyzed for a total of 7 mycotoxins, only DON could be found in 18 samples at levels <176–8100 ng/g (Ertl et al. 2007). Mosch et al. (2015) analyzed dust samples from 8 Bavarian buildings without mold infestation for 8 mycotoxins and could not detect mycotoxins above the limit of quantification in any of them (DON <13 ng/g and OTA <3 ng/g). In winter and summer of 2016 and 2017, airborne fungi and house dust were collected in 2 dwellings of the village, which had been flooded, and one house in a control village in Croatia (Jakšić et al. 2021). A total of 75 metabolites were detected in dust with higher levels in winter and in the flooded village. Highest median concentrations in winter were 10,885 ng/g for cyclopenol, 2264 ng/g for viridicatol, and 1739 ng/g for meleagrin as Penicillum

Table 6.4 Concentrations of mycotoxins in dust in ng/g (modified and extended after Fromme et al. 2016)

References	N	*N* > LOQ	Range	Comments
Aflatoxin				
Ertl et al. (2007)	200	0	<8.6	Germany; control building
Mosch et al. (2015)	8		<1.3	Germany; control building
Beauvericin				
Täubel et al. (2011)	4	4	0.05–1.4	Finland; 9 dwellings with moisture damage and 2 public buildings
Lindemann et al. (2022)	26	9	<4.6–186	Germany; 6 with present mold infestion
Satratoxin G				
Bloom et al. (2009b)	7	0	<LOQ	USA; 5 buildings with moisture damage
Bloom et al. (2009a)	18	0	<LOQ	Sweden; 8 buildings with moisture damage
Mosch et al. (2015)	8		<13	Germany; control building
Satratoxin H				
Bloom et al. (2009b)	7	0	<LOQ	USA; 5 buildings with moisture damage
Bloom et al. (2009a)	18	0	<LOQ	Sweden; 8 buildings with moisture damage
Mosch et al. (2015)	8		<7.0	Germany; control building
Sterigmatocystin				
Bloom et al. (2007)	8	1	17	Sweden; 4 buildings with moisture damage
Bloom et al. (2009a)	18	0	<LOQ	Sweden; 8 buildings with moisture damage
Bloom et al. (2009b)	7	2	16 and 28	USA; 5 buildings with moisture damage
Engelhart et al. (2002a)	11	2	2.0 and 4.0	Germany; 8 buildings with moisture damage and mold
Ertl et al. (2007)	200	4	<3.2- > 108	Germany; control building
Mosch et al. (2015)	8	0	<0.7	Germany; control building
Jakšić et al. (2021)	6		3.4[a], 3.2[b]	Croatia, 2016/17; control buildings
Lindemann et al. (2022)	26	3	<11–3145	Germany; 6 with present mold infestion
Verrucarol *				
Bloom et al. (2009a)	18	2	0.6 and 1.7	Sweden; 8 buildings with moisture damage
Bloom et al. (2007)	8	2	19 and 43	Sweden; 4 buildings with moisture damage
Saito et al. (2016)	21	8	0.6–36.7	USA, 2006/07; 2 buildings with mold
Trichodermol				
Bloom et al. (2007)	8	2	2.4 and 3.4	Sweden; 4 buildings with moisture damage

(continued)

Table 6.4 (continued)

References	N	*N* > LOQ	Range	Comments
Bloom et al. (2009b)	7	0	<LOQ	USA; 5 buildings with moisture damage
Deoxynivalenol (DON)				
Ertl et al. (2007)	200	18	<176–8100	Germany; control building
Mosch et al. (2015)	8	0	<13	Germany; control building
Jakšić et al. (2021)	6		64[a]; 147[b]	Croatia, 2016/17; control buildings
Gliotoxin				
Ertl et al. (2007)	200	1	<15–110	Germany; control building
Mosch et al. (2015)	8	0	<13	Germany; control building
Trichothecene				
Charpin-Kadouch et al. (2006)	15 9		21.9 ppb 0.31 ppb	France; building with mold France; control building
Ochratoxin A (OTA)				
Kasel et al. (1999)	47	13	<0.5–5.1	Germany; control building
Richard et al. (1999)	1	0	<0.5	USA
Ertl et al. (2007)	200	0	<43	Germany; control building
Mosch et al. (2015)	8	0	<3.0	Germany; control building
Chaetoglobosins				
Täubel et al. (2011)	4	1	3100	Finland; 9 dwellings with moisture damage and 2 public buildings.
Emodin				
Täubel et al. (2011)	4	4	0.24–84	Finland; 9 dwellings with moisture damage and 2 public buildings.
Enniatin A				
Lindemann et al. (2022)	26	2	<35–79.5	Germany; 6 with present mold infestion
Enniatin B				
Täubel et al. (2011)	4	4	0.63–11	Finland; 9 dwellings with moisture damage and 2 public buildings.
Jakšić et al. (2021)	6		1.3[a]; 1.3[b]	Croatia, 2016/17; control buildings
Lindemann et al. (2022)	26	13	<0.3–564	Germany; 6 with present mold infestion
Enniatin B_1				
Lindemann et al. (2022)	26	9	<1.7–42	Germany; 6 with present mold infestion
Citrinin				
Ertl et al. (2007)	200	0	<8.6	Germany; control building
Jakšić et al. (2021)	6		<LOQ[a]; <LOQ[b]	Croatia, 2016/17; control buildings

*Hydrolysis product of macrocyclic trichothecenes; LOQ: limit of quantification

[a] summer

[b] winter

metabolites; 2523 ng/g for hydroxysidonic acid, 2162 ng/g for notoamide derivative, and 1414 ng/g for tryprostatin B, as Aspergillus metabolites. Stachybotrys metabolite stachybotrylactam were found with a median of 435 ng/g. Lindemann et al. (2022) applied a mass spectrometric detection using triple quadrupole after ultrahigh-performance liquid chromatographic (UHPLC-TQMS) method for the analysis of 21 dust samples and were able to quantify especially beauvericin, enniatin B, enniatin B_1, and sterigmatocystin.

6.4 β-D-glucan

The 1,3-β-glucans are a group of high-molecular-weight β-D-glucose polysaccharides. They occur as components in the cell walls of fungi, plants, and some bacteria, among others. The chemical structure is shown in Fig. 6.3. Analytically, glucan is very often measured with the so-called glucan-specific Limulus amoebocyte lysate test (LAL test) and sometimes also with an immunoassay. In addition to measurement in dust and in air, wipe samples of indoor surfaces are often examined (Holst et al. 2020). However, there is as yet no harmonized procedure for any of the aforementioned measurement methods. The choice of sampling technique and sample preparation can also have a considerable influence on the final result and has not yet been standardized.

6.4.1 Health Aspects

In various reviews, the toxicological and epidemiological knowledge is presented in summary form (Rylander 1999; Douwes 2005; Parker et al. 2021). Animal studies revealed inflammatory reactions in the respiratory tract at high doses. Experimental studies in human subjects resulted in inflammatory effects in the upper respiratory tract and an increase in respiratory symptoms when exposed to 1,3-β-D-glucan. Epidemiologic studies of the general population and workplaces found symptom increases (e.g., cough) and increased airway responsiveness. Overall, however, the results are thought to be inconsistent (Douwes 2005). In addition, there is evidence suggesting some protective effect of β-D-glucan in the development of atopy in

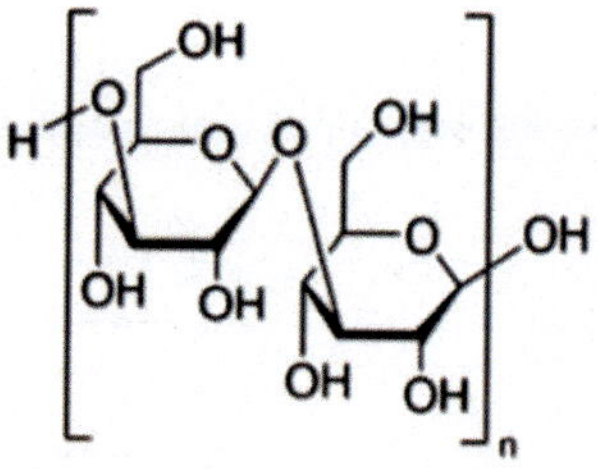

Fig. 6.3 Chemical structure of β-1,3-Glucan

childhood. Because of methodological limitations of the database Parker et al. (2021) derived a health-based occupational exposure limit of 150 ng/m^3 for workers using a LOEL (Lowest Observed Effect Level) based on immune system modulation in guinea pigs.

6.4.2 Occurrence in the Air

So far, only limited measurement results from indoor air in residences are available. In Sweden, all 1,3-β-D-glucan concentrations in 405 buildings were below the limit of quantification of <0.1 ng/m^3, while 20 measurements in 2 schools showed levels of 5.2 ng/m^3 (Rylander 1999).

In the USA, 2 schools were surveyed 5 times per year and geometric mean values of 0.2 ng/m^3 were found in $PM_{2.5}$ in carpeted rooms and 0.5 ng/m^3 in rooms with smooth floor coverings (Foarde and Berry 2004). In homes with water damage or mold infestation, concentrations of 0.06–0.19 ng/m^3 in summer and 0.34–0.52 ng/m^3 in winter have been reported in the USA (Reponen et al. 2007), while in 184 homes with and without possible sources the geometric mean was 1.5 ng/m^3 (Reponen et al. 2010). In a study of two locations in the USA, mean concentrations in $PM_{2.5}$ of 0.006 ng/m^3 (urban) and 0.033 ng/m^3 (rural) were measured in 18 and 3 indoor locations, respectively, in 2001 and 2007 (Menetrez et al. 2009). Again, in 2005 Lee et al. (2006) found geometric mean level of 0.92 ng/m^3 (0.31–9.35 ng/m^3) in 5 homes that had no visible mold growth in Cincinnati, USA. Similarly, Singh et al. (2011) observed mean levels of 1.91 ng/m^3 in airborne dust (PM_1) in 184 indoor residential spaces in the USA, with the highest levels observed in the >1.8 μm (approximately 50%) and < 1.0 μm (approximately 28%) fractions.

6.4.3 Occurrence in Dust

Measurement results are available from very extensive epidemiological studies from Europe. For example, in 1997/1998, a geometric mean (median) of 1711 ng/g 1,3-β-D-glucan (1830 ng/g) was measured in 606 apartments in Erfurt and Hamburg (Gehring et al. 2001). A significant increase in levels was reported in the presence of carpeting, when dogs were present, when more than one person lived permanently in the home, and when less dust was vacuumed. The investigation of 1065 children's rooms in Germany, the Netherlands, and Sweden found median concentrations in dust from 2000 to 2800 ng/g in each country (Giovannangelo et al. 2007). In this study, a significant association with floor covering was observed, with significantly higher levels in carpeted floors. In addition, a geometric mean (median) of 710 ng/g (1300 ng/g) was observed in 1999/2000 when 292 Danish children's rooms were studied (Holst et al. 2015). Concentrations of 1,3-β-D-glucan in this case were positively associated with the presence of carpet in the room, but not with pet ownership, season, smoking behavior, location, and type of building.

In 290 Canadian homes of 4 cities large regional variations of mold glucans were observed (geometric mean: 620000–2,825,000 ng/g) as well as very high concentration, maybe due to the analytical assay used (Sun et al. 2022). Elevated indoor relative humidity and the use of dehumidifiers were associated with mold glucan loadings.

6.5 Viruses

The importance of indoor environments for the spread and transition of viruses, especially those with pandemic potential, has been highlighted much more in recent years. The focus has been on influenza virus, SARS-CoV-1 (Severe Acute Respiratory Syndrome Coronavirus 1), MERS-CoV (Middle East Respiratory Syndrome Coronavirus), and especially SARS-CoV-2 (Severe Acute Respiratory Syndrome Coronavirus 2). Similarly, in the 2020 COVID-19 pandemic caused by SARS-CoV-2, it became clear that the majority of illnesses were mediated by indoor exposure (Qian et al. 2021). No RNA of SARS-CoV-2 could be determined in the outdoor air of two Italian regions heavily affected by COVID-19 (Chirizzi et al. 2021). The viruses considered here have a size between approximately 20 and 300 nm. As a common analytical method in indoor air, real-time polymerase chain reaction (RT-PCR) is often used, e.g., for the determination of viral RNA (Cox et al. 2020). In addition, various methods are basically available for collection on culture media to determine replicable viruses, such as the plaque assay or the endpoint dilution test. Borges et al. (2021) and Silva et al. (2022) gave an overview of the different sampling devices.

6.5.1 Indoor Transmission Conditions

In principle, the following transmission routes must be taken into account indoors, which are shown schematically in Fig. 6.4. In addition to droplet infection, which occurs in the context of close contact with an infected person, exposure to smaller aerosol particles is increasingly being discussed. Secondarily, indirect contact with a surface contaminated with highly virus-laden droplets from an infected person is also a possibility for transmission.

The main starting point of respiratory viruses is considered to be the emission of virus-containing fluid particles, which can be produced during breathing, but especially during talking, coughing, and sneezing (Wei and Li 2016; Morawska et al. 2020; Schijven et al. 2021). Humans always produce a spectrum of particles of different sizes during the aforementioned activities. Depending on the particle diameter, a distinction is made between droplets larger than 5 μm and potentially

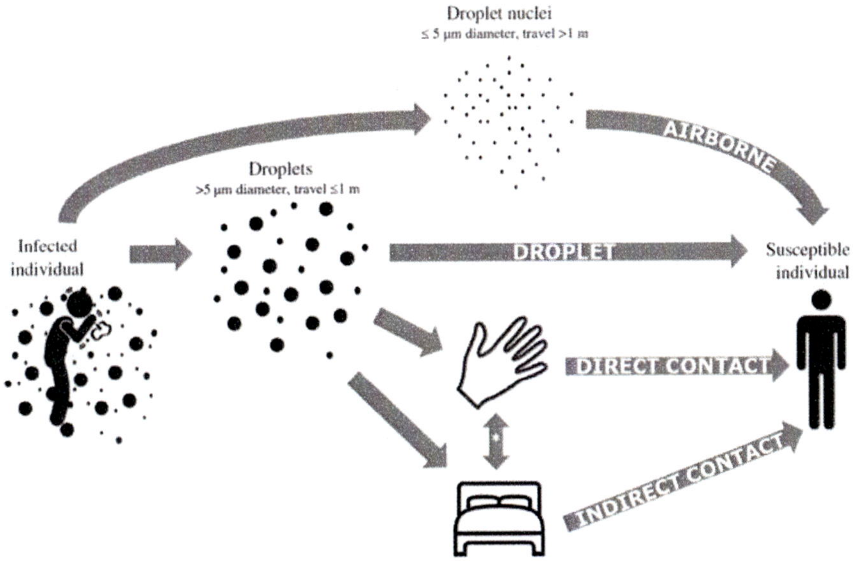

Fig. 6.4 Possible transmission routes for indoor viruses (after Otter et al. 2016)

infectious aerosol particles (so-called droplet nuclei) smaller than 5 µm. While larger droplets quickly sediment onto surfaces such as the ground, small particles can remain airborne for a very long time and may thus pose an exposure risk over a longer period of time (Vuorinen et al. 2020). For example, the time for a particle to sink one meter due to gravity is approximately 3.3 seconds, 5.6 minutes, 9.3 hours, or 39 days for particles with diameters of 100 µm, 10 µm, 1 µm, or 0.1 µm (Tsuda et al. 2013). Exhaled droplets of a size between 60 and 100 µm can be transported approximately 1.5–2 meters in room air, under approximately 1 meter during normal breathing and over approximately 2 meters during coughing (Xie et al. 2007). When there is turbulence in the room, smaller particles in particular are also distributed further (Wei and Li 2015; Liu et al. 2020a). Some recent reviews summarized the knowledge of virus transmission in indoor spaces (Klompas et al. 2021; Ribaric et al. 2022; Dinoi et al. 2022).

During normal breathing, droplets/particles <1 µm dominate, during talking and coughing >1 µm and during loud talking and strong coughing >100 µm (Wei and Li 2016). One explanation for the size differences can be attributed to the different site of droplet formation in the upper or in the deeper regions of the respiratory tract. In an experimental study of 48 healthy subjects, it was shown that particle exhalation is proportionally dependent on volume and can range from 1 to 50 particles (>0.5 µm) per second (Asadi et al. 2019). In this study, it was also observed that some participants emitted an extraordinary amount of particles. Laughter, shouting, and singing also lead to increased formation of aerosols with small particles (Stadnytskyi et al. 2020). In the context of influenza virus infections, the volume of aerosol produced was found to be significantly greater during the phase with symptoms of illness than in the asymptomatic state (Lindsley et al. 2012), but differed only insignificantly between normal breathing and coughing (Lindsley et al. 2016). In SARS-CoV-2, viral spread is possible even in the asymptomatic state, a time window at which there is simultaneously a particularly high viral load in the nasopharynx (Rothe et al. 2020; WHO 2020). Moreover, patients with severe or critical COVID-19 may shed infectious virus for longer periods of time (median: 8 days) compared to what has been reported for in patients with mild COVID-19 (van Kampen et al. 2021).

An important estimation is the secondary attack rate (SAR), defined as the probability of a contact sufficient to infect a subject from an infectious household member to another given household member. In a meta-analysis of 54 relevant studies with 77,758 participants, Madewell et al. (2020) concluded that the SAR in the context of transmission in homes was higher in symptomatic individuals than in asymptomatic individuals, higher in adults than in children, and higher in life partners than in other family members. Overall, estimated mean household secondary attack rate was 7.5% for SARS-CoV-1 and 4.7% for MERS-CoV, both lower than the household secondary attack rate of 16.6% for SARS-CoV-2. A dataset of 7771 contacts of 4921 index cases (positive SARS-CoV-2 PCR-test) indicates that vaccination confers protection against onward transmission of the virus from vaccinated index cases compared to unvaccinated subjects, albeit somewhat less for the Delta than for the Alpha variant (de Gier et al. 2021).

6.5.2 Transmission Possibility in Public and Private Indoor Spaces

In a recent review of 77 scientific papers Dinoi et al. (2022) concluded that average positivity rate (i.e., ratio between positive and total samples) was larger in hospital compared to outdoors and public indoor sites. Moreover, they described that 5 out of 8 studies found no airborne SARS-CoV-2 RNA traces in outdoor air and that airborne transmission is generally unlikely with the exception of crowded areas or sites with close contact to a source.

In an experimental study using a nebulizer, SARS-CoV-1 and SARS-CoV-2 in aerosol were found to have comparable half-lives for viable viruses between 1.1 and 1.2 hours (van Doremalen et al. 2020). In another study in which SARS-CoV-2 was nebulized in a test chamber, the half-life was found to be 2.4 hours (Fears et al. 2020). The authors of this paper report that viable viruses, albeit very few, could still be detected after 16 hours.

Various COVID outbreak events also suggest transmission via aerosols in principle. This relates, for example, to indoor choral singing (Hamner et al. 2020), participation in carnival events (Streek et al. 2020), group gym attendance (Jang et al. 2020), club attendance (Muller et al. 2020), and music events (Furuse et al. 2020). For example, after choral singing (Skagit Valley Chorale) for 2.5 hours, 53 of

61 participants developed COVID disease, and 2 individuals died (Hamner et al. 2020). Using different schlieren imaging methods the spread of breathing air was investigated when playing several wind instruments and singing (Becher et al. 2021). The results showed that the ejected airflow from the examined instruments and singers did not exceed a spreading range of 1.2 m into the room.

A particular problem is the very high interindividual variation in the exhalation of viruses, which can result in one individual (so-called super-spreaders or super-spreading events) secondarily infecting many other individuals. For example, studies in the MERS-CoV pandemic in Korea in 2015 already showed that approximately 80% of the 186 confirmed cases could be traced to only 5 individuals (Kim et al. 2018). This was also described for the SARS-CoV-2 pandemic in 2020 (Yang et al. 2020). Whether a super-spreading event occurs appears to depend on the physiological circumstances of the source individual (e.g., their immune status), environmental conditions (e.g., high numbers of people in confined spaces with low ventilation), and the characteristics of the infectious agent (Stein 2011). A better understanding of these events and anticipation of potential transmission could have a substantial impact on the effectiveness of interventions.

Using two different methods for a school scenario with masked attendants and various ventilation condition Foster and Kinzel (2021) found that mask mandates well-designed HVAC systems and the combination of exposure time with number of occupants is of increased importance compared to physical distancing alone.

The majority of studies on the presence of SARS-CoV-2 in indoor air samples detected viral RNA using different, not standardized methods, namely filter-based samplers, impingers, impactors, cyclones, water-based condensation, and passive sampling (Silva et al. 2022). For example, studies in China, England, Italy, Singapore, Iran, Brazil, and the USA have used RT-PCR to examine indoor air from hospitals where patients were treated with COVID-19. While low concentrations of viral ribonucleic acid (RNA) were found in some of the air samples (Chia et al. 2020; Guo et al. 2020; Santarpia et al. 2020; Zhou et al. 2021; Liu et al. 2020b; Ma et al. 2020; Razzini et al. 2020; Passos et al. 2021), no RNA was detectable in indoor air in other studies (Ding et al. 2021; Faridi et al. 2020; Cheng et al. 2020; Ong et al. 2020; Wu et al. 2020; Vosoughi et al. 2021). Influenza virus and respiratory syncytial virus (RSV) RNA have also been determined in the presence of infected individuals in hospital room air (Lindsley et al. 2010; Shiu et al. 2020). Döhla et al. (2020) were unable to find viral RNA in the indoor air of 21 homes where 43 adults (60% SARS-CoV-2 positive) and 9 children (44% SARS-CoV-2 positive) were sequestered during quarantine.

Although the presence of RNA should be interpreted as the probable presence of a viable virus, information on infectivity of the virus particles are necessary and Silva et al. (2022) performed a systematic review by including 76 qualitative studies. Fourteen assessed infectivity, but only five of them detected viable viruses using impactor, water-based condensation, and cyclone collection methods.

In a hospital room (approximately 25 m^2) occupied by two symptomatic COVID-19 patients, replicable viruses could be detected at a distance of 2 m and 4.8 m (Lednicky et al. 2020). In this case, an endpoint dilution test was performed in addition to RT-PCR and a $TCID_{50}$ (Tissue Culture Infectious Dose) of 6–74 units/l was measured. The $TCID_{50}$ is the pathogen dose at which infection occurs in 50% of tissue cultures. The authors consider that this study gives indication that viable viruses are still present in the aerosol at some distance to the patients. Kitagawa et al. (2022) supported these findings by detecting viable virus in the vicinity of 18 patients with an early infection. On the contrary, Mallach et al. (2021) found in 9.1% virus RNA but viable and replicative virus was not recovered in any samples (n: 138) collected 2 to 3 m from the patient's head in 99 different rooms with COVID-19 positive patients. Again, air samples collected from airborne infection isolation rooms and a community isolation facility housing COVID-19 patients were frequently positive for SARS-CoV-2 nucleic acid with PCR, though PCR-positive air samples were negative on viral culture (Ong et al. 2021). Aerosol sampling was conducted around six hospitalized patients in five rooms in two wards in China in 2020 and SARS-CoV-2 RNA was detected in all samples, and three out of 18 samples confidently indicated the presence of viral replication (Santarpia et al. 2022). Moreover, only three out of 146 PCR-positive samples were observed in one or two meter distance from COVID-19 patients at six acute care hospitals in Canada; none yielded viable virus (Kotwa et al. 2021). Samples collected from the bedroom of a self-isolating university student with mild COVID-19 symptoms and a confirmed presence of Delta variant SARS-CoV-2 strain, viable viruses were found with one sampling method (BioSpot-VIVAS), but not using another sampler from NIOSH (Vass et al. 2022).

Overall, the health significance of these findings cannot yet be conclusively assessed. Linde et al. (2022) detected SARS-CoV-2 RNA in rooms of 10 residents with oropharyngeal swabs in 50.5% of the samples, while in the rooms of residents with negative oropharyngeal swabs only in 3.6% of the samples.

6.5.3 *Possibility of Transmission in Aircraft Cabins*

The high temporal and spatial mobility, which was essentially achieved by the development of flying, always poses a risk for the spread of infectious diseases. Before the corona pandemic around 900 million passengers annually are carried on flights within the European Union (EU) alone.

Regarding potential risks of airline passengers themselves, Leitmeyer and Adlhoch (2016) summarized 14 scientific papers describing tracking in the context of influenza infection on airplanes. Of the 2165 individuals tracked, 163 (7.5%) were identified as secondary cases. A total of 42% of these individuals were seated no more than two rows away from the index cases. Thus, in principle, aerogenic transmission of viruses appears to be possible during flight. Modeling of the importance of airborne transmission pathways concludes for influenza A H1N1 that both the airborne pathway and direct contact are important (Lei et al. 2018).

In a review, the European Centre for Disease Prevention and Control summarized publications that described MERS-CoV infections after previous air travel (ECDC

2020). Overall, according to the ECDC, many studies show methodological weaknesses. Only two events could be truly well substantiated afterward, with follow-up revealing no secondary cases. Therefore, the likelihood of airborne transmission is considered low by ECDC for this virus. Olsen et al. (2003) also describe a flight in which 119 people were exposed to a person infected with SARS-CoV-1. Virus was found in 16 air travelers during follow-up, and two were diagnosed as infected. Mangili and Gendreau (2005) published a comprehensive compilation of what is known about the importance of communicable diseases on airplanes.

The scientific literature related to the 2019 corona pandemic is still limited. A recent systematic review including 18 studies on in-flight SARS-CoV-2 transmission (130 unique flights) concluded that the virus can be transmitted during aircraft travel, but published data do not permit any conclusive assessment of likelihood and extent (Rosca et al. 2021). Overall, secondary attack rate among studies following up >80% of passengers and crew varied between 0 and 8.2%.

During a 15-hour flight—without hygiene measures—2 people had typical COVID-19 symptoms and also tested positive after the flight (Schwartz et al. 2020). Nearby passengers and all other individuals (350 individuals in total) were then followed up. Only 6 individuals developed respiratory symptoms, but all of them tested negative for SARS-CoV-2. During follow-up of a COVID-19 outbreak following a flight in China, the incidence rate (proportion of a population contracting the infection in an epidemic) was 0.44% for persons who had never been to Wuhan but 14% of the persons were from Wuhan. Also, following a flight to New Zealand, 4 of a total of 86 airline passengers tested positive for SARS-CoV-2 (Swadi et al. 2021). Genome sequencing traced the infections to 2 index cases from the same aircraft seated within 2 rows of seats. In a 2-hour domestic flight in Japan, 14 confirmed and 6 probable cases were identified among passengers and flight attendants with a secondary attack rate of 9.7% (Toyokawa et al. 2022). Risk factors for infection included not using a face mask, partial face mask use, and being seated within two rows from the index patient. Using experimentally derived aerosol dispersion data Wang et al. (2021) reported that for a 2-hour flight the average infection probability varies from 0.1% to 2.5% and for a 12-hour flight from 0.8% to 10.8%. If all passengers wear face masks throughout the 12-hour flight, the average infection probability can be reduced by around 73%. In a Chinese study 161 laboratory-confirmed cases from 4492 passengers with suspected COVID-19 infection, and by epidemiological investigation inferred whether these passengers might have been infected during air travel (Zhang et al. 2021). Their findings suggest that the universal use of face masks on the flight, together with the airplane's ventilation system, likely prevented all secondary cases of COVID-19. This was supported by a computational fluid dynamic (CFD) simulation of a Boeing 737–600 cabin with a single infector (a passenger) (Horstman and Rahai 2021).

An IATA summary reported that 18 major airlines reported only 4 suspect transmissions for the period between January and March 2020 (IATA 2020). In 3 cases, the transfer occurred from passenger to flight attendants and in another case between 2 pilots. A further survey of 70 airlines revealed no evidence of recorded passenger-to-passenger transmission. A reduction in possible infection risks during the flight is seen in increased ventilation with a higher proportion of outside air and a flow direction from the cabin roof to the cabin floor, filtration of the air by means of HEPA filters and the wearing of a mouth-nose covering (IATA 2020). Also, the seat back has some barrier function to the person in front. In particular, restricting the movement of people in the cabin is seen as a useful measure. In addition, regular disinfection of cabin surfaces is recommended. IATA published in April 2022 a brief summary of the current medical evidence of the risks related to air travel during COVID-19 pandemic (IATA 2022). EASA/ECDC published an operational guideline to ensure the health and safety of passengers and the aviation personnel by non-pharmaceutical interventions and other measures customized for each stage of the journey as well as risk-based recommendations to minimize the risk of SARS-CoV-2 transmission (EASA/ECDC 2021).

6.5.4 Transmission Through Contaminated Room Surfaces

In addition to the airborne pathway, contaminated surfaces can be considered as a source of infection. For example, Otter et al. (2016) showed in a review paper that under experimental conditions influenza viruses, SARS-CoV-1 and MERS-CoV can remain on dry surfaces for some time, even if in principle an exponential decrease was observed. For example, after experimental nebulization of SARS-CoV-1 and SARS-CoV-2, van Doremalen et al. (2020) described median half-lives of up to 6.8 hours on plastic surfaces. However, the overall virus concentration was low. Various studies in hospital wards with COVID-19 patients showed principally contaminations of various surfaces (Ding et al. 2021; Wu et al. 2020; Zhou et al. 2021; Ma et al. 2020; Razzini et al. 2020). Nevertheless, due to the generally low viral load it remains unclear whether infection is really possible via this route.

6.5.5 Measures to Protect Against Indoor Transmission

The main measure for reducing possible virus transmissions indoors is seen in improved room ventilation (Jiang et al. 2009; Duchaine 2016; Morawska et al. 2020; Mousavi et al. 2021). For rooms with technical ventilation systems, this means increasing the airflow if necessary, but in any case minimizing the amount of recirculated air. For naturally ventilated rooms, regular, sufficient ventilation is an indispensable and effective measure. In addition, spacing people apart, keeping the number of people in the room as low as possible, and wearing masks indoors are important protective measures. In addition to the general hygiene rules, the disinfection of hands and decontamination of room surfaces are also important in the medical sector.

For other measures discussed, such as the use of portable air purifiers, there are so far only studies on the reduction of particles in indoor air, but no meaningful scientific studies on the effectiveness in the context of the transmission of respiratory viruses (Hammond et al. 2021). They concluded in their systematic review that there is an important absence of evidence regarding the effectiveness of a potentially cost-efficient intervention for indoor transmission of respiratory infections, including SARS-CoV-2. A Dutch research tested the removal of aerosols simulated by air-filled soap bubbles in front of a maniken in a 68 m^3 test room under five different airflow rate settings of a portable HEPA filter and different ventilations (Bluyssen et al. 2021). They concluded that the mobile HEPA filter system performed better as compared to the "No ventilation" regime, for some settings even better than all the tested mixing ventilation regimes, while the "Open door and windows" regime (cross natural ventilation) showed the least counted particles. Again, Uhde et al. (2022) tested the decay of a nebulized MS2 bacteriophage solution, acting as a surrogate of viruses like SARS-CoV-2, in a typical classroom (600 m^3). In this case the nebulizer acts as a point source and different ventilation scenarios were tested, with and without a commercial air cleaner (pre-filter, activated charcoal, HEPA filter). All three interventions (A: one window is tilted; B: air cleaners are switched on and the windows are closed; C: the window is open and the air cleaners are switched on) showed on average a 90% reduction of the initial virus concentration within 30 minutes. The German Federal Environment Agency concluded that a central preventive measure is to enhance room and building ventilation in school, and that besides the concentrations of possibly infectious particles, ventilation is neccessary to reduce carbon dioxide concentrations, humidity, and other chemical substances in indoor air as well (Birmili et al. 2021).

In a systematic review of the effectiveness of interventions for SARS-CoV-1, MERS-CoV, and SARS-CoV-2, Chu et al. (2020), considering 44 studies, concluded that the probability of transmission with SARS-CoV-2 was significantly lower at a distance greater than one meter than below one meter (OR: 0.18; 95% CI: 0.09–0.38). Protection also increased with increasing distance.

Wearing a mask was also associated with a significant risk reduction (OR: 0.15; 95% CI: 0.07–0.34), although the association was stronger for N95 masks (equivalent to FFP 3) than for wearing mouth-to-nose coverings (so-called surgical masks). A positive effect was also observed for eye protection in this study. In an experimental study on human head dummies, the tightness of masks to particles loaded with influenza viruses was simulated by PCR in the oral cavity (Lindsley et al. 2012). A tight-fitting surgical mask retained 95% of viral particles and an N95 mask retained over 99%, while a loose-fitting N 95 mask retained only about 65%. The authors emphasize that there are significant differences between different mask types and the passage of air at the lateral edges. In a recent study, surgical masks were shown to decrease droplet emission in individuals who had seasonal coronavirus, influenza, or rhinovirus infections (Leung et al. 2020). An older Cochrane review also concluded that masks can prevent viral distribution, but found no significant difference between surgical masks and N 95 masks (Jefferson et al. 2011). Pan et al. (2021) evaluated the inward (before inhaling) and outward (after exhaling) protection efficiency of 10 different masks and a face shield on a manikin, using NaCl aerosols over the size range of 0.04 μm to >100 μm. At the most penetrating particle size, the vacuum bag, microfiber cloth, and single-layer surgical-type mask had material filtration efficiencies >50%, however, these efficiencies increased rapidly with particle size. Again, a study involving particles generated by coughing human subjects reported an outward protection efficiency of 77% at a distance of 0.3 m from the source for a cloth mask, but only 4% for a face shield (Li et al. 2021).

Face mask or respirator use was assessed among 652 case-participants who had received positive test results for SARS-CoV-2 and 1176 matched control-participants who had received negative test results for SARS-CoV-2 (Andrejko et al. 2022). Always using a face mask or N95/KN95 respirator in indoor public settings was associated with lower adjusted odds of a positive test result compared with never wearing a face mask or respirator in these settings (OR = 0.44; 95% CI = 0.24–0.82). In another study Cowger et al. (2022) used a difference-in-differences analysis to compare the incidence of Covid-19 among students and staff (around 350,000 subjects) in school districts in the greater Boston area during the 2021–2022 school year. Among school districts, the lifting of masking requirements was associated with an additional 44.9 Covid-19 cases per 1000 students and staff during the 15 weeks after the statewide masking policy was rescinded.

6.6 Inhalation Allergens

The large group of inhalation allergens are proteins or protein compounds with a molecular weight between 17 and 30 kDa. For example, the allergen of the cat Fel d 1 (Felis domesticus) has a diameter of 1–5 μm, can easily bind to dust particles, and has the property of floating in the air for a very long time. Two allergens are produced by house dust mites that Der p 1 (Dermatophagoides pteronyssinus) from the European house dust mite and Der f 1 (Dermatophagoides farinae 1) from the American house dust mite. These are cysteine proteases. For these allergens, most information on indoor exposure is available. In addition, allergens from dogs (Can f 1, Canis familiaris), mice (Mus m 1, Mus musculus), rats (Rat n 1, Rattus norvegicus), and cockroaches (Bla g 2, Blattella germanica), for example, can still be detected more frequently indoors.

6.6.1 Health Aspects

Aeroallergens are substances that, after inhalation, are able to trigger hypersensitivity reactions in the immune system and produce sensitization (Farokhi et al. 2018; Akar-Ghibril and Phipatanakul 2020). With further or renewed contact, allergic disease is possible or specific symptoms may be exacerbated (Naclerio et al. 2020). One allergy-based disease may be bronchial asthma. It is a chronic inflammatory disease of the airways characterized by reversible airway obstruction of variable severity and increased airway irritability (hyperreactivity). In addition to genetic factors, the level and duration of exposure may also play a role in pathogenesis.

For example, in the patient-based Global Asthma and Allergy European Network (GA2LEN) study, 28% of participants in 14 European countries had sensitization to cats and 27% to dogs (Heinzerling et al. 2009). The possible relationship between exposure to allergens from fur-bearing animals, particularly cats and dogs, and sensitization or allergic symptoms has been widely studied but is still considered inconsistent (Zahradnik and Raulf 2017). One explanation could lie in the limited exposure data available.

6.6.2 *Occurrence of Cat Allergens in the Air of Residences*

The most important study results in apartments are summarized in Table 6.5. The studies, which often date from the 1990s, show a wide variation in the results. In these studies, they were often below or in the range of the limit of quantification, with a median between <0.02 and 0.40 ng/m^3. In this regard, very low concentrations are described by Munir et al. (2003) using an electrostatic collection method in 9 apartments without a cat (median: 0.001 ng/m^3; range: 0.00007–0.003 ng/m^3), while a median of 3.3 ng/m^3 (0.005–9375 ng/m^3) was found in 9 apartments with a cat. In this study, there was a significant correlation of air content to house dust samples measured in parallel. In 567 French homes, a median of <0.18 ng/m^3 (95th percentile: 2.7 ng/m^3) was found in total dust in 2003–2005, with only 25% of rooms having values above the limit of quantification (Kirchner et al. 2008).

6.6.3 *Occurrence of Cat Allergens in the Air of Schools*

Indoor air measurements of cat allergens in schools have rarely been performed. As summarized in Table 6.5, data are available from Germany, England, Sweden, and the USA.

Fromme et al. (2013) report on a study of 14 classrooms in Munich elementary schools, each sampled for 20 days, and one classroom in Berlin, which was examined on 83 days over the course of the year. Here, in Munich, Fel d 1 in PM_{10} was determined to be above the limit of quantification in 98% of the samples, with a median and 95th percentile of 0.21 ng/m^3 and 0.58 ng/m^3, respectively. Significantly higher levels were observed with increasing number of room occupants and in winter compared to summer. No statistically significant difference was found between rooms with and without a cozy corner.

In England, when rooms of 9 school children were examined, Fel d 1 values of less than 0.5 ng/m^3 were found in 3 samples and 0.79 ng/m^3 in one (Custovic et al. 1996). Almqvist et al. (1999) collected person-related dust from 5 children in 12 classrooms. They reported concentrations of 2.94 ng/m^3 in schools with a high

Table 6.5 Allergens in the air of residences and schools in ng/m^3

References	N	Cat allergen Median (Min-Max)	Mite allergens Median (Min-Max)	Country, year
Residences				
Sakaguchi et al. (1993)	13	<0.02 (<0.02–23)	0.05[a] (<0.02–0.09)*	Japan
Oliver et al. (1995)	45	0.4 (<LOQ-14.7)[b]	<LOQ (<LOQ-2.5)[b]; Der p 1	UK
Bollinger et al. (1996)	40	<2.0 (<2.0–88.5)	–	USA
Custovic et al. (1998)	75	<0.1 (<0.1–1.78)	–	UK
Custovic et al. (1999)	62	<0.16 (<0.16–1.8)	<0.8	UK
Almqvist et al. (1999)	26	0.15 (0.04–0.27)	–	Sweden
Munir et al. (2003)	9	0.001 (0.00007–0.003)	–	Sweden
Kirchner et al. (2008)	567	<0.18 (<0.18–5.8)	–	France, 2003–05
Schools				
Custovic et al. (1996)	4	< 0.5 (<0.5–0.79)	< 0.5 Der p 1	UK
Almqvist et al. (1999)	5	1.54 (0.6–2.9)	–	Sweden
Munir et al. (2003)	7	0.014 (0.0003–0.0004)	–	Sweden
Karlsson et al. (2004)	14	- (0.8–1.5)	–	Sweden, 2002
Foarde and Berry (2004)	2	63.5 and 96.6[c]	<32 and 74[c] *	USA, 2000
Raja et al. (2010)	6	–	- (0.05–0.175) Der p 1	USA, 2007/08
Permaul et al. (2012)	12	1.82[d] (–)	–	USA, 2008/10
Öder et al. (2012)	6	30 (6.8–94.2)	<LOQ	Germany, 2007/08
Fromme et al. (2013)	14	0.21 (0.02–1.15)	0.04 (0.02–0.36) Der p 1 0.04 (0.02–1.28) Der f 1	Germany, 2012

* Sum of Der p1 and Der f1; *LOQ* limit of quantification
[a] mean value
[b] fifth and 95th percentile
[c] room with carpet and with smooth floor covering
[d] geometric mean value

proportion of children with domestic cats and five-fold lower levels when the proportion of children with domestic cats was low. In another Swedish study, an electrostatic collection technique was used to determine an unusually low Fel d 1 level of 0.00032–0.042 ng/m^3 (Munir et al. 2003). In contrast, Karlsson et al.

(2004) reported levels of 0.8–1.5 ng/m^3 (weekly averages) in 25 classrooms in 2 schools in 2002. Cat allergen levels were unaltered in all classes after intervention (e.g., upholstery and plants were removed, bookshelves were replaced with cupboards and regular cleaning was increased).

In the USA, a geometric mean value of 1.82 ng/m^3 was measured in 197 measurements in 12 schools (Permaul et al. 2012). This working group found no correlation between air and house dust levels. A second American study in 2 schools measured a level of 63.5 ng/m^3 in a carpeted room and 96.6 ng/m^3 in one with smooth flooring (Foarde and Berry 2004).

6.6.4 Occurrence of Cat Allergens in Dust

A compilation of previously published study results in residences is given in Table 6.6. In an epidemiological study in Germany, which included the determination of cat allergens in 777 apartments, the median was 260 ng/g (<15–3,680,000 ng/g) and the 90th percentile was 49,000 ng/g in 1995/1997 (Heinrich et al. 2003). In a more recent study of 501 Austrian homes a median of 760 ng/g (5% > 344,000 ng/g) was reported (Stemeseder et al. 2017).

In the USA, comparable results were observed in studies in the 2000s in residential interiors with geometric mean values between 60 and 1500 ng/g Fel d 1. For example, in the comprehensive study by Gent et al. (2012), Fel d 1 was found in 84% of 1211 house dust samples with levels between the limit of quantification and 1,032,000 ng/g (75th percentile: 5100 ng/g). Additionally, in 290 Canadian homes of 4 cities large regional variations were observed (geometric mean: 477–5179 ng/g) (Sun et al. 2022). Household characteristics including presence of carpet in the living room and presence of pet cats were positively associated.

Study results in schools were summarized in Table 6.7. An investigation of 147 German classrooms in 2004/2005 revealed allergen concentrations of 20–45,160 ng/g (median: 460 ng/g; 95th percentile: 3260 ng/g) (Fromme et al. 2008). The allergen content was higher in rooms with carpeted floors than in those with smooth floor coverings. Compared with the above results, the mean measurement results in studies from the USA were essentially in a similar range of 190 and 1700 ng/g (geometric mean values). Levels were significantly higher in schools than in homes (Permaul et al. 2012).

Various studies conclude that the allergen content on carpeted floors is approximately a factor of 1.7 to 6.7 higher than on smooth floor coverings (Dybendal and Elsayed 1994; Patchett et al. 1997; Tortolero et al. 2002; Foarde and Berry 2004). Damp floor cleaning in schools and increased cleaning of other textile furnishings or their complete elimination can significantly reduce levels of cat allergens (Munir et al. 1993; Smedje and Norback 2001). However, it is important to note here that other problems associated with school cleaning such as higher humidity in the room, bacterial problems, VOC from cleaning centers do not occur.

Table 6.6 Allergens in dust from residences in ng/g

Reference	N	Mite allergen Median (Min-Max)	Cat allergen Median (Min-Max)	Country, year
Heinrich et al. (2003)	777	<10 (<10–146,000) Der p1 <10 (<10–190,500) Der p1	260 (<15–3,680,000)	Germany, 1995–97
Arbes et al. (2004)		–	1430 (10–6,000,000)	USA
Breysse et al. (2005)	96	<LOQ (<LOQ-2143) (Der p1) 12 (< LOQ-17632) (Der f1)	525 (<50 < – 553,939)	USA
Kirchner et al. (2008)	567	1600[a] Der p 1 2200[a] Der f 1	–	France, 2003–05
Sheehan et al. (2009)	38	660[a] (<30–26,000) (Der p 1) 80[a] (<30–27,550) (Der f1)	1310[a] (<12–104,000)	USA
Salo et al. (2018)	3655 3911	226 (13144000) (Der p1) 210 (13–142,000) (Der f1)	344 (4–2,944,000)	USA-wide, 2005/6
Cho et al. (2008)	47	<LOQ (<LOQ-26000)	540 (<LOQ-2700000)	USA, 2004/5
Leung et al. (2011)	44	–	70 (50-110[b])	Hong Kong, 2006
Williams et al. (2011)	141	410[a] (Der p 1) 1691[a] (Der f 1)	2010[a]	USA, 2006/7
Gent et al. (2012)	1211	170 (<LOQ-55800) (Der p1) 660 (<LOQ-483800) (Der f1)	1300 (<LOQ-1032000)	USA, 2007–09
Sun et al. (2022)	290	4.5-253[a] (0.3-1088[c]) (Der p1) 73-2068[a] (16-12682[c]) (Der f1)	477-5179[a] (77-64736[c])	Canada, 2005–10
Permaul et al. (2012)	118	20[a] (10–11,010) (Der p1) 80[a] (10–11,770) (Der f1)	60[a] (4–39,234)	USA, 2008–10
Van Dyke et al. (2012)	43	100[a] (<10–980)*	1500[a] (<4–1,170,000)	USA
Stemeseder et al. (2017)	501	30 (Der p 1)	760	Austria, 2013–14
Casley et al. (2018)	40	641 (256–933)[c] (Der p1)	–	UK, 2016/17

LOQ limit of quantification; * Sum of Der p1 and Derf1
[a] geometric mean
[b] interquartile range
[c] interquartile range

Table 6.7 Allergens in dust from schools in ng/g

References	N	Mite allergen Median (Min-Max)	Cat allergen Median (Min-Max)	Country, year
Kim et al. (2005)	23	<200	860[a] (<200–4700)	Sweden, 2000
Foarde & Berry (2004)	2	3400 and 7000	6000 and 4800	USA, 2000
Amr et al. (2003)	12	380[b] (0–11,900)	1660[b] (20–12,200)	USA, 2000/01
Arbes et al. (2005)	89	200[a] (9–70,000) (Der p1) 100[a] (9–70,000) (Der f1)	1430[a] (30–6,000,000)	USA
Perry et al. (2008)	33	600 (<20–22,000)	400 (<20–4300)	USA, 2003
Zhao et al. (2006)	10	–	<100 (<100–177)	China
Van Dijken et al. (2006)	11	<30 (<30–316) (Der p1) <30 (<30–89) (Der f1)	–	The Netherlands, 2004
Fromme et al. (2008)	72 75	–	485 (20–45,160) winter 417 (40–7470) summer	Germany, 2004/05
Sheehan et al. (2009)	46	30[a] (<30–40) (Der p1) 40[a] (<30–70) (Der f)	1040[a] (170–4360)	USA
Permaul et al. (2012)	12	10[a] (10–780) (Der p1) 40[a] (10–1640) (Der f)	190[a] (4–285,000)	USA, 2008–10
Carreiro-Martins et al. (2016)	125	670[c] (460–830) (Der p1)	–	Portugal, 2011
Csobod et al. (2014)	292	<LOQ (<LOQ-337) (Der p 1) <LOQ (< LOQ-167) (Der f 1)	<LOQ (< LOQ-1763)	Europe, 23 countries, 2011/12
Ruggieiri et al. (2019)	12	- (200–840) (Der p1)	–	Italy 2012/13

LOQ limit of quantification
[a] geometric mean
[b] mean
[c] median and 25th and 75th percentiles

6.6.5 *Occurrence of Mite Allergens in the Air of Residences*

To date, there are only a few studies in which mite allergens have been measured in indoor air (Table 6.5). In residential interiors, the Der p 1 allergen was in every case below 0.8 ng/m^3, often much lower or below the limit of quantification (Swanson et al. 1989; de Blay et al. 1991; Sakaguchi et al. 1993, 1996; Custovic et al. 1999). Only one study, using personal measurements to examine the air of 45 homes where asthmatic children lived, determined a 95th percentile of 2.5 ng/m^3 (Oliver et al. 1995).

6.6.6 Occurrence of Mite Allergens in the Air of Schools

The data available regarding indoor school air is also limited and often refers to only a small number of observed rooms (Table 6.5).

The most comprehensive results come from a study of 14 classrooms in Munich elementary schools, each of which was sampled for 20 days, and a classroom in Berlin, which was examined on 83 days over the course of the year (Fromme et al. 2013). In Munich, Der p 1 and Der f 1 were detected in the PM_{10} fraction in only 44% and 16% of the samples, respectively. The 95th percentile values for Der p 1 were 0.22 ng/m^3 (maximum: 0.36 ng/m^3) and for Der f 1 0.16 ng/m^3 (maximum: 1.28 ng/m^3), respectively. No statistically significant difference was found between rooms with and without a cozy corner. In an English study, 9 pupils were examined in the context of personal sampling (Custovic et al. 1996). All results were below the detection limit of 0.5 ng/m^3 for Der p 1.

In the USA, Raja et al. (2010) collected PM_{10} in 3 schools and determined Der p 1 concentrations between 0.05 and 0.18 ng/m^2. In another study, Der p 1 was determined in $PM_{2.5}$ from 2 classrooms in 2000 (Foarde and Berry 2004). A high Der p 1 level was found to be <32.3 ng/m^3 in a room with carpeting and 73.7 ng/m^3 in one with smooth flooring.

6.6.7 Occurrence of Mite Allergens in Dust

A compilation of studies of mite allergens in residences published to date in Europe and the USA is given in Table 6.6. In an epidemiological study from Germany in which, among other things, mite allergens were determined in 777 apartments, the median for Der p 1 in 1995/1997 was <10 ng/g (<10–146,000 ng/g) and the 90th percentile was 1587 ng/g, or for Der f 1 < 10 ng/g (<10–190,500 ng/g) and the 90th percentile were 5722 ng/g and 1291 ng/g, respectively (Heinrich et al. 2003). For dwellings, measurement results are also still available from Europe from two studies conducted in 2003–2005 in 567 indoor spaces in France (Kirchner et al. 2008) and in 2016/2017 in 40 buildings (Casley et al. 2018). In both studies, higher concentrations were observed than in the aforementioned German study.

Additionally, in 290 Canadian homes of 4 cities large regional variations were observed (Der f 1 geometric mean: 73–2068 ng/g) (Sun et al. 2022s). Household characteristics including high bed floor cleaning frequency, indoor relative humidity

>45%, home built before 1990, and presence of pet rodents were partially positively associated.

Overall, it can be seen that mite allergens can be found in relatively low concentrations in public facilities (Table 6.7). As a rule, they are in the same concentration range or somewhat lower than in measurements carried out in parallel in residential interiors. The level of the content depends on the geo-climatic conditions of the investigated area, the structural situation of the investigated object and room, and the furnishings. The largest European study is the so-called SINPHONIE study, which was conducted in 2011/2012 in 23 countries in a total of 292 schools (Csobod et al. 2014). In it, very low concentrations are described compared to the Portuguese study. A significantly higher level was found in community facilities with textile flooring compared to smooth flooring (Dybendal and Elsayed 1994; Zock and Brunekreef 1995; Engelhart et al. 2002b; Tortolero et al. 2002; Amr et al. 2003; Abramson et al. 2006; Salo et al. 2009).

6.6.8 *Occurrence of Other Allergens in the Indoor Air*

In the 2003–2005 survey of 567 indoor sites in France, Can f 1 was determined in only 10% of the homes with a median of <1.0 ng/m^3 and a 95th percentile of 1.6 ng/m^3 (Kirchner et al. 2008). In samples collected during 2 weeks in bedrooms of 20 UK homes concentrations ranged from <0,03–2.76 ng/m^3 (Can f 1), <0.005–0.005 ng/m^3 (Mus m 1), and all values <0.01 (Rat n 1). In another study, a geometric mean of 1.80 ng/m^3 (0.39–39.9 ng/m^3) could be found for Mus m 1 and 1.17 ng/m^3 (0.67–4.91 ng/m^3) for Can f 1 in 197 measurements in 12 schools (Permaul et al. 2012). While there was a significant correlation between air and dust concentrations for the mouse allergen, this was not observed for the dog allergen.

6.6.9 *Occurrence of Other Allergens in Dust*

Can f 1 was determined in 111 homes in Sweden with a median of 320 ng/g (<10–332,000 ng/g) and in 108 homes in Estonia of 750 ng/g (20–592,000 ng/g) (Böttcher et al. 2003). In 92 measurements in 23 school classrooms, Kim et al. (2005) found a median of 750 ng/g (<200–6200 ng/g) in Sweden in 2000. Results were similar for 84 measurements in 3 Polish daycare centers with a geometric mean of 970 ng/g (10–10,390 ng/g) (Cyprowski et al. 2013). Lower levels were observed in 10 Chinese schools with a median of<200 ng/g (<200–527 ng/g) (Zhao et al. 2006). In a more recent study of 501 Austrian homes a median of 60 ng/g Can f 1 (5% > 344,000 ng/g) was reported (Stemeseder et al. 2017).

Comparable concentrations were found in about 7000 homes in the USA as part of the representative NHANES (National Health and Nutrition Examination Survey)

(Salo et al. 2018). Here, 2005–2006 Can f 1 levels were 842 ng/g and the range was from 13 to 1,216,000 ng/g. Also, Van Dyke et al. (2012) determined a geometric mean of 9080 n/g Can f 1 (200–833,000 ng/g) in 43 homes in Denver. A geometric mean of 80 ng/g (10–99,420 ng/g) and 30 ng/g (10–74,480 ng/g) was obtained from 229 measurements in 12 classrooms and 118 measurements in homes conducted in 2008–2010 in the USA, respectively (Permaul et al. 2012).

For the mouse allergen Mus m 1, only few study results in house dust are available so far. For example, Muti et al. (2012) reported a median of only 5.24 ng/g in 30 French homes. In contrast, 229 measurements in mainly 12 classrooms and 118 measurements in homes conducted in 2008–2010 in the USA yielded geometric means of 650 ng/g (1–544,370 ng/g) and 100 ng/g (2–82,560 ng/g), respectively (Permaul et al. 2012). Also, Salo et al. (2018) published results of a representative study in approximately 7000 homes in 2005–2006 in the USA with a geometric mean of 23 ng/g, with a wide concentration range between 2 and 134,000 ng/g.

For cockroach allergen Bla g 2, geometric means in homes of <200 ng/g (<200–900 ng/g) and 1100 ng/g (200–272,000 ng/g) are described in two US studies (Permaul et al. 2012; Salo et al. 2018). In 3 Polish nurseries, the geometric mean was 10 ng/g (4–680 ng/g) (Cyprowski et al. 2013). 115 Dust samples collected 2006 in Hong Kong showed a median of 29.6 ng/g Bla g 2 (range: 17.4–46.6 ng/g) (Leung et al. 2011).

6.7 Endotoxins

Endotoxins are substances that can be found in the outer cell wall of gram-negative bacteria. They consist of lipopolysaccharides (LPS), which are responsible for the biological effects. The occurrence of LPS is directly related to the occurrence of gram-negative bacteria or their cell fragments. Despite structural similarity, differences in action potential have been described for individual bacterial species. Analytically, endotoxins or LPS are determined with the so-called Limulus Amoebocyte Lysate assay (LAL assay). However, there is no harmonized measurement method so far. The choice of sampling method is also expected to have a significant influence on the final results.

In a recent review, Salonen et al. (2016) compiled indoor endotoxin loading and concentrations. They consider the age of the house, the type and frequency of cleaning, location in rural or urban areas, presence of textile or smooth floor coverings, and the presence of dogs and cats in the room as major factors influencing indoor levels. An overview of occurrence and fate in the environment is given by Rasuli et al. (2022).

6.7.1 Health Aspects

It is assumed that the interaction of LPS with receptors in the respiratory tract, especially also in the alveoli, leads to inflammatory effects that are triggered and maintained by an increased release of proinflammatory cytokines, reactive oxygen species (ROS), and lysosomal enzymes (DECOS 2010). Systemic effects could also be due to the cytokines produced pulmonarily, which are subsequently released into the circulation.

Based on experimental and epidemiological studies, mainly in polluted agricultural workplaces, symptoms such as dry cough, dyspnea, and reduced lung function as well as fever and fatigue have been described in acute exposure (DECOS 2010). After several hours, bronchoconstriction and joint and head pain may also occur. A representative study in the USA showed a synergistic interaction between endotoxins in house dust and air pollutants such as $PM_{2.5}$ and nitrogen oxides (Mendy et al. 2019). Liebers et al. (2020) summarized studies published in the last 10 years on endotoxin measurement and health effects due to endotoxin in occupational settings.

For general residential interiors, no values for the assessment of endotoxins in indoor air are available to date. A scientific working group of the Health Council of the Netherlands has derived a value of 90 EU/m^3 for workplaces to protect against acute and chronic effects (DECOS 2010).

6.7.2 Occurrence in the Air of Residences

The results published in the scientific literature to date are summarized in Table 6.8. In Europe, medians of 0.59 EU/m^3 (<0.005–17,740 EU/m^3) and 0.61 EU/m^3 (0.018–14,820 EU/m^3) could be found in two measurement periods when 162 children's rooms were studied over the course of a year in Paris in 2003 (Dassonville et al. 2008). In 100 urban and 50 rural indoor environments and 50 farms in France, 1.8 EU/m^3 and 0.9 EU/m^3 and 1.9 EU/m^3 (mean values) were observed in winter (Barnig et al. 2013). The results did not reveal any significant differences between the study areas. Frankel et al. (2012) collected 127 samples in 5 Danish homes over the course of a year and found concentrations ranging from 0.078 to 8.32 EU/m^3. The medians were 1.24 EU/m^3 in spring, 1.48 EU/m^3 in summer, 0.88 EU/m^3 in autumn, and 0.86 EU/m^3 in winter. Except for spring, indoor concentrations were higher than outdoor concentrations.

Studies from North America show a wide range of results, which can range from 0.009 to 4.2 EU/m^3 for the mean/median content. Concentrations of 0.02–19.82 EU/m^3 with a median of 0.64 EU/m^3 were measured in 15 homes in the USA in 1995/1996 (Park et al. 2000). In Canada, a geometric mean of 0.49 EU/m^3 was determined when 332 children's rooms were surveyed in 1997–2003 (Dales et al. 2006). Reponen et al. (2010) reported a geometric mean of 4.2 EU/m^3 (3.2–5.3

Table 6.8 Endotoxins in the air of residences and schools in EU/m³

Reference	N	Mean (min-max)	Country, year
Residences			
Menetrez et al. (2009)	18	0.009 (0.002–0.01)	USA, 2001
Dassonville et al. (2008)	162	0.51/0.58[a] (<0.05–17,700)	France, 2003
Delfino et al. (2015)	45	0.61[b] (0.002–25.3)	USA, 2004
Horick et al. (2006)	82	0.81 (0.23–5.87)	USA
Chen & Hildemann (2009)	40	0.5[c] (0.1–1.6) $PM_{2.5}$ 1.4[c] (0.2–6.0) PM_{10}	USA, 2005
Wheeler et al. (2011)	146	0.71 (0.06–4.1) summer 0.22 (0.02–1.45) winter	Canada, 2007
Menetrez et al. (2009)	10	0.051 (0.024–0.08)	USA, 2007
Mazique et al. (2011)	85	0.05[b] (0.001–1.68)	USA, 2007/8
Pavilonis et al. (2013)	115	0.32 (0.01–4.52)	USA, 2007–11
Reponen et al. (2010)	184	4.2	USA
Hyvärinen et al. (2006)	5	0.43[c] (0.05–3.99)	Finland, 2008
Barnig et al. (2013)	150	0.09–0.19 (–)	France
Bari et al. (2014)	74	0.41 (0.005–53) summer 0.12 (0.002–12) winter	Canada, 2010
Frankel et al. (2012)	5	1.48 (0.32–38)	Denmark, 2010/11
Escobedo et al. (2014)	30	0.05	USA, 2014
Yoda et al. (2017)	55	0.13[c] (0.01–0.57)	Japan, 2014
Ramagopal et al. (2014)	75	1.0[b] (0.09–16)	USA
Bose et al. (2016)	84	0.06[b] (0.01–0.36)[d]	USA
Schools			
Holst et al. (2015)	20	9.34[c] (2.83–225)[e]	Denmark, 1999/2000
Foarde & Berry (2004)	2	0.3 and 0.9[f]	USA, 2000
Yen et al. (2020b)	120	1.4 (0.02–8.13)	Taiwan, 2010–12
Fromme et al. (2013)	14	15.3[b] (0.5–84)	Germany, 2012
Liu et al. (2018)	165[g]	72 (37.9–97.6)	China, 2015

[a] geometric mean with two samplings
[b] median
[c] geometric mean
[e] tenth and 99th percentile
[d] interquartile range
[f] room with carpet and with smooth floor covering
[g] different rooms in auniversity campus

EU/m³) in 184 children's rooms in the USA in a study conducted in 2001–2003. A total of 356 samples were examined in 45 individuals over an extended period of time in the USA in 2003/2004 (Delfino et al. 2015). The median in this study was 0.61 EU/m³ (0.002–25.3 EU/m³).

In 26 and 50 Canadian residential interiors in 2010, the median was found to be 0.12 EU/m³ (0.002–12 EU/m³) in winter and 0.41 EU/m³ (0.005–53 EU/m³) in summer (Bari et al. 2014). The indoor and outdoor air content was about a factor of 4 higher in summer compared to winter. Mendy et al. (2020) reported air endotoxin concentration of 7.2 EU/m^3 and 6.2 EU/m^3, respectively, in two groups of a total of 89 homes in rural areas in the USA.

In an urban city in Japan concentrations from 0.01 to 0.57 EU/m^3 (geometric mean: 0.13 EU/m^3) were reported at weekdays in the air of 55 residences in 2014 (Yoda et al. 2017). Yen et al. (2020b) found a median (range) of 0.67 EU/m^3 (0.02–8.13 EU/m^3) in 120 housing samples collected in Taiwan in 2010–2012.

6.7.3 *Occurrence in the Air of Public Facilities*

To date, endotoxins have rarely been determined in the indoor air of community facilities, so results are only available from some studies (see Table 6.8).

In Germany, endotoxins were investigated in the air of 15 Munich schools, each on 20 consecutive days, and in one Berlin school over the course of a year (83 sampling days) (Fromme et al. 2013). In the Munich facilities, concentrations in PM_{10} and in the inhalable dust fraction ranged from 0.5 to 84.1 EU/m^3 (median: 15.3 EU/m^3) and from 0.03 to 115 EU/m^3 (median: 8.4 EU/m^3), respectively. The annual course of the concentrations in the Berlin school is shown in Fig. 6.5.

In 20 Danish classrooms, a geometric mean of 9.34 EU/m^3 and a 90th percentile of 23.4 EU/m^3 were found (Holst et al. 2015). In addition, Roda et al. (2011) studied the air in 28 Paris kindergartens over 24 hours in each of two seasons. Concentrations with an annual geometric mean of 1.8 EU/m^3 were found. The geometric mean

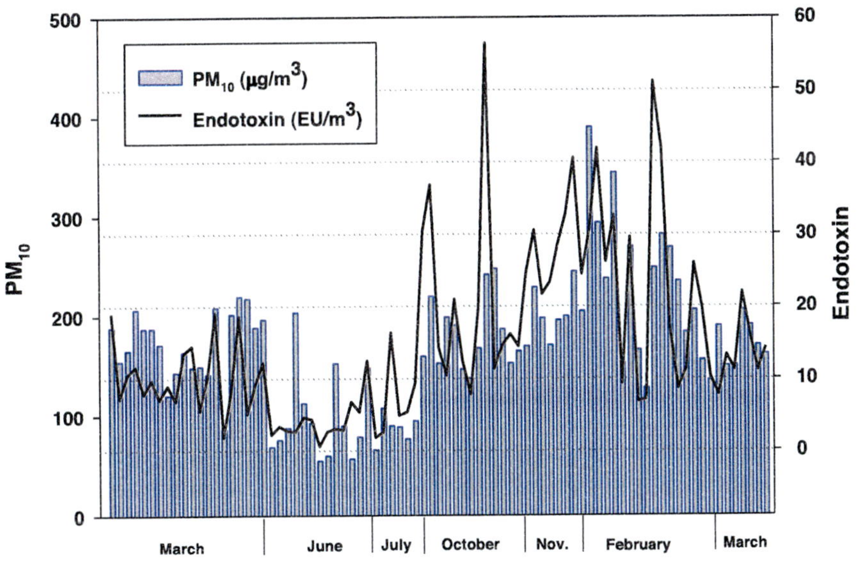

Fig. 6.5 Time course of PM_{10} and endotoxin concentrations in a classroom (after Fromme et al. 2013)

values in winter (1.4 EU/m^3) were not significantly different from those in summer (1.6 EU/m^3).

In the first study from the USA, endotoxin was determined in $PM_{2.5}$ in 2 school rooms, 5 times over 2 hours during the school year (Foarde and Berry 2004). A geometric mean of 0.3 EU/m^3 was found in a room with textile flooring and 0.9 EU/m^3 in a room with smooth flooring. In the second study, a mean content of 0.3 EU/m^3 in PM_{10} was measured in 6 schools (Raja et al. 2010). In Taiwan, median airborne endotoxin concentration of 0.67 EU/m^3 (0.02–8.13 EU/m^3) was reported in 120 schools in 2010–2012 (Yen et al. 2020a).

6.7.4 Occurrence in Dust of Residences

A study of 777 German homes in 1995–1997 showed a wide range of endotoxin levels from 1.5–1476 EU/mg, with a median exposure of 27.8 EU/mg (Heinrich et al. 2003). In the Dutch birth cohort study PIAMA, geometric mean levels ranging from 12.6 EU/mg to 15.3 EU/mg were reported in the living room of 854 subjects for the different groups of participants (Gehring et al. 2020). Moniruzzaman et al. (2012) detected concentrations of 0.14–942 EU/mg (median: 5.7 EU/mg) in bedroom dust and 0.48–188 EU/mg (median: 5.3 EU/mg) in residential dust in 390 Swedish residential interiors in 2001–2002. The Finnish LUKAS2 birth cohort showed a geometric mean of 17.6 EU/mg in 2004–2005 (Casas et al. 2013). Winter levels of 3800 ng/g and 3600 ng/g and 6100 ng/g (mean values) were observed in 100 urban and 50 rural indoor environments and 50 farms in France (Barnig et al. 2013). The results showed no significant differences between urban and rural areas, but from both aforementioned locations to dust levels in farms. Other studies also indicate that endotoxin levels in the living area of children who lived on farms were almost twice as high (81.8 EU/mg) as those who lived in the same village but not on farms (44.9 EU/mg) (Waser et al. 2004).

In the 2005–2006 National Health and Nutrition Examination Survey, a representative study across the USA, concentrations of 0.024–454,500 EU/mg (median 19 EU/mg; 95th percentile: 149 EU/mg) were measured in 6963 house dust samples (Thorne et al. 2005). Significant predictors of higher endotoxin exposure were low income, age of participants, cats and dogs in the home, smokers, and textile floor coverings. Oluwole et al. (2018) determined endotoxins in dust from 102 Canadian homes of persons with mild and severe bronchial asthma in 2015–2016 with geometric mean values of 58.7 and 49.2 EU/mg, respectively. Additionally, in 290 Canadian homes of 4 cities large regional variations were observed (geometric mean: 8.1–205 EU/mg) (Sun et al. 2022). Presence of carpet in the living room was positively associated in all cities.

In a sub-cohort of the Japan Environment and Children's Study (JECS) in a total of 4188 dust samples dust from children's mattresses was analyzed and a lower median of 17.0 EU/mg (inter-quartile range: 11.0–30.0 EU/mg) was observed (Kojima et al. 2022). Shinohara et al. (2023) collected dust from 37 long-term uninhabited and 17 inhabited houses in Japan in 2017/18 and found high median concentrations of 12 EU/mg and 8.9 EU/mg. In a study performed in Vietnam geometric means of 108 EU/mg (6.0–1537 EU/mg) were reported in dust samples from 100 living rooms and 124 EU/mg (4.4–6720 EU/mg) in 100 bedrooms (Tran et al. 2023). Higher levels were observed in kitchen with 149 EU/mg.

6.7.5 Occurrence in Dust of Schools

An investigation of 147 classrooms in Bavarian schools revealed median endotoxin levels in PM_{10} of 19.7 EU/mg dust (6.7–154 EU/mg) in winter and 32.2 EU/mg (9.7–219 EU/mg) in summer (Fromme et al. 2008). Concentrations were significantly higher in the winter measurement period than in the summer, regardless of the floor covering. In a Danish study in 64 school classrooms, mean concentrations were 12.7 EU/mg in schools with no apparent mold problems, 29.0 EU/mg in those with moderate mold contamination, and 27.9 EU/mg in schoolrooms with high mold contamination (Meyer et al. 2011). In another study in 2 schools in Denmark, endotoxin concentrations in dust samples were 16.8 EU/mg and 14.3 EU/mg, respectively (Allermann et al. 2006).

An American study in two schools with different floor coverings found a geometric mean of 43 EU/mg for textile flooring and 51 EU/mg for smooth flooring (Foarde and Berry 2004). In a study conducted in South Carolina at three schools (Fox et al. 2005), endotoxin concentrations were found to be higher in dust deposited on the floor in classrooms that were regularly used than in classrooms that were not used. The authors concluded that skin particles were a possible source of the gram-negative bacteria levels.

References

Abramson SL, Turner-Henson A, Anderson L, Hemstreet MP, Tang S, Bartholomew LK, Joseph CLM, Tyrrell S, Clark NM, Ownby D (2006) Allergens in school settings: results of environmental assessments in 3 city school systems. J Sch Health 76:246–249

Afanou AK, Straumfors A, Eduard W (2019) Fungal aerosol composition in moldy basements. Indoor Air 29:780–790

Akar-Ghibril N, Phipatanakul W (2020) The indoor environment and childhood asthma. Curr Allergy Asthma Rep 20:43

Al Hallak M, Verdier T, Bertron A, Roques C, Bailly JD (2023) Fungal contamination of building materials and the aerosolization of particles and toxins in indoor air and their associated risks to health: a review. Toxins (Basel) 15:175

Allermann L, Wilkins CK, Madsen AM (2006) Inflammatory potency of dust from the indoor environment and correlation to content of NAGase and fungi. Toxicol In Vitro 20:1522–1531

Almqvist C, Larsson PH, Egmar AC, Hedrén M, Malmberg P, Wickman M (1999) School as a risk environment for children allergic to cats and a site for transfer of cat allergen to homes. J Allergy Clin Immunol 103:1012–1017

Amr S, Bollinger ME, Myers M, Hamilton RG, Weiss SR, Rossman M, Osborne L, Timmins S, Kimes DS, Levine ER, Blaisdell CJ (2003) Environmental allergens and asthma in urban elementary schools. Ann Allergy Asthma Immunol 90:34–40

Andrejko KL, Pry JM, Myers JF, Fukui N, DeGuzman JL, Openshaw J, Watt JP, Lewnard JA, Jain S (2022) Effectiveness of face mask or respirator use in indoor public settings for prevention of SARS-CoV-2 infection. MMWR 71:212–216

Araki A, Kanazawa A, Kawai T, Eitaki Y, Morimoto K, Nakayama K, Shibata E, Tanaka M, Takigawa T, Yoshimura T, Chikara H, Saijo Y, Kishi R (2012) The relationship between exposure to microbial volatile organic compound and allergy prevalence in single-family homes. Sci Total Environ 423:18–26

Arbes Jr SJ, Cohn RD, Yin M, Muilenberg ML, Friedman W, Zeldin DC (2004) Dog allergen (Can f 1) and cat allergen (Fel d 1) in US homes: results from the National Survey of Lead and Allergens in Housing. J Allergy Clin Immunol 114:111–117

Arbes SJ, Sever M, Mehta J, Collette N, Thomas B, Zeldin DC (2005) Exposure to indoor allergens in day-care facilities: results from 2 North Carolina counties. J Allergy Clin Immunol 116:133–139

Asadi S, Wexler AS, Cappa CD, Barreda S, Bouvier NM, Ristenpart WD (2019) Aerosol emission and superemission during human speech increase with voice loudness. Sci Rep 9:2348

Augustyniuk-Kram A, Dmowska E (2013) Spectrum and concentration of culturable fungi in house dust from flats in Warsaw, Poland. Aerosol Air Qual Res 13:1438–1447

Bari MA, MacNeill M, Kindzierski WB, Wallace L, Héroux M-E, Wheeler AJ (2014) Predictors of coarse particulate matter and associated endotoxin concentrations in residential environments. Atmos Environ 92:221–230

Barnig C, Reboux G, Roussel S, Casset A, Sohy C, Dalphin JC, de Blay F (2013) Indoor dust and air concentrations of endotoxin in urban and rural environments. Lett Appl Microbiol 56:161–167

Bartlett KH, Kennedy SM, Brauer M, Van Netten C, Dill B (2004) Evaluation and a predictive model of airborne fungal concentrations in school classrooms. Ann Occup Hyg 48:547–554

Becher L, Gena AW, Alsaad H, Richter B, Spahn C, Voelker C (2021) The spread of breathing air from wind instruments and singers using schlieren techniques. Indoor Air 31:1798–1814

Bennett JW, Klich M (2003) Mycotoxins. Clin Microbiol Rev 16:497–516

Birmili W, Selinka H-C, Moriske H-J, Daniels A, Straff W (2021) Ventilation concepts in schools for the prevention of transmission of highly infectious viruses (SARSCoV-2) by aerosols in indoor air. Bundesgesundheitsbl 64:1570–1580

Bloom E, Bal K, Nyman E, Must A, Larsson L (2007) Mass spectrometry-based strategy for direct detection and quantification of some mycotoxins produced by Stachybotrys and Aspergillus spp. in indoor environments. Appl Environ Microbiol 73:4211–4217

Bloom E, Nyman E, Must A, Pehrson C, Larsson L (2009a) Molds and mycotoxins in indoor environments – a survey in water-damaged buildings. J Occup Environ Hyg 6:671–678

Bloom E, Grimsley LF, Pehrson C, Lewis J, Larsson L (2009b) Molds and mycotoxins in dust from water-damaged homes in New Orleans after hurricane Katrina. Indoor Air 19:153–158

Bluyssen PM, Ortiz M, Zhang D (2021) The effect of a mobile HEPA filter system on 'infectious' aerosols, sound and air velocity in the SenseLab. Build Environ 188:107475

Bollinger ME, Eggleston PA, Flanagan E, Wood RA (1996) Cat antigen in homes with and without cats may induce allergic symptoms. J Allergy Clin Immunol 97:907–914

Borges JT, Nakada LYK, Maniero MG, Guimarães JR (2021) SARS-CoV-2: a systematic review of indoor air sampling for virus detection. Environ Sci Pollut Res Int 28:40460–40473

Bos LDJ, Sterk PJ, Schultz MJ (2013) Volatile metabolites of pathogens: a systematic review. PLoS Pathog 9:e1003311

Bose S, Rivera-Mariani F, Chen R, Williams D, Belli A, Aloe C, McCormack MC, Breysse PN, Hansel NN (2016) Domestic exposure to endotoxin and respiratory morbidity in former smokers with COPD. Indoor Air 26:734–742

Böttcher MF, Björkstén B, Gustafson S, Voor T, Jenmalm MC (2003) Endotoxin levels in Estonian and Swedish house dust and atopy in infancy. Clin Exp Allergy 33:295–300

Brasel TL, Douglas DR, Wilson SC, Straus DC (2005) Detection of airborne Stachybotrys chartarum macrocyclic trichothecene mycotoxins on particulates smaller than conidia. Appl Environ Microbiol 71:114–122

Breysse PN, Buckley TJ, Williams DA, Beck CM, Jo S-J, Merriman B, Kanchanaraksa S, Swartz LJ, Callahan KA, Butz AM, Rand CS, Diette GB, Krishnan JA, Moseley AM, Curtin-Brosnan J, Durkin NB, Eggleston PA (2005) Indoor exposures to air pollutants and allergens in the homes of asthmatic children in inner-city Baltimore. Environ Res 98:167–176

Carreiro-Martins P, Papoila AL, Caires I, Azevedo S, Cano MM, Virella D, Leiria-Pinto P, Teixeira JP, Rosado-Pinto J, Annesi-Maesano I, Neuparth N (2016) Effect of indoor air quality of day care centers in children with different predisposition for asthma. Ped Allergy Immunol 27:299–306

Casas L, Tischer C, Wouters IM, Valkonen M, Gehring U, Doekes G, Torrent M, Pekkanen J, Garcia-Esteban R, Hyvärinen A, Heinrich J, Sunyer J (2013) Endotoxin, extracellular polysaccharides, and β(1-3)-glucan concentrations in dust and their determinants in four European birth cohorts: results from the HITEA project. Indoor Air 23:208–218

Casley LS, Godec T, Logan JG, Pearce JC, Smith HMP, Stewart SA, Dewhirst SY (2018) How clean is your house? A study of house dust mites, allergens and other contents of dust samples collected from households. Int J Environ Health Res 28:341–357

Charpin-Kadouch C, Maurel FR, Queralt J, Ramadour M, Dumon H, Garans M, Botta A, Charpin D (2006) Mycotoxin identification in moldy dwellings. J Appl Toxicol 26:475–479

Chen Q, Hildemann LM (2009) Size-resolved concentrations of particulate matter and bioaerosols inside versus outside of homes. Aerosol Sci Technol 43:699–713

Cheng VCC, Chen JHK, Yip CCY, Wong S-C, Chuang VWM, Tsang OTY, Sridhar S, Chan JFW, Ho P-L, Yuen K-Y (2020) Escalating infection control response to the rapidly evolving epidemiology of the coronavirus disease 2019 (COVID-19) due to SARS-CoV-2 in Hong Kong. Infect Control Hosp Epidemiol 41:493–498

Chia PY, Coleman KK, Tan YK, Ong SWX, Gum M, Lau SK, Lim XF, Lim AS, Sutjipto S, Lee PH, Son TT, Young BE, Milton DK, Gray GC, Schuster S, Barkham T, De PP, Vasoo S, Chan M, Ang BSP, Tan BH, Leo YS, Ng OT, Wong MSY, Marimuthu K (2020) Detection of air and surface contamination by SARS-CoV-2 in hospital rooms of infected patients. Nat Commun 11:2800

Chirizzi D, Conte M, Feltracco M, Dinoi A, Gregoris E, Barbaro E, La Bella G, Ciccarese G, La Salandra G, Gambaro A, Contini D (2021) SARS-CoV-2 concentrations and virus-laden aerosol size distributions in outdoor air in north and south of Italy. Environ Int 146:106255

Cho SJ, Ramachandran G, Grengs J, Ryan AD, Eberly LE, Adgate JL (2008) Longitudinal evaluation of allergen and culturable fungal concentrations in inner-city households. J Occup Environ Hyg 5:107–118

Choi H, Schmidbauer N, Bornehag CG (2017) Volatile organic compounds of possible microbial origin and their risks on childhood asthma and allergies within damp homes. Environ Int 98: 143–151

Chu DK, Akl EA, Duda S, Solo K, Yaacoub S, Schünemann HJ (2020) Physical distancing, face masks, and eye protection to prevent person-to-person transmission of SARS-CoV-2 and COVID-19: a systematic review and meta-analysis. Lancet 395:1973–1987

Cowger TL, Murray EJ, Clarke J, Bassett MT, Ojikutu BO, Sánchez SM, Linos N, Hall KT (2022) Lifting universal Mmsking in schools – Covid-19 incidence among students and staff. N Engl J Med 387:1935–1946

Cox J, Mbareche H, Lindsley WG, Duchaine C (2020) Field sampling of indoor bioaerosols. Aerosol Sci Technol 54:572–584

Csobod E, Annesi-Maesano I, Carrer P, Kephalopoulos S, Madureira J, Rudnai P, de Oliveira Fernandes E (2014) SINPHONIE. Schools indoor pollution and Health. Final Report, EUR 26738. https://publications.jrc.ec.europa.eu/repository/handle/JRC91160

Custovic A, Green R, Taggart SC, Smith A, Pickering CA, Chapman MD, Woodcock A (1996) Domestic allergens in public places. II: Dog (Can f1) and cockroach (Bla g 2) allergens in dust and mite, cat, dog and cockroach allergens in the air in public buildings. Clin Exp Allergy 26: 1246–1252

Custovic A, Simpson A, Pahdi H, Green RM, Chapman MD, Woodcock A (1998) Distribution, aerodynamic characteristics, and removal of the major cat allergen Fel d 1 in British homes. Thorax 53:33–38

Custovic A, Simpson B, Simpson A, Hallam C, Craven M, Woodcock A (1999) Relationship between mite, cat, and dog allergens in reservoir dust and ambient air. Allergy 54:612–616

Cyprowski M, Buczyńska A, Szadkowska-Stańczyk I (2013) Indoor allergens in settled dust from kindergartens in city of Lódz, Poland. Int J Occup Med Environ Health 26:890–899

Dales R, Miller D, Ruest K, Guay M, Judek S (2006) Airborne endotoxin is associated with respiratory illness in the first 2 years of life. Environ Health Perspect 114:610–614

Dassonville C, Demattei C, Vacquier B, Bex-Capelle V, Seta N, Momas I (2008) Indoor airborne endotoxin assessment in homes of Paris newborn babies. Indoor Air 18:480–487

De Blay F, Heymann PW, Chapman MD, Platts-Mills TAE (1991) Airborne dust mite allergens: comparison of group II allergens with group I mite allergen and cat-allergen Fe/d I. J Allerg Clin Immunol 88:919–926

De Gier B, Andeweg S, Backer JA, RIVM COVID-19 surveillance and epidemiology team, Hahné SJ, van den Hof S, de Melker HE, Knol MJ (2021) Vaccine effectiveness against SARS-CoV-2 transmission to household contacts during dominance of Delta variant (B.1.617.2), the Netherlands. Euro Surveill 26:2100977

DECOS (2010) Endotoxins. Health-based recommended occupational exposure limit. The Hague, The Netherlands. https://www.healthcouncil.nl/documents/advisory-reports/2010/07/15/endotoxins-health-based-recommended-occupational-exposure-limit

Degen GH, Partosc F, Muñoz K, Gundert-Remy U (2017) Daily uptake of mycotoxins – TDI might not be protective for nursed infants. Toxicol Lett 277:69–75

Delfino RJ, Staimer N, Tjoa T, Gillen DL (2015) Relations of exhaled nitric oxide and FEV_1 to personal endotoxin exposure in schoolchildren with asthma. Occup Environ Med 72:830–836

Ding Z, Qian H, Xu B, Huang Y, Miao T, Yen HL, Xiao S, Cui L, Wu X, Shao W, Song Y, Sha L, Zhou L, Xu Y, Zhu B, Li Y (2021) Toilets dominate environmental detection of severe acute respiratory syndrome coronavirus 2 in a hospital. Sci Total Environ 753:141710

Dinoi A, Feltracco M, Chirizzi D, Trabucco S, Conte M, Gregoris E, Barbaro E, La Bella G, Ciccarese G, Belosi F, La Salandra G, Gambaro A, Contini D (2022) A review on measurements of SARS-CoV-2 genetic material in air in outdoor and indoor environments: Implication for airborne transmission. Sci Total Environ 809:151137

Döhla M, Wilbring G, Schulte B, Kümmerer BM, Diegmann C, Siba E, Richter E, Haag A, Engelhart S, Eis-Hübinger AM, Exner M, Streeck H, Schmithausen RM (2020) SARS-CoV-2 in environmental samples of quarantined households. medRxiv preprint https://doi.org/10.1101/2020.05.28.20114041

Douwes J (2005) (1 fi 3)-b-D-glucans and respiratory health: a review of the scientific evidence. Indoor Air 15: 160–169

Du C, Li B, Yu W (2021) Indoor mould exposure: Characteristics, influences and corresponding associations with built environment—a review. J Build Eng 35:101983

Duchaine C (2016) Assessing microbial decontamination of indoor air with particular focus on human pathogenic viruses. Am J Infect Control 44:S121–S126

Dybendal T, Elsayed S (1994) Dust from carpeted and smooth floors. VI. Allergens in homes compared with those in schools in Norway. Allergy 49:210–216

EASA/ECDC (European Union Aviation Safety Agency, European Centre for Disease Prevention and Control) (2021) COVID-19 Aviation Health Safety Protocol Operational guidelines for the management of air passengers and aviation personnel in relation to the COVID-19 pandemic. Issue No: 03. https://www.ecdc.europa.eu/sites/default/files/documents/Joint%20EASA-ECDC-Aviation-Health-Safety-Protocol-issue-3-17-June-2021.pdf

ECDC (European Centre for Disease Prevention and Control) (2020) Risk assessment guidelines for infectious diseases transmitted on aircraft (RAGIDA). Middle East Respiratory Syndrome Coronavirus (MERS-CoV). Stockholm, Sweden. https://www.ecdc.europa.eu/en/publications-data/risk-assessment-guidelines-infectious-diseases-transmitted-aircraft-ragida-middle

Eduard W (2009) Fungal spores: a critical review of the toxicological and epidemiological evidence as a basis for occupational exposure limit setting. Crit Rev Toxicol 39:799–864

EFSA (2020) Mykotoxins. https://www.efsa.europa.eu/en/topics/topic/mycotoxins

Engelhart S, Loock A, Skutlarek D, Sagunski H, Lommel A, Färber H, Exner M (2002a) Occurrence of toxigenic Aspergillus versicolor isolates and sterigmatocystin in carpet dust from damp indoor environments. Appl Environ Microbiol 68:3886–3890

Engelhart S, Bieber T, Exner M (2002b) House dust mite allergen levels in German day-care centers. Int J Hyg Environ Health 205:453–457

Ertl H, Portner C, Plegge V, Teutenberg T, Türk J, Butte W (2007) Quantification of 7 mycotoxins in house dust by HPLC-MS/MS and MSn screening for mold markers as well as other toxins. Poster

Escobedo LE, Champion WM, Li N, Montoya LD (2014) Indoor air quality in Latino homes in Boulder, Colorado. Atmos Environ 92:69–75

Fakunle AG, Jafta N, Naidoo RN, Smit LAM (2021) Association of indoor microbial aerosols with respiratory symptoms among under-five children: a systematic review and meta-analysis. Environ Health 20:77

Fan L, Han X, Wang X, Li L, Gong S, Qi J, Li X, Ge T, Liu H, Ye D, Cao Y, Liu M, Sun Z, Su L, Yao X, Wang X (2021a) Levels, distributions and influential factors of residential airborne culturable bacteria in 12 Chinese cities: multicenter on-site survey among dwellings. Environ Res 212(Pt C):113425

Fan L, Wang J, Yang Y, Yang W, Zhu Y, Zhang Y, Li L, Li X, Yan X, Yao X, Wang L, Wang X (2021b) Airborne culturable fungi under general living scenario: on-site investigation in 12 typical cities, China. Environ Int 155:106669

Faridi S, Niazi S, Sadeghi K, Naddafi K, Yavarian J, Shamsipour M, Jandaghi NZS, Sadeghniiat K, Nabizadeh R, Yunesian M, Momeniha F, Mokamel A, Hassanvand MS, Mokhtari AT (2020) A field indoor air measurement of SARSCoV-2 in the patient rooms of the largest hospital in Iran. Sci Total Environ 725:138401

Farokhi A, Heederik D, Smit LAM (2018) Respiratory health effects of exposure to low levels of airborne endotoxin – a systematic review. Environ Health 17:14

Fears AC, Klimstra WB, Duprex P, Hartman A, Weaver SC, Plante KS, Mirchandani D, Plante JA, Aguilar PV, Fernández D, Nalca A, Totura A, Dyer D, Kearney B, Lackemeyer M, Bohannon JK, Johnson R, Garry RF, Reed DS, Roy CJ (2020) Persistence of severe acute respiratory syndrome coronavirus 2 in aerosol suspensions. Emerg Infect Dis 26:2168–2171

Fiedler K, Schütz E, Geh S (2001) Detection of microbial volatile organic compounds (MVOCs) produced by moulds on various materials. Int J Hyg Environ Health 204:111–121

Foarde K, Berry M (2004) Comparison of biocontaminant levels associated with hard vs. carpet floors in nonproblem schools: results of a year long study. J Expo Anal Environ Epidemiol 14: S41–S48

Foster A, Kinzel M (2021) Estimating COVID-19 exposure in a classroom setting: a comparison between mathematical and numerical models. Phys Fluids (1994) 33:021904

Fox A, Harley W, Feigley C, Salzberg D, Toole C, Sebastian A, Larsson L (2005) Large particles are responsible for elevated bacterial marker levels in school air upon occupation. J Environ Monit 7:450–456

Frankel M, Beko G, Timm M, Gustavsen S, Hansen EW, Madsen AM (2012) Seasonal variations of indoor microbial exposures and their relation to temperature, relative humidity, and air exchange rate. Appl Environ Microbiol 78:8289–8297

Fromme H, Heitmann D, Dietrich S, Schierl R, Körner W, Kiranoglu M, Zapf A, Twardella D (2008) Air quality in schools – classroom levels of carbon dioxide (CO_2), volatile organic compounds (VOC), aldehydes, endotoxins and cat allergen. Gesundheitswesen 70:88–97. [in German]

Fromme H, Bischof W, Dietrich S, Lahrz T, Schierl R, Schwegler U (2013) Airborne allergens, endotoxins, and particulate matter in elementary schools, results from Germany (LUPE 2). J Occup Environ Hyg 10:573–582

Fromme H, Gareis M, Völkel W, Gottschalk C (2016) Overall internal exposure to mycotoxins and their occurrence in occupational and residential settings – an overview. Int J Hyg Environ Health 219:143–165

Furuse Y, Sando E, Tsuchiya N, Miyahara R, Yasuda I, Ko YK, Saito M, Morimoto K, Imamura T, Shobugawa Y, Nagata S, Jindai K, Imamura T, Sunagawa T, Suzuki M, Nishiura H, Oshitani H (2020) Clusters of coronavirus disease in communities, Japan, January–April 2020. Emerg Infect Dis 26:2176–2179

Gehring U, Douwes J, Doekes G, Koch A, Bischof W, Fahlbusch B, Richter K, Wichmann H-E, Heinrich J (2001) β(1→3)-glucan in house dust of German homes: housing characteristics, occupant behavior, and relations with endotoxins, allergens, and molds. Environ Health Perspect 109:139–144

Gehring U, Wijga AH, Koppelman GH, Vonk JM, Smit HA, Brunekreef B (2020) House dust endotoxin, asthma and allergic sensitization through childhood into adolescence. Clin Exp Allergy 50:1055–1064

Gent JF, Kezik JM, Hill ME, Tsai E, Li DW, Leaderer BP (2012) Household mold and dust allergens: exposure, sensitization and childhood asthma morbidity. Environ Res 118:86–93

Ghosh B, Lal H, Srivastava A (2015) Review of bioaerosols in indoor environment with special reference to sampling, analysis and control mechanisms. Environ Int 85:254–272

Giovannangelo MECA, Gehring U, Nordling E, Oldenwening M, van Rijswijk K, de Wind S, Hoek G, Heinrich J, Bellander T, Brunekreef B (2007) Levels and determinants of β(1→3)-glucans and fungal extracellular polysaccharides in house dust of (pre-)schoolchildren in three European countries. Environ Int 33:9–16

Gottschalk C, Bauer J, Meyer K (2008) Detection of satratoxin G and H in indoor air from a water-damaged building. Mycopathologia 166:103–107

Guo ZD, Wang ZY, Zhang SF, Li X, Li L, Li C, Cui Y, Fu RB, Dong YZ, Chi XY, Zhang MY, Liu K, Cao C, Liu B, Zhang K, Gao YW, Lu B, Chen W (2020) Aerosol and surface distribution of severe acute respiratory syndrome coronavirus 2 in hospital wards, Wuhan, China. Emerg Infect Dis 26:1583–1591

Haines SR, Hall EC, Marciniak K, Misztal PK, Goldstein AH, Adams RI, Dannemiller KC (2021) Microbial growth and volatile organic compound (VOC) emissions from carpet and drywall under elevated relative humidity conditions. Microbiome 9:209

Hammond A, Khalid T, Thornton HV, Woodall CA, Hay AD (2021) Should homes and workplaces purchase portable air filters to reduce the transmission of SARS-CoV-2 and other respiratory infections? A systematic review. PLoS One 16:e0251049

Hamner L, Dubbel P, Capron I, Ross A, Jordan A, Lee J, Lynn J, Ball A, Narwal S, Russell S, Patrick D, Leibrand H (2020) High SARS-CoV-2 attack rate following exposure at a choir practice – Skagit County, Washington, March 2020. MMWR 69:606–610

Heinrich J, Hölscher B, Douwes J, Richter K, Koch A, Bischof W, Fahlbusch B, Kinne RW, Wichmann HE (2003) Reproducibility of allergen, endotoxin and fungi measurements in the indoor environment. J Expo Anal Environ Epidemiol 13:152–160

Heinzerling LM, Burbach GJ, Edenharter G, Bachert C, Bindslev-Jensen C, Bonini S, Bousquet J, Bousquet-Rouanet L, Bousquet PJ, Bresciani M, Bruno A, Burney P, Canonica GW, Darsow U, Demoly P, Durham S, Fokkens WJ, Giavi S, Gjomarkaj M, Gramiccioni C, Haahtela T, Kowalski ML, Magyar P, Muraközi G, Orosz M, Papadopoulos NG, Röhnelt C, Stingl G, Todo-Bom A, von Mutius E, Wiesner A, Wöhrl S, Zuberbier T (2009) GA(2)LEN skin test study I: GA(2)LEN harmonization of skin prick testing: novel sensitization patterns for inhalant allergens in Europe. Allergy 64:1498–1506

Holme JA, Øya E, Afanou AKJ, Øvrevik J, Eduard W (2020) Characterization and pro-inflammatory potential of indoor mold particles. Indoor Air 30:662–681

Holst G, Høst A, Doekes G, Meyer HW, Madsen AM, Sigsgaard T (2015) Determinants of house dust, endotoxin, and β-(1→3)-D-glucan in homes of Danish children. Indoor Air 25:245–259

Holst GJ, Pørneki A, Lindgreen J, Thuesen B, Bønløkke J, Hyvärinen A, Elholm G, Østergaard K, Loft S, Brooks C, Douwes J, Linneberg A, Sigsgaard T (2020) Household dampness and microbial exposure related to allergy and respiratory health in Danish adults. Eur Clin Respir J 7:1706235

Horick N, Weller E, Milton DK, Gold DR, Li R, Spiegelman D (2006) Home endotoxin exposure and wheeze in infants: correction for bias due to exposure measurement error. Environ Health Perspect 114:135–140

Horstman R, Rahai H (2021) A risk assessment of an airborne disease inside the cabin of a passenger airplane. SAE Int J Adv Curr Prac Mobility 3:1263–1271

Hyvärinen A, Roponen M, Tiittanen P, Laitinen S, Nevalainen A, Pekkanen J (2006) Dust sampling methods for endotoxin – an essential, but underestimated issue. Indoor Air 16:20–27

IATA (International Air Transport Association, Medical Advisory Group) (2020) Restarting aviation following COVID-19. Medical evidence for various strategies being discussed as at 25 May 2020. https://www.iata.org/contentassets/f1163430bba94512a583eb6d6b24aa56/covid-medical-evidence-for-strategies-200525.pdf

IATA (2022) COVID-19: A summary of current medical evidence relevant to air travel. Version 13–28 April 2022. IATA Medical Advisor/Medical Advisory Group. https://www.iata.org/globalassets/iata/programs/covid/restart/covid-public-health-meausures-evidence-doc.pdf

Jakšić D, Sertić M, Kifer D, Kocsubè S, Mornar Turk A, Nigović B, Šarkanj B, Krska R, Sulyok M, Šegvić KM (2021) Fungi and their secondary metabolites in water-damaged indoors after a major flood event in eastern Croatia. Indoor Air 31:730–744

Jang S, Han SH, Rhee J-Y (2020) Cluster of coronavirus disease associated with fitness dance classes, South Korea. Emerg Infect Dis 26:1917–1920

Jarvis BB, Miller J (2005) Mycotoxins as harmful indoor air contaminants. Appl Microbiol Biotechnol 66:367–372

Jefferson T, Del Mar CB, Dooley L, Ferroni E, Al-Ansary LA, Bawazeer GA (2011) Physical interventions to interrupt or reduce the spread of respiratory viruses. Cochrane Database Syst Rev 7

Jiang Y, Zhao B, Li X, Yang X, Zhang Z, Zhang Y (2009) Investigating a safe ventilation rate for the prevention of indoor SARS transmission: an attempt based on a simulation approach. Build Simul 2:281–289

Jovanovic S, Felder-Kennel A, Gabrio T, Kouros B, Link B, Maisner V, Piechotowski I, Schick KH, Schrimpf M, Weidner U, Zollner I, Schwenk. (2004) Indoor fungi levels in homes of children with and without allergy history. Int J Hyg Environ Health 207:369–378

Karlsson A-S, Renström A, Hedrén M, Larsson K (2004) Allergen avoidance does not alter airborne cat allergen levels in classrooms. Allergy 59:661–667

Kasel U, Wichmann G, Bleck M (1999) Ochratoxin A in house dust. Umweltmed Forsch Prax 4: 301–303. [in German]

Kim JL, Elfman L, Mi Y, Johansson M, Smedje G, Norbäck D (2005) Current asthma and respiratory symptoms among pupils in relation to dietary factors and allergens in the school environment. Indoor Air 15:170–182

Kim Y, Ryu H, Lee S (2018) Agent-based modeling for super-spreading events: a case study of MERS-CoV transmission dynamics in the Republic of Korea. Int J Environ Res Public Health 15:2369

Kirchner S, Derbez M, Duboudin C, Elias P, Gregoire A, Lucas J-P, Pasquier N, Ramalho O, Weiss N (2008) Indoor air quality in French dwellings. Proceedings Indoor Air 2008, Copenhagen, Denmark

Kitagawa H, Nomura T, Kaiki Y, Kakimoto M, Nazmul T, Omori K, Shigemoto N, Sakaguchi T, Ohge H (2022) Viable SARS-CoV-2 detected in the air of hospital rooms of COVID-19 patients with early infection. Int J Infect Dis 7:S1201-9712(22)00589-6

Klompas M, Milton DK, Rhee C, Baker MA, Leekha S (2021) Current insights into respiratory virus transmission and potential implications for infection control programs: a narrative review. Ann Intern Med 174:1710–1718

Kojima R, Shinohara R, Kushima M, Horiuchi S, Otawa S, Miyake K, Yokomichi H, Akiyama Y, Ooka T, Yamagata Z (2022) Exposure to House Dust Mite allergen and endotoxin in early life and sensitization and allergicrRhinitis: the JECS. Int J Environ Res Public Health 19:14796

Korpi A, Järnberg J, Pasanen A-L (2009) Microbial volatile organic compounds. Crit Rev Toxicol 39:139–193

Kotwa JD, Jamal AJ, Mbareche H, Yip L, Aftanas P, Barati S, Bell NG, Bryce E, Coomes E, Crowl G, Duchaine C, Faheem A, Farooqi L, Hiebert R, Katz K, Khan S, Kozak R, Li AX, Mistry HP, Mozafarihashjin M, Nasir JA, Nirmalarajah K, Panousis EM, Paterson A, Plenderleith S, Powis J, Prost K, Schryer R, Taylor M, Veillette M, Wong T, Zhong XZ, Mc Arthur AG, Mc Geer AJ, Mubareka S (2021) Surface and air contamination with SARS-CoV-2 from hospitalized COVID-19 patients in Toronto, Canada, March–May 2020. J Infect Dis Nov 27: jiab578

Laußmann D, Eis D, Schleibinger H (2004) Comparison of mycological and chemical-analytical laboratory methods for the detection of mold infestations in indoor spaces. Bundesgesundheitsbl 47:1078–1094. [in German]

Lednicky JA, Lauzardo M, Fan ZH, Jutla A, Tilly TB, Gangwar M, Usmani M, Shankar SN, Mohamed K, Eiguren-Fernandez A, Stephenson CJ, Alam MM, Elbadry MA, Loeb JC, Subramaniam K, Waltzek TB, Cherabuddi K, Morris Jr JG, Wu CY (2020) Viable SARS-CoV-2 in the air of a hospital room with COVID-19 patients. Int J Infect Dis 100:476–482

Lee T, Grinshpun SA, Martuzevicius D, Adhikari A, Crawford CM, Luo J, Reponen T (2006) Relationship between indoor and outdoor bio-aerosols collected with a button inhalable aerosol sampler in urban homes. Indoor Air 16:37–47

Lei H, Li Y, Xiao S, Lin C-H, Norris SL, Wei D, Hu Z, Ji S (2018) Routes of transmission of influenza A H1N1, SARS CoV, and norovirus in air cabin: comparative analyses. Indoor Air 28: 394–403

Leitmeyer K, Adlhoch C (2016) Review article: influenza transmission on aircraft – a systematic literature review. Epidemiology 27:743–751

Leung TF, Wong YS, Chan IH, Yung E, Sy HY, Lam CW, Wong GW (2011) Domestic exposure to aeroallergens in Hong Kong families with asthmatic children. Pediatr Pulmonol 46:632–639

Leung NHL, Chu DKW, Shiu EYC, Chan KH, McDevitt JJ, Hau BJP, Yen HL, Li Y, Ip DKM, Peiris JSM, Seto WH, Leung GM, Milton DK, Cowling BJ (2020) Respiratory virus shedding in exhaled breath and efficacy of face masks. Nat Med 26:676–680

Li L, Niu M, Zhu Y (2021) Assessing the effectiveness of using various face coverings to mitigate the transport of airborne particles produced by coughing indoors. Aerosol Sci Technol 55:332–312

Liebers V, Brüning T, Raulf M (2020) Occupational endotoxin exposure and health effects. Arch Toxicol 94:3629–3644

Linde KJ, Wouters IM, Kluytmans JAJW, Kluytmans-van den Bergh MFQ, Pas SD, Geurtsvan Kessel CH, Koopmans MPG, Meier M, Meijer P, Raben CR, Spithoven J, Tersteeg-Zijderveld MHG, Heederik DJJ, Dohmen W, COCON Consortium (2022) Detection of SARS-CoV-2 in air and on surfaces in rooms of infected nursing home residents. Ann Work Expo. Health 7: wxac056

Lindemann V, Schmidt J, Cramer B, Humpf HU (2022) Detection of mycotoxins in highly matrix-loaded house-dust samples by QTOF-HRMS, IM-QTOF-HRMS, and TQMS: advantages and disadvantages. Anal Chem 94:4209–4217

Lindsley WG, Blachere FM, Davis KA, Pearce TA, Fisher MA, Khakoo R, Davis SM, Rogers ME, Thewlis RE, Posada JA, Redrow JB, Celik IB, Chen BT, Beezhold DH (2010) Distribution of airborne influenza virus and respiratory syncytial virus in an urgent care medical clinic. Clin Infect Dis 50:693–698

Lindsley WG, Pearce TA, Hudnall JB, Davis KA, Davis SM, Fisher MA, Khakoo R, Palmer JE, Clark KE, Celik I, Coffey CC, Blachere FM, Beezhold DH (2012) Quantity and size distribution of cough-generated aerosol particles produced by influenza patients during and after illness. J Occup Environ Hyg 9:443–449

Lindsley WG, Blachere FM, Beezhold DH, Thewlis RE, Noorbakhsh B, Othumpangat S, Goldsmith WT, McMillen CM, Andrew ME, Burrell CN, Noti JD (2016) Viable influenza A virus in airborne particles expelled during coughs versus exhalations. Influenza Other Respir Viruses 10: 404–413

Liu H, Zhang Z, Wen N, Wang C (2018) Determination and risk assessment of airborne endotoxin concentrations in a university campus. J Aerosol Sci 115:146–157

Liu F, Qian H, Luo Z, Wang S, Zheng X (2020a) A laboratory study of the expiratory airflow and particle dispersion in the stratified indoor environment. Build Environ 180:106988

Liu Y, Ning Z, Chen Y, Guo M, Liu Y, Gali NK, Sun L, Duan Y, Cai J, Westerdahl D, Liu X, Xu K, Ho KF, Kan H, Fu Q, Lan K (2020b) Aerodynamic analysis of SARS-CoV-2 in two Wuhan hospitals. Nature 582:557–560

Lu R, Pørneki AD, Lindgreen JN, Li Y, Madsen AM (2021) Species of fungi and pollen in the PM_1 and the inhalable fraction of indoor air in homes. Atmos 12:404

Ma J, Qi X, Chen H, Li X, Zhang Z, Wang H, Sun L, Zhang L, Guo J, Morawska L, Grinshpun SA, Biswas P, Flagan RC, Yao M (2020) Exhaled breath is a significant source of SARS-CoV-2 emission. medRxiv preprint https://doi.org/10.1101/2020.05.31.20115154

Madewell ZJ, Yang Y, Longini IM, Halloran ME, Dean NE (2020) Household transmission of SARS-CoV-2: a systematic review and meta-analysis. JAMA Netw Open 3:e2031756

Madureira J, Paciencia I, Rufo JC, Pereira C, Teixeira JP, de Oliveira Fernandes E (2015a) Assessment and determinants of airborne bacterial and fungal concentrations in different indoor environments: homes, child day-care centres, primary schools and elderly care centres. Atmos Environ 109:139–146

Madureira J, Paciencia J, Rufo J, Ramos E, Barros H, Teixeira JP, de Oliveira Fernandes E (2015b) Indoor air quality in schools and its relationship with children's respiratory symptoms. Atmos Environ 118:145–156

Mallach G, Kasloff SB, Kovesi T, Kumar A, Kulka R, Krishnan J, Robert B, McGuinty M, den Otter-Moore S, Yazji B, Cutts T (2021) Aerosol SARS-CoV-2 in hospitals and long-term care homes during the COVID-19 pandemic. PLoS One 16:e0258151

Mangili A, Gendreau MA (2005) Transmission of infectious diseases during commercial air travel. Lancet 365:989–996

Matysik S, Herbarth O, Mueller A (2009) Determination of microbial volatile organic compounds (MVOCs) by passive sampling onto charcoal sorbents. Chemosphere 76:114–119

Mazique D, Diette GB, Breysse PN, Matsui EC, McCormack MC, Curtin-Brosnan J, Williams DL, Peng RD, Hansel NN (2011) Predictors of airborne endotoxin concentrations in inner city homes. Environ Res 111:614–617

Mendes A, Aelenei D, Luísa Papoila A, Carreiro-Martins P, Aguiar L, Pereira C, Neves P, Azevedo S, Cano M, Proença C, Viegas J, Silva S, Mendes D, Neuparth N, Teixeira JP (2014) Environmental and ventilation assessment in child day care centers in Porto: The Envirh Project. J Toxicol Environ Health Part A 77:931–943

Mendy A, Wilkerson J, Salo PM, Weir CH, Feinstein L, Zeldin DC, Thorne PS (2019) Synergistic association of house endotoxin exposure and ambient air pollution with asthma outcomes. Am J Respir Crit Care Med 200:712–720

Mendy A, Metwali N, Perry SS, Chrischilles EA, Wang K, Thorne PS (2020) Household endotoxin reduction in the Louisa Environmental Intervention Project for rural childhood asthma. Indoor Air 30:88–97

Menetrez MY, Foarde KK, Esch RK, Schwartz TD, Dean TR, Hays MD, Cho SH, Betancourt DA, Moore SA (2009) An evaluation of indoor and outdoor biological particulate matter. Atmos Environ 43:5476–5483

Meyer HW, Suadicani P, Nielsen PA, Sigsgaard T, Gyntelberg F (2011) Moulds in floor dust – a particular problem in mechanically ventilated rooms? A study of adolescent schoolboys under the Danish moulds in buildings program. Scand J Work Environ Health 37:332–340

Misztal PK, Lymperopoulou DS, Adams RI, Scott RA, Lindow SE, Bruns T, Taylor JW, Uehling J, Bonito G, Vilgalys R, Goldstein AH (2018) Emission factors of microbial volatile organic compounds from environmental bacteria and fungi. Environ Sci Technol 52:8272–8282

Moniruzzaman S, Hägerhed Engman L, James P, Sigsgaard T, Thorne PS, Sundell J, Bornehag CG (2012) Levels of endotoxin in 390 Swedish homes: determinants and the risk for respiratory symptoms in children. Int J Environ Health Res 22:22–36

Morawska L, Tang JW, Bahnfleth W, Bluyssen PM, Boerstra A, Buonanno G, Cao J, Dancer S, Floto A, Franchimon F, Haworth C, Hogeling J, Isaxon C, Jimenez JL, Kurnitski J, Li Y, Loomans M, Marks G, Marr LC, Mazzarella L, Melikov AK, Miller S, Milton DK, Nazaroff W, Nielsen PV, Noakes C, Peccia J, Querol X, Sekhar C, Seppänen O, Tanabe SI, Tellier R, Tham KW, Wargocki P, Wierzbicka A, Yao M (2020) How can airborne transmission of COVID-19 indoors be minimised? Environ Int 142:105832

Mosch C, Gottschalk C, Gareis M, Völkel W, Fromme H (2015) Mycotoxins in house dust. Report of the Bayerischen Landesamtes für Gesundheit und Lebensmittelsicherheit. München, Germany. [in German]

Moularat M, Hulin M, Robine E, Annesi-Maesano I, Caillaud D (2011) Airborne fungal volatile organic compounds in rural and urban dwellings. Detection of mould contamination in 94 homes determined by visual inspection and airborne fungal volatile organic compounds method. Sci Total Environ 409:2005–2009

Mousavi ES, Kananizadeh N, Martinello RA, Sherman JD (2021) COVID-19 Outbreak and hospital air quality: a systematic review of evidence on air filtration and recirculation. Environ Sci Technol 55:4134–4147

Muller N, Kunze M, Steitz F, Saad NJ, Mühlemann B, Beheim-Schwarzbach JI, Schneider J, Drosten C, Murajda L, Kochs S, Ruscher C, Walter J, Zeitlmann N, Corman VM (2020) Severe acute respiratory syndrome coronavirus 2 outbreak related to a nightclub, Germany, 2020. Emerg Infect Dis 27:645–648

Munir AK, Einarsson R, Schou C, Dreborg SK (1993) Allergens in school dust. I. The amount of the major cat (Fel d I) and dog (Can f I) allergens in dust from Swedish schools is high enough to probably cause perennial symptoms in most children with asthma who are sensitized to cat and dog. J Allergy Clin Immunol 91:1067–1074

Munir AKM, Einarsson R, Dreborg S (2003) Variability of airborne cat allergen, Fel d1, in a public place. Indoor Air 13:353–358

Muti D, Purohit A, Dazy A, Verot A, de Blay F (2012) Mouse (Mus m1) and rat (Rat n1) allergen levels in dust from private and public houses in Strasbourg, France are lower than houses in the U.S.A. Eur Ann Allergy Clin Immunol 44:93–95

Naclerio R, Ansotegui IJ, Bousquet J, Canonica GW, D'Amato G, Rosario N, Pawankar R, Peden D, Bergmann KC, Bielory L, Caraballo L, Cecchi L, Cepeda SAM, Chong Neto HJ, Galán C, Gonzalez Diaz SN, Idriss S, Popov T, Ramon GD, Ridolo E, Rottem M, Songnuan W, Rouadi P (2020) International expert consensus on the management of allergic rhinitis (AR) aggravated by air pollutants: impact of air pollution on patients with AR: Current knowledge and future strategies. World Allergy Organ J 13:100106

Neumann H-D (2011) Air quality and ventilation in schools. Gefahrst Reinh Luft 71:495–497. [in German]

Neumann H-D, Buxtrup M, Sander A, Klus K (2018) Molds in sediment dust from classrooms. Gefahrst Reinh Luft 78:301–315. [in German]

Nevalainen A, Täubel M, Hyvärinen A (2015) Indoor fungi: companions and contaminants. Indoor Air 25:125–156

Öder S, Weichenmeier I, Dietrich S, Schober W, Pusch G, Jörres RA, Schierl R, Nowak D, Fromme H, Behrendt H, Buters J (2012) Toxicity and elemental composition of particulate matter from outdoor and indoor air of elementary schools in Munich, Germany. Indoor Air 22: 148–158
Oliver J, Birmingham K, Crewes A, Weeks J, Carswell F (1995) Allergen levels in airborne and surface dust. Int Arch Allergy Immunol 107:452–453
Olsen SJ, Chang HL, Cheung TY, Tang AFY, Fisk TL, Ooi SPL, Kuo H-W, Jiang DD-S, Chen K-T, Lando J, Hsu K-H, Chen T-J, Dowell SF (2003) Transmission of severe acute respiratory syndrome on aircraft. N Engl J Med 349:2416–2422
Oluwole O, Rennie DC, Senthilselvan A, Dyck R, Afanasieva A, Kirychuk S, Katselis G, Lawson JA (2018) The association between endotoxin and beta-(1→3)-D-glucan in house dust with asthma severity among schoolchildren. Respir Med 138:38–46
Ong SWX, Tan YK, Chia PY, Lee TH, Ng OT, Wong MSY, Marimuthu K (2020) Air, surface environmental, and personal protective equipment contamination by Severe Acute Respiratory Syndrome Coronavirus 2 (SARS-CoV-2) from a symptomatic patient. JAMA 323:1610–1612
Ong SWX, Tan YK, Coleman KK, Tan BH, Leo YS, Wang DL, Ng CG, Ng OT, Wong MSY, Marimuthu K (2021) Lack of viable severe acute respiratory coronavirus virus 2 (SARS-CoV-2) among PCR-positive air samples from hospital rooms and community isolation facilities. Infect Control Hosp Epidemiol 42:1327–1332
Otter JA, Donskey C, Yezli S, Douthwaite S, Goldenberg SD, Weber DJ (2016) Transmission of SARS and MERS coronaviruses and influenza virus in healthcare settings: the possible role of dry surface contamination. J Hospital Infect 92:235–250
Pan J, Harb C, Leng W, Marr LC (2021) Inward and outward effectiveness of cloth masks, a surgical mask, and a face shield. Aerosol Sci Technol 55:718–733
Park JH, Spiegelman DL, Burge HA, Gold DR, Chew GL, Milton DK (2000) Longitudinal study of dust and airborne endotoxin in the home. Environ Health Perspect 108:1023–1028
Parker JA, Boles C, Buerger AN, Fung ES, Maier A (2021) Derivation of an occupational exposure limit for β-glucans. Regul Toxicol Pharmacol 123:04959
Passos RG, Silveira MB, Abrahão JS (2021) Exploratory assessment of the occurrence of SARS-CoV-2 in aerosols in hospital facilities and public spaces of a metropolitan center in Brazil. Environ Res 195:110808
Patchett K, Lewis S, Crane J, Fitzharris P (1997) Cat allergen (Fel d 1) levels on school children's clothing and in primary school classrooms in Wellington, New Zealand. J Allergy Clin Immunol 100:755–759
Pavilonis BT, Anthony TR, O'Shaughnessy PT, Humann MJ, Merchant JA, Moore G, Thorne PS, Weisel CP, Sanderson WT (2013) Indoor and outdoor particulate matter and endotoxin concentrations in an intensely agricultural county. J Expo Sci Environ Epidemiol 23:299–305
Permaul P, Hoffman E, Fu C, Sheehan W, Baxi S, Gaffin J, Lane J, Bailey A, King E, Chapman M, Gold D, Phipatanakul W (2012) Allergens in urban schools and homes of children with asthma. Pediatr Allergy Immunol 23:543–549
Perry TT, Vargas PA, Bufford J, Feild C, Flick M, Simpson PM, Hamilton RG, Jones SM (2008) Classroom aeroallergen exposure in Arkansas head start centers. Ann Allergy Asthma Immunol 100:358–363
Polizzi V, Delmulle B, Adams A, Moretti A, Susca A, Picco AM, Rosseel Y, Kindt R, Van Bocxlaer J, De Kimpe J, Van Peteghema C, De Saeger S (2009) JEM Spotlight: fungi, mycotoxins and microbial volatile organic compounds in mouldy interiors from water-damaged buildings. J Environ Monit 11:1849–1858
Qian H, Miao T, Liu L, Zheng X, Luo D, Li Y (2021) Indoor transmission of SARS-CoV-2. Indoor Air 31:639–645
Rai S, Singh DK, Kumar A (2021) Microbial, environmental and anthropogenic factors influencing the indoor microbiome of the built environment. J Basic Microbiol 61:267–292
Raja S, Xu Y, Ferro AR, Jaques PA, Hopke PK (2010) Resuspension of indoor aeroallergens and relationship to lung inflammation in asthmatic children. Environ Int 36:8–14

Ramagopal M, Wang Z, Black K, Hernandez M, Stambler AA, Emoekpere OH, Mainelis G, Shalat SL (2014) Improved exposure characterization with robotic (PIPER) sampling and association with children's respiratory symptoms, asthma and eczema. J Expo Sci Environ Epidemiol 24: 421–427

Rasuli L, Dehghani MH, Aghaei M, Mahvi AH, Mubarak NM, Karri RR (2022) Occurrence and fate of bacterial endotoxins in the environment (air, water, wastewater) and remediation technologies: an overview. Chemosphere 303(Pt 2):135089

Razzini K, Castrica M, Menchetti L, Maggi L, Negroni L, Orfeo NV, Pizzoccheri A, Stocco M, Muttini S, Balzaretti CM (2020) SARS-CoV-2 RNA detection in the air and on surfaces in the COVID-19 ward of a hospital in Milan, Italy. Sci Total Environ 742:140540

Reponen T, Seo S-C, Grimsley F, Lee T, Crawford C, Grinshpun SA (2007) Fungal fragments in moldy houses: a field study in homes in New Orleans and Southern Ohio. Atmos Environ 41: 8140–8149

Reponen T, Singh U, Schaffer C, Vesper S, Johansson E, Adhikari A, Grinshpun SA, Indugula R, Ryan P, Levin L, Lemasters G (2010) Visually observed mold and moldy odor versus quantitatively measured microbial exposure in homes. Sci Total Environ 408:5565–5574

Rering CC, Gaffke AM, Rudolph AB, Beck JJ, Alborn HAT (2020) A comparison of collection methods for microbial volatiles. Front Sustain Food Syst 4:598967

Ribaric NL, Vincent C, Jonitz G, Hellinger A, Ribaric G (2022) Hidden hazards of SARS-CoV-2 transmission in hospitals: a systematic review. Indoor Air 32:e12968

Richard JL, Plattner RD, May J, Liska SL (1999) The occurrence of ochratoxin A in dust collected from a problem household. Mycopathologia 146:99–103

Roda C, Barral S, Ravelomanantsoa H, Dusseaux M, Tribout M, LeMoullec Y, Momasa I (2011) Assessment of indoor environment in Paris child daycare centers. Environ Res 111:1010–1017

Rosca EC, Heneghan C, Spencer EA, Brassey J, Plüddemann A, Onakpoya IJ, Evans DH, Conly JM, Jefferson T (2021) Transmission of SARS-CoV-2 associated with aircraft travel: a systematic review. J Travel Med 28:taab133

Rothe C, Schunk M, Sothmann P, Bretzel G, Froeschl G, Wallrauch C, Zimmer T, Thiel V, Janke C, Guggemos W, Seilmaier M, Drosten C, Vollmar P, Zwirglmaier K, Zange S, Wölfel R, Hoelscher M (2020) Transmission of 2019-nCoV infection from an asymptomatic contact in Germany. N Engl J Med 382:970–971

Ruggieri S, Longo V, Perrino C, Canepari S, Drago G, L'Abbate L, Balzan M, Cuttitta G, Scaccianoce G, Minardi R, Viegi G, Cibella F (2019) Indoor air quality in schools of a highly polluted south Mediterranean area. Indoor Air 29:276–290

Ryan TJ, Beaucham C (2013) Dominant microbial volatile organic compounds in 23 US homes. Chemosphere 90:977–985

Rylander R (1999) Indoor air-related effects and airborne (1-3)-P3-D-glucan. Environ Health Perspect 107(Suppl 3):501–503

Sahlberg B, Gunnbjörnsdottir M, Soon A, Jogi R, Gislason T, Wieslander G, Janson C, Norback D (2013) Airborne molds and bacteria, microbial volatile organic compounds (MVOC), plasticizers and formaldehyde in dwellings in three North European cities in relation to sick building syndrome (SBS). Sci Total Environ 444:433–440

Saito R, Park J-H, LeBouf R, Green BJ, Park Y (2016) Measurement of macrocyclic trichothecene in floor dust of water-damaged buildings using gas chromatography/tandem mass spectrometry—dust matrix effects. J Occup Environ Hyg 13:442–450

Sakaguchi M, Inouye S, Irie T, Miyazawa H, Watanabe M, Yasueda H, Shida T, Nitta H, Chapman MD, Schou C, Aalberse RC (1993) Airborne cat (Fel d I), dog (Can f I), and mite (Der I and Der II) allergen levels in the homes of Japan. J Allergy Clin Immunol 92:797–802

Sakaguchi M, Inouye S, Sasaki R, Hashimoto M, Kobayashi C, Yasueda H (1996) Measurement of airborne mite allergen exposure in individual subjects. J Allergy Clin Immunol 97:1040–1044

Salo PM, Sever ML, Zeldin DC (2009) Indoor allergens in school and day care environments. J Allergy Clin Immunol 124:185–192

Salo PM, Wilkerson J, Rose KM, Cohn RD, Calatroni A, Mitchell HE, Sever ML, Gergen PJ, Thorne PS, Zeldin DC (2018) Bedroom allergen exposures in US households. J Allergy Clin Immunol 141:1870–1879

Salonen H, Duchaine C, Létourneau V, Mazaheri M, Laitinen S, Clifford S, Mikkola R, Lappalainen S, Reijula K, Morawska L (2016) Endotoxin levels and contribution factors of endotoxins in resident, school, and office environments—a review. Atmos Environ 142:360–369

Santarpia JL, Rivera DN, Herrera V, Morwitzer MJ, Creager H, Santarpia GW, Crown KK, Brett-Major D, Schnaubelt E, Broadhurst MJ, Lawler JV, Reid SP, Lowe JJ (2020) Transmission potential of SARS-CoV-2 in viral shedding observed at the University of Nebraska Medical Center. medRxiv 2020.03.23.20039446

Santarpia JL, Herrera VL, Rivera DN, Ratnesar-Shumate S, Reid SP, Ackerman DN, Denton PW, Martens JWS, Fang Y, Conoan N, Callahan MV, Lawler JV, Brett-Major DM, Lowe JJ (2022) The size and culturability of patient-generated SARS-CoV-2 aerosol. J Expo Sci Environ Epidemiol 32:706–711

Sauni R, Uitti J, Jauhiainen M, Kreiss K, Sigsgaard T, Verbeek JH (2013) Remediating buildings damaged by dampness and mould for preventing or reducing respiratory tract symptoms, infections and asthma (Review). Evid Based Child Health 8:944–1000

Schijven J, Vermeulen LC, Swart A, Meijer A, Duizer E, de Roda Husman AM (2021) Exposure assessment for airborne transmission of SARS-CoV-2 via breathing, speaking, coughing and sneezing. Environ Health Perspect 129:47002

Schleibinger H, Laussmann D, Bornehag CG, Eis D, Rueden H (2008) Microbial volatile organic compounds in the air of moldy and mold-free indoor environments. Indoor Air 18:113–124

Schuchardt S, Strube A (2009) Microbial volatile organic compounds in moldy interiors: A long-term climate chamber study. J Basic Microbiol 53:532–538

Schwartz KL, Murti M, Finkelstein M, Leis JA, Fitzgerald-Husek A, Bourns L, Meghani H, Saunders A, Allen V, Yaffe B (2020) Lack of COVID-19 transmission on an international flight. CMAJ 192:E410

Sheehan WJ, Rangsithienchai PA, Muilenberg ML, Rogers CA, Lane JP, Ghaemghami J, Rivard DV, Otsu K, Hoffman EB, Israel E, Gold DR, Phipatanakul W (2009) Mouse allergens in urban elementary schools and homes of children with asthma. Ann Allergy Asthma Immunol 102: 125–130

Shinohara N, Hashimoto K, Kim H, Yoshida-Ohuchi H (2023) Fungi, mites/ticks, allergens, and endotoxins in different size fractions of house dust from long-term uninhabited houses and inhabited houses. Build Environ 229:109918

Shiu EYC, Huang W, Ye D, Xie Y, Mo J, Li Y, Cowling BJ, Yang Z, Leung NHL (2020) Frequent recovery of influenza A but not influenza B virus RNA in aerosols in pediatric patient rooms. Indoor Air 30:805–815

Silva PG, Branco PTBS, Soares RRG, Mesquita JR, Sousa SIV (2022) SARS-CoV-2 air sampling: a systematic review on the methodologies for detection and infectivity. Indoor Air 32:e13083

Singh U, Reponen T, Cho KJ, Grinshpun SA, Adhikari A, Levin L, Indugula R, Green BJ (2011) Airborne endotoxin and β-D-glucan in PM_1 in agricultural and home environments. Aerosol Air Qual Res 11:376–386

Smedje G, Norbäck D (2001) Irritants and allergens at school in relation to furnishings and cleaning. Indoor Air 11:127–133

Stadnytskyi V, Bax CE, Bax A, Anfinrud P (2020) The airborne lifetime of small speech droplets and their potential importance in SARS-CoV-2 transmission. Proc Natl Acad Sci 117:11875–11877

Stein RA (2011) Super-spreaders in infectious diseases. Int J Infect Dis 15:e510–e513

Stemeseder T, Schweidler B, Doppler P, Klinglmayr E, Moser S, Lueftenegger L, Himly M, Lang R, Zumbach J, Oostingh GJ, Hawranek T, Bathke AC, Gadermaier G (2017) Exposure to indoor allergens in different residential settings and its influence on IgE sensitization in a geographically confined Austrian cohort. PLoS One 12:e0168686

Streek H, Schulte B, Kümmerer B, Richter E, Höller T, Fuhrmann C, Bartok E, Dolscheid R, Berger M, Wessendorf L, Eschbach-Bludau M, Kellings A, Schwaiger A, Coenen M, Hoffmann P, Stoffel-Wagner B, Nothen MM, Eis-Hubinger A-M, Exner M, Schmithausen RM, Schmid M, Hartmann G (2020) Infection fatality rate of SARS-CoV-2 infection in a German community with a super-spreading event. Nat Commun 11:5829

Sun L, Miller JD, Van Ryswyk K, Wheeler AJ, Héroux ME, Goldberg MS, Mallach G (2022) Household determinants of biocontaminant exposures in Canadian homes. Indoor Air 32: e12933

Swadi T, Geoghegan JL, Devine T, McElnay C, Sherwood J, Shoemack P, Ren X, Storey M, Jefferies S, Smit E, Hadfield J, Kenny A, Jelley L, Sporle A, McNeill A, Reynolds GE, Mouldey K, Lowe L, Sonder G, Drummond AJ, Huang S, Welch D, Holmes EC, French N, Simpson CR, de Ligt J (2021) Genomic evidence of in-flight transmission of SARS-CoV-2 despite predeparture testing. Emerg Infect Dis 27:687–693

Swanson MC, Campbell AR, Klauck MJ, Reed CF (1989) Correlation between levels of mite and cat allergens in settled and airborne dust. J Allergy Clin Immunol:776–783

Täubel M, Sulyok M, Vishwanath V, Bloom E, Turunen M, Järvi K, Kauhanen E, Krska R, Hyvärinen A, Larsson L, Nevalainen A (2011) Co-occurrence of toxic bacterial and fungal secondary metabolites in moisture-damaged indoor environments. Indoor Air 21:368–375

Thorn RMS, Greenman J (2012) Microbial volatile compounds in health and disease conditions. J Breath Res 6:024001

Thorne PS, Kulhánková K, Yin M, Cohn R, Arbes SJ, Zeldin DC (2005) Endotoxin exposure is a risk factor for asthma: the national survey of endotoxin in U.S. Housing. Am J Respir Crit Care Med 172:1371–1377

Tortolero SR, Bartholomew LK, Tyrrell S, Abramson SL, Sockrider MM, Markham CM, Whitehead LW, Parcel GS (2002) Environmental allergens and irritants in schools: a focus on asthma. J Sch Health 72:33–38

Toyokawa T, Shimada T, Hayamizu T, Sekizuka T, Zukeyama Y, Yasuda M, Nakamura Y, Okano S, Kudaka J, Kakita T, Kuroda M, Nakasone T (2022) Transmission of SARS-CoV-2 during a 2-h domestic flight to Okinawa, Japan, March 2020. Influenza Other Respir Viruses 16: 63–71

Tran NT, Nguyen DKT, Tran TTT, Hauglustaine J-M, Michel O, Bouland C (2023) Indoor environmental parameters are related to house typology in the South of Vietnam. ResearchSquare

Tsuda A, Henry FS, Butler JP (2013) Particle transport and deposition: basic physics of particle kinetics. Compr Physiol 3:1437–1471

UBA (2019) Guideline on the prevention, detection and remediation of moulds in buildings. Berlin. https://www.umweltbundesamt.de/sites/default/files/medien/1410/publikationen/191022_uba_schimmelleitfaden_engl_bf.pdf

Uhde E, Salthammer T, Wientzek S, Springorum A, Schulz J (2022) Effectiveness of air-purifying devices and measures to reduce the exposure to bioaerosols in school classrooms. Indoor Air 32: e13087

Unfallkasse (Unfallkasse Nordrhein-Westfalen) (2014) Healthy air in schools – Part 2. Assessment of CO_2 concentration and thermal comfort in classrooms. Düsseldorf. [in German]

Van Dijken F, van Bronswijk JEMH, Sundell J (2006) Indoor environment and pupils'health in primary schools. Build Res Inform 34:437–446

Van Doremalen N, Bushmaker T, Morris DH, Holbrook MG, Gamble A, Williamson BN, Tamin A, Harcourt JL, Thornburg NJ, Gerber SI, Lloyd-Smith JO, de Wit E, Munster VJ (2020) Aerosol and surface stability of SARS-CoV-2 as compared with SARS-CoV-1. New Engl J Med 382: 1564–1567

Van Dyke MW, Martyny JW, Marola J, Ramamoorthy P, Ridder A, Harbeck RJ, Rose CS (2012) Efficacy of occupant-collected dust samples in the evaluation of residential allergen and fungal lvels. J Occup Environ Hyg 9:14–24 van Kampen JJA, van de Vijver DAMC, Fraaij PLA, Haagmans BL, Lamers MM, Okba N, van den Akker JPC, Endeman H, Gommers DAMPJ, Cornelissen JJ, Hoek RAS, van der Eerden MM, Hesselink DA, Metselaar HJ, Verbon A, de Steenwinkel JEM, Aron GI, van Gorp ECM, van Boheemen S, Voermans JC, Boucher CAB, Molenkamp R, Koopmans MPG, Geurtsvankessel C, van der Eijk AA (2021) Duration and key determinants of infectious virus shedding in hospitalized patients with coronavirus disease-2019 (COVID-19). Nat Commun 12:267

Vass WB, Lednicky JA, Shankar SN, Fan ZH, Eiguren-Fernandez A, Wu CY (2022) Viable SARS-CoV-2 Delta variant detected in aerosols in a residential setting with a self-isolating college student with COVID-19. J Aerosol Sci 165:106038

Vosoughi M, Karami C, Dargahi A, Jeddi F, Jalali KM, Hadisi A, Haghighi SB, Dogahe HP, Noorimotlagh Z, Mirzaee SA (2021) Investigation of SARS-CoV-2 in hospital indoor air of COVID-19 patients' ward with impinger method. Environ Sci Pollut Res Int 28:50480–50488

Vuorinen V, Aarnio M, Alava M, Alopaeus V, Atanasova N, Auvinen M, Balasubramanian N, Bordbar H, Erästö P, Grande R, Hayward N, Hellsten A, Hostikka S, Hokkanen J, Kaario O, Karvinen A, Kivistö I, Korhonen M, Kosonen R, Kuusela J, Lestinen S, Laurila E, Nieminen HJ, Peltonen P, Pokki J, Puisto A, Råback P, Salmenjoki H, Sironen T, Österberg M (2020) Modelling aerosol transport and virus exposure with numerical simulations in relation to SARS-CoV-2 transmission by inhalation indoors. Saf Sci 130:104866

Walser SM, Gerstner DG, Brenner B, Bünger J, Eikmann T, Janssen B, Kolb S, Kolk A, Nowak D, Raulf M, Sagunski H, Sedlmaier N, Suchenwirth R, Wiesmüller G, Wollin KM, Tesseraux I, Herr CE (2015) Evaluation of exposure – response relationships for health effects of microbial bioaerosols – a systematic review. Int J Hyg Environ Health 18:577–589

Walser SM, Brenner B, Heinze S, Szewzyk R, Wolter E, Herr CEW (2017) Environmental health relevance of airbornemicroorganisms in ambient and indoor air. Bundesgesundheitsbl 60:618–624. [in German]

Wang Z, Galea ER, Grandison A, Ewer J, Jia F (2021) Inflight transmission of COVID-19 based on experimental aerosol dispersion data. J Travel Med 28:taab023

Waser M, Schierl R, von Mutius E, Maisch S, Carr D, Riedler J, Eder W, Schreuer M, Nowak D, Braun-Fahrländer C (2004) Determinants of endotoxin levels in living environments of farmers' children and their peers from rural areas. Clin Exp Allergy 34:389–397

Wei J, Li Y (2015) Enhanced spread of expiratory droplets by turbulence in a cough jet. Build Environ 93:86–96

Wei J, Li Y (2016) Airborne spread of infectious agents in the indoor environment. Am J Infect Control 44:S102–S108

Wheeler AJ, Dobbin NA, Lyrette N, Wallace L, Foto M, Mallick R, Kearney J, Van Ryswyk K, Gilbert NL, Harrison I, Rispler K, Héroux M-E (2011) Residential indoor and outdoor coarse particles and associated endotoxin exposures. Atmos Environ 45:7064–7071

WHO (2009) WHO guidelines for indoor air quality: dampness and mould. WHO Regional Office for Europe, Copenhagen, Denmark. https://www.who.int/airpollution/guidelines/dampness-mould/en/

WHO (2020) Transmission of SARS-CoV-2: implications for infection prevention precautions. Scientific brief. 09 July 2020. https://www.who.int/news-room/commentaries/detail/transmission-of-sars-cov-2-implications-for-infection-prevention-precautions

Williams AH, Smith JT, Hudgens EE, Rhoney S, Ozkaynak H, Hamilton RG, Gallagher JE (2011) Allergens in household dust and serological indicators of atopy and sensitization in Detroit children with history-based evidence of asthma. J Asthma 48:674–684

Wolf M, Lai PS (2020) Indoor microbial exposures and chronic lung disease: from microbial toxins to the microbiome. Clin Chest Med 41:777–796

Wu F, Groopman JD, Pestka JJ (2014) Public health impacts of foodborne mycotoxins. Annu Rev Food Sci Technol 5:351–372

Wu S, Wang Y, Jin X, Tian J, Liu J, Mao Y (2020) Environmental contamination by SARS-CoV-2 in a designated hospital for coronavirus disease 2019. Am J Infect Control 48:910–914

Xie X, Li Y, Chwang ATQ, Ho PL, Seto WH (2007) How far droplets can move in indoor environments – revisiting the Wells evaporation–falling curve. Indoor Air 17:211–225

Yang X, Dong N, Chan EW-C, Chen S (2020) Genetic cluster analysis of SARS-CoV-2 and the identification of those responsible for the major outbreaks in various countries. Emerg Microbes Infect 9:1287–1299

Yen YC, Yang CY, Ho CK, Yen PC, Cheng YT, Mena KD, Lee TC, Chen PS (2020a) Indoor ozone and particulate matter modify the association between airborne endotoxin and schoolchildren's lung function. Sci Total Environ 705:135810

Yen YC, Yang CY, Wang TN, Yen PC, Ho CK, Mena KD, Lee TC, Chen KS, Lin YC, Chen PS (2020b) Household airborne endotoxin associated with asthma and allergy in elementary school-age children: a case-control study in Kaohsiung, Taiwan. Environ Sci Pollut Res Int 27:19502–19509

Yoda Y, Tamura K, Shima M (2017) Airborne endotoxin concentrations in indoor and outdoor particulate matter and their predictors in an urban city. Indoor Air 27:955–964

Zahradnik E, Raulf M (2017) Respiratory allergens from furred mammals: environmental and occupational exposure. Vet Sci 4:38

Zhang J, Qin F, Qin X, Li J, Tian S, Lou J, Kang X, Lian H, Niu S, Zhang W, Chen Y (2021) Transmission of SARS-CoV-2 during air travel: a descriptive and modelling study. Ann Med 53:1569–1575

Zhao ZH, Elfman L, Wang ZH, Zhang Z, Norbäck D (2006) A comparative study of asthma, pollen, cat and dog allergy among pupils and allergen levels in schools in Taiyuan city, China, and Uppsala, Sweden. Indoor Air 16:404–413

Zhou J, Otter JA, Price JR, Cimpeanu C, Garcia DM, Kinross J, Boshier PR, Mason S, Bolt F, Holmes AH, Barclay WS (2021) Investigating SARS-CoV-2 surface and air contamination in an acute healthcare setting during the peak of the COVID-19 pandemic in London. Clin Infect Dis 73:e1870–e1877

Zock JP, Brunekreef B (1995) House dust mite allergen levels in dust from schools with smooth and carpeted classroom floors. Clin Exp Allergy 25:549–553

Chapter 7
Inorganic Gases

Contents

Abstract The focus of this chapter are inorganic gases like the air quality indicator carbon dioxide (CO_2), carbon monoxide (CO), nitrogen dioxide (NO_2), ammonia (NH_3), sulfur dioxide (SO_2), and ozone (O_3). They are presented individually in detail. In addition to a brief description of the chemical-physical characteristics and the use of the substance, an overview of main health risks and, in particular, references to further scientific literature are given. In the foreground, however, are extensive presentations of the occurrence of the substances in indoor air and in sedimented dust.

Keywords Inorganic gases · NO_2 · Ozone · CO_2 · Ammonia

7.1 Carbon Dioxide (CO_2)

Carbon dioxide (CO_2) is an odorless and colorless gas that is a natural component of atmospheric air (approx. 0.03% by volume, equivalent to 300 ppm). It has become known because of its climatic effect in outdoor air, as it promotes the "greenhouse effect" by absorbing infrared radiation in the CO_2 molecule. In addition to humans themselves, various combustion processes, such as burning candles, operating oil and gas lights, heaters and gas stoves, and tobacco smoking are possible indoor sources of CO_2. Leaky chimneys and soil gases can also lead to indoor pollution where appropriate.

H. Fromme, *Indoor Air Quality*, https://doi.org/10.1007/978-3-031-40078-0_7

Table 7.1 Classification of indoor air quality according to DIN EN 13779:2007–09 (DIN EN 2007)

Indoor air category	Description	Increase in CO_2 concentration compared to outside air (ppm)	Absolute CO_2 concentration in indoor air (ppm)	Ventilation rate (l/s person)	Outdoor air flow rate (m^3/h person)
IDA 1	High indoor air quality	≤ 400	≤ 800	> 15	>54
IDA 2	Average indoor air quality	> 400–600	> 800–1000	10–15	>36–54
IDA 3	Moderate indoor air quality	> 600–1000	> 1000–1400	6–10	>22–36
IDA 4	Low indoor air quality	> 1000	> 1400	< 6	<22

The table contains the specifications of DIN EN 13779 in columns 1–3 as well as 5 and 6. Column 4 shows absolute CO_2 concentrations in the indoor air for a CO_2 outdoor air concentration of 400 ppm as an example

However, the most important source of CO_2 indoors is the human being himself. Depending on the type of activity and physical exertion (activity), an adult person exhales between approx. 10 and 200 liters of CO_2 per hour. For the hygienic assessment of indoor spaces, Max Josef von Pettenkofer therefore already proposed appropriate hygiene standards in the late nineteenth century (Pettenkofer 1858). The so-called Pettenkofer number of 1000 ppm CO_2, named after him, characterizes an air-hygienically acceptable range in this context. In addition, there are now other values available. For example, according to the old ventilation standard DIN 1946 (Part 2), a value of 1500 ppm should not be exceeded, which was used to assess the deterioration of the indoor air due to odorous substances and evaporation from people (DIN 2019). However, it is generally recommended only as a hygienic indoor air guideline value for recreation rooms without the special release of working substances. It was originally developed for rooms with ventilation systems where sedentary or light activity is performed. In the meantime, the above-mentioned standard has been replaced by DIN EN 13779, which classifies the indoor air quality of occupied rooms in non-residential buildings based on the CO_2 content (reduced by the CO_2 outdoor air concentrations). Four quality categories are distinguished (see Table 7.1). DIN EN 13779 refers to rooms in which smoking is not permitted and contamination is mainly caused by human metabolism (DIN EN 2007). However, no justification, not even a health justification, is given for the categorization made. In addition, the German Committee on Indoor Guideline Values has developed so-called hygienic guideline values to be used for assessment in classrooms of schools (Ad-hoc-AG 2008). Here, a distinction is made between a hygienically harmless range of <1000 ppm CO_2 and a hygienically unacceptable range of

>2000 ppm CO_2. If the harmless range is exceeded, further organizational, ventilation, or structural measures must be taken, because ventilation alone does not permanently lower the CO_2 content. In the range in between, ventilation measures should be taken and ventilation plans drawn up.

From the known data of human CO_2 production, it can be calculated that an adult person can increase the carbon dioxide content in a room with a volume of around 35 m^3 from the background level (300 ppm) to approximately 1000 ppm CO_2 in one hour. Accordingly, the same amount of fresh air must be supplied to the room every hour to keep the amount of CO_2 produced by a person in the concentration range below 1000 ppm. This results in a fresh air rate of 9 liters/second and person. To achieve the quality criterion of "moderate" indoor air quality according to DIN EN 13779 (increase in CO_2 concentration compared to outdoor air between 600 and 1000 ppm), an outdoor air volume flow rate of fan-assisted ventilation systems of 22–36 m^3/h per student is required (see Table 7.1). ASHRAE 62.1–2019 recommends standards to specify minimum ventilation rates and other measures to provide indoor air quality that is acceptable for humans and that minimizes health effects (ASHRAE 2019a). Again, ASHRAE 62.2–2019 defines minimum requirements for mechanical and natural ventilation systems to provide acceptable indoor air (ASHRAE 2019b).

Villanueva et al. (2022) summarized the different methods for CO_2 sampling and analysis and their advantages and limitations.

7.1.1 Health Aspects

An essential physiological function of carbon dioxide in the organism is that it regulates respiration via the chemoreceptors in the organism, which reflexively excite the respiratory center in the brainstem. Elevated CO_2 concentrations in the inhaled air therefore subsequently increase the respiratory rate and the respiratory volume. Even in the range of 1000–2000 ppm CO_2 in the breath, oxygen transport in the blood decreases due to the slight shift in blood pH (Azuma et al. 2018). This can lead to headaches, drowsiness, and a reduction in performance. Effects described in the scientific literature on the human organism include an increase in respiratory rate, a change in blood pH, and a decrease in physical performance. At concentrations above 30,000–40,000 ppm, headaches, dizziness, ringing in the ears, palpitations may occur. Only at concentrations above 100,000 ppm do more serious health disorders occur (Azuma et al. 2018).

It has been described relatively consistently in the scientific literature that reduced ventilation in buildings is associated with an increase in CO_2 and non-specific health symptoms or reduced performance/attention (Seppänen and Fisk 2004; Erdmann and Apte 2004; Wargocki and Wyon 2006; Twardella et al. 2012; Petersen et al. 2016; Maula et al. 2017; Künn et al. 2019; Du et al. 2020). Overall, based on the available studies, it is assumed that it is not possible to designate a CO_2 concentration above which adverse health effects occur in classrooms (HCN 2010).

Nevertheless, using a within-subjects counterbalanced study 31 volunteers were exposed in a naturally ventilated office room during short time (<60 min) to 830 ppm and 2700 ppm CO_2 and cognitive performance, subjective and neurophysiological parameters were evaluated (Snow et al. 2019). They reported a lack of an expected performance improvement in executive function and cognitive flexibility parameters using two test batteries when CO_2 was higher and this absence can occur without changes to SBS symptoms or perceived air quality and could not be explained by other factors.

7.1.2 *Occurrence Indoors*

Carbon dioxide is an easily determined parameter in indoor air that can be used as an indicator of indoor air quality. It depends on the number of people in the room, their physical activity, and the current ventilation conditions.

A special exposure situation exists in common rooms of, e.g., schools and daycare centers since here many people meet in a relatively small space and the ventilation conditions are often inadequate. Figure 7.1 shows an example of the course of the minute-by-minute individual measurements in a typical classroom. It is clear that the CO_2 concentration increases continuously in the course of the teaching

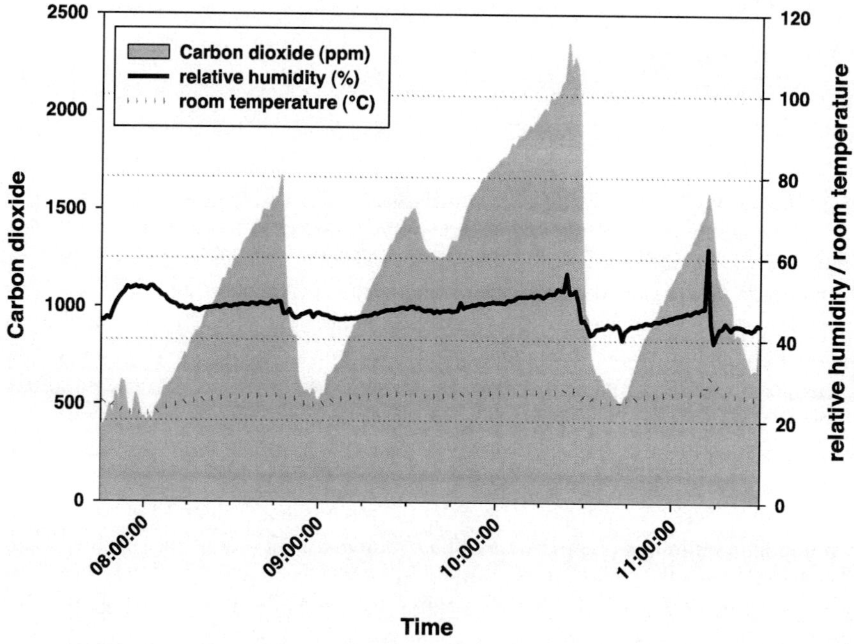

Fig. 7.1 Typical course of carbon dioxide (CO_2) levels and various room climate parameters during teaching time in classrooms

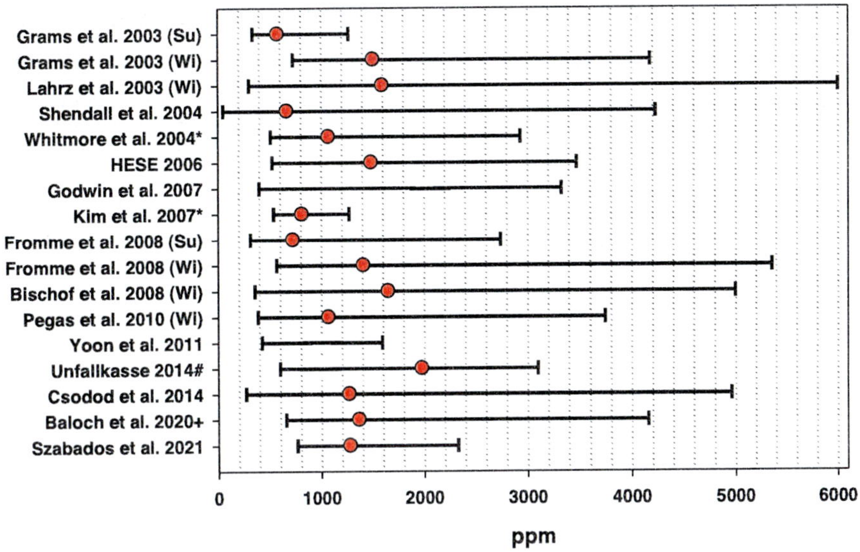

Fig. 7.2 Minimum, median, and maximum carbon dioxide (CO_2) levels in classrooms (*Wi* winter, *Su* summer; * arithmetic mean; + tenth and 99th percentile; # 95th percentile)

time depending on the number of people using the room and that a certain drop in the CO_2 concentrations is only observed when the room is ventilated during the breaks. However, it is also evident that it cannot be assumed that intensive ventilation exclusively during the long break can always lower the room air content below, for example, 1500 ppm or into a hygienically satisfactory range. The situation is particularly critical in the winter months, as ventilation is particularly infrequent during this period.

The German working group on ventilation assumes that, according to the current state of science and technology, confirmed by numerous measurements and simulation calculations, window ventilation alone is not capable of meeting the requirements for indoor air quality (defined via the CO_2 content) throughout the year (UBA 2017a). Therefore, the aforementioned working group recommends that to ensure health protection, it is necessary to plan for additional demand-controlled technical ventilation for schools in addition to optimally designed windows. This is to be done using a detailed ventilation concept, which is to be taken into account already during planning, but also during later use. It is recommended that the required personal ventilation rate be implemented as so-called hybrid ventilation, i.e. as a combination of mechanical basic ventilation and additional ventilation via windows as required.

There are many results in the scientific literature where high CO_2 content in indoor school air has been used as an indicator of inadequate fresh air supply (summarized, e.g., in Daisey et al. 2003; Salthammer et al. 2016).

Exemplary measurement results from European classrooms are summarized in Fig. 7.2. Some very extensive studies have investigated CO_2 in school classrooms. For example, in the Health Effects of the School Environment (HESE) study in

2004–2005, a median of 1490 ppm (range: 525–3475 ppm) was observed in 48 classrooms of 21 schools (HESE 2006). In addition, the Schools Indoor Pollution and Health: Observatory Network in Europe (SINPHONIE) study examined a total of 340 classrooms in 23 European countries in 2010–2012 (Csobod et al. 2014) with a median of 1257 ppm (range: 269–4960 ppm. Moreover, Szabados et al. (2021) investigated the Transnational Adaption Actions for Integrated Indoor Air Quality Management Project (InAirQ) for 5 consecutive days in 64 primary schools in the Czech Republic, Hungary, Italy, and Slovenia in 2017–2018. They found a median CO_2 value of 1284 ppm (range: 767–2328 ppm). A mean indoor air concentration of 200 to 1900 ppm CO_2 (median: 500 ppm) was observed as relatively good indoor air quality in common rooms of 70 Berlin kindergartens from 2000 to 2001 (Fromme et al. 2002). Again, in 2004/2005, 92 classrooms in Munich and the surrounding area were examined in winter and 76 in summer (Fromme et al. 2008). The median CO_2 content ranged from 598 to 4172 ppm in winter and was significantly lower between 480 and 1875 ppm in summer. It was found that a larger number of users, a smaller room size, and a smaller room volume were significantly associated with the deterioration of air quality in terms of an increase in CO_2. When only the time when students are present in the classroom is considered, they are exposed to CO_2 levels of >1500 ppm for a median of about 59% (winter) and 0% (summer) of class time, and to CO_2 levels of >1000 ppm for a median of about 90% (winter) and 20% (summer) of class time. The CO_2 concentrations and the percentage of exceedance during class time of the German studies are summarized in Table 7.2.

Similar results were reported in 64 elementary and middle schools in the USA in 2003 with means during school days between 553 and 1522 ppm (Godwin and Batterman 2007) and in 71 classrooms of 17 preschools in Korea with means of 847 ppm in urban settings and 607 ppm in rural regions (Yoon et al. 2011). In Beijing, China, CO_2 was measured in 33 classrooms in 2017 to 2018, with mean daily concentrations in classrooms with mechanical ventilation (MV) and natural ventilation (NV) of 953 ppm and 1274 ppm in transitional seasons, 937 ppm and 1785 ppm in summer, and 1093 ppm and 1448 ppm in winter, respectively (Cai et al. 2021). The proportions of CO_2 concentrations > 1000 ppm when the classrooms were occupied were 40% and 60% in MV and NV classrooms. Although, the maximum CO_2 concentration in NV classrooms occasionally exceeded 5000 ppm. In the nationwide Chinese Indoor Environment and Health Surveillance 2018 a total of 66 classrooms in primary schools were measured and a median of 600 ppm (range: 100–5500 ppm) was observed (Zhu et al. 2021).

Overall, the available CO_2 results, especially during the winter period, indicate a frequently unsatisfactory indoor air quality in schools, which must be considered unacceptable from a hygienic point of view.

Table 7.2 Carbon dioxide concentration and exceedance percentages during teaching time in classrooms in German schools (according to Ad-hoc-AG 2008)

Source	Number	Median (Min-Max)	% Exceedance > 1000 ppm during class time	% Exceedance > 2000 ppm during class time	Measuring period	Comments
Fromme et al. (2008)	90	1412 (367–5359)	82	30	Winter 2004/2005	Measurements during class time; per
	75	728 (314–2742)	29	5	Summer 2005	Classroom 1 full day of classes
Lahrz et al. (2003)	39	1600 (300–6000)	80	17	Winter 2002/2003	Measurements during class time; 1 full class day per classroom
Bischof et al. (2008)	220	1654 (355–4998)	86	32	Winter 2005/2006	Measurement during teaching time, one school hour per classroom, windows closed during measurement
Gabrio and Volland (2007)	19	1459 (431–4380)	79	23	Winter 2005/2007	Evaluation of instructional time, per class-room 1 whole
	18	820 (304–3554)	36	3	Summer 2006	Teaching day (from measurements of whole school days)
Grams et al. (2003)	36	1510 (730–4177)	89[a]		Winter 2001/2002	Summer and winter measurements were taken at the same 7 schools
	22	581 (339–1270)	32[a]		Summer 2001/2002	Measurements were typically taken over approximately 48 hours/classroom

[a] Percent exceedance at 1500 ppm

7.2 Carbon Monoxide (CO)

Carbon monoxide (CO) is a colorless, odorless, tasteless, non-irritant gas. It is produced naturally and anthropogenically in all incomplete biomass combustion processes. Important sources in indoor air, in addition to the—usually small—contribution from outdoor air, are cooking or heating with fossil fuels and tobacco smoking. In addition, defective or improperly operated heating systems as well as ornamental and heating fireplaces in living spaces can lead to significant loads. Open indoor fireplaces are a significant source of CO. Villanueva et al. (2022) reported different methods for CO sampling and analysis, reviewing their advantages and limitations.

Conversion between different common units: 1 ppm = 1.145 mg/m^3 and 1 mg/m^3 = 0.873 ppm.

7.2.1 *Health Aspects*

Carbon monoxide can be rapidly absorbed via the alveoli and, due to its 20 to 300 times higher affinity for hemoglobin than oxygen, leads to a reduced oxygen transport capacity and a reduced delivery of oxygen to the peripheral tissues. Depending on the concentration level and exposure duration, headaches, dizziness, fatigue, angina pectoris, confusion, cardiac arrhythmias, apathy, and finally unconsciousness occur (Kinoshita et al. 2020). In Germany, about 75,000 hospital treatments and 3000 deaths must be expected annually (Kaiser and Schaper 2012). While poisoning cases were on the decline until 2008, they have been rising continuously since 2009. In the 28 EU Member States a total of 140,490 CO-related deaths were reported with an annual death rate of 2.2 of 100,000) within the reporting period from 1980 to 2008 (Braubach et al. 2013).

An increasing problem is the exposure of hookah smokers. Due to the high levels of carbon monoxide produced during the smoking process, cases of poisoning are observed time and again (Fromme and Schober 2015). It must be assumed that, in addition, many cases of poisoning remain undetected or are not attributed to smoking a shisha.

Individuals with pre-existing lung and cardiovascular conditions are particularly sensitive to higher CO concentrations (Barn et al. 2018). In this context, pregnant women and fetuses must also be considered a risk group for CO effects.

Experimental subject studies and a large number of epidemiological studies showed acute and chronic effects of the cardiovascular system at a mean content between 0.6 and 10.9 mg/m^3 (WHO 2010, 2021). In light of this, the WHO recommends that the following time-graded guideline levels should be used: 15 minutes—100 mg/m^3, 1 hour—35 mg/m^3, 8 hours—10 mg/m^3, and 24 hours—4 mg/m^3.

7.2.2 *Occurrence in Indoor Air*

In the absence of the aforementioned indoor sources, very low concentrations, often in the range of the limit of quantification, can be expected in indoor air. The WHO assumes mean levels between 0.23 and 2.1 mg/m^3 in indoor living spaces (WHO 2010). A mean level of 0.92 mg/m^3 was observed in 12 Portuguese indoor non-smokers bedrooms (Canha et al. 2019). In another study, where personal exposure was determined over 48 hours in 5 European cities, geometric mean values between 0.45 and 2.17 mg/m^3 were shown (Georgoulis et al. 2002). There was a statistically significant correlation with the respective outdoor concentration. In South Korea mean values of 0.11 mg/m^3 (0.11–0.13 mg/m^3) were reported in 40 living and bedrooms in 2010 (Choi et al. 2021), while in 60 bedrooms in Taiwan means of 0.45 mg/m^3 (0–1.43 mg/m^3) using 12-hour measurements and 0.32 mg/m^3 (0–1.28 mg/m^3) in 24 hours, respectively, were found in 2017–2018 (Chen et al. 2021). Within the 2018 National Chinese Indoor Environment and Health Surveillance, 30 classrooms in primary schools were surveyed and medians of 0.44 mg/m^3 (range: 0.02–1.61 mg/m^3) were found (Zhu et al. 2021).

In the presence of indoor sources or in connection with accidents, indoor air levels of several hundred mg/m^3 are possible. Significantly higher concentrations also occur when smoking tobacco products. For example, mean levels in hookah establishments ranged from 8.0 to 20.6 mg/m^3, reaching maximum levels of up to 137 mg/m^3 in North America and 303 mg/m^3 in Arabic hookah cafes (Fromme and Schober 2015). In a German study in which a hookah was smoked for 4 hours in a typical indoor environment, mean concentrations of 55 mg/m^3 could be found (Fromme et al. 2009). Clearly higher levels from 1.7 mg/m^3 to 126 mg/m^3 were observed in 384 buildings in a town in Ethiopia if biomass fuel was the major energy source and if incensing or smoking was common (Gizaw and Teka 2020). High CO concentrations ranging from <6–150 mg/m^3 (median: 40 mg/m^3) could also be measured in indoor environments where gasoline-powered go-karts are used (Wolf et al. 2018).

7.3 Nitrogen Dioxide (NO_2)

Pure nitrogen dioxide is a reddish-brown gas. It has an oxidizing and fire-promoting effect, is heavier than air, and has a pungent odor. Nitrogen oxides are emitted during all types of combustion processes, primarily nitrogen monoxide (NO), which is subsequently oxidized in the atmosphere primarily through reaction with ozone to nitrogen dioxide (NO_2), which is more significant for health. Typical outdoor sources are natural sources (e.g., volcanism, vegetation fires, microbiological transformations in soil) and anthropogenic sources in the deliberate combustion of nitrogen-containing material (e.g., fossil fuels in motorized road transport) (US EPA 2017). Indoors, heating with fossil fuels, cooking with gas, cigarette smoke, and candle fires can contribute significantly to exposure. According to estimates, in 2015, 38% of NO_2 emissions across Germany were caused by transport, 25% by the energy sector, 15% by industry and commerce, 11% by agriculture, and 11% by households/small consumers and other minor activities (UBA 2017b). In the USA, using data of the National Emissions Inventory, substantial reductions of NOx emissions were observed in the last decades, but to day mobile source-related emissions still dominate the emissions. Highway vehicles are the largest source NO_X emissions, contributing 34% of the total emissions (US EPA 2016, 2017).

Conversion between different common units: 1 ppm = 1.882 mg/m^3 and 1 mg/m^3 = 0.531 ppm.

7.3.1 Health Aspects

The health effects of nitrogen dioxide have been studied in numerous clinical studies under controlled exposure conditions, in epidemiological studies, and also in animal experiments (HC 2016a; US EPA 2016; Huangfu and Atkinson 2020; Dominski et al. 2021). Inhaled nitrogen dioxide reaches the lower respiratory tract where it reacts with substances in the mucosa. Long-term exposure to elevated concentrations of nitrogen dioxide can cause emphysema-like remodeling of lung tissue. In controlled studies, evidence of alteration in airway responsiveness and lung immune response and inflammatory response has been observed in humans, particularly in asthmatics (Hesterberg et al. 2009; Brown 2015).

Epidemiological studies indicate associations between NO_2 (as an indicator of road traffic emissions) in outdoor air and chronic obstructive pulmonary disease (COPD). An increase in respiratory disease with hospitalization of children, moderate and severe asthma attacks in children and adults, or allergic rhinitis have been described (Nhung et al. 2017; Orellano et al. 2017; Zhang et al. 2018a).

Various scientific organizations assume that a causal relationship between chronically elevated NO_2 exposure and cardiovascular effects, diabetes, reproductive and developmental toxic effects, all-cause mortality, and cancer is possible (HC 2016a; US EPA 2016). However, long-term epidemiological studies are difficult to assess regarding the specific effect of NO_2, as part of the many traffic-related pollutants.

7.3.2 Occurrence in the Air of Residences

The level of NO_2 in indoor air mainly depends on the NO_2 concentration in outdoor air, ventilation rate, and human activities indoors like smoking, cooking, and combustion heating. Table 7.3 summarizes the results of NO_2 levels in indoor air of residences. In studies since 2000, mean values ranged from 9–67 μg/m^3. A clear

Table 7.3 Nitrogen dioxide concentrations in the air of residences in μg/m^3

References	*N*	Mean	Min-Max	sampling time	Country, year
Gallelli et al. (2002)	89	25	–	1 week	Italy, 2000
Kornartit et al. (2010)	30	15	7.8–22	1 week; winter	UK, 2000
	30	25	21–29	1 week; summer	UK, 2001
Stranger et al. (2007)	18	33	12–65	1 week	Belgium, 2001/02
Offermann (2009)	29	5.7	2.6–22.6[a]	1 day	USA, 2002–05
Walker et al. (2017)	16	7.9	2.1–26.1	1 week	USA
Wichmann et al. (2010)	18	9	3–16	2 week	Sweden, 2003/04
Valero et al. (2009)	8	36[b]	16–67	48 hours; urban	Spain, 2003/04
	20	32[b]	18–48	48 hours; rural	
Sahsuvaroglu et al. (2009)	20	27.1	6.8–96.4	72 hours	Canada, 2003/04
Michikawa et al. (2014)	65	23.0–67.0[c]	– –	1 week	Japan, 2003–05
Underhill et al. (2015)	16[d] 18[e]	16.0 11.7	10.2–27.5 4.5–20.1	1 week	Peru
Bozkurt et al. (2015)	15	37	–	24 hours; summer	Turkey, 2006
	15	57	–	24 hours; winter	
Physick et al. (2011)	62	31.6	12.6–45.7	Four times for 2 days	Australia, 2006/07
Zechmeister et al. (2020)	20	21.4	11.4–36.9	1 month	Spain, 2008
Lee et al. (2013)	151[d] 150[e]	39.9 43.9	– –	5 weekdays	Korea, 2008/09
Schembari et al. (2013)	54	39	15–118	1 week	Spain, 2008/09
Meier et al. (2015)	80	10[b]	–	1 week	Switzerland, 2011/12
Dėdelė and Miškinytė (2016)	12	–	9–36	2 weeks	Lithuania, 2014
Lee et al. (2016)	20	71.3	33.3–171	48 hours	Korea, 2013–2015
Zusman et al. (2021)	287	19.9	1.9–218	2 weeks	USA, 1014–16
Singer et al. (2020)	66	10.9	2.1–22.6[f]	1 week	USA, 2011–17
Jung et al. (2021)	4623	60	n.a.-1600	1 week	Japan, 2016/17
Vu et al. (2022)	81	14.2	0.2–136		UK, 2016–18
Chan et al. (2020)	70	11.6	0.9–996[f]	1 week	USA, 2016/18
Wohlgemuth et al. (2020)	27	24	7.9–43[g]		Switzerland, 2018

[a] tenth and 90th percentiles
[b] median
[c] medians of annual means
[d] summer
[f] calculated from a figure
[e] winter
[g] means of the buildings

temporal trend in indoor air concentrations is not evident, but a decrease in peak concentrations over time was reported.

In a recent study in the USA, mean concentrations of 5.8 μg/m^3 were reported in 66 indoor environments in 2011–2017 (Singer et al. 2020). In another nationwide study 4843 and 4623 homes were investigated two times as part of the Japan Environment and Children's Study (JECS) from 2014–2016 and 2016–2017 (Jung et al. 2021). A median (95th percentile) of 28.2 μg/m^3 (659 μg/m^3) was observed. Significantly higher levels can be observed indoors whenever specific sources such as gas stoves, open fires, gas stoves, oil and kerosene lamps, tobacco smoking or go-karts are present (Cotterill and Kingham 1997; Cyrys et al. 2000; Wolf et al. 2018). High mean concentrations of NO_2 of 231 μg/m^3 could also be found in the indoor air of taxicabs making a total of 499 trips through Paris in 2019/2020 (Hachem et al. 2020). In central London, the taxi driver's average NO_2 exposure while working (40 work shifts) was nearly twice as high for diesel taxicab drivers (101.9 μg/m^3) compared with electric drivers (55.3 μg/m^3), respectively (Bos et al. 2021).

Various studies showed a dependence of indoor air levels on the proximity to a road or with outdoor NO_2 levels, especially in the absence of effective indoor sources (Janssen et al. 2001; Kousa et al. 2001; Gauvin et al. 2001; Rotko et al. 2001; Gallelli et al. 2002; Emenius et al. 2004; Bellander et al. 2012; Hannam et al. 2013; Cibella et al. 2015; Mehmood et al. 2015; Lee et al. 2016; Majd et al. 2019; Vu et al. 2022). Indoor/outdoor air (I/O) NO_2 ratios of 0.6 to 1.1 were found in these studies. In an extensive German study from 1995/1996, an I/O ratio of 0.8 was observed in summer and 0.5 in winter (Cyrys et al. 2000). In the before mentioned very large JECS study median I/O ratios at two time points were 1.4 and 1.2, respectively. If indoor sources with high NO_2 release are present and there is little ventilation, the indoor-to-outdoor ratio may well rise into the range of 2.8 (Levy et al. 1998; Gallelli et al. 2002). In a recent review Hu and Zhao (2020) summarized the results of 218 publications and gave an overview of the major influencing factors.

Several studies showed a strong correlation between the person-related fraction of exposure obtained by staying outdoors and the stationary concentrations measured in outdoor air (Piechocki-Minguy et al. 2006; Yazar et al. 2011). Additionally, there is evidence from a study performed in four cities in Korea that personal exposure was significantly correlated with residential indoor concentration (Lee et al. 2013). In this study, the personal exposure estimated by residential indoor, workplace indoor and outdoor levels explained 38% and 41% of the measured personal exposure variance in summer and winter, respectively. Due to the high temporal and spatial changes in NO_2 levels, a more personal exposure assessment would in principle be helpful (Van Roosbroeck et al. 2008). One starting point could be small-scale measurements in outdoor air with model-based estimation of indoor air pollution. For example, in 43 rounds of measurements by bicycle through a German city center in February 2018, a mean value in outdoor air of 25 μg/m^3 was measured in a park-like area (urban background), while it was 60 μg/m^3 and 100 μg/m^3 in two road areas heavily polluted by traffic (Samad and Vogt 2021). The study shows the large spatial variability of the measurement results and thus probably also of the indoor air levels.

7.3.3 *Occurrence in the Air of Public Facilities*

The results of classrooms in schools and daycare centers are given in Table 7.4. Overall, similar mean concentrations were found compared to residential interiors. A recent review of the exposure in schools and offices is provided by Salonen et al. (2019).

During the academic years 2015–2017, very high levels of NO_2 was measured in 220 classrooms seasonally in the USA, excluding summer (Kabirikopaei et al. 2021). In contrast, low levels were observed in 10 German schools in 2019 as part of the Europe-wide HEAL (Health and Environment Alliance) study with concentrations from 5 to 22 $\mu g/m^3$ (mean: 10.6 $\mu g/m^3$) (HEAL 2019) as well as in 6 schools in rural Spain in 2013 (mean: 8.9 $\mu g/m^3$) (Villanueva et al. 2018).

Outdoor concentration is the single largest predictor for indoor concentrations explaining around 50% of NO_2 indoor air concentrations in classrooms during school hours in 32 primary and secondary schools in Hong Kong (Chen et al. 2021). Again, Matthaios et al. (2022) summarized data of 309 classrooms of 74 inner-city schools in the USA and that outdoor air was the most important predictor explaining 34.1% of the indoor NO_2, school-based predictors explained 34.2% of NO_2 levels, whereas classroom-based predictors explained 29.3% of NO_2 concentrations, respectively.

In 32 primary and secondary schools in Hong Kong a median I/O ratio of 0.69 was observed (Chen et al. 2021), while Verriele et al. (2016) found a median I/O ratio of 0.8 in 10 recently built energy-efficient French schools. The authors explain the higher ratios with the location of the objects in an area with low outdoor air concentrations.

7.4 Ammonia (NH_3)

Ammonia (NH_3, CAS No.: 7664-41-7) is a colorless nitrogenous gas with a very distinctive and pungent odor. It occurs naturally in the environment as part of the nitrogen cycle and microbial degradation processes. It was estimated that human production of fixed nitrogen as NH_3 is now estimated to be 140 million N per year, an amount that is similar to non-anthropogenic sources (ATSDR 2004).

For Germany, it is estimated that in 2013 agriculture is the main source of ammonia in outdoor air with a share of about 95%. More than 70% of the total emissions come from livestock farming with a share of 52% for cattle farming, 20% for pig farming, and 9% for poultry farming (UBA 2023). The share of mineral fertilizer and digestate application in total ammonia emissions is 25%.

Exposure to ammonia inside buildings occurs due to emissions directly from humans through exhaled breath and sweat, from the use of consumer surface cleaning products, which typically contain low levels of ammonia (between 5 and 10%), by tobacco smoke, cooking by-products and concrete (Ampollini et al. 2019;

Table 7.4 Nitrogen dioxide concentrations in the air of schools and daycare centers in μg/m^3

References	*N*	Mean	Min-Max	sampling time	Country, year
Stranger et al. (2007)	27	57	14–158	1 week	Belgium, 2002/03
Wichmann et al. (2010)	6	17	3–47	2 weeks	Sweden, 2003/04
Wichmann et al. (2010)	10	12	2–25	2 weeks	Sweden, 2003/04
Zhang et al. (2011)	23	501	35.3–63.8	1 week	China, 2006
Gaffin et al. (2018)	218	21	8.1–55.9	1 week	USA, 2008–13
Norback et al. (2019)	32	43	25.5–53.3	1 weeks	China, 2010
Zhang et al. (2014)	32	43	22.2–53.3	1 week	China, 2010–12
Rosbach et al. (2016)	17	19	10–30	1 week	Netherlands, 2011/12
Guerriero et al. (2016)	5	25	9.1–41.2	1 week	UK, 2011/12
Jovanovic et al. (2014)	3	15	8–22	1 week	Serbia, 2012
Rivas et al. (2014)	39	30	5–69	1 week	Spain, 2012/13
Ruggieri et al. (2019)	12	55-99[a]	–	48 hours	Italy, 2012/13
Mehmood et al. (2015)	27	19.0–137	10–376	1 week	Pakistan
Verriele et al. (2016)	10	15.3	–	School hours, 4.5 days	France
Villanueva et al. (2018)	6	8.9	1.4–29	1 week	Spain, 2013, rural
	6	15	7.5–23	1 week	Spain, 2013, urban
	6	13	5.4–21	1 week	Spain, 2013, industrial
Branco et al. (2015)	4	6–136	1–189	1 hour	Portugal, 2013
Kabirikopaei et al. (2021)	220	196[b]	33.0–319	35 hours, unoccupied	USA, 2015–17
Majd et al. (2019)	32	24-34[c]	3.4–129	School hours for 5 days	USA, 2015–17
Che et al. (2021)	76	32	–	7 hours	Hong Kong
Matthaios et al. (2022)	74[d]	21	–	1 week	USA, 2008–13 + 2015–19
HEAL (2019)	10	11	5–22	4 weeks	Germany, 2019

[a]median
[b] annual mean
[c] three measurement periods
[d] 74 schools with 309 classrooms

Villanueva et al. 2022). For example, Li et al. (2020) measured the ammonia emissions using four volunteers under different conditions in a 22.5 m^3 test chamber and reported significantly increased emission rates at higher temperature and more skin exposure. They additionally summarized the average emission rates in mg/hour/

person reported before in 9 studies for nose breath (0.016–0.046), mouth breath (0.11–0.5), dermal (0.41–5.9), and whole body (0.4–5.2). Since ammonia (0.01–0.6%) occurs naturally in cured tobacco leaves and is now being used as an additive to enhance flavor and increase the amount of free nicotine in cigarette smoke, it can be detected in tobacco smoke. Using the 3R4F reference cigarette mean total ammonia level was 12.3 μg per cigarette under ISO smoking conditions (Watson et al. 2016). Sidestream smoke to mainstream smoke distribution ratios (SS/MS ratios) were 37–64 using four types of cigarettes (Sakuma et al. 1984).

A systematic description of the sampling and analysis techniques ammonia in indoor air is presented by Villanueva et al. (2022). Nazaroff and Weschler (2020) gave a comprehensive overview of indoor ammonia emissions concentrations.

Under the pressure and temperature conditions near the ground, a conversion factor of 1 $\mu g/m^3$ = 1.43 ppb was described.

7.4.1 Health Aspects

Estimates of odor thresholds for ammonia vary from 0.04–103 ppm and near these levels, exposed persons can experience annoyance and believe the odor to be a nuisance (OEHHA 1999). Gaseous ammonia causes irritation of the skin and mucous membranes of the upper respiratory tract and the eye while high concentrations lead to conjunctivitis, laryngitis, and pulmonary edema. (ATSDR 2004). Due to its high water solubility ammonia dissolves in moisture on the before mentioned body surfaces forming ammonium hydroxide, which causes liquefaction necrosis of the tissues as well as saponification of cell membrane lipids, resulting in cell disruption and cell death (ATSDR 2004). Various occupational cohort studies of farmers working in livestock buildings reported an association between exposure to pollutants, including ammonia, and an increase in respiratory symptoms like bronchial reactivity/hyperresponsiveness, inflammation, cough, wheezing, or shortness of breath and a decrease in pulmonary function (ATSDR 2004). Nevertheless, the importance of these findings is still questionable due to, e.g., co-exposure to other pollutants and endotoxin.

7.4.2 Occurrence in Indoor Air

Table 7.5 shows the results of ammonia measurements in the indoor air of dwellings and in public buildings such as schools and daycare centers.

Tuomainen et al. (2001) investigated one building constructed in the conventional way (control building) and the other by following the instructions of the Finnish Classification of Indoor Climate, Construction and Finishing Materials (case building) before occupants moved in and after a 5-month occupancy. The study showed that good indoor air quality and lower ammonia concentrations can be achieved by

Table 7.5 Indoor and outdoor ammonia concentration in residences and schools and daycare centers in ppb

References	*N*	Indoor		Outdoor	Sampling conditions	Country, Year
		Mean	Min-max	Mean		
Residences						
Brauer et al. (1991)	5 6	19.3 8.1	14.3–41.1 0.25–20.1	1.13 1.87	Winter Summer	US
Atkins and Lee (1993)	10 10	53.4 45.5	18.6–110 22.9–68.6	– –	Living rooms Bedrooms	UK
Tidy and Cape (1993)	5	24.2	7–63	–	Winter	UK, 1991/92
Leaderer et al. (1999)	49 9 148	32.1 27.5 38.0	1.5–93.0 2.0–49.5 0.13–234	2.8 2.8 1.6	Summer, homes with AC Summer homes without AC Winter, without kerosene heater	USA, 1994–98
Suh et al. (1992)	24	21.7 [a]	n.g.-58.4	0.3[a]		USA
Tuomainen et al. (2001)	12[b] 7[b]	21 69	– –	–	Residences in case buildings Residences in control buildings	Finland
Fischer et al. (2003)	1	15.4–21.2	–	12.4–13.6	Unoccupied, three measurement periods	USA, 2000/01
Järnström et al. (2006)	14	60–61	–	–	Three measurement periods	Finland, 1999–03
Spengler et al. (1993)	10	20.1	13.9–29.6	–		USA, 1990
Suzuki et al. (2020)	47	104	19–229	21	Living rooms of newly houses	Japan, 2014–16
Uchiyama et al. (2015)	602	23 53	n.g.-501 n.g.-1430	1.4 14.2	Winter Summer	Japan, 2012–14 Japan, 2012/13
Schools and common areas						
Liang and Waldman (1992)	3	10, 31, 56	24, 42, 114[c]	3.1[d]	Institutional settings for elderly and childcare; summer, daytime	USA, 1989

(continued)

Table 7.5 (continued)

References	*N*	Indoor		Outdoor	Sampling conditions	Country, Year
		Mean	Min-max	Mean		
Gomzi (1999)	2	77	6–613	–	Winter industrial area	Croatia, 1990/91
	1	90	3–217		Winter rural area	
	2	40	0–603		Summer industrial area	
	1	43	0–137		Summer rural area	
Meininghaus et al. (2003)	1	446	–	–	School	France

n.g. not given
[a]geometric mean
[b] after 5-month occupancy
[c] 90th percentile
[d] around 4 km away

careful design, choice of proper materials and equipment, and construction of high quality. Again, Järnström et al. (2006) analyzed ammonia in 14 newly finished buildings and after 6 and 12 months in Finland and reported levels between 29 and 86 ppb in the newly finished buildings. The concentrations correlated with the season, with 86–129 ppb (relative humidity >50%) in the 6–12-month-old buildings during the summer period but only 29–72 ppb in the winter period (relative humidity <40%). In Japan, 47 new houses were investigated two weeks after completion and means of 104 ppb in the living rooms and 109 ppb in the bedrooms were observed (Suzuki et al. 2020).

In a large study from Japan the mean ammonia concentration increased with indoor temperature and tended to be higher in the houses with pets (Uchiyama et al. 2015); e.g., the house with the highest ammonia concentration of 1430 ppb contained six dogs and one cat. Furthermore, Sun et al. (2021) reviewed 56 studies from China 1980–2019 and found overall median concentrations (range of means) in residences of 300 ppb (43–14,800 ppb), in offices of 372 ppb (157–4876 ppb), and in school buildings of 215 ppb (29–601 ppb). Higher concentrations were found in the severe cold zone and the regions with higher economic level.

Overall, concentrations in indoor air are generally higher than in outdoor air, and indoor air levels can be higher by a factor of 16–23 in winter (Brauer et al. 1991; Uchiyama et al. 2015) and by a factor of 3.6 in summer (Brauer et al. 1991; Uchiyama et al. 2015; Suzuki et al. 2020). For person-related measurements in 24 children, the geometric mean values of 56 ppb compared to 22 ppb were about twice as high as for stationary measurements in their living rooms (Suh et al. 1992).

Measurement results in the indoor air of schools and other common areas have hardly been published in the scientific literature so far. In the late 1980s, for example, three institutional settings for elderly and childcare were investigated, with only slight differences between daytime and nighttime measurements, but significantly higher levels inside than in the outside air (Liang and Waldman 1992). In France, three classrooms of a rural school were investigated in which the users complained of non-specific health effects. In all of the rooms, there was an equal and high measurement result of 446 ppb (Meininghaus et al. 2003). A more extensive study comes from Croatia, in which air was analyzed at three sampling points each over 24 hours in two schools in an industrial region and in a rural school (Gomzi 1999).

So far, only a few results for office buildings are available. In six offices of five modern buildings ammonia was analyzed in former Yugoslavia with means between 55–239 ppb (maximum: 402 ppb) in winter and 60–187 ppb (maximum:242 ppb) in summer (Sisovic et al. 1987). The mean indoor-to-outdoor ratios in winter and summer were 4.9–10.5 and 2.7–6.7, respectively. In addition, Salonen et al. (2009) obtained 47 air samples from 14 office buildings in Finland in 2001–2006 and reported a median of 17 ppb and a range between 1 and 70 ppb. Of the 88% of buildings for which ventilation information was provided, 95% were equipped with a mechanical supply and exhaust ventilation system.

Some studies have investigated the known influencing factors and their contribution to indoor air concentration. Thus, Risner and Conner (1991) reported that mean concentrations obtained from the smoking of four cigarettes in a 28 m^3 room was 153 ppb. Tidy and Cape (1993) found the highest mean values of 93 ppb and 57 ppb in rooms of two living rooms of smokers, 7 ppb and 8 ppb in unoccupied rooms, and 14 ppb in a room of non-smokers.

As part of the HOMEChem study real-time concentration of ammonia indoors was measured with a mean background concentration in an unoccupied test room of 32 ppb, with further increase during cooking to 62 ppb (maximum: 1390 ppb), cleaning to 73 ppb (maximum: 1592 ppb), and occupancy activities to 51 ppb (maximum: 99 ppb) (Ampollini et al. 2019). Furthermore, concentrations were strongly influenced by indoor temperatures and heating, ventilation, and air conditioning (HVAC) operation.

7.5 Sulfur Dioxide (SO_2)

Sulfur dioxide (SO_2, CAS No.: 7446-09-5) is a colorless, pungent-smelling, water-soluble gas that is mainly emitted during combustion processes of fossil fuels such as coal and oil by oxidation of the sulfur contained in the product. The main natural sources are volcanoes, forest fires, and wildfires (HC 2016b). Industrially, sulfur dioxide is used in large quantities as a captive intermediate in the production of sulfuric acid and in the pulp and paper industry (ATSDR 1998). Domestic burning of coal or biomass for cooking and heating, practiced by many persons worldwide, is an important source of indoor SO_2 as well as unvented kerosene heaters, gas appliances, and oil furnaces (Brauer et al. 2002; Seow et al. 2016).

Zhong et al. (2020) modeled a global SO_2 emissions inventory for 1960–2014 and describe a peak in emissions in Europe and North America in the early 1980s and a significant decline thereafter, while in Asia the peak was not reached until the mid-2000s and then emissions declined slowly. In Asia, power generation and industry were the predominant contributors, accounting for 50% and 42% of the emission in 2014, respectively. Overall, the study results suggest that the predominant emission contributor has shifted from developed countries in the early 1960s (61%) to developing countries at present (83%).

The conversion factors at 20 °C and 1013 hPa were: 1 ppb = 2.66 μg/m^3 and 1 μg/m^3 = 0.38 ppb.

7.5.1 Health Aspects

Humans can be exposed to SO_2 mainly via inhalation and the gas is absorbed in the upper respiratory tract. The odor threshold ranges from 800 to 2660 μg/m^3, mainly around 1330 μg/m^3 (HC 2016b). Exposure to higher concentrations of sulfur dioxide may cause irritation to the eyes, nose, and throat. In contact with water sulfurous acid was formed which accounts for the irritant effects to eyes and mucous membranes (Arts et al. 2006). Arts et al. (2006) summarized the results of 16 studies using healthy volunteers and irritation of the upper airways and eyes was reported at exposure levels of 2660 μg/m^3 and higher in combination with exercise, while such effects were not observed at levels of 2000 μg/m^3 and below. Moreover, lung function changes were also seen at levels of 2660 μg/m^3 and higher, although at 2660 μg/m^3 these changes were found almost exclusively in combination with physical exercise.

The US EPA concludes in his 2009 integrated science assessment that the controlled human exposure, epidemiologic, and toxicological evidence support a causal relationship between respiratory morbidity and short-term (5-minutes to 24-hours) exposure to SO_2 and that all asthmatics are at high risk of experiencing adverse respiratory effects from SO_2 exposure (US EPA 2009). On behalf of the WHO, results of epidemiologic studies were summarized in two systematic reviews (Zheng et al. 2021; Orellano et al. 2021). The authors found for sulfur dioxide exposure and asthma hospital admissions and emergency room visits a meta-analytic effect estimate of RR = 1.010 (95% CI: 1.001–1.020) per 10 μg/m^3, on sulfur dioxide and daily mortality a meta-analytic effect estimate of RR = 1.0059 (95% CI: 1.0046–1.0071) per 10 μg/m^3, and for respiratory mortality a meta-analytic effect estimate of RR = 1.0067 (95% CI: 1.0025–1.0109) per 10 μg/m^3 sulfur dioxide (WHO 2021).

7.5.2 Occurrence in Indoor Air

Brauer et al. (2002) reviewed the studies before 1990 in North America, Europe, and Asia and reported mean SO_2 levels from 0.5–32 μg/m^3. For example, Spengler et al. (1979) presented the results of 1 year's indoor monitoring for SO_2 in six US communities with widely varying outdoor levels and found mean concentrations from 1 to 22 μg/m^3 (range: 0–26 μg/m^3).

A large study was performed by Leaderer et al. (1999) in the USA in 1994–1998 in 281 homes. In the summer sampling period mean SO_2 concentrations of 2.4 μg/m^3 and 0.8 μg/m^3 were found in non-AC homes and AC homes, respectively, while the means in winter were 2.2 μg/m^3 in nonkerosene homes and 43 μg/m^3 in homes where kerosene heaters were used. Using passive monitoring techniques 7-day average concentrations of SO_2 in 25 homes in two Canadian communities resulted in medians of 0.5 μg/m^3 and 1.4 μg/m^3 (range: 0.2–5.2 μg/m^3) (Kindzierski and Sembaluk 2001). Sarnat et al. (2006) assessed 4-hr personal exposures to SO_2 for 10 non-smoking adults in the USA and found mean values of 4 μg/m^3 in summer and 1.9 μg/m^3 in fall 2000.

Chao (2001) investigated 10 apartments in Hong Kong in 1997 with concentrations ranging from 2.61 to 10.44 μg/m^3 with an average of 6.30 μg/m^3. Average indoor concentrations in the living rooms of four households in China in 2002 with the combustion of either coal–biomass briquettes or low-grade coal varied from 50 to 960 μg/m^3 while in the kitchen the average levels ranged from 280 to 7070 μg/m^3 (Isobe et al. 2005). In two rural Chinese regions with a high incidence of malignant and non-malignant lung diseases 163 homes were investigated in 2008/09 and measurement results differed by fuel type, stove design, and smoky coal source (Seow et al. 2016). The study indicates that homes burning fuel in unventilated stoves and firepits had the highest rate of SO_2 detection. Very low concentrations were reported in a study performed 2010–2012 in 120 living rooms in a highly industrialized city in Taiwan with a median of 0 μg/m^3 and 90th percentile of 189 μg/m^3 (Yen et al. 2020).

In Australia the mean maximum hourly average was 1436 μg/m^3 (range: 0–18,000 μg/m^3) for SO_2 in 75 homes using mainly paraffins for heating and cooking as a fuel (Bailie et al. 1999). The highest concentrations depending on the income levels of users and types of fuels used were reported for 550 homes in South Africa in 1995/96 with average hourly concentrations between 8000 and 39,000 μg/m^3 in living rooms and 11,000 and 90,000 μg/m^3 in the cooking areas (Sanyal and Maduna 2000).

In 11 daycare centers in Thailand in 2014/15 a mean SO_2 level of 16 μg/m^3 was reported (Siwarom et al. 2017). Higher levels were observed in 34 daycare centers in a Chinese city with an average of 40.5 μg/m^3 (range: 5.0–85.8 μg/m^3) (Zhang et al. 2018b).

In the absence of specific indoor sources such as—open—heating and cooking with fossil fuels, the indoor concentrations are significantly below the levels in the outdoor air. The observed average ratios of the indoor and outdoor concentrations of SO_2 were 0.11 and 0.14 (Kindzierski and Sembaluk 2001), 0.38 (Zhang et al. 2018b), and 0.78 (Chao 2001).

7.6 Ozone (O_3)

Ozone (O_3) is a colorless to bluish, characteristically pungent-smelling gas whose molecules consist of three oxygen atoms. In the stratosphere, at an altitude of between 15 and 30 kilometers, the so-called ozone layer absorbs the sun's ultraviolet radiation and thus prevents negative effects on the earth's biosphere. In the troposphere, where human activities mainly take place, there is an influx from the stratosphere due to vertical air movements in spring, especially in the northern hemisphere (Zhang et al. 2019). However, the most significant source of tropospheric or ground-level ozone is its secondary formation in the presence of high-energy sunlight from reaction with various non-methane volatile organic compounds (VOC) or oxides of nitrogen (NO, NO_2) (WHO 2021). Higher ozone concentrations can be found in areas of low anthropogenic emissions. Major sources include vehicular exhaust, fugitive evaporation of gasoline and other gaseous fuels (e.g., natural gas and propane), biomass and fossil fuel combustion, and industrial solvent use (Zhang et al. 2019). Ground-level ozone concentrations are subject to very large temporal and spatial fluctuations caused by the complex processes of formation and depletion. For example, typically diurnal pattern with high concentrations during afternoon and early evening hours were observed.

For outdoor air, Directive 2008/50/EC of the European Parliament and of the Council of 21 May 2008 on ambient air quality and cleaner air for Europe sets a maximum daily eight-hour average value for ozone of 120 $\mu g/m^3$ to protect human health. This value may not be exceeded on a maximum of 25 days per year. In addition, an information threshold of 180 $\mu g/m^3$, above which the population must be informed, and an alert threshold of 240 $\mu g/m^3$ apply to the one-hour value. The existing primary and secondary US National Ambient Air Quality Standards (NAAQS), established in 2015, are 140 $\mu g/m^3$ as the fourth-highest daily maximum 8-hour concentration, averaged across three consecutive years. In 2012, National Ambient Air Quality Standards of 100 $\mu g/m^3$ (special regions like national parks) and 160 $\mu g/m^3$ (all other areas) (daily, 8-hour maximum) were introduced by the Chinese Ministry of Environmental Protection (Yang et al. 2020).

In a summary of results from 20 European countries in 2021 the maximum daily eight-hour average value of 120 $\mu g/m^3$ was met in only 19% of the stations. The short-term WHO Air Quality Guideline of 100 $\mu g/m^3$ was exceeded in 92% of all the reporting stations in 2021 (Targa et al. 2022).

Different personal and ambient ozone measurement devices were used like nitrate-coated active or passive samplers, electrochemical or ultraviolet absorption monitors as well as organic vapor diffusion-based ozone monitoring systems using charcoal pads (summary in Kim and Rohr 2021). Due to the limited number of central ozone monitors and different personal behavior patterns, it must be taken into account that the usual stationary ambient ozone measurements, which are often used as a surrogate for determining personal exposure, may be a source of bias, especially in epidemiologic studies (Kim and Rohr 2021).

Under the pressure and temperature conditions near the ground, conversion factors of 1 part per million (ppm) = 1.9957 mg/m^3 and 1 mg/m^3 = 0.5011 ppm were described.

7.6.1 *Health Aspects*

After inhalation, ozone reacts with lipids, proteins, and antioxidants in the epithelial lining fluid of the respiratory tract, creating secondary oxidation products (US EPA 2020). Meanwhile, in a present meta-analysis, an overall positive association of short-term exposures to O_3 with a biomarker of systematic inflammation (C-reactive protein, CRP) was reported including 38 epidemiological studies (Xu et al. 2022). Nevertheless, the O_3 systemic mechanisms are still a matter of research.

US EPA concluded that there is a "causal relationship" between short-term ozone exposure and respiratory effects in controlled human studies (e.g., airway responsiveness, respiratory symptoms, and respiratory tract inflammation) and epidemiologic studies (e.g., asthma exacerbation, chronic obstructive pulmonary disease exacerbation, respiratory infection, as well as hospital admissions and emergency department visits) (US EPA 2020). Moreover, there is a "likely to be causal relationship" between long-term ozone exposure and respiratory effects like new-onset asthma, respiratory symptoms in children with asthma, and respiratory mortality. In addition, there is some but limited evidence that exposure to ozone is likely to induce metabolic effects and can affect the cardiovascular and nervous systems, reproduction and development, and mortality. Further reviews of the short- and long-term health effects are available (e.g., Bell et al. 2014; Atkinson et al. 2016; Niu et al. 2022; Sun et al. 2022; Holm and Balmes 2022).

For example, a recent review of short-term effects determined from 67 studies reported that higher levels of maximal 8-hour or average 24-hour concentrations of O_3 were associated with significantly increased risks of asthma-related exacerbations (RR = 1.014 (95% CI: 1.008–1.020), defined as emergency room visits and hospital admissions (Zheng et al. 2021). Moreover, Bell et al. (2014) reported significant lower short-term mortality risks for younger persons compared to older persons in a review and meta-analysis of 167 relevant publications.

The systematic review of long-term effects by Huangfu and Atkinson (2020) on ozone and all non-accidental mortality reported a meta-analytic effect estimate of RR (relative risk) = 1.01 (95% CI: 1.00–1.02) per 10 μg/m^3 increase and on respiratory mortality of RR = 1.02 (95% CI: 0.99–1.05) in peak-season average of daily maximum 8-hour mean ozone concentrations, assuming a linear relationship. These estimates were used to assess the WHO air quality guideline levels (WHO 2021). The WHO stated that in the case of ozone, it is customary to calculate daily maximum of 8-hour mean concentrations rather than 24-hour averages because of the strong diurnal variation in ozone concentration. On the other hand, using pooled

Table 7.6 Ozone concentration in indoor air and indoor-to-outdoor concentration ratios (I/O) in residences in μg/m³

References	Number	Min-Max	Median	Mean	I/O *	Country, year
Stranger et al. (2007)	19	–	–	3.2	0.11	Belgium, 2001/02
Bozkurt et al. (2015)	15	–	–	9.3	–	Turkey, 2006 summer
Bozkurt et al. (2015)	15	–	–	2.9	–	Turkey, 2006/07, winter
Yen et al. (2020)	120	2.0–30[b]	10.0	–	–	Taiwan, 2010/11
Uchiyama et al. (2015)	602	NG-27	1.1	1.7	0.05	Japan, 2012–14 (winter)
Uchiyama et al. (2015)	602	NG-62	7.2	10	0.29	Japan, 2012/13 (summer)
Kalimeri et al. (2017)	16	0.4–14.1	–	4.1	0.11	Greece, 2013
Jung et al. (2021)	4843	<LOQ-60	4.0	–	0.10[a]	Japan, 2014–17
Barkjohn et al. (2021)	43	10–36	12.0	16.0	0.22	China, 2017

* using mean values; *NG* not given
[a] using median values
[b] tenth and 90th percentiles; *NG* not given

data from 15 cohorts from 11 European countries the ELAPSE (Effects of Low-Level Air Pollution: A Study in Europe) study reported an inverse associations between O_3 and all causes of death examined (Stafoggia et al. 2022). Nevertheless, the authors concluded that, due to the small variation in O_3 exposure, their results were less suited to assessing associations with O_3 than those based in other locations (e.g., Canada or the USA) where greater variability in exposures has been reported.

7.6.2 Occurrence in the Air

Overall, most indoor ozone originates from ambient air, but some indoor sources like ozone generators, laser printers, photocopy machines, electrostatic precipitators, and air ionizers must be considered (Weschler 2000; Guo et al. 2019). Nevertheless, filtration systems for reducing indoor air pollution now exist for some products. In a recent review Nazaroff and Weschler (2022) summarized the measurements in indoor air of residences, schools, and offices and found the following most important influencing factors of indoor-to-outdoor concentration ratio: air-exchange rates, ozone loss rate on fixed indoor surfaces, and in densely occupied spaces, ozone loss rate on human occupants.

Tables 7.6 and 7.7 summarize the results of measurements in residences and in schools and daycare centers. Overall, ozone concentrations in indoor air were low compared with the outdoor air and were often near or below the detection limit. Mean concentrations in school common areas ranged from 0.8 to 74 μg/m³, with the

Table 7.7 Ozone indoor air concentration and indoor-to-outdoor concentration ratios (I/O) in schools and daycare centers in μg/m^3

References	Number	Min-Max	Median	Mean	I/O*	Country, year
Mi et al. (2006)	30	1.1–7.0	–	5.3	0.25	China, 2000, winter
Stranger et al. (2008)	27 (162)	<LOD-4.0	–	1.5	0.10	Belgium, 2002 winter, urban
Stranger et al. (2008)	27 (162)	<LOD-1.8	–	0.8	0.30	Belgium, 2002, winter, suburban
Stranger et al. (2008)	27 (162)	<LOD-9.9	–	3.0	0.10	Belgium, 2003, summer, urban
Stranger et al. (2008)	27	<LOD-64	–	2.1	0.04	Belgium, 2003, summer, suburban
HESE (2006)	41	3.0–49	6.9	14.3	–	Italy, Norway, Sweden, Denmark, France, 2004/5
Zuraimi et al. (2007)	59	–	–	73.6	0.80	Singapore
Karthikeyan et al. (2007)	(28)	0.1–12.0	3.0	3.5	0.40	Singapore, 2005
Bozkurt et al. (2015)	3	–	–	12.0	0.21	Turkey, 2006, summer
Bozkurt et al. (2015)	3	–	–	2.3	0.32	Turkey, 2006/07, winter
Mullen et al. (2011)	6	–	–	12.0	0.32	USA, 2008
Demirel et al. (2014)	64	6.2–55.1	18.0	18.6	0.21	Turkey, 2008/09
Verriele et al. (2016)	10	–	3.9	3.1	0.08	France
Norback et al. (2019)	32	2.9–44	–	8.6	0.48	China, 2010
Csobod et al. (2014)	340	<LOQ-142	3.0	8.0	–	Europe, 2011/12, schools
Csobod et al. (2014)	25	<LOQ-16	6.0	6.0	–	Europe, 2011/12, kindergartens
Jovanovic et al. (2014)	1 (3/5)	8.8–15.9	–	15.5	0.07	Serbia, 2012
Branco et al. (2015)	3	<LOQ-53		13–24	0.33	Portugal, 2013
Branco et al. (2019)	50	15.0[a]	7.8	–	–	Portugal, 2013–16
Hwang and Park (2020)	8	10–18	11.0	12.2	–	Korea, 2014/15
Kabirikopaei et al. (2021)	220	0.3–53	–	14.4	–	USA, 2015–17
Che et al. (2021)	(76)	–	–	28	0.58	Hong Kong, 2017/18

* using mean values; + only naturally ventilated buildings; *LOD* limit of detection, *LOQ* limit of quantification

[a] interquartile range

lowest concentrations found in winter in 27 Belgian schools and the highest in 59 schools in Singapore. Comparable results with mean values between 1.7 and 16 μg/m^3 were observed in homes. Salonen et al. (2018) summarize the results from different studies performed in school and office indoor environments.

Generally, indoor air O_3 concentrations are lower in winter than in summer (e.g., Stranger et al. 2008). For example, ozone indoor concentration in winter was 5.9 times lower than that in summer in 602 houses throughout Japan (Uchiyama et al. 2015).

Karthikeyan et al. (2007) observed that naturally ventilated daycare centers have significantly higher indoor O_3 concentrations than those that are mechanically ventilated. Again, those located in close proximity to traffic emissions or in densely populated regions had lower O_3 concentrations indoors compared to rural areas (Liu et al. 1993; Karthikeyan et al. 2007; Demirel et al. 2014).

References

Ad-hoc-AG (2008) Health assessment of carbon dioxide in indoor air. Bundesgesundheitsbl Gesundheitsforsch Gesundheitsschutz 51:1358–1369. [in German]

Ampollini L, Katz EF, Bourne S, Tian Y, Novoselac A, Goldstein AH, Lucic G, Waring MS, DeCarlo PF (2019) Observations and contributions of real-time indoor ammonia concentrations during HOMEChem. Environ Sci Technol 53:8591–8598

Arts JH, de Heer C, Woutersen RA (2006) Local effects in the respiratory tract: relevance of subjectively measured irritation for setting occupational exposure limits. Int Arch Occup Environ Health 79:283–298

ASHRAE (2019a) ASHRAE 62.1–2019. Ventilation for Acceptable Indoor Air Quality

ASHRAE (2019b) ASHRAE 62.2–2019. Ventilation and Acceptable Indoor Air Quality in Residential Buildings

Atkins DHF, Lee DS (1993) Indoor concentrations of ammonia and the potential contribution of humans to atmospheric budgets. Atmos Environ 27A:1–7

Atkinson RW, Butland BK, Dimitroulopoulou C, Heal MR, Stedman JR, Carslaw N, Jarvis D, Heaviside C, Vardoulakis S, Walton H, Anderson HR (2016) Long-term exposure to ambient ozone and mortality: a quantitative systematic review and meta-analysis of evidence from cohort studies. BMJ Open 23:e009493

ATSDR (2004) Toxicological profile for ammonia. https://www.atsdr.cdc.gov/toxprofiles/tp126.pdf

Azuma K, Kagi N, Yanagi U, Osawa H (2018) Effects of low-level inhalation exposure to carbon dioxide in indoor environments: a short review on human health and psychomotor performance. Environ Int 121:51–56

Bailie RS, Pilotto LS, Ehrlich I, Mbuli S, Truter R, Terblanche P (1999) Poor urban environments: use of paraffin and other fuels as sources of indoor air pollution. J Epidemiol Comm Health 53: 585–586

Barkjohn KK, Norris C, Cui X, Fang L, Zheng T, Schauer JJ, Li Z, Zhang Y, Black M, Zhang JJ, Bergin MH (2021) Real-time measurements of PM2.5 and ozone to assess the effectiveness of residential indoor air filtration in Shanghai homes. Indoor Air 31:74–87

Barn P, Giles L, Héroux ME, Kosatsky T (2018) A review of the experimental evidence on the toxicokinetics of carbon monoxide: the potential role of pathophysiology among susceptible groups. Environ Health 17:13

Bell ML, Zanobetti A, Dominici F (2014) Who is more affected by ozone pollution? A systematic review and meta-analysis. Am J Epidemiol 180:15–28

Bellander T, Wichmann J, Lind T (2012) Individual exposure to NO_2 in relation to spatial and temporal exposure indices in Stockholm, Sweden: the INDEX study. PLoS One 7:e39536

Bischof W, Bebersdorf J, Koch A (2008) Survey of the indoor air hygiene situation in Erfurt schools. Personal communication. [in German]

Bos B, Lim S, Hedges M, Molden N, Boyle S, Mudway DI, Barratt DB (2021) Taxi drivers' exposure to black carbon and nitrogen dioxide in electric and diesel vehicles: a case study in London. Environ Res 195:110736

Bozkurt Z, Dogan G, Arslanbas D, Pekey B, Pekey H, Dumanoglu Y, Bayram A, Tuncel G (2015) Determination of the personal, indoor and outdoor exposure levels of inorganic gaseous pollutants in different microenvironments in an industrial city. Environ Monit Assess 187:590

Branco PT, Nunes RA, Alvim-Ferraz MC, Martins FG, Sousa SI (2015) Children's exposure to indoor air in urban nurseries—Part II: Gaseous pollutants' assessment. Environ Res 142:662–670

Branco PTBS, Alvim-Ferraz MCM, Martins FG, Sousa SIV (2019) Quantifying indoor air quality determinants in urban and rural nursery and primary schools. Environ Res 176:108534

Braubach M, Algoet A, Beaton M, Lauriou S, Héroux ME, Krzyzanowski M (2013) Mortality associated with exposure to carbon monoxide in WHO European Member States. Indoor Air 23: 115–125

Brauer M, Koutrakis P, Keeler GJ, Spengler JD (1991) Indoor and outdoor concentrations of inorganic acidic aerosols and gases. J Air Waste Manag Assoc 41:171–181

Brauer M, Henderson SB, Kirkham T, Lee KS, Rich K, Teschke K (2002) Review of the health risks associated with nitrogen dioxide and sulfur dioxide in indoor air. https://open.library.ubc.ca/soa/cIRcle/collections/facultyresearchandpublications/52383/items/1.0048208

Brown JS (2015) Nitrogen dioxide exposure and airway responsiveness in individuals with asthma. Inhal Toxicol 27:1–14

Cai C, Sun Z, Weschler LB, Li T, Xu W, Zhang Y (2021) Indoor air quality in schools in Beijing: field tests, problems and recommendations. Build Environ 205:108179

Canha N, Lage J, Coutinho JT, Alves C, Almeida SM (2019) Comparison of indoor air quality during sleep in smokers and non-smokers' bedrooms: a preliminary study. Environ Pollut 249: 248–256

Chan W R, Kim Y-S, Less BB, Singer BC, Walker IS (2020) Ventilation and indoor air quality in new California homes with gas appliances and mechanical ventilation. California Energy Commission. Publication number: CEC-500-2020-023. https://www.energy.ca.gov/sites/default/files/2021–05/CEC-500-2020-023.pdf

Chao CYH (2001) Comparison between indoor and outdoor air contaminant levels in residential buildings from passive sampler study. Build Environ 36:999–1007

Che W, Li ATY, Frey HC, Tang KTJ, Sun L, Wei P, Hossain MS, Hohenberger TL, Leung KW, Lau AKH (2021) Factors affecting variability in gaseous and particle microenvironmental air pollutant concentrations in Hong Kong primary and secondary schools. Indoor Air 31:170–187

Chen JK, Wu C, Su TC (2021) Positive association between indoor gaseous air pollution and obesity: an observational study in 60 households. Int J Environ Res Public Health 18:11447

Choi HS, Suh MJ, Hong SC, Kang JW (2021) The association between the concentration of heavy metals in the indoor atmosphere and atopic dermatitis symptoms in children aged between 4 and 13 years: a pilot study. Children (Basel) 8:1004

Cibella F, Cuttitta G, Della Maggiore R, Ruggieri S, Panunzi S, De Gaetano A, Bucchieri S, Drago G, Melis MR, La Grutta S, Viegi G (2015) Effect of indoor nitrogen dioxide on lung function in urban environment. Environ Res 138:8–16

Cotterill A, Kingham S (1997) Nitrogen dioxide in the home: cooking, double glazing or outdoor air? Indoor Built Environ 6:344–349

Csobod E, Annesi-Maesano I, Carrer P, Kephalopoulos S, Madureira J, Rudnai P, de Oliveira Fernandes E (2014) SINPHONIE. Schools indoor pollution and Health. Final Report, EUR 26738. https://publications.jrc.ec.europa.eu/repository/handle/JRC91160

Cyrys J, Heinrich J, Richter K, Wölke G, Wichmann HE (2000) Sources and concentrations of indoor nitrogen dioxide in Hamburg (West Germany) and Erfurt (East Germany). Sci Total Environ 250:51–62

Daisey JM, Angell WJ, Apte MG (2003) Indoor air quality, ventilation and health symptoms in schools: an analysis of existing information. Indoor Air 13:53–64

Dėdelė A, Miškinytė A (2016) Seasonal variation of indoor and outdoor air quality of nitrogen dioxide in homes with gas and electric stoves. Environ Sci Pollut Res Int 23:17784–17792

Demirel G, Ozden O, Döğeroğlu T, Gaga EO (2014) Personal exposure of primary school children to BTEX, NO_2 and ozone in Eskişehir, Turkey: relationship with indoor/outdoor concentrations and risk assessment. Sci Total Environ 473–474:537–548

DIN (2019) DIN 1946-6:2019–12. Ventilation and air conditioning – Part 6: Ventilation for residential buildings – General requirements, requirements for design, construction, commissioning and handover as well as maintenance. Beuth Verlag, Berlin. https://www.beuth.de/en/standard/din-1946-6/314483915

DIN EN (2007) DIN EN 13779:2007–09. Ventilation for non-residential buildings – Performance requirements for ventilation and room-conditioning systems. Beuth Verlag, Berlin. https://www.beuth.de/en/standard/din-en-13779/94054008

Dominski FH, Lorenzetti Branco JH, Buonanno G, Stabile L, Gameiro da Silva M, Andrade A (2021) Effects of air pollution on health: a mapping review of systematic reviews and meta-analyses. Environ Res 201:111487

Du B, Tandoc MC, Mack ML, Siegel JA (2020) Indoor CO_2 concentrations and cognitive function: a critical review. Indoor Air 00:1–16

Emenius G, Vartengren M, Korsgaard J, Nordvall L, Pershagen G, Wickman M (2004) Building characteristics, indoor air quality and recurrent wheezing in very young children (BAMSE). Indoor Air 14:34–42

Erdmann CA, Apte MG (2004) Mucous membrane and lower respiratory building related symptoms in relation to indoor carbon dioxide concentrations in the 100-building BASE dataset. Indoor Air Suppl 8:127–134

Fischer ML, Littlejohn D, Lunden MM, Brown NJ (2003) Automated measurements of ammonia and nitric acid in indoor and outdoor air. Environ Sci Technol 37:2114–2119

Fromme H, Schober W (2015) Waterpipes and e-cigarettes: impact of alternative smoking techniques on indoor air quality and health. Atmos Environ 2015(106):429–441

Fromme H, Lahrz T, Piloty M, Pfeiler P, Honigmann I, Gebhardt H, Oddoy A (2002) Priority program - substances of health concern in public facilities in Berlin. Report of the Institut für Lebensmittel, Arzneimittel und Tierseuchen. Berlin. [in German]

Fromme H, Heitmann D, Dietrich S, Schierl R, Körner W, Kiranoglu M, Zapf A, Twardella D (2008) Air quality in schools – classroom levels of carbon dioxide (CO_2), volatile organic compounds (VOC), aldehydes, endotoxins and cat allergen. Gesundheitswesen 70:88–97. [in German]

Fromme H, Dietrich S, Heitmann D, Dressel H, Diemer J, Schulz T, Jörres RA, Berlin K, Völkel W (2009) Indoor air contamination during a waterpipe (narghile) smoking session. Food Chem Toxicol 47:1636–1641

Gabrio T, Volland G (2007) Personal communication

Gaffin JM, Hauptman M, Petty CR, Sheehan WJ, Lai PS, Wolfson JM, Gold DR, Coull BA, Koutrakis P, Phipatanakul W (2018) Nitrogen dioxide exposure in school classrooms of inner-city children with asthma. J Allergy Clin Immunol 141:2249–2255.e2

Gallelli G, Orlando P, Perdelli F, Panatto D (2002) Factors affecting individual exposure to NO_2 in Genoa (northern Italy). Sci Total Environ 287:31–36

Gauvin S, Le Moullec Y, Bremont F, Momas I, Balducci F, Ciognard F, Poilve MP, Zmirou D (2001) Relationships between nitrogen dioxide personal exposure and ambient air monitoring measurements among children in three French metropolitan areas: VESTA study. Arch Environ Health 56:336–341

Georgoulis LB, Hänninen O, Samoli E, Katsouyanni K, Künzli K, Polanska L, Bruinen de Bruin Y, Alm S, Jantunen M (2002) Personal carbon monoxide exposure in five European cities and its determinants. Atmos Environ 36:963–974

Gizaw Z, Teka Z (2020) Correlates of indoor concentration of carbon monoxide in residential buildings in Gondar Town, Northwest Ethiopia. Environ Health Insights 14: 1178630220978234

Godwin C, Batterman S (2007) Indoor air quality in Michigan schools. Indoor Air 17:109–121

Gomzi M (1999) Indoor air and respiratory health in preadolescent children. Atmos Environ 33: 4081–4086

Grams H, Hehl O, Dreesman J (2003) Breathing easy in schools. Investigation results and modeling approaches to indoor air quality in schools. Gesundheitswesen 64:447–456. [in German]

Guerriero C, Chatzidiakou L, Cairns J, Mumovic D (2016) The economic benefits of reducing the levels of nitrogen dioxide (NO_2) near primary schools: the case of London. J Environ Manag 181:615–622

Guo C, Gao Z, Shen J (2019) Emission rates of indoor ozone emission devices: a literature review. Build Environ 158:302–318

Hachem M, Saleh N, Bensefa-Colas L, Momas I (2020) Determinants of ultrafine particles, black carbon, nitrogen dioxide, and carbon monoxide concentrations inside vehicles in the Paris area: PUF-TAXI study Indoor Air Dec 22

Hannam K, McNamee R, De Vocht F, Baker P, Sibley C, Agius R (2013) A comparison of population air pollution exposure estimation techniques with personal exposure estimates in a pregnant cohort. Environ Sci Process Impacts 15:1562–1572

HC (2016a) Human health risk assessment for ambient nitrogen dioxide. http://publications.gc.ca/site/eng/9.816718/publication.html

HC (2016b) Human health risk assessment for sulphur dioxide. Ottawa, Canada. https://publications.gc.ca/collections/collection_2016/sc-hc/H144-29-2016-eng.pdf

HCN (Health Council of the Netherlands) (2010) Indoor air quality in primary schools and the value of carbon dioxide as an indicator of air quality. The Hague. Netherlands. https://www.healthcouncil.nl/binaries/healthcouncil/documents/advisory-reports/2010/04/29/indoor-air-quality-in-primary-schools/advisory-report-indoor-air-quality-in-primary-schools.pdf

HEAL (Health and Environment Alliance) (2019) Healthy air, healthier children. Report. https://www.env-health.org/wp-content/uploads/2019/06/HEAL-Healthy-air-children_EU.pdf

HESE (Health Effects of School Environment) (2006) Final scientific report prepared for the Health & Consumer Protection Directorate General. Siena, Italy. http://ec.europa.eu/health/ph_projects/2002/pollution/fp_pollution_2002_frep_04.pdf

Hesterberg TW, Bunn WB, McClellan RO, Hamade AK, Long CM, Valberg PA (2009) Critical review of the human data on short-term nitrogen dioxide (NO_2) exposures: evidence for NO_2 no-effect levels. Crit Rev Toxicol 39:743–781

Holm SM, Balmes JR (2022) Systematic review of ozone effects on human lung function, 2013 through 2020. Chest 161:190–201

Hu Y, Zhao B (2020) Relationship between indoor and outdoor NO_2: a review. Build Environ 180: 106909

Huangfu P, Atkinson R (2020) Long-term exposure to NO_2 and O_3 and all-cause and respiratory mortality: a systematic review and meta-analysis. Environ Int 144:105998

Hwang SH, Park WM (2020) Indoor air concentrations of carbon dioxide (CO_2), nitrogen dioxide (NO_2), and ozone (O_3) in multiple healthcare facilities. Environ Geochem Health 42:1487–1496

Isobe Y, Yamada K, Wang Q, Sakamoto K, Uchiyama I, Mizoguchi T, Zhou Y (2005) Measurement of indoor sulfur dioxide emission from coal–biomass briquettes. Water Air Soil Pollut 163: 341–353

Janssen NAH, van Vliet PHN, Aarts F, Harssema H, Brunekreef B (2001) Assessment of exposure to traffic related air pollution of children attending schools near motorways. Atmos Environ 35: 3875–3884

Järnström H, Saarela K, Kalliokoski P, Pasanen A-L (2006) Reference values for indoor air pollutant concentrations in new, residential buildings in Finland. Atmos Environ 40:7178–7191

Jovanovic M, Vucicevic B, Turanjanin V, Zivkovi M, Spasojevic V (2014) Investigation of indoor and outdoor air quality of the classrooms at a school in Serbia. Energy 77:42–48

Jung CR, Nishihama Y, Nakayama SF, Tamura K, Isobe T, Michikawa T, Iwai-Shimada M, Kobayashi Y, Sekiyama M, Taniguchi Y, Yamazaki S (2021) Indoor air quality of 5,000 households and its determinants. Part B: Volatile organic compounds and inorganic gaseous pollutants in the Japan Environment and Children's study. Environ Res 197:111135

Kabirikopaei A, Lau J, Nord J, Bovaird J (2021) Identifying the K-12 classrooms' indoor air quality factors that affect student academic performance. Sci Total Environ 786:147498

Kaiser G, Schaper A (2012) Acute carbon monoxide poisoning Notfall Rettungsmed 15: 429–435. [in German]

Kalimeri KK, Bartzis JG, Saraga DE (2017) Commuters' personal exposure to ambient and indoor ozone in Athens. Greece Environments 4:53

Karthikeyan S, Perumal SV, Balasubramanian R, Zuraimi MS, Tham KW (2007) Determination of ozone in outdoor and indoor environments using nitrite-impregnated passive samplers followed by ion chromatography. J Air Waste Manag Assoc 57:974–980

Kim CS, Rohr AC (2021) Review and analysis of personal ambient ozone measurements. J Air Waste Manage Assoc 71:1333–1346

Kindzierski W, Sembaluk S (2001) Indoor outdoor relationship of SO2 concentrations in a rural and an urban community of Alberta. Canadian J Civil Eng 28:163–169

Kinoshita H, Türkan H, Vucinic S, Naqvi S, Bedair R, Rezaee R, Tsatsakis A (2020) Carbon monoxide poisoning. Toxicol Rep 20:169–173

Kornartit C, Sokhi RS, Burton MA, Ravindra K (2010) Activity pattern and personal exposure to nitrogen dioxide in indoor and outdoor microenvironments. Environ Int 36:36–45

Kousa A, Monn C, Rotko T, Alm S, Oglesby L, Jantunen M (2001) Personal exposures to NO_2 in the Expolisstudy: relation to residential indoor, outdoor and workplace concentrations in Basel, Helsinki and Prague. Atmos Environ 35:3405–3412

Künn S, Palacios J, Pestel N (2019) Indoor air quality and cognitive performance. IZA DP No. 12632

Lahrz T, Piloty M, Oddoy A, Fromme H (2003) Substances of health concern in public facilities in Berlin. Studies on indoor air quality in Berlin schools. Report of the Institute for Food, Drugs and Animal Diseases, Department of Environmental and Health Protection. Berlin. [in German]

Leaderer BP, Naeher L, Jankun T, Balenger K, Holford TR, Toth C, Sullivan J, Wolfson JM, Koutrakis P (1999) Indoor, outdoor, and regional summer and winter concentrations of PM_{10}, $PM_{2.5}$, SO_4^{2-}, H^+, NH_4^+, NO_3^-, NH_3, and nitrous acid in homes with and without kerosene space heaters. Environ Health Perspect 107:223–231

Lee K, Yeom J, Yoon C, Yang W, Son B-S, Jeon JM, Yu S-D, Park C-H (2013) Seasonal and geographic effects on predicting personal exposure to nitrogen dioxide by time-weighted microenvironmental model. Atmos Environ 67:143–149

Lee JY, Ryu SH, Kim CH, Bae G-N (2016) Indoor-to-outdoor pollutant concentration ratio modeling of CO_2, NO_2, and lung-deposited nanoparticles. Atmos Pollut Res 7:664–670

Levy JI, Lee K, Spengler JD, Yanagisawa Y (1998) Impact of residential nitrogen dioxide exposure on personal exposure: an international study. JAWMA 48:553–560

Li M, Weschler CJ, Bekö G, Wargocki P, Lucic G, Williams J (2020) Human ammonia emission rates under various indoor environmental conditions. Environ Sci Technol 54:5419–5428

Liang CSK, Waldman JM (1992) Indoor exposures to acidic aerosols at child and elderly care facilities. Indoor Air 2:196–207

Liu LJ, Koutrakis SP, Suh HH, Mulik JD, Burton M (1993) Use of personal measurements for ozone exposure assessment: a pilot study. Environ Health Perspect 101:318–324

Majd E, McCormack M, Davis M, Curriero F, Berman J, Connolly F, Leaf P, Rule A, Green T, Clemons-Erby D, Gummerson C, Koehler K (2019) Indoor air quality in inner-city schools and its associations with building characteristics and environmental factors. Environ Res 170:83–91

Matthaios VN, Kang CM, Wolfson JM, Greco KF, Gaffin JM, Hauptman M, Cunningham A, Petty CR, Lawrence J, Phipatanakul W, Gold DR, Koutrakis P (2022) Factors influencing classroom exposures to fine particles, black carbon, and nitrogen dioxide in inner-city schools and their implications for indoor air quality. Environ Health Perspect 130:47005

Maula H, Hongisto V, Naatula V (2017) The effect of low ventilation rate with elevated bioeffluent concentration on work performance, perceived indoor air quality, and health symptoms. Indoor Air 27:1141–1153

Mehmood T, Ali Z, Noor N, Sidra S, Nasir ZA, Colbeck I (2015) Measurementof NO_2 indoor and outdoor concentrations in selected public schools of Lahore unsing passive sampler. J Anim Plant Sci 25(suppl. 2):681–686

Meier R, Eeftens M, Phuleria HC, Ineichen A, Corradi E, Davey M, Fierz M, Ducret-Stich RE, Aguilera I, Schindler C, Rochat T, Probst-Hensch N, Tsai MY, Künzli N (2015) Differences in indoor versus outdoor concentrations of ultrafine particles, $PM_{2.5}$, PM absorbance and NO_2 in Swiss homes. J Expo Sci Environ Epidemiol 25:499–505

Meininghaus R, Kouniali A, Mandin C, Cicolella A (2003) Risk assessment of sensory irritants in indoor air – a case study in a French school. Environ Int 28:553–557

Mi YH, Norbäck D, Tao J, Mi YL, Ferm M (2006) Current asthma and respiratory symptoms among pupils in Shanghai, China: influence of building ventilation, nitrogen dioxide, ozone, and formaldehyde in classrooms. Indoor Air 16:454–464

Michikawa T, Nakai S, Nitta H, Tamura K (2014) Validity of using annual mean particulate matter concentrations as measured at fixed site in assessing personal exposure: an exposure assessment study in Japan. Sci Total Environ 466-467:673–680

Mullen NA, Bhangar S, Hering SV, Kreisberg NM, Nazaroff WW (2011) Ultrafine particle concentrations and exposures in six elementary school classrooms in northern California. Indoor Air 21:77–87

Nazaroff WW, Weschler CJ (2020) Indoor acids and bases. Indoor Air 30:559–644

Nazaroff WW, Weschler CJ (2022) Indoor ozone: concentrations and influencing factors. Indoor Air 32:e12942

Nhung NTT, Amini H, Schindler C, Kutlar Joss M, Dien TM, Probst-Hensch N, Perez L, Künzli N (2017) Short-term association between ambient air pollution and pneumonia in children: a systematic review and meta-analysis of time-series and case-crossover studies. Environ Pollut 230:1000–1008

Niu Z, Duan Z, Yu H, Xue L, Liu F, Yu D, Zhang K, Han D, Wen W, Xiang H, Qin W (2022) Association between long-term exposure to ambient particulate matter and blood pressure, hypertension: an updated systematic review and meta-analysis. Int J Environ Health Res 4:1–16

Norback D, Li T, Bai X, Li C, Zhao Z, Zhang X (2019) Onset and remission of rhinitis among students in relation to the home and school environment-A cohort study from Northern China. Indoor Air 29:527–538

OEHHA (Office of Environmental Health Hazard Assessment) (1999) Acute RELs and toxicity summaries using the previous version of the Hot Spots Risk Assessment guidelines. https://oehha.ca.gov/media/downloads/crnr/appendixd2final.pdf

Offermann F (2009) Ventilation and Indoor Air Quality in New Homes. California Energy Commission Report CEC-500-2009-005

Orellano P, Quaranta N, Reynoso J, Balbi B, Vasquez J (2017) Association of outdoor air pollution with the prevalence of asthma in children of Latin America and the Caribbean: a systematic review and meta-analysis. J Asthma 6:1–13

Orellano P, Reynoso J, Quaranta N (2021) Short-term exposure to sulphur dioxide (SO2) and all-cause and respiratory mortality: a systematic review and meta-analysis. Environ Int 150: 106434

Petersen S, Jensen KL, Pedersen AL, Rasmussen HS (2016) The effect of increased classroom ventilation rate indicated by reduced CO_2 concentration on the performance of schoolwork by children. Indoor Air 26:366–379
Pettenkofer M (1858) About the air exchange in residential buildings. Cotta Verlag, München. [in German]
Physick W, Powell J, Cope M, Boast K, Lee S (2011) Measurements of personal exposure to NO_2 and modelling using ambient concentrations and activity data. Atmos Environ 45:2095–2102
Piechocki-Minguy A, Plaisance H, Schadkowski C, Sagnier I, Saison JY, Gallo JC, Guillermo R (2006) A case study of personal exposure to nitrogen dioxide using a new high sensitive diffusive sampler. Sci Total Environ 366:55–64
Risner CH, Conner JM (1991) Collection of ammonia in indoor air by means of a weak cation exchange cartridge. Environ Toxicol Chem 10:1417–1423
Rivas I, Viana M, Moreno T, Pandolfi M, Amato F, Reche C, Bouso L, Àlvarez-Pedrerol M, Alastuey A, Sunyer J, Querol X (2014) Child exposure to indoor and outdoor air pollutants in schools in Barcelona, Spain. Environ Int 69:200–212
Rosbach J, Krop E, Vonk M, van Ginkel J, Meliefste C, de Wind S, Gehring U, Brunekreef B (2016) Classroom ventilation and indoor air quality-results from the FRESH intervention study. Indoor Air 26:538–545
Rotko T, Kousa A, Alm S, Jantunen M (2001) Exposures to nitrogen dioxide in EXPOLIS Helsinki: microenvironment, behavioral and sociodemographic factors. J Expo Anal Environ Epidemiol 11:216–233
Ruggieri S, Longo V, Perrino C, Canepari S, Drago G, L'Abbate L, Balzan M, Cuttitta G, Scaccianoce G, Minardi R, Viegi G, Cibella F (2019) Indoor air quality in schools of a highly polluted south Mediterranean area. Indoor Air 29:276–290
Sahsuvaroglu T, Su JG, Brook J, Burnett R, Loeb M, Jerrett M (2009) Predicting personal nitrogen dioxide exposure in an elderly population: integrating residential indoor and outdoor measurements, fixed-site ambient pollution concentrations, modeled pollutant levels, and time-activity patterns. J Toxicol Environ Health A 72:1520–1533
Sakuma H, Kusama M, Yamaguchi K, Matsuki T, Sugawara S (1984) The distribution of cigarette smoke components between mainstream and sidestream smoke. Beiträge zur Tabakforschung Int 12:199–209
Salonen HJ, Pasanen A-L, Lappalainen SK, Riuttala HM, Tuomi TM, Pasanen PO, Bäck BC, Reijula KE (2009) Airborne concentrations of volatile organic compounds, formaldehyde and ammonia in Finnish office buildings with suspected indoor air problems. J Occup Environ Hyg 6:200–209
Salonen H, Salthammer T, Morawska L (2018) Human exposure to ozone in school and office indoor environments. Environ Int 119:503–514
Salonen H, Salthammer T, Morawska L (2019) Human exposure to NO_2 in school and office indoor environments. Environ Int 130:104887
Salthammer T, Uhde E, Schripp T, Schieweck A, Morawska L, Mazaheri M, Clifford C, He C, Buonanno G, Querol X, Viana M, Kumar P (2016) Children's well-being at schools: impact of climatic conditions and air pollution. Environ Int 94:196–210
Samad A, Vogt U (2021) Mobile air quality measurements using bicycle to obtain spatial distribution and high temporal resolution in and around the city center of Stuttgart. Atmos Environ 244: 117915
Sanyal DK, Maduna ME (2000) Possible relationship between indoor pollution and respiratory illness in an Eastern Cape community. South Afr J Sci 96:94–96
Sarnat SE, Coull BA, Schwartz J, Gold DR, Suh HH (2006) Factors affecting the association between ambient concentrations and personal exposures to particles and gases. Environ Health Perspect 114:649–654
Schembari A, Triguero-Mas M, de Nazelle A, Dadvand P, Vrijheid M, Cirach M, Martinez D, Figueras F, Querol X, Basagaña X, Eeftens M, Meliefste K, Nieuwenhuijsen MJ (2013)

Personal, indoor and outdoor air pollution levels among pregnant women. Atmos Environ 64: 287–295

Seow WJ, Downward GS, Wei H, Rothman N, Reiss B, Xu J, Bassig BA, Li J, He J, Hosgood HD, Wu G, Chapman RS, Tian L, Wei F, Caporaso NE, Vermeulen R, Lan Q (2016) Indoor concentrations of nitrogen dioxide and sulfur dioxide from burning solid fuels for cooking and heating in Yunnan Province, China. Indoor Air 26:776–783

Seppänen OA, Fisk WJ (2004) Summary of human responses to ventilation. Indoor Air Suppl 7: 102–118

Singer BC, Chan WR, Kim Y-S, Offermann FJ, Walker IS (2020) Indoor air quality in California homes with code-required mechanical ventilation. Indoor Air 30:885–899

Sisovic A, Sega K, Kalinic N (1987) Indoor/outdoor relationship of ammonia concentration in selected office buildings. Sci Total Environ 61:73–77

Siwarom S, Puranitee P, Plitponkarnpim A, Manuyakorn W, Sinitkul R, Arj-Ong Vallipakorn S (2017) Association of indoor air quality and preschool children's respiratory symptoms. Asian Pac J Allergy Immunol 35:119–126

Spengler JD, Ferris BG, Dockery DW, Speizer FE (1979) Sulfur dioxide and nitrogen dioxide levels inside and outside homes and the implications on health effects research. Environ Sci Technol 13:1276–1280

Spengler JD, Brauer M, Samet JM, Lambert WE (1993) Nitrous acid in Albuquerque, New Mexico, homes. Environ Sci Technol 27:841–845

Stafoggia M, Oftedal B, Chen J, Rodopoulou S, Renzi M, Atkinson RW, Bauwelinck M, Klompmaker JO, Mehta A, Vienneau D, Andersen ZJ, Bellander T, Brandt J, Cesaroni G, de Hoogh K, Fecht D, Gulliver J, Hertel O, Hoffmann B, Hvidtfeldt UA, Jöckel KH, Jørgensen JT, Katsouyanni K, Ketzel M, Kristoffersen DT, Lager A, Leander K, Liu S, Ljungman PLS, Nagel G, Pershagen G, Peters A, Raaschou-Nielsen O, Rizzuto D, Schramm S, Schwarze PE, Severi G, Sigsgaard T, Strak M, van der Schouw YT, Verschuren M, Weinmayr G, Wolf K, Zitt E, Samoli E, Forastiere F, Brunekreef B, Hoek G, Janssen NAH (2022) Long-term exposure to low ambient air pollution concentrations and mortality among 28 million people: results from seven large European cohorts within the ELAPSE project. Lancet Planet Health 6:e9–e18

Stranger M, Potgieter-Vermaak SS, Van Grieken R (2007) Comparative overview of indoor air quality in Antwerp, Belgium. Environ Int 33:789–797

Stranger M, Potgieter-Vermaak SS, Van Grieken R (2008) Characterization of indoor air quality in primary schools in Antwerp, Belgium. Indoor Air 18:454–463

Suh HH, Spengler JD, Koutrakis P (1992) Personal exposures to acid aerosols and ammonia. Environ Sci Technol 26:2507–2517

Sun C, Hong S, Cai G, Zhang Y, Kan H, Zhao Z, Deng F, Zhao B, Zeng X, Sun Y, Qian H, Liu W, Mo J, Guo J, Zheng X, Su C, Zou Z, Li H, Huang C (2021) Indoor exposure levels of ammonia in residences, schools, and offices in China from 1980 to 2019: a systematic review. Indoor Air 31:1691–1706

Sun HZ, Yu P, Lan C, Wan MWL, Hickman S, Murulitharan J, Shen H, Yuan L, Guo Y, Archibald AT (2022) Cohort-based long-term ozone exposure-associated mortality risks with adjusted metrics: a systematic review and meta-analysis. Innovation (Camb) 20:100246

Suzuki N, Nakaoka H, Eguchi A, Hanazato M, Nakayama Y, Tsumura K, Takaguchi K, Takaya K, Todaka E, Mori C (2020) Concentrations of formic acid, acetic acid, and ammonia in newly constructed houses. Int J Environ Res Public Health 17:1940

Szabados M, Csákó Z, Kotlík B, Kazmarová H, Kozajda A, Jutraz A, Kukec A, Otorepec P, Dongiovanni A, Di Maggio A, Fraire S, Szigeti T (2021) Indoor air quality and the associated health risk in primary school buildings in Central Europe – The InAirQ study. Indoor Air 31: 989–1003

Targa J, Ripoll A, Banyuls L, González A, Soares J (2022) Status report of air quality in Europe for year 2021, using validated and up-to-date data. (Eionet Report – ETC/HE 2022/3). https://www.eionet.europa.eu/etcs/etc-he/products/etc-he-products/etc-he-reports/etc-he-report-2022-3-status-report-of-air-quality-in-europe-for-year-2021-using-validated-and-up-to-date-data

Tidy G, Cape JN (1993) Ammonia concentrations in houses and public buildings. Atmos Environ 27A:2235–2237

Tuomainen M, Pasanen A-L, Tuomainen A, Liesivuori J, Juvonen P (2001) Usefulness of the Finnish classification of indoor climate, construction and finishing materials: comparison of indoor climate between two new blocks of flats in Finland. Atmos Environ 35:305–313

Twardella D, Matzen W, Lahrz T, Burghardt R, Spegel H, Hendrowarsito L, Frenzel AC, Fromme H (2012) Effect of classroom air quality on students' concentration: results of a cluster-randomized cross-over experimental study. Indoor Air 22:378–387

UBA (2017a) Requirements for ventilation concepts in buildings. Recommendations of the Working Group on Ventilation (AK Lüftung) at the German Federal Environment Agency. Dessau. [in Gernman] https://www.umweltbundesamt.de/sites/default/files/medien/1410/publikationen/uba_empfehlungspapier_lueftung_unterrichtsgebaeude_final_bf.pdf

UBA (2017b) Air pollutant emissions in Germany. Nitrogen oxides (NOx). German Federal Environment Agency. https://www.umweltbundesamt.de/daten/luft/luftschadstoff-emissionen-in-deutschland. [in German]

UBA (2023) Ammonia, dust and odour emissions. https://www.umweltbundesamt.de/en/topics/soil-agriculture/ecological-impact-of-farming/ammonia-dust-odour-emissions

Uchiyama S, Tomizawa T, Tokoro A, Aoki M, Hishiki M, Yamada T, Tanaka R, Sakamoto H, Yoshida T, Bekki K, Inaba Y, Nakagome H, Kunugita N (2015) Gaseous chemical compounds in indoor and outdoor air of 602 houses throughout Japan in winter and summer. Environ Res 137:364–372

Underhill LJ, Bose S, Williams DL, Romero KM, Malpartida G, Breysse PN, Klasen EM, Combe JM, Checkley W, Hansel NN (2015) Association of roadway proximity with indoor air pollution in a peri-urban community in Lima, Peru. Int J Environ Res Public Health 12:13466–13481

US EPA (2009) Risk and exposure asessment to support the Review of the SO2 Primary National Ambient Air Quality Standards. https://www3.epa.gov/ttn/naaqs/standards/so2/data/200908SO2REAFinalReport.pdf

US EPA (2016) Integrated Science Assessment for Oxides of Nitrogen Health Criteria (2016 Final Report). EPA/600/R-15/068. NC; USA. https://cfpub.epa.gov/ncea/isa/recordisplay.cfm?deid=310879

US EPA (2017) Policy Assessment for the Review of the Primary National Ambient Air Quality Standards for Oxides of Nitrogen. https://www.epa.gov/naaqs/policy-assessment-review-primary-national-ambient-air-quality-standards-oxides-nitrogen

US EPA (2020) Integrated science assessment for ozone and related photochemical oxidants. Center for Public Health and Environmental Assessment. https://www.epa.gov/isa/integrated-science-assessment-isa-ozone-and-related-photochemical-oxidants

Valero N, Aguilera I, Llop S, Esplugues A, de Nazelle A, Ballester F, Sunyer J (2009) Concentrations and determinants of outdoor, indoor and personal nitrogen dioxide in pregnant women from two Spanish birth cohorts. Environ Int 35:1196–1201

Van Roosbroeck S, Li R, Hoek G, Lebret E, Brunekreef B, Spiegelman D (2008) Traffic-related outdoor air pollution and respiratory symptoms in children: the impact of adjustment for exposure measurement error. Epidemiology 19:409–416

Verriele M, Schoemaecker C, Hanoune B, Leclerc N, Germain S, Gaudion V, Locoge N (2016) The MERMAID study: indoor and outdoor average pollutant concentrations in 10 low-energy school buildings in France. Indoor Air 26:702–713

Villanueva F, Tapia A, Lara S, Amo-Salas M (2018) Indoor and outdoor air concentrations of volatile organic compounds and NO_2 in schools of urban, industrial and rural areas in Central-Southern Spain. Sci Total Environ 622–623:222–235

Villanueva F, Ródenas M, Ruus A, Saffell M, Marta F (2022) Sampling and analysis techniques for inorganic air pollutants in indoor air. Appl Spectr Rev 57:531–557

Vu TV, Stewart GB, Kitwiroon N, Lim S, Barratt B, Kelly FJ, Thompson R, Smith RB, Toledano MB, Beevers SD (2022) Assessing the contributions of outdoor and indoor sources to air quality in London homes of the SCAMP cohort. Build Environ 222:109359

Walker IS, Kim Yang-Seon, Singer BC, Chan WR (2017) Assessing occupant and outdoor air impacts on indoor air quality in new California homes. Proc AIVC Conference. https://escholarship.org/content/qt2xg7s5nv/qt2xg7s5nv.pdf

Wargocki P, Wyon DP (2006) Research report on effects of HVAC on student performance. ASHRAE Jl 48:23–26

Watson CV, Feng J, Valentin-Blasini L, Stanelle R, Watson CH (2016) Method for the determination of ammonia in mainstream cigarette smoke using ion chromatography. PLoS One 11: e0159126

Weschler CJ (2000) Ozone in indoor environments: concentration and chemistry. Indoor Air 10: 269–288

WHO (2010) Guidelines for indoor air quality: selected pollutants. World Health Organization, Regional Office for Europe, Copenhagen (http://www.euro.who.int/en/what-we-do/health-topics/environmental-health/air-quality/publications/2010/who-guidelines-for-indoor-air-quality-selected-pollutants)

WHO (2021) WHO global air quality guidelines Particulate matter ($PM_{2.5}$ and PM_{10}), ozone, nitrogen dioxide, sulfur dioxide and carbon monoxide. https://www.who.int/publications/i/item/9789240034228

Wichmann J, Lind T, Nilsson MAM, Belland T (2010) $PM_{2.5}$, soot and NO_2 indoor outdoor relationships at homes, pre-schools and schools in Stockholm, Sweden. Atmos Environ 44: 4536–4544

Wohlgemuth L, McLagan D, Flückiger B, Vienneau D, Osterwalder S (2020) Concurrently measured concentrations of atmospheric mercury in indoor (household) and outdoor air of Basel, Switzerland. Environ Sci Technol Lett 14:234–239

Wolf J, Berlin K, Fembacher L, Heitmann D, Matzen W, Nitschke L, Sysoltseva M, Fromme H (2018) Air quality in indoor go-kart facilities in Germany. Indoor Air 28:950–962

Xu Z, Wang W, Liu Q, Li Z, Lei L, Ren L, Deng F, Guo X, Wu S (2022) Association between gaseous air pollutants and biomarkers of systemic inflammation: a systematic review and meta-analysis. Environ Pollut 292(Pt A):118336

Yang G, Liu Y, Li X (2020) Spatiotemporal distribution of ground-level ozone in China at a city level. Sci Rep 29:7229

Yazar M, Bellander T, Merritt AS (2011) Personal exposure to carcinogenic and toxic air pollutants in Stockholm, Sweden: a comparison over time. Atmos Environ 45:2999–3004

Yen YC, Yang CY, Ho CK, Yen PC, Cheng YT, Mena KD, Lee TC, Chen PS (2020) Indoor ozone and particulate matter modify the association between airborne endotoxin and schoolchildren's lung function. Sci Total Environ 705:135810

Yoon C, Lee K, Park D (2011) Indoor air quality differences between urban and rural preschools in Korea. Environ Sci Pollut Res Int 18:333–345

Zechmeister HG, Rivera M, Köllensperger G, Marrugat J, Künzli N (2020) Indoor monitoring of heavy metals and NO2 using active monitoring by moss and Palmes diffusion tubes. Environ Sci Eur 32:156

Zhang X, Zhao Z, Nordquist T, Norback D (2011) The prevalence and incidence of sick building syndrome in Chinese pupils in relation to the school environment: a two-year follow-up study. Indoor Air 21:462–471

Zhang X, Li F, Zhang L, Zhao Z, Norback D (2014) A longitudinal study of sick building syndrome (SBS) among pupils in relation to SO_2, NO_2, O_3 and PM_{10} in schools in China. PLoS One 9: e112933

Zhang Z, Wang J, Lu W (2018a) Exposure to nitrogen dioxide and chronic obstructive pulmonary disease (COPD) in adults: a systematic review and meta-analysis. Environ Sci Pollut Res Int 25: 15133–15145

Zhang X, Fan Q, Bai X, Li T, Zhao Z, Fan X, Norbäck D (2018b) Levels of fractional exhaled nitric oxide in children in relation to air pollution in Chinese day care centres. Int J Tuberc Lung Dis 22:813–819

Zhang J, Wei Y, Fang Z (2019) Ozone pollution: a major health hazard worldwide. Front Immunol 10:2518

Zheng XY, Orellano P, Lin HL, Jiang M, Guan WJ (2021) Short-term exposure to ozone, nitrogen dioxide, and sulphur dioxide and emergency department visits and hospital admissions due to asthma: a systematic review and meta-analysis. Environ Int 150:106435

Zhong Q, Shen H, Yun X, Chen Y, Ren Y, Xu H, Shen G, Du W, Meng J, Li W, Ma J, Tao S (2020) Global sulfur dioxide emissions and the driving forces. Environ Sci Technol 54:6508–6517

Zhu YD, Li X, Fan L, Li L, Wang J, Yang WJ, Wang L, Yao XY, Wang XL (2021) Indoor air quality in the primary school of China-results from CIEHS 2018 study. Environ Pollut 291: 118094

Zuraimi MS, Tham KW, Chew FT, Ooi PL (2007) The effect of ventilation strategies of child care centers on indoor air quality and respiratory health of children in Singapore. Indoor Air 17:317–327

Zusman M, Gassett AJ, Kirwa K, Barr RG, Cooper CB, Han MK, Kanner RE, Koehler K, Ortega VE, Paine R, Paulin L, Pirozzi C, Rule A, Hansel NN, Kaufman JD (2021) Modeling residential indoor concentrations of $PM_{2.5}$, NO_2, NOx, and secondhand smoke in the Subpopulations and Intermediate Outcome Measures in COPD (SPIROMICS) Air study. Indoor Air 31:702–716

Chapter 8
Radon and Metals

Contents

Abstract The focus of this chapter are on radon, the substance with the highest carcinogenic potential indoors, and metals (especially heavy metals). They are presented individually in detail. In addition to a brief description of the chemical–physical characteristics and the use of the substance, an overview of main health risks and, in particular, references to further scientific literature are given. In the foreground, however, are extensive presentations of the occurrence of the substances in indoor air and in sedimented dust.

Keywords Radon · Radon reduction · Heavy metals · Lead · Mercury

8.1 Radon

Radon is a naturally occurring, odorless, tasteless, and colorless radioactive noble gas. Radon is formed as an intermediate product in the decay series of uranium-238, uranium-235, and thorium-232 from the parent nuclide radium (ICRP 2014). Of the various isotopes of radon, radon-222 is the most stable, with a half-life of about 3.8 days. Due to their short half-lives (seconds), exposure to radon-219 and radon-220 in ambient and indoor air is generally less significant (ICRP 2014). In general, the term radon is always understood simplistically as radon-222.

In the soil, the occurrence and dispersion of radon are strongly influenced by the soil properties like porosity and permeability as well as the weather and can vary considerably from region to region (Cinelli et al. 2019). All rocks contain uranium in small amounts between 1 and 3 parts per million (ppm), but some types of rocks like light-colored volcanic rocks, granites, dark shales, sedimentary rocks that contain

H. Fromme, *Indoor Air Quality*, https://doi.org/10.1007/978-3-031-40078-0_8

phosphate, and metamorphic rocks derived from these rocks have higher than average uranium contents. Due to the mechanical breakdown of rocks uranium reached the surface soils. Additionally, radon can dissolve in groundwater and thus reach drinking water or is released into the atmosphere with the other soil gases.

A particular exposure pathway is the transfer of soil gas through leaks in the building envelope to indoor spaces, especially basements. This process depends on the nature of the subsurface and the temporal variability of radon outgassing, the nature of the building fabric (e.g., impermeability), the pressure ratios between indoor and outdoor air, the ventilation conditions in the rooms, and the time of year (usually higher indoor concentrations in winter) (Schubert et al. 2018). In contrast, building materials are in the majority of cases not a significant source.

In a nationwide survey of different Canadian indoor environments, based on contamination levels and residence times, it was found for children and adults that about 90% of the total exposure occurs through residence in dwellings (Chen 2019).

Inexpensive passive monitoring dosimeters for long-term measurements and active (electrical) devices for short-term measurements are available for indoor analysis (IAEA I2014). While active systems give only a momentary impression of the exposure situation, measurements with passive detectors are performed over 3–12 months and thus allow a valid estimation of long-term indoor exposure. Commonly used passive devices are track-etch detectors (alpha track dositmeters) that use plastic (polyallyl-diglycol carbonate, PADC; cellulose nitrate, LR-115 and CN 85; polycarbonate, Makrofol) as detector material or activated charcoal as well as Electrets, which are passive detectors that use an electrostatically charged disk (Cinelli et al. 2019).

8.1.1 Health Aspects

As an inert gas, radon affects the surfaces of the lungs after inhalation and is exhaled unchanged. The decay of radon produces radiation in the form of alpha radiation, which can damage biological tissues. Studies show a clear link between the level of radon exposure and the development of lung cancer. Thus, in a joint evaluation of 13 studies from 9 European countries, including 7148 cases and about 14,200 controls, it was concluded that indoor radon is responsible for about 9% of lung cancer deaths (Darby et al. 2005). The United Nations Scientific Committee on the Effects of Atomic Radiation (UNSCEAR) has provided a detailed and authoritative update of its reviews of the epidemiology and dosimetry of radon (Harrison 2021).

The Handbook on Indoor Radon of WHO (2009) evaluated the health effects of exposure in dwellings and has come to the following conclusions:

- Epidemiological studies confirm that radon in homes increases the risk of lung cancer in the general population. Other health effects of radon have not been demonstrated consistently.

- The proportion of all lung cancers linked to radon is estimated to lie between 3% and 14%, depending on the average radon concentration in the country and the method of calculation.
- There is no known threshold concentration below which radon exposure presents no risk. Even low concentrations of radon can result in a small increase in the risk of lung cancer.

For Germany, calculations by Menzler et al. (2008) yield a proportion of lung cancer deaths attributable to indoor radon of 5% (95% confidence interval: 1.7–12.6%). The WHO estimates that long-term exposure over 30 years can be expected to result in a 16% increase in risk for each increase of 100 Bq/m^3 using a linear dose-response relationship (WHO 2010). For lifetime exposure, this results in an excess lung cancer risk for non-smokers and current smokers of 0.6×10^{-5} per Bq/m^3 and 15 x 10^{-5} per Bq/m^3, respectively. A recent review and meta-analysis of Cheng et al. (2021) included relevant studies between 1990 and 2020 found adjusted excess relative risks per 100 Bq/m^3 of radon of 15% for never-smokers and 9% for ever-smokers. In another meta-analysis that included a total of 8 ecological, 9 case-control, and 15 ecological-cohort studies, the overall pooled relative risk increase for each 100 Bq/m^3 of radon dose was 1.12 (95% CI: 1.0184–1.2363) (Moon and Yoo 2021).

After tobacco smoke, radon is thus the most significant carcinogenic lung cancer risk in many countries. The World Health Organization's cancer agency has classified radon as a clear human carcinogen in Group 1 (IARC 1988).

8.1.2 *Occurrence in the Air of Residences*

In Europe, the Joint Research Centre (JRC) published the European Indoor Radon Map, which reports the arithmetic means over 10 km × 10 km grid cells of annual indoor radon concentration in ground-floor rooms in 35 countries in 2018 (Cinelli et al. 2019). The results are given in Fig. 8.1. For all the countries participating, 34% of the non-empty cells have an arithmetic mean above 100 Bq/m^3 and 4% of them above 300 Bq/m^3. In the Czech Republic, more than 90% of the means of all non-empty cells exceed 100 Bq/m^3, while none of the cells in the Netherlands have a mean value above this level. More in detail, in Fig. 8.2, district-specific concentration of about 14,000 full-year measurements in about 7000 German households are compiled, which were carried out on behalf of the Federal Office for Radiation Protection (BfS) from 2019 to 2021 (Petermann et al. 2023). On this basis a geometric mean of 41 Bq/m^3 (95th percentile: 180 Bq/m^3) has been found, with approximately 10.5 million (12.5% of the population) above 100 Bq/m^3 and 1.9 million (2.2%) above 300 Bq/m^3. In Austria, a total of 8460 rooms were investigated in a National Survey between 1992 and 2001 and annual mean values of 102 Bq/m^3 (median: 61 Bq/m^3; maximum: 11000 Bq/m^3) were observed (Maringer et al. 2008), while a recent Austrian study conducted in 2013 to 2019 mean levels of 166 Bq/m^3

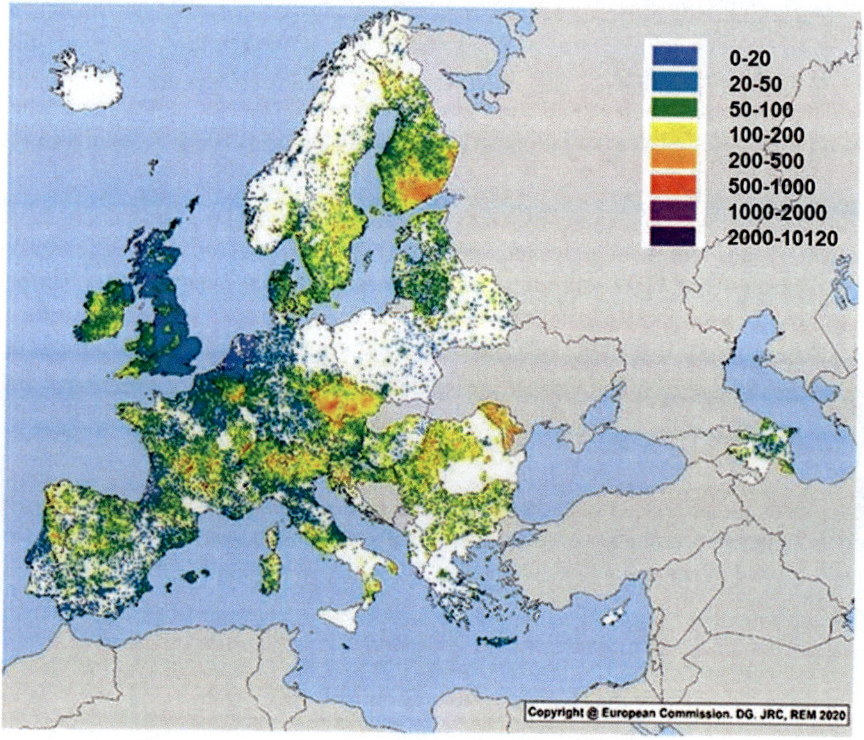

Fig. 8.1 European indoor radon map (arithmetic means of Bq/m³ over 10 km × 10 km grid cells of annual indoor radon concentration in ground-floor rooms) (Cinelli et al. 2019)

(median: 99 Bq/m^3) were reported using about 50,000 measurements (Gruber et al. 2021). Kropat et al. (2014) report an annual average of about 152 Bq/m^3 in 5671 residential interiors for Switzerland as part of the summary of data from 30 years. In a study in which 4 living rooms of houses were partially examined over one year, significantly higher concentrations could be found with a median (90th percentile) of 69 Bq/m^3 (206 Bq/m^3) in the first floor of energy-efficient refurbished buildings ($N = 114$) than in 105 comparison buildings with a median of 40 μg/m^3 (71 Bq/m^3) (Meyer 2019). In contrast, no difference was seen in this study in 35 low-energy houses (passive houses) compared to 41 comparison buildings. Here, the medians were 39 Bq/m^3 (passive houses) and 41 Bq/m^3 (comparison group), respectively. Using ca. 80,000 measurements in the Swiss Radon Database (1994 to 2017) median radon concentrations were higher in single family homes (95.1 Bq/m^3) compared to apartments (85.1 Bq/m^3) and levels decreased with increasing floor of dwelling (from median 121.1 Bq/m^3 in the inhabited basements to 57.6 Bq/m^3 on the third floor and above) (Vienneau et al. 2021).

To explore a hotspot area in the USA, around 1,800,000 measurements from 1988 to 2018 were conducted with mean concentration of 165 Bq/m^3 (Kellenbenz and Shakya 2021). Geology may be the primary factor influencing indoor concentrations as demonstrated by the radon levels in different geographical regions. In 26 houses

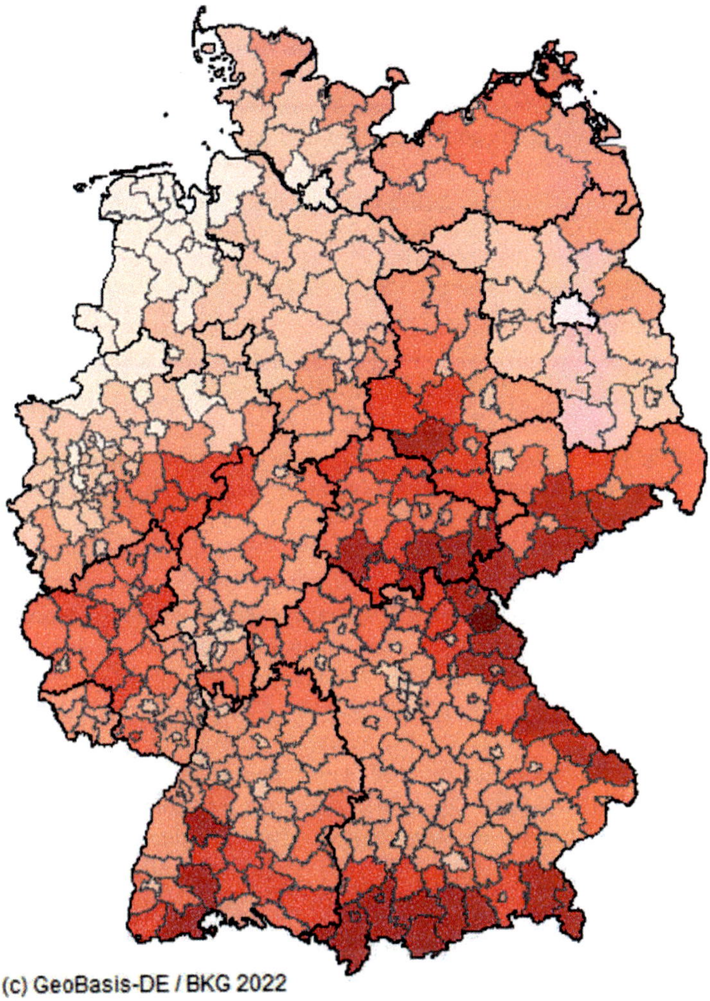

Fig. 8.2 District-related indoor air radon concentration based on 90th percentiles in Germany (according to Petermann et al. 2023)

of non-smokers in the Boston metropolitan area indoor air was sampled two times for 5 days over two seasons, a non-heating and heating season in 2017–2018 (Kang et al. 2020). Radon averaged 48.0 Bq/m^3 for the non-heating season, while it averaged 62.8 Bq/m^3 for the heating season and over three times higher levels were seen in basements compared to family rooms.

A systematic review of indoor radon in 20450 dwellings in China from 2000 to 2020 observed a mean (median) level of 59.3 Bq/m^3 (40.6 Bq/m^3) (Su et al. 2022a). The indoor radon concentration was related to seasons, climate regions, ventilation, decoration, soil and outdoor air concentration and followed an increasing time trend. In China, 24-hour average levels of 30.4 Bq/m^3 were reported in an investigation of

21 bedrooms in winter 2019 with higher concentration in the first floor (42.2 Bq/m^3) compared to the sixth (34.2 Bq/m^3), and twelfth floor 26.0 Bq/m^3 (Li et al. 2021). Results of investigations conducted in 3900 houses in Japan from 2007 to 2010 (Suzuki et al. 2010) and in 599 residences in South Korea from 2015 to 2016 (Park et al. 2018) observed arithmetic mean indoor concentrations of 14.3 Bq/m^3 and 70.8 Bq/m^3 (6.5–572 Bq/m^3), respectively. A systematic review of 31 studies of residential radon in Central and South America found that the prevalence of exposure to radon was uneven, both among the different countries and within individual countries (Giraldo-Osorio et al. 2020).

8.1.3 *Occurrence in the Air of Public Facilities*

In a recent review, Zhukovsky et al. (2018) compiled the results of 63 studies in daycare centers and schools worldwide. Radon concentration in schools and kindergartens is 1.5 times higher than worldwide average value estimated by UNSCEAR using results of national indoor radon surveys. Estimated number of objects with radon concentration $>$ 300 Bq/m^3 in children's institutions is about 1.5%, that is 5 times higher than in the dwellings. The authors explained the elevated indoor radon concentrations in children's institutions in comparison with dwellings by low ventilation and that schools and especially kindergartens are often located at ground level.

In Germany, a total of 172 schools were investigated over 100 days each in 2015/2016, and 48 were found to have at least one occupied room exceeding 300 Bq/m^3 and 128 schools with at least one value exceeding 100 Bq/m^3 (Fesenbeck et al. 2017). In Upper Austria, an annual median of 75 Bq/m^3 (maximum: 3500 Bq/m^3) was found in 633 daycare centers in 1992–2001 (Maringer et al. 2008). In the 1990s, an extensive study was conducted in 35,500 rooms of schools and daycare centers in the Czech Republic, where the annual median was 177 Bq/m^3 and the geometric mean was 124 Bq/m^3 (Fojtikova and Navratilova 2014). For Switzerland, an annual mean of 158 Bq/m^3 was measured in 1010 schools over the past 30 years (Kropat et al. 2014). In 525 rooms of 39 kindergartens and primary schools in southern Italy in 2020/21 radon concentrations were measured and ranged from 11 to 1416 Bq/m^3, with a geometric mean of 77 Bq/m^3 (Loffredo et al. 2022). In this study, total 30% of the classrooms were above 100 Bq/m^3 and 3% above 300 Bq/m^3. Ivanova et al. (2014) reported a geometric mean of 101 Bq/m^3 (9–1415 Bq/m^3) in the indoor air of 512 rooms in Bulgarian kindergarten investigated during three months in 2013. Moreover, Ivanova et al. (2022) measured radon in 437 rooms of 55 Bulgarian schools in 2019/20 and found a geometric mean of 114 Bq/m^3 (19.6–544 Bq/m^3); 30% of the measured buildings exceeded 300 Bq/m^3. In addition, a total of 115 schools were studied in 23 countries in Europe as part of the 2011–2012 SINPHONIE (Schools Indoor Pollution and Health: Observatory Network in Europe) study, and 114 Bq/m^3 (median) and 316 Bq/m^3 (90th percentile) were observed in 4066 measurements (Baloch et al. 2020). In 59 primary school buildings of five Central European countries (Czech Republic, Hungary, Italy, Poland, and

Slovenia) investigated in 2017/18 an arithmetic mean of 98 Bq/m^3 was found (9–507 Bq/m^3) (Szabados et al. 2021).

A systematic review of indoor radon in Chinese studies from 2000 to 2020 observed a mean (median) level of 54.2 Bq/m^3 (39.0 Bq/m^3) in 209 schools (Su et al. 2022a). Annual medians of 42 Bq/m^3 (maximum: 620 Bq/m^3) were found in 180 kindergarten in Russia in 2013–2016 (Onishchenko et al. 2017). Poulin et al. (2012) investigated 65 schools in Canada in 2010 and reported mean levels of 56 Bq/m^3 (15–663 Bq/m^3). Geometric mean radon concentrations in 2289 classrooms in 66 of Utah's public schools were 31.4 Bq/m^3 (14.8–673 Bq/m^3) and thirty-seven (2%) of the classrooms and 20% of the schools had levels at or above the US EPA recommended action level of 148 Bq/m^3 (Davis et al. 2020).

8.1.4 Measures to Reduce Exposure

The IRCP recommends several radon mitigation techniques to reduce convection and diffusion of radon from the soil beneath a building (IRCP 2014). This could be done by strengthening the air tightness of the shell of the building and reversing the air pressure between indoor spaces and the ground, as well as diluting indoor radon with fresh outside air (ICRP 2014). Studies worldwide have investigated a variety of construction solutions, radon mitigation, and remediation systems with different levels of effectiveness summarized by Khan et al. (2019). They stated that sub-slab or sump depressurization system (SSDS) and active ventilation technique were found more effective in achieving a significant and sustained radon reduction than the passive methods such as sealing, membrane, block and beam, simple ventilation, or filtration. Moreover, radon testing in indoor air should be done periodically over the lifetime of the building. Additionally the review indicates that in high radon areas and for larger buildings, effective mitigation of indoor radon can be achieved with a combination of house-specific long-term stable mitigation techniques rather than using any single measure, particularly in the extreme cold countries.

In a systematic review Ajrouche et al. (2017) assessed the effects of radon mitigation on lung cancer in studies conducted in several countries. For example, in the USA 25% radon-related lung cancer deaths could be prevented if all homes above 148 Bq/m^3 were effectively remediated and in Canada 52% of the radon-induced lung cancer deaths nationwide can be prevented per year if all homes with radon above 100 Bq/m^3 were remediated to outdoor levels or 28% if all indoor levels are below the Canadian action level of 200 Bq/m^3. Again using data from 2017 in Finland, a reduction of radon exposure to 100 Bq m^3 action level would eliminate approximately 30% of radon-attributable lung cancer cases (Kurkela et al. 2023).

Although it has been known for a long time that a reduction in indoor concentrations of radon leads to a significant decrease in cancer risk, systematic measurements and, if necessary, mitigation techniques are still often implemented hesitantly. For example, Denu et al. (2019) reported that only 31% of Wisconsin landlords and 35% of Wisconsin school districts have tested their schools for radon and 60% of the homeowners with elevated radon level mitigated, while 47% of schools did nothing in response to a high radon tests.

In Canada qualitative interviews with homeowners and tenants aimed to identify the reasons behind the gap between risk awareness and actual adoption of preventive action and found that having mere cognitive risk awareness is not enough to prompt preventive action and additional affective awareness motivates residents to act (Khan et al. 2021). They concluded that the radon health communication program would be more effective through addressing both these aspects of risk perception along with plausible regulations and necessary incentives.

8.2 Metals

Metals are ubiquitously present as a natural component of the earth's crust. In addition to this natural occurrence, exposure via anthropogenic activities such as mining, processing, and large-scale use in the context of many industrial productions must also be taken into account. These have led to a, in some cases spatially particularly high, contamination of the environmental media (Tchounwou et al. 2012, Barthi and Sharma 2022). For example the use of leaded gasoline (as tetraethyl lead) since the 1920s resulted in a comprehensive environmental contamination and thereby via inhalation to human exposure worldwide. After the ban of lead between the 1980s and 2000s decreasing concentrations were observed in ambient and indoor air as well as in human blood.

In the following, we will briefly discuss the occurrence and use of some important metals such as arsenic, cadmium, lead, and nickel (see Table 8.1) as well as mercury (IARC 2012; NTP 2012; US EPA 2022; OSHA 2023; ECHA 2023).

There are three major groups of **arsenic** compounds: Inorganic arsenic of the trivalent forms like arsenic trioxide, sodium arsenite, and arsenic trichloride and of the pentavalent forms like arsenic pentoxide, arsenic acid, and arsenates as well as organic arsenic like arsanilic acid, methylarsonic acid, and arsenobetaine. Anthropogenic sources are estimated to account for nearly 24,000 tons of arsenic emitted to the global atmosphere per year, while 7900 tons come from natural sources.

Cadmium (Cd) hydroxide is used in electrodes for Ni-Cd batteries, cadmium sulfide compounds as pigments in plastic glass, ceramics, rubber, artist colors, and fireworks, cadmium and cadmium alloys in electroplated coatings on metals, and cadmium salts of organic acids were formerly used as heat and light stabilizers for plastics. Global cadmium production, excluding the US production, decreased slightly to an estimated 25,400 tons in 2010. About 85% of cadmium was used in batteries, predominantly in rechargeable nickel-cadmium batteries.

The primary use of **lead** (Pb) is for batteries and for lead-alloys in ammunition, pipes, cable covering, building material, solder, radiation shielding, collapsible tubes, and fishing weights. Lead is also used in ceramic glazes and as a stabilizer in plastics. Organic lead compounds like tetramethyl lead and tetraethyl lead were used as an antiknock additive in gasoline but were banned in different countries

Table 8.1 Major sources and brief overview of health effects (modify from Rehman et al. 2018; IARC 2012)

Name	Major sources	Major uses	Health effects	IARC evaluation
Arsenic	Smelting, fossil fuel burning, agricultural pesticides, industrial waste	Pharmaceuticals, wood preservatives, agricultural chemicals, and applications in the mining, metallurgical, glass-making, and semiconductor industries	Arsenicosis, psychological effects, decreased mental performance, hypertension, cardiovascular disease risk, carotid atherosclerosis and diabetes mellitus, cancer of the lung, urinary bladder, and skin	Arsenic and inorganic arsenic compounds in group 1. Dimethylarsinic acid and monomethylarsonic acid in group 2B. Arsenobetaine and other organic arsenic compounds in group 3.
Cadmium	Incineration fumes (from cadmium containing products), phosphate fertilizers, color pigments, PVC products, cigarette smoking	Nickel-cadmium batteries, pigments, coatings and plating, stabilizers for plastics, non-ferrous alloys, semiconductors and photovoltaic devices	Neurodegenerative disorders, end-stage renal disease, breast cancer, prostate cancer, demineralization of bones, diabetes.	Cadmium and cadmium compounds in group 1
Lead	Mining, fossil fuel burning, manufacturing of lead - acid batteries, oxide synthesis for paint, and pigments		Neurotoxic effects on intelligence, decreased memory, hemolytic anemia, cardiovascular diseases, reproductive toxicity, lung cancer, bladder cancer	Lead in group 2B Inorganic lead compounds in group 2A Organic lead compounds in group 3
Nickel	Chemical industries, food processing industries, forest fires, volcanic emissions, incineration of waste, combustion of coal	Electroplating, ceramics, pigments, production of ceramics, alloy steel and stainless steel, nickel-cadmium batteries, in petrochemical metallurgical industry	Allergic contact dermatitis, oral hypersensitivity and risk of gingival hyperplasia, oral cancer, skin cancer, lung cancer, asthma, bronchitis, reproductive toxicity, carcinogenesis	Nickel compounds in group 1

[a] Overall evaluation: Group 1: carcinogenic to humans; Group 2A: probably carcinogenic to humans; Group 2B possibly carcinogenic to humans; Group 3 not classifiable as to their carcinogenicity to humans; Group 4: probably not carcinogenic to humans

since decades. In 2018, US production of lead was estimated at 1.3 million metric tons.

Nickel (Ni) salts are used in electroplating, ceramics, pigments, and as intermediates (e.g., catalysts, formation), nickel oxide is used in nickel catalysts in the ceramics industry, in the manufacture of alloy steel and stainless steel, in the manufacture of nickel salts for specialty ceramics, and in the manufacture of

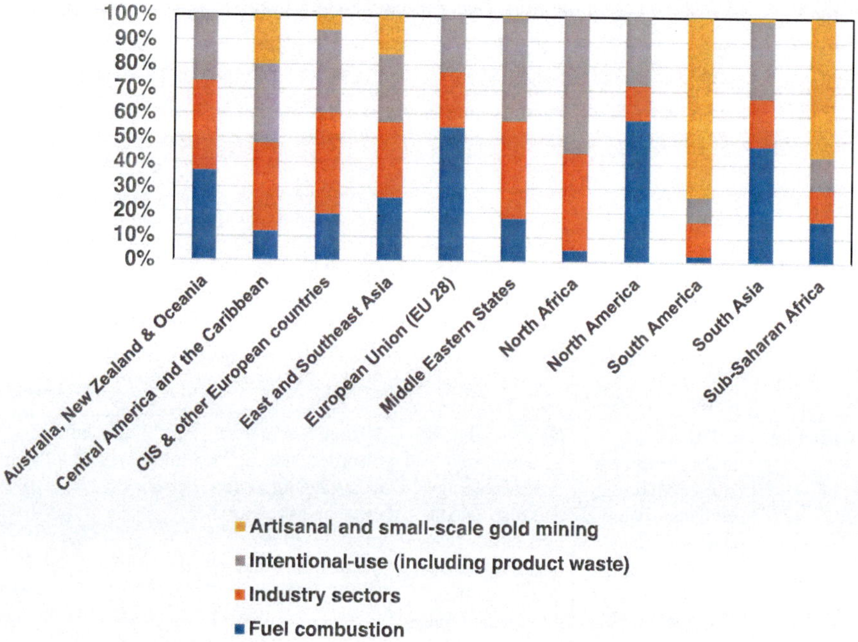

Fig. 8.3 Share of different sectors to the total mercury emission in different regions (modified from UNEP 2019)

nickel-cadmium batteries, and nickel–metal–hydride batteries, while nickel sulfide is used as a catalyst in the petrochemical industry or as an intermediate in the metallurgical industry. The overall worldwide use in 2006 of primary nickel was 1.40 million tons. Estimates for the emission of nickel into the atmosphere from natural sources were 30 million kg/year in the early 1990s.

Mercury (Hg) occurrence in the environment results from anthropogenic and natural sources in three different forms: Elemental (Hg0), inorganic compounds (Hg_2^{++} and Hg^{2+}), and organic like alkylmercury compounds. Exposure occurs during mining, production, and transportation of mercury, as well as mining and refining of gold and silver ores. Mercury is commonly found in thermometers, manometers, barometers, gauges, valves, switches, batteries and was used as amalgam for dentistry, preservatives, heat transfer technology, pigments, catalysts, and lubricating oils. Exposure to methylmercury most commonly occurs via contaminated fish and shellfish. A global inventory of mercury emissions to air from anthropogenic sources in 2015 quantifies global emissions at about 2220 tons and the majority of these emissions occur in Asia (49%), followed by South America (18%), sub-Saharan Africa (16%), Europe (9.1%), and North America (1.8%) (UNEP 2019). Emissions associated with artisanal and small-scale gold mining account for almost 38% of the global total amounts. In Fig. 8.3 the share of different sectors to the total emission of that regions is given.

8.2.1 Health Aspects

Of particular health significance are the heavy metals, especially lead, cadmium, copper, chromium, and zinc, as well as the semimetal arsenic. Heavy metals are generally referred to as those metals which possess a specific density of more than 5 g/cm^3 (Jaishankar et al. 2014). They affect the environment and various biochemical and physiological functions of living organisms in a dose-dependent manner and serious health effects are observed for many of these metals at higher concentrations. A large number of reviews on the topic are available from scientific organizations, but also in the more recent literature (e.g., EFSA 2010; US DHHS 2012; Jaishankar et al. 2014; Nordberg et al. 2015; Pagano et al. 2019; Chen et al. 2019; CLS 2019; Kumar et al. 2020; Barthi and Sharma 2022).

In the following, we will briefly discuss the health effects of some important metals such as arsenic, cadmium, lead, and nickel (see also Table 8.1) as well as mercury (IARC 2012; NTP 2012; US EPA 2022; OSHA 2023; ECHA 2023):

Chronic high exposure to **arsenic** leads to arsenical keratinosis, darkened patchy skin hyperpigmentation, acne, cardiovascular disease and hypertension, and peripheral neuropathy. In addition, there is sufficient evidence in humans of carcinogenicity in the context of mixed exposure to inorganic arsenic compounds. In this case, they can cause cancer of the lung, urinary bladder, and skin. There is also a positive associations between exposure to arsenic and inorganic arsenic compounds and cancer of the kidney, liver, and prostate.

High exposure to **cadmium** can result to kidney, bone, and lung diseases. The main critical effects include increased excretion of low-molecular-weight proteins in the urine due to proximal tubular cell damage of the kidney and an increased risk of osteoporosis. An increased risk of lung cancer has also been reported following inhalation exposure and a positive association has been observed between exposure to cadmium and cadmium compounds and cancer of the kidney and of the prostate.

Lead accumulates in the skeleton and can be mobilized from bones e.g. during pregnancy and lactation. Chronic exposure are associated with impaired kidney function, cardiovascular effects (e.g., hypertension), nervous system and neurobehavioral effects, cognitive dysfunction later in life, reduced fetal growth, and decreased glomerular filtration rate. Exposure in utero and early childhood with the impairment of neurodevelopment in children is the most critical effect. Within the scope of very high exposure situation anemia, peripheral neuropathy, interstitial kidney fibrosis, and severe abdominal cramping were observed. ECHA as well as IARC considers that the epidemiological data are inconsistent as regards increased risk of cancer from inorganic lead compounds.

Nickel is the most common contact allergen. After a preceding sensitization phase an allergy can develop in the form of allergic contact dermatitis. Chronic inhalation exposure to nickel in humans also results in respiratory effects, including a typical form of asthma, decreased lung function, and bronchitis. At high doses, there is evidence from animal studies of reproductive toxicity of nickel salts taken orally or by inhalation. There is sufficient evidence in humans for the carcinogenicity of nickel compounds and nickel metal to cause cancers of the lung, nasal cavity, and paranasal sinuses.

Mercury exposure at high levels can harm the brain, heart, kidneys, lungs, and immune system of people of all ages. Main mercury target after long-term exposure is the nervous system with emotional changes (e.g., nervousness, irritability, and mood swings), neuromuscular changes (e.g., weakness and muscle atrophy), and headache. Mercury became known as part of extensive contamination in Japan (Minamata) by a chemical factory that discharged methylmercury, a by-product of acetaldehyde production, into a bay for decades. Subsequently, adults developed a range of neurotoxic effects, including blurred vision, hearing impairment, olfactory and gustatory disorders, ataxic gait, hand clumsiness, speech disorders, and somato-sensory and psychiatric disorders. Severe mental and motor development disorders have been observed in affected fetuses.

8.2.2 *Occurrence in the Air of Residences*

In Vienna, for example, mean values for lead and cadmium in $PM_{2.5}$ of 24 ng/m^3 (PM_{10}: 28 ng/m^3) and 0.6 ng/m^3 (PM_{10}: 0.5 ng/m^3) were measured in 3 buildings (Komarnicki 2005). The water-soluble fractions of the total metal content in $PM_{2.5}$ were 8% for lead and 29% for cadmium. In 4 Portuguese homes in 2017/2018, mean concentrations in $PM_{2.5–10}$ were 1.6 ng/m^3 for lead, 1.4 ng/m^3 for manganese, 0.3 ng/m^3 for chromium, and 25.9 ng/m^3 for aluminum (Martins et al. 2020). In 159 $PM_{2.5}$ samples collected in a small city in Malta in 2018–2019 levels of 1.9 ng/m^3 for lead, 0.03 ng/m^3 for cadmium, 0.26 ng/m^3 for arsen, and 35 ng/m^3 for aluminum were reported (Camilleri et al. 2022).

In the USA, a total of 235 homes were sampled at 3 seasons in 1999 and medians of 2.4 ng/m^3 (lead), 0.12 ng/m^3 (cadmium), 1.2 ng/m^3 (chromium), 6.5 ng/m^3 (zinc), 1.5 ng/m^3 (manganese), and 0.21 ng/m^3 (antimony) were found (Adgate et al. 2007). Comparable concentration ranges were observed in 2005–2006, except for chromium, in a Canadian of 121 homes (Rasmussen et al. 2018). Here, medians of 1.4 ng/m^3 (lead), 0.08 ng/m^3 (cadmium), 8.6 ng/m^3 (chromium), 13.0 ng/m^3 (zinc), 1.9 ng/m^3 (manganese), and 0.23 ng/m^3 (antimony) were determined, while in an investigation of 340 homes in the USA in 2006–2010 medians of 1.07 ng/m^3 for lead, 0.33 ng/m^3 for chromium, 4.35 ng/m^3 for zinc, 0.48 ng/m^3 for manganese, 9.73 ng/m^3 for aluminum, and 0.76 ng/m^3 for antimony were reported (Matthaios et al. 2021).

Huang et al. (2014) quantified the concentrations of metals in seven indoor $PM_{2.5}$ samples in Guangzhou, China, urban area in 2010. They found high medians of 52.9 ng/m^3 (lead), 0.04 ng/m^3 (cadmium), 85.4 ng/m^3 (chromium), 166 ng/m^3 (zinc), 54 ng/m^3 (manganese), and 2.4 ng/m^3 (antimony). As part of the Japan Environment and Children's Study (JECS) 37 air samples were analyzed in 2017 and the following medians were reported: 14 ng/m^3 for lead, <0.4 ng/m^3 for cadmium, and $<$ 7.0 ng/m^3 for manganese, and $<$ ng/m^3 for selenium (Ma et al.

2020). Moreover, Su et al. (2022b) investigated 22 children's bedroom and observed mean concentrations of 5.2 ng/m^3 for chromium, 10.0 ng/m^3 for mangenese, 13.2 ng/m^3 for lead, 4.5 ng/m^3 for vanadium, and 60.4 ng/m^3 for zinc in $PM_{2.5}$ in Taiwan in 2018–2020.

8.2.3 Occurrence of Elemental Mercury in Air

Measurement results of gaseous elemental mercury (Hg°) in indoor living spaces without a suspicion of contamination are rather rare. In an older American study, short-term measurements in 39 children's living rooms showed concentrations of about 10–10,000 ng/m^3 with a median of 50 ng/m^3 (Hudson et al. 1987). In 12 residential and office rooms in New York and Connecticut, levels ranging from 6.6 to 523 ng/m^3 could be found over 3–5 hours with a median of 69 ng/m^3 (Carpi and Chen 2001). In this study it was shown that the content can vary spatially within a living space (factor > 10). In their study, Garetano et al. (2008) reported a mean Hg level of 5 ng/m^3 (95th percentile: 11 ng/m^3) in 38 buildings in New Jersey without known Hg contamination. In Australia, Hg concentrations of 1–100 ng/m^3 (mean: 5 ng/m^3) were measured in 10 properties, with median levels found to be 3 times higher in older homes (<40 years) (Gillett et al. 2010). Wohlgemuth et al. (2020) reported concentrations of 2.0–10.8 ng/m^3 (median: 3.5 ng/m^3) in 27 homes in Switzerland in 2018. As part of the Japan Environment and Children's Study (JECS) 37 air samples were collected in 2017 and only 2.7% were above the LOD of 0.05 ng Hg/m^3 with a maximum of 0.087 ng/m^3 (Ma et al. 2020). Very high indoor exposure has to be expected for people living or working in and around artisanal small-scale gold mining communities especially in Africa and South America. For example, Gyamfi et al. (2022) reported levels from 73 to >50,000 ng/m^3 (upper instrumental limit) in 316 households where amalgam burning was reported in Ghana in 2018.

Significantly higher elemental mercury level has been reported in school laboratories (about 10–6600 ng/m^3) (Olivero-Verbel et al. 2006), in hospitals (20–13,900 ng/m^3) (Prokopowicz and Mniszek 2005), and in processing areas of fluorescent lamp recycling facility (200–6800 ng/m^3) (Wilson et al. 2017). In another study drilling of dental amalgam generates significant amounts of mercury vapor generally for more than an hour after removal with mean peak concentration of 195,000 ng/m^3 during 21 sessions (Warwick et al. 2019). Metallic mercury or an amalgam can also be released in indoor living spaces when compact fluorescent lamps (CFL) or fluorescent tubes are broken. Salthammer et al. (2012) were able to find concentrations of up to 7820 ng/m^3 in a 24-m^3 chamber and up to 5740 ng/m^3 in an office room for a short time when breaking a mercury-containing CFL. In another study, maximum room air levels between 680 and 10,300 ng/m^3 were measured when Hg-containing CFL were broken with restricted ventilation of the room (air-exchange rate: 0.1 l/h) (Fromme et al. 2012). Both studies show that amalgam-containing CFL lead to lower indoor air levels than mercury-containing ones when broken (maximum 414 ng/m^3 and 140 ng/m^3, respectively). Studies from the USA and Greece in test chambers and indoor environments also show very high concentrations of mercury in the air shortly after CFL breakage, up to the range of several tens of thousands ng/m^3 (Stahler et al. 2008; Sarigiannis et al. 2012). Snow et al. (2021) collected Hg during a fluorescent light crushing event and reported concentrations of 7 ng/m^3 before, 10,400 ng/m^3 during, and 230 ng/m^3 after the experiment. It can be concluded that in particular room air temperature, ventilation conditions, and the size of the surface area occupied by the Hg beads after breakup significantly determine the measurement results.

8.2.4 *Occurrence in the Air of Public Facilities*

To date, only limited studies have investigated classrooms. Viana et al. (2014) found mean concentrations in $PM_{2.5–10}$ of 3.2 ng/m^3 for lead, 0.22 ng/m^3 for cadmium, 0.7 ng/m^3 for chromium, 29.2 ng/m^3 for zinc, 11.8 ng/m^3 for manganese, and 0.04 ng/m^3 for antimony in 39 Spanish classrooms in 2012/2013. In 12 Italian schools in an industrially polluted area, slightly higher levels of 5.5 ng/m^3 (lead), 0.39 ng/m^3 (cadmium), 4.6 ng/m^3 (manganese), and 1.6 ng/m^3 (antimony) were observed in 2012/2013 (Ruggieri et al. 2019). The authors blame outdoor sources for indoor air pollution. In 4 naturally ventilated Portuguese schools, mean concentrations in $PM_{2.5–10}$ in 2017/2018 were 5.1 ng/m^3 for lead, 8.3 ng/m^3 for manganese, 2.6 ng/m^3 for chromium, and 285 ng/m^3 for aluminum (Martins et al. 2020). In 39 schools in Spain and Italy medians of 0.48 ng/m^3 for arsenic, 0.16 ng/m^3 for cadmium, and 2.74 ng/m^3 for nickel were found (Pacitto et al. 2018). Pallarés et al. (2019) studied 7 schools in Spain over a year during class time and found mean levels in PM_{10} of 422–938 ng/m^3 for lead, 55–118 ng/m^3 for cadmium, and 3.6–10.6 ng/m^3 for arsenic. For all elements, the indoor levels were in some cases significantly higher than in the outdoor air. The authors do not give an explanation for the high indoor pollution. Again, in 4 kindergartens from a polluted area in Poland in 2014/15, very high average levels in $PM_{2.5}$ in winter and spring with 38 and 25 ng/m^3 (lead), 15 and 18 ng/m^3 (cadmium), 29 and 22 ng/m^3 (manganese), and 2.6 and 0.8 ng/m^3 (antimony), respectively, were observed (Mainka and Fantke 2022). Moreover, Almeida et al. (2022) analyzed elements in $PM_{2.5}$ in 5 primary schools in Portugal 2017/18 and found mean levels of 13 ng/m^3 for lead, 2.6 ng/m^3 for chromium, 8.9 ng/m^3 for manganese, and 2.5 ng/m^3 for vanadium. The indoor-to-outdoor concentration ratios of the before mentioned substances are 1.4, 1.6, 1.6, and 1.1.

Wang et al. (2023) analyzed $PM_{2.5}$ of two schools situated in a highly polluted area in China in 2012 and reported medians of 15 and 30 ng/m^3 for arsenic, 290 and 103 ng/m^3 for lead, 6 and 9 ng/m^3 for nickel, and 15 and 30 ng/m^3 for chromium. The most abundant elements were Ca, K, Fe, S, Zn, and Cl both indoor and outdoor in two schools, which accounted for about 90% of the total elements.

8.2.5 *Lead in the Air of Indoor Firing Ranges*

When shooting with lead-containing ammunition in closed shooting ranges, not only the external exposure situation, but also the internal exposure of the exposed persons is strongly dependent on the type of ammunition used, the weapons used, and the shooting sequence. In addition, ventilation conditions, room geometry and size, and cleaning are important influencing factors. While food is the main exposure pathway for the general population, shooters are also exposed via inhalation, which is an additional exposure that should not be neglected. A recent review paper summarizes the results, also the internal exposure of shooters (Laidlaw et al. 2017).

Stationary or person-related study results on lead exposure in air from indoor shooting ranges are known from Belgium (Dams et al. 1987), the USA (Tripathi et al. 1991; NIOSH, 2009), Sweden (Svensson et al. 1992), Korea (Kang & Park 2017), and Germany (Demmler et al. 2009; Mühle et al. 2010). They found peak values of up to 5000 μg/m^3 in Belgium and mean values of 210 μg/m^3, 190 μg/m^3, 2000 μg/m^3, and 5100 μg/m^3 in 4 studies from the USA The mean values in Sweden were 660 μg/m^3 when shooting with cartridge ammunition and 5 μg/m^3 when shooting with air guns. Higher lead levels were also observed in 3 recent studies: in the USA with 44 and 350 μg/m^3 (stationary, two shooting ranges) (Scott et al. 2012), in the Czech Republic with 72 μg/m^3 in one facility (Lach et al. 2015), and in Korea with a 2-h mean of 400 μg/m^3 (pistol) and 1000 μg/m^3 (rifle) (Wang et al. 2017). In Korea, an 8-hour time-weighted average of 15–387 μg/m^3 was analyzed at an indoor firing range in which the manager had lead poisoning (Kang & Park 2017). In a room shooting facility in Germany, lead levels of 4050 μg/m^3 and 4550 μg/m^3 were measured in the head area of the 4 sport shooters and of 2200 μg/m^3 and 2240 μg/m^3 in a person 1.5 m behind the shooter in 2 exposure experiments over one hour (Mühle et al. 2010). Personal monitoring samples ($n = 8$) were worn by volunteers for an approximate six-hour period while they participated in practice shooting in Australia in 2021 (Alcock et al. 2022). Lead concentrations ranged from 78–178 μg/m^3. During sessions of a nine-day urban combat training of Swedish military personnel with lead ammunition instructors' airborne exposure was higher than that of cadets (geometric mean: 72.0 μg/m^3 vs. 42.9 μg/m^3) (Schenk et al. 2021). Remote-controlled firing under laboratory conditions was conducted in a closed shooting range and resulted in the lowest airborne lead levels using the unleaded ammunition such as the brass and frangible ammunition types (means: 7 and 16 μg/m^3), since their only source of lead is the primer (Greenberg et al. 2022). They concluded that the use of a lead-free or a low-lead ammunition will lead to reduced airborne lead emissions. Again, Weber et al. (2020) measured lead and copper during indoor training activities over a two-year period and found a reduction using lead ammunition of 64 μg/m^3 to 6–22 μg/m^3 using lead-free ammunition. For copper, mean breathing zone concentrations increased 77–130 μg/m^3 with the change to lead-free frangible ammunition.

8.2.6 Occurrence of metals in Dust

The results of studies in residences are summarized in Table 8.2 and in schools and daycare centers in Table 8.3. The highest dust levels were reported for zinc and copper, while the mean lead concentrations were higher in Europe, Australia, and Canada compared to China and Japan. Very high concentrations were reported in kindergartens of the Upper Silesia Region, Poland, typified by the use of fossil fuels for power and heat purposes (Mainka and Fantke 2022).

Within the DustSafe study a total of 2265 samples from 35 countries were collected since 2016 and sieved to 250 µm prior to analysis (Isley et al. 2022). Multiple comparison tests show that Australia, the USA, and New Zealand are more significantly different than all other countries in their individual trace metal concentrations in dust. Australia had high Pb levels (median: 126 µg/g), New Zealand had high As (median: 39 µg/g), and the USA had high Cu and Zn (medians: 305 µg/g and 1380 µg/g), relative to other countries in the study. For example, in the 111 Chinese samples medians of 112 µg/g for Pb, 10 µg/g for As, and 204 µg/g and 849 µg/g for Cu and Zn, respectively. Overall the data show that home age is a significant predictor of As, Cu, Mn, Ni, Pb, and Zn in dust and the presence of a garden and interior and exterior peeling paint was associated with the largest percentage differences in As and Pb concentration.

As part of the Canadian House Dust Study 1025 samples were collected in single family dwellings across 13 cities from 2007 to 2010 (Rasmussen et al. 2013). Significant positive relationships between house age and metal concentration were observed for Pb, Cd, and Zn. On the other hand, the metal concentration in house dust was not higher in smokers' households compared to homes of non--smokers. Nevertheless, the dust loading (mg/m^2) of surfaces was higher in homes of smoker (Rasmussen et al. 2013). Additionally, results of this study showed that concentrations in settled dust correlated significantly with concentrations in indoor PM_{10} for 11 out of 17 studied elements: Ag, Al, As, B, Cu, Fe, Mn, Pb, U, V, and Zn (Rasmussen et al. 2018). Furthermore, five elements showed significant correlations between personal PM_{10} and settled dust: Ag, Al, As, Mn, and Sb. According to an extensive study in France, which examined 1834 rooms in 2008–2009, dust levels are strongly determined by the load on the ground in front of the buildings (Lucas et al. 2014). In this study, dust mostly originating from outside the building was the major contributor to floor dust lead by track-in exterior soil into the residences.

In a study performed in 2020, 38 urban and rural dust samples were collected and concentrations of Pb from China (Liu et al. 2021). Samples from Northern China were significantly higher than those from Southern China and the concentrations of heavy metals from rural areas were higher than those from urban areas, except for copper. Li et al. (2022) summarized the results from 30 peer-reviewed papers published in China during from 1980 to 2020 and found sample-weighted concentrations of 0.89 µg/g for Hg, 10.6 µg/g for Cd, 61.9 µg/g for As, 112 µg/g for Cr, 380 µg/g for Pb, and 643 µg/g for Mn.

Table 8.2 Mean element concentrations in dust of residences (in µg/g)

References	N	As	Cd	Cr	Cu	Mn	Ni	Pb	Zn	Sieve fraction	Country, year
Uhl et al. (2004)	24	–	3.3	87	240	–	–	350	1640	<63 µm	Austria, 2003
Kempken & Butte (2010)*	16	–	1.5	–	–	230	–	78	850	<63 µm	Germany
Zota et al. (2011)	59	4.0	2.7	–	–	121	–	63	678	<125 µm	USA, 2005/06
Kurt-Karakus (2012)	31	–	0.8	55	156	136	263	28	832	<100 µm	Turkey, 2012
Rasmussen et al. (2013)	1025	13.1	6.0	117	279	–	102	210	833	<80 µm	Canada, 2007–10
Doyi et al. (2019)	224	17.6	–	90	272	220	50.9	299	1876	<250 µm	Australia
Plumejeaud et al. (2018)	19	14.0	1.2	67	339	193	70	229	1520	<150 µm	Portugal
Al Hejami et al. (2020)*	67	–	1.7	42	136	58	23	36	386		Canada, 2013
Ali et al. (2021)	20#	2.2	–	–	–	–	–	94	–	<180 µm	Saudi Arabia, urban
	20#	3.1	–	–	–	–	–	11	–	<180 µm	Saudi Arabia, rural
Gustafsson et al. (2018)		–	–	20	120	100	20	50	800	<3.31[a]	Sweden, 2015
Zhao et al. (2022)	33	–	2.0	104	87	–	35	188	568	<100 µm	China, 2018
Liu et al. (2021)	16*	–	1.9	12.4	12.9	–	–	21.8	129	<150 µm	China, rural, 2020
	17*	–	0.4	8.0	10.6	–	–	11.9	70.9	<150 µm	China, urban, 2020
Ma et al. (2020)	37	–	1.3	–	–	151	–	28	–	<250 µm	Japan
Wang et al. (2022)	60[b]	10	0.4	55	35	480	40	30	90	<250 µm	China, 2020

* Median; # levels in ppm
[a]Mass median aerodynamic diameter
[b]Calculated from a figure

Table 8.3 Mean element concentrations in dust of schools or daycare centers in µg/g

References	N	As	Cd	Cr	Cu	Mn	Ni	Pb	Zn	Sieve fraction	Country, year
Darus et al. (2012)	9	–	–	16.9	30.2	–	–	31.2	149	<100 µm	Malysia, 2010
Zhu et al. (2013)	69	15	1.0	82	70	–	26	118	424		China
Chen et al. (2014)		12	–	129	67	547	34	145	400		China
Lu et al. (2014)	48	13.5	–	134	69.3	558	34.1	145	409		China
Al Hejami et al. (2020)[a]	11	–	2.5	39	188	223	47	41	890		Canada, 2014
Mainka & Fantke (2022)	4	82	198	769	294	370	278	415	3266		Poland, winter, 2014/15
	4	15	313	6290	245	355	314	455	3179		Poland, spring, 2014/15
Rehman et al. (2020)	16	6.0	4.1	139	93.1	–	44.7	177	866	<200 µm	Pakistan, 2018
Lovaković et al. (2022)[a]	10	2.4	0.5	39.9	26.0	209	17.4	17.0	196	<500 µm	Croatia, 2020–21

[a] Median

Results in which particle fraction the highest lead concentration could be found are not consistent. While in a study of 376 house dust samples in Australia arsenic, copper, lead, and zinc were found especially in the 90–150 μm fraction and chromium and manganese in the sieve fraction <45 μm (Doyi et al. 2020), there is also evidence that lead is more likely to be found in the lower sieve fractions (Bright et al. 2006; Kempken and Butte 2010; Ikegami et al. 2014). Kempken and Butte (2010) reported that when screened to different particle sizes, the highest concentrations resulted in the smallest fraction. For example, it was 36 mg/kg for lead in the dust fraction <2 mm, but 78 mg/kg in the fraction <64 μm.

References

Adgate JL, Mongin SJ, Pratt GC, Zhang J, Field MP, Ramachandran G, Sexton K (2007) Relationships between personal, indoor, and outdoor exposures to trace elements in $PM_{2.5}$. Sci Total Environ 386:21–32

Ajrouche R, Ielsch G, Cléro E, Roudier C, Gay D, Guillevic J, Laurier D, Le Tertre A (2017) Quantitative health risk assessment of indoor radon: a systematic review. Radiat Prot Dosim 177:69–77

Al Hejami A, Davis M, Prete D, Lu J, Wang S (2020) Heavy metals in indoor settled dusts in Toronto, Canada. Sci Total Environ 703:134895

Alcock R, Wajrak M, Oosthuizen J (2022) Assessment of the effectiveness of ventilation controls in managing airborne and surface lead levels at a newly commissioned indoor shooting range. Int J Environ Res Public Health 19:11711

Ali N, Eqani SAMAS, Nazar E, Alhakamy NA, Rashid MI, Shahzad K, Zeb J, Shen H, Ismail IMI, Albar HMSA (2021) Arsenic and lead in the indoor residential settings of different socio-economic status; assessment of human health risk via dust exposure. Environ Sci Pollut Res Int 28:13288–13299

Almeida SM, Faria T, Martins V, Canha N, Diapouli E, Eleftheriadis K, Manousakas MI (2022) Source apportionment of children daily exposure to particulate matter. Sci Total Environ 835: 155349

Baloch RM, Maesano CN, Christoffersen J, Banerjee S, Gabriel M, Csobod É, de Oliveira FE, Annesi-Maesano I (2020) Indoor air pollution, physical and comfort parameters related to schoolchildren's health: data from the European SINPHONIE study. Sci Total Environ 739: 139870

Barthi R, Sharma R (2022) Effect of heavy metal: an overview. Mat Today Proc 51:880–885

Bright DA, Richardson GM, Dodd M (2006) Do current standards of practice in Canada measure what is relevant to human exposure at contaminated sites? I: a discussion of soil particle size and contaminant partitioning in soil. Hum Ecol Risk Assess 12:591–605

Camilleri R, Vella AJ, Harrison RM, Aquilina NJ (2022) Source apportionment of indoor $PM_{2.5}$ at a residential urban background site in Malta. Atmos Environ 278:119093

Carpi A, Chen Y-F (2001) Gaseous elemental mercury as an indoor air pollutant. Environ Sci Technol 35:4170–4173

Chen J (2019) Risk assessment for radon exposure in various indoor environments. Radiat Prot Dosim 185:143–150

Chen H, Lu X, Li LY (2014) Spatial distribution and risk assessment of metals in dust based on samples from nursery and primary schools of Xi'an, China. Atmos Environ 88:172–182

Chen QY, DesMarais T, Costa M (2019) Metals and mechanisms of carcinogenesis. Ann Rev Pharmacol Toxicol 59:537–554

Cheng ES, Egger S, Hughes S, Weber M, Steinberg J, Rahman B, Worth H, Ruano-Ravina A, Rawstorne P, Yu XQ (2021) Systematic review and meta-analysis of residential radon and lung cancer in never-smokers. Eur Respir Rev 30:200230

Cinelli G, De Cort M, Tollefsen T (eds) (2019) European Atlas of Natural Radiation, Publication Office of the European Union. Luxembourg. https://publications.jrc.ec.europa.eu/repository/handle/JRC116795

CLS (Columbia Law School Health Justice Advocacy) (2019) Literature review: overview of childhood lead poisoning and its health effects. https://web.law.columbia.edu/sites/default/files/microsites/clinics/health-advocacy/literature_review_health_effects_final_july_2019_1.pdf

Dams R, Vandecasteele C, Desmet B, Helsen M, Nagels M, Vermeir G, Yu ZQ (1987) Element concentrations in the air of an indoor shooting range. Sci Total Environ 77:1–13

Darby S, Hill D, Auvinen A, Barros-Dios JM, Baysson H, Bochicchio F, Deo H, Falk R, Forastiere F, Hakama M, Heid I, Kreienbrock L, Kreuzer M, Lagarde F, Mäkeläinen I, Muirhead C, Oberaigner W, Pershagen G, Ruano-Ravina A, Ruosteenoja E, Rosario AS, Tirmarche M, Tomásek L, Whitley E, Wichmann HE, Doll R (2005) Radon in homes and risk of lung cancer: collaborative analysis of individual data from 13 European case-control studies. BMJ 330:223

Darus FM, Nasir RA, Sumari SM, Ismail ZS, Omara NA (2012) Heavy metals composition of indoor dust in nursery schools building. Proc Soc Behav Sci 38:169–175

Davis EA, Ou JY, Chausow C, Verdeja MA, Divver E, Johnston JD, Beard JD (2020) Associations between school characteristics and classroom radon concentrations in Utah's public schools: a project completed by university environmental health students. Int J Environ Res Public Health 17:5839

Demmler M, Nowak D, Schierl R (2009) High blood lead levels in recreational indoor-shooters. Int Arch Occup Environ Health 82:539–542

Denu RA, Maloney J, Tomasallo CD, Jacobs NM, Krebsbach JK, Schmaling AL, Perez E, Bersch AJ, LeCaire TJ, Meiman JG, Malecki KM, LoConte NK (2019) Survey of radon testing and mitigation by Wisconsin residents, landlords, and school districts. WMJ 118:169–176

Doyi INY, Isley CF, Soltani NS, Taylor MP (2019) Human exposure and risk associated with trace element concentrations in indoor dust from Australian homes. Environ Int 133(Pt A):105125

Doyi INY, Strezov V, Isley CF, Yazdanparast T, Taylor MP (2020) The relevance of particle size distribution and bioaccessibility on human health risk assessment for trace elements measured in indoor dust. Sci Total Environ 733:137931

ECHA (2023) Different scientific reports. Helsinki, Finland. Searched online

EFSA (2010) Scientific Opinion on Lead in Food. EFSA Panel on Contaminants in the Food Chain (CONTAM). EFSA J 8:1570

Fesenbeck I, Frank G, Naber C, Wilhelm C (2017) Radon in Baden-Württemberg schools. Final report. https://www.lubw.baden-wuerttemberg.de/documents/10184/701474/Abschlussbericht+Radon+in+Schulen+KIT/097bd07c-251b-4377-9f90-6d12961197e3 [in German]

Fojtikova I, Navratilova RK (2014) Influence of energy-saving measures on the radon concentration in some kindergartens in the Czech Republic. Radiat Prot Dosim 160:149–153

Fromme H, Büscher O, Matzen W, Drasch G, Roscher E, Nitschke L (2012) Indoor air contamination after the breakage of mercury-containing compact fluorescent lamps (CFLs). Gefahrstoffe-Reinhaltung der Luft 71:215–220. [in German]

Garetano G, Stern AH, Robson M, Gochfeld M (2008) Mercury vapor in residential building common areas in communities where mercury is used for cultural purposes versus a reference community. Sci Total Environ 397:131–139

Gillett RW, Keywood MD, Galbally IE (2010) Indoor project – Persistent organic pollutants and metals. Report prepared for Commonwealth of Australia, Canberra

Giraldo-Osorio A, Ruano-Ravina A, Varela-Lema L, Barros-Dios JM, Pérez-Ríos M (2020) Residential radon in central and South America: a sytematic review. Int J Environ Res Public Health 17:4550

Greenberg NB, Meyer R, Derazne ES (2022) The integrated effect of various small arm weapons and ammunition types on the airborne lead levels. Air Qual Atmos Health 15:1913–1921

Gruber V, Baumann S, Wurm G, Ringer W, Alber O (2021) The new Austrian indoor radon survey (ÖNRAP 2, 2013-2019): Design, implementation, results. J Environ Radioact 233:106618

Gustafsson Å, Krais AM, Gorzsás A, Lundh T, Gerde P (2018) Isolation and characterization of a respirable particle fraction from residential house-dust. Environ Res 161:284–290

Gyamfi O, Sorenson PB, Darko G, Ansah E, Bak JL (2022) Human health risk assessment of exposure to indoor mercury vapour in a Ghanaian artisanal small-scale gold mining community. Chemosphere 241:125014

Harrison JD (2021) Lung cancer risk and effective dose coefficients for radon: UNSCEAR review and ICRP conclusions. J Radiol Prot 41(2)

Huang M, Wang W, Chan CY, Cheung KC, Man YB, Wang X, Wong MH (2014) Contamination and risk assessment (based on bioaccessibility via ingestion and inhalation) of metal(loid)s in outdoor and indoor particles from urban centers of Guangzhou, China. Sci Total Environ 479–480:117–124

Hudson PJ, Vogt RL, Brondum J, Witherell L, Myers G, Paschal DC (1987) Elemental mercury exposure among children of thermometer plant workers. Pediatrics 79:935–938

IAEA (International Atomic Energy Agency) (2014) Radiation protection and safety of radiation sources: International basic safety standards. General Safety Requirements Part 3. Vienna, Austria. https://www-pub.iaea.org/MTCD/Publications/PDF/Pub1578_web-57265295.pdf

IARC (1988) Man-made mineral fibres and radon. IARC Monographs on the evaluation of the carcinogenic risks to humans, Vol. 43. https://publications.iarc.fr/Book-And-Report-Series/Iarc-Monographs-On-The-Identification-Of-Carcinogenic-Hazards-To-Humans/Man-Made-Mineral-Fibres-And-Radon-1988

IARC (2012) Arsenic, metals, fibres, and dusts. IARC Monographs on the evaluation of carcinogenic risks to humans 100 C: 219–309. Lyon, France. https://publications.iarc.fr/Book-And-Report-Series/Iarc-Monographs-On-The-Identification-Of-Carcinogenic-Hazards-To-Humans/Arsenic-Metals-Fibres-And-Dusts-2012

ICRP (2014) Radiological protection against radon exposure. ICRP Publication 126. Ann. ICRP 43. https://journals.sagepub.com/doi/pdf/10.1177/ANIB_43_3

Ikegami M, Yoneda M, Tsuji T, Bannai O, Morisawa S (2014) Effect of particle size on risk assessment of direct soil ingestion and metals adhered to children's hands at playgrounds. Risk Anal 34:1677–1687

Isley CF, Fry KL, Liu X, Filippelli GM, Entwistle JA, Martin AP, Kah M, Meza-Figueroa D, Shukle JT, Jabeen K, Famuyiwa AO, Wu L, Sharifi-Soltani N, Doyi INY, Argyraki A, Ho KF, Dong C, Gunkel-Grillon P, Aelion CM, Taylor MP (2022) International analysis of sources and human health risk associated with trace metal contaminants in residential indoor dust. Environ Sci Technol 56:1053–1068

Ivanova K, Stojanovska Z, Tsenova M, Badulin V, Kunovska B (2014) Measurement of indoor radon concentration in kindergartens in Sofia, Bulgaria. Radiat Prot Dosim 162:163–166

Ivanova K, Chobanova N, Kunovska B, Djounova J, Stojanovska Z (2022) Exposure due to indoor radon in Bulgarian schools. Aerosol Air Qual Res 22:220279

Jaishankar M, Tseten T, Anbalagan N, Mathew BB, Beeregowda KN (2014) Toxicity, mechanism and healtheffects of some heavy metals. Interdiscip Toxicol 7:60–72

Kang KW, Park WJ (2017) Lead poisoning at an indoor firing range. J Korean Med Sci 32:1713–1716

Kang CM, Liu M, Garshick E, Koutrakis P (2020) Indoor particle alpha radioactivity origins in occupied homes. Aerosol Air Qual Res 20. https://doi.org/10.4209/aaqr.2020.01.0037

Kellenbenz KR, Shakya KM (2021) Spatial and temporal variations in indoor radon concentrations in Pennsylvania, USA from 1988 to 2018. J Environ Radioact 233:106594

Kempken B, Butte W (2010) Concentrations of lead, cadmium, manganese and zinc in fractions of household dust. Gefahrst Reinh Luft 70:98–102. [in German]

Khan SM, Gomes J, Krewski DR (2019) Radon interventions around the globe: a systematic review. Heliyon 5:e01737

Khan SM, Gomes J, Chreim S (2021) A mixed methods population health approach to explore radon-induced lung cancer risk perception in Canada. Cancer Control 28:10732748211039764

Komarnicki GJ (2005) Lead and cadmium in indoor air and the urban environment. Environ Pollut 136:47–61

Kropat G, Bochud F, Jaboyedoff M, Laedermann JP, Murith C, Palacios M, Baechler S (2014) Major influencing factors of indoor radon concentrations in Switzerland. J Environ Radioact 129:7–22

Kumar A, Kumar A, MMS C-P, Chaturvedi AK, Shabnam AA, Subrahmanyam G, Mondal R, Gupta DK, Malyan SK, Kumar SS, Khan SA, Yadav KK (2020) Lead toxicity: Health hazards, influence on food chain, and sustainable remediation approaches. Int J Environ Res Public Health 17:E2179

Kurkela O, Nevalainen J, Pätsi SM, Kojo K, Holmgren O, Auvinen A (2023) Lung cancer incidence attributable to residential radon exposure in Finland. Radiat Environ Biophys 62:35–49

Kurt-Karakus PB (2012) Determination of heavy metals in indoor dust from Istanbul, Turkey: estimation of the health risk. Environ Int 50:47–55

Lach K, Steer B, Gorbunov B, Micka V, Muir RB (2015) Evaluation of exposure to airborne heavy metals at gun shooting ranges. Ann Occup Hyg 59:307–323

Laidlaw MA, Filippelli G, Mielke H, Gulson B, Ball AS (2017) Lead exposure at firing ranges – a review. Environ Health 16:34

Li X, Ke H, Ouyang C, Yu X, Liu Y, Wang F, Li W (2021) Investigation of the indoor 222Rn and 220Rn levels in the residential environment and estimation of the annual effective radiation dose for ordinary residents. PLoS One 16:e0253463

Li X, Fan L, Wang X, Yang Y, Zhu Y, Han X, Li L, Ge T, Liu H, Qi J, Gong S, Zhang Q, Guo W, Su L, Yao X, Wang X (2022) Characteristics, distribution, and children exposure assessment of 13 metals in household dust in China: a big data pilot study. Indoor Air 32:e12943

Liu B, Huang F, Yu Y, Li X, He Y, Gao L, Hu X (2021) Heavy metals in indoor dust across China: occurrence, sources and health risk assessment. Arch Environ Contam Toxicol 81(1):67–76

Loffredo F, Opoku-Ntim I, Meo G, Quarto M (2022) Indoor radon monitoring in kindergarten and primary schools in south Italy. Atmos 13:478

Lovaković BT, Jagić K, Dvoršćak M, Klinčić D (2022) Trace elements in indoor dust-Children's health risk considering overall daily exposure. Indoor Air 32:e13104

Lu X, Zhang X, Li LY, Chen H (2014) Assessment of metals pollution and health risk in dust from nursery schools in Xi'an, China. Environ Res 128:27–34

Lucas JP, Bellanger L, Le Strat Y, Le Tertre A, Glorennec P, Le Bot B, Etchevers A, Mandin C, Sébille V (2014) Source contributions of lead in residential floor dust and within-home variability of dust lead loading. Sci Total Environ 470–471:768–779

Ma C, Iwai-Shimada M, Tatsuta N, Nakai K, Isobe T, Takagi M, Nishihama Y, Nakayama SF (2020) Health risk assessment and source apportionment of mercury, lead, cadmium, selenium, and manganese in Japanese women: an adjunct study to the Japan Environment and Children's Study. Int J Environ Res Public Health 17:2231

Mainka A, Fantke P (2022) Preschool children health impacts from indoor exposure to $PM_{2.5}$ and metals. Environ Int 160:107062

Maringer FJ, Kaineder H, Nadschläger E, Sperker S (2008) Standards and experience in radon measurement and regulation of radon mitigation in Austria. Appl Radiat Isot 66:1644–1649

Martins V, Faria T, Diapouli E, Manousakas MI, Eleftheriadis K, Viana M, Almeida SM (2020) Relationship between indoor and outdoor size-fractionated particulate matter in urban microenvironments: levels, chemical composition and sources. Environ Res 183:109203

Matthaios VN, Liu M, Li L, Kang C-M, Vieira CLZ, Gold DR, Koutrakis P (2021) Sources of indoor $PM_{2.5}$ gross α and β activities measured in 340 homes. Environ Res 197:111114

Menzler S, Piller G, Gruson M, Rosario AS, Wichmann HE, Kreienbrock L (2008) Population attributable fraction for lung cancer due to residential radon in Switzerland and Germany. Health Phys 95:179–189

Meyer W (2019) Impact of constructional energy-saving measures on radon levels indoors. Indoor Air 29:680–685

Moon J, Yoo H (2021) Residential radon exposure and leukemia: a meta-analysis and dose-response meta-analyses for ecological, case-control, and cohort studies. Environ Res 202: 111714

Mühle P, Schierl R, Nowak D (2010) Increased lead intake in indoor shooting ranges. Gefahrst Reinh Luft 70:464–468. [in German]

NIOSH (National Institute for Occupational Safety and Health) (2009) Preventing occupational exposures to lead and noise at indoor firing ranges. https://www.cdc.gov/niosh/docs/2009-136/default.html

Nordberg GF, Fowler BA, Nordberg M (eds) (2015) Handbook on the toxicology of metals.4th edn. Academic Press

NTP (2012) NTP Monograph on health effects of low-level lead. https://ntp.niehs.nih.gov/ntp/ohat/lead/final/monographhealtheffectslowlevellead_newissn_508.pdf

Olivero-Verbel J, Ropero-Vega J, Ortiz-Rivera W, Vera-Ospina P, Torres-Fuentes N, Montoya-Rodriguez N (2006) Air mercury levels in a pharmaceutical and chemical sciences school building. Bull Environ Contam Toxicol 76:1038–1043

Onishchenko A, Malinovsky G, Vasilyev A, Zhukovsky M (2017) Radon measurements in kindergartens in Ural region (Russia). Radiat Prot Dosim 177:112–115

OSHA (Occupational Safety & Health Administration) (2023). https://www.osha.gov/

Pacitto A, Stabile L, Viana M, Scungio M, Reche C, Querol X, Alastuey A, Rivas I, Álvarez-Pedrerol M, Sunyer J, van Drooge BL, Grimalt JO, Sozzi R, Vigo P, Buonanno G (2018) Particle-related exposure, dose and lung cancer risk of primary school children in two European countries. Sci Total Environ 616-617:720–729

Pagano G, Thomas PJ, Di Nunzio A, Trifuoggi M (2019) Human exposures to rare earth elements: present knowledge and research prospects. Environ Res 171:493–500

Pallarés S, Gómez ET, Martínéz A, Jordán MM (2019) The relationship between indoor and outdoor levels of PM_{10} and its chemical composition at schools in a coastal region in Spain. Heliyon ee2270

Park TH, Kang DR, Park SH, Yoon DK, Lee CM (2018) Indoor radon concentration in Korea residential environments. Environ Sci Pollut Res Int 25:12678–12685

Petermann E, Bossew P, Hoffmann B (2023) Estimating national indoor radon exposure at a high spatial resolution – improvements by a machine learning based probabilistic approach. European Geosciences Union (EGU) General Assembly 2023, Vienna, Austria, 23–28 April 2023

Plumejeaud S, Reis AP, Tassistro V, Patinha C, Noack Y, Orsière T (2018) Potentially harmful elements in house dust from Estarreja, Portugal: characterization and genotoxicity of the bioaccessible fraction. Environ Geochem Health 40:127–144

Poulin P, Leclerc JM, Dessau JC, Deck W, Gagnon F (2012) Radon measurement in schools located in three priority investigation areas in the province of Quebec, Canada. Radiat Prot Dosim 151: 278–289

Prokopowicz A, Mniszek W (2005) Mercury vapour determination in hospitals. Environ Monit Assess 104:147–154

Rasmussen PE, Levesque C, Chénier M, Gardner HD, Jones-Otazo H, Petrovic S (2013) Canadian House Dust Study: population-based concentrations, loads and loading rates of arsenic, cadmium, chromium, copper, nickel, lead, and zinc inside urban homes. Sci Total Environ 443: 520–529

Rasmussen PE, Levesque C, Chenier M, Gardner HD (2018) Contribution of metals in resuspended dust to indoor and personal inhalation exposures: Relationships between PM_{10} and settled dust. Build Environ 143:513–522

Rehman K, Fatima F, Waheed I, Akash MSH (2018) Prevalence of exposure of heavy metals and their impact on health consequences. J Cell Biochem 119:157–184

Rehman A, Liu G, Yousaf B, Zia-ur-Rehman M, Ali MU, Rashid MS, Farooq MR, Javed Z (2020) Characterizing pollution indices and children health risk assessment of potentially toxic metal (oid)s in school dust of Lahore, Pakistan. Ecotoxicol Environ Saf 190:110059

Ruggieri S, Longo V, Perrino C, Canepari S, Drago G, L'Abbate L, Balzan M, Cuttitta G, Scaccianoce G, Minardi R, Viegi G, Cibella F (2019) Indoor air quality in schools of a highly polluted south Mediterranean area. Indoor Air 29:276–290

Salthammer T, Uhde E, Omelan A, Lüdecke A, Moriske H-J (2012) Estimating human indoor exposure to elemental mercury from broken compact fluorescent lamps (CFLs). Indoor Air 22: 289–298

Sarigiannis DA, Karakitsios SP, Antonakopoulou MP, Gotti A (2012) Exposure analysis of accidental release of mercury from compact fluorescent lamps (CFLs). Sci Total Environ 435–436:306–315

Schenk L, Wingfors H, Skoog B, Forsgard N, Nyberg C, Taube F (2021) Exposures to lead during urban combat training. Int J Hyg Environ Health 235:113773

Schubert M, Musolff A, Weiss H (2018) Influences of meteorological parameters on indoor radon concentrations (^{222}Rn) excluding the effects of forced ventilation and radon exhalation from soil and building materials. J Environ Radioact 192:81–85

Scott EE, Pavelchak N, DePersis R (2012) Impact of housekeeping on lead exposure in indoor law enforcement shooting ranges. J Occup Environ Hyg 9:D45–D51

Snow MA, Feigis M, Lei YD, Mitchell CPJ, Wania F (2021) Development, characterization, and testing of a personal passive sampler for measuring inhalation exposure to gaseous elemental mercury. Environ Int 146:106264

Stahler D, Ladner S, Jackson H. (2008) Maine Compact Fluorescent Lamp Study. Report of the Maine Department of Environmental Protection, Maine, USA. https://www.maine.gov/dep/homeowner/cflreport.html

Su C, Pan M, Zhang Y, Kan H, Zhao Z, Deng F, Zhao B, Qian H, Zeng X, Sun Y, Liu W, Mo J, Guo J, Zheng X, Sun C, Zou Z, Li H, Huang C (2022a) Indoor exposure levels of radon in dwellings, schools, and offices in China from 2000 to 2020: a systematic review. Indoor Air 32: e12920

Su HJ, Jung CC, Wang JH, Chen NT, Chang WH (2022b) Estimations of infiltration factors of diurnal $PM_{2.5}$ and heavy metals in children's bedrooms. Indoor Air 32:e13037

Suzuki G, Yamaguchi I, Ogata H, Sugiyama H, Yonehara H, Kasagi F, Fujiwara S, Tatsukawa Y, Mori I, Kimura S (2010) A nation-wide survey on indoor radon from 2007 to 2010 in Japan. J Radiat Res 51:683–689

Svensson B-G, Schiitz A, Nilsson A, Skerfving S (1992) Lead exposure in indoor firing ranges. Int Arch Occup Environ Health 64:219–221

Szabados M, Csákó Z, Kotlík B, Kazmarová H, Kozajda A, Jutraz A, Kukec A, Otorepec P, Dongiovanni A, Di Maggio A, Fraire S, Szigeti T (2021) Indoor air quality and the associated health risk in primary school buildings in Central Europe – The InAirQ study. Indoor Air 31: 989–1003

Tchounwou PB, Yedjou CG, Patlolla AK, Sutton DJ (2012) Heavy metals toxicity and the environment. EXS 101:133–164

Tripathi RK, Sherertz PC, Llewellyn GC, Armstrong CW (1991) Lead exposure in outdoor firearm instructors. Am J Public Health 81:753–755

Uhl M, Hohenblum P, Scharf S, Trimbacher C (2004) House dust - an indicator of indoor pollution. Report BE-258, Umweltbundesamt Österreich. https://www.umweltbundesamt.at/fileadmin/site/publikationen/BE258.pdf [in German]

UNEP (2019) Global Mercury Assessment 2018. Geneva, Switzerland. https://www.unep.org/resources/publication/global-mercury-assessment-2018

US DHHS (2012) U.S. Department of Health and Human Services. National Toxicology Program monograph on health effects of low-level lead. https://www.semanticscholar.org/paper/

NTP-monograph-on-health-effects-of-low-level-lead./f765e1f6fc8e630289ce71515a5bf9b01 8e6495b
US EPA (2022) Basic Information about Mercury. https://www.epa.gov/mercury/basic-information-about-mercury
Viana M, Rivas I, Querol X, Alastuey A, Sunyer J, Álvarez-Pedrerol M, Bouso L, Sioutas C (2014) Indoor/outdoor relationships and mass closure of quasi-ultrafine, accumulation and coarse particles in Barcelona schools. Atmos Chem Phys 14:4459–4472
Vienneau D, Boz S, Forlin L, Flückiger B, de Hoogh K, Berlin C, Bochud M, Bulliard JL, Zwahlen M, Röösli M (2021) Residential radon – comparative analysis of exposure models in Switzerland. Environ Pollut 271:116356
Wang J, Li H, Bezerra ML (2017) Assessment of shooter's task-based exposure to airborne lead and acidic gas at indoor and outdoor ranges. J Chem Health Saf:14–21
Wang B, Gao F, Li Y, Lin C, Cheng H, Duan X (2022) Assessment of children's metal exposure via handwipe, outdoor soil and indoor dust and their associations with blood biomarkers. Int J Environ Res Public Health 19:14614
Wang R, Ding X, Wang J, Dong Z, Xu H, Ma G, Gao B, Song H, Yang M, Cao J (2023) Trace elements in outdoor and indoor $PM_{2.5}$ in urban schools in Xi'an, Western China: characteristics, sources identification and health risk assessment. Environ Geochem Health 45:1027–1044
Warwick D, Young M, Palmer J, Ermel RW (2019) Mercury vapor volatilization from particulate generated from dental amalgam removal with a high-speed dental drill – a significant source of exposure. J Occup Med Toxicol 14:22
Weber AK, Bannon DI, Abraham JH, Seymour RB, Passman PH, Lilley PH, Parks KK, Braybrooke G, Cook ND, Belden AL (2020) Reduction in lead exposures with lead-free ammunition in an advanced urban assault course. J Occup Environ Hyg 17:598–610
WHO (2009) WHO handbook on indoor radon: a public health perspective. Geneva, Switzerland. https://apps.who.int/iris/bitstream/handle/10665/44149/9789241547673_eng.pdf; jsessionid=9FF350618319D0694B3B7F24E1EF39B5?sequence=1
WHO (2010) Guidelines for indoor air quality: selected pollutants. World Health Organization, Regional Office for Europe, Copenhagen (http://www.euro.who.int/en/what-we-do/health-topics/environmental-health/air-quality/publications/2010/who-guidelines-for-indoor-air-quality-selected-pollutants)
Wilson E, Lafferty JS, Thiboldeaux R, Tomasallo C, Grajewski B, Wozniak R, Meiman J (2017) Occupational mercury exposure at a fluorescent lamp recycling facility – Wisconsin, 2017. Morb Mortal Wkly Rep 67:763–766
Wohlgemuth L, McLagan D, Flückiger B, Vienneau D, Osterwalder S (2020) Concurrently measured concentrations of atmospheric mercury in indoor (household) and outdoor air of Basel, Switzerland. Environ Sci Technol Lett 14:234–239
Zhao P, Liu X, Yin L, Tian D, Wang X (2022) Levels, distribution characteristics, and sources of heavy metals in indoor dust in Shijiazhuang, China. Environ Monit Assess 194:857
Zhu Z, Sun G, Bi X, Li Z, Yu G (2013) Identification of trace metal pollution in urban dust from kindergartens using magnetic, geochemical and lead isotopic analyses. Atmos Environ 77:9–15
Zhukovsky M, Vasilyev A, Onishchenko A, Yarmoshenko I (2018) Review of idoor radon concentrations in schools and kindergartens. Rad Prot Dosimetry 181:6–10
Zota AR, Schaider LA, Ettinger AS, Wright RO, Shine JP, Spengler JD (2011) Metal sources and exposures in the homes of young children living near a miningimpacted superfund site. J Expo Sci Environ Epidemiol 21:495–505

Chapter 9
Smoking

Check for updates

Contents

Abstract The focus of this chapter are smoking and the use of smoking techniques like tobacco smoking, shisha smoking, e-cigarette vaping, and use of illicit drugs (e.g., cannabis). They are presented individually in detail. In addition to a brief description of the chemical–physical characteristics and the use of the substance, an overview of main health risks and, in particular, references to further scientific literature are given. In the foreground, however, are extensive presentations of the occurrence of the substances in indoor air.

Keywords Tobacco · Shisha · Vaping · e-cigarette · Illicit drugs

9.1 Smoking Tobacco

Despite all prevention efforts, tobacco smoking is widespread in many countries and represents an extraordinary health problem. The TackSHS-Survey that included 11,876 participants, representative of the population aged ≥15 years in 12 European countries in 2017–2018, reported that 25.9% of participants were current smokers (31.0% of men and 21.2% of women), while 16.5% were former smokers (Gallus et al. 2021a). Smoking prevalence ranged in this study from 18.9% in Italy to 37.0% in Bulgaria. It decreased with increasing age, level of education, and self-rated household economic level. Even though smoking rates have been declining in recent years, especially in the adolescent and young adult groups, not only further prevention efforts but also targeted legal regulations are needed. Among the US

H. Fromme, *Indoor Air Quality*, https://doi.org/10.1007/978-3-031-40078-0_9

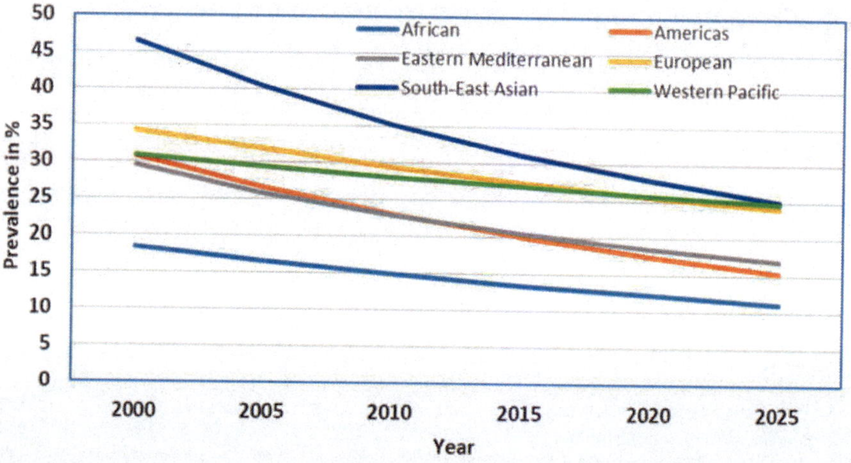

Fig. 9.1 Trends in current tobacco use among people aged ≥15 years (modified from WHO 2019)

adults in 2019, 16.7% (estimated 40.8 million) used any combustible tobacco product (Cornelius et al. 2020). On a global scale, WHO (2019) presented a report in the age-standardized tobacco use prevalence rates in all WHO regions (see Fig. 9.1).

With respect to indoor environments, two exposure options are of particular importance:

Since tobacco smoke contains several thousand substances, some of which are very harmful to health, passive smoke exposure can occur indoors due to the smoker's exhaled sidestream smoke (approx. 15–20%) and the substances produced during smoke breaks by the smoldering of the tobacco (80–85%). This is referred to as SHS (Second-hand Smoke) or ETS (Environmental Tobacco Smoke).

In contrast, the term Third-hand Smoke (THS) is used when components of tobacco smoke accumulate on surfaces of the room, furnishings, and people present during smoking (e.g., in his hair) and are remobilized from there (Jacob et al. 2017; Sheu et al. 2020). These are complex reactions with accumulation in the organic boundary layers of the surfaces and release to the gas phase or dust and aging processes of the aerosol (Sleiman et al. 2014; Wu et al. 2022). Reactions with oxidizing substances such as ozone or nitrogen oxides and the reactive formation of secondary aerosols have also been described (Sleiman et al. 2010). While, for example, ozone- and OH radical-driven oxidation is a major pathway in indoor air nitrosation of nicotine by nitrous acid (HONO) emitted from combustion sources (including smoking) produced tobacco-specific nitrosamines on indoor surfaces.

As a THS working definition (Jacob et al. 2017):

- THS constituents may **remain** adsorbed to surfaces and dust particles, often penetrating deep into materials such as wallboard or upholstery.
- As they persist they may **react** with atmospheric oxidants to yield potentially harmful by-products.

– Both precursors and by-products may be **reemitted** back to the gas phase, and airborne particles that initially deposited onto indoor surfaces may be **re-suspended**.

9.1.1 Health Aspects

Tobacco in any form kills and sickens millions of people every year and around eight million people died from a tobacco-related disease in 2017 (WHO 2019). For smoking tobacco, multiple effects have long been clearly documented for the active smoker. The World Health Organization estimates that 22% of all cancer-related deaths and 12% of all deaths worldwide are caused by tobacco smoking (WHO 2012). In Germany in 2010, about 23% of all new cancer cases in men and 8% in women, and thus a total of about 72,000 cancer cases, were attributable to smoking (Wienecke et al. 2014). When looking at age-standardized new cancer and death rates for tumors of the upper respiratory and digestive organs and the urinary tract for the period 1995 to 2014/2015, declining rates were seen in men for all cancers considered except esophageal cancer in Germany (Wienecke and Kraywinkel 2018). However, disease rates among women increased for all tobacco-associated cancers except urinary tract malignancies during this time.

Additionally, health risks are substantially for persons exposed via secondhand smoke (SHS). A very extensive scientific literature is now available on the health effects of SHS exposure, providing clear evidence in particular for an increased risk with regard to acute and chronic obstructive respiratory and cardiovascular diseases and the development of lung cancer (IARC 2004, 2009; US DHHS 2006; Benowitz 2010; Oberg et al. 2011; Cao et al. 2015; Ni et al. 2020). The US Centers for Disease Control and Prevention stated that there is no risk-free level of exposure to SHS numerous health problems in infants and children, including more frequent and severe asthma attacks, respiratory infections, ear infections, and sudden infant death syndrome (SIDS) (CDC 2020a). Moreover, during pregnancy more than 1000 infant deaths annually in the USA are caused by SHS.

The World Health Organization's cancer agency has also classified SHS as a human carcinogen (IARC 2004). Globally, SHS exposure is one of the leading preventable causes of disease and death, with approximately 600,000 nonsmokers also dying from its effects each year (WHO 2011). Estimates suggest that 40–50% of children currently still live in households where at least one person smokes (Faber et al. 2016).

A particular exposure situation with tobacco smoke exists in public facilities, such as pubs, restaurants, bars, discotheques, where smoking is not prohibited (Fromme et al. 2009a). Meanwhile, there are results from several reviews that a consistent smoking ban protects non smokers and is associated with an improvement in the health situation of the population (Bolte et al. 2009, 2015; Lee et al. 2014, 2019; Frazer et al. 2016). Also, the fear that smokers will smoke more indoors in their homes after a smoking ban in restaurants, possibly exposing children more, has not been confirmed (Liang et al. 2016; Nanninga et al. 2018). A systematic review tried to identify facilitators and barriers that influenced smokers' compliance with smoking bans in public places (Zhou et al. 2016). They reported that at the individual level, nicotine dependence, insufficiency of tobacco-related knowledge, and the negative attitudes toward smoking bans were the most commonly identified barriers; at the interpersonal level, the smoking behaviors of people around, close relatives, and friends' approval were the main barriers; and at the organizational level, the main barriers were inefficient implementation of the bans and the inconvenience of the designative smoking areas.

9.1.2 Occurrence in the Air of Residences

Secondhand smoke is a complex mixture of thousands of substances that are produced due to the combustion and smoldering processes and can be detected in the gaseous or particulate phase. The following substances are of particular importance: nicotine, 4-methylnitrosamine-1-(3-pyridyl)-1-butanol (NNK), acetaldehyde, acrolein, benzene, benzo[*a*]pyrene as an indicator of polycyclic aromatic hydrocarbons (PAH), 1,3-butadiene, carbon monoxide, and formaldehyde (WHO 2000). The chemical composition of the mainstream and sidestream smoke is qualitatively similar but differs significantly in their absolute and relative amounts. Table 9.1 shows the ratios of the major carcinogens. Since quantification of individual substances is analytically laborious, indicators of tobacco smoke exposure were sought early on. For the field of human biomonitoring, the nicotine metabolites cotinine and *trans-3′-hydroxycotinines,* in particular, but also mercapturic acids (e.g., of benzene or butadiene) come into question (Schettgen et al. 2008; Benowitz et al. 2009). In indoor air, studies show that in addition to nicotine itself, both 3-ethenylpyridine and 2,5-dimethylfuran (possibly also 1,3-butadiene in some cases) are very good indicators of indoor tobacco smoke exposure (Alonso et al. 2010; Aquilina et al. 2010). Higher $PM_{2.5}$ levels may also indicate indoor tobacco smoke exposure. Several past

Table 9.1 Emitted amount of carcinogens (group 1 according to IARC) and ratio of sidestream to mainstream smoke when smoking a cigarette (according to IARC 2004)

Substance	Quantity per cigarette	Ratio of sidestream smoke to mainstream smoke
4-Aminobiphenyl	20.8–31.8 ng	5.41
Arsenic	3.5–26.5 ng	1.51
Benzo[*a*]pyrene	51.8–94.5 ng	3.22
Benzene	70.7–134.3 µg	1.07
1,3-Butadiene	81.3–134.7 µg	1.30
Cadmium	122–265 ng	1.47
Formaldehyde	540.4–967.5 µg	14.78
NNK[a]	50.7–75.7 ng	0.40

[a]4-(N-nitrosmethylamine)-1-(3-pyridyl)-1-butanone

reviews have summarized the very extensive measurements of chemical constituents of tobacco smoke (NCI 1999).

Data for individual substances from some more recent studies are compiled below. The nicotine content as a geometric mean was significantly lower in 17 dwellings where there was no smoking (0.32 μg/m^3) than in dwellings with smoking (2.6 μg/m^3) (Matt et al. 2004). In a compilation of 4 studies between 2009–2014, a mean level between 1.0 and 2.1 μg nicotine/m^3 was found in indoor environments where smokers live (Rosen et al. 2015). In the USA, a median of 14.3 μg $PM_{2.5}$/m^3 could be observed in 15 smokers' homes and 5.1 μg/m^3 in nonsmokers' indoor environments (Russo et al. 2015). Immediately during smoking, $PM_{2.5}$ levels of up to 1000 μg/m^3 can be detected with continuous measurement devices. In 7 nonsmoking and 5 smoking bedrooms, with no smoking activity during sampling, significant differences were nevertheless observed in Portugal between mean concentrations of carbon monoxide (1.6 to 0.21 ppm), formaldehyde (109 to 52 μg/m^3), and $PM_{2.5}$ (61.2 to 8.9 μg/m^3) (Canha et al. 2019). Typical tobacco plant constituents also include metals and rare earths, which are absorbed into the plant through the roots. An Italian study examined 73 homes and found significantly higher levels of $PM_{2.5}$ and cadmium, as well as cerium, lanthanum, and thallium, in smokers' interiors than in non-smokers' homes (Drago et al. 2018).

9.1.3 Occurrence in the Air of Hospitality Venues

The pollution in bars, restaurants, discotheques, and comparable establishments was high before the introduction of smoking bans. For the particulate fraction $PM_{2.5}$, measurement results in these indoor spaces before and after the introduction of smoking bans or restrictions are summarized in Table 9.2. Before the regulations, mean $PM_{2.5}$ concentrations between 35 and 465 μg/m^3 and peak concentrations of 1250 μg/m^3 could be found. In a German study, for example, medians of 1014 μg PM_{10}/m^3 and 869 μg $PM_{2.5}$/m^3 were found in 10 discotheques and 210 μg/m^3 (PM_{10}) and 195 μg/m^3 ($PM_{2.5}$) in 18 bars (Bolte et al. 2008). From this study, exemplary time courses of measurements in the air of gastronomic facilities in Bavaria, Germany, before the smoking ban are shown in Fig. 9.2. Overall, there is evidence of a significant reduction in the finer particle fraction of between 68 and 97%, mostly above 90%, which can be achieved simply by implementing consistent non-smoker protection in these establishments. On the other hand, only partial spatial separation of a smoking area in restaurants cannot contribute to a significant reduction of the pollution situation (Bohac et al. 2013; Ward et al. 2013). For example, in a 2010/2011 study in Vienna, $PM_{2.5}$ was found to average 321 μg/m^3 in 134 establishments where smoking occurred, 68 μg/m^3 in non-smoking areas that were not fully separated structurally, and only 10 μg/m^3 in non-smoking establishments (Neuberger et al. 2013).

Results on measurements of the particle number concentration of ultrafine particles are limited. In a US study, median levels of 13,000 and 17,000 particles/cm^3,

Table 9.2 Change in mean indoor air pollution in restaurants and similar facilities before and after the introduction of smoking bans

N, Place	Parameter	Concentration before ban	Concentration after ban	Dimension	Reduction in %	Place, measurement conditions	Reference
Particulate matter (PM)							
14 bars and restaurants	$PM_{2.5}$	412	27.0	µg/m³	93	New York, USA; stationary	Travers et al. (2004)
29 bars and restaurants	$PM_{2.5}$	206	14.0	µg/m³	93	Massachusetts, USA; inpatient for approx. 30 min	Connolly et al. (2005)
6 bars, 1 casino, 1 bar in an indoor swimming pool	PM_4	231	24.5	µg/m³	89	Delaware, USA; stationary	Repace (2004)
7 pubs	PM_4	179	7.7	µg/m³	96	Boston, USA; inpatient over 20–70 min	Repace et al. (2006)
6 bars	$PM_{2.5}$	95	26	µg/m³	73	Trieste, Italy	Tominz et al. (2006)
	PM_{10}	175	107	µg/m³	39		
17 bars	$PM_{2.5}$	151	11	µg/m³	93	Texas, USA; steady state over 30–90 min	Waring and Siegel (2007)
42 bars	$PM_{2.5}$	35.5	5.8	µg/m³	84	Dublin, Ireland	Goodman et al. (2007)
	PM_{10}	72.1	45.5	µg/m³	37		
72 bars	$PM_{2.5}$	202	28.0	µg/m³ [a]	86	Scotland; personal	Semple et al. (2007)
10 bars, restaurants	$PM_{2.5}$	304	9	µg/m³	97	Kentucky, USA; stationary during one day	Hahn et al. (2006)
8 restaurants	$PM_{2.5}$	90	20	µg/m³	78	Kentucky, USA; stationary during one day	Lee et al. (2007)
13 bars and restaurants	Total dust	275	77	µg/m³	72	Oslo, Norway	Skogstad et al. (2006)
40 bars, fast food restaurants	$PM_{2.5}$	119	38.2	µg/m³	68	Rome, Italy; stationary	Valente et al. (2007)

Restaurants, game rooms, pubs	UFP	76,956	38,079	particles/cm^3	51	over 8 h	
8 Casinos and casino restaurants	$PM_{2.5}$	43.4	3.4	µg/m^3	92	Colorado, USA	TPEG (2008)
Organic substances							
2 pubs[b]	Benzene	4.83	0.54	µg/m^3	89	Dublin, Ireland; stationary	McNabola et al. (2006)
	1,3-Butadiene	4.15	0.22	µg/m^3	95	over 4 h	
26 bars	Benzene	18.8	3.7	µg/m^3	80	Dublin, Ireland	Goodman et al. (2007)
3 bars	Benzene	31	<8.0*	µg/m^3	>75	Texas, USA; stationary over 60 min	Waring and Siegel (2007)
10 restaurants	Nicotine	2.0	0.1	µg/m^{3b}	95	Northern Italy; passive over 1 week	Gorini et al. (2008)
15 Discos and pubs	Nicotine	35.2	0.01	µg/m^{3b}	99	Northern Italy; personal monitoring	Gorini et al. (2008)
13 bars and restaurants	Nicotine	28.0	0.6	µg/m^3	98	Oslo, Norway	Skogstad et al. (2006)
3 bars	Nicotine	19.0	0.25	µg/m^{3b}	99	Northern Italy; passive over 1 week	Gorini et al. (2008)
20 bars	Nicotine	35.5	5.9	µg/m^{3b}	83	Ireland, stationary over 7–10 hours	Mulcahy et al. (2005)
6 bars, 1 casino, 1 bar in an indoor swimming pool	PAH	175	4.6	ng/m^3	97	Delaware, USA; stationary	Repace (2004)
7 pubs	PAH	65.1	6.3	ng/m^3	90	Boston, USA; inpatient over 20–70 min	Repace et al. (2006)

(continued)

Table 9.2 (continued)

N, Place	Para-meter	Concentration before ban	Concentration after ban	Dimension	Reduction in %	Place, measurement conditions	Reference
Carbon monoxide							
17 bars	CO	5.9	2.5	ppm	58	Texas, USA; steady state over 30–90 min	Waring and Siegel (2007)
7 pubs	CO	2.5	1.2	ppm	50	Boston, USA; inpatient over 20–70 min	Repace et al. (2006)

[a]Geometric mean; [b]Median; UFP: ultrafine particles; *:Limit of quantification

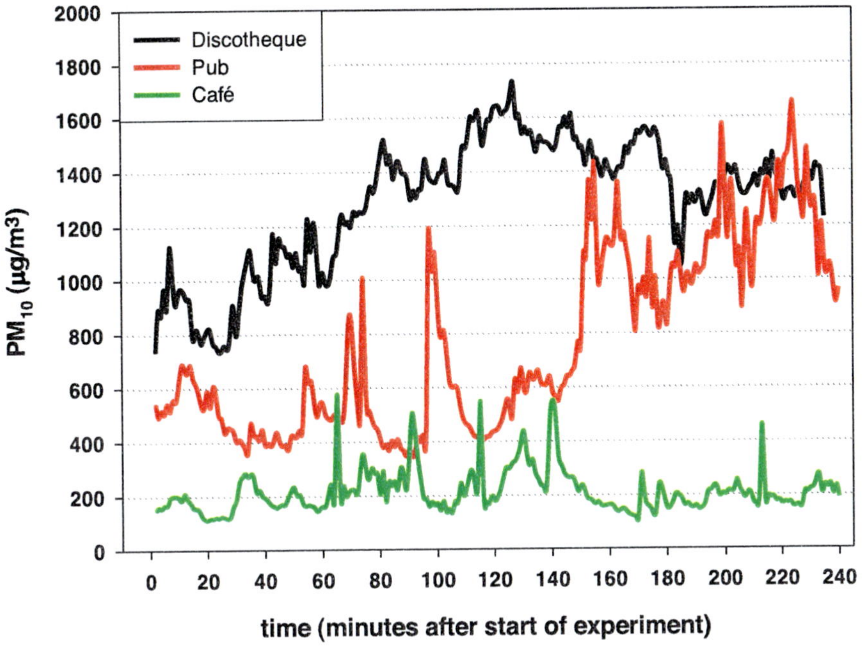

Fig. 9.2 Time course of $PM_{2.5}$ in three hospitality venues in Germany (modified from Bolte et al. 2008)

(in non-smoking rooms) and 109,000 and 103,000 particles/cm^3 (in smoking rooms) were observed in four restaurants (Milz et al. 2007). In a German study, medians of 29,084–221,000 particles/cm^3 were measured in 4 cafés/restaurants, of 119,088–147,444 particles/cm^3 in 2 bars, and 101,309–289,928 particles/cm^3 in 7 discotheques (Bolte et al. 2008). Similarly, in Vienna, measurements in 59 smoking establishments showed a median of 66,011 particles/cm^3 and 7408 particles/cm^3 in 16 non-smoking establishments (Neuberger et al. 2013).

Other tobacco-typical substances can also be found in hospitality venues if smoking is taking place. For example, Bolte et al. (2008) describe medians of 15–193 µg/m^3 (nicotine), 8–20 µg/m^3 (benzene), 3–10 ng/m^3 (cadmium) 215–375 ng/m^3 (sum of the 16 PAH according to US EPA) in a German study of cafés, restaurants, bars, and discotheques.

To characterize outdoor SHS exposure, airborne nicotine was measured in terraces of 220 bars, cafeterias, and pubs from 11 European countries in 2017–2018 (Henderson et al. 2021). Overall median nicotine concentrations were 0.85 µg/m^3 and increased during nighttime to 1.45 µg/m^3, 2.97 µg/m^3 in enclosed venues, 2.79 µg/m^3 in venues with more than two smokers, and 1.20 µg/m^3 in venues in countries with total indoor smoking bans. Again, Kaplan et al. (2019) investigated indoor and outdoor air of 72 hospitality venues in 11 Turkish cities and found overall $PM_{2.5}$ medians of 95 µg/m^3 indoors and 31 µg/m^3 in patios and 37 µg/m^3 and 9 µg/m^3 if smokers were absent.

9.2 Shisha (Water pipe) Smoke

Shisha smoking is a traditional behavior that is very common, especially in the Middle East, Africa, and South Asia. It is known by the names argileh, narghileh, nargila, hookah, shisha, chicha, gozah, and ghalya. In Europe today, about 5–20 g of maassel (or muassel) is smoked per session, which consists of about 30% tobacco, with a nicotine content of 1.8–6.3 mg/g, and about 70% honey or molasses (Hadidi and Mohammed 2004). This tobacco is often infused with flavors such as apple, cherry, mango, peppermint, or cappuccino. As shown in Fig. 9.3, the tobacco is placed in the tobacco head, covered with a screen or perforated foil, and the charcoal is placed on top. After igniting the charcoal, the tobacco carbonizes and the tobacco smoke is sucked in through a water bath via a mouthpiece and inhaled.

Numerous studies have investigated pollutants in Shisha mainstream smoke and found—similar to cigarette smoke—metals such as aluminum, manganese, nickel, cadmium, and lead, organic substances such as phenols, aldehydes, nicotine, aromatic amines, polycyclic aromatic hydrocarbons, flavorings, as well as particulates and carbon monoxide (Schubert et al. 2015). In other studies, various flavoring agents such as vanillin, ethyl vanillin, and benzyl alcohol, which are known contact allergens, were also detected in high concentrations in mainstream smoke (Sepetdjian et al. 2013; Schubert et al. 2013). A "washout" of smoke pollutants in the flow-through water container does not provide protection from these substances, as often assumed by users. Thus, only highly soluble substances are partially retained in the water (Shihadeh 2003). The pollutant contents in the mainstream smoke of a heat pump thus remain high.

The trends of shisha use between 2009 and 2017 was analyzed in 28 EU member states using within the Eurobarometer survey (Filippidis et al. 2019). The prevalence of ever shisha use increased from 11.6% in 2009 to 12.9% in 2017 for subjects older 15 years, while the prevalence of smoking at least once a month was 0.9% in 2017, but higher in Austria (3.6%), Latvia (2.5%), and Belgium (2.0%). For younger respondents aged 15–24 years the prevalence was 27.6% (ever smoking). The German Health Interview and Examination Survey for Children and Adolescents during 2014–2017 found prevalence of current WP use among adolescents of 8.5%, use in the last 12 months of 19.7%, while high prevalence rates were particularly observed among 16- to 17-year-olds with 9.0% and 18.5%, respectively (Klosterhalfen et al. 2020).

Traditionally, smoking a shisha is particularly prevalent in Mediterranean and Arab countries, where 5–15% of adults smoke shisha regularly and 63% of adults report ever having smoked (Akl et al. 2011). In a national representative study in 17 Arab countries (n = 34,788), smoking prevalence among 13–15 year olds was 5.7% in the past 30 days in 2002–2011 (Veeranki et al. 2016).

In the USA, current smoking prevalence was 3.8% among 18–40 year olds (Salloum et al. 2017) and only 1.6% among 12–17 year olds (Trinidad et al. 2017). Nationwide data are also available from Canada for approximately 21,000

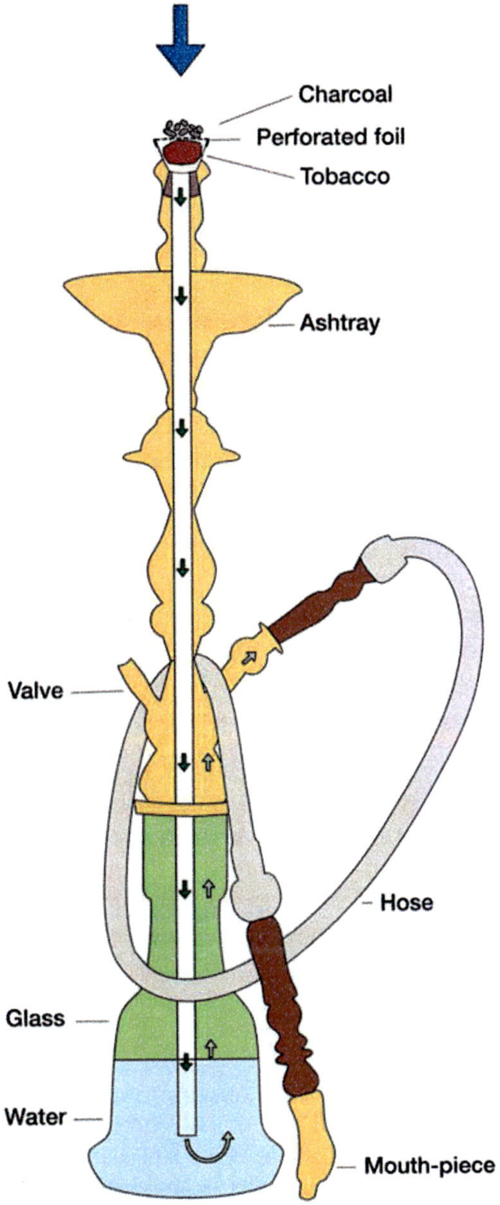

Fig. 9.3 Schematic structure of a shisha (hookah) (from Fromme and Schober 2015)

individuals older than 15 years (Abdullah et al. 2017). Here, the current prevalence was 0.8% and the lifetime prevalence was 8.9%.

In addition to the increasing popularity of hookah, there is often a perception among users that this smoking technique is unlikely to pose any health risks, or in any case fewer health risks than cigarette smoking (Martinasek et al. 2017). Also, the fact that shisha smoking is seen as a shared group experience and contributes to relaxation leads to a higher acceptance in the population compared to cigarette smoking, especially among adolescents and young adults (Minaker et al. 2015).

The current scientific literature, especially on the occurrence, exposure of critical environmental media, but also on health effects, has been summarized in some reviews (Fromme and Schober 2015; Bhatnagar et al. 2019; Adetona et al. 2021; Darawshy et al. 2021; Ghaffari et al. 2021).

9.2.1 Importance of the So-called Tobacco-Free Smoking Products

Due to the discussion on the health effects of tobacco smoking and increasing bans/ restrictions, the so-called tobacco-free smoking products are also increasingly used. They are often labeled “herbal” or “herbal” and are supposed to consist only of “unproblematic” natural substances. These are a mixture of molasses (sugarcane syrup), herbs and flavorings, offered, for example, by suppliers such as SOEX (India), Hydro (USA), Texas Hookah (USA), and Al-Fakher Tobacco Company (United Arab Emirates), especially via the Internet.

There is limited data to date on the extent to which “tobacco-free” shisha are used in the general population. In a study of 3447 college students in the USA, 44% reported ever having smoked a TP, and of these, 37% reported ever having smoked an “herbal” TP (Sutfin et al. 2014).

An experimental study using a smoking machine that mimicked the smoking behavior of 33 shisha smokers examined pollutants in mainstream smoke (Shihadeh et al. 2012). There were no significant differences in the levels of carbon monoxide (CO), tar, total dust, polycyclic aromatic hydrocarbons (PAH), and volatile aldehydes between shisha tobacco and “tobacco-free” smoking products. Hammal et al. (2015) compared the contaminant levels in the mainstream and sidestream smoke of 3 “tobacco-free” smoking products with a classic WP tobacco. They found higher levels of CO (237–286 mg vs. 197 mg vs), tar (1080–2350 mg vs. 909 mg), and formaldehyde (41–111 μg vs. 36 μg), but comparable levels of PAH in the mainstream smoke in the “tobacco-free” products compared to the shisha tobacco, for example. Comparable results were reported for sidestream smoke, with the difference that formaldehyde levels were lower and PAH concentrations higher in the “tobacco-free” products. Cobb et al. (2012) studied 32 healthy shisha smokers who smoked a tobacco-containing product once and a “tobacco-free” product once for 120 min. Exhaled CO (31 ppm vs. 12 ppm) was higher in the group smoking “tobacco-free” products, while—as expected—no nicotine appeared in the plasma of this group.

Even if the so-called tobacco-free smoking products are used for shisha, nicotine intake can be circumvented, but not the other pollutants of mainstream or sidestream smoke.

9.2.2 Health Aspects

Various studies have described acute effects for shisha smokers and individuals exposed to secondhand smoke, especially due to the high content of carbon monoxide (CO). Repeatedly, cases of poisoning leading to hospitalization have been reported in this context (Fromme 2019; Verweij et al. 2019; Dorey et al. 2020; Cazes et al. 2021; Adetona et al. 2021). A summary of different studies is given in Table 9.3. In these cases, individuals showed typical signs of poisoning for CO such as headache, dizziness, nausea, vomiting, confusion, drowsiness, and even unconsciousness. Many cases of poisoning likely go undiagnosed or are not attributed to shisha smoking at all. Other acute effects of shisha smoking are mainly related to the cardiovascular system with effects such as an increase in blood pressure, heart rate, and vascular resistance as well as a decrease in heart rate variability. A review of acute and long-term health effects has been published (Raad et al. 2011; El-Zaatari et al. 2015; Bhatnagar et al. 2019; Al Ali et al. 2020; Darawshy et al. 2021; Adetona et al. 2021). An association between hookah smoking and infectious diseases is discussed, but cannot be confidently inferred from the studies available to date. Since shisha smoke also contains a variety of carcinogenic substances, several systematic reviews conclude that there is an association between exposure and forms of cancer of various localizations (e.g., lung and esophageal cancer) (Akl et al. 2010; El-Zaatari et al. 2015; Aljadani et al. 2020). However, the methodological quality of the studies is considered to be often low.

9.2.3 Occurrence in the Air

Carbon monoxide is a typical gas produced during the smoldering of hookah tobacco and emitted into the indoor air of homes and especially shisha cafés. The average level in these hospitality is between 7 and 18 ppm, reaching maximum levels between 120 ppm in North America and 265 ppm in Arab shisha cafes (Fromme and Schober 2015). Average concentrations of 48 ppm could also be found in a German study in which a shisha was smoked in a room (Fromme et al. 2009b). The typical time course of CO concentrations in indoor air is shown in Fig. 9.4. In twelve UK shisha cafes mean CO concentration of 6.4 ppm was found, higher than concentrations measured within the vicinity of the shisha premises (CO: 0.7 ppm) (Gurung et al. 2016). In an Iranian study performed from 2017 to 2018 in 14 shisha cafés reported mean levels at working days and weekends of 7.1 ppm (4.4–11.4 ppm) and 7.5 ppm (1.4–14.2 ppm), respectively (Masjedi et al. 2019). In 11 rooms located in Dubai, UAE, with shisha exposue and 10 non-smoking rooms mean levels of 11.0 ppm and 1.5 ppm were observed, respectively (Weitzman et al. 2016). Moreover, Shearston et al. (2023) analyzed CO in 12 New York homes where only shishas were smoked and found a mean of 8.4 ppm (0.5–19.2 ppm).

Results on particle mass in the indoor air of shisha cafés show a wide range, depending on room size, ventilation conditions, and number of active smokers. In

Table 9.3 Casuistry of CO poisoning associated with hookah use

Reference	Year	Country	Number	Age	Main symptoms[a]	CO-Hb (%)
Lim et al. (2009)	2009	Singapoore	1	19	Dizziness	27.8
Uyanýk et al. (2011)	2009	Turkey	1	25	Headache, vomiting, dizziness	28.7
Cavus et al. (2010)	2010	Turkey	1	25	Unconsciousness	31.1
Arziman et al. (2011)	2011	Turkey	5	17–27	3 × nausea, dizziness, 1 × unconsciousness	11.4–24.3
La Fauci et al. (2012)	2011	Italy	1	16	Confusion	24.0
Höjer and Enghag (2011)	2011	Sweden	3	15–28	Headache, dizziness, vomiting, unconsciousness	21–32
Clarke et al. (2012)	2012	UK	12	17–38	Headache, dizziness	2–25
Ashurst et al. (2012)	2012	USA	1	21	Vomiting, unconsciousness	15.3
Karaca et al. (2013)	2013	Turkey	1	20	Unconsciousness	31.1
Ozkan et al. (2013)	2013	Turkey	1	19	Unconsciousness	32.7
Bens et al. (2013)	2013	The Netherlands	3		Dizziness, headache	5.7–22
Villano et al. (2014)	2014	USA	2	18, 20	Dizziness, headache, vomiting, syncope,	23 und 19
von Rappard et al. (2014)	2014	Germany	4	17–21	3 × unconsciousness, 1 × asymptomatic	16.7–29.6
Misek and Patte (2014)	2014	USA	1	24	Nausea, headache, chest tightness, general weakness	33.8
Medford et al. (2015)	2015	USA	1	55	Dyspnea, chest tightness, syncope	29.0
Wang et al. (2015)	2015	Australia	1	20	Headache, nausea	25.4
Veen (2016)	2016	The Netherlands	3	15–28	Headache, nausea, vomiting, shortness of breath, syncope	25–39.2
Ateş et al. (2016)	2016	Turkey	1	27	Nausea, syncope	32.6
Maalem et al. (2019)		Saudi Arabia	1	43	Headache, nausea, confusion, body weakness	38.8
Van Ussel and Klink (2021)		Belgium	1	13	Dizziness, headache, tonic episode with urine incontinence	25.3

[a]Not all symptoms always occurred at the same time

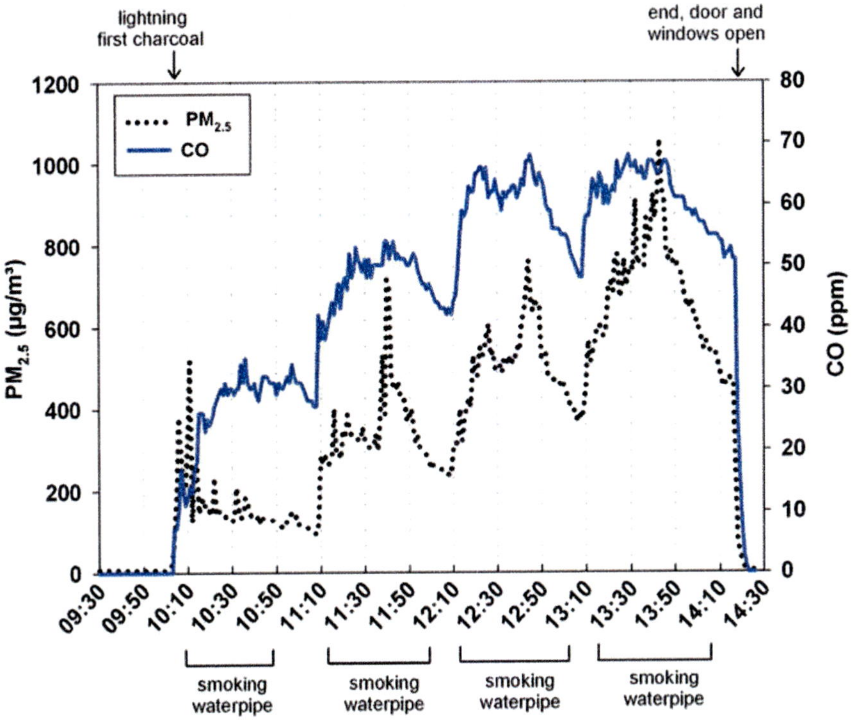

Fig. 9.4 Results of continuously measured $PM_{2.5}$ and CO concentrations during the four water pipe smoking sessions (from Fromme et al. 2009b)

these studies, even the mean values of $PM_{2.5}$ ranged between 136 and 1745 µg/m^3 (Fromme and Schober 2015). The highest levels were found in 13 Pakistani bars and 1419 µg/m^3 in 12 Canadian cafés, with a mean of 1745 µg/m^3 (Zaidi et al. 2011; Zhang et al. 2013). Also, 263 µg/m^3 and 406 µg/m^3 $PM_{2.5}$ were observed in test rooms (Maziak et al. 2008) or under more experimental conditions (see also Fig. 9.4) (Fromme et al. 2009b). PM_{10} concentrations were measured infrequently and yielded mean values between 365 and 432 µg/m^3, showing that especially the fine particles had a large contribution to the mass-based particle pollution of the air (Fromme and Schober 2015). Median concentrations of 231 µg/m^3 (maximum 1759 µg/m^3) were reported in a Spanish study in 2019 collected $PM_{2.5}$ samples from 20 shisha cafés (Feliu et al. 2020). In 11 homes in Dubai, UAE, with shisha exposure and 10 non-smoking rooms mean levels of 429 µg/m^3 and 93 µg/m^3 were observed, respectively. In a more recent study, Shearston et al. (2023) collected $PM_{2.5}$ in 12 New York homes where only a shisha was smoked and reported mean concentration of 309 µg/m^3 (6.3–1992 µg/m^3) in rooms in which a shishas was used. In this study, an average value of 15 µg/m^3 was observed in rooms (n = 15) with no smoking activity. Little information is available on indoor exposure to ultrafine particles. For example, in a living room where a shisha was smoked for 4 hours, mean particle number concentrations of 319,000 particles/cm^3 were found, with particles having a mean diameter of 18 nm (Fromme et al. 2009b). Markowicz et al. (2014) found a mean of 190,000 particles/cm^3 in an exposure chamber. Comparable concentrations can also be observed in cigarette smoking (Markowicz et al. 2014; Bolte et al. 2008).

Moon et al. (2015) studied 46 establishments where shishas were smoked. They found a median level of 136 ng/m^3 (1.3–1963 ng/m^3) for total particle-bound PAH. In a study where a hookah was smoked in a typical indoor room, concentrations of 270 ng/m^3 could be found for the sum of PAH and 140 ng/m^3 in a control room (Fromme et al. 2009b). Naphthalene alone accounted for 110 ng/m^3. The 6 PAH classified as probable carcinogens by the US Environmental Protection Agency (B2-PAH) were 4.9 ng/m^3 in the aforementioned study. Markowicz et al. (2014) studied only the particle-bound PAH in a 21 m^3 test chamber and found comparable results averaging about 7.5 ng/m^3 for the B2-PAH. Additionally, Rostami et al. (2019) reported mean concentrations of PAH of 5915 ng/m^3, 2404 ng/m^3, and 7305 ng/m^3 for cafés with fruit flavored tobacco, regular tobacco, and the both, respectively.

Nicotine can also be measured in the indoor air of shisha cafés when smoking a water pipe. Here, the medians in various studies in the USA, Canada, and Turkey ranged from 1.4 to 10.1 μg/m^3 and the maximum values ranged from 1.9 to 39 μg/m^3. Nicotine levels also increased from <0.05 to 18 μg/m^3 in a typical residential indoor environment where a hookah was smoked for 4 hours (Fromme et al. 2009b). In an Iranian study investigated 14 shisha cafés from 2017 to 2018 and mean nicotine levels at working days and weekends of 5.1 μg/m^3 (1.4–10.3 μg/m^3) and 6.6 μg/m^3 (3.7–12.6 μg/m^3), respectively (Masjedi et al. 2019). The number of "active water pipe heads" had the maximum impact on generation of pollutants inside the cafés. Median nicotine concentrations of 1.15 μg/m^3 (maximum 8.16 μg/m^3) were reported in a Spanish study in 2019 collected samples from 20 shisha cafés (Feliu et al. 2020).

The tobacco-specific nitrosamine 4-(methylnitrosamino)-1-(3-pyridyl)-1-butanone (NNK) has only been detected in indoor air from 46 facilities where shisha was smoked, with mean levels of 1.9 ng/m^3 (0.2–10.4 ng/m^3) (Moon et al. 2015).

There are hardly any results available on the exposure of indoor rooms to VOC in the context of shisha smoking. Fromme et al. (2009b) report benzene levels of 15.0 μg/m^3 and for the sum of volatile compounds, so-called TVOC value (total volatile organic compounds), of 1800 μg/m^3 in a room in which a shisha was smoked experimentally for 4 hours. In an Iranian study investigated 81 shisha cafés very high means of 4960 μg/m^3 for benzene, 4860 μg/m^3 for toluene, 4380 μg/m^3 for ethylbenzene, and 6690 μg/m^3 for xylene were reported (Hazrati et al. 2015). Again, Heydari et al. (2020) found 6680 μg/m^3 for benzene, 3520 μg/m^3 for toluene, 3110 μg/m^3 for ethylbenzene, and 5710 μg/m^3 for xylene in 15 shisha cafés at weekend session and when regular tobacco was used in Iran in 2018/2019. Additionally, the mean concentrations of benzene (7.19 vs. 0.82 μg/m^3), toluene (1.62 vs. 0.3 μg/m^3), and xylenes (2.9 vs. 0.31 μg/m^3) were considerably higher in the indoor air of 30 houses where shisha was smoked compared to 30 non-smoking houses in 2020 (Tabatabaei et al. 2021).

Naddafi et al. (2019) observed mean concentrations of formaldehyde and acetaldehyde of 151 and 60 μg/m^3 as well as 282 and 108 μg/m^3 in Iranian shisha cafés with fruit flavored tobacco and regular tobacco, respectively. Another Iranian study investigated 14 shisha cafés from 2017 to 2018 and mean formaldehyde levels at working days and weekends of 549 μg/m^3 (121–931 μg/m^3) and 618 μg/m^3 (181–1110 μg/m^3), respectively, while the levels in non-smoking cafés the mean levels were 46 and 57 μg/m^3 (Masjedi et al. 2019).

In principle, improved natural or mechanical ventilation with a high proportion of fresh air can reduce the concentrations of problematic substances in the indoor air of, e.g., shisha cafés. However, the extent of the reduction seems to be limited and depends significantly on the quality of the ventilation and other structural conditions (Seidenberg et al. 2019; Rostami et al. 2020). The limited influence of ventilation has already been described for smoking cigarettes, etc. indoors.

9.3 Smoking e-Cigarettes and e-Shishas

E-cigarettes and e-shishas (not to be confused with shishas or water pipes) were developed in China in 2003 and are becoming increasingly popular. Other common names are electronic nicotine delivery systems (ENDS) or electronic non-nicotine delivery systems (ENNDS), depending on whether nicotin is also aerolized. They are technical products consisting of a power source (rechargeable battery), an electric heating element (nebulizer), and a cartridge containing the operating liquid (e-liquid) to be vaporized (see Fig. 9.5). By sucking on the mouthpiece, the e-liquid is nebulized under the effect of heat, and the resulting aerosol can be inhaled as visible vapor by the consumer. Since, in contrast to conventional cigarettes, no typical tobacco combustion occurred, e-cigarettes are often offered as a healthier alternative to smoking. The e-liquids vaporized in e-cigarettes and e-shishas are a mixture of different chemicals such as the carrier substances propylene glycol and glycerin, as well as different flavorings (e.g., vanilla, chocolate, strawberry, cherry) and often nicotine in varying concentrations. Users' exposure to harmful substances is strongly determined by their own behavior. Experienced users inhale more than newcomers,

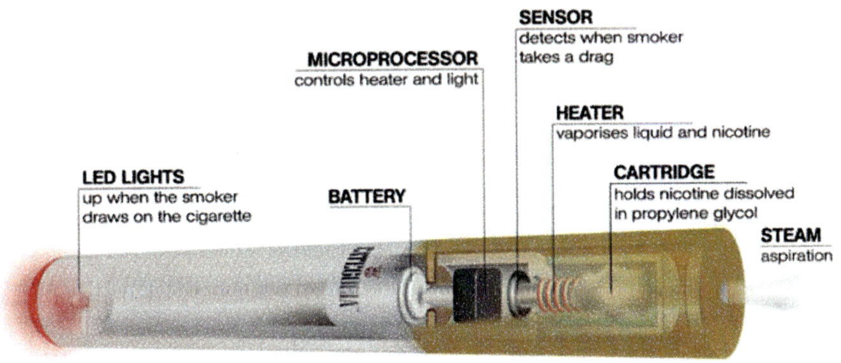

Fig. 9.5 Schematic structure of an e-cigarette

and more powerful e-cigarettes, e.g., second-generation e-cigarettes, or devices with individually selectable vapor production (e.g., advanced personal vaporizers) result in higher exposure. The liquids used can be varied, complicating a health assessment (Glantz and Bareham 2018).

The TackSHS (tackling secondhand tobacco smoke and e-cigarette emissions) that included 11,876 participants, representative of the population aged ≥15 years in 12 European countries in 2017–2018, reported that 2.4% of the subjects currently use an e-cigarette, ranging from 0.6% in Spain to 7.2% in England (Gallus et al. 2021b). Of the 272 users, 52.6% were dual users (i.e., users of both electronic and conventional cigarettes) and 58.8% used liquids with nicotine. A recent meta-analysis of data from 26 national surveys from 69 countries and territories between 2016 and 2020 showed that the pooled prevalence among those younger than 20 years old was 17.2% of ever ENDS/ENNDS vapers (any lifetime use) and 7.8% for current use in past 30 days (Yoong et al. 2021). The prevalence was highest in high-income geografical regions.

A survey among experts from all countries of the World Health Organization (WHO) European Region in 2018 concluded that 48 countries (48.3%) had legislation on e-cigarette use at the national level but only a few of the legislation protect bystanders in indoor settings (Amalia et al. 2022).

9.3.1 Health Aspects

Health risks for smokers are dose dependent with regard to propylene glycol (irritation of mucous membranes and respiratory tract, influence on lung function), ethylene glycol (irritant to mucous membranes, effects on heart, lungs and kidneys), nicotine (physical and psychological dependence, damaging to fertility, suspected of causing cancer), tobacco-type nitrosamines (carcinogenic in humans), formaldehyde (irritation of mucous membranes, affecting lung function), and acetaldehyde/acrolein (irritating to skin and mucous membranes, risk factor for chronic lung disease) (Fromme and Schober 2015). In addition, health risks of flavoring, fragrance, and preservatives in e-cigarettes need to be considered. In addition, cross-sectional and longitudinal studies suggest that e-cigarette smoking promotes initiation of tobacco cigarette smoking (summary in Glantz and Bareham 2018).

Due to the uncertainties that often exist regarding the composition of vaporized liquids, unpredictable health effects are likely. For example, an outbreak (EVALI, e-cigarette, or vaping, product use-associated lung injury) occurred in the summer of 2019 in the USA, in which 2807 e-cigarette smokers had to be hospitalized with severe signs of illness (CDC 2020b). By February 2020, a total of 68 of them had died. Ambulatory treatment figures are not available. Patients presented respiratory (shortness of breath 86%, cough 80%, chest pain 49%), gastrointestinal (nausea 71%, vomiting 70%), and general symptoms (fever 79%, physical discomfort 75%), and hypoxemia was present in 77% (Cao et al. 2020). The most likely cause was tetrahydrocannabinol (THC)-containing liquids that contained vitamin E acetate (α-tocopheryl acetate) as an additive or thickener.

Several articles and reviews on smokers' exposure and potential health effects, particularly on the respiratory and cardiovascular systems, have now been published (Vardavas et al. 2012; Gennimata et al. 2012; Marini et al. 2014; Glantz and Bareham 2018; Li et al. 2020; Buchanan et al. 2020). In a recent review Snoderly et al. (2021) summarized, that, due to the presence of toxicants in e-liquid composition, adverse effects in animal models, association with acute lung injury and cardiovascular disease, and ability to modulate different cell populations in the lung and blood toward pro-inflammatory phenotypes appeared. Moreover, Abbott et al. (2023) reviewed 78 studies and summarized thar e-cigarette use, although weaker than cigarette smoking, is significantly associated with adverse effects on the cardio-respiratory and immune system and periodontal health.

9.3.2 *Occurrence in Indoor Air*

In contrast to conventional tobacco cigarettes, no typical sidestream smoke is produced as in the case of cigarette smoldering; only the exhaled components of the inhaled vapor enter the indoor air. Basically, the indoor exposure situation depends on the type of operating liquid, the type of e-cigarette, and the user behavior.

Chamber experiments show that liquid particles can form from supersaturated propylene glycol during the exhalation phase (Schripp et al. 2013). Recent study results of $PM_{2.5}$ and ultrafine particles (UFP) in indoor air are summarized in Table 9.4. Emission measurements in test chambers, some with a very extensive substance spectrum, are also available (Schripp et al. 2013). In a study from Germany, in which three vaporizers smoked several liquids in one room, mean $PM_{2.5}$ concentrations of 197 μg/m^3 and particle number concentrations (PNC) between 48,620 and 88,386 particles/cm^3 (medians) were found, with the particle distribution showing a maximum at a diameter of 24–36 nm (Schober et al. 2014). In the other studies, the content also ranges widely for both particulate mass and PNC, but always well above the background content in the rooms. In a Spanish study, maximum PNC of 47,000 particles/cm^3 was observed in indoor air during 3 smoking sessions over one hour (Protano et al. 2017). Significant differences were found in both $PM_{2.5}$ and PNC with e-cigarette type (Melstrom et al. 2017; Volesky et al. 2018). In an additional experimental study in two confined settings, a room and a car, two bystanders were exposed to medians of 21 μg/m^3 (pre-exposure 8 μg/m^3) and 16 μg/m^3 (pre-exposure 7 μg/m^3) during a 30-minute ad libitum use of an e-cigarette (Amalia et al. 2022).

The exposure of indoor air to individual VOC has been reported on various occasions. For example, Schober et al. (2014) found mean values for propylene glycol of 199 μg/m^3 (background: <0.04 μg/m^3), for glycerol of 72.7 μg/m^3 (background: <0.04 μg/m^3), and for nicotine of 2.2 μg/m^3 (background: <0.04 μg/m^3). Formaldehyde, benzene, and acrolein were not elevated above background levels. Czogala et al. (2014) also found concentrations ranging from

Table 9.4 $PM_{2.5}$ and ultrafine particles (UFP) in indoor environments during e-cigarette use

	$PM_{2.5}$ (µg/m³)		UFP (particles × 10³/cm³)		
Reference	Median (min–max)	Median background	Mean value (min–max)	Mean value background	Background
Shearston et al. (2021)	997[g] (1.1–9537)	15.1[g] (0.9–180)	–	–	15 homes with e-cigarette use; 15 control homes
Li et al. (2021)	276[g]	–	28	–	6 so-called vape stores
Son et al. (2020)	1660[g] (30–3490)	80[g] (20–190)	–	–	5 so-called vape stores
Nguyen et al. (2019)[a]	10.5–2528e	1.4–52.9[e,f]	14.9–112[e]	16.0–48.8[e,f]	6 so-called vape stores
Nguyen et al. (2019)[b]	5.5–585[e]	2.4–11.6[e,f]	10.4–56.6[e]	5.5–23.9[e,f]	
Melstrom et al. (2017)[c]	35 (2–19,961)	6 (1–228)	31.2 (1.8–239)	2.0 (1.4–4.7)	3 vapers over 2 hours
Melstrom et al. (2017)[d]	515 (7–19,972)	14 (5–390)	25.2 (3.4–225)	3.3 (2.1–4.1)	
Soule et al. (2017)	311–819[e]	1.9–15.5[e]	–	–	2 days of an e-cigarette event
Volesky et al. (2018)	290 [g]	1.8 [g]	7.2	1.4	1 vaper and 3 different liquids
Zhao et al. (2017)	188[g]	8.0[g]	24.8	6.4	13 vapers
Schober et al. (2014)	18–561[e]	6	48.6–88.4[e]	4.4	3 vapers, different liquids, over 2 hours
Czogala et al. (2014)	152[g]	32.4[g]	–	–	5 vapers over 7 minutes
Ruprecht et al. (2017)	–	–	8.7[g]	–	2–3 vapers

[a]Much business; [b]Little business; [c]Disposable; [d]Tank type; [e]Range of individual medians; [f]Outdoor air readings; [g]Mean

0.22 µg/m³ (detection limit) to 3.3 µg/m³ when e-cigarettes were vaporized. In five homes with vaporizers (>2 hours per day), significantly higher nicotine levels were found at 0.11 µg/m³ (median) than in the rooms of 24 controls (0.01 µg/m³) (Ballbè et al. 2014).

Measurement results on PAHs in indoor air are hardly available. Schober et al. (2014) found for the sum of the 16 US EPA PAH an approx. 30–90% higher content during vaporizing compared to the background exposure, whereby the

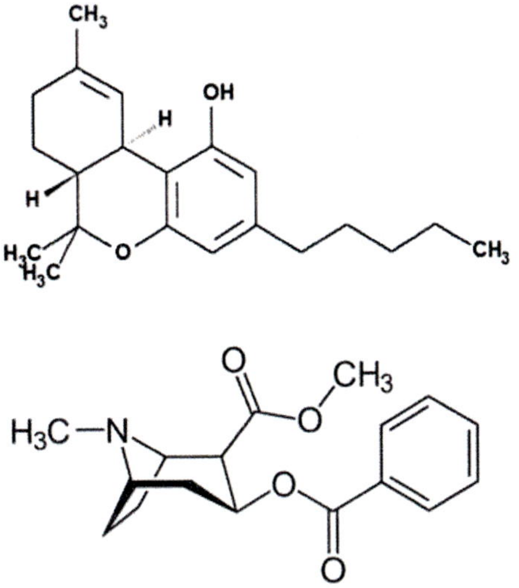

Fig. 9.6 Chemical structure of tetrahydrocannabinol (above) and cocaine (below)

concentrations of the more volatile PAHs such as naphthalene, acenaphthene, fluorene, and phenanthrene were determined. For the 7 PAHs classified as carcinogenic by the US EPA, the mean level during vaporization was 147 ng/m^3 (background exposure: 123 ng/m^3).

9.4 Smoking Illicit Drugs

Worldwide, around 275 million people used illicit drugs in 2020 which resulted in 18 million years of healthy life lost, mostly due to opioids (UNODC 2021). The UNODC report further summarized that roughly 200 million people used cannabis, 20 million people used cocaine, and 62 million people used opioids for non-medical purposes, and 27 million people used amphetamine type stimulants in 2019.

Based on data from 26 EU countries that undertook a survey between 2015 and 2020, last year cannabis use among inhabitants aged 15–34 is estimated at 15.4% (ranging from 3.4% to 21.8%) and when only 15- to 24-year-olds are considered, the past-year prevalence is 19.2% (9.1 million) and 10.3% (4.9 million) in the last month (EMCDDA 2021). In the EU, past-month and past-year prevalence of cannabis use of adults increased by 27% between 2010 and 2019 from 3.1% to 3.9% and from 13.2% to 15.7%, respectively, and 20% of all past-month users reported high-risk use patterns (Manthey et al. 2021). Cross-sectional data from over 100,000 sentinel drug users from 17 countries as part of the Global Drug Survey 2019 showed that 53.6% and 12.2% of the respondents reported past-year or monthly or less cannabis smoking in their homes, respectively, with the highest value of 80.4% in France and 36.9% in Slovakia (Bellettiere et al. 2022). Using data of an internet panel of 4088 US adults aged ≥18 years in 2018, overall 27.4% of the participants reported past-week marijuana SHS exposure in indoor and/or outdoor public areas and 52.4% regarded marijuana SHS as harmful and 81.0% opposed public marijuana smoking (Schauer et al. 2020).

The presence of drugs and their residues have been shown in various environmental media (Zambas-Adams and Honeychurch 2022). Various illicit drugs are also smoked and can thus contaminate indoor spaces and possibly lead to passive smoke exposure of other persons. This applies in particular to cocaine or cocaine cooked up with baking powder (crack), methamphetamines (e.g., crystal meth), and cannabis. Thus, in addition to the frequently illicit smoking of cannabis products, cannabis is also either administered orally or inhaled or smoked after vaporization using an inhaler in the context of medically supervised or accompanied self-therapies. The chemical structure of tetrahydrocannabinol and cocaine is presented in Fig. 9.6. No passive exposure data are yet available when cannabis is used medicinally, what is permitted in many countries. Various studies that determined secondhand smoke exposure of non-smokers to the psychoactive ingredient of cannabis, Δ9-THC (tetrahydrocannabinol), concluded that secondhand smoke also leads to increased internal exposure of those exposed, especially if rooms are poorly ventilated (Mason et al. 1983; Perez-Reyes et al. 1983; Niedbala et al. 2005; Röhrich et al. 2010; Cone et al. 2015; Herrmann et al. 2015). In a review, Zambas-Adams and Honeychurch (2022) describe the analytical basis of the studies and their results in detail.

9.4.1 Health Aspects

The health aspects for the user himself have long been known and comprehensively described in the scientific literature. Epidemiological studies that have investigated an association between secondhand smoke exposure of illicit drugs and health effects are not yet available. Some more experimental studies with a marijuana exposure come to the following conclusions: Röhrich et al. (2010) reported no changes in participants' well-being after exposure to cannabis. In the study by Cone et al. (2015), subjects reported no subjective effects or very few effects compared to active smokers. Nevertheless, a systematic review summarizes the health effects of SHS and THS cannabis exposure and advises caution in light of experience with discussions of secondhand smoke in tobacco smoking (Holitzki et al. 2017).

9.4.2 Occurrence in the Air

Studies using a smoking machine show that the sidestream smoke from marijuana cigarettes contains, on average, slightly more aromatics and PAH, significantly more NOx (2-fold), ammonia (2.6-fold), and hydrogen cyanide (eight-fold) compared to tobacco cigarettes, but fewer tobacco-related nitrosamines, lead, cadmium, and mercury (Moir et al. 2008).

Cecinato et al. (2014b) studied three schools and four homes in Rome at different times of the year in 2011/2012 with no known illicit drug use. They found concentrations in schools in winter (spring) of <0.02–0.05 ng/m^3 (<0.02 ng/m^3) for cannabidiol (CBD), 0.44–1.54 ng/m^3 (0.08–0.16 ng/m^3) for cannabicyclol (CBL), and 0.20–0.28 ng/m^3 (<0.02–0.01 ng/m^3) for Δ^9-THC. In apartments, winter (spring) levels were < 0.02–0.09 ng/m^3 (<0.02–0.03 ng/m^3) for CBD, 0.31–0.65 ng/m^3 (0.09–0.50 ng/m^3) for CBL, and 0.48–1.02 ng/m^3 (<0.02–0.03 ng/m^3) for Δ^9-THC. Cocaine levels ranged from 0.04 to 0.18 ng/m^3 in schools and from 0.05–0.24 ng/m^3 in homes. In 2013, the same study group found concentrations of 0.16–0.45 ng/m^3 for CBD, 1.05–3.17 ng/m^3 for CBL, 4.80–9.12 ng/m^3 for Δ^9-THC, and 0.06–0.22 ng/m^3 for cocaine in three residential interiors (Cecinato et al. 2014a). In one bar, levels were significantly higher at 4.7 ng/m^3 (CBD), 2.5 ng/m^3 (CBL), 14.3 ng/m^3 (Δ^9-THC), and 0.33 ng/m^3 (cocaine) in this study. In a study performed in South Africa in 2018–2019, 49 children were given a Fresh Air wristband (passive sampler) to wear for three days and Δ^9-THC could be quantified in 16% of the samples with a median of 0.32 ng/m^3 (DeLay et al. 2022).

In contrast, concentrations of 8.2–1387 μg/m^3 Δ^9-THC were found when smoking marijuana cigarettes in a 2.5 m^3 storage room, with Δ^9-THC content in the product ranging from 0.01 to 1.25% (Chou et al. 2007).

As with smoking tobacco, smoking marijuana also releases particles. When particle number concentration (diameter 0.5–2.5 μm) was measured in 33 smoking rooms, higher levels were found compared to 149 non-smoking homes (Klepeis et al. 2017). Similarly, in California residences, 7-day mean values of 10,437 particles/cm^3 could be found for particle number concentration (diameter: 0.5–2.5 μm) over 7 days in rooms ($N = 29$) where only marijuana was smoked, but only 6560 particles/cm^3 in non-smoking rooms ($N = 163$) (Posis et al. 2019). Ott et al. (2021) conducted 24 individual smoking activities with over-the-counter joints in a 43-m^3 room. The CBD content of the joints ranged from 0–1.5%, and the Δ9-THC content ranged from 8.6–27.6% (mean: 17.7%). During smoking activities, a peak mean $PM_{2.5}$ concentration of 540 μg/m^3 (153–809 μg/m^3) was measured in this study, while the 24-h mean was 52 μg/m^3 (background: 3.2 μg/m^3). Cheng et al. (2022) conducted 35 experiments for spatial measurement at 1, 2, and 3 m horizontal distances from the marijuana smoker in a current smoker's residential spaces. The smoker who performed prescribed 5-min smoking and vaping activities. The mean 5-min $PM_{2.5}$ concentrations at 1 m were higher and more variable for indoor vaping with 1330 μg/m^3 than for indoor smoking with 870 μg/m^3, however, the levels of indoor vaping decreased more noticeably with distance than for indoor smoking. Three experiments with pre-rolled marijuana joints smoked in the living room produced 5-hour mean $PM_{2.5}$ concentrations of 59.8 μg/m^3, 108.3 μg/m^3, and 96 μg/m^3 (Ott et al. 2022). The authors concluded that secondhand $PM_{2.5}$ concentrations from smoking a marijuana joint in the home were found to be 4–7 times greater than secondhand $PM_{2.5}$ concentrations from smoking a tobacco cigarette. In another study, levels of $PM_{2.5}$ were measured before, during, and after 8 cannabis bong smoking sessions in one 20-m^2 household living room (Nguyen and Hammond 2022). During the first 10 minutes of smoking, mean $PM_{2.5}$ concentrations increased to 410 μg/m^3, after 15 minutes to 570 μg/m^3, after 30 minutes to 1000 μg/m^3, and went as high as 2500 μg/m^3 in one session.

In 8 experimental trials in an indoor room where methamphetamine was produced, concentrations between 79 and 5500 μg/m^3 (mean: 1524 μg/m^3) were found in the area of the production site and between 3 and 4200 μg/m^3 (mean: 1283 μg/m^3) in an adjacent but spatially connected area (Martyny et al. 2007). In another experiment, when methamphetamine was produced in a 47 m^2 room, levels were determined to be between 520 and 760 μg/m^3, while one day later, without corresponding production activity, they were still 70–210 μg/m^3 (VanDyke et al. 2008). Simulated smoking of methamphetamine in a hotel room with 24 m^2 of floor area found concentrations of 300 and 520 μg/m^3 in two smoke tests, while 1600 μg/m^3 was measured in close proximity to the source (Martyny et al. 2007). The authors conclude that a mean indoor air concentration of 373 μg/m^3 can be expected when smoking 100 mg methamphetamine (purity 91%).

Various authors have reported in experimental trials that methamphetamine released into indoor air precipitates very well on room surfaces, furnishings, and clothing, where it can be detected for weeks (Morrison et al. 2015; Bitter 2017).

References

Abbott AJ, Reibel YG, Arnett MC, Marka N, Drake MA (2023) Oral and systemic health implications of electronic cigarette usage as compared to conventional tobacco cigarettes: a review of the literature. J Dent Hyg 97:21–35

Abdullah P, Costanian C, Khanlou N, Tamim H (2017) Prevalence and characteristics of water-pipe smoking in Canada: results from the Canadian Tobacco Use Monitoring Survey. Public Health 148:102–108

Adetona O, Mok S, Rajczyk J, Brinkman MC, Ferketich AK (2021) The adverse health effects of waterpipe smoking in adolescents and young adults: a narrative review. Tob Induc Dis 19:81

Akl EA, Gaddam S, Gunukula SK, Honeine R, Jaoude PA, Irani J (2010) The effects of waterpipe tobacco smoking on health outcomes: a systematic review. Int J Epidemiol 2010(10):834–857

Akl EA, Gunukula SK, Aleem S, Obeid R, Jaoude PA, Honeine R, Irani J (2011) The prevalence of waterpipe tobacco smoking among the general and specific populations: a systematic review. BMC Public Health 11:244

Al Ali R, Vukadinović D, Maziak W, Katmeh L, Schwarz V, Mahfoud F, Laufs U, Böhm M (2020) Cardiovascular effects of waterpipe smoking: a systematic review and meta-analysis. Rev Cardiovasc Med 21:453–468

Aljadani RH, Algabbani AM, Alamir JA, Alqahtani AS, BinDhim NF (2020) Waterpipe tobacco chemical content, microbial contamination, and genotoxic effects: a systematic review. Int J Toxicol 39:256–262

Alonso M, Godayol A, Anticó E, Sanchez JM (2010) Assessment of environmental tobacco smoke contamination in public premises: significance of 2,5-dimethylfuran as an effective marker. Environ Sci Technol 44:8289–8294

Amalia B, Fu M, Feliu A, Tigova O, Fayokun R, Mauer-Stender K, Fernández E (2022) Regulation of electronic cigarette use in public and private areas in 48 countries within the WHO European Region: a survey to in-country informants. J Epidemiol 32:131–138

Aquilina NJ, Delgado-Saborit JM, Meddings C, Baker S, Harrison RM, Jacob 3rd P, Wilson M, Yu L, Duan M, Benowitz NL (2010) Environmental and biological monitoring of exposures to PAHs and ETS in the general population. Environ Int 36:763–771

Arziman I, Acar YA, Yildirim O, Cinar O, Cevik E, Eyi YE, Kaldirim U (2011) Five cases of carbon monoxide poisoning due to narghile (shisha). Hong Kong Emerg Med 18:254–257

Ashurst JV, Urquhart M, Cook MD (2012) Carbon monoxide poisoning secondary to hookah smoking. J Am Osteopath Assoc 112:686–688

Ateş A, Arikan M, Özgök A (2016) An unusual cause of carbon monoxide poisoning: Narghile smoking. Am J Case Rep 17:660–662

Ballbè M, Martínez-Sánchez JM, Sureda X, Fu M, Pérez-Ortuño R, Pascual JA, Saltó E, Fernández E (2014) Cigarettes vs. e-cigarettes: passive exposure at home measured by means of airborne marker and biomarkers. Environ Res 135:76–80

Bellettiere J, Liles S, Posis AIB, Anuskiewicz B, Tripathi O, Nguyen B, Chavez P, Zhu SH, Park JY, Winstock A, Ferris J (2022) In-home cannabis smoking more prevalent than in-home tobacco smoking among 2019 Global Drug Survey respondents. Addict Behav 125:107130

Benowitz NL (2010) Nicotine addiction. N Engl J Med 362:2295–2303

Benowitz NL, Dains KM, Herrera B, Yu L, Jacob III P (2009) Urine nicotine metabolite concentrations in relation to plasma cotinine during low-level nicotine exposure. Nicotine Tob Res 11: 954–960

Bens BWJ, ter Maaten JC, Ligtenberg JJM (2013) Carbon monoxide poisoning after smoking from a water pipe. Ned Tijdschr Geneeskd 157:A6201

Bhatnagar A, Maziak W, Eissenberg T, Ward KD, Thurston G, King BA, Sutfin EL, Cobb CO, Griffiths M, Goldstein LB, Rezk-Hanna M (2019) Water pipe (hookah) smoking and cardiovascular disease risk: a scientific statement from the American Heart Association. Circulation 139:e917–e936

Bitter JL (2017) The persistence of illicit drug smoke residues and their recovery from common household surfaces. Drug Test Anal 9:603–612

Bohac DL, Hewett MJ, Kapphahn KI, Grimsrud DT, Apte MG, Gundel LA (2013) Secondhand smoke exposure in the nonsmoking section: how much protection? Nicotine Tob Res 15:1265–1272

Bolte G, Heitmann D, Kiranoglu M, Schierl R, Diemer J, Körner W, Fromme H (2008) Exposure to environmental tobacco smoke in German restaurants, pubs and discotheques. J Expo Sci Environ Epidemiol 18:262–271

Bolte G, Kuhn J, Twardella D, Fromme H (2009) Smoking bans in public places: current epidemiologic evidence for population-level cardiovascular health effects. Das Gesundheitswesen 71:140–151. [in German]

Bolte G, Wildner M, Fromme H (2015) Effects of nonsmoker protection legislation on children's tobacco smoke exposure and social inequalities in exposure. Prävention 38:7–10. [in German]

Buchanan ND, Grimmer JA, Tanwar V, Schwieterman N, Mohler PJ, Wold LE (2020) Cardiovascular risk of electronic cigarettes: a review of preclinical and clinical studies. Cardiovasc Res 116:40–50

Canha N, Lage J, Coutinho JT, Alves C, Almeida SM (2019) Comparison of indoor air quality during sleep in smokers and non-smokers' bedrooms: a preliminary study. Environ Pollut 249: 248–256

Cao S, Yang C, Gan Y, Lu Z (2015) The health effects of passive smoking: an overview of systematic reviews based on observational epidemiological evidence. PLoS One 10:e0139907

Cao DJ, Aldy K, Hsu S, McGetrick M, Verbeck G, De Silva I, Feng SY (2020) Review of health consequences of electronic cigarettes and the outbreak of electronic cigarette, or vaping, product use-associated lung injury. J Med Toxicol 16:295–310

Cavus UY, Rehber ZH, Ozeke O, Ilkay E (2010) Carbon monoxide poisoning associated with narghile use. Emerg Med J 27:406

Cazes N, Duchier C, Beaume S, Menot P, Renard A, Peduzzi F (2021) Think about acute shisha carbon monoxide intoxication: that's one small step for the patient, one giant leap for our society. Am J Emerg Med 257:e3–257.e5

CDC (Centers for Disease Control and Prevention) (2020a) Health effects of secondhand Smoke. https://www.cdc.gov/tobacco/data_statistics/fact_sheets/secondhand_smoke/health_effects/index.htm

CDC (Centers for Disease Control and Prevention) (2020b) Outbreak of lung injury associated with the use of e-cigarette, or vaping, products. https://www.cdc.gov/tobacco/basic_information/e-cigarettes/severe-lung-disease.html

Cecinato A, Romagnoli P, Perilli M, Patriarca C, Balducci C (2014a) Psychotropic substances in indoor environments. Environ Int 71:88–93

Cecinato A, Balducci C, Romagnoli P, Perilli M (2014b) Behaviours of psychotropic substances in indoor and outdoor environments of Rome, Italy. Environ Sci Pollut Res Int 21:9193–9200

Cheng K-C, Ott W, Wallace L, Zhud Y, Hildemann L (2022) $PM_{2.5}$ exposure close to marijuana smoking and vaping: a case study in residential indoor and outdoor settings. Sci Total Environ 802:149897

Chou S-L, Ling Y-C, Yang M-H, Pai C-Y (2007) Determination of Δ^9-tetrahydrocannabinol in indoor air as an indicator of marijuana cigarette smoking using adsorbent sampling and in-injector thermal desorption gas chromatography–mass spectrometry. Anal Chim Acta 598: 103–109

Clarke SF, Stephens C, Farhan M, Ward P, Keshishian C, Murray V, Zenner D (2012) Multiple patients with carbon monoxide toxicity from water-pipe smoking. Prehosp Disaster Med 27: 612–614

Cobb CO, Sahmarani K, Eissenberg T, Shihadeh A (2012) Acute toxicant exposure and cardiac autonomic dysfunction from smoking a single narghile waterpipe with tobacco and with a "healthy" tobacco-free alternative. Toxicol Lett 215:70–75

Cone EJ, Bigelow GE, Herrmann ES, Mitchell JM, LoDico C, Flegel R, Vandrey R (2015) Nonsmoker xposure to secondhand cannabis smoke. III. Oral fluid and blood drug concentrations and corresponding subjective effects. J Anal Toxicol 39:497–509

Connolly GN, Carpenter C, Alpert HR, Skeer M, Travers M (2005) Evaluation of the Massachusetts smoke-free workplace law. A preliminary report of the Division of Public Health Practice, Harvard School of Public Health, Tobacco Research Program. http://www.hsph.harvard.edu/php/pri/tcrtp/Smoke-free_Workplace.pdf

Cornelius ME, Wang TW, Jamal A, Loretan CG, Neff LJ (2020) Tobacco product use among adults – United States, 2019. Morb Mortal Wkly Rep 69:1736–1742

Czogala J, Goniewicz ML, Fidelus B, Zielinska-Danch W, Travers MJ, Sobczak A (2014) Secondhand exposure to vapors from electronic cigarettes. Nicotine Tob Res 16:655–662

Darawshy F, Abu Rmeileh A, Kuint R, Berkman N (2021) Waterpipe smoking: a review of pulmonary and health effects. Eur Respir Rev 30:200374

DeLay K, Lin EZ, Koelmel JP, Bornman R, Obida M, Chevrier J, Godri Pollitt KJ (2022) Personal air pollutant exposure monitoring in South African children in the VHEMBE birth cohort. Environ Int 170:107524

Dorey A, Scheerlinck P, Nguyen H, Albertson T (2020) Acute and chronic carbon monoxide toxicity from tobacco smoking. Mil Med 12:e61–e67

Drago G, Perrino C, Canepari S, Ruggieri S, L'Abbate L, Longo V, Colombo P, Frasca D, Balzan M, Cuttitta G, Scaccianoce G, Piva G, Bucchieri S, Melis M, Viegi G, Cibella F (2018) Relationship between domestic smoking and metals and rare earth elements concentration in indoor $PM_{2.5}$. Environ Res 165:71–80

El-Zaatari ZM, Chami HA, Zaatari GS (2015) Health effects associated with waterpipe smoking. Tob Control 24(Suppl 1):i31–i43

EMCDDA (European Monitoring Centre for Drugs and Drug Addiction) (2021) European Drug Report 2021: Trends and Developments. https://www.emcdda.europa.eu/publications/edr/trends-developments/2021_en

Faber T, Been JV, Reiss IK, Mackenbach JP, Sheikh A (2016) Smoke-free legislation and child health. NPJ Prim Care Respir Med 26:16067

Feliu A, Fu M, Russo M, Martinez C, Sureda X, López MJ, Cortés N, Fernández E (2020) Exposure to second-hand tobacco smoke in waterpipe cafés in Barcelona, Spain: an assessment of airborne nicotine and $PM_{2.5}$. Environ Res 184:109347

Filippidis FT, Jawad M, Vardavas CI (2019) Trends and Correlates of Waterpipe use in the European Union: analysis of selected eurobarometer surveys (2009–2017). Nicotine Tob Res 21:469–474

Frazer K, Callinan JE, McHugh J, van Baarsel S, Clarke A, Doherty K, Kelleher C (2016) Legislative smoking bans for reducing harms from secondhand smoke exposure, smoking prevalence and tobacco consumption. Cochrane Database Syst Rev CD005992

Fromme H (2019) Occurrence of polybrominated diphenyl ethers and hexabromocyclododecane. Encyclopedia of Environ Health, 2nd edn. https://doi.org/10.1016/B978-0-12-409548-9.11243-6

Fromme H, Schober W (2015) Waterpipes and e-cigarettes: impact of alternative smoking techniques on indoor air quality and health. Atmos Environ 2015(106):429–441

Fromme H, Kuhn J, Bolte G (2009a) Secondhand smoke in hospitality venues. Exposure, body burden, economic and health aspects in conjunction with smoking bans. Das Gesundheitswesen 71:242–257. [in German]

Fromme H, Dietrich S, Heitmann D, Dressel H, Diemer J, Schulz T, Jörres RA, Berlin K, Völkel W (2009b) Indoor air contamination during a waterpipe (narghile) smoking session. Food Chem Toxicol 47:1636–1641

Gallus S, Lugo A, Liu X, Behrakis P, Boffi R, Bosetti C, Carreras G, Chatenoud L, Clancy L, Continente X, Dobson R, Effertz T, Filippidis FT, Fu M, Geshanova G, Gorini G, Keogan S, Ivanov H, Lopez MJ, Lopez-Nicolas A, Precioso J, Przewozniak K, Radu-Loghin C, Ruprecht A, Semple S, Soriano JB, Starchenko P, Trapero-Bertran M, Tigova O, Tzortzi AS, Vardavas C, Vyzikidou VK, Colombo P, Fernandez E, TackSHS Project Investigators (2021a) Who smokes in Europe? Data from 12 European countries in the TackSHS Survey (2017–2018). J Epidemiol 31:145–151

Gallus S, Lugo A, Stival C, Cerrai S, Clancy L, Filippidis FT, Gorini G, Lopez MJ, López-Nicolás Á, Molinaro S, Odone A, Soriano JB, Tigova O, Van Den Brandt PA, Vardavas CI, Fernandez E; TackSHS Project Investigators (2021b) Electronic cigarette use in 12 European countries. Results from the TackSHS survey. J Epidemiol Nov 13

Gennimata SA, Palamidas A, Kaltsakas G, Tsikrika S, Vakali S, Gratziou C, Koulouris N (2012) Acute effects of e-cigarette on pulmonary function in healthy subjects and smokers. ERS Congress, Vienna

Ghaffari HR, Kamari Z, Dindarloo K, Ghanbarnejad A, Tajvar A, Turki H (2021) The systematic review of environmental pollutants in the indoor air of waterpipe café. Tob Health 1:83–93

Glantz SA, Bareham DW (2018) E-cigarettes: use, effects on smoking, risks, and policy implications. Annu Rev Public Health 39:215–235

Goodman P, Agnew M, McCaffrey M, Paul G, Clancy L (2007) Effects of the Irish smoking ban on respiratory health of bar workers and air quality in Dublin pubs. Am J Respir Crit Care Med 175:840–845

Gorini G, Moshammer H, Sbrogiò L, Gasparrini A, Nebot M, Neuberger M, Tamang E, Lopez MJ, Galeone D, Serrahima E (2008) Italy and Austria before and after study: second-hand smoke exposure in hospitality premises before and after 2 years from the introduction of the Italian smoking ban. Indoor Air 18:328–334

Gurung G, Bradley J, Delgado-Saborit JM (2016) Effects of shisha smoking on carbon monoxide and PM2.5 concentrations in the indoor and outdoor microenvironment of shisha premises. Sci Total Environ 548–549:340–346

Hadidi KA, Mohammed FI (2004) Nicotine content in tobacco used in hubble-bubble smoking. Saudi Med J 25:912–917

Hahn EJ, Rayens MK, York N, Okoli CT, Zhang M, Dignan M, Al-Delaimy WK (2006) Effects of a smoke-free law on hair nicotine and respiratory symptoms of restaurant and bar workers. J Occup Environ Med 48:906–913

Hammal F, Chappell A, Wild C, Kindzierski W, Shihadeh A, Vanderhoek A, Huynh CK, Plateel G, Finegan BA (2015) 'Herbal' but potentially hazardous: an analysis of the constituents and smoke emissions of tobacco-free waterpipe products and the air quality in the cafés where they are served. Tob Control 24:290–297

Hazrati S, Rostami R, Fazlzadeh M (2015) BTEX in indoor air of waterpipe cafés: levels and factors influencing their concentrations. Sci Total Environ 524–525:347–353

Henderson E, Continente X, Fernández E, Tigova O, Cortés-Francisco N, Gallus S, Lugo A, Semple S, Dobson R, Clancy L, Keogan S, Ruprecht A, Borgini A, Tzortzi A, Vyzikidou VK, Gorini G, López-Nicolás A, Soriano JB, Geshanova G, Osman J, Mons U, Przewozniak K, Precioso J, Brad R, López MJ, TackSHS project Investigators. (2021) Secondhand smoke exposure assessment in outdoor hospitality venues across 11 European countries. Environ Res 200:111355

Herrmann ES, Cone EJ, Mitchell JM, Bigelow GE, LoDico C, Flegel R, Vandrey R (2015) Non-smoker exposure to secondhand cannabis smoke II: effect of room ventilation on the physiological, subjective, and behavioral/cognitive effects. Drug Alcohol Depend 151:194–202

Heydari G, Vakilabadi DR, Kermani M, Rayani M, Poureshgh Y, Behroozi M, Fanaei F, Arfaeinia H (2020) Load characteristics and inhalation risk assessment of benzene series (BTEX) pollutant in indoor air of Ghalyan and/or cigarette cafes compared to smoking-free cafes. Environ Pollut Bioavail 32:26–35

Höjer J, Enghag M (2011) Carbon monoxide poisoning caused by water pipe smoking. Clin Toxicol 49:702–703

Holitzki H, Dowsett LE, Spackman E, Noseworthy T, Clement F (2017) Health effects of expo-sure to second- and third-hand marijuana smoke: a systematic review. CMAJ Open 24(5):E814–E822

IARC (2004) Tobacco smoke and involuntary smoking. IARC Monographs on the evaluation of carcinogenic risks to humans, Vol. 83, Lyon, France. https://publications.iarc.fr/_publications/media/download/2636/1567f1ed6fa20d5ef35978ef5b585b63e6101379.pdf

IARC (2009) Evaluating the effectiveness of smoke-free policies. IARC Handbooks of Cancer Prevention, Vol. 13. Lyon, France. https://www.iarc.who.int/wp-content/uploads/2018/07/handbook13.pdf

Jacob 3rd P, Benowitz NL, Destaillats H, Gundel L, Hang B, Martins-Green M, Matt GE, Quintana PJ, Samet JM, Schick SF, Talbot P, Aquilina NJ, Hovell MF, Mao JH, Whitehead TP (2017) Thirdhand smoke: new evidence, challenges, and future directions. Chem Res Toxicol 30:270–294

Kaplan B, Grau-Perez M, Çarkoglu A, Ergör G, Hayran M, Navas-Acien A, Cohen JE (2019) Smoke-free Turkey: evaluation of outdoor areas of public places. Environ Res 175:79–83

Karaca Y, Eryigit U, Aksut N, Turkmen S (2013) Syncope associated with water pipe smoking. BMJ Case Rep Apr 19

Klepeis NE, Bellettiere J, Hughes SC, Nguyen B, Berardi V, Liles S, Obayashi S, Hofstetter CR, Blumberg E, Hovell MF (2017) Fine particles in homes of predominantly low-income families with children and smokers: key physical and behavioral determinants to inform indoor-air-quality interventions. PLoS One 12:e0177718

Klosterhalfen S, Kotz D, Kuntz B, Zeiher J, Starker A (2020) Waterpipe use among adolescents in Germany: prevalence, associated consumer characteristics, and trends (German Health Interview and Examination Survey for Children and Adolescents, KiGGS). Int J Environ Res Public Health 17:7740

La Fauci G, Weiser G, Steiner IP, Shavit I (2012) Carbon monoxide poisoning in narghile (water pipe) tobacco smokers. CJEM 14:57–59

Lee K, Hahn EJ, Riker C, Seithers P (2007) Immediate impact of smoke-free laws on indoor air quality. South Med J 100:885–889
Lee PN, Fry JS, Forey BA (2014) A review of the evidence on smoking bans and incidence of heart disease. Regul Toxicol Pharmacol 70:7–23
Lee PN, Fry JS, Thornton AJ (2019) Updating the evidence relating smoking bans to incidence of heart disease. Regul Toxicol Pharmacol 101:172–186
Li L, Lin Y, Xia T, Zhu Y (2020) Effects of electronic cigarettes on indoor air quality and health Ann Rev. Public Health 41:6.1–6.18
Li L, Nguyen C, Lin Y, Guo Y, Fadel NA, Zhu Y (2021) Impacts of electronic cigarettes usage on air quality of vape shops and their nearby areas. Sci Total Environ 760:143423
Liang LA, Weber A, Herr C, Hendrowarsito L, Meyer N, Bolte G, Nennstiel-Ratzel U, Kolb S, Group GME (2016) Children's exposure to second-hand smoke before and after the smoking ban in Bavaria – a multiple cross-sectional study. Eur J Pub Health 26:969–974
Lim BL, Lim GH, Seow E (2009) Case of carbon monoxide poisoning after smoking shisha. Int J Emerg Med 2:121–122
Maalem R, Alali A, Alqahtani S (2019) Tobacco hookah smoking-induced carbon monoxide poisoning: a case report of non-ambient exposure. Clin Case Rep 7:1178–1180
Manthey J, Freeman TP, Kilian C, López-Pelayo H, Rehm J (2021) Public health monitoring of cannabis use in Europe: prevalence of use, cannabis potency, and treatment rates. Lancet Reg Health Eur 10:100227
Marini S, Buonanno G, Stabile L, Ficco G (2014) Short-term effects of electronic and tobacco cigarettes on exhaled nitric oxide. Toxicol Appl Pharmacol 278:9–15
Markowicz P, Löndahl J, Wierzbicka A, Suleiman R, Shihadeh A, Larsson L (2014) A study on particles and some microbial markers in waterpipe tobacco smoke. Sci Total Environ 499:107–113
Martinasek MP, Haddad LG, Wheldon CW, Barnett TE (2017) Beliefs and attitudes associated with hookah smoking among a United States college population. Respir Care 2017(62):370–379
Martyny JW, Arbuckle SL, McCammon Jr CS, Esswein EJ, Erb N, Van Dyke M (2007) Chemical concentrations and contamination associated with clandestine methamphetamine laboratories. J Chem Health Saf July/August:40–52
Masjedi MR, Taghizadeh F, Hamzehali S, Ghaffari S, Fazlzadeh M, Jafari AJ, Niazi S, Mehrizi EA, Moradi M, Pasalari H, Arfaeinia H (2019) Air pollutants associated with smoking in indoor/outdoor of waterpipe cafés in Tehran, Iran: concentrations, affecting factors and health risk assessment. Sci Rep 9:3110
Mason AP, Perez-Reyes M, McBay AJ, Foltz RL (1983) Cannabinoid concentrations in plasma after passive inhalation of marijuana smoke. J Anal Toxicol 7:172–174
Matt GE, Quintana PJE, Hovell MF, Bernert JT, Song S, Novianti N, Juarez T, Floro J, Gehrman C, Garcia M, Larson S (2004) Households contaminated by environmental tobacco smoke: sources of infant exposures. Tob Control 13:29–37
Maziak W, Rastam S, Ibrahim I, Ward KD, Eissenberg T (2008) Waterpipe-associated particulate matter emissions. Nicotine Tob Res 10:519–523
McNabola A, Broderick B, Johnston P, Gill L (2006) Effects of the smoking ban on benzene and 1,3-butadiene levels in pubs in Dublin. J Environ Sci Health A 41:799–810
Medford MA, Gasier HG, Hexdall E, Moffat AD, Freiberger JJ, Moon RE (2015) Research report: Charcoal type used for hookah smoking influences CO production. Undersea Hyperb Med 42: 375–380
Melstrom P, Koszowski B, Thanner MH, Hoh E, King B, Bunnell R, McAfee T (2017) Measuring $PM_{2.5}$, ultrafine particles, nicotine air and wipe samples following the use of electronic cigarettes. Nicotine Tob Res 19:1055–1061
Milz S, Akbar-Khanzadeh F, Ames A, Spino S, Tex C, Lanza K (2007) Indoor air quality in restaurants with and without designated smoking rooms. J Occup Environ Hyg 4:246–252

Minaker LM, Shuh A, Burkhalter RJ, Manske SR (2015) Hookah use prevalence, predictors, and perceptions among Canadian youth: findings from the 2012/2013 Youth Smoking Survey. Cancer Causes Contr 26:831–838

Misek R, Patte C (2014) Carbon monoxide toxicity after lighting coals at a hookah bar. J Med Toxicol 10:295–298

Moir D, Rickert WS, Levasseur G, Larose Y, Maertens R, White P, Desjardins S (2008) A comparison of mainstream and sidestream marijuana and tobacco cigarette smoke produced under two machine smoking conditions. Chem Res Toxicol 21:494–502

Moon KA, Magid H, Torrey C et al (2015) Secondhand smoke in waterpipe tobacco venues in Istanbul, Moscow, and Cairo. Environ Res 142:568–574

Morrison G, Shakila NV, Parker K (2015) Accumulation of gasphase methamphetamine on clothing, toy, fabrics, and skin oil. Indoor Air 25:405–414

Mulcahy M, Evans DS, Hammond SK, Repace JL, Byrne M (2005) Secondhand smoke exposure and risk following the Irish smoking ban: an assessment of salivary cotinine concentrations in hotel workers and air nicotine levels in bars. Tob Control 14:384–388

Naddafi K, Nabizade R, Rostami R, Reza H, Gaffari HR, Fazlzadeh M (2019) Formaldehyde and acetaldehyde in the indoor air of waterpipe cafés: measuring exposures and assessing health effects. Build Environ 165:106392

Nanninga S, Lhachimi SK, Bolte G (2018) Impact of public smoking bans on children's exposure to tobacco smoke at home: a systematic review and meta-analysis. BMC Public Health 18:749

NCI (National Cancer Institite) (1999) Smoking and Tobacco Control Monograph No. 10. https://cancercontrol.cancer.gov/brp/tcrb/monographs/10/index.html

Neuberger M, Moshammer H, Schietz A (2013) Exposure to ultrafine particles in hospitality venues with partial smoking bans. J Expo Sci Environ Epidemiol 23:519–524

Nguyen PK, Hammond SK (2022) Fine particulate matter exposure from secondhand cannabis bong smoking. JAMA Netw Open 5:e224744

Nguyen C, Li L, Sen CA, Ronquillo E, Zhu Y (2019) Fine and ultrafine particles concentrations in vape shops. Atmos Environ 211:159–169

Ni Y, Shi G, Qu J (2020) Indoor $PM_{2.5}$, tobacco smoking and chronic lung diseases: a narrative review. Environ Res 181:108910

Niedbala RS, Kardos KW, Fritch DF, Kunsman KP, Blum KA, Newland GA, Waga J, Kurtz L, Bronsgeest M, Cone EJ (2005) Passive cannabis smoke exposure and oral fluid testing. II. Two studies of extreme cannabis smoke exposure in a motor vehicle. J Anal Toxicol 29:607–615

Oberg M, Jaakkola MS, Woodward A, Peruga A, Pruss-Ustun A (2011) Worldwide burden of disease from exposure to second-hand smoke: a retrospective analysis of data from 192 countries. Lancet 377:139–146

Ott WR, Zhao T, Cheng K-C, Wallace LA, Hildemann LM (2021) Measuring indoor fine particle concentrations, emission rates, and decay rates from cannabis use in a residence. Atmos Environ 10:100106

Ott WR, Wallace LA, Cheng K-C, Hildemann LM (2022) Measuring $PM_{2.5}$ concentrations from secondhand tobacco vs. marijuana smoke in 9 rooms of a detached 2-story house. Sci Total Environ 52:158244

Ozkan S, Ozturk T, Ozmen Y, Durukan P (2013) Case report syncope associated with carbon monoxide poisoning due to narghile smoking. Case Reports Emerg Med:2–4

Perez-Reyes M, Di Guiseppi S, Mason AP, Davis KH (1983) Passive inhalation of marihuana smoke and urinary excretion of cannabinoids. Clin Pharmacol Ther 34:36–41

Posis A, Bellettiere J, Liles S, Alcaraz J, Nguyen B, Berardi V, Klepeis NE, Hughes SC, Wu T, Hovell MF (2019) Indoor cannabis smoke and children's health. Prev Med Rep 14:100853

Protano C, Manigrasso M, Avino P, Vitali M (2017) Second-hand smoke generated by combustion and electronic smoking devices used in real scenarios: ultrafine particle pollution and age-related dose assessment. Environ Int 107:190–195

Raad D, Gaddam S, Schunemann HJ, Irani J, Jaoude PA, Honeine R, Akl EA (2011) Effects of water-pipe smoking on lung function. A systematic review and meta-analysis. Chest 139:764–774
Repace J (2004) Respirable particles and carcinogens in the air of Delaware hospitality venues before and after a smoking ban. J Occup Environ Med 46:887–905
Repace J, Hyde JN, Brugge D (2006) Air pollution in Boston bars before and after a smoking ban. BMC Public Health 6:266
Röhrich J, Schimmel I, Zörntlein S, Becker J, Drobnik S, Kaufmann T, Kuntz V, Urban R (2010) Concentrations of Δ^9-tetrahydrocannabinol and 11-nor-9-carboxytetrahydrocannabinol in blood and urine after passive exposure to cannabis smoke in a coffee shop. J Anal Toxicol 34:196–203
Rosen LJ, Myers V, Winickoff JP, Kott J (2015) Effectiveness of interventions to reduce tobacco smoke pollution in homes: a systematic review and meta-analysis. Int J Environ Res Public Health 18:16043–16059
Rostami R, Zarei A, Saranjam B, Ghaffari HR, Hazrati S, Poureshg Y, Fazlzade M (2019) Exposure and risk assessment of PAHs in indoor air of waterpipe cafés in Ardebil, Iran. Build Environ 155:47–57
Rostami R, Naddafi K, Arfaeinia H, Nazmara S, Fazlzadeh M, Saranjam B (2020) The effects of ventilation and building characteristics on indoor air quality in waterpipe cafés. J Expo Sci Environ Epidemiol 30:805–813
Ruprecht AA, De Marco C, Saffari A, Pozzi P, Mazza R, Veronese C, Angellotti G, Munarini E, Ogliari AC, Westerdahl D, Hasheminassab S, Shafer MM, Schauer JJ, Repace J, Sioutas C, Boffi R (2017) Environmental pollution and emission factors of electronic cigarettes, heat-not-burn tobacco products, and conventional cigarettes. Aerosol Sci Technol 51:674–684
Russo ET, Hulse TE, Adamkiewicz G, Levy DE, Bethune L, Kane J, Reid M, Shah SN (2015) Comparison of indoor air quality in smoke-permitted and smoke-free multiunit housing: findings from the Boston Housing Authority. Nicotine Tob Res 17:316–322
Salloum RG, Thrasher JF, Getz KR, Barnett TE, Asfar T, Maziak W (2017) Patterns of waterpipe tobacco smoking among U.S. young adults, 2013–2014. Am J Prev Med 52:507–512
Schauer GL, Tynan MA, Marynak K (2020) Self-reported exposure to, perceptions about, and attitudes about public marijuana smoking among US adults, 2018. Addiction 115:1320–1329
Schettgen T, Musiol A, Kraus T (2008) Simultaneous determination of mercapturic acids derived from ethylene oxide (HEMA), propylene oxide (2-HPMA), acroleine (3-HPMA), acrylamide (AAMA) and N,N-dimethylformamide (AMCC) in human urine using liquid chromatography/tandem mass spectrometry. Rapid Comm Mass Spectrom 22:2629–2638
Schober W, Szendrei K, Matzen W, Osiander-Fuchs H, Heitmann D, Schettgen T, Jörres RA, Fromme H (2014) Use of electronic cigarettes (e-cigarettes) impairs indoor air quality and increases FeNO levels of e-cigarette consumers. Int J Hyg Environ Health 217:628–637
Schripp T, Markewitz D, Uhde E, Salthammer T (2013) Does e-cigarette consumption cause passive vaping? Indoor Air 23:25–31
Schubert J, Luch A, Schulz TG (2013) Waterpipe smoking: analysis of the aromaprofile of flavored waterpipe tobaccos. Talanta 115:665–674
Schubert J, Müller FD, Schmidt R, Luch A, Schulz TG (2015) Waterpipe smoke: source of toxic and carcinogenic VOCs, phenols and heavy metals? Arch Toxicol 89:2129–2139
Seidenberg AB, Orlan EN, Travers MJ, Sutfin EL (2019) Air quality and presence of air ventilation systems inside waterpipe cafés in North Carolina. Tob Control 28:356–358
Semple S, Maccalman L, Naji AA, Dempsey S, Hilton S, Miller BG, Ayres JG (2007) Bar workers' exposure to second-hand smoke: the effect of Scottish smoke-free legislation on occupational exposure. Ann Occup Hyg 51:571–580
Sepetdjian E, Halim RA, Salman R, Jaroudi E, Shihadeh A, Saliba NA (2013) Phenolic compounds in particles of mainstream waterpipe smoke. Nicotine Tob Res 15:1107–1112
Shearston JA, Eazor J, Lee L, Vilcassim MJR, Reed TA, Ort D, Weitzman M, Gordon T (2021) Effects of electronic cigarettes and hookah (waterpipe) use on home air quality. Tob Control May 21: tobaccocontrol-2020-056437

Shearston JA, Eazor J, Lee L, Vilcassim MJR, Reed TA, Ort D, Weitzman M, Gordon T (2023) Effects of electronic cigarettes and hookah (waterpipe) use on home air quality. Tob Control 32:36–41

Sheu R, Stönner C, Ditto JC, Klüpfel T, Williams J, Gentner DR (2020) Human transport of thirdhand tobacco smoke: a prominent source of hazardous air pollutants into indoor nonsmoking environments. Sci Adv 6:eaay4109

Shihadeh A (2003) Investigation of mainstream smoke aerosol of the argileh water pipe. Food Chem Toxicol 41:143–152

Shihadeh A, Salman R, Jaroudi E, Saliba N, Sepetdjian E, Blank MD, Cobb CO, Eissenberg T (2012) Does switching to a tobacco-free waterpipe product reduce toxicant intake? A crossover study comparing CO, NO, PAH, volatile aldehydes, "tar" and nicotine yields. Food Chem Toxicol 50:1494–1498

Skogstad M, Kjaerheim K, Fladseth G, Gjølstad M, Daae HL, Olsen R, Molander P, Ellingsen DG (2006) Cross shift changes in lung function among bar and restaurant workers before and after implementation of a smoking ban. Occup Environ Med 63:482–487

Sleiman M, Destaillats H, Smith JD, Liu CL, Ahmed M, Wilson KR, Gundel LA (2010) Secondary organic aerosol formation from ozone-initiated reactions with nicotine and secondhand tobacco smoke. Atmos Environ 44:4191–4198

Sleiman M, Logue JM, Luo W, Pankow JF, Gundel LA, Destaillats H (2014) Inhalable constituents of thirdhand tobacco smoke: chemical characterization and health impact considerations. 2014. Environ Sci Technol 48:13093–13101

Snoderly HT, Nurkiewicz TR, Bowdridge EC, Bennewitz MF (2021) E-Cigarette use: device market, study design, and emerging evidence of biological consequences. Int J Mol Sci 22: 12452

Son Y, Giovenco DP, Delnevo C, Khlystov A, Samburova V, Meng Q (2020) Indoor air quality and passive e-cigarette aerosol exposures in vape-shops. Nicotine Tob Res 22:1772–1779

Soule EK, Maloney SF, Spindle TR, Rudy AK, Hiler MM, Cobb CO (2017) Electronic cigarette use and indoor air quality in a natural setting. Tob Control 26:109–112

Sutfin EL, Song EY, Reboussin BA, Wolfson M (2014) What are young adults smoking in their hookahs? A latent class analysis of substances smoked. Addict Behav 39:1191–1196

Tabatabaei Z, Baghapour MA, Hoseini M, Fararouei M, Abbasi F, Baghapour M (2021) Assessing BTEX concentrations emitted by hookah smoke in indoor air of residential buildings: health risk assessment for children. J Environ Health Sci Eng 19:1653–1665

Tominz R, Poropat C, Bovenzi M (2006) Changes in PM_{10} and $PM_{2.5}$ air levels in bars after the enforcement of the smoking ban in the Italian legislation. Epidemiol Prev 30:325–333

TPEG (Tobacco Program Evaluation Group) (2008) Smoking pollution in gaming venues. Report prepared for the Colorado Department of Public Health and Environment

Travers MJ, Cummings KM, Hyland A, Repace J, Babb J, Pechacek T, Caraballo R (2004) Indoor air quality in hospitality venues before and after implementation of a Clean Indoor Air Law – Western New York, 2003. MMWR 12:1038–1041

Trinidad DR, Pierce JP, Sargent JD, White MM, Strong DR, Portnoy DB, Green VR, Stanton CA, Choi K, Bansal-Travers M, Shi Y, Pearson JL, Kaufman AR, Borek N, Coleman BN, Hyland A, Carusi C, Kealey S, Leas E, Noble ML, Messer K (2017) Susceptibility to tobacco product use among youth in wave 1 of the population Assessment of tobacco and health (PATH) study. Prev Med 101:8–14

UNODC (United Nations Office on Drugs and Crime) (2021) World Drug Report 2021. https://www.unodc.org/unodc/en/data-and-analysis/wdr2021.html

US DHHS (2006) U.S. Department of Health and Human Services. The health consequences of involuntary exposure to tobacco smoke: A Report of the Surgeon General. Atlanta, GA, USA. https://www.cdc.gov/tobacco/data_statistics/sgr/2006/index.htm

Uyanık B, Arslan ED, Akay H, Erçelik E, Tez M (2011) Narghile (hookah) smoking and carboxyhemoglobin levels. J Emerg Med 40:679

Valente P, Forastiere F, Bacosi A, Cattani G, Di Carlo S, Ferri M, Figà-Talamanca I, Marconi A, Paoletti L, Perucci C, Zuccaio P (2007) Exposure to fine and ultrafine particles from secondhand smoke in public placet bifore and after the smoking ban, Italy 2005. Tobacco Contr 16:312–317

Van Ussel I, Klink D (2021) Carbon monoxide intoxication due to waterpipe smoking as cause of a seizure in an adolescent: a case report. Belgium J Paediatr 23:82–83

VanDyke M, Erb N, Arbuckle S, Martyny J (2008) A 24-hour study to investigate persistent chemical exposures associated with clandestine methamphetamine laboratories. J Occup Environ Hyg 6:82–89

Vardavas CI, Anagnostopoulos N, Kougias M, Evangelopoulou V, Connolly GN, Behrakis PK (2012) Short-term pulmonary effects of using an electronic cigarette: impact on respiratory flow resistance, impedance, and exhaled nitric oxide. Chest 141:1400–1406

Veen M (2016) Carbon monoxide poisoning caused by water pipe smoking: a case series. J Emerg Med 51:41–44

Veeranki SP, Alzyoud S, Dierking L, Kheriallah K, Mzayek F, Pbert L, Ward KD (2016) Associations of adolescents' cigarette, waterpipe, and dual tobacco use with parental tobacco use. Nicotine Tob Res 18:879–884

Verweij BGF, Rood PPM, Schuit SCE, Bouwhuis MG (2019) Waterpipe smoking: not as innocent as it may seem. Neth J Med 77:156–159

Villano J, Minns A, Kreshak A, Tomaszewski C (2014) From hookah to hyperbarics: carbon monoxide toxicity due to hookah smoking. Clin Toxicol 52:728

Volesky KD, Maki A, Scherf C, Watson L, Van Ryswyk K, Fraser B, Weichenthal SA, Cassol E, Villeneuve PJ (2018) The influence of three e-cigarette models on indoor fine and ultrafine particulate matter concentrations under real-world conditions. Environ Pollut 243:882–889

Von Rappard J, Schönenberger M, Bärlocher L (2014) Accidental carbon monoxide intoxication after hookah use. Dtsch Ärztebl 2014(111):674–679. [in German]

Wang LW, He EY, Ghosh D, Day RO, Jones GR, Subbiah RN, Holloway CJ (2015) Severe arbon monoxide poisoning from waterpipe smoking: a public health concern. Med J Aust 202:446–447

Ward M, Currie LM, Kabir Z, Clancy L (2013) The efficacy of different models of smoke-free laws in reducing exposure to second-hand smoke: a multi-country comparison. Health Policy 110: 207–213

Waring MS, Siegel JA (2007) An evaluation of the indoor air quality in bars before and after a smoking ban in Austin, Texas. J Expo Sci Environ Epidem 17:260–268

Weitzman M, Yusufali AH, Bali F, Vilcassim MJR, Gandhi S, Peltier R, Nadas A, Sherman S, Lee L, Hong Z, Shearston J, Park SH, Gordon T (2016) Effects of hookah smoking on indoor air quality in homes. Tob Control 26:586–591

WHO (2000) Air Quality Guidelines for Europe. WHO Regional Publications, European Series No. 91. https://www.euro.who.int/en/publications/abstracts/air-quality-guidelines-for-europe

WHO (2011) Report on the global tobacco epidemic 2011: Warning about the dangers of tobacco. Geneva, Switzerland. http://apps.who.int/iris/bitstream/10665/44616/1/9789240687813_eng.pdf

WHO (2012) Global Report: Mortality attributable to tobacco. Genua, Italy. http://apps.who.int/iris/bitstream/handle/10665/44815/9789241564434_eng.pdf;jsessionid=AF1625CB7879EA708B4477F7CA60D9EC?sequence=1

WHO (2019) Global report on trends in prevalence of tobacco use 2000–2025. Third edition. https://www.who.int/publications/i/item/who-global-report-on-trends-in-prevalence-of-tobacco-use-2000-2025-third-edition

Wienecke A, Kraywinkel K (2018) Tobacco-Associated Cancers in Germany – Development of Incidence and Mortality since 1995. Bundesgesundheitsbl 61:1399–1406. [in German]

Wienecke A, Barnes B, Lampert T, Kraywinkel K (2014) Changes in cancer incidence attributable to tobacco smoking in Germany, 1999–2008. Int J Cancer 134:682–691

Wu J-X, Lau ATY, Xu Y-M (2022) Indoor secondary pollutants cannot be ignored: third-hand smoke. Toxics 10:363

Yoong SL, Hall A, Leonard A, McCrabb S, Wiggers J, Tursan d'Espaignet E, Stockings E, Gouda H, Fayokun R, Commar A, Prasad VM, Paul C, Oldmeadow C, Chai LK, Thompson B, Wolfenden L (2021) Prevalence of electronic nicotine delivery systems and electronic non-nicotine delivery systems in children and adolescents: a systematic review and meta-analysis. Lancet Public Health 6:e661–e673

Zaidi SMA, Moin O, Khan JA (2011) Second-hand smoke in indoor hospitality venues in Pakistan. Int J Tuberc Lung Dis 15:972–977

Zambas-Adams P, Honeychurch KC (2022) Analytical approaches and trends in the determi-nation of psychoactive drugs in air. Science 4:1

Zhang B, Haji F, Kaufman P, Muir S, Ferrence R (2013) Enter at your own risk': a multimethod study of air quality and biological measures in Canadian waterpipe cafes. Tob Control 24:175–181

Zhao T, Nguyen C, Lin C-H, Middlekauff HR, Peters K, Moheimani R, Guo Q, Zhu Y (2017) Characteristics of secondhand electronic cigarette aerosols from active human use. Aerosol Sci Technol 51:1368–1376

Zhou L, Niu L, Jiang H, Jiang C, Xiao S (2016) Facilitators and barriers of smokers' compliance with smoking bans in public places: a systematic review of quantitative and qualitative literature. Int J Environ Res Public Health 13:1228

Chapter 10
Indoor Risk Assessment and Guideline Values

Contents

Abstract Quantitative risk assessment is a multi-stage process at the end of which there is a qualitative and quantitative health assessment of the risks for the affected population, which can then form the basis of risk management. In addition to knowledge of the exposure situation to substances, the hazard potential must be known and knowledge of the dose–response relationship must be available. In this chapter basic information and key principles of the risk assessment process are given.

When applying assessment values for indoor air, it is important to consider their respective technical derivation and justification as well as their legal classification. Up to now, only a limited number of legally binding (mandatory) assessment values exist for indoor air, while significantly more toxicologically derived guideline values like the US EPA reference concentrations, the World Health Organization air quality guidelines, and the indoor air guideline values of the German Committee for Indoor Guideline Values. Furthermore, reference values, which represent the background exposure that existed at the time they were established, can be used for evaluation. This chapter gives a brief overview of the available assessment values.

Keywords Quantitative risk assessment · QRA · Health · Risk management · Guideline values · Reference values

H. Fromme, *Indoor Air Quality*, https://doi.org/10.1007/978-3-031-40078-0_10

10.1 Basics of Risk Assessment

The concept of risk, in this case the health risk, is particularly important and must be concretized. In insurance industry, risk is understood to be the product of the amount of damage with the probability of the damage occurring, while in technology it is used to describe a situation that lies in a range between safety (acceptable risks) and dangerousness (technically no longer acceptable). These definitions cannot be used directly to describe health risks. From a medical point of view, risk can be understood as the conditional probability of the occurrence of an adverse health effect in the presence of exposure to a noxious agent. It can be expressed as the product of the hazardousness of a substance (totality of adverse effects) with exposure to that substance:

Risk = Hazard × Exposure

In addition to the risk-relevant properties inherent in a substance, there must therefore always be a corresponding exposure in order for a health risk to occur. The exposure itself is influenced by the level, frequency, and duration with which a person is exposed to a substance.

Quantitative risk assessment, following the US National Research Council, has been understood since the 1980s as a multi-stage process at the end of which there is a qualitative and quantitative health assessment of the risks for the affected population, which can then form the basis of risk management. Quantitative risk assessments represent procedures that, on a scientific-technical basis and incorporating heuristic-pragmatic assumptions (conventions), aim to derive protective values based on the probability of the occurrence of adverse health effects, depending on the level of exposure (AUH 1995).

Indoor health assessment follows the aforementioned principles, which are also applied to the risk assessment of other intake pathways such as outdoor air or food. Generally, the process is shown in Fig. 10.1. According to this, the following essential steps are to be considered after clarification of the underlying question:

- The hazard potential (qualitative effect properties) must be known (hazard identification). This is, of course, always dependent on the state of the knowledge in science; as an example, endocrine effects of foreign substances, which have only been increasingly investigated and taken into account since the beginning of the 1990s. However, such "new" effects do not necessarily represent the most toxicologically relevant endpoints.
- Knowledge of the dose-response relationship must be available. Here, it is important to consider above all the effects that occur even at low exposures. Assessment or guideline values can then be derived on the basis of this knowledge.
- Valid measurements or at least prognostic statements on external and/or internal exposure must be available. Where appropriate, other potentially significant uptake pathways must also be considered as part of the assessment.

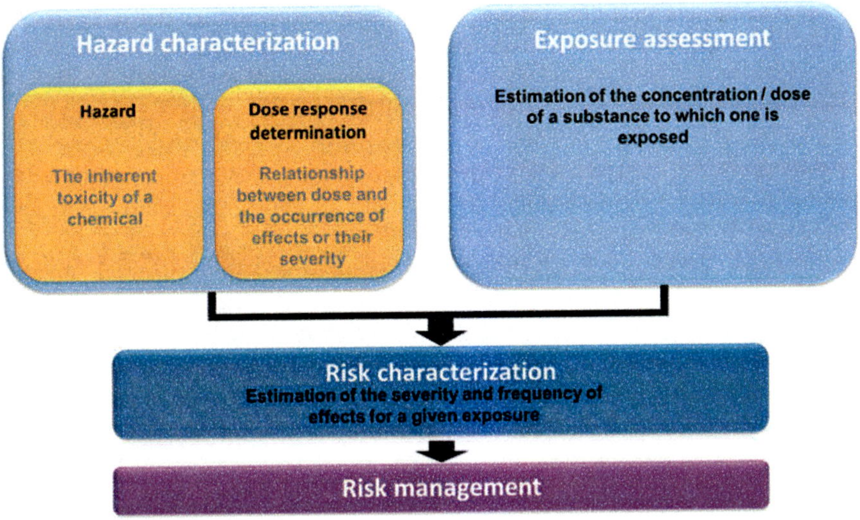

Fig. 10.1 Steps and process of the quantitative risk assessment

In quantitative risk assessment, the exposure is compared with suitable assessment values, i.e. ultimately with the dose-response curve. This can be used to determine whether or to what extent health hazards exist or have to be expected.

To estimate noncancer hazards for a single air toxic the US EPA calculated a so-called hazard ratio (HR) (NATA 2018). The HR is defined as the ratio of the potential exposure to a substance and the level at which no adverse effects occurred:

$$HQ = \frac{C}{GV}$$

where C is the estimated (long-term) inhalation exposure concentration for a specific air toxic and GV the corresponding guide value or toxicological reference concentration for that air toxic. A hazard quotient of 1 or lower means adverse noncancer effects are unlikely, and thus can be considered to have negligible hazard. For HQ greater than 1, the potential for adverse effects increases, but we do not know by how much. Chronic noncancer hazards for multiple air toxics were estimated by summing chronic noncancer HQ for individual air toxics that cause similar adverse health effects (affect the same organ or organ system).

To estimate cancer risks of a population the inhalation unit risk (IUR) was often used, taken into account that the risk is directly proportional to dose and that there is no threshold for the carcinogenic effect. The IUR is defined as the upper-bound excess lifetime cancer risk estimated to result from continuous exposure to an agent at a concentration of 1 μg/m^3 in air, over a 70-year lifetime and assuming a daily inhalation rate of about 20 m^3/day (NATA 2018). As a basis, the US EPA proposed that linear extrapolation of carcinogenic risk in the low-dose region of the curve is a reasonable approach for estimating risk at exposures typical for the general population. The estimated incremental lifetime cancer risk (ILCR) due to exposure to a specific air toxic is:

$$\text{ILCR} = C \text{ x IUR}$$

where C is the estimate of long-term inhalation exposure concentration for a specific air pollutant (in for example μg/m^3) and IUR the corresponding inhalation unit risk (in 1/[μg/m^3]). For example, if the UR is 5×10^{-6}, and the daily lifetime exposure is 10 μg/m^3, the ILCR is 50×10^{-6} or 50 excess tumors per one million exposed person.

If the carcinogen has a mutagenic mode of action, ATSDR applies an additional age-dependent adjustment factors (ADAF) (from birth to <2 years: factor 10; from 3 to <16 years: factor 3) to the abovementioned cancer risk equation for the increased susceptibility of children in that case (ATSDR 2022).

Furthermore, in addition to exposure to carcinogens via indoor air, it is important to consider other uptake pathways such as outdoor air, food, and drinking water. Only in this way can a comprehensive picture of the concrete importance of each pathway be established and overall prioritizations made. Against this background, Setton et al. (2013), for example, have developed indicators for the exposure of the population to carcinogens that allow not only a comparison of different substances and intake pathways with each other, but also a prioritization. They estimate lifetime average daily intake using population characteristics combined with concentrations in outdoor air, indoor air, dust, drinking water, and food and beverages from existing monitoring databases or comprehensive literature reviews and multiplied it by cancer potency factors from different scientific institutions. They summarized, that based on data representing average conditions circa 2006, a total of 18 carcinogen-exposure pathways had potential lifetime excess cancer risks greater than 1 per million (1×10^{-6}), based on varying data quality (Fig. 10.2). Carcinogens with moderate to high data quality and lifetime excess cancer risk greater than 1 per million included radon, formaldehyde, diesel engine exhaust, benzene, 1,3-butadiene, and asbestos in indoor air.

The result of a quantitative risk assessment is usually placed once again in a larger context, in which other aspects besides health aspects also come into play. These include, in particular, social and general societal aspects, which above all must be assessed politically. Finally, on the basis of such a risk assessment, a decision is made on how to deal with any hazards (risk management). It must be used to take all necessary measures in a targeted and consistent manner. In particular, the need for and scope of measures must be critically scrutinized time and again. In addition, aspects of risk communication must be taken into account throughout the entire process, since they often have a significant influence on the further course of an indoor problem. The most important prerequisites for successful target-group-oriented risk communication and the basis are confidence-building, transparency, and credibility. In the run-up, clarity must be established about the task and objectives of risk communication. However, a realistic, technically sound and, in particular, comprehensible assessment can also reduce fears in the population or among the

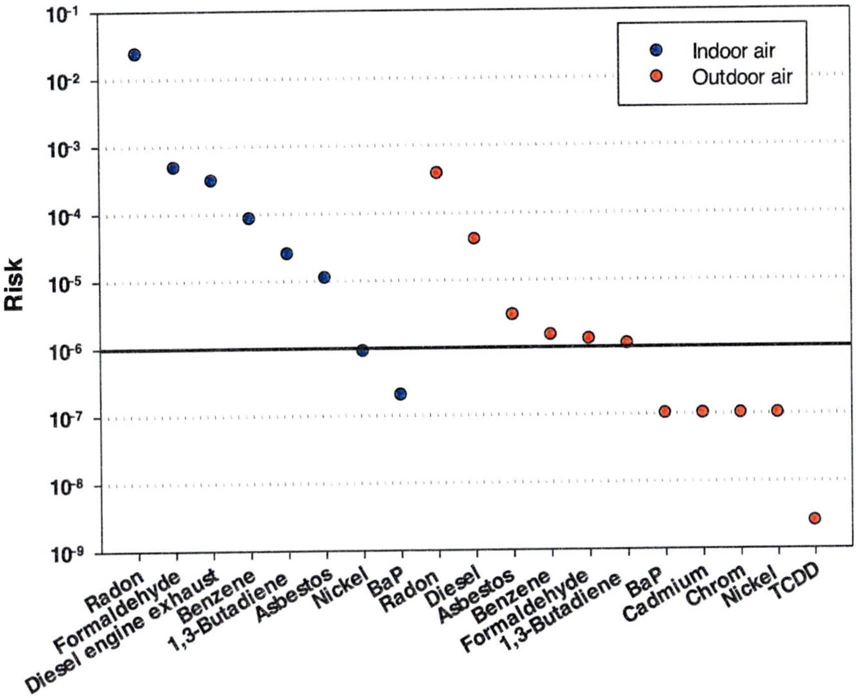

Fig. 10.2 Estimates of lifetime excess cancer risk in indoor and outdoor air in Canada (modified after Setton et al. 2013)

users of space and increase the degree of approval for the implementation of measures.

10.2 Key Points and Process of an Health Evaluation

Figure 10.3 shows a schematic diagram of the assessment procedure in the event of a problem with indoor pollutants. In a first step, it should be clarified which legal bases are to be used for the assessment. In particular, a distinction must be made to indoor spaces (or workplaces) in which an activity involving a hazardous substance is carried out. In this case, the requirements of the Occupational Exposure Limits (OEL) in the USA or the Maximum Occupational Concentrations (MAK value) in Germany set limits. These hazardous substance-related values, on the other hand, are not to be applied directly to the indoor environment such as residences, offices, or schools. The definition of indoor spaces to which the following assessment scheme is to be applied can be found in the introduction.

In a further step, it is necessary to describe the objectives of the investigation and a question that is as concrete as possible. On this basis, the scope of measurement

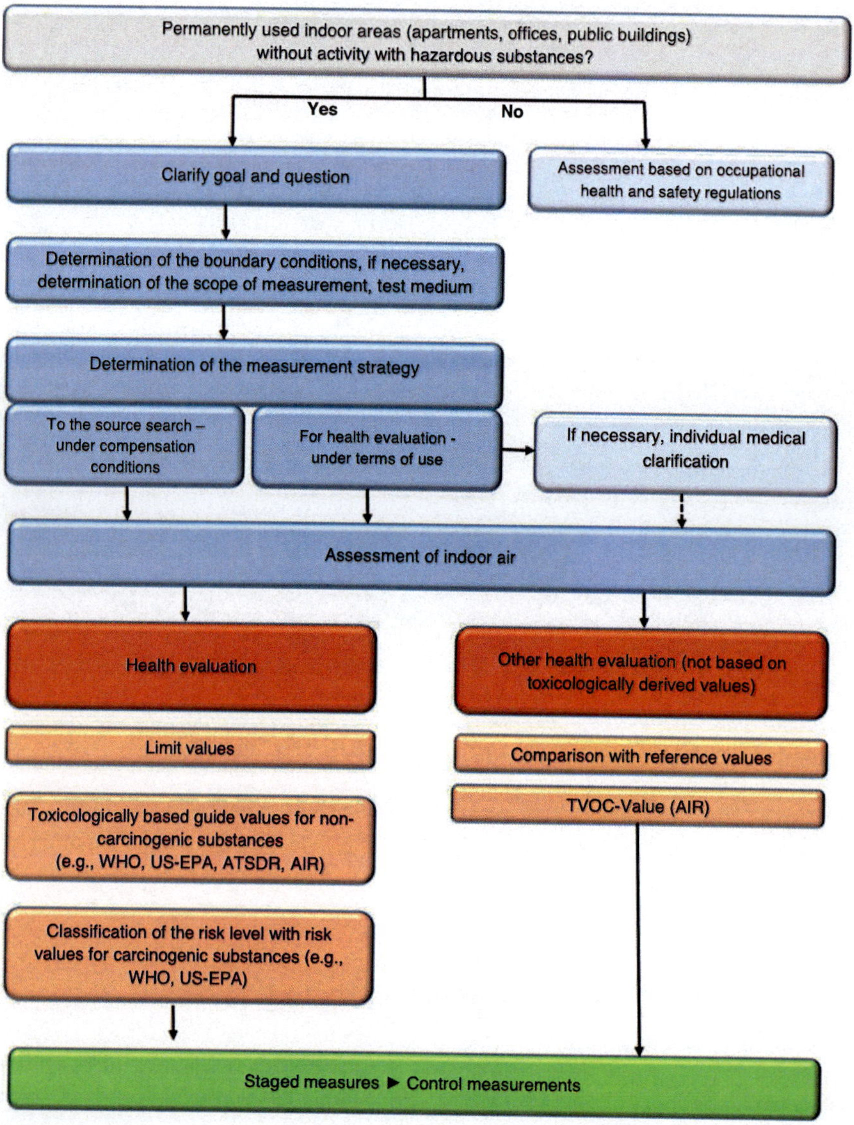

Fig. 10.3 Flow chart for the evaluation of indoor air pollutants (TVOC: total volatile organic compounds, levels set by the German Committee on Indoor Air Guide Values, AIR)

can then be clarified and a targeted measurement strategy can be defined. In order to be able to carry out a subsequent assessment properly, e.g. on the basis of the guideline values set by WHO and other scientific institutions, it is important to distinguish between the search for sources indoors and a health-related assessment, because different sampling strategies are required, for example, in Germany (Ad-hoc-AG 2014). In the first case, measurements must be made under compensation conditions, in the second case under conditions of use (see definitions in Sect. 1.5.1). In the presence of health complaints attributed to the pollutant, it may also be necessary to conduct an individual medical examination and consultation. The actual assessment is then possible on the basis of limit values or toxicologically justified guideline and reference values. If none of the above-mentioned assessment values are available for the substance to be evaluated, at least an assessment of the indoor pollution can be made in comparison with statistically determined reference values.

Carcinogenic substances are a special case. They can be simplified into the group of genotoxic (primary) carcinogens, which interact directly with DNA themselves or through their metabolites (e.g., through mutations, adduct formation), or non-genotoxic (secondary) carcinogens, which exert their effect indirectly, e.g., via oxidative stress, inhibition of topoisomerase or of DNA repair enzymes (Cartus and Schrenk 2017). For the first group, it is generally accepted that no effect threshold can be assumed, so that in this case a risk reduction should take place in the sense of the ALARA principle (As Low As Reasonably Achievable). For the second group, on the other hand, the existence of an effect threshold can be assumed. If the carcinogenic effect is the most sensitive endpoint, the evaluation should be based on this, otherwise on the most sensitive toxicological endpoint that has been shown in technically reliable studies.

For the general, not occupationally exposed population, there is as yet no definition of a tolerable or acceptable risk to carcinogenic substances that is accepted by society as a whole. Some institutions and authorities have made statements about an acceptable level of risk:

- The World Health Organization drinking water guideline specifies that guideline values are conservatively presented as the concentrations in drinking-water associated with an estimated upper-bound excess lifetime cancer risk of 10^{-5} (WHO 2011). But member states may consider that a different level of hypothetical risk is more appropriate to their circumstances, and values relating to risks of 10^{-4} or 10^{-6} additional cancer cases over a lifetime of exposure may be determined.
- The US Agency for Toxic Substances and Disease Registry recommends that if the estimated cancer risk is greater than 10^{-6}, health assessor should conduct an in-depth toxicological effects analysis.
- Although there is no EU legislation setting the "tolerable" risk level for carcinogens, European Chemicals Agency recommends a cancer risk level of 10^{-6} could be seen as indicative tolerable risks levels of the general population (ECHA 2012).

Since many genotoxic substances have their source in outdoor air or occur in food, risk reduction measures are primarily useful in these areas—and not in indoor air. Examples would include reducing diesel engine emissions or benzene emissions from fuel combustion.

If no risk-related guideline values for carcinogenic substances are available, a rough classification of the risk level of the measured concentrations in comparison with the risk estimators of the WHO (see Table 10.2) or those of the US Environmental Protection Agency could be considered.

10.3 Values for Assessing Indoor Air Contamination

When applying assessment values for indoor air, it is important to consider their respective technical derivation and justification as well as their legal classification. Generally, it is necessary to differentiate between non-occupational indoor spaces and occupational/industrial indoor workplaces. For the latter, numerous guideline values are recommended or mandatory. While the occupational values refer to a lifetime working and eight-hour work shift, the guide values for the general population in residences and public buildings refer normally to a lifelong period (24 hours a day, seven days a week) and also apply to susceptible subsets of the population like children and people with pre-existing diseases.

The following values can be distinguished:

- Only a limited number legally binding (mandatory) assessment values exist for indoor air.
- Toxicologically derived guideline values are characterized by the fact that they are based on sound and appropriate knowledge of toxic effects of the substance in question; they often include safety margins to protect susceptible subsets of the population as well.
- Other guideline values or precautionary values are usually set lower than toxicologically based values and are also intended to prospectively avoid, or at least reduce, the exposure of humans or other objects of protection.
- Reference values represent the background exposure that existed at the time they were established. As a rule, they do not provide any information about the degree of health risk. If exposures are in the range of the background values, they probably do not pose an increased risk compared to the risk of the general public.

10.3.1 Mandatory Values (Limit Values)

Mandatory limits on the concentrations of substances in the indoor air are generally rare. For these values, there is a legally defined obligation to comply with the values. The measurements must be carried out under precisely defined sampling and examination conditions.

10.3.1.1 Mandatory Values in Europe

In France, mandatory values for long-term exposure were recommended for formaldehyde (30 μg/m^3 since 2015 and 10 μg/m^3 since 2023) as well as for benzene (2 μg/m^3) in 2011 (ANSES 2021).

In Germany, four indoor contaminants are currently regulated.

Based on the Ordinance on the Emission Limitation of Volatile Halogenated Organic Compounds (second BImSchV) of 1990 a value of <0.1 mg/m^3 was specified for perchloroethylene (tetrachloroethene) (Anonymous 2001). It applies in a non-premises room used for human occupancy or in an adjacent food establishment. This regulation applies to indoor spaces adjacent to surface treatment facilities, dry cleaning and textile cleaning facilities, and extraction facilities (e.g., for animal carcass disposal). If an indoor air concentration of more than 0.1 mg/m^3 is detected, determined as an average value over a period of 7 days, which is attributable to the operation of an adjacent plant, the operator of this plant must take measures within 6 months to ensure that an indoor air concentration of 0.1 mg/m^3 is not exceeded.

As part of the general building regulation requirements from the building codes of the federal states, there are so-called introduced technical building regulations and guidelines for the implementation of building regulation requirements. These are mandatory under building regulations if they have been introduced as technical building regulations. This is the case in most federal states, but not in all, with the guidelines on polychlorinated biphenyls, pentachlorophenol, and asbestos.

- For polychlorinated biphenyls (PCB), the following value range applies on the basis of an assessment by the former German Federal Health Office, which was incorporated into the building law (e.g., BStMI 2012):
  - Indoor air concentrations <300 ng PCB/m^3 are to be considered tolerable in the long term (precautionary value).
  - At 300–3000 ng PCB/m^3, sources must be traced and eliminated with regard to proportionality, or at least a reduction in PCB concentrations (e.g., through regular ventilation and thorough cleaning and dust removal from rooms) must be sought (target value: <300 ng/m^3).
  - At >3000 ng PCB/m^3 control analyses are required immediately. If the value is confirmed, measures to reduce the indoor air concentration of PCBs must be taken immediately, depending on the exposure, to avoid health risks in these rooms. The remediation measures must be suitable to effectively reduce the PCB intake. The target value here is also less than 300 ng PCB/m^3 air (remediation guide value).

PCBs in this case are understood as the sum of the 6 indicator congeners (PCB 28, 52, 101, 138, 153, and 180) multiplied by factor 5.

- For pentachlorophenol (PCP), the following regulations have been introduced into building law (e.g., BStMI 2012):

- In recreation rooms, a possible health hazard is to be assumed if the indoor air concentration to be expected as an annual average is above 1 μg PCP/m^3 air.
- In the case of dwellings or other rooms in which people regularly spend more than 8 hours a day over a longer period of time and in which exposure via dust and food, etc. is to be expected due to the use, such as in daycare centers or homes, however, a health hazard is already possible if the indoor air concentration to be expected as an annual average is below 1 μg PCP/m^3 air but above 0.1 μg PCP/m^3 air and at the same time there is a PCP contamination in the blood of more than 70 μg PCP/l (serum) or in the urine of more than 40 μg PCP/l.

• Within the framework of the asbestos guideline, the following provisions have been made in building law (e.g., BStMI 2012):

The guideline applies within the framework of assessment and remediation in the presence of weakly bound asbestos products. These are asbestos products with a bulk density of less than 1000 kg/m^3. An assessment of the urgency of remediation is made by means of a predefined questionnaire as part of the site inspection. When monitoring the success of remediation, it must be demonstrated that the following two conditions have been met:

- The asbestos fiber concentration with fiber lengths $L \geq 5$ μm, fiber diameters $D < 3$ μm, and a length-to-width ratio >3:1 is calculated from the number of fibers observed on the filter. Each measured value must be less than 500 fibers/m^3.
- The upper limit of the 95% confidence interval of the asbestos fiber concentration calculated from the number of asbestos fibers with a fiber length $L \geq 5$ μm, a fiber diameter $D < 3$ μm, and a fiber length-to-width ratio of L: $D > 3$:1 according to the Poisson distribution must be below 1000 fibers/m^3.

10.3.1.2 Mandatory Values in Asia

In Taiwan, the Indoor Air Quality Management Act was enacted in 2011 to improve indoor air quality and to protect public health (Tsai 2019). The recommended standards for any closed or semi-closed residential and public building are given in Table 10.1.

In Korea the Public Sanitation Act promulgated in 1986 and revised in 2005 specified the standard of pollutants in the public facility (Jeong 2006) (see Table 10.1). The Health Act in School by the Ministry of Education and Human Resources assessed air quality standards for classrooms and attached rooms in schools (Yang et al. 2019) (see Table 10.1).

10.3.2 Toxicologically Based Guideline Values

Although guideline values are not legally binding, they are sometimes treated as equivalent to mandatory values, especially if they are toxicologically based, since

Table 10.1 Indoor air quality guidelines in Taiwan, Korea, and China

	Taiwan		Korea (public buildings)		Korea (facilities for sensitive population)	China	
Substance	Value	Time	Value	Time	Value	Value	Time
Carbon dioxide (CO_2) ppm	1000	8 h	1000	1 h	1000	0.1%	24 h
Carbon monoxide (CO) (ppm)	9	8 h	25	1 h	19	8.7	1 h
Nitrogen dioxide (NO_2) (μg/m^3)	–		–		94	240	1 h
Sulfur dioxide (SO_2) (μg/m^3)	–		–		–	500	1 h
Ammonia (NH_3) (μg/m^3)	–		–		–	200	1 h
Formaldehyde (μg/m^3)	75	1 h	120	8 h	100	100	1 h
TVOC	0.56 ppm	1 h	–		400 μg/m^3	600 μg/m^3	8 h
Bacterial (CFU/m^3)	1500	Ceiling	–		800	2500	
Fungi (CFU/m^3)	1000	Ceiling	–		–	–	
PM_{10} (μg/m^3)	75	24 h	150	24 h	100	150	24 h
$PM_{2.5}$ (μg/m^3)	35	24 h	–		–	–	–
Ozone (O_3) (ppm)	0.06	8 h	–		0.06	300	1 h
Benzo[*a*]pyrene (ng/m^3)	–		–		–	1.0	24 h
Radon (Bq/m^3)	–		–		–	400	annual

CFU Colony-forming unit, *PM* Particulate matter

they are often used by courts of law, for example, to justify their decisions. Exceeding these values must result in correspondingly effective measures.

10.3.2.1 Guideline Values of the World Health Organization (WHO)

Guideline values for toxic air pollutants were first established in 1987 by a WHO working group on the basis of toxicological knowledge at that time, taking protective factors into account (WHO 2000, 2006, 2010, 2021). They are not legally binding limit values, but background information intended, for example, to provide citizens and authorities with guidance on assessing health risks from air pollution. They refer to the general population exposed to a long-term exposure situation of outdoor air—including indoor air. In the majority of cases, they do not apply to short-term or peak concentrations (e.g., in the event of accidents). In 2016 the WHO started to update the air quality guidelines and recommendations on new levels are formulated, together with interim targets, for the following pollutants: $PM_{2.5}$, PM_{10}, ozone, nitrogen dioxide, sulfur dioxide, and carbon monoxide (WHO 2021). The evidence-informed derivation of each AQG level and an indication of the reduction in health risk associated with the achievement of consecutive interim targets can be found. All assessments have been proposed on the occasion of expert meetings of the WHO, which are summarized in Table 10.2.

10.3.2.2 Guideline Values in France

Since 2004, the French Agency for Food, Environmental and Occupational Health and Safety developed Indoor Air Quality Guidelines Values (IAQG) (ANSES 2021). They were established after summarizing the toxicity data to humans and animals, proven for consistency, and following exposure through inhalation, taking into account associated durations of exposure (acute, sub-chronic, or chronic) (Mandin et al. 2009). The IAQG proposed by ANSES are accompanied by an analysis of the measurement methods, guidance on the sampling strategy, and a proposal of information enabling the health benefits associated with compliance with the IAQG. On the basis of the information to the critical health effect 12 substances were assessed (see Table 10.3).

10.3.2.3 Guideline Values in Belgium

In Belgium, target and intervention guidance values for indoor air have been established for 23 substances, mainly volatile organic compounds (VOC), NO_2, $PM_{2.5}$, and asbestos (De Brouwere et al. 2020). The target guidance value of a chemical corresponds to a concentration where negative impact on health of the occupants due to the indoor environment is not expected, while the intervention guidance value is a measurable concentration of a chemical, physical, or biological factor in the indoor environment requiring a remediation action because the concentrations might provoke a health risk for the occupants of the indoor environment. For each substance, a document on the basis of the existing guidance values was created and the most appropriate value was used at the end.

10.3.2.4 Guideline Values in the UK

In the UK a working group reviewed existing scientific evidence of health-based guidelines proposed by other countries and organizations, to identify which existing guidelines could be adopted for individual VOC (Shrubsole et al. 2019). They did not aim to carry out a full review of the toxicological evidence, to produce new guidelines but they selected the most appropriate existing health-based guideline

Table 10.2 Guideline values of the World Health Organization for non-carcinogenic substances (WHO 2000, 2006, 2010, 2021)

Substance	Guide value µg/m³	Exposure duration
Lead	0.5	1 year
Cadmium	0.005	1 year
Dichloroethane, 1,2-	700	1 year
Dichloromethane	3000	24 h
	450	1 week
Formaldehyde	100[a]	30 min
Carbon monoxide (CO)	100,000[b]	15 min
	35,000[b]	1 h
	10,000	8 h
	4000	24 h
Sulfur dioxide (SO_2)	500	10 min
	40	24 h
Nitrogen dioxide (NO_2)	10	1 year
	25	24 h
	200	1 h
Carbon disulfide	100	24 h
	20[c]	30 min
Hydrogen sulfide	150	24 h
	7[c]	30 min
Manganese	0.15	1 year
Naphthalene	0.01	1 year
Ozone	60	Peak season [d]
	100	8 h
PM_{10}	15	1 year
	45	24 h
$PM_{2.5}$	5	1 year
	15	24 h
Mercury (inorganic)	1	1 year
Styrene	260	1 week
	70[c]	30 min
Tetrachloroethene (perchloroethylene)	250	1 year
Toluene	260	1 week
Vanadium	1	24 h
Benzo[*a*]pyrene	1.2 ng/m³ ≙ 1/10000 0.12 ng/m³ ≙ 1/100000 0.012 ng/m³ ≙ 1/1000000	Additional lifetime risk
Benzene	17 µg/m³ ≙ 1/10000 1.7 µg/m³ ≙ 1/100000 0.17 µg/m³ ≙ 1/1000000	Additional lifetime risk
Trichloroethane	230 µg/m³ ≙ 1/10000 23 µg/m³ ≙ 1/100000 2.3 µg/m³ ≙ 1/1000000	Additional lifetime risk

[a]Also protects against long-term exposure and against carcinogenic effects; [b]Not more than once per day; [c]protection against odor nuisance; [d]Average of daily maximum 8-hour mean O_3 concentration in the six consecutive months with the highest six-month running-average O_3 concentration

Table 10.3 Indoor air quality guidelines in France (ANSES 2021)

Substance	Year of publication	Type of value	IAQGs	Recommended measurement methods
Carbon monoxide (CO)	2007	Short-term IAQG: – For exposure of 8 hours. – For exposure of 1 hour. – For exposure of 30 minutes. – For exposure of 15 minutes	10 mg/m³ 30 mg/m³ 60 mg/m³ 100 mg/m³	–
Benzene	2008	Short-term IAQG: for exposure of 1–14 days	30 µg/m³	–
		Intermediate IAQG: for exposure from 14 days to 1 year	20 µg/m³	–
		Long-term IAQG: for exposure of >1 year	10 µg/m³	–
		Long-term IAQG: for lifelong exposure corresponding to a risk level of 10^{-6}	0.2 µg/m³	–
		Long-term IAQG: for lifelong exposure corresponding to a risk level of 10^{-5}	2 µg/m³	–
Naphthalene	2009	Long-term IAQG: for exposure of >1 year	10 µg/m³	–
Trichloroethylene	2009	Intermediate IAQG: for exposure from 14 days to 1 year	800 µg/m³	Sampling by passive diffusion with carbon disulfide desorption and GC/FID or GC/MS analysis
		Long-term IAQG: for lifelong exposure corresponding to a risk level of 10^{-6}	2 µg/m³	Sampling by passive diffusion through a tube with thermal desorption followed by GC/FID or GC/MS analysis
		Long-term IAQG: for lifelong exposure corresponding to a risk level of 10^{-5}	20 µg/m³	
Tetrachloroethylene	2010	Short-term IAQG: for exposure of 1–14 days	1380 µg/m³	Active sampling by pumping through an activated charcoal tube with carbon disulfide desorption and

(continued)

Table 10.3 (continued)

Substance	Year of publication	Type of value	IAQGs	Recommended measurement methods
				GC/FID or GC/MS analysis
		Long-term IAQG: for exposure of >1 year	250 µg/m³	Sampling by passive diffusion through an activated charcoal tube with carbon disulfide desorption followed by GC/FID or GC/MS analysis
Particulate matter* ($PM_{2.5}$ and PM_{10})	2010	No IAQGs proposed	–	–
Hydrocyanic acid (HCN)	2011	No short-term IAQGs proposed	–	–
Nitrogen dioxide (NO_2)	2013	Short-term IAQG: for exposure of 1 hour	200 µg/m³	Direct measuring by chemiluminescence. Sampling by pumping over a coated medium and analysis by spectrophotometry or ion chromatography.
		Long-term IAQG: for exposure of >1 year	20 µg/m³	Sampling by passive diffusion and analysis by spectrophotometry or ion chromatography
Acrolein	2013	Short-term IAQG: for exposure of 1 hour	6.9 µg/m³	Canister sampling, preconcentration and gas chromatography analysis, and mass spectrometry detection
		Long-term IAQG: for exposure of >1 year	0.8 µg/m³	No measurement method proposed: development and validation needed
Acetaldehyde	2014	Short-term IAQG: for exposure of 1 hour	3000 µg/m³	Sampling by pumping over a coated medium, solvent desorption and analysis by liquid chromatography using a UV/visible detector
		Long-term IAQG: for exposure of >1 year	160 µg/m³	No measurement method proposed: development and validation needed

(continued)

Table 10.3 (continued)

Substance	Year of publication	Type of value	IAQGs	Recommended measurement methods
Ethylbenzene	2016	Short-term IAQG: for exposure of 24 hours	22,000 μg/m³	Active sampling with a sorbent tube, solvent desorption and gas chromatography analysis, flame ionization detection
		Long-term IAQG: for exposure of >1 year	1500 μg/m³	
Formaldehyde	2018	Short-term IAQG. To be complied with repeatedly and continuously throughout the day	100 μg/m³	Active sampling on a DNPH coated silica gel in a sampling tube—Determination by liquid chromatography using a UV/visible detector or Passive sampling on a DNPH/H_3PO_4-coated badge (DSD-DNPH cartridge)—Determination by liquid chromatography using a UV/visible detector
Toluene	2018	IAQG. To be complied with for short-term and long-term measurement	20,000 μg/m³	Active sampling with a sorbent tube, solvent desorption and gas chromatography analysis, coupled with either flame ionization detection or mass spectrometry detection

values (HBGV) for inhalation of 11 organic substances and propose these as UK IAQ guidelines.

10.3.2.5 Guideline Values in Germany

10.3.2.5.1 Toxicologically Based Guideline Values

The indoor air guideline values (Table 10.4) are derived by the German Committee for Indoor Guideline Values (AIR) (former German Working Group on Indoor Air Guidelines) (Ad-hic-AG 2012; Fromme et al. 2019). The background to the committee's activities is § 3 of the building regulations of the federal states, according to which installations must be arranged, erected, altered, and maintained in such a way that public safety and order, in particular life, health and the natural foundations of life, are not endangered. In addition, according to § 13, structural installations must

Table 10.4 Guideline values for indoor air in Germany (UBA 2022)

Name	CAS No.	Year[a]	GV II	GV I	Unit	Remarks[b]
Aldehydes						
Formaldehyde	50-00-0	2016	–	0.10	mg/m^3	
Acetaldehyde	75-07-0	2013	1.0	0.10	mg/m^3	
2-Furaldehyde	98-01-1	2011	0.10	0.010	mg/m^3	
Benzaldehyde	100-52-7	2010	0.20	0.020	mg/m^3	V
$\sum C_4–C_{11}$ Aldehydes	Various	2009	2.0	0.10	mg/m^3	G
Aliphatic hydrocarbons						
$\sum C_9–C_{14}$-Alkanes/Isoalkanes	Various	2005	2.0	0.20	mg/m^3	G
Alcohols						
1-Propanol	71-23-8	2022	46	14	mg/m^3	
Methanol	67-56-1	2022	40	13	mg/m^3	60 min
2-Propanol	67-63-0	2021	45	22	mg/m^3	
Propan-1,2-diol	57-55-6	2017	0.60	0.060	mg/m^3	
1-Butanol	71-36-3	2014	2.0	0.70	mg/m^3	
2-Ethylhexanol	104-76-7	2013	1.0	0.10	mg/m^3	V, S
Benzyl alcohol	100-51-6	2010	4.0	0.40	mg/m^3	
Aromatic hydrocarbons						
$\sum C_7–C_8$ Alkyl benzenes	Various	2016	1	1	–	G, R
Toluene	108-88-3	2016	3.0	0.30	mg/m^3	
$\sum$ Xylenes	Various	2015	0.80	0.10	mg/m^3	G
$\sum$ Naphthalene and naphthalene-like substances[c]	Various	2013	30	10	$\mu g/m^3$	G
Ethylbenzene	100-41-4	2012	2.0	0.20	mg/m^3	
$\sum C_9–C_{15}$ Alkyl benzenes	Various	2012	1.0	0.10	mg/m^3	G
$\sum$ Cresols	Various	2012	50	5.0	$\mu g/m^3$	G
Phenol	108-95-2	2011	0.20	0.020	mg/m^3	
Styrene	100-42-5	1998	0.30	0.030	mg/m^3	
Carboxylic acids						
Methanoic acid	64-18-6	2023	1.0	0.51	mg/m^3	
Ethanoic acid	64-19-7	2023	3.7	1.3	mg/m^3	
Propanoic acid	79-09-4	2023	1.6	0.78	mg/m^3	
Esters						
Methyl methacrylate	80-62-6	2021	2.1	1.1	mg/m^3	
Ethyl acetate	141-78-6	2014	6.0	0.60	mg/m^3	
Tris(2-chloroethyl) phosphate (TCEP)	115-96-8	2002	50	5.0	$\mu g/m^3$	

(continued)

Table 10.4 (continued)

Name	CAS No.	Year[a]	GV II	GV I	Unit	Remarks[b]
Glycols/Glycol ethers/Glycol esters						
2-Phenoxyethanol	122-99-6	2018	0.10	0.030	mg/m^3	
Ethylene glycol monomethyl ether (EGME)	109-86-4	2013	0.20	0.020	mg/m^3	
Diethylene glycol methyl ether (DEGME)	111-77-3	2013	6.0	2.0	mg/m^3	V
Diethylene glycol dimethyl ether (DEGDME)	111-96-6	2013	0.30	0.030	mg/m^3	
Ethylene glycol monoethyl ether (EGEE)	110-80-5	2013	1.0	0.10	mg/m^3	
Ethylene glycol monoethyl ether acetate (EGEEA)	111-15-9	2013	2.0	0.20	mg/m^3	V
Diethylene glycol monoethyl ether (DEGEE)	111-90-0	2013	2.0	0.70	mg/m^3	V
Ethylene glycol butyl ether (EGBE)	111-76-2	2013	1.0	0.10	mg/m^3	
Ethylene glycol butyl ether acetate (EGBEA)	112-07-2	2013	2.0	0.20	mg/m^3	V
Diethylene glycol butyl ether (DEGBE)	112-34-5	2013	1.0	0.40	mg/m^3	V
Ethylene glycol hexyl ether (EGHE)	112-25-4	2013	1.0	0.10	mg/m^3	
Propylene glycol methyl ether (2PG1ME)	107-98-2	2013	10	1.0	mg/m^3	
Dipropylene glycol monomethyl ether (D2PGME)	34,590-94-8	2013	7.0	2.0	mg/m^3	V, S
Propylene glycol monoethyl ether (2PG1EE)	1569-02-4	2013	3.0	0.30	mg/m^3	
Propylene glycol 1-tert-butyl ether (2PG1tBE)	57,018-52-7	2013	3.0	0.30	mg/m^3	
Default-value: Glycol ether	Various	2013	0.050[e]	0.0050[e]	ppm	V
∑ Glycol ethers	Various	2013	1	1	–	R
Halogenated hydrocarbons						
Tetrachloroethene	127-18-4	2017	1.0	0.10	mg/m^3	
2-Chloropropane	75-29-6	2015	8.0	0.80	mg/m^3	
Dichloromethane	75-09-2	1997	2.0	0.20	mg/m^3	24 h
Pentachlorophenol (PCP)	87-86-5	1997	1.0	0.10	μg/m^3	
Ketones						
Acetophenone	98-86-2	2022	220	66	μg/m^3	
Acetone	67-64-1	2021	160	53	mg/m^3	
Methyl isobutyl ketone	108-10-1	2013	1.0	0.10	mg/m^3	

(continued)

Table 10.4 (continued)

Name	CAS No.	Year[a]	GV II	GV I	Unit	Remarks[b]
Terpenes						
∑ Monocyclic monoterpenes (limonene)	Various	2010	10	1.0	mg/m³	S, L
∑ Bicyclic terpenes (α-pinene)	Various	2003	2.0	0.20	mg/m³	L
Others						
Benzothiazole	95-16-9	2020	–	15	μg/m³	V
Nitrogen dioxide (NO_2)	10,102-44-0	2018	0.25	0.080	mg/m³	60 min
2-Butanone oxime	96-29-7	2015	60	20	μg/m³	
1-Methyl-2-pyrrolidone	872-50-4	2014	1.0	0.10	mg/m³	
∑ Cyclic dimethylsiloxanes D_3–D_6	Various	2011	4.0	0.40	mg/m³	G
∑ Diisocyanates	Various	2000	d	d	mg/m³	G
Mercury (as metallic vapor)	7439-97-6	1999	0.35	0.035	μg/m³	

[a]Year of publication in Bundesgesundheitsblatt
[b]Remarks: G (details on substance spectrum see in the respective publications); L (guiding substance); R (∑Ri = Ci/RWi); S (value refers to stereoisomers mixtures as for single stereoisomers); V (preliminary); times given for averaging periods deviate from the usual long-term value
[c]These guide values can provisionally also be used as a sum value for the total group of bi- and tricyclic aromatic hydrocarbons. Indoor air measurement of tricyclic aromatic hydrocarbons should be restricted to the occurrence of directly emitting building products such as asphalt floor coverings
[d]The working group did not consider the specification of an RW II for diisocyanates (DI) to be reasonable (see explanation in the publication): The initially higher concentration in the room air during the processing of coatings and adhesives containing diisocyanates quickly decreases and, after completion of the curing process, continuous exposure is not to be expected. In general, good ventilation should be used when processing products containing DI
[e]For compounds with insufficient data, a default guideline value II and I were proposed on the basis of statistical evaluations of the data for all glycol ethers. For the overall assessment of all glycol ethers and glycol esters found in indoor air, the ratio of the concentration of each compound divided by its guideline value is formed and added to a sum guideline value. The cumulative guide values are considered to be complied with if their value is less than 1 in each case

be arranged, constructed, and fit for use in such a way that no hazards or unacceptable nuisances are caused by water, moisture, plant and animal pests or other chemical, physical, or biological influences.

The derivations are based on the available toxicological knowledge in the current scientific literature and are determined in a comprehensible manner using a basic scheme (see Fig. 10.4). As far as possible, an attempt is made to derive two guideline values for a substance in order to enable a three-range system for staged assessments and measures.

The following definitions underlie the derivations of the two values:

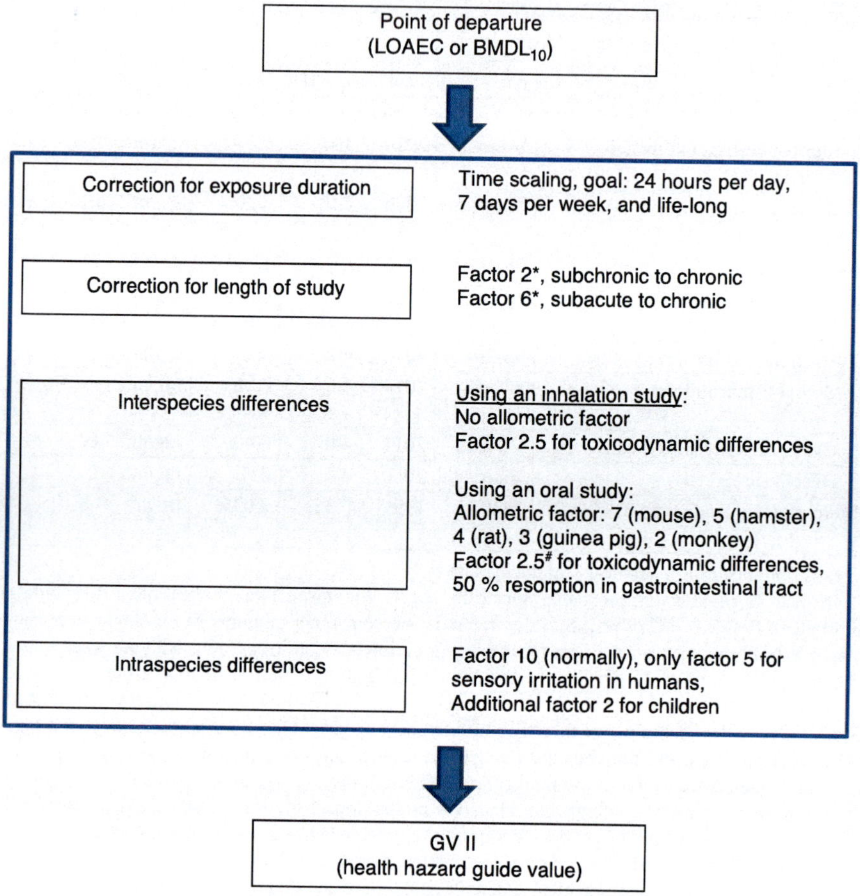

Fig. 10.4 Scheme of the German Committee for Indoor Guideline Values to assess indoor contaminants (after Fromme et al. 2019). *This factor may be omitted if a developmental toxicity study is the starting point; # only if only one or no study in primates is available; LOAEL: Lowest Observed Adverse Effect Level; LOAEC: Lowest Observed Adverse Effect Concentration; $BMDL_{10}$: Benchmark Dose Lower Confidence Limit 10%

Guideline Value II (GV II)

The GV II is an effect-related, justified value based on current toxicological and epidemiological knowledge of the effect threshold of a substance with the introduction of uncertainty factors. It represents the concentration of a substance at which immediate action is required if it is reached or exceeded, since this situation is likely to pose a health hazard, especially to sensitive persons who stay permanently in the rooms. The need for action is to be understood as an immediate need for testing, e.g. with regard to remediation decisions to reduce exposure. Closure of the rooms may therefore be necessary.

The application of guide values as assessment standards requires the performance of a measurement under normal conditions of use. The exceeding of GV II should immediately be combined with a control measurement under usual conditions of use and—as far as possible and reasonable—a determination of the internal exposure of the room occupants.

Guideline Value I (GV I)
The guideline value I is the concentration of a substance in indoor air at which, according to current knowledge, no adverse health effects are to be expected, even with lifelong exposure. Exceeding the limit is associated with hygienically undesirable exposure above the usual level. For precautionary reasons, action is also required in the concentration range between GV I and GV II. GV I can serve as a remediation target value. It should not be "exhausted," but undercut if possible.

10.3.2.5.2 Other German Guideline Values

The Committee establishes guideline values whenever there is systematic, practical experience that the probability of complaints or adverse health effects increases with increasing concentration, but the current state of knowledge is not sufficient to derive toxicologically based guideline values.

Dioxin-like Polychlorinated Biphenyls (dl-PCB)

For this group of PCB, an assessment was carried out in 2007, in which the committee follows the following approach, which has also been introduced internationally (Ad-hoc-AG 2007b). In the context of assessment and regulation, the concentration data of dioxin-like PCBs are usually expressed in toxicity equivalents (TEQ). For this purpose, the levels in air are multiplied by a toxicity equivalency factor (TEF) defined for the individual congener. A TEF value takes into account the ratio of this substance to the most toxicologically significant substance, 2,3,7,8-tetrachlorodibenzo-p-dioxin (2,3,7,8-TCDD). For 2,3,7,8-TCDD, the value is set to 1 in accordance with the convention.

Against this background, the committee has derived a test value of 5 pg PCB-TEQ per cubic meter of indoor air. Since the measurement of dioxin-like PCBs is laborious, a method is presented to estimate the sum of PCB TEQ based on PCB indoor air concentrations determined by standard analytical methods.

Two cases are distinguished:

1. If there are clearly joint compounds with PCB whose chlorination level is lower than Clophen A60. In this case, the TEQ value is well below 5 pg/m^3 if the total PCB guideline value of 3 μg/m^3 is met. Total PCB in this case means the sum of the 6 indicator PCB (PCB 28, 52, 101, 138, 153, and 180) multiplied by 5 (excluding PCB 118). At indoor air concentrations above 3 μg/m^3 for total PCBs, exposure-reducing measures should be considered. At concentrations below this, the ventilation behavior must be checked and improved if necessary.

2. If ceiling tiles containing Clophen-A50 or -A60 have been installed and if mainly highly chlorinated Clophenes occur as PCB sources, the test value of 5 pg PCB-TEQ may already be exceeded above 1 μg/m^3 total PCB. In view of the limited data available, these estimates are associated with considerable uncertainties. In this case, the working group therefore recommends estimating the TEQ value by including the dioxin-like lead congener PCB 118. At levels below 0.01 μg PCB 118/m^3, a TEQ value of 5 pg TEQ/m^3 is met in indoor air, regardless of the PCB source type. At indoor air concentrations above 0.01 μg PCB 118/m^3, exposure-reducing measures should be considered immediately. At concentrations below this, the ventilation behavior must be checked and improved if necessary.

Carcinogenic Substances

The Committee on Indoor Guideline Values has presented recommendations for the evaluation of carcinogenic substances (AIR 2015). The starting point for the final assessment is ECHA's risk assessment guidance, which sets a so-called indicative cancer risk of 10^{-6} for the general population in the absence of legal regulations (ECHA 2012). If the commonly occurring concentration of a carcinogenic contaminant in indoor air, represented by its reference value, falls below a theoretical cancer risk of 1:1,000,000 (10^{-6}) exposed persons assuming lifetime exposure, it seems disproportionate to the committee, given the current state of knowledge, to take exposure-reducing measures at such a concentration. In this case, the committee designates the concentration associated with a theoretical cancer risk of 10^{-6} as the risk-based guideline level. However, if the commonly occurring concentration of a carcinogenic contaminant in indoor air, represented by its reference level, exceeds a theoretical cancer risk of 10^{-6} assuming lifetime exposure, the committee does not designate a risk-based guideline level in such a case but uses the reference level (the 95th percentile of a sample that is as representative as possible for Germany) as the interim guideline level. The committee emphasizes that this is a purely technical value above which exposure-reducing measures should generally be taken.

So far, assessment values have been developed for 5 substances. A risk-based guideline value for trichloroethene of 20 μg/m^3 and interim guideline values for benzene of 4.5 μg/m^3, for 1,2-dichloroethane of 1.0 μg/m^3, for benzo[*a*]pyrene of 0.80 ng/m^3, and for vinyl chloride of 2.3 μg/m^3.

Carbon Dioxide in Indoor Air

The derived guideline values for carbon dioxide in indoor air based on the evaluation of current intervention studies from the point of view of health and hygiene and has recommended appropriate measures (Ad-hoc-AG 2008), which are summarized in Table 10.5.

They are based on the current CO_2 concentrations (instantaneous values) determined during the measurement (at a representative location in the room) with a

Table 10.5 Hygienic assessment of carbon dioxide (CO_2) concentrations in indoor air (Ad-hoc-AG 2008)

Level (ppm)	Hygienic evaluation	Recommendations
<1000	Hygienically harmless	No further measures
1000–2000	Hygienically conspicuous	Take ventilation measure (increase outside air volume flow or air exchange). If ventilation measures and an improvement in ventilation behavior do not show the desired success, a ventilation plan with defined responsibilities should be drawn up and implemented. The implementation of a control measurement to verify the effectiveness of the ventilation measures should be considered. In rooms with a very large number of people and limited room volume, care should be taken to ensure that the room is thoroughly ventilated before it is used, so that the indoor air largely corresponds to the outdoor air quality. It is also advisable to ensure adequate air exchange even before a CO_2 concentration of 1000 ppm is reached.
>2000	Hygienically unacceptable	If a value of 2000 ppm CO_2 is exceeded, ventilation is required. The aim should be to keep the CO_2 concentration below 1000 ppm. If the CO_2 concentration is not permanently below 2000 ppm despite increased ventilation and if higher ventilation frequencies lead to considerable disturbances (e.g., of the teaching process in a classroom), further organizational, ventilation or constructional measures are required. These include, for example, checking the ventilation of the room, reducing the number of people in the room, or installing technically controlled ventilation (e.g., a CO_2-controlled window opening or a mechanical room ventilation system).

sufficient temporal resolution (averaging period usually <2 min). The measurement strategy follows the standard VDI 4300-9 (VDI 2003).

TVOC (Total Volatile Organic Compounds) Values

In order to take account of the fact that indoor air contains a large number of organic compounds and that health complaints are often expressed even when existing guideline values for individual indoor air pollutants are complied with, evaluation standards for the assessment of indoor air quality were developed with the aid of the sum of volatile organic compounds (TVOC). TVOC are defined as the sum of compounds that can be analytically detected on a deactivated non-polar column in the elution range between n-hexane and n-hexadecane. It must be emphasized, however, that these TVOC values are not toxicologically based according to the basic scheme. However, they can be used for exposure characterization, source tracing, and in risk assessment as screening parameters for possible sensory irritation. The committee specifically notes that although there is evidence that the likelihood of discomfort reactions and adverse health effects increases with increasing TVOC concentrations, the health risk does not depend on the sum of the VOC but crucially on the nature and proportions of individual compounds. Table 10.6 shows the individual stages of the TVOC scheme and the recommendations derived from them (Ad-hoc-AG 2007a).

10.3.2.6 Guidelines for Radon in Europe

The following assessments by national and international institutions can be used for indoor radon exposure:

In the EU, Directive 2013/59/EURATOM of December 5, 2013, established 300 Bq/m^3 as the action level for dwellings and workplaces.

For example, this was transposed into German law, for example, by the Radiation Protection Act (StrlSchG) and the Radiation Protection Ordinance (StrlSchV), respectively. The StrlSchG establishes regulations for the protection of humans and, as far as the long-term protection of human health is concerned, the environment from the harmful effects of ionizing radiation. Protection against radon is regulated in paragraphs 121–130. Since radon exposure is higher in some areas due to geogenic conditions, radon precautionary areas must be designated by the federal states by the end of 2020. The federal government has prepared a radon action plan that is updated regularly (BMUV 2019). In addition, the authorities are obliged to inform the population about exposure to radon in occupied rooms and the associated health risks.

Section 124 of the StrlSchG establishes a reference value for the radon activity concentration in the air of occupied rooms averaged over the year of 300 Bq/m^3. This value serves as a benchmark for the adequacy of protective measures.

In the designated radon precautionary areas, the following regulations apply for protection against radon (BfS 2020):

- For private, already existing residential buildings, owners can take voluntary measures. There is no statutory obligation to do so.
- For private new buildings, the building owner has the obligation to prevent radon from entering the building to a large extent by structural measures.

Outside of radon precautionary areas, it is assumed that the usual protection against soil moisture is sufficient and that no special measures need to be implemented in new construction.

The German Radiation Protection Commission (SSK) made the following recommendation in 2005 (SSK 2005):

The European pooling study has provided evidence of an increase in the risk of lung cancer during prolonged residence (30 years) in dwellings at radon concentrations lower than 250 Bq/m^3. In view of the well-established results of the European study, the range below 250 Bq/m^3 should also be considered when making decisions about concrete measures to reduce radon concentrations. The SSK states that for the group of lifelong non-smokers, for whom the results cannot be influenced by

Table 10.6 Hygienic evaluation of the sum of VOC (TVOC) in indoor air (according to Ad-hoc-AG 2007a)

Stage/ Concentration	Hygienic evaluation	Issues to be clarified	Recommendations
Stage 1: ≤0.3 mg/m³	Hygienically harmless, usually no complaints	Are there any guideline values that have been exceeded?	No further measures
Stage 2: >0.3–1 mg/m³	Hygienically still harmless, as long as no guideline values for individual substances or substance groups are exceeded. In individual cases, complaints or odor perceptions, e.g. after minor renovation work or new furnishings in recent weeks.	Are there any reference value exceedances? Are there any conspicuous reference value exceedances? Are the room climatic conditions perfect?	Ventilate sufficiently, especially after renovation work. Determine VOC sources (e.g., walk-through of the room). Check the use of plaster and cleaning agents. Follow-up measurements to check if guideline values are exceeded under conditions of use.
Stage 3: >1–3 mg/m³	Hygienically conspicuous. Use in rooms that are regularly used is only acceptable for a limited period (<12 months). Within approx. 6 months, TVOC concentration should be reduced significantly below the TVOC value measured initially. Cases with complaints or odor perceptions, e.g. after major renovation work.	Are there any reference value exceedances? Are there any conspicuous reference value exceedances? Are the room climatic conditions (air exchange, temperature, humidity) perfect?	Check immediately if reference values are exceeded by re-measurement under conditions of use and take into account the information on reference values in the evaluation. Check conspicuous reference value exceedances for health relevance. In any case: search for sources and check ventilation behavior: ventilate intensively and, if necessary, define usage and ventilation conditions. Control measurement or remeasurement recommended after about one month (under conditions of use). If the TVOC concentration is still above 1 mg/m3 after 12 months despite the efforts described, adequate remediation measures must be included in further planning.
Stage 4: >3–10 mg/m³	Hygienically questionable. Use in rooms that are regularly used only acceptable for a limited period (< 1 month).	Are there any reference value exceedances? Are there any conspicuous reference value exceedances?	Check immediately if reference values are exceeded by re-measurement under conditions of use and

(continued)

Table 10.6 (continued)

Stage/ Concentration	Hygienic evaluation	Issues to be clarified	Recommendations
	TVOC concentration should be reduced below 3 mg/m^3 within one month. Cases with accumulation of complaints or odor perceptions, e.g. after major renovation work.	Are the room climatic conditions (air exchange, temperature, humidity) perfect?	take into account the information on reference values when evaluating. Check conspicuous reference value exceedances for health relevance. Toxicological evaluation of individual substances or substance groups required. In any case: carry out source search and ventilate intensively and, if necessary, determine conditions of use and ventilation and arrange for suitable minimization measures. A possibly necessary stay is only tolerable with a time restriction per day over a maximum period to be specified by the health authority (per day by the hour/time-limited). Control measurement or re-measurement recommended after approx. 1 month (under conditions of use). If the TVOC concentration is still above 3 mg/m3 after 1 month despite the efforts described, adequate remediation measures must be included in further planning.
Stage 5: >10 mg/m^3	Hygienically unacceptable. Avoid use of room if possible. At most, a stay is permissible per day by the hour/for a limited period of time. At values above 25 mg/m^3, use of the room must be avoided. The TVOC concentration should be reduced to below 3 mg/	Are there any conspicuous reference value exceedances? Are there any reference value exceedances? Are the room climatic conditions (air exchange, temperature, humidity) perfect?	Check immediately if reference values are exceeded by re-measurement under conditions of use and take into account the information on reference values in the evaluation. Check conspicuous reference value exceedances for health

(continued)

Table 10.6 (continued)

Stage/ Concentration	Hygienic evaluation	Issues to be clarified	Recommendations
	m^3 within one month. Usually complaints and odor perceptions, e.g. after misuse, accidents.		relevance. Toxicological evaluation of individual substances or substance groups required. In any case: search for sources and ventilate intensively, determine conditions of use and ventilation and arrange for suitable minimization measures. A possibly necessary stay is only tolerable with a time restriction per day over a maximum period to be specified by the health authority (per day by the hour/time-limited). Control measurement or re-measurement within one month (under conditions of use). If, as a result of minimization measures, the TVOC concentration falls below 10 mg/m^3 in the period under consideration, but a concentration of 3 mg/m^3 continues to be exceeded, the recommendations for measures as under stage 4 apply. If, after one month, the TVOC concentration continues to be above 10 mg/m^3 despite the efforts described, the room should not be used and adequate remediation measures must be initiated.

smoking, there is a significant increase in risk already from the concentration interval of 100–199 Bq/m^3.

In 2007, Health Canada established the national guideline for radon in indoor air of 200 Bq/m^3 as an action level. The guideline recommends that corrective action should be undertaken if the average annual radon level exceeds 200 Bq/m^3 in the normal occupancy area of a building, take corrective action sooner, the higher the radon level is, and the corrective action should reduce the radon concentration as much as is practicable. Again, the construction of new buildings should use techniques that minimize radon entry.

On the basis of the Indoor Radon Abatement Act from 1988 the U.-S. Environmental Protection Agency recommends remediation at a maximum of 4 picocuries/liter (pCi/l) of radon in air of dwellings and schools, with the caveat that radon concentrations below this level still carry a risk and in many cases are reducible (ATSDR 2010). This value is equivalent to approximately 148 Bq/m^3.

In its summary of scientific data, the World Health Organization (WHO) has recommended a reference value of 100 Bq/m^3 in indoor air to minimize health risks. If this level cannot be achieved under country-specific conditions, the reference value of 300 Bq/m^3 should not be exceeded. This concentration corresponds approximately to an exposure of 10 mSv per year (WHO 2009, 2010).

In 2014, the International Atomic Energy Agency (IAEA) set an upper national reference level of 300 Bq/m^3 as an annual average as radiation protection standards (IAEA 2014). As part of its requirements (Requirement 50), the IAEA recommends that governments provide information on indoor exposure and associated health effects and establish an action plan to control indoor exposure.

The International Commission on Radiological Protection (ICRP) issued updated recommendations in 2014 (ICRP 2014): it, too, is in a range of 100–300 Bq/m^3 with the recommended reference levels, which would correspond to a dose range of 1–20 mSv per year for population exposure from natural sources. As an upper reference value—to be set nationally—for apartments/houses, the ICRP recommends 300 Bq/m^3, clearly indicating that reference values should be set by national authorities as low as possible in the country.

10.3.2.7 Guideline Values in North America

10.3.2.7.1 National Values in the USA

The US Environmental Protection Agency (US EPA) operates the Integrated Risk Information System (IRIS), which comprehensively characterizes the health risks of a large number of environmental chemicals. For many pollutants, a so-called Reference Concentration (RfC) has been derived in relation to non-carcinogenic effects. It describes a concentration level below which serious health effects are not to be expected, even in the case of lifelong exposure. It takes into account both effects at the point of entry (in this case, in the respiratory tract) and systemic effects, and explicitly includes sensitive groups such as children. A list for the individual pollutants and often a substance dossier listed under IRIS can be found at: https://www.epa.gov/iris.

In addition, the ATSDR (Agency for Toxic Substances and Disease Registry), in coordination with US EPA, sets the so-called Minimal Risk Level (MRL) for priority hazardous substances. In this case, taking into account the available toxicological studies and epidemiological knowledge, MRL are set for acute, intermediate, and chronic exposure conditions. Acute exposure means 1–14 days, intermediate means 15–364 days, and chronic means 365 days or longer. MRLs are developed using a safety factor from the toxicological value at which no adverse effect was observed any longer (called the no-observed-adverse-effect level, NOAEL). The level of the safety factor results, for example, from the transfer of results from animal experiments to humans, from the different sensitivity of organisms, and from the uncertainty of the basic toxicological data. Carcinogenic effects are generally not considered in the derivation.

By definition, the MRL are also values below which no substance-related negative health effects are to be expected—even in the case of daily intake. An updated list of the 164 substances currently assessed for inhalation (including a detailed substance dossier) can be found at: https://www.atsdr.cdc.gov/mrls/mrllist.asp.

10.3.2.7.2 Values in California

The California Environmental Protection Agency's (Cal-EPA) Office of Environmental Health Hazard Assessment (OEHHA) has established so-called Acute Reference Exposure Levels (A-RELs) and Chronic Reference Exposure Levels (C-RELs) for the non-carcinogenic effects of pollutants, which can provide guidance on how to evaluate pollutants at different indoor exposure durations.

Acute Reference Exposure Levels (A-REL) describe an exposure level below which adverse health effects in the general population are not expected in the case of short-term exposure over one hour. A current publication of all A-REL (in some cases also values for an 8-hour exposure) can be found at the following Internet address: https://oehha.ca.gov/air/general-info/oehha-acute-8-hour-and-chronic-reference-exposure-level-rel-summary.

Chronic Reference Exposure Levels (C-REL) are general air reference levels that describe a concentration level below which adverse health effects are not expected to occur from long-term (lifetime) exposure. They are based, taking into account safety factors, on the most sensitive endpoint of an effect described in the scientific literature. They apply to the general population, with all population groups (including children) considered. A current publication of all C-RELs can be found at the following web address: https://oehha.ca.gov/air/general-info/oehha-acute-8-hour-and-chronic-reference-exposure-level-rel-summary.

10.3.2.7.3 Values in Canada

The Canadian health authority Health Canada has drawn up the so-called Residential Indoor Air Quality Guidelines which recommend exposure limits for contaminants that affect indoor air quality (HC 2020). These are the following VOC: Acetaldehyde (24 hours: 280 $\mu g/m^3$; 1 hour: 1420 $\mu g/m^3$), Acrolein (24 hours: 0.44 $\mu g/m^3$; 1 hour: 38 $\mu g/m^3$), Benzene (as low as possible), formaldehyde (8 hours: 50 $\mu g/m^3$; 1 hour: 123 $\mu g/m^3$), naphthalene (24 hours: 10 $\mu g/m^3$), and toluene (24 hours: 2300 $\mu g/m^3$; 8 hours: 15000 $\mu g/m^3$). Values were also derived for carbon dioxide (24 hours: 1000 ppm), carbon monoxide (24 hours: 10 ppm; 1 hour: 25 ppm), nitrogen dioxide (24 hours: 20 $\mu g/m^3$; 1 hour: 170 $\mu g/m^3$), ozone (8 hours: 40 $\mu g/m^3$), and radon (exposure limit: 200 Bq/m^3). With regard to fine particulate matter HC stated: keep indoor levels of $PM_{2.5}$ as low as possible and use a stovetop fan (which exhausts outdoors) while cooking and do not allow smoking indoors. New guidelines are posted as more contaminants are assessed.

In addition, the so-called Indoor Air Reference Levels (IRL) were compiled for supportive assessment on the basis of values set by other organizations, mostly the US or California environmental agencies (see Table 10.7).

10.3.2.8 Guideline Values in Asia

10.3.2.8.1 Values in China

In 1988, the Ministry of Health of the People's Republic of China issued a set of hygienic standards for public places, in which concentration limits of carbon monoxide (CO), carbon dioxide (CO_2), inhalable particulate matter (PM_{10}), and bacteria were stipulated (Hao et al. 2019). Revised versions of these standards were promulgated in 1996. The first national indoor air quality standard (GB/T 18883-2002) came into effect in 2003 (see Table 10.1; Hao et al. 2019; Niu et al. 2015).

10.3.2.8.2 Values in Japan

In Japan, toxicologically based indoor air quality guidelines have been established for 13 chemicals since 1997 with the last revision in 2012 (see Table 10.8) (Azuma et al. 2020). On the basis of chronic toxicity studies with long-term exposure (except formaldehyde) the guidelines are acceptable values for indoor air concentrations of these chemicals. The guideline values should be applied to all indoor spaces, including housing, offices, medical facilities, welfare facilities, schools, public facilities, and transportation facilities, but exclude specific spaces such as industrial plants.

Table 10.7 Indoor Air Reference Levels (IARL) of Health Canada in μg/m^3 (HC 2020)

Substance (CAS No.)	IARL
1,3-Butadiene (106-99-0)	1.7
1,4-Dichlorobenzene (106-46-7)	60
2-Butoxyethanol (111-76-2)	11,000
2-ethoxyethanol (110-80-5)	70
3-Chloropropene (107-05-1)	1.0
Acetone (67-64-1)	70,000
Acrolein (107-02-8)	0.35
Aniline (62-53-3)	1.0
Carbon tetrachloride (56-23-5)	1.7
Chloroform (67-66-3)	300
Cyclohexane (110-82-7)	6000
Dichloromethane (75-09-2)	600
Epichlorohydrin (106-89-8)	1.0
Ethylbenzene (100-41-4)	2000
Ethylene oxide (75-21-8)	0.002
Isopropyl alcohol (67-63-0)	7000
Isopropylbenzene (98-82-8)	400
Methyl ethyl ketone (78-93-3)	5000
Methyl isobutyl ketone (108-10-1)	3000
Propionaldehyde (123-38-6)	8.0
Propylene oxide (75-56-9)	2.7
Styrene (100-42-5)	850
Tetrachloroethylene (127-18-4)	40
Toluene diisocyanate (26471-62-5)	0.008
Xylene, mixture (1330-20-7)	100

10.3.3 Reference Values

As a convention, the reference value is determined as the 95th percentile of the distribution of the contamination levels in a reference population (Poulsen and Christensen 1997). An assessment based on reference values only provides information on the exposure relative to a reference group. Exceeding a reference value does not automatically imply a health risk to the user. Generally, reference values represent the recent exposure in a population, preferably obtained from a representative basis. Therefore, reference values must be regularly updated (Fromme et al. 2019).

If a reference value is exceeded, the plausibility of the dataset should be checked. This verification step should include analytical conditions and method validation, building characteristics, and the sampling procedure. Next, the health relevance of

Table 10.8 Indoor air quality guidelines in Japan in μg/m^3 (Azuma et al. 2020)

Chemicals	Guideline value
Formaldehyde	100
Toluene	260
Xylene	200
p-Dichlorobenzene	240
Ethylbenzene	3800
Styrene	220
Chlorpyrifos	1 for adults: 0.1 for children
Di-*n*-butyl phthalate	17
Tetradecan	330
Di-(2-ethylhexyl) phthalate)	100
Diazinon	0.29
Acetaldehyde	48
Fenobucarb	33
Nonanal	Interim value: 41 (0.007)
TVOC	400 (tentative target value)

the concentrations should be assessed on the basis of the available toxicological data. If the contamination concentration poses a health risk, the reason and source of the contamination must be found. The strength and cause of the emission source must also be determined. The necessity and magnitude of the recommended response should be assessed based on the collected information. The results of indoor air measurements are strongly influenced by sampling strategy, ventilation, and climatic factors; thus, specific recommendations for the standardization of sampling procedures and analytical conditions are essential.

In Germany, for example, recent reference values for the indoor air of residences are available from the German Environmental Survey performed by the German Environment Agency in 2014–2017 (Bethke et al. 2021). In this study, representative of the German population, 639 indoor living spaces where children usually spend time were examined. The statistical parameters are explained in the chapter on VOC and aldehydes and summarized in Tables 3.3 and 3.4. Additionally, the results of plasticizers, organophosphate flame retardants, and musk compounds from this study are summarized in Table 10.9 (Daniels et al. 2021).

Table 10.9 Substances in dust of residences in the German Environmental Health Survey 2014–2017 (in mg/kg) (after Daniels et al. 2021)

Substance	N	% ≥ LOQ	Median	90th percentile	95th percentile	Maximum
Phthalates						
DMP	646	7	<0.70	<0.70	0.70	352.26
DEP	646	57	<0.70	2.42	5.26	175.61
DnBP	646	100	7.58	25.20	44.67	423.71
DiBP	646	100	8.69	44.32	89.51	414.93
BzBP	646	61	2.95	35.04	88.45	890.99
DCHP	646	23	<4.00	4.29	10.52	258.16
DIHP	643	93	5.40	215.55	429.76	2498.11
DEHP	646	100	152.97	542.96	769.69	7382.43
DINP	643	97	193.34	1260.62	1815.36	9951.22
DIDP	643	96	30.60	92.04	140.78	2781.74
DPHP	607	29	8.89	23.68	38.59	265.98
DIUP	643	63	<4.07	4.28	8.40	80.05
Alternative plasticizers						
DBA	646	0	<3.00	<3.00	<3.00	<3.00
DEHA	646	57	<1.00	3.82	6.81	60.55
DINA	626	17	<3.84	<3.84	<3.84	6.19
DEHAZ	626	8	<1.38	<1.38	<1.38	3.23
DEHS	646	9	<3.32	<3.32	5.39	45.14
ATBC	646	98	9.94	44.68	79.73	766.45
DEHT	646	99	82.94	471.74	831.63	5985.88
DINCH	626	97	17.26	64.01	125.87	1756.80
TOTM	646	20	<5.00	<5.00	8.68	28.19
Organophosphate flame retardant						
TnBP	643	18	<0.90	<0.90	< 0.90	23.15
TCEP	643	26	<3.97	<3.97	5.53	39.68
TCPP	643	77	2.42	9.68	19.98	178.44
TTP	643	28	<4.43	<4.43	< 4.43	30.14
TBEP	643	93	6.45	45.18	73.17	584.81
DPEHP	643	25	<1.56	<1.56	2.02	150.35
TEHP	643	24	< 0.80	<0.80	0.93	26.94
Musk fragrances						
Musk xylene	646	0	<8.00	<8.00	<8.00	34.11
Musk Ketone	646	0	<6.00	<6.00	<6.00	79.03
Tonalid	646	6	<1.55	<1.55	<1.55	17.64
Galaxolide	646	7	<13.00	<13.00	<13.00	147.04
Galaxolide lactone	646	9	<4.50	<4.50	5.37	24.52
Iso E Super	646	34	<8.00	11.11	22.30	112.23

LOQ limit of quantification

References

Ad-hoc-AG (2007a) Evaluation of indoor air contaminants by means of reference and guideline values. Bundesgesundheitsbl 50:990–1005. [in German]

Ad-hoc-AG (2007b) Health evaluation of dioxin-like polychlorinated biphenyls in indoor air. Bundesgesundheitsbl – Gesundheitsforsch – Gesundheitsschutz 50:1–12. [in German]

Ad-hoc-AG (2008) Health assessment of carbon dioxide in indoor air. Bundesgesundheitsbl Gesundheitsforsch Gesundheitsschutz 51:1358–1369. [in German]

Ad-hoc-AG (2014) Determination and assessment of chemical contamination of the air of indoor workplaces (without activity with hazardous substances). Joint communication of the Working Group on Air Analyses of the Permanent Senate Commission for the Examination of Harmful Working Substances of the German Research Foundation and the German Working Group on Indoor Air Guidelines. Bundesgesundheitsbl 57:1002–1018. [in German]

AIR (German Committee on Indoor Guidelines) (2015) Health evaluation of indoor air carcinogens: first addendum to the basic scheme. Bundesgesundheitsbl 58:769–773. [in German]

Anonymous (2001) Second Ordinance on the Implementation of the Federal Immission Control Act (Ordinance on the Emission Limitation of Volatile Halogenated Hydrocarbons – 2. BImschV) vom 10.12.1990. BGBl. I S. 2694) last modified 21.8.2001 (BGBl. I S. 2209). [in German]

ANSES (2021) Indoor Air Quality Guidelines (IAQGs). https://www.anses.fr/en/content/indoor-air-quality-guidelines-iaqgs

ATSDR (2010) What are the standards and regulations for environmental radon levels? https://www.atsdr.cdc.gov/csem/radon/standards.html#epa_guidelines

ATSDR (2022) Public health assessment guidance manual (PHAGM). Calculating hazard quotients and cancer risk estimates. https://www.atsdr.cdc.gov/pha-guidance/conducting_scientific_evaluations/epcs_and_exposure_calculations/hazardquotients_cancerrisk.html

AUH (Ausschuss für Umwelthygiene) (1995) Standards for exposure assessment. Report. Behörde für Arbeit, Gesundheit und Soziales der Hansestadt Hamburg. http://www.apug.de/archiv/pdf/Expositionsabschaetzung_Laender_1995.pdf. [in German]

Azuma K, Jinno H, Tanaka-Kagawa T, Sakai S (2020) Risk assessment concepts and approaches for indoor air chemicals in Japan. Int J Hyg Environ Health 225:113470

Bethke R et al (2021) Indoor volatile organic compounds in the German Environmental Survey for Children and Adolescents (GerES, 2014–2017). Manuscript in preparation

BfS (2020) Radon precautionary areas in Germany. https://www.bfs.de/EN/topics/ion/environment/radon/maps/precautionary-areas.html;jsessionid=F29B8B2AB58A2D484D7B411327FE20E9.1_cid349

BMUV (German Federal Ministry for the Environment, Nature Conservation, Nuclear Safety and Consumer Protection) (2019) Radon action plan for the long-term reduction of radon exposure. https://www.bmuv.de/publikation/radon-action-plan

BStMI (2012) (Bavarian State Ministry of the Interior) Enforcement of Art. 3 § 2 Sentence 1 of the Bavarian Building Code (BayBO); list of technical rules introduced as technical building regulations (including asbestos, PCB, PCP guidelines): Last amended by notice from 30th of November 2012. [in German]

Cartus A, Schrenk D (2017) Current methods in risk assessment of genotoxic chemicals. Food Chem Toxicol 106:574–582

Daniels A et al (2021) Phthalates, substitutes for phthalates, flame retardants and aromatic musk compounds in house dust in the German Environmental Survey for Children and Adolescents (GerES, 2014–2017). Manuscript in preparation

De Brouwere K, Bautmans B, Benoy S, Geerts L, Stranger M, Van Holderbeke M (2020) Establishing target and intervention guidance values for indoor air in dwellings and publicly accessible buildings: The Flemish approach. Int J Hyg Environ Health 230:113579

ECHA (2012) Guidance on information requirements and chemical safety assessment. Chapter R.8: Characterisation of dose [concentration]-response for human health. Version: 2.1, November

2012. https://www.echa.europa.eu/documents/10162/13632/information_requirements_r8_en.pdf
Fromme H, Debiak M, Sagunski H, Röhl C, Kraft M, Kolossa-Gehring M (2019) The German approach to regulate indoor air contaminants. Int J Hyg Environ Health 222:347–354
Hao J, Zhu T, Fan X (2019) Indoor air pollution and its control in China. In: Pluschke P, Schleibinger H (eds) Indoor Air Pollution. Handbook of environmental chemistry. Springer
HC (2020) Residential indoor air quality guidelines. https://www.canada.ca/en/health-canada/services/air-quality/residential-indoor-air-quality-guidelines.html
IAEA (International Atomic Energy Agency) (2014) Radiation protection and safety of radiation sources: International basic safety standards. General Safety Requirements Part 3. Vienna, Austria. https://www-pub.iaea.org/MTCD/Publications/PDF/Pub1578_web-57265295.pdf
ICRP (2014) Radiological protection against radon exposure. ICRP Publication 126. Ann. ICRP 43. https://journals.sagepub.com/doi/pdf/10.1177/ANIB_43_3
Jeong JY (2006) Recently issues on Indoor air quality in Korea. https://www.semanticscholar.org/paper/Recently-issues-on-Indoor-air-quality-in-Korea-Jeong/edb27018fc8586ed8754df50d5afc8af5d226d51
Mandin C, Bonvallot N, Kirchner S, Keirsbulck M, Alary R, Cabanes PA, Dor F, Moullec YL, Mullot J-U, Peel AE, Rousselle C (2009) Development of French indoor air quality guidelines. Clean 37:494–499
NATA (2018) Technical Support Document. EPA's 2014 National Air Toxics Assessment. https://www.epa.gov/national-air-toxics-assessment/2014-nata-technical-support-document
Niu X, Guinot B, Cao J, Xu H, Sun J (2015) Particle size distribution and air pollution patterns in three urban environments in Xi'an, China. Environ Geochem Health 37:801–812
Poulsen OM, Christensen HJM (1997) Calculation and application of covarage intervals for biological reference values. Pure Appl Chem 69:1601–1611
Setton E, Hystad P, Poplawski K, Cheasley R, Cervantes-Larios A, Keller CP, Demers PA (2013) Risk-based indicators of Canadians' exposures to environmental carcinogens. Environ Health 12:15.
Shrubsole C, Dimitroulopoulou S, Foxall K, Gadeberg B, Doutsi A (2019) IAQ guidelines for selected volatile organic compounds (VOCs) in the UK. Build Environ 165:106382
SSK (German Commission on Radiological Protection) (2005) Lung cancer risk from residential radon exposures. Recommendations of the Radiation Protection Commission. http://www.ssk.de/de/werke/2005/kurzinfo/ssk0504.htm) [in German]
Tsai WT (2019) An overview of health hazards of volatile organic compounds regulated as indoor air pollutants. Rev Environ Health 34:81–89
UBA (2022) Indoor air guide values of the German Committee on Indoor Air Guide Values. https://www.umweltbundesamt.de/en/topics/health/commissions-working-groups/german-committee-on-indoor-air-guide-values#german-committee-on-indoor-air-guide-values-air
VDI (Verein Deutscher Ingenieure) (2003) Measurement of indoor air pollutants – measurement strategy for carbon dioxide (CO_2). VDI 4300 Blatt 9. Beuth, Berlin, Germany. [in German]
WHO (2000) Air Quality Guidelines for Europe. WHO Regional Publications, European Series No. 91. https://www.euro.who.int/en/publications/abstracts/air-quality-guidelines-for-europe
WHO (2006) Air quality guidelines. Global Update 2005. WHO Regional Office for Europe, Copenhagen (http://www.euro.who.int/en/what-we-do/health-topics/environment-and-health/air-quality/publications/pre2009/air-quality-guidelines.-global-update-2005.-particulate-matter,-ozone,-nitrogen-dioxide-and-sulfur-dioxide)
WHO (2009) WHO handbook on indoor radon: a public health perspective. Geneva, Switzerland. https://apps.who.int/iris/bitstream/handle/10665/44149/9789241547673_eng.pdf;jsessionid=9FF350618319D0694B3B7F24E1EF39B5?sequence=1
WHO (2010) Guidelines for indoor air quality: selected pollutants. World Health Organization, Regional Office for Europe, Copenhagen (http://www.euro.who.int/en/what-we-do/health-topics/environmental-health/air-quality/publications/2010/who-guidelines-for-indoor-air-quality-selected-pollutants)

WHO (2011) Report on the global tobacco epidemic 2011: Warning about the dangers of tobacco. Geneva, Switzerland. http://apps.who.int/iris/bitstream/10665/44616/1/9789240687813_eng.pdf

WHO (2021) WHO global air quality guidelines Particulate matter ($PM_{2.5}$ and PM_{10}), ozone, nitrogen dioxide, sulfur dioxide and carbon monoxide. https://www.who.int/publications/i/item/9789240034228

Yang J, Seo JH, Jeong NN, Sohn JR (2019) Effects of legal regulation on indoor air quality in facilities for sensitive populations – a field study in Seoul, Korea. Environ Manag 64:344–352